Artificial Intelligence and Causal Inference

Chapman & Hall/CRC Machine Learning & Pattern Recognition

Introduction to Machine Learning with Applications in Information Security
Mark Stamp

A First Course in Machine Learning
Simon Rogers, Mark Girolami

Statistical Reinforcement Learning: Modern Machine Learning Approaches
Masashi Sugiyama

Sparse Modeling: Theory, Algorithms, and Applications
Irina Rish, Genady Grabarnik

Computational Trust Models and Machine Learning
Xin Liu, Anwitaman Datta, Ee-Peng Lim

Regularization, Optimization, Kernels, and Support Vector Machines
Johan A.K. Suykens, Marco Signoretto, Andreas Argyriou

Machine Learning: An Algorithmic Perspective, Second Edition
Stephen Marsland

Bayesian Programming
Pierre Bessiere, Emmanuel Mazer, Juan Manuel Ahuactzin, Kamel Mekhnacha

Multilinear Subspace Learning: Dimensionality Reduction of Multidimensional Data
Haiping Lu, Konstantinos N. Plataniotis, Anastasios Venetsanopoulos

Data Science and Machine Learning: Mathematical and Statistical Methods
Dirk P. Kroese, Zdravko Botev, Thomas Taimre, Radislav Vaisman

Deep Learning and Linguistic Representation
Shalom Lappin

Artificial Intelligence and Causal Inference
Momiao Xiong

For more information on this series please visit: https://www.routledge.com/Chapman–HallCRC-Machine-Learning–Pattern-Recognition/book-series/CRCMACLEAPAT

Artificial Intelligence and Causal Inference

Momiao Xiong

CRC Press
Taylor & Francis Group
Boca Raton London New York

CRC Press is an imprint of the
Taylor & Francis Group, an **informa** business
A CHAPMAN & HALL BOOK

First edition published 2022
by CRC Press
6000 Broken Sound Parkway NW, Suite 300, Boca Raton, FL 33487-2742

and by CRC Press
4 Park Square, Milton Park, Abingdon, Oxon, OX14 4RN

CRC Press is an imprint of Taylor & Francis Group, LLC

Library of Congress Cataloging-in-Publication Data

Names: Xiong, Momiao, author.
Title: Artificial intelligence and causal inference / Momiao Xiong.
Description: First edition. | Boca Raton : CRC Press, 2022. | Includes bibliographical references and index.
Identifiers: LCCN 2021040988 (print) | LCCN 2021040989 (ebook) | ISBN 9780367859404 (hardback) | ISBN 9781032193281 (paperback) | ISBN 9781003028543 (ebook)
Subjects: LCSH: Artificial intelligence. | Causation. | Inference.
Classification: LCC Q335 .X56 2022 (print) | LCC Q335 (ebook) | DDC 006.3/1--dc23/eng/20211130
LC record available at https://lccn.loc.gov/2021040988
LC ebook record available at https://lccn.loc.gov/2021040989

ISBN: 978-0-367-85940-4 (hbk)
ISBN: 978-1-032-19328-1 (pbk)
ISBN: 978-1-003-02854-3 (ebk)

DOI: 10.1201/9781003028543

Typeset in Minion
by KnowledgeWorks Global Ltd.

Contents

Preface

Artificial intelligence (AI) is the ability of the machine or program to think, learn, and make decisions from simulating human behavior and experience. The magnitude of the changes in the science, engineering, medicine, and socioeconomics caused by AI will be bigger than what we can imagine.

AI has made great progress in automatic natural language processing, speech recognition, image-to-image translation, image-to-text translation, causal inference, automation, robotics, intelligent manufacture, agriculture, and medicine. Computers can read legal documents, write scientific report, and make disease diagnosis and medical decisions. Machines can do everything—think, understand, reason, create—and will even make scientific discovery in the near future. AI is transforming humans and machines and making them work together. In human history, we experienced three great technological revolutions: agricultural, industrial, and electronic and computational. Now we are facing the fourth technological revolution – the AI revolution. The current AI revolution is developing rapidly and is unstoppable.

One crucial issue in AI development is to understand mechanisms underlying intelligence, including reasoning, planning, and imagination. Understanding, transfer and generalization are major principles that give rise, to intelligence. One key component for understanding is causal inference. Causal inference includes intervention, domain shift learning, temporal structure, and counterfactual thinking as major concepts to understanding causation. Unfortunately, these essential components of causality are often overlooked by machine learning, which leads to some failure of deep learning. To overcome these limitations, data augmentation, pre-training, self-supervision, and architecture with special functions are often used in deep learning methods (Scholkopf et al. 2021). However, learning association between variables depends on the distribution of the training data. It is difficult to generalize association results to test data, which are outside the distribution of the training data. Therefore, when environments change and distribution shifts, the interventions and counterfactuals allow modeling of distribution shifts, which will lead to success of generalization.

The principle of the module and compositionality is an essential component of human cognition (Russin et al. 2021). Understanding the complex object or process consists of three stages: (1) identifying the subcomponents, (2) investigating the contents and meaning of the subcomponents, and (3) combining the relevant subcomponents into the whole object or entire process using structural information. The structural causal model (SCM) decomposes the complex object or process into several modules and learns their structural knowledge. Each module may correspond physical causal mechanisms and many modules have similar functions across different tasks. A new task can be accomplished by a combination of a few typical modules.

Intervention and counterfactuals are two basic approaches to causal inference from observational data. Intervention, in general, is the language for the experiment. We use intervention to infer the effect or learn the response of the system under intervention. The counterfactual is the imagined effect or response of the system under the imagined action, i.e., the counterfactual is the imagined consequence of the alternative action.

Causation is a subtle concept. The data format is an essential part of causal inference. There are two types of data: observational and experimental. Intervention is defined for experiments. However, experiments are often expensive, time consuming, and even unethical and infeasible. In practice, only observational data are available. For observational data, both intervention and counterfactuals are unobservable. They can only be imagined, but cannot be observed. Only partial knowledge of the true cause-effect is available.

Therefore, causal inference from observational data needs some assumptions and causal models. The first widely used model that is based on intervention is the SCM. Intuitively, the causal models imply that changes in cause will definitely induce changes in the effect. SCM consists of (1) an effect variable and a set of causal variables, (2) a set of functional structural equations, and (3) a set of independent noise variables. The interventions include (1) changing the functional form, (2) changing the exogenous variables or noise variables, which cannot be explained by the models, and (3) setting the function to a constant. The SCM defines a causal process (or a mechanism) that maps a set of parents (endogenous variables) and the exogenous variables to an effect (or response variables). The SCM represents the data-generating process. The SCM implies the independent mechanism, i.e., the conditional distribution of effect, given cause, is independent of the distribution of cause.

To fully uncover the causal mechanism, we need to consider alternative interventions and imaginary consequences under alternative interventions. These imaginary consequences or results are called counterfactual outcomes, which may correspond to modified mechanisms. Similar to the SCM, there are two types of counterfactual causal reasoning. One type of the counterfactual causal analysis considers two variables: treatment (cause) and effect.

Counterfactual outcome is taken as a potential outcome. The treatment variable can take binary, categorical, and continuous values. The difference between the factual and counterfactual outcomes can be taken as the measure of causal effect. The second type of counterfactual analysis uses the SCM. After the SCM is identified and estimated, we can estimate counterfactual results under the alternative intervention using the SCM.

AI and causal inference involve (1) using AI techniques as major tools for causal analysis and (2) applying the causal concepts and causal analysis methods to solving AI problems. The purpose of this book is to fill the gap between the AI and modern causal analysis.

The classical causal analysis has several limitations. Its first limitation is to assume a linear functional form. The second limitation is to assume no confounders or observed confounders. The third limitation has difficulty dealing with high-dimensional data. The fourth limitation is unable to infer the causal network with large size. The fifth limitation is lack of language or paradigm to composite the smaller modular causal model into the large causal model. In this book, we introduce the AI-based causal inference methods to overcome these limitations.

AI provides rich tools for function representation. In addition to deep feedforward neural networks, convolutional networks and recurrent neural networks (RNNs), the variational autoencoder (VAE), generative adversary network (GAN), or conditional GAN can be used to model nonlinear functions and nonlinear stochastic processes. Therefore, the VAEs and GANs (CGANs) are used to infer the causal-effect relations between two variables and estimate the individualize effect of binary, categorical, and continuous treatment.

The classical statistical methods for construction of causal networks are formulated as a combinatorial optimization problem. It is well known that the combinatorial optimization problem is an NP-hard problem. Solving the combinatorial optimization problem requires heavy computations. Therefore, we are unable to construct large causal networks. Recently, the construction of causal networks is formulated as a continuous optimization problem and is much easier to solve.

A key issue in classical causal analysis is a confounding problem. The VAE, BiCoGAN, other generative implicit models, attention mechanism, and transformer can map the observed variables to the latent space, which provides good approximations to the unobserved confounders. Therefore, the VAE, BiCoGAN, other generative implicit models, attention mechanism, and transformer can be used for causal analysis with unobserved confounders.

The classical statistical methods can only model linear SCM with the scalar feature of each node. The VAE can model quasi-nonlinear SCM with multiple features of each node. The VAE and matrix normal distribution can be used to formulate the construction of causal networks with multiple features of each node as a continuous optimization problem. The VAE can be used to infer the large causal networks with multiple features of each node.

To completely construct functional causal networks with both linear and nonlinear functions, we need to use multiple AI techniques. First, we use BiCoGAN to infer the functional network but with partial directed and cyclic network. Then, the transformer and attention techniques are used to encode and decode directed acyclic graph (DAG). Finally, reinforcement learning (RL) is used to reformulate the discrete optimization problem into a continuous problem. Searching optimal policy in the RL improves the score function of the DAG. Iteratively repeating the RL search, we can obtain the causal network with the best reward. Therefore, the RL with BiCoGAN, transformer, and attention techniques can be used to construct large functional causal networks.

With the help of multiple AI techniques, we introduce the AI-based methods for estimation of individualized treatment effect in the presence of network interference. The book introduces application of the VAE, neural differential equations, RNN, and RL to estimate counterfactual outcomes. The book also covers large-scale nonlinear mediation causal analysis and nonlinear instrumental variable analysis.

To meet core deep learning challenges of robustness, generalizability, bias, and explainability, and build human-like machine intelligence, in the book we develop causal network-based deep learning methods. There are two approaches. For one approach the pipeline is described as follows. First, we construct causal networks from the original data such as images, text variables, and genomic data. Then, map the constructed networks into the latent space and extract features from the latent variables. Finally, deep neural networks use the extracted features for prediction.

The second approach is graph classification (Nguyen et al. 2020). Given a set of graphs and their labels, we learn embedding for each graph to predict its label. Graph networks including undirected and directed graphs update the vector representation of each node by recursively aggregating and transforming the vector representation of its neighbors. Then, we use a polling function to obtain the vector representation of the entire graphs, which are input to the neural networks for classification.

Applications of causal inference to RL for deconfounding and counterfactual estimation are also included in the book. We consider both observed and unobserved confounders. We introduce *do*-calculus as a tool for adjusting the observed confounders to remove spurious association and return true causation. To deal with unobserved confounders, we first collect many covariates, e.g., images, text data, genomics, geographic data, and lab test results, and then use VAE to map these covariates into the latent space. We develop causal models to infer the causal relations among the latent variables, and action and reward variables that simultaneously affect both action and reward variables. Finally, we adjust for each identified confounders in the RL.

To design alternative actions and predict counterfactual rewards or outcomes in the future under a sequence of alternative actions in the RL, we introduce the SCM and BiCOGAN for estimation of the causal mechanisms and counterfactual outcomes. Then, we introduce data augmentation by counterfactual reasoning, which provides additional information for future reward prediction in RL.

The book is organized into eight chapters. The following is a description of each chapter.

Chapter 1, "Deep Neural Networks", covers (1) three types of neural networks, (2) dynamic approach to deep learning, and (3) optimal control for deep learning. The power of the neural networks comes from their deep architectures. Deep learning has a potential to improve human lives and our works and is a major component of AI. The deep neural networks are classified as three basic neural networks: feedforward neural networks, convolutional neural networks (CNN), and recurrent neural networks (RNNs). The coverage of three types of neural networks in this book will provide a foundation for application of AI to causal inference. Differential equations including ordinary differential equations, stochastic differential equations, and partial differential equations can be used to design the neural networks and estimate their parameters. This book will introduce differential equations for residual networks (ResNet), reversible networks (RevNet), wide residual networks, generative adversarial networks (GANs), and generative latent function time-series models. Estimation problems in neural networks can be formulated as an optimal control problem where the parameters are taken as control signals. In this book, we formulate deep learning as an optimal control problem and use Pontryagin's Maximum Principle to derive necessary conditions for optimality and develop training algorithms for parameter estimation.

Chapter 2, "Gaussian Processes and Learning Dynamic for Wide Neural Networks", includes (1) linear models for learning in neural networks, (2) Gaussian processes, and (3) wide neural networks as a Gaussian process. Alternative to deep neural networks, this book covers wide neural networks where the number of hidden units in a fully connected layer or the number of channels in a convolutional layer is infinite. A goal of neural networks is to identify unknown functions that map inputs to outputs. The book includes the linear model and Gaussian process that generates distributions over functions as powerful tools for analysis of wide neural networks and discusses how to define the Gaussian process for the single-layer neural network and multiple layer neural networks.

Chapter 3, "Deep Generative Models", covers (1) variational inference, (2) variational autoencoders, (3) convolutional variational autoencoders, and (4) graphic convolutional variational autoencoders. Variational inference (VI) is a powerful method for representation learning that maps high-dimensional data to low-dimensional latent space. We first take a statistical approach to fully understand VI. Then, we introduce VAEs. The VAE is a type of generative models. The generative models attempt to find interesting patterns, cluster, statistical correlations, and causal structures of the data. Finally, we introduce two types of very useful VAEs: convolutional variational autoencoder (CVAE) and graphic CVAE. CVAE can be used to extract the important features of two- or three-dimensional data such as the captured videos or images, which reduce the dimension of the datasets while keeping the most of valuable information. Graphic data analysis has broad applications including node clustering and classification, link prediction, and molecular and drug design. Convolutional graph neural networks that redefine the "convolution" for graphs have been developed following the recurrent graph neural networks. The convolutional graph neural networks are divided into the spectral-based approaches and the spatial-based approaches. In addition, many alternative methods for graph neural networks such as graph autoencoders (GAEs) and spatial-temporal graph neural networks, have been developed in the past few years. In general, graph neural networks can be classified into recurrent graph neural networks, convolutional graph neural network, GAEs, and spatial-temporal graph neural networks. This book will focus on GAEs, and specifically graphic CVAEs.

Chapter 4, "Generative Adversarial Networks (GAN)", moves to another generative implicit model – GAN models. This chapter covers (1) generative adversarial networks, (2) seven types of GANs, and (3) generative implicit networks for causal inference with measured and unmeasured confounders. The GANs that can produce large numbers of samples of data distribution without precise modeling and learn deep representations have become one of the most promising advances in AI in the past decade. The GANs can now (1) generate text, images, image-to-image translation, clinical data, and even music that is indistinguishable as fake to human generated, (2) translate from voice to face, (3) be applied to survival analysis, (4) be applied to regression, (5) be applied to time series generation, (6) be applied to signal processing and speech recognition, (7) be applied to hypothesis testing (Bellot and van der Schaar2019), and (8) provide tools for causal inference. In this chapter, we will introduce the main concepts and the theoretic models of GANs, their computational algorithms, the evaluation metrics, Wasserstein GAN, seven types of widely used other GANs: conditional GAN, bidirectional conditional GAN, adversarial autoencoder, adversarially regularized GAE, graph variational GAN, deep convolutional GAN, and multi-agent GAN, and generative implicit networks. The application of the GANs to regression, hypothesis testing, and time series will be discussed.

Chapter 5, "Deep Learning for Causal Inference", focuses on the applications of the AI to causal inferences. This chapter covers (1) functional additive models for causal inference, (2) VAE and GAN for causal analysis, (3) a general framework for formulation of causal inference into continuous optimization, (4) VAE and graph neural network models for learning structural causal models among observed variables, (5) latent causal models, (6) causal mediation analysis, (7) confounding, and (8) linear and nonlinear instrumental variable models.

The problems for causal discovery can be divided into two classes: bivariate causal discovery and causal network discovery. There are many methods for inferring causal relationships. A popular method that can be used for both bivariate causal discovery and causal network discovery is properly constrained functional SCMs, including linear and nonlinear functions. The functional SCMs can be used to model the observational variables and latent variables. The neural network is a universal approximation to any function. In this chapter, we will introduce deep learning as a general framework for causal inference. Two types of generative models, GANs and VAEs, will be explored as major tools for both bivariate causal discovery and large causal network reconstruction with discrete and continuous variables. Mediation analysis is to study how independent variables (causes, interventions) influence the outcomes (effects, outcomes) through intermediate variables (mediators). Univariate, multivariate, cascade unobserved mediation, and application of VAE to cascade unobserved mediation will be discussed. Confounders that affect both an intervention and its outcome obscure the real effect of the causes. Confounders will cause bias. It is a major obstacle to making valid causal inferences from observational data. This chapter introduces VAE with latent variable models and linear and nonlinear instrumental variables for causal inference with unobserved confounders.

Chapter 6, "Causal Inference in Time Series", covers (1) four concepts of causality for multiple time series: Granger causality, Sims causality, intervention causality, and structural causality, (2) statistical methods for Granger causality inference in time series, and (3) nonlinear structural equation models for causal inference on multivariate time series. An essential difference between time series and cross-sectional data is that the time series data have temporal order, but cross-sectional data do not have any order. As a consequence, the causal inference methods for cross-sectional data, which were discussed in the previous chapters, cannot be directly applied to time series data. In this chapter we introduce causal inference methods in time series. First, we introduce four basic concepts of causality for multiple time series: intervention, structural, Granger, and Sims causality. Then we focus on investigation of two major causal graphical models: Granger graphical and dynamic direct acyclic graphical (DAG) models. Nonlinear structural equation models for causal inference on time series which are implemented by neural networks are discussed.

Chapter 7, "Deep Learning for Counterfactual Inference and Treatment Effect Estimation", covers (1) potential outcome framework and counterfactual causal inference, (2) combine deep learning with classical treatment effect estimation methods, (3) counterfactual variational autoencoders, (4) variational autoencoders for survival analysis, (5) VAE causal survival analysis, (6) VAE-Cox model for survival analysis, (7) time series causal survival analysis, (8) a general GAN model for estimation of ITE with discrete outcome and any type of treatment, and (9) adversarial variational autoencoder-generative adversarial network (AVAE-GAN) for estimation of treatment effect in the presence of unmeasured confounders. Historically, structural equation models, potential outcome frameworks, and counterfactuals developed relatively independently in different fields, but they can be unified using interventional queries with *do*-calculus. This allows methods and algorithms developed within one framework to be easily applied to one another, and also allows predictions about the consequences of intervening upon (rather than merely observing) the variables, and provides a method of evaluating counterfactual claims. The SCMs and counterfactual causal inferences can be unified by *do*-calculus. In Chapter 5, we introduce the SCMs and application of VAE and GAN to the SCMs. In this chapter, we will investigate applications of the VAE and GAN to the counterfactuals and treatment effect estimation.

In causal inference, several methods for dealing with unobserved confounding, including instrumental variables, proxy (or surrogate) variables, network structure, and multiple causes have been developed. The counterfactual VAE (CFVAE) is a remarkable work of application of artificial intelligence (AI) to estimation of ITE in the presence of unobserved confounding.

Causal survival analysis from observational data raises two great challenges. The first challenge is presence of confounders that affect both the treatment and survival time. The second challenge is the censoring problem where we only know that an event has not occurred up to a certain point in time and do not know the exact time when time-to-event takes place. Recently, the neural network-based causal methods use the learned representation to balance distributions across treatment and control groups. Therefore, in this chapter, we introduce the VAE-based causal survival analysis methods, which utilize the balanced (latent) representation learning to predict counterfactual survival outcomes and estimate ITE in observational studies.

Treatment response is heterogeneous. However, the classical methods treat the treatment response as homogeneous and estimate the average treatment effects. Therapy should be offered personally to ensure that the right therapy is offered to "the right patient at the right time". Consequently, alternative to calculating the average effect of an intervention over a population, we should estimate individualized treatment effects (ITEs). To achieve this, the book covers the GAN-based models for unbiasedly estimating ITE in the absence of unobserved confounding, including vanilla GAN, the counterfactual-GAN (cGAN), and residual GAN. The book also covers the integration models of VAE and GAN for estimating ITE in the presence of unmeasured confounders, including one GAN and two VAEs with measurable proxy for the latent confounders and GAN with adversarial balancing-based representation for ITE estimation in the presence of unmeasured confounders.

Chapter 8, "Reinforcement Learning and Causal Inference", covers (1) basic reinforcement learning theory, (2) approximate function and approximate dynamic programming, (3) value-based methods – Q-learning and deep Q-learning, (4) policy gradient methods, (5) actor-critic methods, (6) temporal difference method, (7) Sarsa, (8) actor-critic and eligibility trace method, (9) deconfounding reinforcement learning, (10) counterfactuals and reinforcement learning, and (11) reinforcement learning for inferring causal networks. The purpose of this chapter is to stimulate investigation of applications of causal inference to RL and application of RL to causal discovery. To achieve this, we first cover the core part of the RL. Application of causal inference to RL is referred to as causal RL. The current application of structural model to RL is to use SCM to deconfound RL for observational data. The counterfactual can be used as an alternative causal inference framework which is incorporated into RL to remove spurious return. The widely used traditional methods for construction of causal networks are based on score functions. Inferring causal network is reduced to a combinatorial optimization problem to search the causal network with the best score. However, finding the optimal combinatorial problem solution is NP-hard. In Chapter 5, we introduced the recent works on formulating acyclic constraint in term of smooth continuous function and transforming the combinatorial optimization problem to a continuous optimization problem. However, these methods can only infer linear or quasi-nonlinear causal networks. Fortunately, using a combination of RL and graph encoding-decoding algorithms, we also can transform the problem of construction of nonlinear functional causal networks into a continuous optimization problem. Specifically, we use BiCoGAN to infer functional SCM and its graph realization. To enforce acyclicity, exponential smooth constraint is also introduced. Then, we use the recent development in transformer and attention to encoder graphs into continuous latent space. Finally, we use RL to search the functional causal network with the best score in the continuous latent space via searching optimal policy. Therefore, attention networks and RL provide a second approach to transforming a discrete optimization problem for inferring the causal network to a continuous optimization problem. This opens a new way for construction of causal networks with large size.

Overall, this book systematically investigates the application of AI to causal inference and causal inference for improving AI algorithms. This book sets the basis and analytical platforms for further research in this challenging and rapidly changing field. The expectation is that the presented concepts, methods, computational algorithms, and analytic platforms in the book will stimulate the integration of AI and causal inference, and facilitate training next-generation scientists to become armed with both solid AI and causal inference knowledge.

I am deeply grateful to my colleagues and collaborators Eric Boerwinkle, Li, Jin, Wei Lin, and others whom I have worked with for many years. I would especially like to thank my former and current students and postdoctoral fellows for their strong dedication to the research and scientific contributions to the book: Qiyang Ge, Zhixin Hu, Shenying Fang, Li Luo,

Yuanyuan Liu, Tao Xu, and Zhouxuan Li. Finally, I must thank my editor, David Grubbs, for his encouragement and patience during the process of creating this book.

REFERENCES

Bellot, A. and van der Schaar, M. (2019). Conditional independence testing using generative adversarial networks. arXiv:1907.04068.

Russin, J., Fernandez, R., Palangi, H., Rosen, E., Jojic, N., Smolensky, P. and Gao, J. (2021). Compositional processing emerges in neural networks solving math problems. arXiv:2105.08961.

Schölkopf, B., Locatello, F., Bauer, S., Ke, N. R., Kalchbrenner, N., Goyal, A. and Bengio, Y. (2021). Toward causal representation learning, Proceedings of the IEEE - Advances in Machine Learning and Deep Neural Networks 109: 612–634.

CHAPTER 1

Deep Neural Networks

1.1 THREE TYPES OF NEURAL NETWORKS

Deep neural networks are inspired by the human nervous system and the structure of the brain (Shrestha and Mahmood 2019). Neural Networks have found broad applications ranging from autonomous vehicles to new medical imaging tools. The power of the neural networks comes from their deep architectures. The deep learning has a potential to improve human lives and our works. The neural networks share the similar architectures. They consist of nonlinear processing units (nodes) organized in multiple layers. The nodes in each layer are connected to nodes in adjacent layers, but not connected with the nodes in the same layers. The deep neural networks are classified into three basic neural networks: feedforward neural networks, convolutional neural networks (CNNs), and recurrent neural networks (RNNs).

1.1.1 Multilayer Feedforward Neural Networks

1.1.1.1 Architecture of Feedforward Neural Networks

Multilayer feedforward neural networks transfer information in just one direction from input to output via multiple hidden layers. The networks do not have any cycles. A typical neural network consists of interconnected neurons (processing units) and implements a nonlinear mapping. Given input data $X=[x_1,\ldots,x_p]^T$ to the neural network, the output $Y=[y_1,\ldots,y_q]^T$ of the neural network, is a nonlinear function of X, computed along the forward pass path:

$$Y=\begin{bmatrix} y_1 \\ \vdots \\ y_q \end{bmatrix}=\begin{bmatrix} \sigma_1(X,W) \\ \vdots \\ \sigma_q(X,W) \end{bmatrix}, \tag{1.1}$$

where $\sigma_k(X,W)$ is a nonlinear function and W are weights (parameters). Equation (1.1) can be written in a matrix form:

$$Y=\sigma(X,W), \tag{1.2}$$

where σ is a vector that performs element-wise nonlinear mapping. Neural networks are a universal approximates of nonlinear functions which can learn complex nonlinear functions from the data.

Consider a network with p input neurons, q output neurons, and L layers of interconnected neurons (one input layer, one output layer, and $L-1$ hidden layers) (Figure 1.1). The weight for the connection from the kth neuron in the input layer to the jth neuron in the first layer is denoted by $w_{jk}^{(1)}$. The bias of the jth neuron in the first layer is denoted by $b_j^{(1)}$. The weighted input to the jth neuron and activation of the jth neuron in the first layer are denoted by $z_j^{(1)}$ and $a_j^{(1)}$, respectively. Then, the weighted input $z_j^{(1)}$ from the data input layer to the first layer and activation $a_j^{(1)}$ of the jth neuron in the first layer are given by

$$z_j^{(1)}=\sum_{k=1}^{p} w_{jk}^{(1)}x_k+b_j^{(1)},\ j=1,\ldots,m_1, \tag{1.3}$$

and

$$a_j^{(1)}=\sigma\left(z_j^{(1)}\right), \tag{1.4}$$

respectively, where σ is a nonlinear activation function. We often take a sigmoid function $\sigma(z)=\frac{1}{1+e^{-z}}$ as an activation function or take rectified linear activation functions $\sigma(z)=\max(0,z)$ as an activating function.

DOI: 10.1201/9781003028543-1

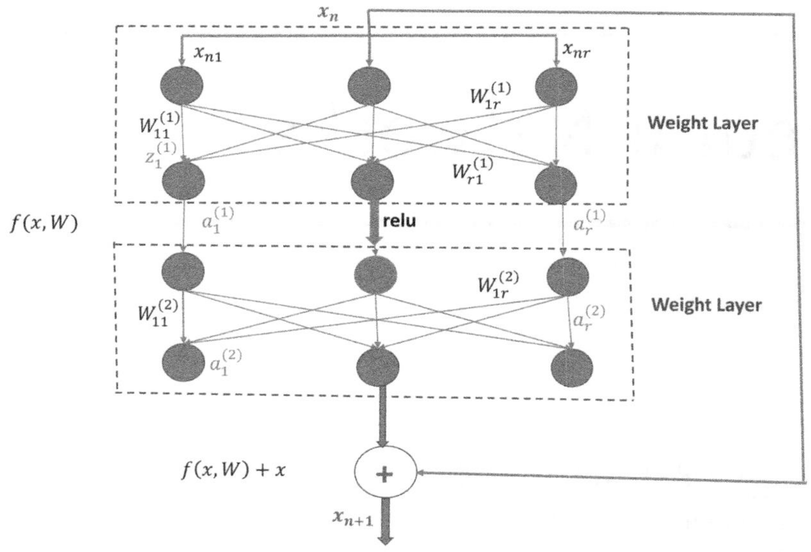

FIGURE 1.1 Architecture of multilayer feedforward neural networks.

Equations (1.3) and (1.4) can be written in a matrix form:

$$Z^{(1)} = W^{(1)}X + b^{(1)}, \tag{1.5}$$

$$a^{(1)} = \sigma\left(Z^{(1)}\right), \tag{1.6}$$

where

$$Z^{(1)} = \begin{bmatrix} z_1^{(1)} \\ \vdots \\ z_{m_1}^{(1)} \end{bmatrix}, W^{(1)} = \begin{bmatrix} w_{11}^{(1)} & \cdots & w_{1p}^{(1)} \\ \vdots & \vdots & \vdots \\ w_{m_1 1}^{(1)} & \cdots & w_{m_1 p}^{(1)} \end{bmatrix},$$

$$X = \begin{bmatrix} x_1 \\ \vdots \\ x_p \end{bmatrix}, b^{(1)} = \begin{bmatrix} b_1^{(1)} \\ \vdots \\ b_{m_1}^{(1)} \end{bmatrix}, a^{(1)} = \begin{bmatrix} a_1^{(1)} \\ \vdots \\ a_{m_1}^{(1)} \end{bmatrix},$$

$$\sigma\left(Z^{(1)}\right) = \begin{bmatrix} \sigma\left(z_1^{(1)}\right) \\ \vdots \\ \sigma\left(z_{m_1}^{(1)}\right) \end{bmatrix}.$$

Similarly, for the lth layer, we denote the weight for the connection from the kth neuron in the $(l-1)^{th}$ layer to the jth neuron in the lth layer by $w_{jk}^{(l)}$, the bias of the jth neuron in the lth layer by $b_j^{(l)}$. The weighted input to the jth neuron and activation of the jth neuron in the lth layer are denoted by $z_j^{(l)}$ and $a_j^{(l)}$, respectively. Equations (1.3) and (1.4) are, in general, expressed as

$$z_j^{(l)} = \sum_{k=1}^{m_{l-1}} w_{jk}^{(l)} a_k^{(l-1)} + b_j^{(l)},\ j = 1,\ldots, m_l, \tag{1.7}$$

$$a_j^{(l)} = \sigma\left(z_j^{(l)}\right). \tag{1.8}$$

Finally, for the output layer, we have

$$z_j^{(L)} = \sum_{k=1}^{m_{L-1}} w_{jk}^{(L)} a_k^{L-1} + b_j^{(L)},\ j = 1,\ldots, m_L, \tag{1.9}$$

$$a_j^{(L)} = \sigma\left(z_j^{(L)}\right). \tag{1.10}$$

Similarly, in a matrix form, we have

$$Z^{(l)} = W^{(l)}a^{(l-1)} + b^{(l)}, \tag{1.11}$$

$$a^{(l)} = \sigma\left(z^{(l)}\right), l = 1,\ldots, L. \tag{1.12}$$

The activation of the jth neuron in the output layer is the composition of several activation functions in the previous layers and is given by

$$a_j^L = \sigma\left(\sum_{k=1}^{m_{L-1}} w_{jk}^{(L)} \sigma\left(\sum_{u=1}^{m_{L-2}} w_{km}^{(L-1)} \sigma(\ldots) + b_k^{(L-1)}\right) + b_j^{(L)}\right). \tag{1.13}$$

1.1.1.2 Loss Function and Training Algorithms

Neural networks can be viewed as a general class of nonlinear functions from a vector X of input variables to a vector Y of output variables. Our goal is to approximate output variables as accurately as possible using neural networks. Given a set of input variables $\{x_1,\ldots,x_N\}$ and output variables $\{y_1,\ldots,y_N\}$, a cost function for measuring the approximation error when the output is continuous is defined as

$$C(W)=\sum_{n=1}^{N}C_n(x_n,W),$$

where

$$C_n(x_n,W)=\frac{1}{2N}\left\|y_n-a^L(x_n,W)\right\|^2, \tag{1.14}$$

where N is the number of training samples, W are weights in the network and $a^L(x_n,W)$ is the vector of activations output from the network when input data x_n is given.

When we deal with classification problems, softmax function should be used as the activation function of the output layer. The output of the softmax function is equivalent to a probability distribution of the classes; it measures the probability of the truth of classes. If q classes are studied, the softmax function is then defined as

$$\hat{y}_j=\sigma\left(z_j^{(L)}\right)=\frac{e^{z_j^{(L)}}}{\sum_{i=1}^{q}e^{z_i^{(L)}}},\ j=1,\ldots,q.$$

The cross entropy will be used as a loss function and is defined as

$$C_n(x_n,W)=-\sum_{j=1}^{q}y_j\log\hat{y}_j(n), \tag{1.15}$$

$$C(W)=\sum_{n=1}^{N}C_n(x_n,W),$$

where $\hat{y}_j(n)$ is the output of the network when input data are x_n.

It is well known that

$$-\sum_{j=1}^{q}y_j\log\hat{y}_j=-\sum_{j=1}^{q}y_j\log\hat{y}_j\frac{y_j}{y_j}$$

$$=-\sum_{j=1}^{q}y_j\log y_j+\sum_{j=1}^{q}y_j\log\frac{y_j}{\hat{y}_j}$$

$$=H(Y)+D_{KL}\left(Y\,||\,\hat{Y}\right), \tag{1.16}$$

where D_{KL} is Kullback-Leibler (KL) distance that measures the closeness between two distribution: the distribution of the observed classes and the distribution of the predicted classes (the output of the neural networks). Minimizing loss function is equivalent to minimizing the distance between the observed distribution and output distribution of the classes.

The weights W can be estimated by minimizing

$$\min_W C(W)=\frac{1}{N}\sum_{n=1}^{N}C_n(x_n,W). \tag{1.17}$$

The algorithms for solving minimization problem (1.17) include gradient descent, stochastic gradient descent, momentum, and Levenberg-Marquardt algorithm (Shrestha and Mahmood 2019).

A popular algorithm to minimize $C(W)$ is gradient descent methods. The idea is to update the weights along the direction of fastest descent of $C(W)$ and stop once the gradient becomes zero. We first work on a single training example and then work on the whole dataset by summarizing all training examples. The gradient descent algorithm for updating weights is

$$W^{k+1}=W^k-\eta_k\frac{\partial C_n(x_n,W)}{\partial W}, \tag{1.18}$$

where $\eta_k\in R_+$ is the learning rate.

The gradient $\frac{\partial C_n(x_n,W)}{\partial W}$ is computed by the backpropagation algorithm. Since $C_n(x_n,w)$ is a complicated composite function of weights w, a key for computing $\frac{\partial C_n(x_n,w)}{\partial w}$ is a chain rule.

We first examine how the changes in z_j^L cause changes in cost function $C_n(x_n,w)$. We define the error rate δ_j^L in the jth neuron of the output layer as

$$\delta_j^L=\frac{\partial C(x_n,W)}{\partial z_j^l}. \tag{1.19}$$

Using equations (1.14) and (1.19), we obtain

$$\delta_j^L=\left(a_j^L-y_{nj}\right)\frac{\partial a_j^L}{\partial z_j^L}$$

$$\delta_j^L = \left(a_j^L - y_{nj}\right)\frac{\partial a_j^L}{\partial z_j^L} = \left(a_j^L - y_{nj}\right)\sigma'\left(z_j^L\right). \quad (1.20)$$

To write equation (1.20) in a vector form, we introduce the Hadamard product that is defined as the element-wise product of the two vectors:

$$u \otimes v = \begin{bmatrix} u_1 \\ \vdots \\ u_m \end{bmatrix} \otimes \begin{bmatrix} v_1 \\ \vdots \\ v_m \end{bmatrix} = \begin{bmatrix} u_1 v_1 \\ \vdots \\ u_m v_m \end{bmatrix}. \quad (1.21)$$

Equation (1.19) can then be rewritten as

$$\delta^L = \left(a^L - y_n\right) \otimes \sigma'\left(z^L\right), \quad (1.22)$$

where

$$a^L = \begin{bmatrix} a_1^L \\ \vdots \\ a_{m_L}^L \end{bmatrix}, y_n = \begin{bmatrix} y_{n1} \\ \vdots \\ y_{nm_L} \end{bmatrix}, \text{ and } \sigma'\left(z^L\right) = \begin{bmatrix} \sigma'\left(z_1^L\right) \\ \vdots \\ \sigma'\left(z_{m_L}^L\right) \end{bmatrix}.$$

Next step for calculation of cost function gradient is to move the error backward from the last layer to the input layer through the network. From equations (1.13) and (1.17), we know that

$$C\left(x_n, w\right) = C\left(z_1^{(l+1)}, \ldots, z_{m_{l+1}}^{(l+1)}\right), \quad (1.23)$$

where

$$z_k^{(l+1)} = w_{kj}^{(l+1)} a_j^{(l)} + \ldots,\ k = 1, \ldots, m_{l+1},\ a_j^{(l)} = \sigma\left(z_j^{(l)}\right). \quad (1.24)$$

Using chain rule, we obtain

$$\delta_j^{(l)} = \frac{\partial C}{\partial z_j^{(l)}} = \sum_{k=1}^{m_{l+1}} \frac{\partial C}{\partial z_k^{(l+1)}} \frac{\partial z_k^{(l+1)}}{\partial z_j^{(l)}} = \sum_{k=1}^{m_{l+1}} \delta_k^{(l+1)} \frac{\partial z_k^{(l+1)}}{\partial z_j^{(l)}}. \quad (1.25)$$

Again, using chain rule and equation (1.24) gives

$$\frac{\partial z_k^{(l+1)}}{\partial z_j^{(l)}} = w_{kj}^{(l+1)} \frac{\partial a_j^{(l)}}{\partial z_j^{(l)}} = w_{kj}^{(l+1)} \sigma'\left(z_j^{(l)}\right). \quad (1.26)$$

Substituting equation (1.26) into equation (1.25), we obtain

$$\delta_j^{(l)} = \sum_{k=1}^{m_{l+1}} \delta_k^{(l+1)} w_{kj}^{(l+1)} \sigma'\left(z_j^{(l)}\right), l = L-1, \ldots, 1. \quad (1.27)$$

Equation (1.27) recursively calculates the error $\delta_j^{(l)}$. Equation (1.27) can be rewritten in a matrix form as

$$\delta^{(l)} = \left(\left(w^{(l+1)}\right)^T \delta^{(l+1)}\right) \otimes \sigma'\left(z^{(l)}\right). \quad (1.28)$$

Now we calculate the rate of change $\frac{\partial C}{\partial b_j^{(l)}}$ of the cost with respect to any bias in the network. Since the cost is a function of $z_j^{(l)}$, by definition and chain rule, we have

$$\frac{\partial C}{\partial b_j^{(l)}} = \frac{\partial C}{\partial z_j^{(l)}} \frac{\partial z_j^{(l)}}{\partial b_j^{(l)}} = \delta_j^{(l)} \frac{\partial z_j^{(l)}}{\partial b_j^{(l)}}. \quad (1.29)$$

Using equation (1.9) gives

$$\frac{\partial z_j^{(l)}}{\partial b_j^{(l)}} = 1. \quad (1.30)$$

Substituting equation (1.30) into equation (1.29), we obtain

$$\frac{\partial C}{\partial b_j^{(l)}} = \delta_j^{(l)}. \quad (1.31)$$

Equation (1.31) can be rewritten in a matrix form as

$$\frac{\partial C}{\partial b} = \delta. \quad (1.32)$$

This shows that the error $\delta_j^{(l)}$ is exactly equal to the rate of change $\frac{\partial C}{\partial b_j^{(l)}}$ of the cost with respect to bias.

Finally, we calculate the rate of change of the cost with respect to any weight in the network $\frac{\partial C}{\partial w_{jk}^{(l)}}$.

It is clear that the cost involves the weight $w_{jk}^{(l)}$ via the following function:

$$z_j^{(l)} = \sum_{k=1}^{m_{l-1}} w_{jk}^{(l)} a_k^{(l-1)} + b_j^{(l)}, \ldots.$$

Therefore, using chain rule, we obtain

$$\frac{\partial C}{\partial w_{jk}^{(l)}} = \frac{\partial C}{\partial z_j^{(l)}} \frac{\partial z_j^{(l)}}{\partial w_{jk}^{(l)}} = \delta_j^{(l)} a_k^{(l-1)}. \quad (1.33)$$

Now consider the cross entropy loss function (1.15). It flows from equation (1.15) that

$$\delta_j^{(L)} = \frac{\partial C_n(x_n, W)}{\partial z_j^{(L)}} = -\frac{y_j}{\hat{y}_j} \frac{\partial \hat{y}_j}{\partial z_j^{(L)}} = -\frac{y_j}{\hat{y}_j} \dot{\sigma}\left(z_j^L\right). \quad (1.34)$$

Define

$$\left(\frac{y}{\hat{y}}\right) = \begin{bmatrix} \frac{y_1}{\hat{y}_1} \\ \vdots \\ \frac{y_q}{\hat{y}_q} \end{bmatrix}.$$

Then, equation (1.34) can be rewritten in a matrix form:

$$\delta^{(L)} = -\left(\frac{y}{\hat{y}}\right) \odot \dot{\sigma}\left(Z^{(L)}\right). \quad (1.35)$$

Other formulas are the same as that for the sum of square loss function discussed earlier.

1.1.2 Convolutional Neural Network

CNN is the neural network of choice for image recognition, computer visions, and natural language processing (Shrestha and Mahmood 2019; Henaff et al. 2015). CNNs are widely used to identify faces, objects, and traffic signs. The CNNs were first proposed by Yann LeCun in the 1980s (Dickson 2020). The successful application of the CNNs is to recognize handwritten digits in banking and postal services. The feature that the CNNs needed large samples and compute resources to work efficiently limited their wide applications. For a quite long time, the CNNs remained on the sidelines of image recognition and artificial intelligence (Dickson 2020). In 2012, AlexNet designed by Alex Krizhevsky won the ImageNet Large Scale Visual Recognition Challenge and achieved a top-five error of 15.3%, more than 10.8% points lower than that of the second one (Krizhevsky et al. 2017; AlexNet 2020). The CNN opens a new era of deep learning and artificial intelligence. Next we introduce the four main operations in a CNN: convolution, nonlinearity (ReLU), pooling, and fully connected layer. These four operations are basic components of the CNN.

1.1.2.1 Convolution

Convolution is to extract features from the input image. Convolution preserves the spatial relationship between pixels in its learning the image features and is implemented by filters (or "kernel" or "feature detector"). A filter is a two-dimensional matrix of values. For example, we have a 3×3 filter:

$$\begin{bmatrix} -1 & 0 & 1 \\ 1 & 0 & 2 \\ 2 & 1 & 0 \end{bmatrix}.$$

As an example of convolution, we consider a 4×4 image:

$$\begin{bmatrix} 40 & 0 & 0 & 30 \\ 50 & 0 & 30 & 1 \\ 60 & 20 & 0 & 70 \\ 10 & 0 & 0 & 90 \end{bmatrix}.$$

To perform convolution, we first overlay the filter in the top left corner of the image and then perform element-wise multiplication between the overlapping image values and filter values.

The first element of our output image after convolution is calculated as follows:

$$-1 \times 40 + 50 \times 1 + 60 \times 2 + 0 \times 0 + 0 \times 0$$
$$+ 20 \times 1 + 0 \times 1 + 30 \times 2 + 0 \times 0 = 210.$$

Similarly, we obtain the rest of the following output image matrix:

$$\begin{bmatrix} 210 & 52 \\ 90 & 141 \end{bmatrix}.$$

The output matrix is called the "Convolved Feature" or "Activation Map" or the "Feature Map". The filters are designed to detect the patterns of the imaging. To detect different patterns, we need to design different filters. Figure 1.2 shows that different values of the filter matrix will produce different convolved feature maps from the same image. The parameters in the filters are the parameters in the CNNs and will be automatically learned from the data during the training

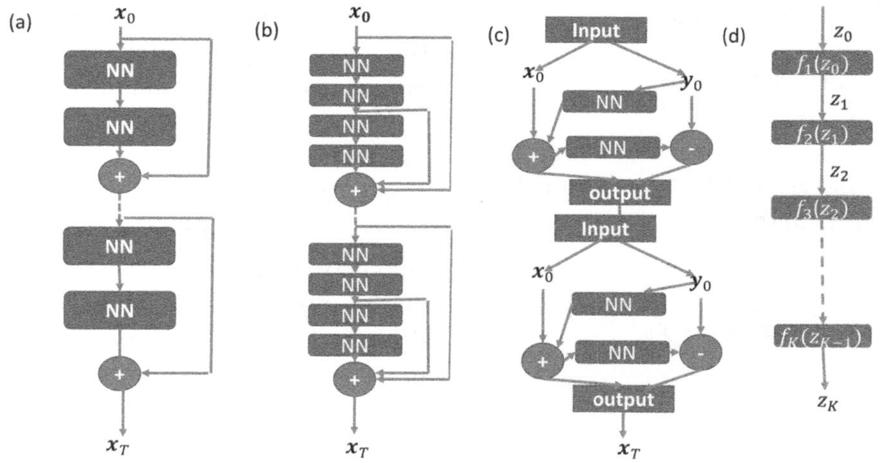

FIGURE 1.2 Convolution operation.

process. However, we still need to design the number of filters, the filter sizes and the sizes of the feature maps. If we want to extract more features of the images, we need to specify the more number of filters. Three parameters: depth, stride, and zero-padding determine the size of the feature maps.

Depth. The depth of the feature maps is defined as the number of feature maps. One filter generates one feature map. Therefore, the depth of the feature maps corresponds to the number of filters.

Stride Size. The stride size is defined as the number of elements (pixels) with which we slide the filter over the input matrix at each step. A larger stride size leads to a smaller output size due to small number of applying filters (Britz 2015). Figure 1.3 plots stride sizes of 1 and 2 of the filter applied to a one-dimensional input.

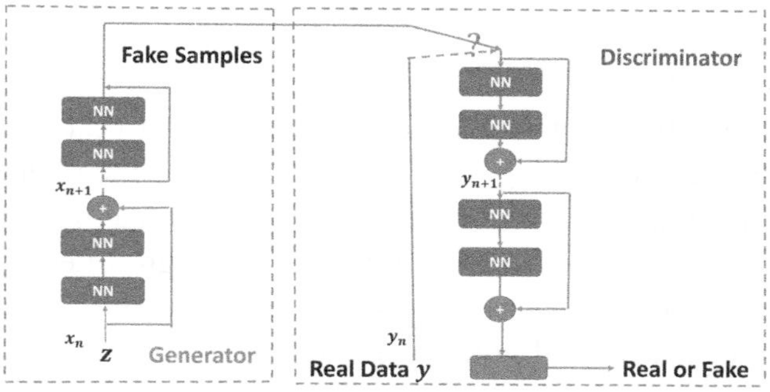

FIGURE 1.3 Convolution stride size. Up: Stride size 1. Bottom: Stride size 2.

Zero-padding. We often want that the output image has the same size as the input image. To achieve this, we can pad the input matrix with zeros around the border, which allows to apply the filter to bordering elements of the input matrix. Padding zero to the border of the input matrix will control the size of the feature maps.

Now we are ready to compute the size of the convolutional layer in one dimension (feature map). Let W be the input data size (one dimension of the input matrix), F be the size of the filter, S be the stride size, P be the number of padding zeros and m be the size of the output (feature map). The size of the output can be calculated as follows:

$$m = \frac{W - F + 2P}{S} + 1. \tag{1.36}$$

Example 1.1: Figure 1.3

For the up feature map, we observe $W = 7$, $F = 3$, $p = 1$, $S = 1$. Then, it follows from equation (1.36) that

$$m = \frac{7 - 3 + 2}{1} + 1 = 7$$

i.e., the size of the feature map is the same as that of the input matrix.

For the bottom feature map, similarly, we obtain $m = 4$.

1.1.2.2 Nonlinearity (ReLU)

An additional nonlinear transformation is used after every convolution operation. Rectified Linear Unit (ReLU) that is defined as $\max(0, x)$ is often used nonlinear transform function. ReLU is an element-wise operation (applied per pixel). All negative pixel values in the feature map are replaced by zero. Other nonlinear transformation functions include tanh or sigmoid function. However, in many scenarios, the ReLU performs better than other nonlinear transformation functions.

1.1.2.3 Pooling

Since the sizes of the input image that also leads to the high dimension of the feature maps are often large, we need to reduce the dimensions of the feature maps while preserving information contained in the feature maps. Spatial Pooling (also called subsampling or downsampling) is usually used to reduce the dimensionality of each feature map but retains the important information in the feature maps. Similar convolution of the filter, to perform pooling, we slide the window (pooling window) of a certain size with a stride over the feature map. Spatial Pooling includes Max, Average, and Sum.

1.1.2.3.1 Max Pooling Max pooling is to take the largest element from the rectified feature map within the pooling window as the output of max pooling. We slide the window with the specified stride over the rectified feature map to obtain the output of each max pooling within the window. The final output elements of pooling form a pooling layer.

1.1.2.3.2 Sum Pooling Instead of taking the maximum of the elements, we can take sum of all elements in the pooling window. The generated pooling layer is called sum-pooling layer.

1.1.2.3.3 Average Pooling Similarly, instead of taking maximum or sum of the elements, we take the average of the elements in the pooling window.

Figure 1.4 shows three pooling operations with 6×6 input matrix, 3×3 pooling window and stride 3. It shows that pooling operation reduced 6×6 dimensions to 2×2 dimensions. Many papers report that in practice Max Pooling performs better than sum and average pooling.

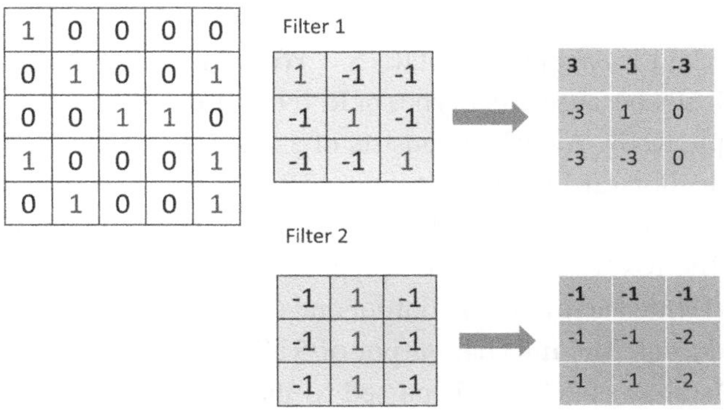

FIGURE 1.4 Three types of pooling.

1.1.2.4 Fully Connected Layers

Convolution, pooling, and nonlinear transformation operations extract the features from the image. We are analyzing these features independently. To jointly use these features, we need to combine them. Furthermore, if our task is classification, we need to convert matrix of features into one dimension for final classification. The fully connected layer that is a traditional multi-layer perceptron (MLP) will be used to implement these tasks. In other words, fully connected layers take the output of the previous layers and convert them into a single vector that can be taken as an input to the next layer. Fully connected layers consist of several layers: the first fully connected layer, hidden fully connected layers, and output fully connected layer.

1.1.3 Recurrent Neural Networks

Unlike feedforward neural networks and CNNs, a recurrent neural network (RNN) is a class of networks that form a directed graph along a temporal sequence. The RNNs are able to model temporal sequential data and to investigate their dynamic behavior (Salehinejad et al. 2018). RNNs are widely used in natural language processing and language modeling (Mikolov et al. 2011), speech recognition (Bourlard and Morgan 2012), translation, video analysis (Venugopalan et al. 2014), and time series forecasting (Petneházi 2019). In this section, we will introduce standard RNN, long short-term memory (LSTM) network (Sherstinsky 2018), convolutional RNN (Keren and Schuller 2016), and space-temporal neural networks (Delasalles et al. 2019).

1.1.3.1 Simple RNN

1.1.3.1.1 Architecture of RNN We begin with a simple RNN because other types of RNN are extension of simple RNN. An RNN consists of a sequence of identical feedforward neural networks. The feedforward neural network is a core of the RNN and called "RNN cells". The RNN consists of two types of inputs and outputs: (1) internal input and output and (2) external input and output. The internal output of RNN can be viewed as "system state" that is passed to the next timestep. An RNN cell receives a prior internal state and a current external input and generates a current internal state and an external current output.

Consider a sequence of data $\{\ldots, X_{t-1}, X_t, X_{t+1}, \ldots\}$, where $X_t = [x_t^1, \ldots, x_t^k]^T$ is a k-dimensional vector. A basic structure of RNN has three layers: input, recurrent hidden, and output layers (Figure 1.5(a)) (Salehinejad et al. 2018). Assume that the dimension of the hidden state space is m and $h_t = [h_t^1, \ldots, h_t^m]^T$ is an m-dimensional state vector. The data X_t is inputted into the input layer. The linear transformation $W_{xh}X_t$ of the data X_t is then sent to the hidden layer, where W_{xh} is an $m \times k$ dimensional matrix. The hidden layer receives information from the input layer and hidden layer at the previous time point. The state is determined by the following nonlinear transformation of its received information:

$$h_t = f_h(W_{hh}h_{t-1} + W_{xh}X_t + b_h), \tag{1.37}$$

where W_{hh} is an $m \times m$ dimensional weight matrix that connects the previous state to the current state, and

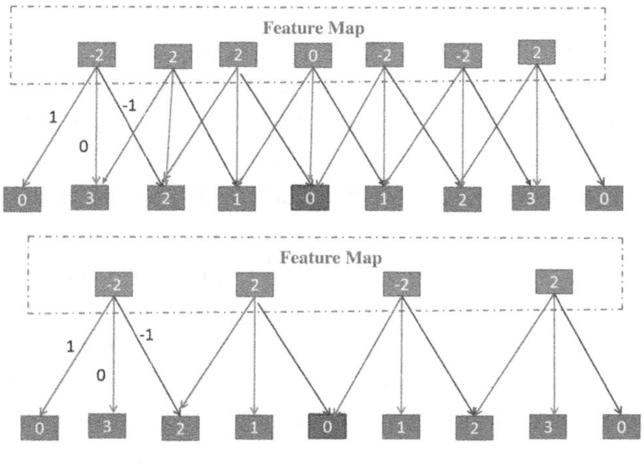

FIGURE 1.5 Basic structure of RNN, architecture of a general RNN.

$b_h = [b_h^1, \ldots, b_h^m]^T$ is an m-dimensional bias vector that corrects the bias, and f_h is a element-wise nonlinear activation function and is often defined as the following "tanh" function:

$$\tanh(x) = \frac{e^x - e^{-x}}{e^x + e^{-x}}.$$

The neurons in hidden layer are connected to the output layer via a $q \times m$ dimensional weight matrix W_{hz}. Let $Z_t = [z_t^1, \ldots, z_t^q]^T$ be a q-dimensional output vector. The output (target) vector is determined by

$$Z_t = f_o(W_{hz}h_t + b_o), \tag{1.38}$$

where f_o is an element-wise activation function and b_o is the bias vector of the output neurons. If the outputs are continuous, the activation functions are sigmoid function defined as $\sigma(x) = \frac{1}{1+e^{-x}}$ or ReLU activation function defined as $\sigma(x) = \max(x, 0)$. If the output variables are discrete, then the activation function is softmax, defined as

$$Z_t = softmax(\alpha) = \begin{bmatrix} \sigma(\alpha_1) \\ \vdots \\ \sigma(\alpha_q) \end{bmatrix}, \tag{1.39}$$

where $\sigma(\alpha_i) = \frac{e^{\alpha_i}}{\sum_{j=1}^{q} e^{\alpha_j}}$.

The architecture of the unfolded RNN through time is shown in Figure 1.5(b). The input information is transmitted through recurrent hidden layer. The neurons of the hidden layer summarize all information about the past states. Finally, we use the neurons of the output layers to make accurate prediction at the output layer.

Example 1.2: Forward Pass of RNN

The data for two time points are given below:

$$W_{xh} = \begin{bmatrix} 0.29 & 0.85 & 0.65 & 0.41 \\ 0.90 & 0.75 & 0.54 & 0.28 \\ 0.54 & 0.12 & 0.65 & 0.55 \end{bmatrix},$$

$$W_{hh} = 0.43, \; b_h = \begin{bmatrix} 0.57 \\ 0.57 \\ 0.57 \end{bmatrix},$$

$$W_{hz} = \begin{bmatrix} 0.37 & 0.97 & 0.83 \\ 0.40 & 0.27 & 0.67 \\ 0.65 & 0.10 & 0.33 \\ 0.92 & 0.35 & 0.21 \end{bmatrix},$$

$$X_1 = \begin{bmatrix} 1 \\ 0 \\ 0 \\ 0 \end{bmatrix}, X_2 = \begin{bmatrix} 0 \\ 1 \\ 0 \\ 0 \end{bmatrix}.$$

First we use equation (1.37) to calculate the hidden state h_0. The procedures are given as follow:

Set the initial state

$$h_0 = \begin{bmatrix} 0 \\ 0 \\ 0 \\ 0 \end{bmatrix}.$$ Then, we have $W_{hh}h_{t-1}$

$$= W_{hh}h_0 = \begin{bmatrix} 0 \\ 0 \\ 0 \end{bmatrix}, W_{xh}X_1 = \begin{bmatrix} 0.29 \\ 0.90 \\ 0.54 \end{bmatrix} \text{ and}$$

$$\beta = W_{hh}h_0 + W_{xh}X_1 + b_h = \begin{bmatrix} 0.86 \\ 1.47 \\ 1.11 \end{bmatrix},$$

$$\text{which implies } h_1 = \begin{bmatrix} 0.70 \\ 0.90 \\ 0.80 \end{bmatrix}.$$

Next we calculate h_2. We first calculate

$$W_{hh}h_1 = 0.43 * \begin{bmatrix} 0.70 \\ 0.90 \\ 0.80 \end{bmatrix} = \begin{bmatrix} 0.301 \\ 0.387 \\ 0.344 \end{bmatrix}$$

$$\text{and } W_{xh}X_2 = \begin{bmatrix} 0.85 \\ 0.75 \\ 0.12 \end{bmatrix}, \text{ which implies}$$

$$\beta = W_{hh}h_1 + W_{xh}X_2 + b_h = \begin{bmatrix} 1.72 \\ 1.71 \\ 1.03 \end{bmatrix}$$

$$\text{and } h_2 = \begin{bmatrix} 0.938 \\ 0.937 \\ 0.774 \end{bmatrix}. \text{ Now we obtain}$$

$$\alpha = W_{hz}h_2 = \begin{bmatrix} 0.37 & 0.97 & 0.83 \\ 0.40 & 0.27 & 0.67 \\ 0.65 & 0.10 & 0.33 \\ 0.92 & 0.35 & 0.21 \end{bmatrix} \begin{bmatrix} 0.938 \\ 0.937 \\ 0.774 \end{bmatrix}$$

$$= \begin{bmatrix} 1.91 \\ 1.15 \\ 1.55 \\ 0.80 \end{bmatrix}.$$ Finally, we obtain the output

$$Z_2 = softmax(\alpha) = \begin{bmatrix} 0.40 \\ 0.19 \\ 0.28 \\ 0.13 \end{bmatrix}.$$

1.1.3.1.2 Loss Function Let Y_t be the observed output at the time t and Z_t be the corresponding target. The loss function is defined as

$$L(Y,Z) = \sum_{t=1}^{T} L(Y_t, Z_t), \tag{1.40}$$

where if Y_t is real numbers then $L(Y_t, Z_t)$ is defined as $L(Y_t, Z_t) = \frac{1}{2}\|Y_t - Z_t\|^2$; if Y_t is frequencies, then the loss function $L(Y_t, Z_t)$ is defined as a cross entropy: $L(Y_t, Z_t) = -\sum_k Y_t^k \log Z_t^k$.

1.1.3.1.3 Parameter Estimation and Backpropagation Algorithm A number of algorithms for parameter estimation in the RNN, such as gradient descent, extended Kalman filter (Puskorius and Feldkamp 1994), expectation and maximization (Ma and Ji 1998), and Levenberg-Marquardt algorithm (Chan and Szeto 1999), have been developed. This section will focus on the backpropagation algorithm. Backpropagation algorithm is a gradient descent method. Tensor calculus (Kochman 2017) is a powerful tool for deriving partial derivatives of complicated functions and is briefly summarized in Appendix 1A.

A general recursive procedure for using gradient descent method to update the weights of the RNN is

$$W^{(k+1)} = W^{(k)} - \eta \frac{\partial L}{\partial W}. \tag{1.41}$$

Now a key issue for parameter estimation is how to calculate the gradient of the objective function with parameters $\frac{\partial L}{\partial W}$. The parameters can be scalar variables, vectors, and matrices. To provide a unified language for partial derivative calculations, we use tensor to represent scalar variables, vectors, and matrices. A scalar, vector, and matrix are called a tensor of zeroth, first order, and second order, respectively. Definitions of gradient of a tensor with respect to tensor and their calculations are summarized in Appendix 1A.

Using equations (1.37), (1.38), (1.40), and chain rule in Section 1A2.4, we obtain

$$\frac{\partial L(Y,Z)}{\partial W_{hh}}=\sum_{t=1}^{T}\frac{\partial L(Y_t,Z_t)}{\partial Z_t}\frac{\partial Z_t}{\partial h_t}\frac{\partial h_t}{\partial W_{hh}}. \tag{1.42}$$

First, we calculate $\frac{\partial L(Y_t,Z_t)}{\partial Z_t}$. If loss function is sum of square of errors, we obtain

$$\frac{\partial L(Y_t,Z_t)}{\partial Z_t}=Z_t-Y_t. \tag{1.43}$$

Next we calculate $\frac{\partial Z_t}{\partial h_t}$. It is clear that Z_t is a nonlinear composite function of h_t. Let $u_t=W_{hz}h_t+b_o$. Then, using chain rule, we obtain

$$\frac{\partial Z_t}{\partial h_t}=\frac{\partial Z_t}{\partial u_t}\frac{\partial u_t}{\partial h_t}. \tag{1.44}$$

Using equation (A54), we obtain

$$\frac{\partial u_t}{\partial h_t}=W_{hz}. \tag{1.45}$$

Substituting equation (1.45) into equation (1.44), we obtain

$$\frac{\partial Z_t}{\partial h_t}=\frac{\partial f_0(u_t)}{\partial u_t}W_{hz}. \tag{1.46}$$

Next we calculate $\frac{\partial h_t}{\partial W_{hh}}$. Using chain rule and equation (1.37), we obtain

$$\frac{\partial h_t}{\partial W_{hh}}=\frac{\partial f_h}{\partial W_{hh}}+\frac{\partial f_h}{\partial h_{t-1}}\frac{\partial h_{t-1}}{\partial W_{hh}}. \tag{1.47}$$

Let $a_t=\frac{\partial h_t}{\partial W_{hh}}$, $b_t=\frac{\partial f_h}{\partial W_{hh}}$ and $c_t=\frac{\partial f_h}{\partial h_{t-1}}$. Then, equation (1.47) can be written as

$$a_t=b_t+c_ta_{t-1}. \tag{1.48}$$

Recursively, we obtain

$$\begin{aligned}a_t&=b_t+c_tb_{t-1}+c_tc_{t-1}a_{t-2}\\&=b_t+c_tb_{t-1}+c_tc_{t-1}b_{t-2}+c_tc_{t-1}c_{t-2}a_{t-3}\\&=b_t+\sum_{i=1}^{t-1}\prod_{j=i+1}^{t}c_jb_i.\end{aligned} \tag{1.49}$$

Therefore, we have

$$\frac{\partial h_t}{\partial W_{hh}}=\frac{\partial f_h}{\partial W_{hh}}+\sum_{i=1}^{t-1}\prod_{j=i+1}^{t}\frac{\partial f_h}{\partial h_{j-1}}\frac{\partial f_h}{\partial W_{hh}}, \tag{1.50}$$

which shows that we can use the chain rule to compute $\frac{\partial h_t}{\partial W_{hh}}$ recursively.

Let

$$Z_t=W_{hz}h_t+b_o. \tag{1.51}$$

Then, using equation (A57), we obtain

$$\frac{\partial Z_t}{\partial W_{hz}}=h_t^T. \tag{1.52}$$

Using equations (1.40), (1.43), and (1.52), we obtain

$$\frac{\partial L(Y,Z)}{\partial W_{hz}}=\sum_{t=1}^{T}(Z_t-Y_t)h_t^T. \tag{1.53}$$

Let

$$u_t=W_{hh}h_{t-1}+W_{xh}X_t+b_h. \tag{1.54}$$

It follows from chain rule and equation (A37) that

$$\begin{aligned}b_t&=\frac{\partial f_h(W_{hh}h_{t-1}+W_{xh}X_t+b_h)}{\partial W_{hh}}\\&=\frac{\partial f_h(u_t)}{\partial u_t}\frac{\partial u_t}{\partial W_{hh}}=\frac{\partial f_h(u_t)}{\partial u_t}h_{t-1}^T.\end{aligned} \tag{1.55}$$

Using equation (A54), we obtain

$$\frac{\partial u_t}{\partial h_{t-1}}=W_{hh}. \tag{1.56}$$

Then, $h_t = f_h(u_t)$. Using chain rule and equation (1.55), we obtain

$$c_t = \frac{\partial f_h\left(W_{hh}h_{t-1} + W_{xh}X_t + b_h\right)}{\partial h_{t-1}} = \frac{\partial f_h(u_t)}{u_t}\frac{\partial u_t}{\partial h_{t-1}} = \frac{\partial f_h(u_t)}{u_t}W_{hh}. \quad (1.57)$$

Combining equations (1.49), (1.50), (1.55), and (1.57), we obtain

$$a_t = \frac{\partial h_t}{\partial W_{hh}} = \frac{\partial f_h(u_t)}{\partial u_t}h_{t-1}^T + \sum_{j=1}^{t-2}\left(\prod_{k=j+2}^{t}\frac{\partial f_h(u_k)}{\partial u_k}W_{hh}\right)\frac{\partial f_h(u_{j+1})}{\partial u_{j+1}}h_j^T, \quad (1.58)$$

where $u_t = W_{hh}h_{t-1} + W_{xh}X_t + b_h$.

If f_h is a linear equation, i.e.,

$$f_h\left(W_{hh}h_{t-1} + W_{xh}X_t + b_h\right) = W_{hh}h_{t-1} + W_{xh}X_t + b_h,$$

Then equation (1.58) is reduced to

$$a_t = \frac{\partial h_t}{\partial W_{hh}} = \sum_{j=1}^{t-1}W_{hh}^{t-j-1}h_j^T. \quad (1.59)$$

If f_h is a tanh function, then

$$\frac{\partial f_h(u_t)}{\partial u_t} = \begin{bmatrix} \frac{4}{\left(e^{u_t^1}+e^{-u_t^1}\right)^2} & \cdots & 0 \\ \vdots & \vdots & \vdots \\ 0 & \cdots & \frac{4}{\left(e^{u_t^K}+e^{-u_t^K}\right)^2} \end{bmatrix} = \operatorname{diag}\left(4\left(e^{u_t^l} + e^{-u_t^l}\right)^{-2}\right). \quad (1.60)$$

Equation (1.58) is reduced to

$$a_t = \frac{\partial h_t}{\partial W_{hh}} = \operatorname{diag}\left(4\left(e^{u_t^l} + e^{-u_t^l}\right)^{-2}\right)h_{t-1}^T + \sum_{j=1}^{t-2}\left(\prod_{k=j+2}^{t}W_{hh}\right)\operatorname{diag}\left(4\left(e^{u_k^l} + e^{-u_k^l}\right)^{-2}\right) \operatorname{diag}\left(4\left(e^{u_{j+1}^l} + e^{-u_{j+1}^l}\right)^{-2}\right)h_j^T. \quad (1.61)$$

If the loss function is sum of square errors, then $Z_t = W_{hz}h_t + b_o$. Using equation (1.45), we obtain

$$\frac{\partial Z_t}{\partial h_t} = W_{hz}. \quad (1.62)$$

Therefore, combining equations (1.42), (1.43), (1.61), and (1.62), we obtain

$$\frac{\partial L(Y,Z)}{\partial W_{hh}} = \sum_{t=1}^{T}W_{hz}\left(Z_t - Y_t\right)\left[\operatorname{diag}\left(4\left(e^{u_t^l} + e^{-u_t^l}\right)^{-2}\right)h_{t-1}^T + \sum_{j=1}^{t-2}\left(\prod_{k=j+2}^{t}W_{hh}\right)\operatorname{diag}\left(4\left(e^{u_k^l} + e^{-u_k^l}\right)^{-2}\right) \operatorname{diag}\left(4\left(e^{u_{j+1}^l} + e^{-u_{j+1}^l}\right)^{-2}\right)h_j^T\right] \quad (1.63)$$

If loss function is a cross entropy, then equation (1.42) can be rewritten as

$$\frac{\partial L(Y,Z)}{\partial W_{hh}} = \sum_{t=1}^{T}\frac{\partial L(Y_t,Z_t)}{\partial u_t}\frac{\partial u_t}{\partial h_t}\frac{\partial h_t}{\partial W_{hh}}.$$

In Appendix 1B, we can show that

$$\frac{\partial L(Y_t,Z_t)}{\partial u_t} = Z_t - Y_t.$$

Then, equation (1.63) still holds for cross entropy loss function.

Similarly, we calculate $\frac{\partial L}{\partial W_{xh}}$. What we need to change is to calculate $\frac{\partial f_h(u_t)}{\partial W_{xh}}$ instead of calculating $\frac{\partial f_h(u_t)}{\partial W_{hh}}$.

Using equation (1.52), we obtain

$$\frac{\partial f_h(u_t)}{\partial W_{xh}} = \frac{\partial f_h(u_t)}{\partial u_t}\frac{\partial u_t}{\partial W_{xh}} = \frac{\partial f_h(u_t)}{\partial u_t}X_t^T. \quad (1.64)$$

Similar to deriving equation (1.58), we obtain

$$\frac{\partial h_t}{\partial W_{xh}} = \frac{\partial f_h(u_t)}{\partial u_t}X_t^T + \sum_{j=1}^{t-2}\left(\prod_{k=j+2}^{t}\frac{\partial f_h(u_k)}{\partial u_k}W_{hh}\right)\frac{\partial f_h(u_{j+1})}{\partial u_{j+1}}X_j^T. \quad (1.65)$$

Therefore, we obtain

$$\frac{\partial L(Y,Z)}{\partial W_{xh}} = \sum_{t=1}^{T} W_{hz}(Z_t - Y_t) \left[\begin{array}{l} \operatorname{diag}\left(4\left(e^{u_t^l} + e^{-u_t^l}\right)^{-2}\right) X_t^T \\ + \sum_{j=1}^{t-2} \left(\operatorname{diag}\left(4\left(e^{u_k^l} + e^{-u_k^l}\right)^{-2}\right) \right. \\ \left. \prod_{k=j+2}^{t} W_{hh} \operatorname{diag}\left(4\left(e^{u_{j+1}^l} + e^{-u_{j+1}^l}\right)^{-2}\right) X_j^T \right) \end{array} \right].$$

In summary, the gradients of loss function with respect to weights in the RNN are given by

$$\frac{\partial L(Y,Z)}{\partial W_{hz}} = \sum_{t=1}^{T} (Z_t - Y_t) h_t^T, \tag{1.66}$$

$$\begin{aligned} \frac{\partial L(Y,Z)}{\partial W_{hh}} = \sum_{t=1}^{T} W_{hz}(Z_t - Y_t) \Bigg[& \operatorname{diag}\left(4\left(e^{u_t^l} + e^{-u_t^l}\right)^{-2}\right) h_{t-1}^T \\ & + \sum_{j=1}^{t-2} \left(\prod_{k=j+2}^{t} W_{hh} \right) \operatorname{diag}\left(4\left(e^{u_k^l} + e^{-u_k^l}\right)^{-2}\right) \\ & \operatorname{diag}\left(4\left(e^{u_{j+1}^l} + e^{-u_{j+1}^l}\right)^{-2}\right) h_j^T \Bigg] \end{aligned} \tag{1.67}$$

$$\frac{\partial L(Y,Z)}{\partial W_{xh}} = \sum_{t=1}^{T} W_{hz}(Z_t - Y_t) \left[\begin{array}{l} \operatorname{diag}\left(4\left(e^{u_t^l} + e^{-u_t^l}\right)^{-2}\right) X_t^T \\ + \sum_{j=1}^{t-2} \left(\begin{array}{l} \operatorname{diag}\left(4\left(e^{u_k^l} + e^{-u_k^l}\right)^{-2}\right) \prod_{k=j+2}^{t} W_{hh} \\ \operatorname{diag}\left(4\left(e^{u_{j+1}^l} + e^{-u_{j+1}^l}\right)^{-2}\right) X_j^T \end{array} \right) \end{array} \right]. \tag{1.68}$$

$$\frac{\partial L(Y,Z)}{\partial b_h} = \sum_{t=1}^{T} W_{hz} \left[\begin{array}{l} \operatorname{diag}\left(4\left(e^{u_t^l} + e^{-u_t^l}\right)^{-2}\right) \\ + \sum_{j=1}^{t-2} \left(\prod_{k=j+2}^{t} \operatorname{diag}\left(4\left(e^{u_k^l} + e^{-u_k^l}\right)^{-2}\right) W_{hh} \right) \\ \operatorname{diag}\left(4\left(e^{u_{j+1}^l} + e^{-u_{j+1}^l}\right)^{-2}\right) (Z_t - Y_t) \end{array} \right], \tag{1.69}$$

$$\frac{\partial L(Y,Z)}{\partial b_0} = \sum_{t=1}^{T} \frac{\partial f_0(u_t)}{\partial u_t} (Z_t - Y_t). \tag{1.70}$$

1.1.3.2 Gated Recurrent Units

In the previous section we investigated how the parameters in the RNN are estimated by gradient methods. In practice, the gradients may vanish or explode due to long product of matrices. In training RNN we often need to store vital early information, skip irrelevant information, and reset internal state representation. The gated recurrent units (GRU) that control how to use the present input and previous memory for producing the current state can meet these requirements (Zaremba et al. 2014; Dey and Salem 2017).

The essential function of the GRU is to gate the hidden states via new introduced resetting and updating gates. The outputs of reset and update gates are vectors whose components take values in the interval $[0, 1]$. The inputs to the hidden state are the previous state and new states. The reset and update gates attempt to assign the appropriate weights to the previous and new states. The final hidden state is a convex linear combination of the previous and current hidden states. The architecture of the GRU is shown in Figure 1.6.

Let $x_t \in R^d$ be the external d-dimensional input vector and h_t, r_t, z_t be the l-dimensional hidden state vector, output vectors of reset gate and update gate at the time t, respectively. The output of a gate is given by a fully connected layer with a sigmoid as its activation function. The reset gate and update gate are, respectively, defined as

$$r_t = \sigma\left(W_{rx} x_t + W_{rh} h_{t-1} + b_r\right), \tag{1.71}$$

$$z_t = \sigma\left(W_{zt} x_t + W_{zh} h_{t-1} + b_z\right), \tag{1.72}$$

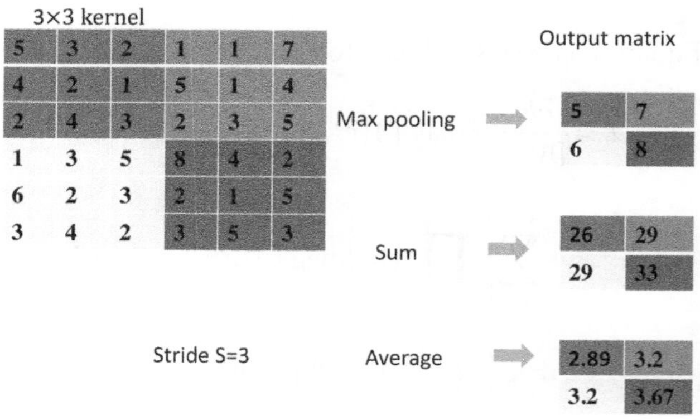

FIGURE 1.6 Architecture of gated recurrent units.

where W_{rx}, W_{rh}, W_{zt}, and W_{zh} are weight matrices, b_r and b_z are biases, σ is an element-wise nonlinear transformation that often uses a sigmoid function.

Next we discuss how to integrate the reset gate with a regular latent state updating information. Recall that in the simple RNN, the regular latent state update is given by

$$h_t = \tanh\left(W_{hh}h_{t-1} + W_{hx}x_t + b_h\right), \quad (1.73)$$

where W_{hh}, W_{hx} are weight matrices and b_h is a bias.

To reduce the influence of the previous states, we can multiply the previous state by r_t element-wise. When r_t is close to 1, we only use the previous hidden state information and keep the previous hidden state unchanged. Similarly, when r_t is close to 0, the previous hidden state information is skipped, we only use the current input information to upstate the current hidden state. The candidate hidden state is calculated by

$$\tilde{h}_t = \tanh\left(W_{hh}\left(r_t \odot h_{t-1}\right) + W_{hx}x_t + b_h\right), \quad (1.74)$$

where $\odot$ represents the element-wise multiplication.

Finally, we incorporate the effect of the update gate into update of the hidden state at the time t;

$$h_t = z_t \odot h_{t-1} + \left(1 - z_t\right) \odot \tilde{h}_t. \quad (1.75)$$

Equation (1.74) shows that when the update gate z_t is close to 1, we ignore the current information from x_t and retain the old state. However, when z_t is close to 0, the candidate hidden state $\tilde{h}_t$ will be taken as the new hidden state h_t.

In summary, the GRU model is given as follows.

1.1.3.2.1 The GRU RNN Model

$$r_t = \sigma\left(W_{rx}\, x_t + W_{rh}h_{t-1} + b_r\right),$$
$$z_t = \sigma\left(W_{zt}x_t + W_{zh}h_{t-1} + b_z\right),$$
$$\tilde{h}_t = \tanh\left(W_{hh}\left(r_t \odot h_{t-1}\right) + W_{hx}x_t + b_h\right),$$
$$h_t = z_t \odot h_{t-1} + \left(1 - z_t\right) \odot \tilde{h}_t.$$

1.1.3.3 Long Short-Term Memory (LSTM)

The LSTM attempts to preserve long-term useful information and skip short-term irrelevant information. Both GRU and LSTM share many common properties. To solve the problem of long-term dependencies, Hochreiter and Schmidhuber (1997) developed long short-term memory (LSTM). Because they have remarkable learning properties, the LSTMs have been widely applied to many areas, including speech recognition (He and Droppo 2016), dynamic trajectory prediction and time series forecasting (Altché and Fortelle 2017) and correlation analysis (Mallinar and Rosset 2018). The LSTM has become one of the focuses of deep learning.

Next I will introduce the basic principles and major components of the LSTM. The architecture of the LSTM is shown in Figure 1.7. The LSTM is inspired by logic gates of a computer. Gated memory cells are major components of the LSTM. A number of gates are used to control a memory cell. We first introduce input, forget, and output gates.

1.1.3.3.1 Input Gates, Forget Gates, and Output Gates Information inputting into the LSTM is the input x_t at the current time point and hidden state h_{t-1} at the previous time point. The key to LSTMs is the cell state. The LSTM does have the ability to remove or add information to the cell state. Removing and adding operations are carefully regulated by the structures, which we call gates. The gates consist of a fully connected layer with sigmoid activation function and a point-wise multiplication operation. The output of the gate which takes values in the interval $[0, 1]$ determines how much information going through. A value of zero indicates no information through, while a value of one indicates that let all information through.

We first introduce forget gate layer. The forget gate decides what information is thrown away from the cell state. Let f_t be the output vector of the forget gate. The input to the forget gate is the current input vector x_t and the hidden state h_{t-1}. The output of the forget gate is defined as

$$f_t = \sigma\left(W_{fx}x_t + W_{fh}h_{t-1} + b_f\right), \quad (1.76)$$

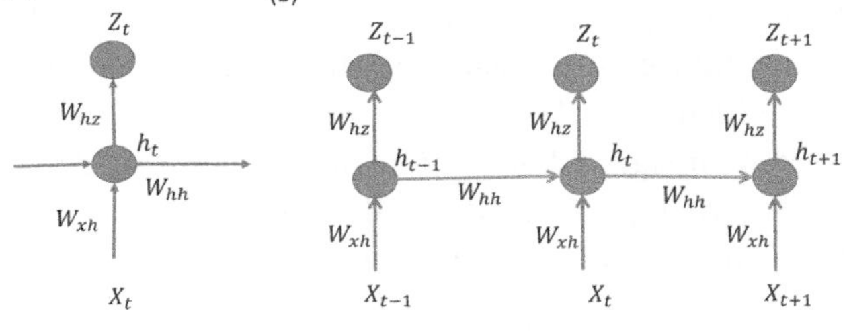

FIGURE 1.7 Architecture of the LSTM.

where W_{fx}, W_{fh} are weight matrices and b_f is a bias vector.

Next we discuss input gate layer. The input gate layer determines what new information stores in the cell state. Again, the input gate layer is a fully connected layer with sigmoid activation function. Let i_t be the output vector of the input gate, which is defined as

$$i_t = \sigma\left(W_{ix}x_t + W_{ih}h_{t-1} + b_i\right), \quad (1.77)$$

where W_{ix}, W_{ih} are weight matrices and b_i is a bias vector.

1.1.3.3.2 Candidate Memory Cell Now we introduce candidate memory cell which determines now much new data should be taken into account. The output of the candidate memory cell is defined as

$$\tilde{c}_t = \tanh\left(W_{cx}x_t + W_{ch}h_{t-1} + b_c\right), \quad (1.78)$$

where W_{cx}, W_{ch} are weight matrices and b_c is a bias vector.

1.1.3.3.3 Memory Cell Memory cell determines how much information in the old state should be memorized and retained and how much new information should be added to the current hidden state in the cell. The previous old cell state should be updated via candidate memory cell, input gate, and forget gate. The new cell state c_t at the time point t should be updated by

$$c_t = f_t \odot c_{t-1} + i_t \odot \tilde{c}_t, \quad (1.79)$$

where $\odot$ denotes element-wise multiplication.

Equation (1.79) shows that when the output of forget gate f_t is close to 1 and the output of input gate is simultaneously close to 0, then the state c_{t-1} of the past memory cell will be stored and passed to the current cell state. This mechanism allows to alleviate the vanishing gradient problem and retains the dependencies in the sequence data.

1.1.3.3.4 Output Gate and Update Hidden State Finally we introduce output gate and discuss how to update the hidden state. The output gate is a fully connected layer with sigmoid activation function, which determines what parts of the cell state will be output. Formally, the output gate is defined as

$$o_t = \sigma\left(W_{ox}x_t + W_{oh}h_{t-1} + b_o\right), \quad (1.80)$$

where W_{ox}, W_{oh} are weight matrices and b_o is a bias vector.

Hidden state h_t can be updated using information from the output gate o_t and cell state c_t as follows:

$$h_t = o_t \odot \tanh\left(c_t\right), \quad (1.81)$$

$$z_t = \operatorname{softmax}\left(W_{zh}h_t + b_z\right), \quad (1.82)$$

where W_{zh} is a weight matrix and b_z is a bias vector.

We update the hidden state h_t by tanh transformation of cell state, multiplied by the output of the sigmoid output gate. When the output gate is 1, we pass all memory information in the cell state through to the predictor. In contrast, when the output is 0, we keep all the information only within the cell.

The Models for the LSTM are summarized by the following equations.

1.1.3.3.5 The LSTM Models The LSTM can be expressed succinctly in the following discrete dynamic and vector equations:

$$f_t = \sigma\left(W_{fx}x_t + W_{fh}h_{t-1} + b_f\right) \quad (1.83)$$

$$i_t = \sigma\left(W_{ix}x_t + W_{ih}h_{t-1} + b_i\right) \quad (1.84)$$

$$\tilde{c}_t = \tanh\left(W_{cx}x_t + W_{ch}h_{t-1} + b_c\right) \quad (1.85)$$

$$c_t = f_t \odot c_{t-1} + i_t \odot \tilde{c}_t \quad (1.86)$$

$$o_t = \sigma\left(W_{ox}x_t + W_{oh}h_{t-1} + b_o\right) \quad (1.87)$$

$$h_t = o_t \odot \tanh\left(c_t\right) \quad (1.88)$$

$$z_t = \operatorname{softmax}\left(W_{zh}h_t + b_z\right) \quad (1.89)$$

1.1.3.4 Applications of RNN to Modeling and Forecasting of Dynamic Systems

Dynamic systems widely exist in all area of science such as engineering, public health, medicine, economics, and finance. System modeling and forecasting are very important for prediction of the dynamic behavior of the system and its control.

Consider a general nonlinear state space model for a dynamic system:

$$h_t = f_h\left(h_{t-1}, u_t, \theta_t\right) + v_t, \quad (1.90)$$

$$y_t = f_Y\left(h_t, u_t, \beta_t\right) + w_t, \quad (1.91)$$

where $h_t \in R^d$ is a state vector, $u_t \in R^l$ is an input or control vector, $Y_t \in R^q$ is an output vector, $\theta_t \in R^k$, β_t

are vectors of parameters, $v_t \in R^d$, and $w_t \in R^q$ are vectors of noises with

$$E[v_t]=0,\ E[w_t]=0 \text{ and } E\left\{\begin{bmatrix} v_t \\ w_t \end{bmatrix}\begin{bmatrix} v_t^T & w_t^T \end{bmatrix} = \begin{bmatrix} Q & S \\ S^T & R \end{bmatrix}\right\}\delta_{ts}.$$

Equation (1.90) is state transition equations and (1.91) is observation equations.

In the case of deterministic identification, we assume $v_t = 0$ and $w_t = 0$. In neural network implementation, we only consider the deterministic case. State transition equation (1.90) is implemented by an RNN and observation equation is implemented by a feedforward neural network (NN). The architecture of recurrent state space model (RSSM) is shown in Figure 1.8(a). Specifically, equation (1.90) can be implemented by

$$h_t = f_h\left(W_{hh}h_{t-1} + W_{xh}\begin{bmatrix} y_t \\ \vdots \\ y_{t-k+1} \end{bmatrix} + W_{uh}u_t + b_h\right), \quad (1.92)$$

where k is the number of lags, W_{hh}, W_{xh} are weight matrices, and b_h is a bias parameter. Equation (1.91) can be implemented by a two-layer MLP:

$$Y_{t+1} = \sigma_2\left(W_{zh}^2\sigma_1\left(W_{zh}^1 h_t + W_{zu}u_t + b_1\right) + b_2\right), \quad (1.93)$$

or

a linear transformation

$$Y_{t+1} = W_{zh}h_t + W_{zu}u_t + b_1, \quad (1.94)$$

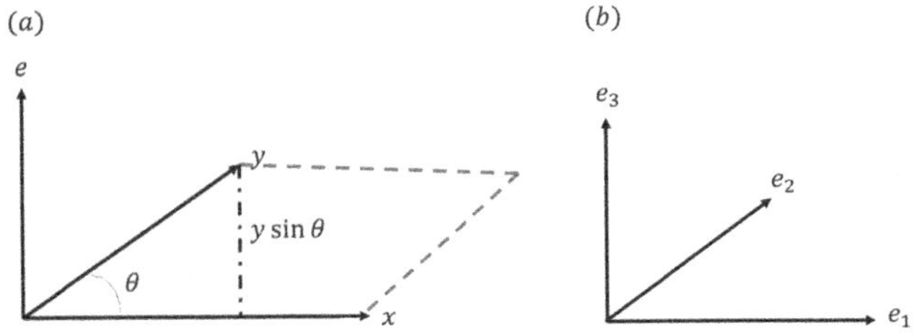

FIGURE 1.8 (a) Architecture of state space RNN, (b) architecture of recurrent state space model with hidden intervention variable.

where σ_1, σ_2 are nonlinear activation functions such as ReLU, W_{zh}^2, W_{zh}^1, W_{zh}, W_{zu} are weight matrices and b_1, b_2 are bias parameters.

1.1.3.5 Recurrent State Space Models with Autonomous Adjusted Intervention Variable

The widely used epidemiological model for predicting the transmission dynamic trajectory of outbreak of infectious diseases is the susceptible-exposed-infected-recovered (SEIR) model which is a mathematical compartmental model based on the average behavior of a population under study. The entire population is divided into four groups (compartments): susceptible group S, exposed group E, infected group I, and recovered group R. The SEIR for modeling the dynamics of epidemic system is given by

$$\frac{dS(t)}{dt} = -\beta I(t)\frac{S(t)}{N}, \quad (1.95)$$

$$\frac{dE(t)}{dt} = \beta I(t)\frac{S(t)}{N} - \sigma E(t), \quad (1.96)$$

$$\frac{dI(t)}{dt} = \sigma E(t) - rI(t), \quad (1.97)$$

$$\frac{dR(t)}{dt} = rI(t), \quad (1.98)$$

where $N = S(t) + E(t) + I(t) + R(t)$ is assumed a constant across the time, β, σ, and r are the exposure, infection, and recovery rates, respectively.

To further simplify the model, we remove the exposure group as intermediate group of infection and assume direct transition from susceptible to infected. The simplified susceptible-infected-recovered (SIR) model is given by

$$\frac{dS(t)}{dt} = -\beta I(t)\frac{S(t)}{N}, \quad (1.99)$$

$$\frac{dI(t)}{dt} = \beta I(t)\frac{S(t)}{N} - rI(t), \quad (1.100)$$

$$\frac{dR(t)}{dt} = rR(t), \quad (1.101)$$

where β and r are infection and recovery rate, respectively. The reproduction number R is defined as

$$R = \frac{\beta}{r}. \quad (1.102)$$

The classical SEIR and SIR models have several serious limitations. First, both SEIR and SIR models assume that the parameters β, σ, and r are constants. Second, if we assume that the parameters of the SEIR and SIR are time dependent, the small sample sizes do not allow the robust parameter estimation. The number of parameters in the SEIR and SIR are 3 and 2, respectively. The observed quantities have only two variables $I(t)$ and $R(t)$. At the fixed time, we only have one sample and two observed variables. In this case, the variance of the estimators will be large. Third, the parameters in the model are hypothesized. The relationships between many public health intervention measures such as the rate of the virus test, travel and public transportation restriction, social distancing cancelling large group gatherings, mandatory quarantine, and school closures and model parameters are not explicitly represented. The changes in the public intervention measures cannot explicitly change the parameters in the models. The intervention measure changes cause changes in the model parameters only through changed observations $I(t)$ and $R(t)$. Fourth, the SEIR and SIR models assume a homogeneous population which is evenly mixed. Fifth, an epidemic system is a control dynamic system. The dynamics of the epidemic system are in the constant changes due to changes in intervention strategies. Sixth, models are not unique and the models with different parameters can coexist for a single system. Some parameters in the epidemiological models are unidentified (Roda et al. 2020).

To make the reproduction number to be time varying, Dandekar and Barbastathis (2020) proposed a following model:

$$\frac{dS(t)}{dt} = -\beta I(t)\frac{S(t)}{N}, \tag{1.103}$$

$$\frac{dI(t)}{dt} = \beta I(t)\frac{S(t)}{N} - \left(r + Q(t)\right)I(t), \tag{1.104}$$

$$\frac{dR(t)}{dt} = rI(t), \tag{1.105}$$

$$\frac{dT(t)}{dt} = Q(t)I(t), \tag{1.106}$$

where

$Q(t)$ is a newly defined variable, quarantine strength, and calculated as follows:

$$\begin{aligned} Q(t) &= \sigma(W_k\sigma\left(W_{k-1}\ldots\sigma\left(W_1 u\right)\right) = NN(W,u), \\ u &= \left(S(t), I(t), R(t), T(t)\right), \end{aligned} \tag{1.107}$$

$W = \left(W_k \ldots W_1\right)$ are weigh matrices and σ is a activation function in the feedforward neural networks.

A time varying reproduction number is defined as

$$R_t = \frac{\beta}{r + Q(t)}. \tag{1.108}$$

The quarantine strength is the variable that is synthesized from the observed $I(t)$, $R(t)$, and unobserved $S(t)$. The MIT model still has three limitations. First, the model still assumes that the infection parameter β is constant across the time. Second, the model assumes that the quarantine strength affects only the recovery rate. However, the quarantine strength also affects the infection rate in the reality. Third, the model still assumes that initial values of $S(t), I(t), R(t), T(t)$ are known. In practice, these values are unknown and guessed subjectively.

Public intervention measure can be decomposed into two parts. One part is due to the large rate of the virus testing, social distance, and quarantine measured, for example, by Google mobility index. Second part is due to autonomous isolation. When an infected individual is detected, the infected individual will be immediately isolated. Then, more and more infected individuals are identified, more and more infected individuals are isolated. The infection and recovery parameters β and r are a function of the infected individuals $I(t)$. The second part intervention cannot be measured from outside system and cannot be inputted to the system as covariates. Therefore, we define a new variable, that is called an intervention measure a_t, to account for the second unmeasured intervention part. The intervention measure a_t is a state variable and should be estimated. The RSSM with intervention measure as a state variable is given by

$$h_t = f_h\left(h_{t-1}, a_{t-1}, y_t, \ldots, y_{t-k+1}, u_t, \theta\right), \tag{1.109}$$

$$a_t = g_i\left(h_{t-1}, a_{t-1}, y_t, \ldots, y_{t-k+1}, u_t, \beta\right), \tag{1.110}$$

$$y_{t+1} = f_y\left(h_t, a_t, u_t, \varphi\right). \tag{1.111}$$

The architecture of the RSSM for implementing equations (1.109)–(1.111) is shown in Figure 1.8(b). Equation (1.108) can be implemented by the following RNN:

$$\begin{bmatrix} h_t \\ a_t \end{bmatrix} = F_h\left(\begin{bmatrix} W_{hh}^1 \\ W_{hh}^2 \end{bmatrix} h_{t-1} + \begin{bmatrix} W_{yh}^1 \\ W_{yh}^2 \end{bmatrix}\begin{bmatrix} y_t \\ \vdots \\ y_{t-k+1} \end{bmatrix} + \begin{bmatrix} W_{uh}^1 \\ W_{uh}^2 \end{bmatrix} u_t + \begin{bmatrix} b_h^1 \\ b_h^2 \end{bmatrix}\right), \quad (1.112)$$

where the parameters θ and β are represented by the weight matrices and bias vectors.

Equation (1.112) can be implemented by a two-layer MLP:

$$Y_{t+1} = \sigma_2\left(W_{zh}^2\sigma_1\left(W_{zh}^1 h_t + W_{za}a_t + W_{zu}u_t + b_1\right) + b_2\right), \quad (1.113)$$

or

a linear transformation

$$Y_{t+1} = W_{zh}h_t + W_{za}a_t + W_{zu}u_t + b_1, \quad (1.114)$$

where σ_1, σ_2 are nonlinear activation functions such as ReLU, W_{zh}^2, W_{zh}^1, $W_{za}W_{zh}$, W_{zu} are weight matrices and b_1, b_2 are bias parameters.

1.2 DYNAMIC APPROACH TO DEEP LEARNING

1.2.1 Differential Equations for Neural Networks

Differential equations including ordinary differential equation (Simmon 2016), stochastic differential equations (Särkkä 2012), and partial differential equations (Miersemann 2012) can be used to design the neural networks and estimate their parameters (Xiong and Wang 1992; Chang et al. 2017; Haber and Ruthotto 2017; Haber et al. 2017; Haber and Ruthotto 2017; Long et al. 2017; Lu et al. 2017; Chen et al. 2018). In this section, we will introduce differential equations for residual networks (ResNet), reversible networks (RevNet) (Chang et al. 2018), wide residual networks (Zagoruyko and Komodakis 2017), generative adversarial networks (GANs) (Grover et al. 2018), and generative latent function time-series model (Kingma and Welling 2014; Choi et al. 2016; Toderici et al. 2017). These will lay foundation for optimal control, stability analysis, and robust control of dynamic systems for modeling deep neural networks.

1.2.2 Ordinary Differential Equations for ResNets

A residual neural network that discretizes a dynamic system is called a difference neural network (He et al. 2015, 2016; Chang et al. 2017, 2018; Haber et al. 2017). Consider a residual neural network with two layers (Figure 1.9(a)). Two-layer neural network can be either feedforward neural network or CNN. Let $x_n = \left[x_{n1},\ldots,x_{nr}\right]^T$ be the r-dimensional input vector of variables and $x_{n+1} = \left[x_{n+1\,1},\ldots,x_{n+1\,r}\right]^T$ be the output of the residual network. Let

$$W^{(1)} = \begin{bmatrix} W_{11}^{(1)} & \cdots & W_{1r}^{(1)} \\ \vdots & \vdots & \vdots \\ W_{r1}^{(1)} & \cdots & W_{rr}^{(1)} \end{bmatrix}, \quad b^{(1)} = \begin{bmatrix} b_1^{(1)} \\ \vdots \\ b_r^{(1)} \end{bmatrix}$$

$$W^{(2)} = \begin{bmatrix} W_{11}^{(2)} & \cdots & W_{1r}^{(2)} \\ \vdots & \vdots & \vdots \\ W_{r1}^{(2)} & \cdots & W_{rr}^{(2)} \end{bmatrix}, \quad b^{(2)} = \begin{bmatrix} b_1^{(2)} \\ \vdots \\ b_r^{(2)} \end{bmatrix}$$

be two weight matrices. Let $\sigma = [\sigma,\ldots,\sigma]^T$ be a vector of ReLU function, i.e., $\sigma(x) = \max(0,x)$.

Then, the output x_{n+1} is given by

$$\begin{aligned} x_{n+1} &= x_n + W^{(2)}\sigma\left(W^{(1)}x_n + b^{(1)}\right) + b^{(2)} \\ &= x_n + f\left(x_n, W, b\right), \end{aligned} \quad (1.115)$$

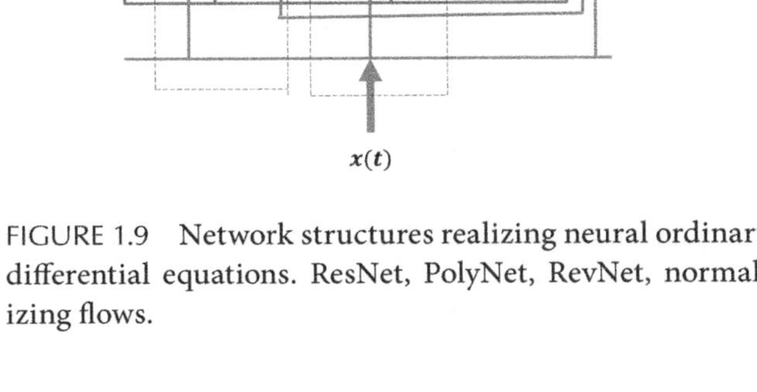

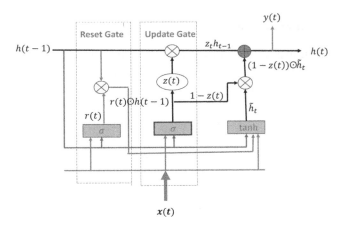

FIGURE 1.9 Network structures realizing neural ordinary differential equations. ResNet, PolyNet, RevNet, normalizing flows.

where

$$f(x_n,W)=\begin{bmatrix} f_1(x_n,W) \\ \vdots \\ f_r(x_n,W) \end{bmatrix}=W^{(2)}\sigma\left(W^{(1)}x_n+b^{(1)}\right)+b^{(2)}.$$

In the limit, equation (1.115) is transformed to ordinary differential equation (ODE):

$$\frac{dx(t)}{dt}=f\left(x(t),\theta\right), \tag{1.116}$$

where $\theta=(W,b)$ and $f\left(x(t),\theta\right)=W^{(2)}\sigma\left(W^{(1)}x(t)+b^{(1)}\right)+b^{(2)}$.

Since function $f\left(x(t),\theta\right)$ is specified by a neural network, equation (1.116) is called a neural ODE (NODE). A node can be realized by a cascade of neural networks (Figure 1.9 (a)) and defines a dynamic system. In the defined dynamic system, the input of the neural networks x_0 is the initial value of the dynamic system, a residual network block with two layers defines function $f\left(x(t),\theta\right)$ and the final solution X_T of the dynamic system (1.116) is the output of the output layer in the cascade of neural networks.

Recursively, equation (1.115) can be written as the summation of the outputs of all preceding residual functions:

$$x_{n+1}=x_0+\sum_{i=0}^{n} f\left(x_i,\theta_i\right). \tag{1.117}$$

Example 1.3: PolyNet

Figure 1.9(b) shows the structure of PolyNet (Zhang et al. 2017). Each residual block can be modeled by

$$x_{n+1}=x_n+F\left(x_n\right)+F\left(F\left(x_n\right)\right).$$

If the block is a feedforward neural network, then the function $F(x)$ is

$$F(X)=W^{(2)}\sigma\left(W^{(1)}x+b^{(1)}\right)+b^{(2)}.$$

Equation (1.116) for PolyNet is then reduced to

$$\frac{dx(t)}{dt}=f\left(x(t),\theta\right),$$

where

$$f\left(x(t),\theta\right)=F(x)+F\left(F(x)\right).$$

If the residual block is the feedforward neural network, then

$$f\left(x(t),\theta\right)=W^{(2)}\sigma\left(W^{(1)}x+b^{(1)}\right)+b^{(2)}$$
$$+W^{(2)}\sigma\left(W^{(1)}\left(W^{(2)}\sigma\left(W^{(1)}x+b^{(1)}\right)+b^{(2)}\right)+b^{(1)}\right)+b^{(2)}. \tag{1.118}$$

1.2.3 Ordinary Differential Equations for Reversible Neural Networks

1.2.3.1 Stability of Dynamic Systems

Reversible neural networks allow the simulation of the dynamic going from the final time to the initial time, and vice versa (Chang et al. 2017). The reversible neural networks are related to the stability of the neural networks or dynamic systems. Instable of the dynamic systems can be caused by (1) perturbation of the initial state or (2) the perturbation of the input. There are three major types of definition of stability: (1) Poisson definition of stability, (2) Lyapunov definition of stability, and (3) asymptotic stability. Poisson stability means that if the dynamic system is perturbed by small changes in input, after a while, the system will return to an arbitrarily small neighborhood of the initial point $x_0=x\left(t_0\right)$. Poisson stability is important, but cannot characterize the behavior of neighboring trajectories, initially close to $x_0(t)$. To overcome this limitation, the Lyapunov stability is proposed. The trajectory $x_0(t)$. is said to be Lyapunov stable if the initial state $x_0=x\left(t_0\right)$ is perturbed, after while, the system $x(t)$ will return to the neighborhood of the original trajectory $x_0(t)$, i.e., $\left\|x(t)-x_0(t)\right\|<\varepsilon$ for small $\varepsilon>0$. If the system $\left\|x(t)-x_0(t)\right\|\to 0$, as $t\to\infty$ after perturbation vanishes, then the system is called asymptotically stable.

If $f(x^*,t)=0$ for all $t\geq 0$ then x^* is called an equilibrium point of a dynamic system. Without loss of generality, we assume that $x^*=0$ is the equilibrium point of the dynamic system. Consider a nonlinear dynamic system:

$$\frac{dx}{dt}=f(x,t),\ x^*=0 \text{ is the equilibrium point.} \tag{1.119}$$

Assume that $f(x,t)$ can be expanded by a Taylor expansion:

$$f(x,t)=J(t)x+R(x), \tag{1.120}$$

where

$$J(t)=\frac{\partial f}{\partial x^T}|_{x^*=0} \text{ and } \frac{\partial f}{\partial x^T}=\begin{bmatrix} \frac{\partial f_1}{\partial x_1} & \cdots & \frac{\partial f_1}{\partial x_r} \\ \vdots & \vdots & \vdots \\ \frac{\partial f_r}{\partial x_1} & \cdots & \frac{\partial f_r}{\partial x_r} \end{bmatrix}.$$

Substituting equation (1.120) into equation (1.119), we obtain

$$\frac{dx}{dt}=J(t)x+R(x). \tag{1.121}$$

The first Lyapunov criterion for stability can be stated as the theorem 1.1.

Theorem 1.1: First Lyapunov Criterion for Stability

1. If all the eigenvalues of the Jacobin matrix $J(t)$ have negative real part, then the equilibrium point x^* is asymptotically stable and is not related with $R(x)$;
2. If at least one of the eigenvalues of the matrix $J(t)$ has positive real part, then the equilibrium point x^* is unstable and is not related with $R(x)$;
3. If at least one of the eigenvalues of the matrix $J(t)$ has zero real part, while all the other eigenvalues have negative real part, then it is not possible to conclude anything about the stability of the equilibrium point.

Example 1.4

Consider the nonlinear dynamic system:

$$\frac{dx_1}{dt}=2x_1-x_1x_2$$
$$\frac{dx_2}{dt}=x_1x_2-x_2.$$

The equilibrium points are $x_{e1}^*=[0,0]$ and $x_{e2}^*=[1,2]$. The Jacobian matrix is

$$J=\begin{bmatrix} 2-x_2 & -x_1 \\ x_2 & x_1-1 \end{bmatrix}.$$

The eigenvalues of the Jacobin matrix J at $x_{e1}^*=[0,0]$ are $\lambda_1=2, \lambda_2=-1$. The system at $x_{e1}^*=[0,0]$ is unstable. The eigenvalues of the Jacobin matrix J at $x_{e2}^*=[1,2]$ are $\lambda_1=\sqrt{2}i$, $\lambda_2=-\sqrt{2}i$. It is not possible to conclude anything about the stability of the equilibrium point $x_{e2}^*=[1,2]$.

1.2.3.2 Second Method of Lyapunov

In this section, we introduce the second method or direct method of Lyapunov that allows to determine the stability properties of the dynamic system without directly integrating differential equation.

We first define the Lyapunov function.

Definition 1.1: Lyapunov Function

If a function $V(x): R^n \to R$ satisfies the following conditions:

1. $V(x)=0$ if and only if $x=0$;
2. $V(x)>0$ if and only if $x>0$;
3. $\frac{dV(x)}{t}=\left(\frac{\partial V(x)}{\partial x}\right)^T f(x,t)\le 0$ for all $x\neq 0$.

Now we are ready to introduce the basic Lyapunov Theorem.

Theorem 1.2: Basic Lyapunov Theorem

1. If Lyapunov function exists, then the system is stable in the sense of Lyapunov.
2. If Lyapunov function exists and $\frac{dV(x)}{t}<0$, then the system is asymptotically stable, i.e., every trajectory of the dynamic system converges to zero as $t\to\infty$.
3. If $V(x)$ is positive $(V(0)=0, V(x)>0$, for all $x>0)$ and $\frac{dV(x)}{t}>0$ then the system is unstable.

Example 1.5

Consider a dynamic system:

$$\frac{dx_1}{dt}=x_2,$$
$$\frac{dx_2}{dt}=-x_1-x_2.$$

It is clear that $x_1=0$ and $x_2=0$ are the equilibrium points. The Jacobian matrix is

$$J=\begin{vmatrix} 0 & 1 \\ -1 & -1 \end{vmatrix}.$$

Its eigenvalues are $x_1 = \frac{-1+\sqrt{3}i}{2}$ and $x_2 = \frac{-1-\sqrt{3}i}{2}$. Since the real part of all eigenvalues is negative, then the system is asymptotically stable.

Now consider a quadratic function

$$V(x) = \frac{1}{2}\left[(x_1 + x_2)^2 + 2x_1^2 + x_2^2\right].$$

It is clear that $V(0) = 0$, $V(x) > 0$, for all $x \neq 0$. Now calculate its derivative with respect to time:

$$\frac{dV(x)}{dt} = \begin{bmatrix} (x_1 + x_2)\frac{dx_1}{dt} + 2x_1\frac{dx_1}{dt} \\ +(x_1 + x_2)\frac{dx_2}{dt} + x_2\frac{dx_2}{dt} \end{bmatrix}$$

$$= (x_1 + x_2)\left(\frac{dx_1}{dt} + \frac{dx_2}{dt}\right) + 2x_1\frac{dx_1}{dt} + x_2\frac{dx_2}{dt}$$

$$= -x_1(x_1 + x_2) + 2x_1x_2 - x_2(x_1 + x_2)$$

$$= -(x_1^2 + x_2^2) < 0.$$

This shows that the system is asymptotically stable. This example shows that when the first method of Lyapunov cannot make conclusion, the second method of Lyapunov determines that the system is asymptotically stable.

Example 1.6: Stability Analysis of ResNet

Consider the ODE for the ResNet with a single layer (Haber et al. 2017):

$$\frac{dx}{dt} = \sigma(Wx + b).$$

Its Jacobian matrix is

$$J = \frac{\partial \sigma}{\partial x^T} = \begin{bmatrix} \dot{\sigma}(W_1 x + b)W_1 \\ \vdots \\ \dot{\sigma}(W_r x + b)W_r \end{bmatrix}$$

$$= \begin{bmatrix} \dot{\sigma}(W_1 x + b) & \cdots & 0 \\ \vdots & \vdots & \vdots \\ 0 & \cdots & \dot{\sigma}(W_r x + b) \end{bmatrix} \begin{bmatrix} W_1 \\ \vdots \\ W_r \end{bmatrix}$$

$= \text{diag}(\dot{\sigma})W$, where

$$\text{diag}(\dot{\sigma}) = \begin{bmatrix} \dot{\sigma}(W_1 \boldsymbol{x} + b) & \cdots & 0 \\ \vdots & \vdots & \vdots \\ 0 & \cdots & \dot{\sigma}(W_r \boldsymbol{x} + b) \end{bmatrix},$$

$W = \begin{bmatrix} W_1 \\ \vdots \\ W_r \end{bmatrix}$ and we use the equality $\frac{\partial \boldsymbol{x}}{\partial x^T} = I$.

We can show that the sign of eigenvalues of the matrix J is the same as that of eigenvalues of the weight matrix W (Exercise 1.2). Therefore, the stability of the ResNet depends on the eigenvalues of the weight matrix W. Therefore, without restriction, the eigenvalues of the weight matrix can be positive, negative, and imaginary. Therefore, the ResNet can be stable, unstable, and ill-posed. For example, consider the following three matrices:

$$W_+ = \begin{bmatrix} 1 & -1 \\ 0 & 2 \end{bmatrix}, W_- = \begin{bmatrix} -1 & 0 \\ 5 & -2 \end{bmatrix} \text{ and } W^* = \begin{bmatrix} 0 & -4 \\ 4 & 0 \end{bmatrix}.$$

It is clear that the eigenvalues of W_+ are 1 and 2, the system is unstable, the eigenvalues of W_- are –1 and –2, the system is stable and the eigenvalues of W^* are $2i$ and $-2i$ and the learning of the ResNet under this condition is ill-posed.

1.2.3.2.1 Krasofskii Method for Construction of Lyapunov Function There are no general methods for construction of Lyapunov function. In practice, we often assume that Lyapunov function has a quadratic form. Krasofskii method provides a practical method for construction of Lyapunov function (Sastry 1999).

1.2.3.2.2 Krasofskii Method Consider the dynamic system:

$$\frac{dx}{dt} = f(x) \text{ in } R^r.$$

If there exists $P, Q \in R^{r \times r}$, two positive definite matrices, such that

$$P\frac{\partial f}{\partial x^T} + \left(\frac{\partial f}{\partial x^T}\right)^T P = -Q.$$

Then, $V(x)=f^T(x)Pf(x)$ is a Lyapunov function and $x^*=0$ is globally asymptotically stable. Specifically, we can assume $P=Q=I$. Then, if $F(x)=\frac{\partial f}{\partial x^T}+\frac{\partial f^T}{\partial x}$ is negative definite, then, $V(x)=f(x)^T f(x)$ is a Lyapunov function and $x^*=0$ is globally asymptotically stable.

Proof

We outline a brief proof. In fact, we have

$$\frac{dV(x)}{dt}=\left(\frac{\partial f(x)}{\partial x^T}f(x)\right)^T Pf(x)+f^T(x)P\frac{\partial f(x)}{\partial x^T}f(x)$$

$$=f^T(x)\left[P\frac{\partial f}{\partial x^T}+\left(\frac{\partial f}{\partial x^T}\right)^T P\right]f(x)$$

$$=-f^T(x)f(x)<0.$$

1.2.3.3 Lyapunov Exponent

To quantify the Lyapunov stability, we use the Lyapunov exponent that measures the rate of separation of $x(t)$ and $x_0(t)$.

Let $\delta x_0=x(0)-x_0(0)$ be the initial separation between two trajectories and $\delta x(t)=x(t)-x_0(t)$ be the divergence between two trajectories at the time t. We want to measure the rate

$$\log\frac{\|\delta x(t)\|}{\|\delta x_0\|}. \quad (1.122)$$

Recall that the dynamic system is defined by

$$\frac{dx}{dt}=f(x,t), \quad (1.123)$$

$$x(t_0)=x_0.$$

Let $J(t)=\frac{\partial f}{\partial x^T}$ be the Jacobian matrix of the nonlinear function $f(x,t)$. Define the transition matrix

$$\frac{d\Phi(t,t_0)}{dt}=J(t)\Phi(t,t_0),\ \Phi(t_0,t_0)=I. \quad (1.124)$$

Assume that singular valued decomposition (SVD) of the transition matrix is

$$\Phi(t,t_0)=W\Lambda V^T.$$

It is clear that

$$\Phi(t,t_0)^T\Phi(t,t_0)=V\Lambda^2 V$$

and v_i is the eigenvector of the matrix $\Phi(t,t_0)^T\Phi(t,t_0)$. Now we define the Lyapunov exponent.

Definition 1.2: Lyapunov Exponent

The Lyapunov exponent of the dynamic system (1.123) is defined as

$$\lambda_i=\lim_{t\to\infty}\frac{1}{t-t_0}\log\|\Phi(t,t_0)v_i(t_0)\|,\ i=1,2,\ldots,r \quad (1.125)$$

or

the eigenvalues of the matrix Λ, where

$$\Lambda=\lim_{t\to\infty}\frac{1}{2(t-t_0)}\log\left[\Phi(t,t_0)^T\Phi(t,t_0)\right]. \quad (1.126)$$

The largest of λ_i $(i=1,\ldots,r)$ is called the maximal Lyapunov exponent or defined as

$$\lambda=\lim_{\to\infty}\lim_{\delta x_0\to 0}\frac{1}{t}\log\frac{\|\delta x(t)\|}{\|\delta x(0)\|}. \quad (1.127)$$

1.2.3.4 Reversible ResNet

It is well known that a two-layer neural network can approximate any monotonically-increasing continuous function (Hornik 1991). Chang et al. (2017) proposed two-layer reversible deep residual neural networks to improve stability and generalization of the ResNet. Consider two-layer reversible ResNet (Fig 1.9c):

$$\frac{d\mathbf{x}}{dt}=\left(W^{(1)}(t)\right)^T\sigma\left(W^{(1)}(t)\mathbf{y}(t)+b_1(t)\right)$$

$$\frac{d\mathbf{y}}{dt}=-\left(W^{(2)}(t)\right)^T\sigma\left(W^{(2)}(t)\mathbf{x}(t)+b_2(t)\right),$$

or

$$\begin{bmatrix}\frac{d\mathbf{x}}{dt}\\ \frac{d\mathbf{y}}{dt}\end{bmatrix}=\begin{bmatrix}\left(W^{(1)}(t)\right)^T & 0\\ 0 & -\left(W^{(2)}(t)\right)^T\end{bmatrix}\sigma\left(\begin{bmatrix}0 & W^{(1)}(t)\\ W^{(2)}(t) & 0\end{bmatrix}\begin{bmatrix}\mathbf{x}\\ \mathbf{y}\end{bmatrix}+\begin{bmatrix}b_1(t)\\ b_2(t)\end{bmatrix}\right), \quad (1.128)$$

where $W^{(1)}(t)$ and $W^{(2)}(t)$ are weight matrices, and $b_1(t)$ and $b_2(t)$ are vectors of bias.

The Jacobian matrix of dynamic system (1.128) is given by

$$J=\begin{bmatrix} \left(W^{(1)}(t)\right)^T & 0 \\ 0 & -\left(W^{(2)}(t)\right)^T \end{bmatrix} \operatorname{diag}(\dot{\sigma})\begin{bmatrix} 0 & W^{(1)}(t) \\ W^{(2)}(t) & 0 \end{bmatrix}, \quad (1.129)$$

where

$$\operatorname{diag}(\dot{\sigma})=\begin{bmatrix} \dot{\sigma} & \cdots & 0 \\ \vdots & \vdots & \vdots \\ 0 & \cdots & \dot{\sigma} \end{bmatrix} \text{ and using } \frac{\partial\begin{bmatrix} x \\ y \end{bmatrix}}{\partial\begin{bmatrix} x^T & y^T \end{bmatrix}}=I.$$

Chang et al. (2017) showed that all eigenvalues of J are imaginary and RevNet is stable and well-posed.

1.2.3.5 Residual Generative Adversarial Networks

Residual neural networks can be used design both generator and discriminator in generative adversarial networks (GANs) (Mehrotra and Dukkipati 2017; Shen and Liu 2017; Zhao et al. 2018). Figure 1.10 shows the structure of residual generative adversarial networks (RGANs), where NN denotes the neural network and both generator and discriminator consist of residual neural networks.

Let $x_n = z$. Then, in the generator, each residual block has two layers of neural networks. The output x_{n+1} and input x_n have the following relationship:

$$x_{n+1}=x_n+W^{(2)}\sigma\left(W^{(1)}x_n+b^{(1)}\right)+b^{(2)}. \quad (1.130)$$

In the limit, we obtain

$$\frac{dx(t)}{dt}=f\left(x(t),\theta\right), \quad (1.131)$$

where $\theta=(W,b)$ and $f\left(x(t),\theta\right)=W^{(2)}\sigma\left(W^{(1)}x(t)+b^{(1)}\right)+b^{(2)}$.

Similarly, we can define the following differential equation for the ResNet in the discriminator:

$$\frac{dy(t)}{dt}=g\left(y(t),\phi\right), \quad (1.132)$$

where $\phi=(W,b)$ and $g\left(y(t),\phi\right)=W^{(4)}\sigma\left(W^{(3)}y(t)+b^{(3)}\right)+b^{(4)}$.

Now we investigate the loss function for the RGANs. The loss function can be defined as

$$\min_{\theta}\max_{\phi} V(D,G)=E_{y\sim P_{data}(y)}\left[\log D(y,\phi)\right]+E_{x\sim P_x(x)}\left[\log\left(1-D\left(G(x,\theta)\right)\right)\right], \quad (1.133)$$

where $D(y,\phi)$ represents the probability that y came from the data rather than faked sample and is calculated as the output of the ResNet for the discriminator, and $G(x,\theta)$ denotes a transformation function and is calculated as the output of the ResNet for the generator.

The expectations in equation (1.133) will be approximated by their sampling formula. The optimization problem (1.133) can be decomposed into the following two problems. The first problem is to assume that m noise sample points $\left\{x^{(1)},\ldots,x^{(m)}\right\}$ and data points $\left\{y^{(1)},\ldots,y^{(m)}\right\}$ are sampled. Then, the optimization

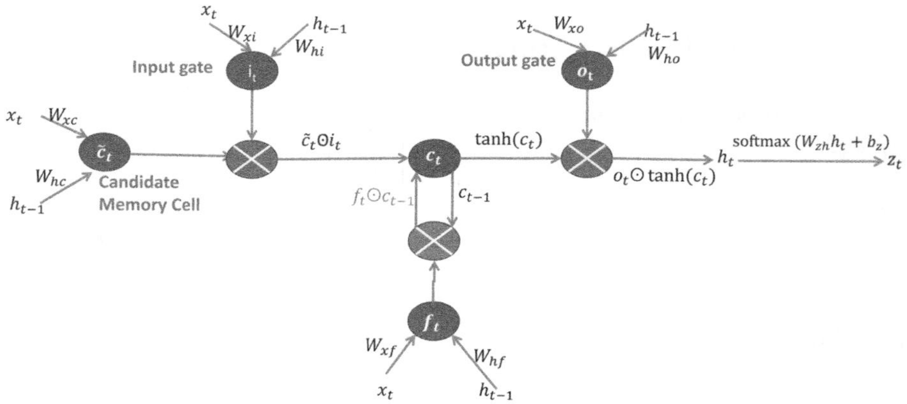

FIGURE 1.10 Structure of residual generative adversarial networks.

problem (1.133) can be first approximated by the following problem:

$$\max_{\phi} V(D, G_*) = \frac{1}{m}\sum_{i=1}^{m}\begin{bmatrix}\log D(y^{(i)}, \phi) \\ + \log\left(1 - D(G(x^{(i)}, \theta_*), \phi)\right)\end{bmatrix}$$

$$\max_{\phi} V(D, G_*) = \frac{1}{m}\sum_{i=1}^{m}\begin{bmatrix}\log Y_T\left(y^{(i)}, \phi\right) \\ + \log\left(1 - Y_T\left(G_T\left(x^{(i)}, \theta_*\right), \phi\right)\right)\end{bmatrix}, \quad (1.134)$$

where $Y_T\left(y^{(i)}, \phi\right)$ is the final value of solution to the ODE (1.132) (discriminator) with initial values $y^{(i)}$ and input ϕ (parameters), $Y_T\left(G\left(x^{(i)}, \theta_*\right), \phi\right)$ is the final value of the solution to the ODE (1.132) with initial values $G_T\left(x^{(i)}, \theta_*\right)$ and input ϕ, and $G_T\left(x^{(i)}, \theta_*\right)$ is the final values of the solution to the ODE (1.131) (generator) with initial values $x^{(i)}$ and input θ_*.

The second problem is to assume that m noise sample points $\left\{x^{(1)}, \ldots, x^{(m)}\right\}$ are sampled. Then, we solve the following optimization problem:

$$\min_{\theta} C(G) = \frac{1}{m}\sum_{i=1}^{m}\log\left(1 - Y_T\left(G_T\left(x^{(i)}, \theta\right), \phi_*\right)\right), \quad (1.135)$$

where $G_T\left(x^{(i)}, \theta\right)$ is the final values of the solution (output) to the ODE (1.131) (generator) with initial values $x^{(i)}$ and input θ, and $Y_T\left(G\left(x^{(i)}, \theta\right), \phi_*\right)$ is the final value of the solution to the ODE (1.132) with initial values $G_T\left(x^{(i)}, \theta\right)$ and input ϕ_*.

1.2.3.6 Normalizing Flows

Consider K invertible mappings (invertible transformations) f_k, i.e., $f_k : R^d \to R^d$, $k = 1, \ldots, K$ with inverse f_k^{-1} such that $f_k^{-1} \circ f_k = f_k^{-1}\left(f_k(z)\right) = z$. Let $p(z_0)$ be the distribution of the random variable z_0. The random variable z_K is obtained by successively transforming a random variable z_0 via a sequence of invertible transformation f_k (Figure 1.9(d)):

$$z_K = f_K \circ f_{K-1} \circ \ldots \circ f_2 \circ f_1(z_0).$$

Now we first calculate the distribution $z_1 = f_1(z_0)$. Using the change of variable theorem, we obtain

$$p(z_1) = p(z_0)\left|\frac{\partial f_1}{\partial z^T}\right|^{-1}, \quad (1.136)$$

where $\frac{\partial f}{\partial z^T}$ is the Jacobian matrix of invertible function and $|.|$ denotes the determinant of the matrix.

Similarly, we have $p(z_2) = p(z_1)\left|\frac{\partial f_2}{\partial z^T}\right|^{-1} = p(z_0)\left|\frac{\partial f_1}{\partial z^T}\right|^{-1}\left|\frac{\partial f_2}{\partial z^T}\right|^{-1}$.

Therefore, using the same argument, we obtain

$$\log p(z_K) = \log p(z_0) - \sum_{k=1}^{K}\log\left|\frac{\partial f_k}{\partial z_{k-1}}\right|. \quad (1.137)$$

Using the similar argument, we obtain

$$\log p(z(t+1)) = \log p(z(t)) - \log\frac{\partial f}{\partial z(t)}. \quad (1.138)$$

1.3 OPTIMAL CONTROL FOR DEEP LEARNING

1.3.1 Mathematic Formulation of Optimal Control

Estimation problem in neural networks can be formulated as an optimal control problem where the parameters are taken as control signals. The optimal control problem consists of three parts: (1) boundary conditions: $x(t_0) = x_0$ and $x(t_f) = x_T$, where t_0 and t_f are the initial time and terminal time, respectively, or the constraint set $S = \left\{x \mid g\left(x, t_f\right) = 0\right\}$; (2) control function $\theta(t) \in R^r$ that satisfies constraints $\theta(t) \in U_r$; and (3) the cost function $l(\theta(.))$ (Appendix 1.C).

Consider a neural ODE for a neural network:

$$\frac{dx}{dt} = f(x, t, \theta), \quad (1.139)$$

where $x = [x_1, \ldots, x_m]^T$ is a vector to represent the state of dynamic system, $f = [f_1, \ldots, f_m]^T$ is a vector of nonlinear functions, t is time, and $\theta = [\theta_1, \ldots, \theta_r]^T$ is a vector of parameters in the neural networks or the control function in the dynamic system. Our goal is to choose (x, θ) to minimize or maximize a cost function subject to boundary conditions and constraints of control function.

Cost functional can be in general defined in three different forms that can be converted from one to another.

1. *Integral form (Lagrange form)*
   Let $L : R^m \times R \times R^r \to R$ be a smooth function and $J(x)$ is a performance index or an objective

function. Then, the integral type of performance index is defined as

$$J(\theta)=\int_{t_0}^{t_f} L(x(t),t,\theta(t))dt. \tag{1.140}$$

2. *Terminal value form (Mayer form)*
Mathematic representation of Mayer form is defined as

$$J(\theta)=\Phi\left(x\left(t_f\right),t_f,\theta\left(t_f\right)\right), \tag{1.141}$$

where $\Phi: R^m \times R \times R^r \to R$ is a function.

3. *General form (Bolza form)*
Combining *Lagrange form* and *Mayer form,* we obtain a general form of performance index or objective function of optimal control defined as

$$J(\theta)=\Phi\left(x\left(t_f\right),t_f\right)+\int_{t_0}^{t_f} L(x(t),t,\theta(t))dt. \tag{1.142}$$

These three forms can be converted to each other.

A general nonlinear optimal control problem for the parameter estimation of neural network can be formulated as follows:

$$\begin{aligned}
&\min_{\theta} J(\theta)=\Phi\left(x\left(t_f\right),t_f\right)+\int_{t_0}^{t_f} L(x(t),t,\theta(t))dt\\
&\text{s.t. } \frac{dx}{dt}=f(x,t,\theta),\\
&x(t_0)=x_0,\ \Psi\left(x\left(t_f\right),t_f\right)=0,\\
&\theta(t)\in U_r,\\
&g(x(t),\theta(t),t)\geq 0,\ \left[\dot{Z}(t)\right]^2=g,\ Z(t_0)=0,\\
&\dot{w}(t)=\theta(t),\ w(t_0)=0.
\end{aligned} \tag{1.143}$$

1.3.2 Pontryagin's Maximum Principle

The Pontryagin's Maximum Principle can be used to solve the optimal control problem (Pontryagin 1987) (Appendix 1C). The Pontryagin's Maximum Principle gives necessary conditions for solving the optimal control problem. Before introducing the Pontryagin's Maximum Principle, we define a Hamiltonian function:

$$H(x,\lambda,u,t)=L(x,u,t)+\lambda^T f(x,u,t), \tag{1.144}$$

where λ is a multiplier vector or adjoint vector. Now we introduce the Pontryagin's Maximum Principle.

Theorem 1.3: Pontryagin's Maximum Principle

The Pontryagin's Maximum Principle gives a first order necessary condition for optimality. Let $\tilde{u}(t)$ be an optimal control, $\tilde{x}(t)$ be its corresponding optimal trajectory, and $\tilde{\lambda}(t)$ be an adjacent vector. Then, $\tilde{u}(t)$, $\tilde{x}(t)$, and $\tilde{\lambda}(t)$ must satisfy the following conditions:

1. Nominal equations

$$\frac{d\tilde{x}}{dt}=\frac{\partial}{\partial\lambda}H\left(\tilde{x}(t),\tilde{\lambda}(t),\tilde{u}(t),t\right), \tag{1.145}$$

$$\frac{d\tilde{\lambda}}{dt}=-\frac{\partial}{\partial x}H\left(\tilde{x}(t),\tilde{\lambda}(t),\tilde{u}(t),t\right), \tag{1.146}$$

2. Optimal control $\tilde{u}(t)$ must be minimizer of the Hamiltonian function on the optimal trajectory $\tilde{x}(t)$:

$$H\left(\tilde{x}(t),\tilde{\lambda}(t),\tilde{u}(t),t\right)\leq \min_{u\in\Omega} H\left(\tilde{x}(t),\tilde{\lambda}(t),u(t),t\right), \tag{1.147}$$

where Ω is a set of admissible control.

If Hamiltonian function is differentiable, then the optimal condition for $\tilde{u}(t)$ is

$$\frac{\partial}{\partial u(t)}H\left(\tilde{x}(t),\tilde{\lambda}(t),u(t),t\right)=0. \tag{1.148}$$

3. Terminal condition:

$$\left\{H\left(\tilde{x}(t),\tilde{\lambda}(t),\tilde{u}(t),t\right)+\frac{\partial\Phi}{\partial t_f}+\frac{\partial\Psi^T}{\partial t_f}v\right\}\bigg|_{t_f}=0. \tag{1.149}$$

4. Transversality condition of adjacent vector $\tilde{\lambda}(t)$:

$$\tilde{\lambda}\left(t_f\right)=\left\{\frac{\partial\Phi}{\partial x}+\frac{\partial\Psi^T}{\partial x}v\right\}\bigg|_{t_f}. \tag{1.150}$$

5. Boundary conditions for the optimal trajectory:

$$x(t_0)=x_0, \tag{1.151}$$

$$\Psi\left(x\left(t_f\right),t_f\right)=0. \tag{1.152}$$

1.3.3 Optimal Control Approach to Parameter Estimation

Learning representation of neural networks can be formulated as learning dynamic systems (Ayed et al. 2019). Neural ODEs and optimal controls are major tools for dynamic approach to neural networks (Chen et al. 2018; Li and Hao 2018; Dupont et al. 2019; Liu et al. 2019; Quaglino et al. 2019). In this section, we formulate deep learning as an optimal control problem and use Pontryagin's Maximum Principle to derive necessary conditions for optimality and develop training algorithms for parameter estimation.

In Section 1.2, a neural network can be formulated as a dynamic system:

$$\frac{dx}{dt} = f(x,\theta,t), \quad x(t_0) = x_0, \tag{1.153}$$

where the parameters θ are used as control signals.

The terminal time of the systems is denoted by t_f. Our goal is to optimize a scalar-valued loss function:

$$L\left(x(t_f)\right) = L\left(\int_{t_0}^{t_f} f(x,\theta,t)dt\right). \tag{1.154}$$

Let

$$Y(t) = \begin{bmatrix} x(t) \\ \theta \\ t \end{bmatrix}.$$

Define an augmented dynamic system (Chen et al. 2018):

$$\dot{Y}(t) = \begin{bmatrix} \dot{x}(t) \\ \dot{\theta} \\ \dot{t} \end{bmatrix} = F(Y,\theta,t) = \begin{bmatrix} f(x,\theta,t) \\ 0 \\ 1 \end{bmatrix}. \tag{1.155}$$

Let $\lambda = \begin{bmatrix} \lambda_x & \lambda_\theta & \lambda_t \end{bmatrix}$ be a vector of adjoint states. Assume that a Hamiltonian function is given by

$$H(y,\lambda,\theta,t) = \lambda F(Y,\theta,t) = \begin{bmatrix} \lambda_x & \lambda_\theta & \lambda_t \end{bmatrix} \begin{bmatrix} f(x,\theta,t) \\ 0 \\ 1 \end{bmatrix}. \tag{1.156}$$

Using equations (1.146) and (1.156), we obtain

$$\begin{aligned} \frac{d\lambda}{dt} &= -\frac{\partial H}{\partial Y^T} \\ &= -\begin{bmatrix} \lambda_x & \lambda_\theta & \lambda_t \end{bmatrix} \begin{bmatrix} \frac{\partial f(x,\theta,t)}{\partial x} & \frac{\partial f(x,\theta,t)}{\partial \theta} & \frac{\partial f(x,\theta,t)}{\partial t} \\ 0 & 0 & 0 \\ 0 & 0 & 0 \end{bmatrix}, \end{aligned} \tag{1.157}$$

which implies that

$$\frac{d\lambda_x}{dt} = -\lambda_x \frac{\partial f(x,\theta,t)}{\partial x}, \tag{1.158}$$

$$\frac{d\lambda_\theta}{dt} = -\lambda_x \frac{\partial f(x,\theta,t)}{\partial \theta}, \tag{1.159}$$

$$\frac{d\lambda_t}{dt} = -\lambda_x \frac{\partial f(x,\theta,t)}{\partial t}. \tag{1.160}$$

Connecting adjoint states with the loss function, we define

$$\lambda_x(t) = -\frac{\partial L}{\partial x(t)}, \tag{1.161}$$

$$\lambda_\theta(t) = \frac{\partial L}{\partial \theta(t)}, \tag{1.162}$$

$$\lambda_t(t) = \frac{\partial L}{\partial t(t)}. \tag{1.163}$$

Combing equations (1.159) and (1.162), we obtain

$$\frac{\partial L}{\partial \theta(t)} = -\int_{t_0}^{t_f} \lambda_x(t) \frac{\partial f(x,\theta,t)}{\partial \theta} dt = \int_{t_f}^{t_0} \lambda_x(t) \frac{\partial f(x,\theta,t)}{\partial \theta} dt. \tag{1.164}$$

Similarly, combining equations (1.160) and (1.163), we obtain

$$\frac{\partial L}{\partial t(t_f)} = -\lambda_x(t_f) \frac{\partial f\left(x(t_f),\theta,t_f\right)}{\partial t_f}, \tag{1.165}$$

$$\frac{\partial L}{\partial t(t_0)} = -\int_{t_0}^{t_f} \lambda_x(t) \frac{\partial f(x(t),\theta,t)}{\partial t} dt. \tag{1.166}$$

Equation (1.158) can be easily derived from equation (1.161). Since loss function $L(x(t))$ is a composite function of time t. Using chain rule, we obtain

$$\frac{\partial L}{\partial t}=\left(\frac{\partial L}{\partial x}\right)^T\frac{\partial x}{\partial t}. \quad (1.167)$$

Substituting equation (1.153) into equation (1.167) yields

$$\frac{\partial L}{\partial t}=(\lambda_x)^T f(x,\theta,t). \quad (1.168)$$

Taking derivative with respect to x on both sides of equation (1.168), we obtain

$$\frac{\partial^2 L}{\partial x\partial t}=(\lambda_x)^T\frac{\partial f(x,\theta,t)}{\partial x}. \quad (1.169)$$

However,

$$\frac{\partial^2 L}{\partial x\partial t}=\frac{\partial}{\partial t}\left(\frac{\partial L}{\partial x}\right)=-\frac{d\lambda_x}{dt}. \quad (1.170)$$

Substituting equation (1.170) into equation (1.169), we obtain equation (1.158):

$$\frac{d\lambda_x}{dt}=-\lambda_x\frac{\partial f(x,\theta,t)}{\partial x}.$$

In summary, we derive thee equations for the adjoint vectors and one equation for the dynamic system:

$$\frac{d\lambda_x}{dt}=-\lambda_x^T\frac{\partial f(x(t),t,\theta)}{\partial x}, \quad (1.171)$$

$$\frac{\partial L}{\partial\theta(t)}=\int_{t_f}^{t_0}\lambda_x(t)\frac{\partial f(x,\theta,t)}{\partial\theta}dt, \quad (1.172)$$

$$\frac{\partial L}{\partial t(t_f)}=-\lambda_x(t_f)\frac{\partial f(x(t_f),\theta,t_f)}{\partial t_f}, \quad (1.173)$$

$$\frac{dx}{dt}=f(x,\theta,t), \quad (1.174)$$

$$\frac{\partial L}{\partial t(t_0)}=\int_{t_f}^{t_0}\lambda_x(t)\frac{\partial f(x(t),\theta,t)}{\partial t}dt. \quad (1.175)$$

$$\lambda_x(t_f)=\frac{\partial L}{\partial x(t_f)}. \quad (1.176)$$

The major procedures for solving the four ODEs are summarized in the Algorithm 1.1 (Chen et al. 2018).

Algorithm 1.1

Input of parameters and initial values:
Parameters θ in the dynamic systems, start time t_0, stop time t_f, final state $x(t_f)$, gradient of loss function $\frac{\partial L}{\partial x(t_f)}$.

Step 1: Compute initial values

$$\frac{\partial L}{\partial t_f}=-\left(\frac{\partial L}{\partial x(t_f)}\right)^T\frac{\partial f(x,\theta,t)}{\partial t_f},$$

$$Y_{t_f}=\begin{bmatrix} x(t_f)\\ \frac{\partial L}{\partial x(t_f)}\\ 0\\ \frac{\partial L}{\partial t_f}\end{bmatrix}.$$

Step 2: Solve the equation

$$\dot{Y}=\begin{bmatrix}\dot{x}\\ \dot{\lambda}_x\\ \frac{\partial L}{\partial\theta}\\ \frac{\partial L}{\partial t}\end{bmatrix}=\begin{bmatrix} f(x,\theta,t)\\ -\lambda_x^T\frac{\partial f(x(t),t,\theta)}{\partial x}\\ \int_{t_f}^{t_0}\lambda_x(t)\frac{\partial f(x,\theta,t)}{\partial\theta}dt\\ \int_{t_f}^{t}\lambda_x(t)\frac{\partial f(x(t),\theta,t)}{\partial t}dt\end{bmatrix}.$$

Step 3: Output all gradients

$$\frac{\partial L}{\partial x(t_0)},\frac{\partial L}{\partial\theta},\frac{\partial L}{\partial t_0},\frac{\partial L}{\partial t_f}.$$

1.3.4 Learning Nonlinear State Space Models

1.3.4.1 Joint Estimation of Parameters and Controls

Optimal control approach can be applied to parameter estimation and identification of nonlinear state space models, which can be taken as a new data-driven paradigm for learning dynamical systems (Ayed et al. 2019;). We formulate the problem of learning nonlinear state space models as a continuous-time optimal control problem, where the nonlinear dynamic systems are modeled by neural networks and parameters of the neural network are viewed as control variables.

The nonlinear state transition equations can be modeled by the following ODEs:

$$\frac{dx}{dt} = f(x,u_1,\theta,t), \qquad (1.177)$$
$$x(t_0) = x_0,$$

where $x(t) \in R^d$ is a vector of state variables, $\theta \in R^q$ is a vector of parameters, $u_1(t) \in R^p$ is a vector of control variables, t_0 is an initial time, and x_0 is a vector of initial values of states. The initial value can also be specified by the constraint:

$$x(t_0) = g(Y_{-k}, \hat{x}_0, \theta). \qquad (1.178)$$

The states are partially observed. Consider the following observation equation in the nonlinear state space model:

$$y(t) = H(x, u_1\ \theta, t), \qquad (1.179)$$

where $y(t)$ is a vector of observed variables.

Consider a sum of square of errors as a loss function:

$$J_\theta(y,\hat{y},\theta,u_1) = \int_{t_0}^{t_f} \|y(t) - \hat{y}(x,u_1,\theta,t)^2\, dt, \qquad (1.180)$$

where $\hat{y}(t, u_1, \theta)$ is the output of the system.

The second loss function that accounts for the cost of control signal $u_1(t)$ is defined as

$$J_u^*(u_1) = \int_{t_0}^{t_f} J_u(x,u_1,t)dt. \qquad (1.181)$$

The Lagrangian is given by

$$L(x,\lambda,u_1,\theta) = J_\theta(y,\hat{y},\theta,u_1) + J_u(u_1) + \int_{t_0}^{t_f} \left[\lambda^T f(x,u_1,\theta,t) - \dot{x}\right]dt + \mu^T\left(g(Y_{-k},\hat{x}_0,\theta) - x_0\right). \qquad (1.182)$$

The Hamiltonian function is

$$H(x,\lambda,u_1,\theta,t) = \|y(t) - \hat{y}(x,u_1,\theta,t)\|^2 + J_u(x,u_1,t) + \lambda^T f(x,u_1,\theta,t). \qquad (1.183)$$

Using equation (1.145), we obtain

$$\dot{\lambda} = -\frac{\partial H}{\partial x} = 2\left(\hat{y}(x,u_1,\theta,t) - y(t)\right)\frac{\partial}{\partial x}\hat{y}(x,u_1,\theta,t) + \frac{\partial}{\partial x}J_u(x,u,t) + \lambda^T\frac{\partial}{\partial x}f(x,u_1,\theta,t). \qquad (1.184)$$

Using equation (1.148), we obtain

$$\frac{\partial H}{\partial u_1} = 2\left(\hat{y}(x,u_1,\theta,t) - y(t)\right)\frac{\partial}{\partial u_1}\hat{y}(x,u_1,\theta,t) + \frac{\partial}{\partial u_1}J_u(x,u,t) + \lambda^T\frac{\partial}{\partial u_1}f(x,u_1,\theta,t) = 0$$

or

$$\frac{\partial}{\partial u_1}J_u(x,u,t) = -\begin{bmatrix} 2\left(\hat{y}(x,u_1,\theta,t) - y(t)\right) \\ \frac{\partial}{\partial u_1}\hat{y}(x,u_1,\theta,t) \\ +\lambda^T\frac{\partial}{\partial u_1}f(x,u_1,\theta,t) \end{bmatrix}. \qquad (1.185)$$

Again, using optimality condition, we obtain

$$\frac{\partial L}{\partial \theta} = \frac{\partial J_\theta}{\partial \theta} + \int_{t_0}^{t_f}\lambda^T\frac{\partial f(x,u_1,\theta,t)}{\partial \theta}dt + \mu^T\frac{\partial g(Y_{-k},\hat{x}_0,\theta)}{\partial \theta} = 0, \qquad (1.186)$$

which implies

$$\frac{\partial J_\theta}{\partial \theta} = -\int_{t_0}^{t_f}\lambda^T\frac{\partial f(x,u_1,\theta,t)}{\partial \theta}dt - \mu^T\frac{\partial g(Y_{-k},\hat{x}_0,\theta)}{\partial \theta}, \qquad (1.187)$$

Procedures for learning nonlinear state space models using optimal control approach are summarized in Algorithm 1.2 (Ayed et al. 2019).

Algorithm 1.2

Input: Training samples $\left(Y_{-k}, \hat{x}_0, y_{t_0}, y_{t_0+1},\ldots,y_{t_f}\right)$

Step 1: Initialization $\theta_0, u_1^0(t), t=t_0,\ldots,t_f$.
If the state variables are all observed then

$$x_0 \leftarrow \hat{x}_0$$

else

$$x_0 \leftarrow g\left(Y_{-k}, \hat{x}_0,\theta\right)$$

end

$$m \leftarrow 0$$

Step 2: While not convergence for parameter θ **do**
Solve transition differential equation:

$$\frac{dx}{dt} = f\left(x,u_1^k(t),\theta^k,t\right),$$

$$x(t_0) = x_0, t = t_0,\ldots,t_f.$$

Solve adjoint state equation:

$$\dot{\lambda} \doteq 2\left(\hat{y}\left(x,u_1^k(t),\theta^k,t\right) - y(t)\right)\frac{\partial}{\partial x}\hat{y}\left(x,u_1^k(t),\theta^k,t\right)$$

$$+\frac{\partial}{\partial x}J_u\left(x,u_1^k(t),t\right) + \lambda^T\frac{\partial}{\partial x}f\left(x,u_1^k(t),\theta^k,t\right).$$

Step 3: For $t = t_0 \ldots t_f$

$$l \leftarrow 0$$

Step 4: While not convergence for control signal $u_1(t)$ **do**
Calculate gradient:

$$\frac{\partial}{\partial u_1}J_u\left(x,u_1^l,t\right) = -\begin{bmatrix} 2\left(\hat{y}\left(x,u_1^l(t),\theta^k,t\right) - y(t)\right) \\ \frac{\partial}{\partial u_1}\hat{y}\left(x,u_1^l(t),\theta,t\right) \\ +\lambda^T\frac{\partial}{\partial u_1}f\left(x,u_1^l(t),\theta,t\right) \end{bmatrix}$$

Update $u_1(t)$

$$u_1^{l+1}(t) = u_1^l(t) - \gamma\frac{\partial}{\partial u_1}J_u\left(x,u_1^l,t\right).$$

$$l \leftarrow l+1.$$

end **while** $u_1^m(t) \leftarrow u_1^{l+1}(t)$.

Step 5: Calculate gradient:

$$\left.\frac{\partial J_\theta}{\partial \theta}\right|_{\theta^m} = -\int_{t_0}^{t_f}\lambda^T\frac{\partial f\left(x,u_1^m(t),\theta^m,t\right)}{\partial\theta}dt$$

$$-\mu^T\frac{\partial g\left(Y_{-k}, \hat{x}_0,\theta^m\right)}{\partial\theta}.$$

Update parameters θ:

$$\theta^{m+1} = \theta^m - \alpha\left.\frac{\partial J_\theta}{\partial\theta}\right|_{\theta^m}.$$

$$m \leftarrow m+1.$$

end while
Output the estimated θ and $u_1(t), t = t_0,\ldots,t_f$.

1.3.4.2 Multiple Samples and Parameter Estimation

In the previous section, we consider one sample and joint estimation of the parameters and control signals. In this section, we consider multiple samples and only estimation of parameters.

Let

$$x_t = \begin{bmatrix} x_t^1 \\ \vdots \\ x_t^S \end{bmatrix}, u_t = \begin{bmatrix} u_t^1 \\ \vdots \\ u_t^S \end{bmatrix}, y_t = \begin{bmatrix} y_t^1 \\ \vdots \\ y_t^S \end{bmatrix},$$

$$f(x,u,\theta) = \begin{bmatrix} f\left(x_t^1, u_t^1, \theta\right) \\ \vdots \\ f\left(x_t^S, u_t^S, \theta\right) \end{bmatrix}.$$

Define system state equation:

$$\dot{x}_t = f(x,u,\theta), \quad (1.188)$$

$$x_{t_0} = x_0, \text{ or}$$

$$x_{t_0} = g\left(Y_{-k}, \hat{x}_0,\theta\right) = \begin{bmatrix} g\left(Y_{-k}^1, \hat{x}_0^1, \theta\right) \\ \vdots \\ g\left(Y_{-k}^S, \hat{x}_0^S, \theta\right) \end{bmatrix}. \quad (1.189)$$

and observation equation:

$$y_t = G\left(x_t,u_t,\beta\right) = \begin{bmatrix} G\left(x_t^1, u_t^1,\beta\right) \\ \vdots \\ G\left(x_t^S,u_t^S,\beta\right) \end{bmatrix}. \quad (1.190)$$

Assume that $\theta \in \Omega$, where Ω is an open set. Define a lost functional of the form:

$$J(x,y,u,\theta)=\frac{1}{S}\sum_{s=1}^{S}\int_{t_0}^{t_f}\left\|y_t^s-G\left(x_t^s,u_t^s,\beta\right)\right\|^2 dt. \quad (1.191)$$

Define a Hamiltonian function:

$$H(x,y,u,\theta,\beta,\lambda)=\frac{1}{S}\sum_{s=1}^{S}\left\|y_t^s-G\left(x_t^s,u_t^s,\beta\right)\right\|^2 + \lambda^T f(x,u,\theta), \quad (1.192)$$

where

$$\lambda=\begin{bmatrix}\lambda_1\\ \vdots\\ \lambda_S\end{bmatrix}. \quad (1.193)$$

Necessary conditions for optimality using the Pontryagin's Maximum Principle are

$$\text{State Equation: } \dot{x}^s = f\left(x_t^s,u_t^s,\theta\right). \quad (1.194)$$

Adjoint Equation:

$$\dot{\lambda}_s=-\frac{\partial H}{\partial x_t^s}=\frac{2}{S}\left(G\left(x_t^s,u_t^s,\beta\right)-y_t^s\right)\frac{\partial G\left(x_t^s,u_t^s,\beta\right)}{\partial x_t^s} + \lambda_s^T\frac{\partial f\left(x_t^s,u_t^s,\theta\right)}{\partial x_t^s}. \quad (1.195)$$

Stationarity Condition:

$$\frac{\partial H}{\partial \beta}=\sum_{s=1}^{S}\left(G\left(x_t^s,u_t^s,\beta\right)-y_t^s\right)\frac{\partial G\left(x_t^s,u_t^s,\beta\right)}{\partial \beta}=0, \quad (1.196)$$

$$\frac{\partial H}{\partial \theta}=\lambda^T\frac{\partial f(x,u,\theta)}{\partial \theta}=0. \quad (1.197)$$

Transversality Conditions: $\lambda^*\left(t_f\right)=0$.

The boundary condition: $x_{t_0}=x_0$, or

$$x_{t_0}=g\left(Y_{-k},\hat{x}_0,\theta\right)=\begin{bmatrix}g\left(Y_{-k}^1,\hat{x}_0^1,\theta\right)\\ \vdots\\ g\left(Y_{-k}^S,\hat{x}_0^S,\theta\right)\end{bmatrix}. \quad (1.198)$$

Equations (1.195) and (1.196) that define local optimum solutions can be solved by the following ODE (Liu 1991):

$$\dot{\beta}=\sum_{s=1}^{S}\left(G\left(x_t^s,u_t^s,\dot{\beta}\right)-y_t^s\right)\frac{\partial G\left(x_t^s,u_t^s,\beta\right)}{\partial \beta}, \quad (1.199)$$

$$\dot{\theta}=\lambda^T\frac{\partial f(x,u,\theta)}{\partial \theta}. \quad (1.200)$$

In Summary, we need to solve the following differential equations:

$$\dot{x}^s = f\left(x_t^s,u_t^s,\theta\right),\ x_{t_0}^s=x_0^s \text{ or } x_{t_0}^s = g\left(Y_{-k}^s,\hat{x}_0^s,\theta\right),\ s=1,\ldots,S. \quad (1.201)$$

$$\dot{\lambda}_s=\frac{2}{S}\left(G\left(x_t^s,u_t^s,\beta\right)-y_t^s\right)\frac{\partial G\left(x_t^s,u_t^s,\beta\right)}{\partial x_t^s} + \lambda_s^T\frac{\partial f\left(x_t^s,u_t^s,\theta\right)}{\partial x_t^s},\ \lambda_s^*\left(t_f\right)=0. \quad (1.202)$$

$$\frac{d\beta}{d\tau}=\sum_{s=1}^{S}\left(G\left(x_t^s,u_t^s,\beta\right)-y_t^s\right)\frac{\partial G\left(x_t^s,u_t^s,\beta\right)}{\partial \beta},\ \beta(0)=\beta_0. \quad (1.203)$$

$$\frac{d\theta}{dt}=\lambda^T\frac{\partial f(x,u,\theta)}{\partial \theta},\ \theta(0) = \theta_0, \text{ randomly initiated vlue.} \quad (1.204)$$

1.3.4.3 Optimal Control Problem

If we assume that dynamic system is learned, our goal is to design optimal strategy to make the cost minimum. Consider the following control problem:

$$\dot{x}=f(x,u,t,\theta), \quad (1.205)$$

$$x\left(t_0\right)=x_0, \quad (1.206)$$

$$\Psi\left(x\left(t_f\right).\,t_f\right)=0,\ t_f \text{ is unknown}, \quad (1.207)$$

$$g(x,u)\geq 0. \quad (1.208)$$

Define the cost function:

$$J(u)=\Phi\left(x\left(t_f\right),t_f\right)+\int_{t_0}^{t_f}L(x,u,t)dt. \quad (1.209)$$

Define Hamiltonian function:

$$H(x,u,\lambda,t)=L(x,u,t)+\lambda^T f(x,u,t,\theta). \quad (1.210)$$

The necessary conditions for the optimality are given by

$$\text{State Equation: } \dot{x} = \frac{\partial H}{\partial \lambda} = f(x,u,t,\theta), \tag{1.211}$$

$$\text{Adjoint Equation: } \dot{\lambda} = -\frac{\partial H}{\partial x} = -\frac{\partial}{\partial x}L(x,u,t) - \lambda^T \frac{\partial f(x,u,t,\theta)}{\partial x}, \tag{1.212}$$

$$\text{Optimal Control: } H(x^*,\lambda^*,u^*,t) = \min_{u\in\Omega} H(x^*,\lambda^*,u,t), \tag{1.213}$$

where $\Omega = \{u(t) \mid g(x,u) \geq 0\}$.

$$\text{Transversality Conditions: } \lambda^*(t_f) = \left.\frac{\partial \Phi(x(t_f),t_f)}{\partial x} + \frac{\partial \Psi(x(t_f).\,t_f)^T}{\partial x} v\right|_{t_f}, \tag{1.214}$$

$$\left. H(x^*,\lambda^*,u^*,t) + \frac{\partial \Phi(x(t_f),t_f)}{\partial t_f} + \frac{\partial \Psi(x(t_f).\,t_f)^T}{\partial t_f} v \right|_{t_f} = 0, \tag{1.215}$$

Boundary Conditions:

$$x(t_0) = x_0, \tag{1.216}$$

$$\Psi(x(t_f).\,t_f) = 0. \tag{1.217}$$

SOFTWARE PACKAGE

Code for "Deep Residual Networks" is posted on the website: https://github.com/KaimingHe/deep-residual-networks

PyTorch Implementation of Differentiable ODE Solvers is published in the website: https://github.com/rtqichen/torchdiffeq

A curated, quasi-exhaustive list of state-of-the-art publications and resources about Generative Adversarial Networks (GANs), their applications and codes are posted on the website: https://github.com/GKalliatakis/Delving-deep-into-GANs

The website: https://github.com/georgezoto/Convolutional-Neural-Networks includes the code for convolutional neural networks.

Code for Recurrent Neural Networks is posted on the website: https://github.com/kjw0612/awesome-rnn

Code for the paper "on complex gated recurrent neural networks" is posted on the website: https://github.com/v0lta/Complex-gated-recurrent-neural-networks

Code for LSTM Neural Network for Time Series Prediction can be downloaded from https://github.com/jaungiers/LSTM-Neural-Network-for-Time-Series-Prediction

APPENDIX 1A: BRIEF INTRODUCTION OF TENSOR CALCULUS

In Appendix 1A, we will briefly introduce tensor algebra and tensor calculus which are adapted from the note written by Kochmann (2017). Tensor calculus can significantly simplify the notations of back propagation.

1A1 Tensor Algebra

1A1.1 Vector and Einstein's Summation Convention

Vector, matrix, and tensor can be expressed in terms of bases. First, we consider an $\boldsymbol{n}$-dimensional vector. Suppose that a set of bases in the $\boldsymbol{n}$-dimensional space is given by

$$S = \{e_1, \ldots, e_n\}.$$

Then, each vector x with components $\{x_i, i = 1, \ldots, n\}$ can be represented as

$$x = x_1 e_1 + \cdots + x_n e_n.$$

If we assume that each basis is represented by

$$e_1 = \begin{bmatrix} 1 \\ 0 \\ \vdots \\ 0 \end{bmatrix}, \ldots, e_n = \begin{bmatrix} 0 \\ \vdots \\ 0 \\ 1 \end{bmatrix},$$

then the vector can be expressed as

$$x = \begin{bmatrix} x_1 \\ x_2 \\ \vdots \\ x_n \end{bmatrix}.$$

These are widely used representations. Next, we introduce Einstein's summation convention of the vector basis representation:

$$x = x_i e_i, \tag{1A1}$$

which is equivalent to summation:

$$x = \sum_{i=1}^{n} x_i e_i.$$

In equation (1A1), the summation sign is dropped and it is understood that the index i is repeated n times. The repeated index i is called a dummy index.

To further simplify index notation, we omit the base vectors for every vector quantity. Vector x is often written as

$$x_i, \tag{1A2}$$

which reads in full as $x_i e_i$. For example, a vector equation $y = x$ can be written as

$$y_i = x_i, \tag{1A3}$$

which implies

$$y_1 = x_1, \ldots, y_n = x_n.$$

1A1.2 Inner Product of Two Vectors

The inner product of two vectors is defined as

$$x.y = \|x\|\|y\|\cos\theta, \tag{1A4}$$

where θ is the angle between two vectors x and y. Let δ_{ij} be Kronecker delta defined as

$$\delta_{ij} = \begin{cases} 1 & i = j \\ 0 & i \neq j \end{cases}. \tag{1A5}$$

Using the definition (1A4), the inner product of set of orthonormal bases is

$$e_i . e_j = \delta_{ij}. \tag{1A6}$$

Let $x = \sum_{i=1}^{n} x_i e_i$ and $y = \sum_{j=1}^{n} y_j e_j$. Then, the inner product is

$$x.y = \left(\sum_{i=1}^{n} x_i e_i\right) . \left(\sum_{j=1}^{n} y_j e_j\right) = \sum_{i=1}^{n} x_i y_i,$$

which can be written as

$$x.y = x_i y_i. \tag{1A7}$$

The norm of a vector x is defined as

$$\|x\| = \sqrt{x.x} = \sqrt{x_i x_i}. \tag{1A8}$$

1A1.3 Cross Product of Two Vectors

Cross product of two vectors is defined as (Figure 1.11a)

$$x \times y = \|x\|\|y\| \sin\theta \; e, \tag{1A9}$$

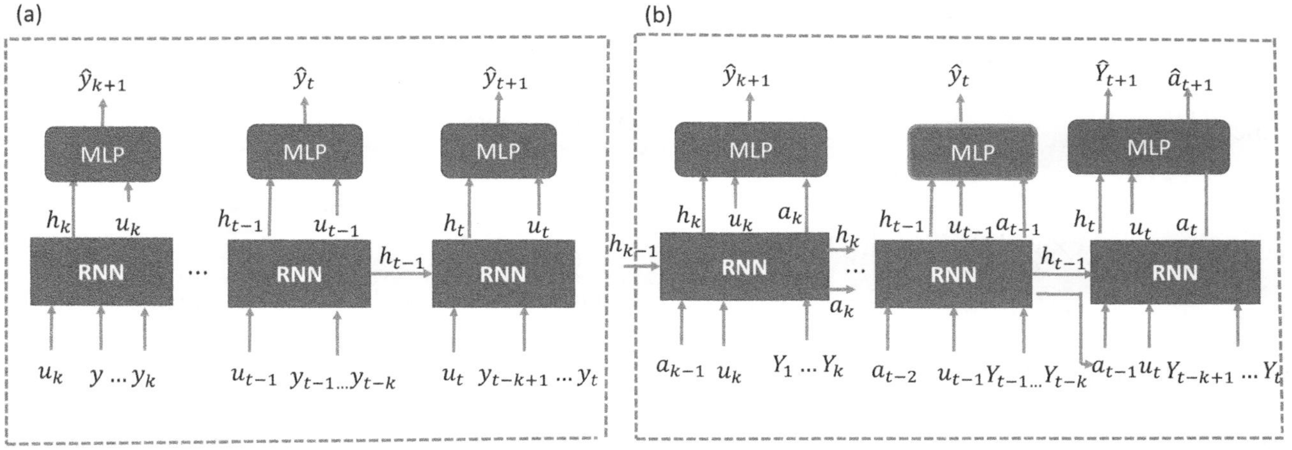

FIGURE 1.11 (a) A geometric representation of cross product of two vectors; (b) a Cartesian system.

where θ is an angle and e is a unit vector perpendicular to the plane spanned by x and y and oriented by the right-hand-rule. The norm of cross product is equal to

$$\|x \times y\| = \|x\|\|y\|\sin\theta,$$

which represents the area of the parallelepiped spanned by vectors x and y.

A set of orthonormal bases (Figure 1.11b) can be represented by cross product as follows:

$$e_1 \times e_2 = e_3,\ e_2 \times e_3 = e_1,\ e_3 \times e_1 = e_2, \quad (1A10)$$

$$e_2 \times e_1 = -e_3,\ e_1 \times e_3 = -e_2,\ e_3 \times e_2 = -e_1, \quad (1A11)$$

$$e_1 \times e_1 = 0,\ e_2 \times e_2 = 0,\ e_3 \times e_3 = 0. \quad (1A12)$$

Equations (1A10)–(1A12) can be rewritten as

$$e_i \times e_j = \varepsilon_{ijk} e_k, \quad (1A13)$$

where ε_{ijk} is defined as

$$\varepsilon_{ijk} = \begin{cases} 1 & i,j,k \text{ is a cycic sequence} \\ -1 & i,j,k \text{ is a anticycic sequence.} \\ 0 & ijk \text{ is a acyclic sequence} \end{cases}$$

Let $x = x_i e_i$ and $y = y_j e_j$. Then, it follows from equation (1A13) that

$$x \times y = x_i y_j e_i \times e_j = \varepsilon_{ijk}\, x_i y_j e_k, \quad (1A14)$$

which implies that

$$(x \times y)_k = \varepsilon_{ijk}\, x_i y_j. \quad (1A15)$$

Using equation (1A14), we obtain

$$x \times y = \sum_{i=1}^{3}\sum_{j=1}^{3}\sum_{k=1}^{3} \varepsilon_{ijk}\, x_i y_j e_k,$$

which implies

$$x \times y = \begin{vmatrix} e_1 & e_2 & e_3 \\ x_1 & x_2 & x_3 \\ y_1 & y_2 & y_3 \end{vmatrix},$$

where $|A|$ denotes the determinant of the matrix A.

1A1.4 Mapping, Tensor, and Matrix

Consider two sets $\mathbb{Q}, \mathcal{H} \in R^d$. Let T be a mapping that assigns to each point $x \in \mathbb{Q}$ a unique point $y = T(x) \in \mathcal{H}$. For the convenience of discussion, we furthermore assume that mapping is linear, i.e.,

$$T(x+y) = T(x) + T(y) \text{ and } T(\alpha x) = \alpha T(x).$$

A scaler is called a zero-order tensor, a vector is called a first-order tensor, and linear transformation is called a second-order tensor or simply called tensor.

Now we study a tensor that acts on the bases vectors $e_i \in R^d$ $(i = 1,\ldots,n)$. Recall that $T(e_i)$ maps e_i to a vector whose components are denoted by $T_{1i}, T_{2i}, \ldots, T_{ni}$. Therefore, we obtain

$$T(e_i) = T_{1i}e_1 + T_{2i}e_2 + \cdots + T_{ni}e_n, \text{ or}$$

$$\begin{bmatrix} T(e_1) \\ \vdots \\ T(e_n) \end{bmatrix} = \begin{bmatrix} T_{11} & \cdots & T_{n1} \\ \vdots & \vdots & \vdots \\ T_{1n} & \cdots & T_{nn} \end{bmatrix} \begin{bmatrix} e_1 \\ \vdots \\ e_n \end{bmatrix}.$$

Since

$$T(e_j) = T_{ij} e_i,$$

then inner product of $T(e_j)$ and e_k is

$$T(e_j).e_k = T_{ij} e_i.e_k = T_{ij}\delta_{ik} = T_{kj}.$$

The tensor can be represented by a matrix. The element T_{ij} represents the ith coordinate of the vector obtained by mapping $T(e_j)$. This shows that matrix $T = (T_{ij})$ defines the components of a second-order tensor T.

Now we consider the action of the second-order tensor T on a vector x:

$$y = y_i e_i = T(x) = T(x_j e_j) = x_j T(e_j) = x_j T_{ij} e_i,$$

which implies

$$y_i = T_{ij} x_j \text{ or } y_i = \sum_{j=1}^{n} T_{ij} x_j \quad (1A16)$$

Equation (1A16) can be written in a matrix form:

$$\begin{bmatrix} y_1 \\ \vdots \\ y_n \end{bmatrix} = \begin{bmatrix} T_{11} & \cdots & T_{1n} \\ \vdots & \vdots & \vdots \\ T_{n1} & \cdots & T_{nn} \end{bmatrix} \begin{bmatrix} x_1 \\ \vdots \\ x_n \end{bmatrix}, \text{ or} \quad (1A17)$$

$$y = Tx. \quad (1A18)$$

1A1.5 Outer Product of Two Vectors

Next we introduce outer product of two vectors or tensor product. The outer product of two vectors $x, y \in R^d$ is denoted by $x \otimes y$. Outer product is a tensor. The action of tensor $x \otimes y$ on a vector z is defined as

$$(x \otimes y)z = x(y.z). \tag{1A19}$$

Equation (1A19) can also be taken as a definition of the outer product of two vectors. We can show that the tensor $T = x \otimes y$ defined by equation (1A19) is a linear mapping. It is clear that

$$\begin{aligned} T(\alpha u+\beta v) &= (x \otimes y)(\alpha u+\beta v) = x\left(y.(\alpha u+\beta v)\right) \\ &= x\left(\alpha(y.u)+\beta(y.v)\right) = \alpha(x(y.u)+\beta\left(x(y.v)\right) \\ &= \alpha(x \otimes y)u+\beta(x \otimes y)v = \alpha T(u)+\beta T(v). \end{aligned}$$

The tensor can also be written in a component form:

$$T = x \otimes y = (x_i e_i) \otimes (y_j e_j) = x_i y_j e_i \otimes e_j, \text{ or} \tag{1A20}$$

$$T = \sum_{i=1}^{n}\sum_{j=1}^{n} x_i y_j e_i \otimes e_j. \tag{1A21}$$

Equation (1A21) implies that the components of the tensor are

$$T_{ij} = x_i y_j,\ i = 1,\ldots,n,\ j = 1,\ldots,n. \tag{1A22}$$

Let

$$x = \begin{bmatrix} x_1 \\ \vdots \\ x_n \end{bmatrix} \text{ and } y = \begin{bmatrix} y_1 \\ \vdots \\ y_n \end{bmatrix}.$$

Then, equation (1A22) can be written as

$$T = xy^T = \begin{bmatrix} x_1 \\ \vdots \\ y_n \end{bmatrix}\begin{bmatrix} y_1 & \cdots & y_n \end{bmatrix} = \begin{bmatrix} x_1 y_1 & \cdots & x_1 y_n \\ \vdots & \vdots & \vdots \\ x_n y_1 & \cdots & x_n y_n \end{bmatrix}. \tag{1A23}$$

Next, we consider action of the second-order tensor T on a vector z in a matrix form. Using equation (1A21) and index notation, we have

$$\begin{aligned} u_i e_i = u = Tz &= (T_{ij} e_i \otimes e_j)(z_k e_k) = T_{ij} z_k (e_i \otimes e_j) e_k \\ &= T_{ij} z_k e_i e_j . e_k = (T_{ij} z_j) e_i, \end{aligned}$$

which implies

$$u_i = T_{ij} z_j \qquad \text{or}$$

$$u = \begin{bmatrix} u_1 \\ \vdots \\ u_n \end{bmatrix} = Tz = \begin{bmatrix} T_{11} & \cdots & T_{1n} \\ \vdots & \vdots & \vdots \\ T_{n1} & \cdots & T_{nn} \end{bmatrix}\begin{bmatrix} z_1 \\ \vdots \\ z_n \end{bmatrix}. \tag{1A24}$$

Now we extend a second-order tensor to an nth-order tensor. We define the tensor of order n as

$$T = T_{ij\ldots n} e_i \otimes e_j \otimes \ldots \otimes e_n$$

where n base vectors are connected by outer products.

Recall that

$$(x \otimes y)z = x(y.z). \tag{1A25}$$

Equation (1A25) can be generalized to cross product:

$$(x \otimes y) \times z = x(y \times z). \tag{1A26}$$

1A1.6 Transpose of Tensor and Trace

Now we first introduce a transpose of a tensor. By definition of inner product, we have

$$x.(Ty) = y.(T^T x). \tag{1A27}$$

Using index notation, we obtain

$$x.(Ty) = x_i T_{ij} y_j = x_j T_{ji} y_i = y_i T_{ij} x_j \text{ and} \tag{1A28}$$

$$y.(T^T x) = y_i (T^T)_{ij} x_j. \tag{1A29}$$

Combining equations (1A27)–(1A29), we obtain

$$(T^T)_{ij} = T_{ji}. \tag{1A30}$$

Equation (1A30) defines the transpose of a tensor. By the similar argument, we can show that

$$\begin{aligned} (x \otimes y)^T &= (x_i y_j e_i \otimes e_j)^T = y_i x_j e_i \otimes e_j \\ &= (y_i e_i)(x_j e_j) = y \otimes x. \end{aligned} \tag{1A31}$$

The trace of a tensor is defined as

$$Tr(x \otimes y) = x.y. \tag{1A32}$$

It follows from equation (1A32) that

$$Tr(x \otimes y) = x_i y_i = T_{ii}, \tag{1A33}$$

where $T = xy^T$.

In other words, the trace of the tensor is the sum of all diagonal entries in the tensor matrix expression.

1A1.7 Multiplication of Tensors

Consider two maps: S and T. Suppose that S and T are represented by

$$S = S_{kl} e_k \otimes e_l \text{ and } T = T_{ij} e_i \otimes e_j.$$

Then, we have

$$\begin{aligned} S(Tu) &= S\left(T\left(u_m e_m\right)\right) = S\left(u_m T\left(e_m\right)\right) \\ &= S\left(u_m T_{ij}\left(e_i \otimes e_j\right) e_m\right) \\ &= S\left(u_m T_{ij} e_i\left(e_j . e_m\right)\right) = S\left(T_{ij} u_j e_i\right) \\ &= T_{ij} u_j S_{kl}\left(e_k \otimes e_l\right) e_i \\ &= T_{ij} u_j S_{kl} e_k\left(e_l . e_i\right) = \left(S_{ki} T_{ij}\right) u_j e_k = (ST) u, \end{aligned}$$

where $(ST)_{kj} = S_{ki} T_{ij}$.

Therefore, the multiplication of two tensors S and T is defined as

$$ST = S_{ik} T_{kj} e_i \otimes e_j, \text{ or} \quad (1A34)$$

$$ST = \sum_k S_{ik} T_{kj} e_i \otimes e_j.$$

Let $S = a \otimes b$ and $T = x \otimes y$. Then, we have the following useful equality:

$$\begin{aligned} ST &= (a \otimes b)(x \otimes y) = ab^T xy^T \\ &= (b.x) ay^T = (b.x)(a \otimes y). \end{aligned}$$

Multiplication of two tensors has another expression:

$$\begin{aligned} (u \otimes v) T &= \left(u_i v_j e_i \otimes e_j\right)\left(T_{kl} e_k \otimes e_l\right) = T_{kl} u_i v_j e_i \otimes e_l e_j . e_k \\ &= T_{jl} v_j u_i e_i \otimes e_l = u_i e_i \otimes T_{jl} v_j e_l = u \otimes \left(T^T v\right). \end{aligned} \quad (1A35)$$

Now consider trace of two tensors. Let $T = u \otimes v = u_i v_k e_i \otimes e_k$ and $S = S_{jl} e_j \otimes e_l$ Then,

$$\begin{aligned} S\, u \otimes v &= \left(S_{jl} e_j \otimes e_l\right)\left(u_i v_k e_i \otimes e_k\right) \\ &= S_{jl} u_i v_k \left(e_j \otimes e_l\right)\left(e_i \otimes e_k\right)\Big) \\ &= S_{jl} u_i v_k e_j \otimes e_k e_l . e_i = S_{ji} u_i v_k e_j \otimes e_k. \end{aligned} \quad (1A36)$$

Therefore, using equation (1A35), we have

$$Tr\left(S\, u \otimes v\right) = S_{ji} u_i v_k \delta_{jk} = v_j S_{ji} u_i = v.Su. \quad (1A37)$$

1A1.8 Inner Product of Two Tensors, Norm of a Tensor

Consider two tensors of the same order with base vectors:

$$e_i \otimes e_j \otimes \ldots \otimes e_m \text{ and } e_k \otimes e_l \otimes \ldots \otimes e_n.$$

The inner product of these two tensors are defined as inner products of all pairs of base vectors:

$$\begin{aligned} &\left(e_i \otimes e_j \otimes \ldots \otimes e_m\right).\left(e_k \otimes e_l \otimes \ldots \otimes e_n\right) \\ &= \left(e_i . e_k\right)\left(e_j . e_l\right) \ldots \left(e_m . e_n\right). \end{aligned} \quad (1A38)$$

Let

$$S = S_{ij \ldots m} e_i \otimes e_j \otimes \ldots \otimes e_m \text{ and } T = T_{kl \ldots n} e_k \otimes e_l \otimes \ldots \otimes e_n.$$

Then, the inner product of two general tensors of the same order is given by

$$S.T = S_{ij \ldots m} T_{kl \ldots n}\left(e_i . e_k\right)\left(e_j . e_l\right) \ldots \left(e_m . e_n\right) = S_{ij \ldots m} T_{ij \ldots m}, \quad (1A39)$$

which is equivalent to

$$S.T = Tr\left(ST^T\right) = Tr\left(S^T T\right). \quad (1A40)$$

Specifically, inner product of two vectors x and, the first-order tensor, is

$$x.y = x_i y_i.$$

Consider two matrices $A = \left(a_{ij}\right)$ and $B = \left(b_{ij}\right)$. Then, the inner product of two matrices is

$$A.B = a_{ij} b_{ij} \quad \text{or}$$

$$A.B = \sum_i \sum_j a_{ij} b_{ij} = Tr\left(AB^T\right). \quad (1A41)$$

The norm of a tensor T is defined as

$$\|T\| = \sqrt{T.T} = \sqrt{Tr\left(TT^T\right)} = \sqrt{\sum_i \sum_j T_{ij}^2}. \quad (1A42)$$

1A2 Tensor Calculus

1A2.1 Tensor-Valued Functions

We consider three types of arguments and three types of function values.

1. Function value is a scalar.
   a. argument is a scalar. $f: R \to R, t \to f(t)$
   b. argument is a vector. $f: R^d \to R, \boldsymbol{x} \to f(\boldsymbol{x})$
2. Function is a vector.
   a. argument is a scalar. $\boldsymbol{F}: R \to R^d, t \to \boldsymbol{F}(t)$
   b. argument is a vector. $\boldsymbol{F}: R^d \to R^d, \boldsymbol{x} \to \boldsymbol{F}(\boldsymbol{x})$
3. Function is a tensor.
   a. argument is a scalar. $\boldsymbol{T}: R \to L(R^d, R^d)$, $t \to \boldsymbol{T}(t)$
   b. argument is a vector. $T: R^d \to L(R^d, R^d)$, $\boldsymbol{x} \to \boldsymbol{T}(\boldsymbol{x})$

1A2.2 Gateaux Derivative

In the classical calculus, we learned three related variation concepts: directional derivatives, partial derivatives, and gradients. Gateaux derivative is a generalization of directional derivative. Let, $h \in R^d$. Then, a Gateaux derivative is defined as

$$D_h f(x) = \lim_{\epsilon \to 0} \frac{f(x+\varepsilon h) - f(x)}{\varepsilon} = \left[\frac{d}{d\varepsilon} f(x+\varepsilon h)\right]_{\varepsilon=0}. \tag{1A43}$$

It is clear that the Gateaux differential is a one-dimensional calculation along a specified direction h. Therefore, we can always use classical one variable calculus as a tool to calculate the Gateaux derivative.

Example 1A.1: Inner Product

Let $f(x) = x.x$. Then,

$$D_h f(x) = \left[\frac{d}{d\varepsilon}(x+\varepsilon h).(x+\varepsilon h)\right]_{\varepsilon=0} = 2x.h.$$

Example 1A.2: Quadratic Function

Let $f(x) = x^T Ax$. Then,

$$D_h f(x) = \left[\frac{d}{d\varepsilon}(x+\varepsilon h)^T A(x+\varepsilon h)\right]_{\varepsilon=0}$$

$$= 2h^T A(x+\varepsilon h)_{\varepsilon=0} = 2h^T Ax.$$

Example 1A.3: Exponential

Let $f(x) = exp(x)$. Then its Gateaux derivative is

$$D_h f(x) = \left[\frac{d}{d\varepsilon} exp(x+\varepsilon h)\right]_{\varepsilon=0}$$

$$= \left[hexp(x+\varepsilon h)\right]_{\varepsilon=0} = hexp(x).$$

1A2.3 Gradient

1A2.3.1 Gradient of a Scalar Function

A gradient of a scale function I defied as a linear mapping which maps each direction h onto its Gateaux derivative $D_h f$:

$$\text{grad} f(x).h = D_h f = \left[\frac{d}{d\varepsilon} f(x+\varepsilon h)\right]_{\varepsilon=0}$$

$$= \left[\frac{\partial f(x+\varepsilon h)}{\partial(x+\varepsilon h)} \cdot \frac{\partial(x+\varepsilon h)}{\partial \varepsilon}\right]_{\varepsilon=0} \tag{1A44}$$

$$= \frac{\partial f}{\partial x}.h = \frac{\partial f}{\partial x_i}.h.$$

However, $\text{grad} f(x).h$ can be written as

$$\text{grad} f(x).h = \left[grad\ f(x)\right]_i h_i. \tag{1A45}$$

Comparing equation (1A45) with equation (1A44), we obtain

$$\left[grad\ f(x)\right]_i = \frac{\partial f}{\partial x_i} \text{ or}$$

$$grad\ f = \frac{\partial f}{\partial x}. \tag{1A46}$$

Example 1A.4: Inner Product Continued

Recall that $f(x) = x.x$. Then,

$$grad\ f = \frac{\partial f}{\partial x} = 2x.$$

For the convenience of discussion, the comma index notation can be used to denote partial derivative as follows:

$$\frac{\partial}{\partial x_i} f = f_{,i} \text{ and } \frac{\partial^2}{\partial x_i \partial x_j} f = f_{,ij}. \tag{1A47}$$

For example, we have

$$\frac{\partial}{\partial x_i}(x.x) = (x.x)_{,i} = 2x_i \text{ and } x_{k,i} = \delta_{ki}.$$

1A2.3.2 Gradient of a Tensor Function

The gradient of a scalar function can be extended to a tensor with high order. In general, a gradient of a tensor function is defined as

$$grad\ f(x) = f_{,j} \otimes e_i. \tag{1A48}$$

Similarly, divergence and curl of a tensor function can be defined as

$$div\ f(x) = f_{,i}.e_i \text{ and } rl\ f(x) = -f_{,i} \times e_i. \tag{1A49}$$

Example 1A.5: Gradient of a Vector Function

Let $\boldsymbol{F}(\boldsymbol{x}): R^d \to R^d$ be a vector function and can be represented by

$$\boldsymbol{F}(\boldsymbol{x}) = F_i(\boldsymbol{x})e_i. \tag{1A50}$$

Then, using equations (1A48) and (1A50), we obtain

$$\text{grad } \boldsymbol{F} = \frac{\partial \boldsymbol{F}}{\partial \boldsymbol{x}_j} \otimes e_j = \frac{\partial F_i(\boldsymbol{x})}{\partial x_j} e_i \otimes e_j, \tag{1A51}$$

which implies that

$$(\text{grad } \boldsymbol{F})_{ij} = \frac{\partial F_i(\boldsymbol{x})}{\partial x_j} = F_{i,j}.$$

grad $\boldsymbol{F}$ can be rewritten as

$$\text{grad } \boldsymbol{F} = \frac{\partial \boldsymbol{F}}{\partial \boldsymbol{x}^T} = \begin{bmatrix} \frac{\partial F_1}{\partial \boldsymbol{x}^T} \\ \vdots \\ \frac{\partial F_d}{\partial \boldsymbol{x}^T} \end{bmatrix} = \begin{bmatrix} \frac{\partial F_1}{\partial x_1} & \cdots & \frac{\partial F_1}{\partial x_d} \\ \vdots & \vdots & \vdots \\ \frac{\partial F_d}{\partial x_1} & \cdots & \frac{\partial F_d}{\partial x_d} \end{bmatrix}.$$

Let $\boldsymbol{F}(\boldsymbol{x}) = \boldsymbol{x}$. Then, grad $\boldsymbol{F} = \boldsymbol{I}$, where I is an identity matrix.

Example 1A.6: Divergence of a Vector Function

Using equation (1A49), we obtain

$$div\ \boldsymbol{F}(\boldsymbol{x}) = \boldsymbol{F}_i.e_i = \frac{\partial F_j e_j}{\partial x_i}.e_i = \frac{\partial F_i}{\partial x_i} = (F_i)_{,i}. \tag{1A52}$$

Example 1A.7: Divergence of a Tensor Function

Let $T: R^d \to L(R^d, R^d)$ be a tensor. Then, using equation (1A48), we obtain

$$div\ T = (T)_{,i}.e_i = \left(T_{jk}e_j \otimes e_k\right)_{,i}.e_i$$

$$= \frac{\partial T_{jk}}{\partial x_i} e_j \otimes e_k.e_i = \frac{\partial T_{ji}}{\partial x_i} e_j = T_{ji,i}e_j.$$

1A2.3.3 Gradient of a Tensor Function with Respect to Tensor Argument

Now we introduce partial derivatives with respect to general tensors of arbitrary order. Let T be any tensor quantity of arbitrary order. Then, gradient of a tensor function with respect to a vector can be calculated by

$$\frac{\partial T}{\partial \boldsymbol{x}} = \text{grad } T = \frac{\partial T}{\partial x_j} \otimes e_j. \tag{1A53}$$

Consider an example. Let $x \in R^d$ be a vector and $T \in L(R^d, R^d)$ be a tensor. Then, gradient of Tx with respect to a vector x is

$$grad\ (Tx) = \frac{\partial Tx}{\partial x_k} \otimes e_k = \frac{\partial T_{ij}x_j e_i}{\partial x_k} \otimes e_k$$

$$= T_{ij}\frac{\partial x_j}{\partial x_k} e_i \otimes e_k = T_{ij}\delta_{jk} e_i \otimes e_k = T_{ij}e_i \otimes e_j,$$

i.e.,

$$\frac{\partial Tx}{\partial x} = T. \tag{1A54}$$

Consider partial derivative of a tensor with respect to a tensor, we assume

$$\frac{\partial T_{ij\ldots n}}{\partial T_{ab\ldots m}} = \delta_{ia}\delta_{jb}\ldots\delta_{nm}. \tag{1A55}$$

Now equation (1A53) for the gradient of a tensor with respect to a vector can be generalized to the gradient of a tensor of arbitrary order, including scalar, vector, and tensor with high order, with respect to tensors of arbitrary order:

$$\frac{\partial(.)}{\partial S} = \frac{\partial T}{\partial S} = \frac{\partial(.)}{\partial S_{ij\ldots n}} \otimes e_i \otimes e_j \otimes \ldots \otimes e_n$$

$$= \frac{\partial T}{\partial S_{ij\ldots n}} \otimes e_i \otimes e_j \otimes \ldots \otimes e_n. \tag{1A56}$$

Applying equation (1A56) to trace of a tensor, we obtain

$$\frac{\partial \mathrm{Tr}(T)}{\partial T} = \frac{\partial \mathrm{Tr}(T)}{\partial T_{ij}} e_i \otimes e_j = \frac{\partial T_{kk}}{\partial T_{ij}} e_i \otimes e_j \qquad (1A57)$$
$$= \delta_{ki}\delta_{kj} e_i \otimes e_j = \delta_{ij} e_i \otimes e_j = \boldsymbol{I}.$$

1A2.4 Chain Rule and Product Rule

Let F, Q be tensors of second order. Then, we calculate the gradient of the product of two tensors F and Q:

$$\frac{\partial (FQ)}{\partial F} = \frac{\partial F_{ij} Q_{kl} \left(e_i \otimes e_j\right)\left(e_k \otimes e_l\right)}{\partial F_{mn}} \left(e_m \otimes e_n\right)$$
$$= \frac{\partial F_{ij} Q_{kl}}{\partial F_{mn}} \left(e_j . e_k\right)\left(e_i \otimes e_l\right)\left(e_m \otimes e_n\right)$$
$$= \delta_{im}\delta_{jn}\delta_{jk} Q_{kl} \left(e_i \otimes e_l\right)\left(e_m \otimes e_n\right)$$
$$= \delta_{im}\delta_{jn}\delta_{jk} Q_{kl} \left(e_l . e_m\right)\left(e_i \otimes e_n\right)$$
$$= \delta_{im}\delta_{jn}\delta_{jk}\delta_{lm} Q_{kl} \left(e_i \otimes e_n\right)$$
$$= Q_{nm} \left(e_m \otimes e_n\right),$$

which implies

$$\frac{\partial (FQ)}{\partial F} = Q^T. \qquad (1A58)$$

The chain rule for tensor calculus is the same to the chain rule for the classical calculus if the scalar numbers in the chain rule for the classical calculus are replaced by tensors of arbitrary orders. I explain chain rule by examples. Let $W : L(R^d, R^d) \rightarrow R$ be a scalar composite function of two tensors $F, Q \in L(R^d, R^d)$. Then, we calculate $\frac{\partial W(FQ)}{\partial F}$ via chain rule and equation (1A58) as follows:

$$\frac{\partial W(FQ)}{\partial F} = \frac{\partial W(FQ)}{\partial (FQ)} \frac{\partial (FQ)}{\partial F} = \frac{\partial W(FQ)}{\partial (FQ)} Q^T. \qquad (1A59)$$

Now consider the second example:

$$\frac{\partial \mathrm{Tr}(A^T A)}{\partial A} = \frac{\partial \mathrm{Tr}(B)}{\partial B} \frac{\partial B}{\partial A} = \frac{\partial B_{nn}}{\partial B_{ij}} \frac{\partial B_{ij}}{\partial A_{kl}} e_k \otimes e_l$$
$$= \delta_{ni}\delta_{nj} \frac{\partial A_{ni} A_{nj}}{\partial A_{kl}} e_k \otimes e_l$$
$$= \delta_{ni}\delta_{nj} \left(\delta_{nk}\delta_{il} A_{nj} + A_{ni}\delta_{nk}\delta_{jl}\right) e_k \otimes e_l$$
$$= 2A_{kl} e_k \otimes e_l,$$

which implies

$$\frac{\partial \mathrm{Tr}(A^T A)}{\partial A} = 2A. \qquad (1A60)$$

Finally, we introduce the product rule for tensor multiplication, including inner, outer, and cross product. Let X, Y, and T be tensors of arbitrary order and $\odot$ denote any tensor multiplication. Then, we have the following product rule:

$$\frac{\partial (X \odot Y)}{\partial T} = \frac{\partial X}{\partial T} \odot Y + X \odot \frac{\partial Y}{\partial T}. \qquad (1A61)$$

APPENDIX 1B: CALCULATE GRADIENT OF CROSS ENTROPY LOSS FUNCTION

Let $u_t = W_{hz} h_t + b_t$,

$$u_t = \begin{bmatrix} u_t^1 \\ \vdots \\ u_t^k \end{bmatrix}, Z_t = \begin{bmatrix} Z_t^1 \\ \vdots \\ Z_t^K \end{bmatrix} \text{ and } Z_t = softmax\left(u_t\right).$$

Softmax function of u_t is defined as

$$Z_t^j = \frac{e^{u_t^j}}{\sum_{i=1}^{K} e^{u_t^i}}. \qquad (1B1)$$

Thus, $\log Z_t^j = u_t^j - \log \sum_{i=1}^{K} e^{u_t^i}$.

Note that

$$Y_t \log Z_t = \sum_{j=1}^{K} Y_t^j \log Z_t^j = \sum_{j=1}^{K} Y_t^j \left(u_t^j - \log \sum_{i=1}^{K} e^{u_t^i} \right), \qquad (1B2)$$

which implies

$$\frac{\partial Y_t \log Z_t}{\partial u_t^l} = \sum_{j \neq l}^{K} Y_t^j \left(- \frac{\exp(u_t^l)}{\sum_{i=1}^{K} \exp(u_t^i)} \right) + Y_t^l \left(1 - \frac{\exp(u_t^l)}{\sum_{i=1}^{K} \exp u_t^i} \right)$$
$$= Y_t^l - \left(\sum_{j=1}^{K} Y_t^j \right) \frac{\exp(u_t^l)}{\sum_{i=1}^{K} \exp u_t^i}. \qquad (1B3)$$

Note that the summation of probability mass is equal to 1, i.e.,

$$\sum_{j=1}^{K} Y_t^j = 1. \qquad (1B4)$$

Substituting equations (1B1) and (1B4) into equation (1B3) yields

$$\frac{\partial Y_t \log Z_t}{\partial u_t^l} = Y_t^l - Z_t^l. \quad (1B5)$$

Therefore,

$$\frac{\partial L(Y_t, Z_t)}{\partial u_t} = -\begin{bmatrix} \frac{\partial Y_t \log Z_t}{\partial u_t^1} \\ \vdots \\ \frac{\partial Y_t \log Z_t}{\partial u_t^K} \end{bmatrix} = Z_t - Y_t. \quad (1B6)$$

APPENDIX 1C: OPTIMAL CONTROL AND PONTRYAGIN'S MAXIMUM PRINCIPLE

In this appendix, we briefly introduce optimal control theory and Pontryagin's Maximum Principle.

1C1 Optimal Control

Consider a dynamic system that can be controlled by some external forces:

$$\dot{x} = f(x(t), u(t)), \quad (1C1)$$

where $x(t) \in R^n$ is an n-dimensional vector of functions, $u(t) \in R^m$ is an m-dimensional vector of control variables, and equation (1C1) is state equation that determine how the state of the system is changed by the external forces.

Let $x(t_0) = x_0$ be the initial value of the system state and $x(t_f)$ be the value of targeted final state, where t_f is the time when the system enters the final state. The target value should satisfy the constraint:

$\Psi(x(t_f), t_f) = 0$, where Ψ is a differentiable function.

The goal of optimal control is to reach $x(t_f)$ with optimal cost, starting with the initial value x_0. The loss function over the entire time period is defined as

$$J(u) = \Phi(x(t_f), t_f) + \int_{t_0}^{t_f} L(x(t), u(t), t)\,dt, \quad (1C2)$$

where $L(x(t), u(t), t)$ is the point loss function of the control at the time t, depending on the state $x(t)$ and control $u(t)$.

To precisely formulate the optimal control problem, we also need to introduce constraints on control variables. Let $g(t) \in R^l$ be an l-dimensional vector of differentiable function. Then, the control variable $u(t)$ should satisfy the following inequality:

$$g(x(t), u(t), t) \geq 0. \quad (1C3)$$

In summary, the optimal control problem can be mathematically defined as

$$\min_u J(u) = \Phi(x(t_f), t_f) + \int_{t_0}^{t_f} L(x(t), u(t), t)\,dt, \quad (1C4)$$

$$\dot{x} = f(x(t), u(t)), \quad (1C5)$$

$$x(t_0) = x_0, \ \Psi(x(t_f), t_f) = 0. \quad (1C6)$$

1C2 Pontryagin's Maximum Principle

The Pontryagin's Maximum Principle can be used to solve the optimal control problem (Pontryagin 1987). The Pontryagin's Maximum Principle gives necessary conditions for solving the optimal control problem. Before introducing the Pontryagin's Maximum Principle, we define a Hamiltonian function:

$$H(x, \lambda, u, t) = L(x, u, t) + \lambda^T f(x, u, t), \quad (1C7)$$

where λ is a multiplier vector or adjoint vector. Now we introduce the Pontryagin's Maximum Principle.

Theorem 1C.1: Pontryagin's Maximum Principle

The Pontryagin's Maximum Principle gives a first order necessary condition for optimality. Let $\tilde{u}(t)$ be an optimal control, $\tilde{x}(t)$ be its corresponding optimal trajectory, and $\tilde{\lambda}(t)$ be an adjacent vector. Then, $\tilde{u}(t)$, $\tilde{x}(t)$, and $\tilde{\lambda}(t)$ must satisfy the following conditions:

1. $$\frac{d\tilde{x}}{dt} = \frac{\partial}{\partial \lambda} H(\tilde{x}(t), \tilde{\lambda}(t), \tilde{u}(t), t), \quad (1C8)$$

   $$\frac{d\tilde{\lambda}}{dt} = -\frac{\partial}{\partial x} H(\tilde{x}(t), \tilde{\lambda}(t), \tilde{u}(t), t), \quad (1C9)$$

2. Optimal control $\tilde{u}(t)$ must be minimizer of the Hamiltonian function on the optimal trajectory $\tilde{x}(t)$:

$$H(\tilde{x}(t), \tilde{\lambda}(t), \tilde{u}(t), t) \leq \min_{u \in \Omega} H(\tilde{x}(t), \tilde{\lambda}(t), u(t), t), \quad (1C10)$$

where Ω is a set of admissible control.

If Hamiltonian function is differentiable, then the optimal condition for $\tilde{u}(t)$ is

$$\frac{\partial}{\partial u(t)} H\left(\tilde{x}(t), \tilde{\lambda}(t), u(t), t\right)=0. \quad \text{(1C11)}$$

3. Terminal condition:

$$\left\{H\left(\tilde{x}(t), \tilde{\lambda}(t), \tilde{u}(t), t\right)+\frac{\partial \Phi}{\partial t_f}+\frac{\partial \Psi^T}{\partial t_f} v\right\}\Bigg|_{t_f}=0. \quad \text{(1C12)}$$

4. Transversality condition of adjacent vector $\tilde{\lambda}(t)$:

$$\tilde{\lambda}\left(t_f\right)=\left\{\frac{\partial \Phi}{\partial x}+\frac{\partial \Psi^T}{\partial x} v\right\}\Bigg|_{t_f}. \quad \text{(1C13)}$$

5. Boundary conditions for the optimal trajectory:

$$x\left(t_0\right)=x_0, \quad \text{(1C14)}$$

$$\Psi\left(x\left(t_f\right), t_f\right)=0. \quad \text{(1C15)}$$

Coronary 1C.1: Typical Boundary Conditions

1. $\Phi\left(x\left(t_f\right), t_f\right)=0$, $x\left(t_f\right)=x_f$ is fixed and t_f is free.
   In this case, we assume that $\Phi\left(x\left(t_f\right), t_f\right)=0$ and $\Psi\left(x\left(t_f\right), t_f\right)=x\left(t_f\right)-x_f=0$, which do not explicitly contain t_f. Thus, we have

$$\frac{\partial \Phi\left(x\left(t_f\right), t_f\right)}{\partial t_f}=0, \frac{\partial \Psi\left(x\left(t_f\right), t_f\right)}{\partial t_f}=0, \text{ and } \frac{\partial}{\partial x} \Psi\left(x\left(t_f\right), t_f\right)=1, \quad \text{(1C16)}$$

which implies

$$H\left(\tilde{x}(t), \tilde{\lambda}(t), \tilde{u}(t), t\right)\Big|_{t_f}=0. \quad \text{(1C17)}$$

Equation (1C17) will determine t_f.

It follows from equations (1C13) and (1C16) that

$$\tilde{\lambda}\left(t_f\right)=v\big|_{t_f}. \quad \text{(1C18)}$$

Equation (1C18) shows that there is no constraint posted on $\tilde{\lambda}\left(t_f\right)$.

2. $\Phi\left(x\left(t_f\right), t_f\right)=0$, $\Psi_i\left(x\left(t_f\right), t_f\right)=h_i\left(x\left(t_f\right)\right)=0$, $i=1, \ldots, n-k$, and t_f is free.
   Under these conditions, we obtain

$$\frac{\partial \Phi\left(x\left(t_f\right), t_f\right)}{\partial t_f}=0, \frac{\partial \Psi\left(x\left(t_f\right), t_f\right)}{\partial t_f}=0,$$

which combines equation (1C12) yields

$$H\left(\tilde{x}(t), \tilde{\lambda}(t), \tilde{u}(t), t\right)\Big|_{t_f}=0. \quad \text{(1C19)}$$

Recall that

$$\frac{\partial \Psi_i\left(x\left(t_f\right), t_f\right)}{\partial x\left(t_f\right)}=\frac{\partial h_i\left(x\left(t_f\right)\right)}{\partial x\left(t_f\right)}, i=1, \ldots, n-k. \quad \text{(1C20)}$$

Using equation (1C13), we obtain

$$\tilde{\lambda}\left(t_f\right)=\left[\begin{array}{ccc}\frac{\partial h_1}{\partial x\left(t_f\right)} & \cdots & \frac{\partial h_{n-k}}{\partial x\left(t_f\right)}\end{array}\right]\left[\begin{array}{c}v_1 \\ \vdots \\ v_{n-k}\end{array}\right]. \quad \text{(1C21)}$$

3. $\Phi\left(x\left(t_f\right), t_f\right)=0$, $x\left(t_f\right)$ and t_f is free.
   Using equations (1C12) and (1C13), we obtain

$$H\big|_{t_f}=0 \text{ and } \lambda\left(t_f\right)=0.$$

4. $\Phi\left(x\left(t_f\right), t_f\right)=0$, $x\left(t_f\right)=h\left(t_f\right)$ or $\Psi\left(x\left(t_f\right)\right)=x\left(t_f\right)-h\left(t_f\right)$, and t_f is free.
   In this case, $\frac{\partial \Phi\left(x\left(t_f\right), t_f\right)}{\partial t_f}=0$ and $\frac{\partial \Psi\left(x\left(t_f\right)\right)}{\partial x}=1$.
   Using equation (1C13), we obtain

$$\tilde{\lambda}\left(t_f\right)=v. \quad \text{(1C22)}$$

Using equation (1C12) yields

$$H\big|_{t_f}=-\frac{\partial \Psi}{\partial t_f} \tilde{\lambda}\left(t_f\right). \quad \text{(1C23)}$$

1C3 Calculus of Variation

Pontryagin's Maximum Principle is a generalization of calculus of variation. Before proving Pontryagin's

Maximum Principle, we introduce calculus of variation, a powerful tool for optimization. Let X be a linear space over R with norm $\|.\|$, define a real-valued functional J on X as a function mapping $J: X \to R$ and denoted as $J(x)$.

Example 1C.1

$$J(x)=\int_0^1 x(t)e^{x(t)}dt,$$

Example 1C.2

$$J(x)=\int_0^1 \left(e^t+\dot{x}\right)^2 dt.$$

Example 1C.3

$$J(y)=\int_0^1 \sqrt{1+\dot{y}(x)^2}\, dx.$$

Example 1C.4

$$J(y)=\int_a^b L(t,y(t),\dot{y}(t))dt.$$

Definition 1C.1: Minimum of Functional

Let $J(x)$ be a functional. A point $\tilde{x}$ be a minimizer of the functional $J(x)$ if $J(\tilde{x}) \le J(x)$ for all $x \in X$.

Definition 1C.2: First Variation (Gateaux Variation, Directional Derivative)

The Gateaux variation (or first variation) of the functional $J(x)$ at $x \in D \subset X$ in the direction h is defined as

$$\delta J(x,h)=\lim_{\varepsilon\to 0}\frac{J(x+\varepsilon h)-J(x)}{\varepsilon} \tag{1C24}$$
$$=\frac{d}{d\varepsilon}J(x+\varepsilon h)\Big|_{\varepsilon=0}.$$

Example 1C.5: Example 1C.1 Continued

It is clear that

$$J(x+\varepsilon h)=\int_0^1 (x+\varepsilon h)e^{x+\varepsilon h}dt.$$

Thus, using the definition of first variation, we obtain

$$\frac{\partial}{\partial\varepsilon}J(x+\varepsilon h)=\int_0^1 \left(he^{x+\varepsilon h}+hxe^{x+\varepsilon h}\right)\Big|_{\varepsilon=0} dt$$
$$=\int_0^1 h(1+x)e^x dt.$$

Example 1C.6: Example 1C.4 Continued, Euler-Lagrange Equation

It follows from example C4 that

$$J(y+\varepsilon h)=\int_a^b L(t,\, y+\varepsilon h,\, \dot{y}+\varepsilon\dot{h})dt.$$

Again, using the definition of first variation and chain rule, we obtain

$$\frac{\partial J(y+\varepsilon h)}{\partial\varepsilon}=\int_a^b\left(\frac{\partial L}{\partial y}h+\frac{\partial L}{\partial\dot{y}}\dot{h}\right)dt$$
$$=\int_a^b \frac{\partial L}{\partial y}hdt+\int_a^b\frac{\partial L}{\partial\dot{y}}\dot{h}dt$$
$$=\int_a^b \frac{\partial L}{\partial y}hd+h\frac{\partial L}{\partial\dot{y}}\Big|_a^b-\int_a^b\frac{d}{dt}\left(\frac{\partial L}{\partial\dot{y}}\right)hdt. \tag{1C25}$$

Since $y(a)$ and $y(b)$ are fixed, then we have

$$h(a)=h(b)=0. \tag{1C26}$$

Substituting equation (1C26) into equation (1C25), we obtain

$$\delta J(y,h)=\frac{\partial J(y+\varepsilon h)}{\partial\varepsilon}=\int_a^b\left(\frac{\partial L}{\partial y}-\frac{d}{dt}\left(\frac{\partial L}{\partial\dot{y}}\right)\right)hdt. \tag{1C27}$$

Taking

$$h=\frac{\partial L}{\partial y}-\frac{d}{dt}\left(\frac{\partial L}{\partial \dot{y}}\right), \tag{1C28}$$

and substituting equation (1C28) into equation (1C27) and letting $\delta J(y,h)=0$, we obtain

$$\frac{\partial L}{\partial y}-\frac{d}{dt}\left(\frac{\partial L}{\partial \dot{y}}\right)=0,$$

which is also called Euler-Lagrange equation.

Lemma 1C.1

Suppose that $\tilde{x}$ is a minimizer of the functional $J(x)$. Then, its first variation $\delta J(x,h)$ should be equal to zero for all direction, i.e.,

$$\delta J(x,h)=0, \text{ for all } h. \tag{1C29}$$

1C4 Proof of Pontryagin's Maximum Principle

First, we use Lagrange multipliers $\lambda \in R^n$, $v \in R^r$, and $\gamma \in R^l$ to transform constrained dynamic optimization problem into the following unconstrainted optimization problem:

$$J(u)=\Phi(x(t_f)+v^T\Psi(x(t_f),t_f)+\int_{t_0}^{t_f}\left\{\begin{matrix} L(x,u,t)+\lambda^T(f(x,u,t)-\dot{x}) \\ +\gamma^T(g(x,u,t)-\dot{Z}^2)\end{matrix}\right\}dt. \tag{1C30}$$

Define function

$$\begin{aligned} &F(x,\dot{x},u,\dot{Z},\lambda,\gamma,t) \\ &=H(x,\lambda,u,t)-\lambda^T\dot{x}+\gamma^T\left[g(x,u,t)-\dot{Z}^2\right], \end{aligned} \tag{1C31}$$

where $H(x,\lambda,u,t)$ is Hamiltonian function defined as

$$H(x,\lambda,u,t)=L(x,u,t)+\lambda^T f(x,u,t). \tag{1C32}$$

Define

$$\dot{w}=u(t).$$

Now equation (1C30) can be rewritten as

$$\begin{aligned} J(u)=&\Phi(x(t_f),t_f)+v^T\Psi(x(t_f),t_f) \\ &+\int_{t_0}^{t_f}F(x,\dot{x},\dot{w},\dot{Z},\lambda,\gamma,t)dt. \end{aligned} \tag{1C33}$$

The first variation of the functional $J(u)$ is

$$\delta J(u)=\delta J_{t_f}+\delta J_x+\delta J_w+\delta J_Z, \tag{1C34}$$

where

$$\begin{aligned} \delta J_{t_f}&=\frac{d}{d\varepsilon}\left[\begin{matrix}\Phi(x(t_f),t_f+\varepsilon\delta t_f)+v^T\Psi\begin{pmatrix}x(t_f),t_f \\ +\varepsilon\delta t_f\end{pmatrix} \\ +\int_{t_0}^{t_f+\varepsilon\delta t_f}F(x,\dot{x},\dot{w},\dot{Z},\lambda,\gamma,t)dt\end{matrix}\right] \\ &=\left[\begin{matrix}\frac{\partial}{\partial t_f}\Phi(x(t_f),t_f)+v^T\frac{\partial}{\partial t_f}\Psi(x(t_f),t_f) \\ +F(x,\dot{x},\dot{w},\dot{Z},\lambda,\gamma,t)\end{matrix}\right]_{t_f}\delta t_f, \end{aligned} \tag{1C35}$$

$$\begin{aligned} \delta J_x&=\frac{d}{d\varepsilon}\left\{\begin{matrix}\Phi(x+\varepsilon\delta x_f,t_f)+\Psi^T(x+\varepsilon\delta x_f,t_f) \\ +\int_{t_0}^{t_f}F(x+\varepsilon\delta x,\dot{x}+\varepsilon\delta\dot{x},\dot{w},\dot{Z},\lambda,\gamma,t)dt\end{matrix}\right\} \\ &=\delta x_f^T\frac{\partial}{\partial x}\{\Phi+v^T\Psi\}\Big|_{t_f}+\int_{t_0}^{t_f}\left[\delta x^T\frac{\partial F}{\partial x}+\delta\dot{x}^T\frac{\partial F}{\partial \dot{x}}\right]dt, \end{aligned} \tag{1C36}$$

Since initial state $x(t_0)$ is fixed, we obtain

$$\delta x\big|_{t_0}=0, \tag{1C37}$$

which implies

$$\delta x^T\frac{\partial F}{\partial \dot{x}}\bigg|_{t_0}=0. \tag{1C38}$$

Using equation (1C38), we obtain

$$\int_{t_0}^{t_f} \delta \dot{x}^T \frac{\partial F}{\partial \dot{x}} dt = \delta x^T \frac{\partial F}{\partial \dot{x}}\Big|_{t_0}^{t_f} - \int_{t_0}^{t_f} \delta x^T \frac{d}{dt}\frac{\partial F}{\partial \dot{x}} = \delta x^T \frac{\partial F}{\partial \dot{x}}\Big|_{t_f} - \int_{t_0}^{t_f} \delta x^T \frac{d}{dt}\frac{\partial F}{\partial \dot{x}} dt. \quad (1C39)$$

Substituting equation (1C39) into equation (1C36), we obtain

$$\delta J_x = \delta x_f^T \frac{\partial}{\partial x}\{\Phi + v^T \Psi\}\Big|_{t_f} + \delta x^T \frac{\partial F}{\partial \dot{x}}\Big|_{t_f} + \int_{t_0}^{t_f} \left[\delta x^T \frac{\partial F}{\partial x} - \delta x^T \frac{d}{dt}\frac{\partial F}{\partial \dot{x}} \right] dt. \quad (1C40)$$

Note that variation of terminal state variable $x(t_f)$ can be expressed as (Figure 1.12)

$$\delta x_f = x(t_f + \delta t_f) - x(t_f) + x(t_f) + \delta x(t_f) - x(t_f) = \dot{x}(t_f)\delta t_f + \delta x(t_f), \quad (1C41)$$

which implies

$$\delta x(t_f) = \delta x_f - \dot{x}(t_f)\delta t_f. \quad (1C42)$$

Substituting equation (1C42) into equation (1C40), we obtain

$$\delta J_x = \delta x_f^T \frac{\partial}{\partial x}\left\{\Phi + v^T \Psi + \frac{\partial F}{\partial \dot{x}}\right\}\Big|_{t_f} - \dot{x}^T \frac{\partial F}{\partial \dot{x}}\Big|_{t_f} \delta t_f + \int_{t_0}^{t_f} \left[\delta x^T \left(\frac{\partial F}{\partial x} - \frac{d}{dt}\frac{\partial F}{\partial \dot{x}} \right) \right] dt. \quad (1C43)$$

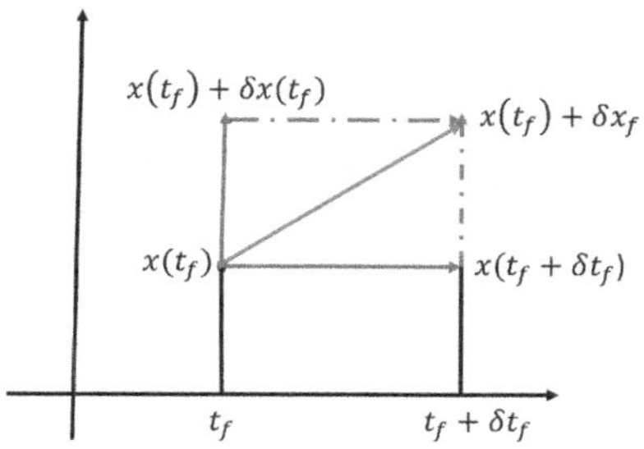

FIGURE 1.12 Scheme of variation of terminal state variable.

Next we calculate δJ_w. Using definition of the first variation, we obtain

$$\delta J_w = \frac{d}{d\varepsilon}\int_{t_0}^{t_f} F(x, \dot{x}, \dot{w} + \varepsilon\delta\dot{w}, \dot{Z}, \lambda, \gamma, t) dt = \int_{t_0}^{t_f} \delta \dot{w}^T \frac{\partial F}{\partial \dot{w}} dt = \delta w^T \frac{\partial F}{\partial \dot{w}}\Big|_{t_0}^{t_f} - \int_{t_0}^{t_f} \delta w^T \frac{d}{dt}\frac{\partial F}{\partial \dot{w}} dt. \quad (1C44)$$

Since $u(t_0)$ is fixed, we have

$$\delta w\big|_{t_0 = 0}. \quad (1C45)$$

Substituting equation (1C45) into equation (1C44) yields

$$\delta J_w = \delta w^T \frac{\partial F}{\partial \dot{w}}\Big|_{t_f} - \int_{t_0}^{t_f} \delta w^T \frac{d}{dt}\frac{\partial F}{\partial \dot{w}} dt. \quad (1C46)$$

Finally, we calculate δJ_Z. Again, using definition of first variation, we obtain

$$\delta J_Z = \frac{d}{d\varepsilon}\int_{t_0}^{t_f} F(x, \dot{x}, \dot{w}, \dot{Z} + \varepsilon\delta\dot{Z}, \lambda, \gamma, t) dt = \int_{t_0}^{t_f} \delta \dot{Z}^T \frac{\partial F}{\partial \dot{Z}} dt = \delta Z^T \frac{\partial F}{\partial \dot{Z}}\Big|_{t_0}^{t_f} - \int_{t_0}^{t_f} \delta Z^T \frac{d}{dt}\frac{\partial F}{\partial \dot{Z}} dt. \quad (1C47)$$

Recall that

$Z(t_0) = 0$, which implies

$$\delta Z^T\big|_{t_0} = 0. \quad (1C48)$$

Substituting equation (1C48) into equation (1C47) leads to

$$\delta J_Z = \delta Z^T \frac{\partial F}{\partial \dot{Z}}\Big|_{t_f} - \int_{t_0}^{t_f} \delta Z^T \frac{d}{dt}\frac{\partial F}{\partial \dot{Z}} dt. \quad (1C49)$$

Combining equations (1C34), (1C35), (1C43), (1C46), and (1C49), we obtain

$$\delta J(u)=\delta t_f\left[\frac{\partial}{\partial t_f}\Phi+\frac{\partial \Psi^T}{\partial t_f}v+F-\dot{x}^T\frac{\partial F}{\partial \dot{x}}\right]\Bigg|_{t_f}$$

$$+\ \delta x_f^T\frac{\partial}{\partial x}\left\{\Phi+v^T\Psi+\frac{\partial F}{\partial \dot{x}}\right\}\Bigg|_{t_f}+\delta w^T\frac{\partial F}{\partial \dot{w}}\Bigg|_{t_f}$$

$$+\ \delta Z^T\frac{\partial F}{\partial \dot{Z}}\Bigg|_{t_f}+\int_{t_0}^{t_f}\left[\delta x^T\left(\frac{\partial F}{\partial x}-\frac{d}{dt}\frac{\partial F}{\partial \dot{x}}\right)\right]dt$$

$$-\int_{t_0}^{t_f}\delta w^T\frac{d}{dt}\frac{\partial F}{\partial \dot{w}}dt-\int_{t_0}^{t_f}\delta Z^T\frac{d}{dt}\frac{\partial F}{\partial \dot{Z}}dt. \tag{1C50}$$

Using the necessary condition $\delta J(u)=0$ of the minimizer of the functional $J(u)$ and arbitrary δt_f, δx_f, δw, δZ, we obtain:

Euler Equations:

$$\frac{\partial F}{\partial x}-\frac{d}{dt}\frac{\partial F}{\partial \dot{x}}=0,\ \frac{d}{dt}\frac{\partial F}{\partial \dot{w}}=0 \text{ and } \frac{d}{dt}\frac{\partial F}{\partial \dot{Z}}=0$$

and transversality conditions:

$$\left[\frac{\partial}{\partial t_f}\Phi+\frac{\partial \Psi^T}{\partial t_f}v+F-\dot{x}^T\frac{\partial F}{\partial \dot{x}}\right]\Bigg|_{t_f}$$

$$=0,\ \frac{\partial}{\partial x}\left\{\Phi+v^T\Psi+\frac{\partial F}{\partial \dot{x}}\right\}\Bigg|_{t_f}$$

$$=0,\ \frac{\partial F}{\partial \dot{w}}\Bigg|_{t_f}=0 \text{ and } \frac{\partial F}{\partial \dot{Z}}\Bigg|_{t_f}=0.$$

EXERCISES

EXERCISE 1.1

Consider the dynamic system:

$$\frac{dx_1}{dt}=2x_2-x_1\left(x_1^2+x_2^2\right)$$

$$\frac{dx_2}{dt}=-x_1-x_2\left(x_1^2+x_2^2\right).$$

Assess its stability.

EXERCISE 1.2

Consider two matrices: $A=\text{diag}\left(a_1,\ldots,a_n\right)$ with $a_i>0, i=1,\ldots,n$ and B. Show that the sign of all eigenvalues of the matrix B as that same of AB. (Hint: $|\lambda I-AB|=|A||\lambda A^{-1}-B|$).

Gaussian Processes and Learning Dynamic for Wide Neural Networks

2.1 INTRODUCTION

In Chapter 1, we have introduced three types of deep neural networks. Despite unprecedented performance of deep neural networks across a wide range of tasks (Devlin et al. 2018), theoretical analysis of deep neural networks including the gradient-based training dynamics of the parameter estimation are intractable (Lee et al. 2019). Alternative to deep neural networks, in this chapter, we introduce wide neural networks where the number of hidden units in a fully connected layer or the number of channels in a convolutional layer is infinite. The wide neural networks have several remarkable features. First, the output of the wide neural networks can be analytically expressed (Borovykh 2019). Second, output dynamics of the evolving wide neural networks are a Gaussian process (GP) (Matthews et al. 2018; Novak et al. 2019). Third, the wide neural networks provide a powerful tool for understanding the network output and its generalization capabilities, which are essential for the success of the neural networks (Li et al. 2019).

In this chapter, I first introduce linear models for learning output of the wide neural networks.

2.2 LINEAR MODELS FOR LEARNING IN NEURAL NETWORKS

2.2.1 Notation and Mathematic Formulation of Dynamics of Parameter Estimation Process

Let $x \in R^{n_0}$ be the input data and $y \in R^k$ be the labels. Assume that N individuals are sampled. Define

$$X^n = \begin{bmatrix} x_1^n \cdots x_{n_0}^n \end{bmatrix}, X = \begin{bmatrix} X^1 \\ \vdots \\ X^N \end{bmatrix},$$

$$Y^n = \begin{bmatrix} Y_1^n \cdots Y_k^n \end{bmatrix} \text{ and } Y = \begin{bmatrix} Y^1 \\ \vdots \\ Y^N \end{bmatrix}.$$

Let $D = \{X, Y\}$ be the training set. Consider a fully-connected feed forward network with L hidden layers and n_l neurons in the lth layer for $i = 1,\ldots,L$, and an output layer with $n_{L+1} = k$ neurons that correspond to the number of labels. Define

$$z^l(x) = \begin{bmatrix} z_1^l & \cdots & z_{n_l}^l \end{bmatrix}, h^l(z) = \begin{bmatrix} h_1^l(z) & \cdots & h_{n_l}^l(z) \end{bmatrix},$$

$$W^{l+1} = \begin{bmatrix} w_{11}^{l+1} & \cdots & w_{1n_{l+1}}^{l+1} \\ \vdots & \cdots & \vdots \\ w_{n_l 1}^{l+1} & \cdots & w_{n_l n_{l+1}}^{l+1} \end{bmatrix}, \text{ and } b^{l+1} = \begin{bmatrix} b_1^{l+1} \\ \vdots \\ b_{n_{l+1}}^{l+1} \end{bmatrix}$$

be the pre- and post-activation function at layer l with input x.

The nonlinear transformation relationships in a feed-forward neural network are given by

$$h^{l+1}(z) = z^l W^{l+1} + b^{l+1} \quad (2.1)$$

$$z^{l+1} = \varphi(h^{l+1}), \quad (2.2)$$

DOI: 10.1201/9781003028543-2

where φ is a nonlinear activation function. The weights and biases can be defined as functions of random variables as follows:

$$W_{ij}^l = \frac{\sigma_\omega}{\sqrt{n_l}}\omega_{ij}^l \text{ and } b_j^l = \sigma_b \beta_j^l, \quad (2.3)$$

where σ_ω^2 and σ_b^2 are variances of weight and bias, ω_{ij}^l, $\beta_j^l \sim N(0,1)$ at initialization. The parametrization methods in equations (2.1)–(2.3) are called the Neural Tangent Kernel (NTK) parameterization.

Let

$$\theta^l = \begin{bmatrix} \omega_{.1}^l \\ \beta_1^l \\ \vdots \\ \omega_{n_l}^l \\ \beta_{n_l}^l \end{bmatrix}, Y^n = f_t^T\left(x^n, \theta(t)\right)$$

$$= \left[\, f_t^1\left(x^n, \theta(t)\right) \cdots f_t^k\left(x^n, \theta(t)\right) \,\right]$$

$$= \left[\, h_1^{L+1}(x^n, \theta(t)) \cdots h_k^{L+1}(x^n, \theta(t)) \,\right], \theta = \begin{bmatrix} \theta^1 \\ \vdots \\ \theta^{L+1} \end{bmatrix},$$

$$\hat{Y}^n = \left[\, \hat{Y}_1^n \cdots \hat{Y}_k^n \,\right] = \hat{f}\left(x^n, \theta\right), \hat{Y} = \begin{bmatrix} \hat{Y}^1 \\ \vdots \\ \hat{Y}^N \end{bmatrix}.$$

Let $l\left(\hat{Y}^n, Y^n\right) = l\left(f_t\left(x^n, \theta_t\right), Y\right) : R^k \times R^k \to R$ be the loss function. Widely used loss functions are mean square error (MSE), which is defined as

$$l\left(\hat{Y}^n, Y^n\right) = \frac{1}{2}\left\|\hat{Y}^n - Y^n\right\|_2^2, \quad (2.4a)$$

and cross entropy loss which is defined as

$$l\left(f_t\left(x^n, \theta_t\right), Y^n\right) = -\sum_{i=1}^{k} \hat{y}_i^n \log y_i^n$$

$$= -\sum_{i=1}^{k} \hat{y}_i^n \log f_i\left(x^n, \theta\right). \quad (2.4b)$$

The loss function for N sample is

$$\mathcal{L}\left(f_t\left(x, \theta_t\right)\right) = \frac{1}{N}\sum_{n=1}^{N} l\left(f_t\left(x^n, \theta_t\right), Y^n\right). \quad (2.5)$$

The lost function is a composite function of parameters θ_t. By chain rule, the gradient of the loss function with respect to θ is given by

$$\nabla_\theta \mathcal{L} = \frac{\partial f_t^T}{\partial \theta}\frac{\partial \mathcal{L}}{\partial f_t}. \quad (2.6)$$

The recursive procedures for updating the parameters θ using gradient descent method are

$$\theta^{t+1} - \theta^t = -\eta\frac{\partial f_t^T}{\partial \theta}\frac{\partial \mathcal{L}}{\partial f_t}, \quad (2.7)$$

where η is a learning rate. Equation (2.7) can be approximated by

$$\dot{\theta}_t = -\eta\frac{\partial f_t^T}{\partial \theta}\frac{\partial \mathcal{L}}{\partial f_t}. \quad (2.8)$$

Define

$$f_t(X) = \begin{bmatrix} f_t(X^1) \\ \vdots \\ f_t(X^N) \end{bmatrix} \text{ and } \theta = \begin{bmatrix} \theta_1 \\ \vdots \\ \theta_{L+1} \end{bmatrix}, \text{ which imply}$$

that

$$\frac{\partial f_t(X)}{\partial \theta^T}\frac{\partial f_t^T(X)}{\partial \theta} = \sum_{l=1}^{L+1}\frac{\partial f_t(X)}{\partial \theta_l^T}\frac{\partial f_t^T(X)}{\partial \theta_l}.$$

By rules of derivative of composite function, we obtain

$$\frac{\partial f_t(X)}{\partial t} = \frac{\partial f_t(X)}{\partial \theta^T}\dot{\theta}_t = -\eta\frac{\partial f_t(X)}{\partial \theta^T}\frac{\partial f_t^T(X)}{\partial \theta}\frac{\partial \mathcal{L}}{\partial f_t(\chi)}$$

$$= -\eta \Pi^t(X, X)\frac{\partial \mathcal{L}}{\partial f_t(\chi)}, \quad (2.9)$$

where

$$\Pi^t(X, X) = \frac{\partial f_t(\chi)}{\partial \theta^T}\frac{\partial f_t^T(\chi)}{\partial \theta}$$

$$= \sum_{l=1}^{L+1}\frac{\partial f_t(X)}{\partial \theta_l^T}\frac{\partial f_t^T(X)}{\partial \theta_l}. \quad (2.10)$$

and is called the tangent kernel at time t.

Example 2.1: Neural Tangent Kernel (NTK) at the First Layer

Now we calculate the NTK $\Pi^t(X, X)$ at time t.

Consider layer $L=1$. It is known that

$$\theta_1 = \begin{bmatrix} W_{.1}^1 \\ \vdots \\ W_{.n_0}^1 \\ b_1 \end{bmatrix}.$$

Then, we have

$$\Pi^t(X^n, X^n) = \frac{\partial f_t(X)}{\partial \theta_1^T}\frac{\partial f_t^T(X)}{\partial \theta_1}$$

$$= \begin{bmatrix} \frac{\partial f_t^1}{\partial \theta_1^T} \\ \vdots \\ \frac{\partial f_t^k}{\partial \theta_1^T} \end{bmatrix} \begin{bmatrix} \frac{\partial f_t^1}{\partial \theta_1} & \cdots & \frac{\partial f_t^k}{\partial \theta_1} \end{bmatrix}$$

$$= \begin{bmatrix} \frac{\partial f_t^1}{\partial \theta_1^T}\frac{\partial f_t^1}{\partial \theta_1} & \cdots & \frac{\partial f_t^1}{\partial \theta_1^T}\frac{\partial f_t^k}{\partial \theta_1} \\ \vdots & \vdots & \vdots \\ \frac{\partial f_t^k}{\partial \theta_1^T}\frac{\partial f_t^1}{\partial \theta_1} & \cdots & \frac{\partial f_t^k}{\partial \theta_1^T}\frac{\partial f_t^k}{\partial \theta_1} \end{bmatrix}$$

$$= \begin{bmatrix} \prod_{11}^t (X^n, X^n) \cdots \prod_{1k}^t (X^n, X^n) \\ \vdots \quad \vdots \quad \vdots \\ \prod_{k1}^t (X^n, X^n) \cdots \prod_{kk}^t (X^n, X^n) \end{bmatrix}.$$

Recall that

$$f_t^i(X^n, \theta) = \frac{\sigma_\omega}{\sqrt{n_0}} X^n W_{.i}^{(1)} + \sigma_b \beta_i^{(0)},$$

$$f_t^j(X^n, \theta) = \frac{\sigma_\omega}{\sqrt{n_0}} \left(W_{.j}^{(0)}\right)^T (X^n)^T + \sigma_b \beta_j^{(0)},$$

Therefore, we have

$$\frac{\partial f_t^i(X^n)}{\partial \theta^T} = \begin{bmatrix} 0 \cdots 0 & \frac{\sigma_\omega}{\sqrt{n_0}} X^n & \sigma_b & 0 \cdots 0 \end{bmatrix},$$

$$\frac{\partial f_t^j(X^n)}{\partial \theta} = \begin{bmatrix} 0 \cdots 0 & \frac{\sigma_\omega}{\sqrt{n_0}} (X^n)^T & \sigma_b & 0 \cdots 0 \end{bmatrix} \text{ and}$$

$$\prod_{ij}^t (X^n, X^n) = \frac{\sigma_\omega^2}{n_0} X^n (X^n)^T + \sigma_b^2.$$

2.2.2 Linearized Neural Networks

The output $f_t(x, \theta_t)$ of neural network is a nonlinear function of parameters. The first order Taylor expansion of the output is given by

$$f_t^{lin}(x, \theta_t) = f_0(x, \theta_0) + \frac{\partial f_0}{\partial \theta^T}\omega_t, \tag{2.11}$$

where $\omega_t = \theta_t - \theta_0$.

It follows from equation (2.8) that

$$\frac{d\omega_t}{dt} = \frac{d\theta_t}{dt} = -\eta \frac{\partial f_t^T}{\partial \theta}\frac{\partial \mathcal{L}}{\partial f_t}. \tag{2.12}$$

Combining equations (2.11) and (2.12), we obtain the derivative of $f_t^{lin}(x, \theta_t)$:

$$\frac{df_t^{lin}(x, \theta_t)}{dt} = \frac{\partial f_0}{\partial \theta^T}\frac{d\omega_t}{dt} = -\eta \frac{\partial f_0}{\partial \theta^T}\frac{\partial f_0^T}{\partial \theta}\frac{\partial \mathcal{L}}{\partial f_t}$$

$$= -\eta \Pi_0(x, x)\frac{\partial \mathcal{L}}{\partial f_t}, \tag{2.13}$$

where $\Pi_0(x, x) = \frac{\partial f_0}{\partial \theta^T}\frac{\partial f_0^T}{\partial \theta}$.

For the MSE loss, we have

$$\frac{\partial l_t(X, \theta_t), Y)}{\partial f_t(X, \theta_t)} = 2\left(f_t(X, \theta_t) - Y\right). \tag{2.14}$$

Substituting equation (2.14) into equation (2.13) and using equation (2.11), we can obtain

$$\frac{d\omega_t}{dt} = -2\eta \frac{\partial f_0^T}{\partial \theta}\left(f_t(X, \theta_t) - Y\right)$$

$$\approx -2\eta \frac{\partial f_0^T}{\partial \theta}\left(f_t^{lin}(X, \theta_t) - Y\right)\Big)$$

$$= -2\eta \frac{\partial f_0^T}{\partial \theta}\left(\frac{\partial f_0}{\partial \theta^T}\omega_t + f_0(x, \theta_0) - Y\right),$$

$$= -2\eta \frac{\partial f_0^T}{\partial \theta}\frac{\partial f_0}{\partial \theta^T}\omega_t - \eta \frac{\partial f_0^T}{\partial \theta}\left(f_0(x, \theta_0) - Y\right)$$

$$= -2\eta \Pi_0(X, X)\omega_t - \eta \frac{\partial f_0^T}{\partial \theta}\left(f_0(X, \theta_0) - Y\right), \tag{2.15}$$

where $\Pi_0(X,X)=\frac{\partial f_0^T}{\partial\theta}\frac{\partial f_0}{\partial\theta^T}$,

and

$$\frac{df_t^{lin}(X,\theta_t)}{dt}=-2\eta\prod_0(X,X)\big(f_t(X,\theta_t)-Y\big)$$
$$\approx-\eta\prod_0(X,X)\big(f_t^{lin}(X,\theta_t)-Y\big)\big)$$
$$=-\eta\prod_0(X,X)f_t^{lin}(X,\theta_t)$$
$$+\eta\prod_0(X,X)Y \quad (2.16)$$

Now we solve ordinary equations (ODEs) (2.15) and (2.16). The solutions to ODEs (2.15) and (2.16) are (Appendix 2B)

$$\omega_t=-\prod_0^{-1}(X,X)\big(I-e^{-\eta\Pi_0(X,X)t}\big)$$
$$\frac{\partial f_0^T}{\partial\theta}\big(f_0(X,\theta_0)-Y\big). \quad (2.17)$$

Similarly, we can prove (Exercise 2.1)

$$f_t^{lin}(X)=\big(I-e^{-\eta\Pi_0(X,X)t}\big)Y+e^{-\eta\Pi_0(X,X)t}f_0(x). \quad (2.18)$$

For any point x, we have

$$f_t^{lin}(x)=\mu_t(x)+\gamma_t(x), \quad (2.19)$$

where

$$\mu_t(x)=\prod_0(x,\chi)\prod_0^{-1}(X,X)\big(I-e^{-\eta\Pi_0(X,X)t}\big)Y \quad (2.20)$$

and

$$\gamma_t(x)=f_0(x)-\prod_0(x,\chi)\prod_0^{-1}(X,X)$$
$$\big(I-e^{-\eta\Pi_0(X,X)t}\big)f_0(X). \quad (2.21)$$

When t go to infinity, we obtain

$$\mu_\infty(x)=\prod_0(x,\chi)\prod_0^{-1}(X,X),$$

$$\gamma_\infty(x)=f_0(x)\left(I-\prod_0(x,\chi)\prod_0^{-1}(X,X)\right)$$, which implies

$$f_\infty^{lin}(x)=f_0(x). \quad (2.22)$$

2.3 GAUSSIAN PROCESSES

2.3.1 Motivation

A goal of neural network is to identify unknown functions that map inputs to outputs. GP that generates distributions over functions is a powerful tool for analysis of wide neural networks. The GP captures a comprehensive picture of relations between inputs and outputs via infinite number of parameters (Schulz et al. 2018).

Let $X\in R^d$ be an input vector and Y be an output variable. The relations between input X and output Y can be statistically described by a conditional distribution of Y, given X. This relation can be decomposed into a systematic and a random component. We assume that the systematic variation is captured by a latent function $f(X)$. Therefore, the output Y can be modeled by

$$Y=f(X)+\varepsilon, \quad (2.23)$$

where ε follows a Gaussian distribution $\varepsilon\sim N(0,\sigma_\varepsilon^2)$ with mean 0 and variance σ_ε^2. If the latent function is a linear function, then the general function model (2.23) is reduced to the linear model:

$$Y=X^TW+\varepsilon, \quad (2.24)$$

where X is a column output vector and W is a column parameter vector.

Viewing the parameters as random variables and taking Bayesian regression approach, we assume that a Gaussian prior over the parameters is $P(W)=N(0,\Sigma)$ and the Gaussian likelihood is given by (Schulz et al. 2018):

$$P(Y|X,W)=N(X^TW,\sigma_\varepsilon^2). \quad (2.25)$$

The posterior distribution of the parameters W, given the observed data $\{X,Y\}$ is given by

$$P(W\mid X,Y)\propto P(W)P(Y\mid X,W)$$
$$\propto\exp\left\{-\frac{1}{2\sigma_\varepsilon^2}(Y-X^TW)^T(Y-X^TW)\right\}$$
$$\exp\left\{-\frac{1}{2}W^T\sum^{-1}W\right\}. \quad (2.26)$$

It can be shown that (Exercise 2.2)

$$\exp\left\{-\frac{1}{2\sigma_\varepsilon^2}\left(Y-X^TW\right)^T\left(Y-X^TW\right)\right\}$$

$$\exp\left\{-\frac{1}{2}W^T\sum^{-1}W\right\}$$

$$\infty \exp\left\{\frac{-1}{2}\left(W-\mu_W\right)^T\left(\frac{XX^T}{\sigma_\varepsilon^2}+\sum^{-1}\right)\left(W-\mu_W\right)\right\}. \tag{2.27}$$

Therefore, we obtain

$$P\left(W|X,Y\right)=N\left(\mu_W,\Lambda\right). \tag{2.28}$$

where

$$\mu_W=\left(\frac{1}{\sigma_\varepsilon^2}XX^T+\sum^{-1}\right)^{-1}XY,$$

$$\Lambda=\left(\frac{XX^T}{\sigma_\varepsilon^2}+\sum^{-1}\right)^{-1}.$$

Next we assume that a new input X_* is given. We want to estimate the posterior distribution of the function value $f_*=X_*^TW$, given the data $\left(X,Y\right)$, i.e.,

$$P(f_*\,|\,X,Y).$$

Since conditional distribution $P(W\,|\,X,Y)$ is Gaussian, the conditional distribution $P(f_*\,|\,X,Y)$ is also Gaussian distribution with the mean

$$\mu_*=X_*^TE\left[W\right]=X_*^T\left(\frac{1}{\sigma_\varepsilon^2}XX^T+\sum^{-1}\right)^TXY \tag{2.29}$$

and the variance

$$\sigma_*^2=X_*^T\Lambda X_*=X_*^T\left(\frac{XX^T}{\sigma_\varepsilon^2}+\sum^{-1}\right)^{-1}X_*. \tag{2.30}$$

In the linear models, if we assume that the prior distributions of the parameters are Gaussian, the posterior distributions of the parameters are also Gaussian and can be analytically computed. The distribution of the new output values corresponding new inputs can also be analytically calculated. However, only few relations in the real world are truly linear (Schulz et al. 2018), we need to extend the above approach from linear regression to nonlinear regression. This motives us to introducing GP in the next section.

2.3.2 Gaussian Process Models

Unlike linear regression where we can introduce Gaussian prior for the parameters, for the nonlinear regression models, we define a Gaussian prior for the nonlinear function. A GP can lead to different paths. These paths are usually referred to realizations of the process. Each realization can be interpreted as a nonlinear function. In theory, a nonlinear function is defined by infinite number of pairs of input and output. However, in the real world, the realization of the nonlinear function is the finite number of pairs of input and output. Therefore, we first define the multivariate Gaussian prior for the nonlinear function with the finite number of pairs of input and output. Consider the nonlinear function $f\left(x\right)$ defined on the following finite set of pairs of input and output:

$$y_i=f\left(x_i\right),\ i=1,\ldots,n. \tag{2.31}$$

We define the following n-variate Gaussian prior over the nonlinear function (2.31):

$$N\left(m(x),\sum_x\right), \tag{2.32}$$

where

$$m(x)=\begin{bmatrix}E\left[f\left(x_1\right)\right]\\ \vdots\\ E\left[f\left(x_n\right)\right]\end{bmatrix}, \tag{2.33}$$

and

$$\sum_x=\begin{bmatrix}E\left[\left(f\left(x_1\right)-m\left(x_1\right)\right)^2 & \cdots & E\left[\left(f\left(x_1\right)-m\left(x_1\right)\right)\left(f\left(x_n\right)-m\left(x_n\right)\right)\right]\\ \vdots & \vdots & \vdots\\ E\left[\left(f\left(x_n\right)-m\left(x_n\right)\right)\left(f\left(x_1\right)-m\left(x_1\right)\right)\right] & \cdots & E\left[\left(f\left(x_n\right)-m\left(x_n\right)\right)^2\right]\end{bmatrix}. \tag{2.34}$$

Now we are ready to define a GP (Schulz et al. 2018).

Definition 2.1: Gaussian Process

A GP is a real valued stochastic process indexed by x and is defined as a collection of any finite number of random variables that follow a joint multivariate Gaussian distribution.

The GP is understood as a distribution over functions $f(x)$ with a continuous time or space domain. GPs have two remarkable properties. First, it is clear from definition that a GP is fully determined by its mean $\mu(x)=E[f(x)]$ and covariance functions $K(x,x')=cov(f(x),f(x'))$. Second, prediction of output for a new input can be done by conditional distribution of the normal distribution, which is a linear function of the observed values. The GP over function $f(x)$ is denoted as

$$f(x) \sim GP(m(x), K(x,x')). \tag{2.35}$$

If the function is defined in equation (2.31). Then, the mean function $m(x)$ and covariance function (kernel) function are defined in equations (2.31) and (2.32), respectively.

A GP is completely determined by a mean function and covariance (kernel) function (Bartels and Hennig 2019). The covariance (kernel) function determines a distribution over functions. The covariance (kernel) function should be positive definite. Here we introduce several covariance (kernel) functions (Duvenaud 2020). The widely used covariance (kernel) function is the squared exponential kernel, or exponentiated quadratic covariance function (also called as the radial basis function (RBF kernel)). It is defined as

$$K_{SE}(x,x') = \sigma_f^2 exp\left\{-\frac{\|x-x'\|^2}{2\tau^2}\right\}, \tag{2.36}$$

where σ_f^2 is referred to as a signal variance and determines the average distance between the function and mean of GP, and τ^2 is a length scale, which determines the length of the "fluctuation" in the function.

Example 2.2 Rational Quadratic Kernel

Rational quadratic kernel is defined as

$$K_{RQ}(x,x') = \sigma^2\left(1+\frac{(x-x')^2}{2\alpha l^2}\right)^{-\alpha}, \tag{2.37}$$

where the parameter α measures the relative weighting of large-scale and small-scale variations. After some algebra, we obtain

$$\left(1+\frac{(x-x')^2}{2\alpha l^2}\right)^{-\alpha} = \left(1+\frac{1}{\frac{2l^2}{(x-x')^2}\alpha}\right)^{-\frac{(x-x')^2}{2l^2}\frac{2l^2}{(x-x')^2}\alpha}$$

$$\rightarrow \exp\left\{-\frac{(x-x')^2}{2l^2}\right\}, \alpha \rightarrow \infty. \tag{2.38}$$

This shows that when, $\alpha \rightarrow \infty$, Rational quadratic kernel converges to the exponentiated quadratic kernel. Adding many exponentiated quadratic kernels with different length scales can generate the rational quadratic kernel.

Exponentiated quadratic kernel and rational quadratic kernel are designed for smooth continuous functions. If functions are discontinuous then either lengthscale will become very small or posterior mean will become zero almost everywhere (Duvenaud 2020).

Example 2.3: Periodic Kernel

A periodic kernel is defined as

$$K_{per} = \sigma^2 \exp\left\{\frac{2sin^2\left(\frac{\pi|x-x'|}{p}\right)}{l^2}\right\}, \tag{2.39}$$

where the parameter p determines the period of the function and l measures the length of wiggles in the function. The periodic kernel is designed to fit the periodic functions.

Example 2.4: Locally Periodic Kernel

Locally periodic kernel is defined as the product of the exponentiated quadratic kernel and periodic kernel:

$$K_{LocalPer}(x,x') = K_{SE}(x,x')K_{per}(x,x')$$

$$= \sigma^2 exp\left\{-\frac{\|x-x'\|^2}{2\tau^2}\right\}\exp\left\{\frac{2sin^2\left(\frac{\pi|x-x'|}{p}\right)}{l^2}\right\}. \tag{2.40}$$

The local periodic kernel is designed to model the functions that are only locally periodic, but with time varying shape of the periodic part of the function.

2.3.3 Gaussian Processes for Regression

Similar to Bayesian regression, GP that models distributions over functions can be used for regression. Given a set of input and output variables as a training dataset, we can define a mean and a kernel function, which determine a GP. GP is then taken as a prior on the regression function. To predict the output of a set of the new input data, we calculate the posterior distribution of the output, given the set of the new input data. The posterior distribution is then used to predict the expected values and probability of the output variables. We first consider the prediction with noise-free observations and then investigate the prediction using noise observation.

2.3.3.1 Prediction with Noise-Free Observations

We assume that the observations are noise free. Consider a training dataset $X=\{x_i, i=1,\dots,n_1\}$ and $Y=\{y_i=f(x_i), i=1,\dots,n_1\}$. Given a new test dataset $X^*=\{x_j^*, j=1,\dots,n_2\}$. We want to predict $Y^*=f(X^*)$. It follows from the GP theory that for the finite number of samples, the outputs (Y, Y^*) follow the following joint Gaussian distribution:

$$\begin{bmatrix} Y \\ Y^* \end{bmatrix} \sim N\left(\begin{matrix} \mu(X) \\ \mu(X^*) \end{matrix} \begin{bmatrix} K(X,X) & K(X,X^*) \\ K(X^*,X) & K(X^*,X^*) \end{bmatrix}\right), \tag{2.41}$$

where

$$\mu(X)=E[f(X)]_{n_1\times 1},\ \mu(X^*)=E[f(X^*)]_{n_2\times 1},$$

$$K(X,X)=E\left[(f(X)-\mu(X))(f(X)-\mu(X))^T\right],$$

$$K(X,X^*)=E\left[(f(X)-\mu(X))(f(X^*)-\mu(X^*))^T\right],$$

$$K(X^*,X)=E\left[(f(X^*)-\mu(X^*))((f(X)-\mu(X))^T\right],$$

$$K(X^*,X^*)=E\left[(f(X^*)-\mu(X^*))(f(X^*)-\mu(X^*))^T\right].$$

Using joint distribution (2.39), we obtain the conditional distribution of Y, given Y, X, and X^*:

$$P(Y^*|Y,X,X^*)=N\left(\mu_{Y^*|Y}, \sum\nolimits_{Y^*|Y}\right), \tag{2.42}$$

where

$$\mu_{Y^*|Y}=\mu(X^*)+K(X^*,X)K(X,X)^{-1}(Y-\mu(X)), \tag{2.43}$$

and

$$\sum\nolimits_{Y^*|Y}=K(X^*,X^*)-(X^*,X)K(X,X)^{-1}K(X,X^*). \tag{2.44}$$

Equation (2.41) shows that the conditional mean $\mu_{Y^*|Y}$ is weighted averages of the observed variables Y with the kernel function K as weights. We can use the conditional mean $\mu_{Y^*|Y}$ of GP to predict outputs Y^* that correspond to the input X^*. Next we discuss the prediction with noise observation.

2.3.3.2 Prediction with Noise Observations

In the previous discussion, we assume that the predictions $f(X)=Y$ are from noiseless observation. Now we assume the noise observations and consider the model:

$$Y=f(X)+\varepsilon, \tag{2.45}$$

where ε follows a Gaussian distribution with variance σ_ε^2. We assume that the GP $GP(m(x), K(X,X))$ as a prior over the function $f(X)$. The outputs Y also follow the Gaussian distribution:

$$N(m(x), K(XX)+\sigma_\varepsilon^2 I). \tag{2.46}$$

Now the outputs (Y, Y^*) follow the following joint Gaussian distribution:

$$\begin{bmatrix} Y \\ Y^* \end{bmatrix} \sim N\left(\begin{matrix} \mu(X) \\ \mu(X^*) \end{matrix} \begin{bmatrix} K(X,X)+\sigma_\varepsilon^2 I & K(X,X^*) \\ K(X^*,X) & K(X^*,X^*) \end{bmatrix}\right). \tag{2.47}$$

The conditional distribution of Y^*, given Y, X, and X^*:

$$P(Y^*|Y, X, X^*) = N\left(\mu_{Y^*|Y}, \sum_{Y^*|Y}\right), \quad (2.48)$$

where

$$\mu_{Y^*|Y} = \mu(X^*) + K(X^*, X)$$

$$\left(K(X,X)+\sigma_\varepsilon^2 I\right)^{-1}\left(Y-\mu(X)\right), \quad (2.49)$$

and

$$\sum_{Y^*|Y} = K(X^*, X^*) - (X^*, X)\left(K(X,X)+\sigma_\varepsilon^2 I\right)^{-1} K(X, X^*). \quad (2.50)$$

For the noise observation, we only need to change the covariance (kernel) functions $K(X,X)$ of the function $f(X)$ to $K(X,X)+\sigma_\varepsilon^2 I$.

2.4 WIDE NEURAL NETWORK AS A GAUSSIAN PROCESS

In this section, we discuss how to derive a GP as a prior for the output of the wide neural networks. When the number of the nodes in the hidden layer approaches infinitely, the central limit theorem ensures that the output of the wide neural network converges to a multivariate Gaussian distribution. The Gaussian distribution will be determined by its mean and covariance (kernel) function. Next we discuss how to define the GP for the single-layer neural networks and multilayer neural networks (Lee et al. 2018).

2.4.1 Gaussian Process for Single-Layer Neural Networks

We first consider the input layer. The pre-activations, $Z_i(X)$, can be expressed as

$$Z_i^0(X) = b_i^0 + \sum_{k=1}^{d_{in}} W_{ik}^0 X_k \text{ and } Z_i^0(X') = b_i^0 + \sum_{k=1}^{d_{in}} W_{ik}^0 X'_k, \quad (2.51)$$

where d_{in} is the dimension of input vector $X \in R^{d_{in}}$.

We assume $W_{ik}^0 \sim N\left(0, \frac{\sigma_w^2}{d_{in}}\right)$, $b_j^0 \sim N(0, \sigma_b^2)$.

Then, it follows from the central limit theorem that

$$\begin{bmatrix} Z_i^0(X) \\ Z_i^0(X') \end{bmatrix} \sim N\left(\mu_Z K(X,X')\right), \quad (2.52)$$

where

$$\mu_Z = \begin{bmatrix} E[Z_i^0(X)] \\ E[Z_i^0(X')] \end{bmatrix} = \begin{bmatrix} 0 \\ 0 \end{bmatrix}, \text{ and}$$

$$K(X,X') = \begin{bmatrix} Var(Z_i^0(X)) & E[Z_i^0(X)Z_i^0(X')] \\ E[Z_i^0(X')Z_i^0(X)] & Var(Z_i^0(X')) \end{bmatrix}. \quad (2.53)$$

It follows from equation (2.51) that

$$Var(Z_i^0(X)) = \sigma_b^2 + var\left(\sum_{k=1}^{d_{in}} W_{ik}^0 X_k\right)$$

$$= \sigma_b^2 + \sum_{k=1}^{d_{in}} \frac{\sigma_w^2}{d_{in}} (X_k)^2 = \sigma_b^2 + \sigma_w^2 X^T X, \quad (2.54)$$

Similarly, we obtain

$$E[Z_i^0(X)Z_i^0(X')] = \sigma_b^2 + \sigma_w^2 X^T X',$$

$$E[Z_i^0(X')Z_i^0(X)] = \sigma_b^2 + \sigma_w^2 X'^T X \text{ and}$$

$$Var(Z_i^0(X')) = \sigma_b^2 + \sigma_w^2 X'^T X'. \quad (2.55)$$

Substituting equations (2.54) and (2.55) into equation (2.53), we obtain

$$K(X,X') = \begin{bmatrix} \sigma_b^2 + \sigma_w^2 X^T X & \sigma_b^2 + \sigma_w^2 X^T X' \\ \sigma_b^2 + \sigma_w^2 X'^T X & \sigma_b^2 + \sigma_w^2 X'^T X' \end{bmatrix}. \quad (2.56)$$

Therefore, GP for the pre-activation function $Z_i(X)$ is

$$Z_i(X) \sim GP(0, K(X,X')). \quad (2.57)$$

Next we consider the first layer. Suppose that the pre-activation function $Z_i^1(X)$ in the first layer is given by

$$Z_i^1(X) = b_i^1 + \sum_{j=1}^{n_1} W_{ij}^1 X_j^1(X), \quad (2.58)$$

where

$$X_j^1(X)=\phi\left(b_j^0+\sum_{j=1}^{d_{in}}W_{jk}^0X_k\right). \tag{2.59}$$

If we consider the collection of $Z_i^1(X^1),\ldots,Z_i^1(X^n)$, then the multidimensional central limit theorem implies that $Z_i^1(X^1),\ldots,Z_i^1(X^n)$ are asymptotically distributed as

$$N\left(\begin{bmatrix}0\\ \vdots\\ 0\end{bmatrix}\begin{bmatrix}K^1(X^1,X^1) & \cdots & K^1(X^1,X^n)\\ \vdots & \vdots & \vdots\\ K^1(X^n,X^1) & \cdots & K^1(X^n,X^n)\end{bmatrix}\right). \tag{2.60}$$

Now we calculate $K(X^i,X^j)$. By definition, we obtain

$$\begin{aligned}K^1(X^i,X^j)&=E\left[Z_i^1(X^i)Z_i^1(X^j)\right]\\ &=E\left[\left(b_i^1+\sum_{j=1}^{n_1}W_{ij}^1X_j^1(X^i)\right)\left(b_i^1+\sum_{j=1}^{n_1}W_{ij}^1X_j^1(X^j)\right)\right]\\ &=\sigma_b^2+\sum_{j=1}^{n_1}\sum_{k=1}^{n_1}E\left[W_{ij}^1W_{ik}^1\right]E\left[X_j^1(X^i)X_k^1(X^j)\right].\end{aligned} \tag{2.61}$$

Using the assumption that weights W_{ij}^1 and W_{ik}^1 are independent, we have

$$E\left[W_{ij}^1W_{ik}^1\right]=\begin{cases}\dfrac{\sigma_w^2}{d_{in}} & j=k\\ 0 & j\neq k\end{cases}. \tag{2.62}$$

Substituting equation (2.62) into equation (2.61) yields

$$\begin{aligned}K^1(X^i,X^j)&=\sigma_b^2+\sigma_w^2E\left[X_j^1(X^i)X_j^1(X^j)\right]\\ &=\sigma_b^2+\sigma_w^2C(X^i,X^j),\end{aligned} \tag{2.63}$$

where for ReLU nonlinear function, $C(X^i,X^j)$ is equal to (Appendix 2B, Cho and Saul 2009)

$$C(X^i,X^j)=\frac{1}{2\pi}\sqrt{\left(\sigma_b^2+\sigma_w^2\|X^i\|^2\right)\left(\sigma_b^2+\sigma_w^2\|X^j\|^2\right)}\left(\sin\theta_{X^i,X^j}^0+\left(\pi-\theta_{X^i,X^j}^0\right)\cos\theta_{X^i,X^j}^0\right),$$

where $\theta_{X^i,X^j}^0=arc\cos\left(\frac{\sigma_b^2+\sigma_w^2\langle X^i,X^j\rangle}{\sqrt{\left(\sigma_b^2+\sigma_w^2\|X^i\|^2\right)\left(\sigma_b^2+\sigma_w^2\|X^j\|^2\right)}}\right)$, $\langle X^i,X^j\rangle$ denotes inner product of X^i and X^j.

2.4.2 Gaussian Process for Multilayer Neural Networks

The derivation of the previous section can be extended to multilayer neural networks via induction. Consider the lth layer. We assume that the pre-activation function Z_i^{l-1} follows a GP $GP(0,K^{l-1}(X^m,X^n))$, where X^m and X^n are the inputs of the mth sample and n^{th} sample, respectively. We also assume that Z_i^{l-1} and Z_j^{l-1} are independent and identically distributed, which implies that $X_j^l(X)$ are independent and identically distributed. The pre-activation function Z_i^l in the lth layer is

$$Z_i^l(X)=b_i^l+\sum_{j=1}^{n_l}W_{ij}^lX_j^l(X),\ X_j^l(X)=\phi\left(Z_j^{l-1}(X)\right). \tag{2.64}$$

Consider a finite collection of $Z_i^1(X^1),\ldots,Z_i^1(X^n)$. Again, multidimensional central limit theorem implies that asymptotic joint distribution of $Z_i^1(X^1),\ldots,Z_i^1(X^n)$ is multivariate Gaussian distribution. By the same argument as that in the previous section,

$$Z_i^l(X)\sim GP\left(0,K^l(X^m,X^n)\right), \tag{2.65}$$

where

$$K^l(X^m,X^n)=\sigma_b^2+\sigma_w^2E_{Z_i^{l-1}\sim GP(0,K^{l-1})}\left[\phi\left(Z_i^{l-1}(X^m)\right)\phi\left(Z_i^{l-1}(X^n)\right)\right]. \tag{2.66}$$

By the previous GP assumption of Z_i^{l-1}, we have

$$\begin{bmatrix}Z_i^{l-1}(X^m)\\ Z_i^{l-1}(X^n)\end{bmatrix}\sim N\left(\begin{matrix}0 & K^{l-1}(X^m,X^m) & K^{l-1}(X^m,X^n)\\ 0 & K^{l-1}(X^n,X^m) & K^{l-1}(X^n,X^n)\end{matrix}\right). \tag{2.67}$$

Therefore, $E_{Z_i^{l-1} \sim GP(0, K^{l-1})}\left[\phi\left(Z_i^{l-1}(X^m)\right)\phi\left(Z_i^{l-1}(X^n)\right)\right]$ is a nonlinear function, denoted as

$$E_{Z_i^{l-1} \sim GP(0, K^{l-1})}\left[\phi\left(Z_i^{l-1}(X^m)\right)\phi\left(Z_i^{l-1}(X^n)\right)\right]$$
$$= F_\phi\left(K^{l-1}(X^m, X^m), K^{l-1}(X^m, X^n), K^{l-1}(X^n, X^n)\right). \quad (2.68)$$

Substituting equation (2.68) into equation (2.66) yields

$$K^l(X^m, X^n) = \sigma_b^2 + \sigma_w^2 F_\phi\left(K^{l-1}(X^m, X^m),\right.$$
$$\left. K^{l-1}(X^m, X^n), K^{l-1}(X^n, X^n)\right). \quad (2.69)$$

In Appendix 2B, we show that for ReLU nonlinear function, we have

$$K^l(X^m, X^n) = \sigma_b^2 + \frac{\sigma_w^2}{2\pi}\sqrt{K^{l-1}(X^m, X^m)K^{l-1}(X^n, X^n)}$$
$$\left(\sin\theta_{X^m,X^n}^{l-1} + \left(\pi - \theta_{X^m,X^n}^{l-1}\right)\cos\theta_{X^m,n}^{l-1}\right),$$

where

$$\theta_{X^m,X^n}^l = arc\cos\left(\frac{\sigma_b^2 + \sigma_w^2\langle X^m, X^n\rangle}{\sqrt{\left(\sigma_b^2 + \sigma_w^2\|X^m\|^2\right)\left(\sigma_b^2 + \sigma_w^2\|X^n\|^2\right)}}\right). \quad (2.70)$$

Now we can use Bayesian regression discussed in Section 2.3.3.2 to predict the output $Z_i^L(X^*)$ of the neural network, given new input X^*. Consider a training dataset $D_{tr} = \{(X^1, Y^1), \ldots, (X^n, Y^n)\}$ and a test dataset $D_{te} = \{X^*\}$. In the earlier section, we shown that the GP for Z_i^L is

$$Z_i^L \sim GP(0, K^L).$$

We assume the model:

$$Y = Z(x) + \varepsilon. \quad (2.71)$$

Using equation (2.47), we obtain

$$\begin{bmatrix} Y \\ Y^* \end{bmatrix} \sim N\left(\begin{bmatrix} 0 \\ 0 \end{bmatrix}\begin{bmatrix} K^L(X,X)+\sigma_\varepsilon^2 I & K^L(X,X^*) \\ K^L(X^*,X) & K^L(X^*,X^*) \end{bmatrix}\right). \quad (2.72)$$

From equations (2.48) and (2.49), the conditional distribution of Y^*, given X, Y, X^*, is

$$Y^* \sim N\left(\mu_{Y^*|Y} \;\; K_{Y^*|Y}\right), \quad (2.73)$$

where

$$\mu_{Y^*|Y} = K^L(X^*, X)\left(K^L(X,X) + \sigma_\varepsilon^2 I\right)^{-1} Y \quad (2.74)$$

and

$$K_{Y^*|Y} = K^L(X^*, X^*) - K^L(X^*, X)$$
$$\left(K^L(X,X) + \sigma_\varepsilon^2 I\right)^{-1} K^L(X, X^*). \quad (2.75)$$

The detailed expressions of equations (2.74) and (2.75) are

$$K^L(X,X) = \begin{bmatrix} K^L(X^1, X^1) & \cdots & K^L(X^1, X^n) \\ \vdots & \vdots & \vdots \\ K^L(X^n, X^1) & \cdots & K^L(X^n, X^n) \end{bmatrix}, \quad (2.76)$$

$$K^L(X, X^*) = \begin{bmatrix} K^L(X^1, X^{*1}) & \cdots & K^L(X^1, X^{*m}) \\ \vdots & \vdots & \vdots \\ K^L(X^n, X^{*1}) & \cdots & K^L(X^n, X^{*m}) \end{bmatrix}, \quad (2.77)$$

$$K^L(X^*, X^*) = \begin{bmatrix} K^L(X^{*1}, X^{*1}) & \cdots & K^L(X^{*1}, X^{*m}) \\ \vdots & \vdots & \vdots \\ K^L(X^{*m}, X^{*1}) & \cdots & K^L(X^{*m}, X^{*m}) \end{bmatrix}, \quad (2.78)$$

and

$$Y = \begin{bmatrix} Y^{*1} \\ \vdots \\ Y^{*m} \end{bmatrix}. \quad (2.79)$$

An essential problem for the success of Gaussian regression is to calculate the kernel functions. For the ReLU function, we introduced analytical formula (2.70) to calculate the kernel function. For other nonlinear functions, we need to use numerical methods to calculate the integral in equation (2.68):

$$F_{mn} = \int_{-\infty}^{\infty} \phi(u_g)\phi(u_h)$$
$$c\exp\left\{-\frac{1}{2}\begin{bmatrix} u_g & u_h \end{bmatrix}\begin{bmatrix} s & sr \\ sr & s \end{bmatrix}\begin{bmatrix} u_g \\ u_h \end{bmatrix}\right. du_g du_h, \quad (2.80)$$

where

$$c = \frac{1}{\int_{-\infty}^{\infty}\int_{-\infty}^{\infty} \exp\left\{\frac{-1}{2}\begin{bmatrix} u_g & u_h \end{bmatrix}\begin{bmatrix} s & sr \\ sr & s \end{bmatrix}\begin{bmatrix} u_g \\ u_h \end{bmatrix}\right\} du_g du_h},$$

$$s = K^l(X^m, X^m) = K^l(X^n, X^n),$$

$$r = \frac{K^l(X^m, X^n)}{\sqrt{K^l(X^m, X^m) K^l(X^n, X^n)}} = \frac{K^l(X^m, X^n)}{S}.$$

Numerical algorithm for kernel function calculation is summarized Algorithm 2.1 (Lee et al. 2018).

Algorithm 2.1: Numerical Calculation of Kernel Function

Step 1: All inputs are normalized to have identical norm.

Step 2: To calculate integral, the pre-activation function $u(Z_i^l)$ is set to be in the interval $u = [-u_{max}, \ldots, u_{max}]$, set variances s in the interval $[0, \ldots, s_{max}]$, and correlations $r = [-1, \ldots, 1]$. The intervals for u, s, and r are divided into n_g, n_v and n_r small intervals, respectively. All intervals and sampling grids are fixed and are reused across data points and layers.

Step 3: Calculate a look up table for F_{mn} in equation (2.68) or (2.80). The total number of elements in the look up table is $n_v n_r$. Using equation (2.80), we calculate the element in the look up table:

$$F_{ij} = \frac{\sum_g \sum_h \phi(u_g)\phi(u_h)\exp\left\{-\frac{1}{2}\begin{bmatrix} u_g & u_h \end{bmatrix}\begin{bmatrix} s_i & s_i r_j \\ s_i r_j & s_i \end{bmatrix}\begin{bmatrix} u_g \\ u_h \end{bmatrix}\right\}}{\sum_g \sum_h \exp\left\{-\frac{1}{2}\begin{bmatrix} u_g & u_h \end{bmatrix}\begin{bmatrix} s_i & s_i r_j \\ s_i r_j & s_i \end{bmatrix}\begin{bmatrix} u_g \\ u_h \end{bmatrix}\right\}}, \tag{2.81}$$

where $u_g, u_h \in = [-u_{max}, \ldots, u_{max}]$, $s_i \in [0, \ldots, s_{max}]$, $r_j \in [-1, \ldots, 1]$.

APPENDIX 2A: RECURSIVE FORMULA FOR NTK CALCULATION

First we denote the pre- and post-activation functions at layer l by $h^l(X)$ and $\alpha^l(X)$, respectively. Their recursive relations in the network are defined as

$$\alpha_0(X, \theta) = X, \tag{2A1}$$

$$h^{l+1}(X, \theta) = \frac{1}{\sqrt{n_l}} W^l \alpha^l(X, \theta) + \beta b^l, \tag{2A2}$$

$$\alpha^l(X, \theta) = \sigma(h^l(X, \theta)), \tag{2A3}$$

where σ is a nonlinear function.

The NTK is defined as

$$\Pi^t(\chi, \chi) = \frac{\partial f_t(\chi)}{\partial \theta^T} \frac{\partial f_t^T(\chi)}{\partial \theta}, \tag{2A4}$$

where

$$\prod^t(\chi, \chi) = \begin{bmatrix} \frac{\partial f_{t1}(X_1, \theta)}{\partial \theta^T} \\ \vdots \\ \frac{\partial f_{t1}(X_n, \theta)}{\partial \theta^T} \\ \vdots \\ \frac{\partial f_{tk}(X_1, \theta)}{\partial \theta^T} \\ \vdots \\ \frac{\partial f_{tk}(X_N, \theta)}{\partial \theta^T} \end{bmatrix} \begin{bmatrix} \frac{\partial f_{t1}(X_1, \theta)}{\partial \theta} \cdots \frac{\partial f_{t1}(X_n, \theta)}{\partial \theta} \cdots \frac{\partial f_{tk}(X_1, \theta)}{\partial \theta} \cdots \frac{\partial f_{tk}(X_N, \theta)}{\partial \theta} \end{bmatrix} \text{ or}$$

$$\prod^t(\chi, \chi) = \begin{bmatrix} \pi_{11}(X_1, X_1) & \cdots & \pi_{11}(X_1, X_N) & \cdots & \pi_{1n_l}(X_1, X_1) & \cdots & \pi_{1k}(X_1, X_N) \\ \vdots & & \vdots\ \vdots\ \vdots & & & \vdots\ \vdots\ \vdots & \\ \pi_{11}(X_N, X_1) & \cdots & \pi_{11}(X_N, X_N) & \cdots & \pi_{1n_l}(X_N, X_1) & \cdots & \pi_{1k}(X_N, X_N) \\ \vdots & \vdots & \vdots & \vdots & \vdots & \vdots & \vdots \\ \pi_{k1}(X_1, X_1) & \cdots & \pi_{k1}(X_1, X_N) & \cdots & \pi_{kk}(X_1, X_1) & \cdots & \pi_{kk}(X_1, X_N) \\ \vdots & \vdots & \vdots & \vdots & \vdots & \vdots & \vdots \\ \pi_{k1}(X_N, X_1) & \cdots & \pi_{k1}(X_N, X_N) & \ldots & \pi_{kk}(X_N, X_1) & \ldots & \pi_{kk}(X_N, X_N) \end{bmatrix},$$

$$\pi_{ij}(X_n, X_m) = \frac{\partial f_{ti}(X_n, \theta)}{\partial \theta^T} \frac{\partial f_{tj}^T(X_m, \theta)}{\partial \theta}.$$

Now we consider layer $L = 1$. Then, $f_t(X, \theta) = h^1(X, \theta)$. Define

$$\prod_1^t(\chi, \chi) = \frac{\partial h^1(\chi, \theta)}{\partial \theta^T} \frac{\partial (h^1(\chi, \theta))^T}{\partial \theta}.$$

Hence, we have

$$f_{ti}(X_n,\theta)=h_i^1(X_n,\theta)=\frac{1}{\sqrt{n_0}}X_n^T\left(W_{i.}^{(0)}\right)^T+\beta b_i^{(0)}, \quad (2A5)$$

$$f_{tj}(X_m,\theta)=h_j^1(X_m,\theta)=\frac{1}{\sqrt{n_0}}W_{j.}^{(0)}X_m+\beta b_j^{(0)}, \quad (2A6)$$

where $W_{i.}^{(0)}=\left[W_{i1}^{(0)}, W_{i2}^{(0)},\ldots,W_{in_0}^{(0)}\right]$.

Therefore, we have

$$\frac{\partial f_{ti}(X_n,\theta)}{\partial\theta^T}=\left[0\cdots0\frac{1}{\sqrt{n_0}}X_n^T\ \beta\ 0\cdots0\cdots0\right] \text{ and} \quad (2A7)$$

$$\frac{\partial f_{tj}(X_m,\theta)}{\partial\theta}=\begin{bmatrix}0\\ \vdots\\ 0\\ 0\\ 0\\ \vdots\\ \frac{1}{\sqrt{n_0}}X_m\\ \beta\\ \vdots\\ 0\end{bmatrix}. \quad (2A8)$$

The row index of the element in the NTK is determined by the index of the neuron in the network, and column index of the element in the NTK is determined by the index of the data samples. We define the element in the NTK as

$$\Pi_{ij}^t(X_n,X_m)=\frac{\partial f_{ti}(X_n,\theta)}{\partial\theta^T}\frac{\partial f_{tj}(X_m,\theta)}{\partial\theta}$$

$$=\begin{cases}\frac{1}{n_0}X_n^TX_m+\beta^2 & i=j\\ 0 & i\neq j\end{cases}, \quad (2A9)$$

where $i=1,\ldots,n_1$, $j=1,\ldots,n_1$, $n=1,\ldots,N$, $m=1,\ldots,N$.

Let

$$\pi^1=\begin{bmatrix}\frac{1}{n_0}X_1^TX_1+\beta^2 & \cdots & \frac{1}{n_0}X_1^TX_N+\beta^2\\ \vdots & \vdots & \vdots\\ \frac{1}{n_0}X_N^TX_1+\beta^2 & \cdots & \frac{1}{n_0}X_N^TX_N+\beta^2\end{bmatrix}.$$

In the first layer, we obtain

$$\Pi_1^t(\chi,\chi)=\pi^1\otimes I_{n_1},$$

where ⊗ denotes Kronecker product and I_{n_1} is a n_1 dimensional identity matrix.

We define

$$\Gamma^1=\tilde{\Gamma}^1\otimes I_{n_1}, \quad (2A10)$$

where $\tilde{\Gamma}^1=\pi^1$.

Next consider layer l. The NTK for layer l is

$$\Pi_l^t(\chi,\chi)=\frac{\partial h^l(\chi,\theta)}{\partial\theta^T}\frac{\partial h^l(\chi,\theta)^T}{\partial\theta}. \quad (2A11)$$

Equation (2A11) can be written as

$$\Pi_l^t(\chi,\chi)=\begin{bmatrix}\Pi_l^t(X_1,X_1) & \cdots & \Pi_l^t(X_1,X_N)\\ \vdots & \ddots & \vdots\\ \Pi_l^t(X_N,X_1) & \cdots & \Pi_l^t(X_N,X_N)\end{bmatrix}, \quad (2A12)$$

where

$$\Pi_l^t(X_n,X_m)=\begin{bmatrix}\pi_{11}^l(X_n,X_m) & \cdots & \pi_{1n_l}^l(X_n,X_m)\\ \vdots & \ddots & \vdots\\ \pi_{n_l1}^l(X_n,X_m) & \cdots & \pi_{n_ln_l}^l(X_n,X_m)\end{bmatrix},$$

$$\pi_{ij}^l(X_n,X_m)=\frac{\partial h_i^l(X_n,\theta)}{\partial\theta^T}\frac{\partial h_j^l(X_m,\theta)^T}{\partial\theta}.$$

It follows from equation (2A2) that

$$h^l(X,\theta)=\frac{1}{\sqrt{n_{l-1}}}W^{l-1}\alpha^{l-1}(X,\theta)+\beta b^{l-1},$$

or

$$h_i^l(X_n,\theta)=\frac{1}{\sqrt{n_{l-1}}}W_{i.}^{l-1}\alpha^{l-1}(X_n,\theta)+\beta b_i^{l-1},\ i=1,\ldots,n_l \quad (2A13)$$

which implies that

$$\frac{\partial h^l(X,\theta)}{\partial \theta^T}=\frac{1}{\sqrt{n_{l-1}}}W^{l-1}\frac{\partial \alpha^{l-1}(X,\theta)}{\partial \theta^T}+\beta\frac{\partial b_i^{l-1}}{\partial \theta^T}+\frac{\partial\left[\frac{1}{\sqrt{n_{l-1}}}W^{l-1}\right]}{\partial \theta^T}$$

$$\frac{\partial h_i^l(X_n,\theta)}{\partial \theta^T}=\frac{1}{\sqrt{n_{l-1}}}W_{i.}^{l-1}\frac{\partial \alpha^{l-1}}{\partial \theta^T}+\frac{1}{\sqrt{n_{l-1}}}\alpha^{l-1}(X_n,\theta)^T\frac{\partial (W_{i.}^{l-1})^T}{\partial \theta^T}+\beta\frac{\partial b_i^l}{\partial \theta^T}. \quad (2A14)$$

Recall equation (2A3)

$$\alpha^{l-1}(X,\theta)=\sigma\left(h^{l-1}(X,\theta)\right) \quad (2A15)$$

and chain rule (1.4.6)

$$\frac{\partial \alpha^{l-1}(X_n,\theta)}{\partial \theta^T}=\frac{\partial \alpha^{l-1}(X_n,\theta)}{\partial h^{l-1}(X_n,\theta)^T}\frac{\partial h^{l-1}(X_n,\theta)}{\partial \theta^T}. \quad (2A16)$$

It follows from equation (2A15) that

$$\frac{\partial \alpha^{l-1}(X_n,\theta)}{\partial h^{l-1}(X_n,\theta)^T}=\begin{bmatrix}\dot{\sigma}\left(h_1^{l-1}(X_n,\theta)\right)\cdots & 0\\ \vdots & \ddots & \vdots\\ 0 & \cdots\dot{\sigma}\left(h_{n_{l-1}}^{l-1}(X_n,\theta)\right)\end{bmatrix}. \quad (2A17)$$

Combining equations (2A16) and (2A17) yields

$$\frac{\partial \alpha^{l-1}(X_n,\theta)}{\partial \theta^T}=\begin{bmatrix}\dot{\sigma}\left(h_1^{l-1}(X_n,\theta)\right)\frac{\partial h_1^{l-1}(X_n,\theta)}{\partial \theta^T}\\ \vdots\\ \dot{\sigma}\left(h_{n_{l-1}}^{l-1}(X_n,\theta)\right)\frac{\partial h_{n_{l-1}}^{l-1}(X_n,\theta)}{\partial \theta^T}\end{bmatrix}. \quad (2A18)$$

Again, using equations (2A13), we obtain

$$\frac{\partial h_j^l(X_m,\theta)^T}{\partial \theta}=\frac{1}{\sqrt{n_{l-1}}}\frac{\partial \alpha^{l-1}(X_m,\theta)^T(W_{i.}^{l-1})^T}{\partial \theta}+\frac{1}{\sqrt{n_{l-1}}}\frac{\partial W_{i.}^{l-1}}{\partial \theta}\alpha^{l-1}(X_m,\theta)+\beta\frac{\partial (b_i^{l-1})^T}{\partial \theta}. \quad (2A19)$$

Similar to equation (2A16), we have

$$\frac{\partial \alpha^{l-1}(X_m,\theta)^T}{\partial \theta}=\frac{\partial h^{l-1}(X_m,\theta)^T}{\partial \theta}\frac{\partial \alpha^{l-1}(X_m,\theta)^T}{\partial h^{l-1}(X_m,\theta)}. \quad (2A20)$$

Using equations (2A15), we obtain

$$\frac{\partial \alpha^{l-1}(X_m,\theta)^T}{\partial h^{l-1}(X_m,\theta)}=\begin{bmatrix}\dot{\sigma}\left(h_1^{l-1}(X_m,\theta)\right)\cdots & 0\\ \vdots & \ddots & \vdots\\ 0 & \cdots\dot{\sigma}\left(h_{n_{l-1}}^{l-1}(X_m,\theta)\right)\end{bmatrix}. \quad (2A21)$$

Thus, combining equations (2A19) and (2A21), we obtain

$$\frac{\partial \alpha^{l-1}(X_m,\theta)^T}{\partial \theta}=\begin{bmatrix}\frac{\partial h_1^{l-1}(X_m,\theta)}{\partial \theta}\dot{\sigma}\left(h_1^{l-1}(X_m,\theta)\right) & \cdots & \frac{\partial h_{n_{l-1}}^{l-1}(X_m,\theta)\dot{\sigma}\left(h_{n_{l-1}}^{l-1}(X_m,\theta)\right)}{\partial \theta}\end{bmatrix}. \quad (2A22)$$

Similar to equation (2A7), we obtain

$$\frac{1}{\sqrt{n_{l-1}}}\alpha^{l-1}(X_n,\theta)^T\frac{\partial (W_{i.}^{l-1})^T}{\partial \theta^T}+\beta\frac{\partial b_i^l}{\partial \theta^T}=\begin{bmatrix}0 & \cdots & 0 & \frac{1}{\sqrt{n_{l-1}}}\alpha^{l-1}(X_n,\theta)^T & \beta & 0 & \cdots & 0 & \cdots & 0\end{bmatrix}. \quad (2A23)$$

Again, similar to equation (2A8), we obtain

$$\frac{1}{\sqrt{n_{l-1}}}\frac{\partial W_{i.}^{l-1}}{\partial \theta}\alpha^{l-1}(X_m,\theta)+\beta\frac{\partial (b_i^{l-1})^T}{\partial \theta}=\begin{bmatrix}0\\ \vdots\\ 0\\ 0\\ 0\\ \vdots\\ \frac{1}{\sqrt{n_{l-1}}}\alpha^{l-1}(X_m,\theta)\\ \beta\\ \vdots\\ 0\end{bmatrix}. \quad (2A24)$$

Recall that $\pi^l_{ij}(X_n, X_m)$ in equation (2A12) is defined as

$$\pi^l_{ij}(X_n, X_m) = \frac{\partial h^l_i(X_n,\theta)}{\partial \theta^T}\frac{\partial h^l_j(X_m,\theta)^T}{\partial \theta}. \quad \text{(2A25)}$$

Using equations (2A18)

$$\frac{\partial \alpha^{l-1}(X_n,\theta)}{\partial \theta^T}\frac{\partial \alpha^{l-1}(X_m,\theta)^T}{\partial \theta}$$

$$= \begin{bmatrix} \dot\sigma\left(h^{l-1}_1(X_n,\theta)\right)\frac{\partial h^{l-1}_1(X_n,\theta)}{\partial \theta^T} \\ \vdots \\ \dot\sigma\left(h^{l-1}_{n_{l-1}}(X_n,\theta)\right)\frac{\partial h^{l-1}_{n_{l-1}}(X_n,\theta)}{\partial \theta^T} \end{bmatrix}$$

$$\begin{bmatrix} \frac{\partial h^{l-1}_1(X_m,\theta)}{\partial \theta}\dot\sigma\left(h^{l-1}_1(X_m,\theta)\right) & \cdots & \frac{\partial h^{l-1}_{n_{l-1}}(X_m,\theta)\dot\sigma\left(h^{l-1}_{n_{l-1}}(X_m,\theta)\right)}{\partial \theta} \end{bmatrix}$$

$$= \begin{bmatrix} \phi^{l-1}_{11}(X_n,X_m) & \cdots & \phi^{l-1}_{1n_{l-1}}(X_n,X_m) \\ \vdots & \ddots & \vdots \\ \phi^{l-1}_{n_{l-1}1}(X_n,X_m) & \cdots & \phi^{l-1}_{n_{l-1}n_{l-1}}(X_n,X_m) \end{bmatrix} = \Phi^{l-1}(X_n,X_m), \quad \text{(2A26)}$$

where

$$\phi^{l-1}_{uv}(X_n,X_m)$$

$$= \dot\sigma\left(h^{l-1}_u(X_n,\theta)\right)\dot\sigma\left(h^{l-1}_v(X_m,\theta)\right)\frac{\partial h^{l-1}_u(X_n,\theta)}{\partial \theta^T}\frac{\partial h^{l-1}_v(X_m,\theta)}{\partial \theta}. \quad \text{(2A27)}$$

It follows from equation (2A12) that

$$\pi^{l-1}_{uv}(X_n,X_m) = \frac{\partial h^{l-1}_u(X_n,\theta)}{\partial \theta^T}\frac{\partial h^{l-1}_v(X_m,\theta)}{\partial \theta}. \quad \text{(2A28)}$$

Combining equations (2A27) and (2A28), we obtain

$$\phi^{l-1}_{uv}(X_n,X_m) = \dot\sigma\left(h^{l-1}_u(X_n,\theta)\dot\sigma\left(h^{l-1}_v(X_m,\theta)\right)\pi^{l-1}_{uv}(X_n,X_m)\right). \quad \text{(2A29)}$$

Define

$$\frac{\partial \alpha^{l-1}(\chi,\theta)}{\partial \theta^T}\frac{\partial \alpha^{l-1}(\chi,\theta)^T}{\partial \theta}$$

$$= \Phi^{l-1}(\chi,\chi) = \begin{bmatrix} \Phi^{l-1}(X_1,X_1), & \cdots & \Phi^{l-1}(X_1,X_N) \\ \vdots & \ddots & \vdots \\ \Phi^{l-1}(X_N,X_1), & \cdots & \Phi^{l-1}(X_N,X_N) \end{bmatrix}. \quad \text{(2A30)}$$

Combining equations (2A23) and (2A24) leads to

$$\left(\frac{1}{\sqrt{n_{l-1}}}\alpha^{l-1}(X_n,\theta)^T\frac{\partial\left(W^{l-1}_{i.}\right)^T}{\partial \theta^T} + \beta\frac{\partial b^l_i}{\partial \theta^T}\right)$$

$$\left(\frac{1}{\sqrt{n_{l-1}}}\frac{\partial W^{l-1}_{j.}}{\partial \theta}\alpha^{l-1}(X_m,\theta) + \beta\frac{\partial\left(b^{l-1}_j\right)^T}{\partial \theta}\right)$$

$$= \begin{cases} \frac{1}{n_{l-1}}\alpha^{l-1}(X_n,\theta)^T\alpha^{l-1}(X_m,\theta) + \beta^2 & i = j \\ 0 & i \neq j \end{cases} \quad \text{(2A31)}$$

Define

$$\gamma^{l-1}(X_n,X_m,\theta) = \frac{1}{n_{l-1}}\alpha^{l-1}(X_n,\theta)^T\alpha^{l-1}(X_m,\theta) + \beta^2, \quad \text{(2A32)}$$

$$\gamma^{l-1}(\chi,\chi) = \begin{bmatrix} \gamma^{l-1}(X_1,X_1) & \cdots & \gamma^{l-1}(X_1,X_N) \\ \vdots & \ddots & \vdots \\ \gamma^{l-1}(X_N,X_1) & \cdots & \gamma^{l-1}(X_N,X_N) \end{bmatrix}, \quad \text{(2A33)}$$

and

$$\Gamma^{l-1}(\chi,\chi) = \gamma^{l-1}(\chi,\chi)\otimes I_{n_{l-1}}. \quad \text{(2A34)}$$

Then,

$$\frac{\partial\left[\frac{1}{\sqrt{n_{l-1}}}W^{l-1}\alpha^{l-1}(\chi,\theta) + \beta b^{l-1}\right]}{\theta^T}$$

$$\frac{\partial\left[\frac{1}{\sqrt{n_{l-1}}}W^{l-1}\alpha^{l-1}(\chi,\theta) + \beta b^{l-1}\right]^T}{\partial \theta} = \Gamma^{l-1}(\chi,\chi). \quad \text{(2A35)}$$

Combining equations (2A11), (2A14), (2A30), and (2A35), we obtain

$$\prod_l^t(\chi,\chi) = \frac{\partial h^l(\chi,\theta)}{\partial \theta^T}\frac{\partial h^l(\chi,\theta)^T}{\partial \theta}$$

$$= \frac{1}{n_{l-1}}\tilde W^{l-1}\Phi^{l-1}(\chi,\chi)\left(\tilde W^{l-1}\right)^T + \Gamma^{l-1}(\chi,\chi), \quad \text{(2A36)}$$

where

$$\tilde{W}^{l-1} = \begin{bmatrix} W^{l-1} \\ \vdots \\ W^{l-1} \end{bmatrix}. \tag{2A37}$$

Let

$$\Delta = \frac{1}{n_{l-1}} \tilde{W}^{l-1} \Phi^{l-1}(\chi, \chi) \left(\tilde{W}^{l-1}\right)^T. \tag{2A38}$$

Then,

$$\Delta = \begin{bmatrix} \tilde{\Delta}_{11} & \cdots & \tilde{\Delta}_{1N} \\ \vdots & \ddots & \vdots \\ \tilde{\Delta}_{N1} & \cdots & \tilde{\Delta}_{NN} \end{bmatrix}, \tag{2A39}$$

where

$$\tilde{\Delta}_{nm} = \frac{1}{n_{l-1}} W^{l-1} \Phi^{l-1}(X_n, X_m) \left(W^{l-1}\right)^T. \tag{2A40}$$

It follows from equation (2A40) that

$$\tilde{\Delta}_{nm} = \begin{bmatrix} \Delta_{11}^{nm} & \cdots & \Delta_{1n_{l-1}}^{nm} \\ \vdots & \ddots & \vdots \\ \Delta_{n_{l-1}1}^{nm} & \cdots & \Delta_{n_{l-1}n_{l-1}}^{nm} \end{bmatrix}$$

$$= \frac{1}{n_{l-1}} \begin{bmatrix} W_{1.}^{l-1} \Phi^{l-1}(X_n, X_m) \left(W_{1.}^{l-1}\right)^T & \cdots & W_{1.}^{l-1} \Phi^{l-1}(X_n, X_m) \left(W_{n_{l-1}.}^{l-1}\right)^T \\ \vdots & \ddots & \vdots \\ W_{n_{l-1}.}^{l-1} \Phi^{l-1}(X_n, X_m) \left(W_{1.}^{l-1}\right)^T & \cdots & W_{n_{l-1}.}^{l-1} \Phi^{l-1}(X_n, X_m) \left(W_{n_{l-1}.}^{l-1}\right)^T \end{bmatrix}. \tag{2A41}$$

Using large number of theorem, we obtain

$$\Delta_{ij}^{nm} = \frac{1}{n_{l-1}} W_{i.}^{l-1} \Phi^{l-1}(X_n, X_m) \left(W_{j.}^{l-1}\right)^T \overset{a.s.}{\rightarrow}$$

$$\frac{1}{n_{l-1}} E\left[W_{i.}^{l-1} \Phi^{l-1}(X_n, X_m) \left(W_{j.}^{l-1}\right)^T \right]. \tag{2A42}$$

By the assumption, we have

$$E\left[\left(W_{j.}^{l-1}\right)^T \right] = 0 \tag{2A43}$$

and

$$Cov\left(\left(W_{j.}^{l-1}\right)^T, W_{i.}^{l-1} \right) = \begin{cases} \sigma_\omega^2 I_{n_{l-1}} & i = j \\ 0 & i \neq j \end{cases}. \tag{2A44}$$

By the formula for the expected value of a quadratic form, we obtain

$$\lambda_{ij}^{nm} = \frac{1}{n_{l-1}} E\left[W_{i.}^{l-1} \Phi^{l-1}(X_n, X_m) \left(W_{j.}^{l-1}\right)^T \right]$$

$$= \begin{cases} \sigma_\omega^2 \frac{1}{n_{l-1}} \mathrm{Tr}\left(\Phi^{l-1}(X_n, X_m)\right) & i = j \\ 0 & i \neq j \end{cases}. \tag{2A45}$$

Combining equations (2A41), (2A42), and (2A45), we obtain

$$\tilde{\Delta}_{nm} \overset{a.s}{\rightarrow} \frac{1}{n_{l-1}} \sigma_\omega^2 \mathrm{Tr}\left(\Phi^{l-1}(X_n, X_m)\right) I_{n_{l-1}}. \tag{2A46}$$

Combining equations (2A39), (2A40), and (2A46) yields

$$\Delta \overset{a.s}{\rightarrow} \Lambda^{l-1}(\chi, \chi) = \Psi^{l-1}(\chi, \chi) \otimes I_{n_{l-1}}, \tag{2A47}$$

where

$$\Psi^{l-1}(\chi, \chi) = \begin{bmatrix} \pi^l(X_1, X_1) & \cdots & \pi^l(X_1, X_n) \\ \vdots & \ddots & \vdots \\ \pi^l(X_N, X_1) & \cdots & \pi^l(X_N, X_N) \end{bmatrix}$$

$$= \frac{1}{n_{l-1}} \sigma_w^2 \begin{bmatrix} \mathrm{Tr}\left(\Phi^{l-1}(X_1, X_1)\right) & \cdots & \mathrm{Tr}\left(\Phi^{l-1}(X_1, X_N)\right) \\ \vdots & \ddots & \vdots \\ \mathrm{Tr}\left(\Phi^{l-1}(X_N, X_1)\right) & \cdots & \mathrm{Tr}\left(\Phi^{l-1}(X_N, X_N)\right) \end{bmatrix}. \tag{2A48}$$

Combining equations (2A36) and (2A47), we obtain

$$\prod_l^t (\chi, \chi) = \Lambda^{l-1}(\chi, \chi) + \Gamma^{l-1}(\chi, \chi),$$

which implies

$$\prod_l^t (\chi, \chi) = \pi^l(\chi, \chi) \otimes I_{n_{l-1}}$$

$$= \Psi^{l-1}(\chi, \chi) \otimes I_{n_{l-1}} + \gamma^{l-1}(\chi, \chi) \otimes I_{n_{l-1}}, \tag{2A49}$$

where

$$\pi^l(\chi, \chi) = \begin{bmatrix} \pi^l(X_1, X_1) & \cdots & \pi^l(X_1, X_n) \\ \vdots & \ddots & \vdots \\ \pi^l(X_N, X_1) & \cdots & \pi^l(X_N, X_N) \end{bmatrix}.$$

If we assume in equation (2A28) that

$$\pi_{uv}^{l-1}(X_n, X_m) = 0, \text{ for } u \neq v. \qquad (2A50)$$

Using large number of theorem, equations (2A29) and (2A30), we obtain

$$\pi^{l-1}(X_n, X_m) = \sigma_\omega^2 \frac{1}{n_{l-1}} \sum_{u=1}^{n_{l-1}} \pi_{uu}^{l-1}(X_n, X_m)$$
$$= \sigma_\omega^2 E\left[\pi_{11}^{l-1}(X_n, X_m)\right]. \qquad (2A51)$$

We assume that

$$\dot{\sigma}\left(h_u^{l-1}(X_n,\theta)\right)\dot{\sigma}\left(h_v^{l-1}(X_m,\theta)\right) \xrightarrow{a.s} E\left[\dot{\sigma}\left(u(X_n)\dot{\sigma}\left(V(X_m)\right)\right)\right]. \qquad (2A52)$$

Using equations (2A26) and (2A29), we obtain

$$\frac{1}{n_{l-1}} \mathrm{Tr}\left(\Phi^{l-1}(X_n, X_m)\right) = \frac{1}{n_{l-1}} \sum_{u=1}^{n_{l-1}} \phi_{uu}^{l-1}(X_n, X_m)$$

$$= \frac{1}{n_{l-1}} \sum_{u=1}^{n_{l-1}} \dot{\sigma}\left(h_u^{l-1}(X_n,\theta)\right)\dot{\sigma}\left(h_v^{l-1}(X_m,\theta)\right)\pi_{uu}^{l-1}(X_n, X_m)$$

$$\xrightarrow{a.s} E\left[\dot{\sigma}\left(u(X_n)\dot{\sigma}\left(V(X_m)\right)\right)\right]\pi^{l-1}(X_n, X_m) \qquad (2A53)$$

Combining equations (2A36)–(2A39), (2A47)–(2A49), and (2A53), we have

$$\pi^l(X_n, X_m) = \gamma^{l-1}(X_n, X_m) + \sigma_\omega^2 E\left[\dot{\sigma}\left(h_u^{l-1}(X_n,\theta)\right)\dot{\sigma}\left(h_v^{l-1}(X_m,\theta)\right)\right] \pi^{l-1}(X_n, X_m). \qquad (2A54)$$

We show that when the number of neurons in the layer l increases to infinite, the output of the network in the layer l will converge to a GP. Recall that the output $h^l(X,\theta)$ of the network layer l is defined as

$$h^l(X,\theta) = \frac{1}{\sqrt{n_{l-1}}} W^{l-1}\alpha^{l-1}(X,\theta) + \beta b^{l-1}.$$

We assume that the weights and bias parameters are i.i.d, and

$$E\left[W^{l-1}\right] = 0, \text{ and}$$

$$E\left[W^{l-1}\left(W^{l-1}\right)^T\right] = n_{l-1}\sigma_w^2 I_{n_{l-1}}.$$

In summary, the recursive formulas for the calculation of the NTK are given as follows (Appendix 2A):

$$\prod_l^t(\chi,\chi) = \pi^l(\chi,\chi) \otimes I_{n_{l-1}}$$
$$= \Psi^{l-1}(\chi,\chi) \otimes I_{n_{l-1}} + \gamma^{l-1}(\chi,\chi) \otimes I_{n_{l-1}}, \; l = 1, \ldots, L+1,$$

$$\prod_l^t(X_n, X_m) = \begin{bmatrix} \pi_{11}^l(X_n, X_m) & \cdots & \pi_{1n_l}^l(X_n, X_m) \\ \vdots & \ddots & \vdots \\ \pi_{n_l 1}^l(X_n, X_m) & \cdots & \pi_{n_l n_l}^l(X_n, X_m) \end{bmatrix},$$

$$\pi_{uv}^l(X_n, X_m) = \frac{\partial h_u^l(X_n,\theta)}{\partial \theta^T} \frac{\partial h_v^l(X_m,\theta)^T}{\partial \theta},$$

$$\Psi^{l-1}(\chi,\chi) = \begin{bmatrix} \pi^l(X_1, X_1) & \cdots & \pi^l(X_1, X_n) \\ \vdots & \ddots & \vdots \\ \pi^l(X_N, X_1) & \cdots & \pi^l(X_N, X_N) \end{bmatrix},$$

$$= \frac{1}{n_{l-1}} \sigma_w^2 \begin{bmatrix} \mathrm{Tr}\left(\Phi^{l-1}(X_1, X_1)\right) & \cdots & \mathrm{Tr}\left(\Phi^{l-1}(X_1, X_N)\right) \\ \vdots & \ddots & \vdots \\ \mathrm{Tr}\left(\Phi^{l-1}(X_N, X_1)\right) & \cdots & \mathrm{Tr}\left(\Phi^{l-1}(X_N, X_N)\right) \end{bmatrix},$$

$$\Phi^{l-1}(X_n, X_m) = \begin{bmatrix} \phi_{11}^{l-1}(X_n, X_m) & \cdots & \phi_{1n_{l-1}}^{l-1}(X_n, X_m) \\ \vdots & \ddots & \vdots \\ \phi_{n_{l-1}1}^{l-1}(X_n, X_m) & \cdots & \phi_{n_{l-1}n_{l-1}}^{l-1}(X_n, X_m) \end{bmatrix},$$

$$\phi_{uv}^{l-1}(X_n, X_m) = \dot{\sigma}\left(h_u^{l-1}(X_n,\theta)\dot{\sigma}\left(h_v^{l-1}(X_m,\theta)\right)\right) \pi_{uv}^{l-1}(X_n, X_m),$$

$$\pi_{uv}^{l-1}(X_n, X_m) = \frac{\partial h_u^{l-1}(X_n,\theta)}{\partial \theta^T} \frac{\partial h_v^{l-1}(X_m,\theta)^T}{\partial \theta}.$$

$$\gamma^{l-1}(\chi,\chi)=\begin{bmatrix} \gamma^{l-1}(X_1,X_1) & \cdots & \gamma^{l-1}(X_1,X_N) \\ \vdots & \ddots & \vdots \\ \gamma^{l-1}(X_N,X_1) & \cdots & \gamma^{l-1}(X_N,X_N) \end{bmatrix},$$

$$\gamma^{l-1}(X_n,X_m,\theta)=\frac{1}{n_{l-1}}\alpha^{l-1}(X_n,\theta)^T\alpha^{l-1}(X_m,\theta)+\beta^2.$$

APPENDIX 2B: ANALYTIC FORMULA FOR PARAMETER ESTIMATION IN THE LINEARIZED NEURAL NETWORKS

Define ODE

$$\frac{d\omega_t}{dt}=-P\omega_t-Q, \tag{2B1}$$

where $P=\eta\Pi_0$, $Q=\eta\frac{\partial f_0^T}{\partial\theta}\left(f_0(x,\theta_0)-Y\right)$.

First, we solve equation

$$\frac{d\omega_t}{dt}=-P\omega_t. \tag{2B2}$$

Equation (2B2) can be transformed to

$$\frac{d\log\omega_t}{dt}=-P. \tag{2B3}$$

Solving ODE (2B3), we obtain

$$\log\omega_t=-Pt+c,$$

or

$$\omega_t=e^{-Pt}C. \tag{2B4}$$

Taking derivative on both sides of equation (2B4), we obtain

$$\frac{d\omega_t}{dt}=\frac{dC}{dt}e^{-Pt}-PCe^{-Pt}=-PCe^{-Pt}-Q, \tag{2B5}$$

which implies

$$e^{-Pt}\frac{dC}{dt}=-Q, \text{ or}$$

$$\frac{dC}{dt}=-e^{Pt}Q. \tag{2B6}$$

Solving equation (2B6), we obtain

$$C=-\int_0^t e^{\eta\Pi_0 d\tau}Q+C_0=-\eta^{-1}\Pi_0^{-1}\left(e^{\eta\Pi_0 t}-I\right)Q+C_0. \tag{2B7}$$

Since $\omega_0=0$, we have $C_0=0$. Therefore, we obtain

$$C=-\eta^{-1}\Pi_0^{-1}\left(e^{\eta\Pi_0 t}-I\right)Q. \tag{2B8}$$

Substituting equation (2B8) into equation (2B4), we obtain

$$\omega_t=-\frac{\partial f_0^T}{\partial\theta}\Pi_0^{-1}\left(I-e^{-\eta\Pi_0 t}\right)\left(f_0(x,\theta_0)-Y\right). \tag{2B9}$$

EXERCISES

EXERCISE 2.1
Please show

$$f_t^{lin}(x)=\left(I-e^{-\eta\Pi_0 t}\right)Y+e^{-\eta\Pi_0 t}f_0(x).$$

EXERCISE 2.2
Please show

$$\exp\left\{-\frac{1}{2\sigma_\varepsilon^2}(Y-X^TW)^T(Y-X^TW)\right\}$$

$$\exp\left\{-\frac{1}{2}W^T\sum^{-1}W\right\}$$

$$\propto\exp\left\{\frac{-1}{2}(W-\mu_W)^T\left(\frac{XX^T}{\sigma_\varepsilon^2}+\sum^{-1}\right)(W-\mu_W)\right\}.$$

CHAPTER 3

Deep Generative Models

3.1 VARIATIONAL INFERENCE

3.1.1 Introduction

Variational inference (VI) is a powerful method for representation learning that maps high-dimensional data to low-dimensional latent space and intends to approximate the posterior distribution (Wainwright and Jordan 2008; Kingma and Welling 2013; Rezende et al. 2014; Ranganath et al. 2016; Nguyen et al. 2017; Blei et al. 2018; Jeremias et al. 2019; Lin et al. 2019; Tang et al. 2019). Variational neural networks have been applied to computer vision, information retrieval, machine translation, image translation, music composition, molecular and chemical design, and speech recognition (Gasulla 2015; Wang et al. 2017; Ban et al. 2018; Salha et al. 2019; Yellapragada and Konkimalla 2019). We take statistical approach to VI to fully understand VI. Most materials in this section are from Blei et al. (2018).

3.1.2 Variational Inference as Optimization

Let $\boldsymbol{x} \in \boldsymbol{R}^D$ be a set of observed variables and $\boldsymbol{z} \in \boldsymbol{R}^d$ be a set of latent variables, where $\boldsymbol{d} \ll \boldsymbol{D}$. Suppose that the observed variables $\boldsymbol{x}$ are generated by the latent variable $\boldsymbol{z}$. Let $\boldsymbol{X} = \left[\boldsymbol{x}_1, \ldots, \boldsymbol{x}_n\right]^T$ and $Z = \left[z_1, \ldots, z_n\right]^T$. Suppose that each $\boldsymbol{z}_i$ is sampled from the prior distribution $\boldsymbol{p}_0\left(\boldsymbol{z}_i\right)$ and each data point $\boldsymbol{x}_i$ is generated from the conditional distribution $\boldsymbol{p}_\theta\left(\boldsymbol{x}_i \mid \boldsymbol{z}_i\right)$ independently (Blei et al. 2018).

Bayesian inference is to compute the posterior distribution of latent variables given observations:

$$p(z|x) = \frac{p(x,z)}{p(x)}, \tag{3.1}$$

where $\boldsymbol{p}(\boldsymbol{x}, \boldsymbol{z})$ is a joint distribution of observed variable $\boldsymbol{x}$ and latent variable $\boldsymbol{z}$, and the denominator is the marginal density $\boldsymbol{p}(\boldsymbol{x})$ of observations and also is referred to as the evidence. The marginal distribution can be calculated by marginalizing out the latent variables from the joint distribution:

$$p(x) = \int p(x,z)\boldsymbol{dz}. \tag{3.2}$$

Since the data dimension is very high and distribution function may be complex, in general, the integral does not have analytical solutions. Computing integral is intractable.

Example 3.1

Consider two unit-variance univariate Gaussian distributions $\boldsymbol{N}(\mu_1, \boldsymbol{1})$ and $\boldsymbol{N}(\mu_2, \boldsymbol{1})$. Assume that $\mu_1 \sim \boldsymbol{N}(\boldsymbol{0}, \sigma^2)$ and $\mu_2 \sim \boldsymbol{N}(\boldsymbol{0}, \sigma^2)$, where μ_1 and μ_2 are independent, σ^2 is a hyperparameter. To generate an observation $\boldsymbol{x}_i$, we define a latent cluster assignment indicator variable $\boldsymbol{c}_i$, using one-hot encoding, where $\boldsymbol{c}_i = [\boldsymbol{1}, \boldsymbol{0}]^T$ indicates an observation $\boldsymbol{x}_i$ comes from the first class, $\boldsymbol{c}_i = [\boldsymbol{0}, \boldsymbol{1}]^T$ indicates an observation $\boldsymbol{x}_i$ comes from the second class, and $\boldsymbol{c}_i \sim$ ***categorical*** $\left(\frac{1}{2}, \frac{1}{2}\right)$. Then, we sample $\boldsymbol{c}_i$ from categorical (1/2, ½) distribution to determine which latent cluster observation comes from and generate $\boldsymbol{x}_i$ from the corresponding Gaussian distribution $\boldsymbol{N}(\boldsymbol{c}_i^T \mu, \boldsymbol{1})$, where $\mu = [\mu_1, \mu_2]^T$. It is clear that the latent variables are $z = \{\mu, \boldsymbol{c}\}$, where $\boldsymbol{c}$ is label

DOI: 10.1201/9781003028543-3

assignment indicator variable, which can take value 0 or 1. The marginal distribution for this example is then given by

$$p(x)=\int p(\mu_1,\mu_2)\big[p(c_1=0)p(x_1|c_1=0,\mu)+p(c_1=1)p(x_1|c_1=1,\mu)\big]+\big[p(c_2=0)p(x_2|c_2=0,\mu)+p(c_2=1)p(x_2|c_2=1,\mu)\big]+\big[p(c_3=0)p(x_3|c_3=0,\mu)+p(c_3=1)p(x_3|c_3=1,\mu)\big]d\mu_1 d\mu_2. \quad (3.3)$$

From equation (3.3), we observe that the term involved in equation (3.3) is 2^3. If the number of clusters for mixture of Gaussians is K, then the time complexity of numerical integrations is $O(K^n)$. Computing marginal distribution $p(x)$ is intractable in practice. Therefore, instead of directly calculating posterior distribution, we approximate posterior distribution with a simpler distribution via VI, which sidesteps the intractability of computing marginal distribution. In other words, we define a set Ω of distributions over the latent variables. Let $q(z)\in\Omega$ be a candidate distribution or variational distribution. The goal of VI is to approximate the posterior distribution $p(z|x)$ by $q(z)$. We use Kullback-Leibler (KL) divergence (relative entropy) to measure the difference between $q(z)$ and $p(z|x)$. The VI is reduced as the following optimization problem:

$$q^*(z)=\underset{q(z)\in\Omega}{\operatorname{argmin}} KL\big(q(z)||p(z|x)\big). \quad (3.4)$$

3.1.3 Variational Bound and Variational Objective

Objective function in equation (3.4) still involves the marginal distribution $p(x)$. To see this, we rewrite:

$$\begin{aligned} KL\big(q(z)||p(z|x)\big) &= E_{q(z)}\log q(z)-E_{q(z)}\log p(z|x)\\ &= E_{q(z)}\log q(z)-E_{q(z)}\log\frac{p(x,z)}{p(x)}\\ &= E_{q(z)}\log q(z)-E_{q(z)}\log p(x,z)+E_{q(z)}\log p(x)\\ &= E_{q(z)}\log q(z)-E_{q(z)}\log p(x,z)+\log p(x). \end{aligned} \quad (3.5)$$

Equation (3.5) shows that the objective function in equation (3.4) implicitly depends on the marginal distribution $p(x)$, which implies that computation of $KL\big(q(z)||p(z|x)\big)$ is intractable. Fortunately, $\min_{q(z)} KL\big(q(z)||p(z|x)\big)$ is equivalent to $\min_{q(z)} KL(q(z)||p(x,z)$. In fact, $\min_{q(z)} KL(q(z)||p(x,z)$ is equivalent to

$$\begin{aligned} &\min_{q(z)} KL\big(q(z)||p(x,z)\big)\\ &=\min_{q(z)} E_{q(z)}\log q(z)-E_{q(z)}\log p(x,z)\\ &=\min_{q(z)}\big\{E_{q(z)}\log q(z)-E_{q(z)}\log p(x,z)+\log p(x)\big\}\\ &=\min_{q(z)} KL\big(q(z)||p(z|x)\big). \end{aligned} \quad (3.6)$$

Let,

$$\mathrm{ELBO}(q)=E_{q(z)}\log p(x,z)-E_{q(z)}\log q(z). \quad (3.7)$$

Combining equations (3.5) and (3.6) yields

$$\log p(x)=KL\big(q(z)||p(z|x)\big)+\mathrm{ELBO}(q), \quad (3.8)$$

which implies

$$\log p(x)\geq \mathrm{ELBO}(q). \quad (3.9)$$

Optimization problem (3.4) can be reduced to

$$q^*(z)=\underset{q(z)\in\Omega}{\operatorname{argmax}}\,\mathrm{ELBO}(q), \quad (3.10)$$

where ELBO is called the evidence lower bound.

Recall that

$$E_{q(z)}\log p(x,z)=E_{q(z)}\log p(z)+E_{q(z)}\log p(x|z). \quad (3.11)$$

Substituting equation (3.11) into equation (3.7) yields

$$\mathrm{ELBO}(q)=E_{q(z)}\log p(x|z)-KL\big(q(z)||p(z)\big). \quad (3.12)$$

When $q(z)=p(z)$, equations (3.9), (3.10), and (3.12) show that the optimization choses values of z that explain the observed data.

3.1.4 Mean-Field Variational Inference

3.1.4.1 A General Framework

The complexity of optimization is largely determined by the complexity of the variational family $\boldsymbol{\Omega}$. In this section, we introduce the mean-field variational family that is a fully factorized distribution (Blei et al. 2018). The mean-field variational family should have two features: (1) accuracy of approximating the posterior distribution and (2) simplicity of the family that leads to tractable approximation. A mean-field approximation assumes that the latent variables are mutually independent and the variational density of each latent variable is determined by a distinct factor. The mean-field variational family for latent variables z is

$$q(z)=\prod_{j=1}^{m} q_j(z_j), \tag{3.13}$$

where $q_j(z_j)$ is a variational factor.

A fully factorized variational distribution allows iteratively performing optimization of each factor, each time updating one latent variable while keeping other latent variables unchanged. Indeed, substituting factorization in equation (3.13) into equation (3.7), we obtain

$$\text{ELBO}(q)=\int q_j(z_j)E_{q(z_{-j})}\log p(x,z_j|z_{-j})dz_j - \int q(z_j)\log q(z_j)dz_j + c_j, \tag{3.14}$$

where z_{-j} denotes the set z excluding z_j and c_j denotes a constant that includes all terms that are constant with respect to z_j in the update of z_j.

Next we need to find $q^*(z_j)$ that minimizes

$$\text{ELBO}(q(z_j))=\int q_j(z_j)E_{q(z_{-j})}\log p(x,z_j|z_{-j})dz_j - \int q(z_j)\log q(z_j)dz_j. \tag{3.15}$$

Using calculus of variations, we obtain (Appendix 3A)

$$q^*(z_j)\propto Exp\left\{E_{q(z_{-j})}\log p(z_j\,|\,z_{-j},x)\right\}, \tag{3.16}$$

or

$$q^*(z_j)\propto Exp\left\{E_{q(z_{-j})}\log p(z_j,z_{-j},x)\right\}. \tag{3.17}$$

Equations (3.16) and (3.17) present algorithms for iteratively updating variational distribution $q(z)$ of each latent variable, which will finally converge the optimal solution. This algorithm is called coordinate ascent variational inference (CAVI) algorithm (Bishop 2006; Blei et al. 2018). It is summarized as follows.

CAVI Algorithm

Step 1: Initialization of variational factors $q_j(z_j)$, $j=1,\ldots,m$.

Calculate $ELBO^{(0)}(q)=E[\log p(x,z)]-E[\log q(z)]$.

Step 2: Set $t \leftarrow t+1$

Step 3: For $j\in\{1,\ldots,m\}$ do

$$q_j(z_j)\leftarrow \exp\left\{E_{q(z_{-j})}\log p(z_j\,|\,z_{-j},x)\right\}$$

End

Step 4: Compute

$$ELBO^{(t)}(q)=E[\log p(x,z)]-E[\log q(z)].$$

Step 5: Check convergence

If $|ELBO^{(t)}(q)-ELBO^{(t-1)}(q)|\le\varepsilon$ then stop, output $q(z)$;

Otherwise, go to Step 2, where ε is a prespecified error.

3.1.4.2 Bayesian Mixture of Gaussians

3.1.4.2.1 Overview As an example of the mean-field variational family, we consider Bayesian mixture of Gaussian. Consider K mixture components and n sampled data points $x=[x_1,\ldots,x_n]^T$. The kth component follows a Gaussian distribution $N(\mu_k,1)$. We further assume that $\mu_k \sim N(\mu_k,\sigma_k^2)$ and they are independent. The mean-field variational family consists of two parts: (1) the distribution on the jth observations's mixture assignment $q(c_j;\varphi_j)$ where c_j is an indicator K-vector, all zeros except for a value 1 in the component corresponding to the cluster the data x_j comes from, and (2) a Gaussian distribution $q(\mu_k:m_k,\sigma_k^2)$ on the kth mixture component's mean parameter. Therefore, the mean-field variational family is of the form:

$$q(z)=q(\mu,c)=\prod_{k=1}^{K}q(\mu_k;m_k,\sigma_k^2)\prod_{j=1}^{n}q(c_j;\varphi_j). \tag{3.18}$$

Next, we define the joint distribution of the observed variable x and latent variable z. By the definition, the joint distribution $p(x,z)$ is given by

$$p(x,z)=p(\mu,c,x)$$
$$=\prod_{k=1}^{K}p(\mu_k)\prod_{j=1}^{n}p(c_j)p(x_j|c_j,\mu), \quad (3.19)$$

where

$$\mu\sim N(0,\sigma^2),\ p(c_j=k)=\frac{1}{K},\ p(x_j|c_j,\mu)$$
$$=\frac{1}{\sqrt{2\pi}}e^{-\frac{(x_j-c_j^T\mu)^2}{2}}.$$

We combine the joint distribution (3.19) and mean-field variational distribution family (3.18) to form the following ELBO for the mixture of Gaussian:

$$ELBO=\sum_{k=1}^{K}E\left[\log p(\mu_k)\right]$$
$$+\sum_{j=1}^{n}\left(E\left[\log p(c_j)\right]+E\left[\log p(x_j|c_j,\mu)\right]\right)$$
$$-\sum_{j=1}^{n}E\left[\log q(c_j)\right]-\sum_{k=1}^{K}E\left[\log q(\mu_k)\right]. \quad (3.20)$$

The CAVI algorithm performs optimization to update each variational parameter in turn. We first study updating the variational cluster assignment factor $q(c)$ and investigate the updating the variational mixture component factor $q(\mu)$.

3.1.4.2.2 Update the Variational Cluster Assignment Recall from equation (3.17) that

$$q^*(c_j)\propto\exp\left\{E_\mu\left[E_{c_{-j}}\left[\log p(\mu,c,x)\right]\right]\right.$$
$$\propto\exp\left\{\log p(c_j)+E\left[\log p(x_j|c_j,\mu)\right]\right\}. \quad (3.21)$$

In equation (3.19), we assume that

$$\log p(c_j)=-\log K,$$

which implies that $\log p(c_j)$ for all possible values of c_j and provides no information for updating cluster assignment distribution. Therefore, we need to only consider the second term in equation (3.21).

Since c_j is an indicator vector, the conditional probability $\log p(x_j|c_j,\mu)$ can be written as

$$p(x_j|c_j,\mu)=\prod_{k=1}^{K}p(x_j|\mu_k)^{c_{jk}}, \quad (3.22)$$

where

$$c_{jk}=\begin{cases}1 & \text{if } x_j \text{ is from the } k^{\text{th}} \text{ cluster}\\ 0 & \text{otherwise}\end{cases}.$$

Taking logarithm on both sides of equation (3.22), we obtain

$$\log p(x_j|c_j,\mu)=\sum_{k=1}^{K}c_{jk}\log p(x_j|\mu_k). \quad (3.23)$$

Taking expectation on both sides of equation (3.23), we obtain

$$\mathrm{E}\left[\log p(x_j|c_j,\mu)\right]=\sum_{k=1}^{K}c_{jk}\mathrm{E}\left[\log p(x_j|\mu_k)\right]. \quad (3.24)$$

It follows from equation (3.19) that

$$\mathrm{E}\left[\log p(x_j|\mu_k)\right]=E\left[-\frac{(x_j-\mu_k)^2}{2}\right]-\frac{1}{2}\log 2\pi. \quad (3.25)$$

Since expectation in equation (3.25) is taken over the parameter of the kth component of the Gaussian distribution and is not taken over the observed variable x_j, we obtain

$$E\left[(x_j-\mu_k)^2\right]=x_j^2-2E[\mu_k]x_j+E[\mu_k^2]. \quad (3.26)$$

Substituting equations (3.25) and (3.26) into equation (3.24) yields

$$E\left[\log p(x_j|c_j,\mu)\right]$$
$$=\sum_{k=1}^{K}c_{jk}\left(E[\mu_k]x_j-\frac{1}{2}E[\mu_k^2]\right)+\text{const.} \quad (3.27)$$

Substituting equation (3.27) into equation (3.12), we obtain

$$q^*(c_j) \propto \exp\left\{\sum_{k=1}^{K} c_{jk}\left(E[\mu_k]x_j - \frac{1}{2}E[\mu_k^2]\right)\right\}. \quad (3.28)$$

Equation (3.28) indicates that calculation requires $E[\mu_k]$ and $E[\mu_k^2]$ for each mixture component. Next we discuss how to calculate the variational density $q^*(\mu_k)$ for the kth mixture component.

3.1.4.2.3 Update the Variational Density of the Mixture Means It follows from equation (3.19) that the joint distribution of the observed variables x and the kth latent mixture component is

$$P(x, \mu_k) = p(\mu_k)\prod_{j=1}^{n} p(x_j \mid c_j, \mu). \quad (3.29)$$

Using equation (3.17), we calculate the coordinate-optimal $q(\mu_k)$ for the mean of the kth latent mixture component:

$$q(\mu_k) \propto \exp\left\{E_{q(\mu_{-k}, c_{-k})} \log P(x, \mu_k)\right\}$$
$$\propto \exp\left\{\log p(\mu_k) + \sum_{j=1}^{n} E_{q(\mu_{-k}, c_{-k})} \log P(x_j \mid c_j, \mu)\right\}. \quad (3.30)$$

Substituting equation (3.17) into equation (3.30) yields

$$q(\mu_k) \propto \exp\left\{\log p(\mu_k) + \sum_{j=1}^{n} E\left[c_{jk} \log p(x_j \mid \mu_k)\right]\right\}$$
$$\propto \exp\left\{\log p(\mu_k) + \sum_{j=1}^{K} E\left[c_{jk}\right] \log p(x_j \mid \mu_k)\right\}. \quad (3.31)$$

Define

$$\varphi_{jk} = E\left[c_{jk}\right]. \quad (3.32)$$

Substituting equation (3.32) into equation (3.31), we obtain

$$q(\mu_k) \propto \exp\left\{\log p(\mu_k) + \sum_{j=1}^{n} \varphi_{jk} \log p(x_j \mid \mu_k)\right\}. \quad (3.33)$$

Recall that

$$p(\mu_k) = \frac{1}{\sqrt{2\pi}} e^{-\frac{\mu_k^2}{2\sigma^2}}, \; p(x_j \mid c_j, \mu) = \frac{1}{\sqrt{2\pi}} e^{-\frac{(x_j - c_j^T \mu)^2}{2}}. \quad (3.34)$$

Substituting equation (3.34) into equation (3.33) yields

$$q(\mu_k) \propto \exp\left\{-\frac{\mu_k^2}{2\sigma^2} + \sum_{j=1}^{n} \varphi_{jk}\left(-\frac{(x_j - \mu_k)^2}{2}\right)\right\} + \text{const},$$
$$\propto \exp\left\{-\frac{\mu_k^2}{2\sigma^2} + \left(\sum_{j=1}^{n} \varphi_{jk} x_j\right)\mu_k - \left(\sum_{j=1}^{n} \varphi_{jk}\right)\frac{\mu_k^2}{2}\right\} + \text{const},$$
$$\propto \exp\left\{\left(\sum_{j=1}^{n} \varphi_{jk} x_j\right)\mu_k - \left(\frac{1}{2\sigma^2} + \frac{\sum_{j=1}^{n} \varphi_{jk}}{2}\right)\mu_k^2\right\} + \text{const}. \quad (3.35)$$

Equation (3.35) shows that μ_k follows a Gaussian distribution with the mean and variance given by

$$m_k = \frac{\sum_{j=1}^{n} \varphi_{jk} x_j}{1/\sigma^2 + \sum_{j=1}^{n} \varphi_{jk}} \text{ and } \sigma_k^2 = \frac{1}{1/\sigma^2 + \sum_{j=1}^{n} \varphi_{jk}}. \quad (3.36)$$

In summary, we introduce coordinate mean-field VI for Bayesian mixture of Gaussians. Variational updates consist of two procedures: (1) update for the cluster assignment:

$$q^*(c_j) \propto \exp\left\{\sum_{k=1}^{K} c_{jk}\left(E[\mu_k]x_j - \frac{1}{2}E[\mu_k^2]\right)\right\} \quad (3.37)$$

and (2) update for the variational mixture-component means and variances:

$$m_k = \frac{\sum_{j=1}^{n} \varphi_{jk} x_j}{1/\sigma^2 + \sum_{j=1}^{n} \varphi_{jk}} \text{ and } \sigma_k^2 = \frac{1}{1/\sigma^2 + \sum_{j=1}^{n} \varphi_{jk}}.$$

3.1.4.2.4 CAVI Algorithms for the Mixture of Gaussians The coordinate mean-field VI algorithms for Bayesian mixture of Gaussians are summarized as follows (Blei et al. 2018):

CAVI algorithms for the mixture of Gaussians.

Data and Parameter Input: $n \times D$ dimensional data matrix, number of components K, variance of prior Gaussian distribution for component means.

Output: Variational densities for the means $q(\mu_k; m_k, \sigma_k^2)$ and data point cluster assignment indicator vector $q(c_j; \varphi_j)$.

Step 1: Initialization. Variational means and variances $m^0 = [m_1^0, \ldots, m_k^0]^T$, $(s^2)^0 = [(\sigma_1^2)^0, \ldots, (\sigma_k^2)^0]^T$ and data point cluster assignment probabilities $\varphi^0 = [(\varphi_1)^0, \ldots, (\varphi_n)^0]^T$. Set $t = 0$.

Step 2: Set $t = t + 1$.

Step 3: For $j \in \{1, \ldots, n\}$ do

Set

$$\varphi_{jk}^t \propto \exp\left\{E\left[\mu_k^{t-1}; m_k^{t-1}, (\sigma_k^2)^{t-1}\right]x_j - E\left[(\mu_k^2)^{t-1}; m_k^{t-1}, (\sigma_k^2)^{t-1}\right]\right\}$$

End

Step 4: For $k \in \{1, \ldots, K)$ do

Set

$$m_k^t \leftarrow \frac{\sum_{j=1}^n \varphi_{jk}^t x_j}{\frac{1}{\sigma^2} + \sum_{j=1}^n \varphi_{jk}^t}$$

$$(\sigma_k^2)^t \leftarrow \frac{1}{\frac{1}{\sigma^{2+\sum_{j=1}^n \varphi_{jk}^t}}}.$$

end

Step 5: Compute ELBO

Compute

$$E\left[\log p(\mu_k^t)\right] = -\frac{(m_k^t)^2}{2\sigma^2}, E\left[\log P(c_j)\right] = -\log K,$$

$$E\left[\log P(x_j \mid c_j^t, \mu^t)\right] = \varphi_{jk}^t x_j m_k^t - \frac{\varphi_{jk}^t (m_k^t)^2}{2},$$

$$E\left[\log q(c_j^t)\right] = -\log K + \sum_{k=1}^K \left[\varphi_{jk}^t m_k^t x_j - \left((\sigma_k^2)^t + (m_k^t)^2\right]\right], \text{ and}$$

$$E\left[\log q(\mu_k^t)\right] = \left(\sum_{j=1}^n \varphi_{jk} x_j\right)\mu_k^t - \left(\frac{1}{2\sigma^2} + \frac{\sum_{j=1}^n \varphi_{jk}}{2}\right)(m_k^t)^2.$$

Compute

$$ELBO^t = \sum_{k=1}^K E\left[\log p(\mu_k^t)\right] + \sum_{j=1}^n \left(E\left[\log p(c_j)\right]\right. + E\left[\log P(x_j \mid c_j^t, \mu^t)\right]\left.\right) - \sum_{j=1}^n E\left[\log q(c_j^t)\right] - \sum_{k=1}^K E\left[\log q(\mu_k^t)\right].$$

Step 6: Check convergence

If $|ELBO^t - ELBO^{t-1}| \le \varepsilon$ then stop and output

$$q(\mu, c) = \prod_{k=1}^K q(\mu_k^t) \prod_{j=1}^n q(c_j^t);$$

else go to Step 2.

We can use the estimated variational density to calculate the posterior distribution and perform a posterior decomposition of the data (Blei et al. 2018). For example, we assume that the posterior of the kth mixture component can be estimated as a Gaussian distribution with means m_k and variance σ_k^2. Therefore, given data points, we can determine their most likely mixture component assignment $c_j = \underset{k}{\text{argmax}}\ \varphi_{jk}$ and estimate cluster mean as variational means m_k.

Bayesian mixture of Gaussian model assume that data come from the corresponding Gaussian $N\left(c_j^T \mu, 1\right)$. The density of new data can be predicted by

$$p\left(x_{new}|x\right)=\frac{1}{K}\sum_{k=1}^{K}\frac{1}{\sqrt{2\pi}}e^{-\frac{(x_{new}-m_k)^2}{2}}. \tag{3.38}$$

3.1.4.3 Mean-Field Variational Inference with Exponential Family

3.1.4.3.1 Exponential Family Complete Conditionals

In Section 3.1.4.2, we investigated Bayesian mixture of Gaussians, proposed a general coordinate-ascent algorithms for optimizing the ELBO, and obtained variational coordinate update in closed form. Gaussian distribution is a specific form of exponential family. We expect that the variational coordinate update for exponential family will also take the simple closed form. In this section, we generalize the variation inference for the Bayesian mixture of Gaussian to exponential family (Amari 1982; Blei et al. 2018).

We assume that each complete conditional follows the distribution in the exponential family:

$$p\left(z_j|z_{-j},x\right)$$
$$=\exp\left\{\eta_j\left(z_{-j},x\right)^T z_j - a\left(\eta_j\left(z_{-j},x\right)\right)\right\}h\left(z_j\right), \tag{3.39}$$

where latent variable z_j is a sufficient statistic, $\eta_j\left(z_{-j},x\right)$ is a function of other latent variables and observed variables that are taken as parameters, and $h\left(z_j\right)$ is a function of latent variable.

Variational exponential family includes Bayesian mixtures of exponential family models with conjugate priors, hierarchical hidden Markov models, Kalman filter models, and Bayesian linear regression.

Using equations (3.13) and (3.16), we can obtain mean-field variational density for the exponential family models:

$$q(z)=\prod_{j=1}^{m}q_j\left(z_j\right),$$

where

$$q_j\left(z_j\right)\propto\exp\left\{E\left[\log p\left(z_j\,|\,z_{-j},x\right)\right]\right\}. \tag{3.40}$$

Taking logarithm on both sides of equation (3.39) and then taking expectation, we obtain

$$\mathrm{E}\left[\log p\left(z_j|z_{-j},x\right)\right]$$
$$=E\left[\eta_j\left(z_{-j},x\right)\right]^T z_j - E\left[a\left(\eta_j\left(z_{-j},x\right)\right)\right]+\log h\left(z_j\right). \tag{3.41}$$

Substituting equation (3.41) into equation (3.40) leads to

$$q_j\left(z_j\right)\propto\exp\left\{\mathrm{E}\left[\eta_j\left(z_{-j},x\right)\right]^T z_j\right.$$
$$\left.-E\left[a\left(\eta_j\left(z_{-j},x\right)\right)\right]+\log h\left(z_j\right)\right\}$$
$$\propto h\left(z_j\right)\exp\left\{\mathrm{E}\left[\eta_j\left(z_{-j},x\right)\right]^T z_j\right\}. \tag{3.42}$$

Equation (3.42) shows that each optimal variational factor is in the same exponential family with the same base function $h(.)$ and log normalizer $a(.)$ as its corresponding complete conditional. Define the variational parameter for the jth variational factor:

$$v_j=E\left[\eta_j\left(z_{-j},x\right)\right]. \tag{3.43}$$

After computing v_j, variational factor can then be easily updated by

$$q_j\left(z_j\right)\propto h\left(z_j\right)\exp\left(v_j\right). \tag{3.44}$$

3.1.4.3.2 Conjugate Prior for Exponential Family

The exponential families that are readily integrated are very useful in practice. Conjugate priors have appealing computational properties, and hence are widely used in practice. In this section, we introduce concept of prior and a general form of conjugate prior for exponential family. Let $p(\theta|x)$ be the posterior distribution of variables θ after data x are observed and $p(\theta)$ be a prior distribution of θ.

Definition 3.1

The prior $P(\theta)$ and posterior distributions $p(\theta|x)$ are called conjugate distributions if they are in the same probability distribution family. The prior $P(\theta)$ is called a conjugate prior for the likelihood function $p(x|\theta)$.

The posterior distribution $p(\theta|x)$ is proportional to the product of prior distribution and likelihood as follows:

$$p(\theta|x) \propto p(\theta)p(x|\theta), \tag{3.45}$$

where $p(x|\theta)$ is a likelihood function. The prior is often found by selecting the prior $p(\theta)$ such that $p(\theta|x)$ and $p(\theta)$ are in the same family. We illustrate this by some examples.

Example 3.2: Bernoulli Distribution and Beta Priors

The likelihood for Bernoulli distribution is

$$p(x|\theta) = \theta^x (1-\theta)^{1-x}. \tag{3.46}$$

To retain the form of a product of powers of θ and $1-\theta$, we need to multiply a distribution with powers of θ and $1-\theta$ similar to

$$p(\theta) \propto \theta^{\alpha_1}(1-\theta)^{\alpha_2}.$$

This suggests that prior distribution is beta distribution. Indeed,

$$\begin{aligned} p(\theta|x) &= B(\alpha_1, \alpha_2)\theta^{\alpha_1-1}(1-\theta)^{\alpha_2-1}\theta^x(1-\theta)^{1-x} \\ &= B(\alpha_1+x, \alpha_2+x)\theta^{\alpha_1+x}(1-\theta)^{\alpha_2+x-1}, \end{aligned}$$

where both $p(\theta|x)$ and $p(\theta)$ are beta distribution.

Example 3.3: A Generic Conjugate Prior for Exponential Family

Now we consider a generic conjugate prior for an exponential family. The exponential family density in canonical form is given by

$$p(x|\theta) = h(x)\exp\{\theta^T T(x) - a(\theta)\}. \tag{3.47}$$

Conjugate prior for exponential family is

$$p(\theta|\gamma) = h(\gamma)\exp\{\theta^T \gamma_1 + \gamma_2(-a(\theta))\}. \tag{3.48}$$

The product of likelihood function and prior is equal to

$$\begin{aligned} p(\theta|\gamma)p(x|\theta) &= p(\theta+x) \propto h(\gamma) \\ &\times\exp\{\theta^T(T(x)+\gamma_1)+(1+\gamma_2)(-a(\theta))\}, \end{aligned} \tag{3.49}$$

which is in the same family as the conjugate prior $p(\theta|\gamma)$.

3.1.4.3.3 Variational Inference for Conditionally Conjugate Models Conditionally conjugate models are an important special case of exponential family. The conditionally conjugate models are characterized by global parameters and local variables. The joint density for the conditionally conjugate models is

$$p(\beta, z, x) = p(\beta)\prod_{j=1}^{n} p(z_j, x_j|\beta), \tag{3.50}$$

where β is a vector global latent variable influencing every date point, and z is a vector of local latent variables whose j^{th} component locally affects the data x_j.

We assume that terms in the model (3.50) are selected to ensure that each complete conditional is the exponential family. To achieve this, we assume that the conditional density of each (x_j, z_j) pair given the global parameters β is

$$p(z_j, x_j|\beta) = h(z_j, x_j)\exp\{\beta^T T(z_j, x_j) - \alpha(\beta)\}. \tag{3.51}$$

Its conjugate prior is

$$p(\beta) = h(\beta)\exp\{\gamma_1^T\beta + \gamma_2(-\alpha(\beta)) - a(\gamma)\}. \tag{3.52}$$

The posterior distribution $p(\beta|x,z)$ is given by

$$p(\beta|z,x) \propto p(\beta)\prod_{j=1}^{n} p(z_j, x_j|\beta). \tag{3.53}$$

Substituting equations (3.51) and (3.52) into equation (3.53) yields the complete conditional of the global variables:

$$p(\beta|z,x) \propto h(\beta)$$

$$\exp\left\{\gamma_1^T\beta + \gamma_2(-\alpha(\beta)) - a(\gamma) + \left(\sum_{j=1}^{n} T(z_j, x_j)\right)^T \beta - n\alpha(\beta)\right\},$$

$$\propto h(\beta)\exp\left\{\left(\gamma_1+\sum_{j=1}^{n}T(z_j,x_j)\right)^T\beta+(\gamma_2+n)(-\alpha(\beta))\right\},$$
$$\propto h(\beta)\exp\left\{[\beta,-\alpha(\beta)]^T\gamma\right\}, \quad (3.54)$$

where

$$\gamma=\left[\gamma_1+\sum_{j=1}^{n}T(z_j,x_j),\gamma_2+n\right]^T. \quad (3.55)$$

Next we study complete conditional of the local variable z_j, given β, x_j and other local variables z_{-j} and other data x_{-j}. However, equation (3.50) shows that local variable z_j is conditionally independent of other local variables z_{-j} and other data x_{-j}, given β, x_j. In other words, we have

$$p(z_j|x_j,\beta,z_{-j},x_{-j})=p(z_j|x_j,\beta). \quad (3.56)$$

In the following discussion, we assume that $p(z_j|x_j,\beta)$ is in an exponential family and given by

$$p(z_j|x_j,\beta)=h(z_j)\exp\left\{\eta(\beta,x_j)^T t(z_j)-\alpha(\eta(\beta,x_j))\right\}. \quad (3.57)$$

Now it is ready to study CAVI for the conditionally conjugate models. We consider two types of parameters: global parameter β and local parameters $z=(z_1,\ldots,z_n)$. Let $q(\beta|\gamma)$ be the variational approximation to posterior distribution of latent global variables and $q(z_j|\varphi_j)$ be the variational approximation to the latent local variables z_j.

Comparing equation (3.42) with equation (3.57) and using equation (3.43), we obtain the local variational update:

$$\varphi_j=E_\gamma\left[\eta(\beta,x_j)\right]. \quad (3.58)$$

Using equation (3.16), we obtain the optimal mean-field variational density for the global latent variables β:

$$q^*(\beta)\propto\exp\left\{E\left[\log P(\beta|z,x)\right]\right\}. \quad (3.59)$$

Substituting equations (3.54) and (3.55) into equation (3.59), we obtain

$$q^*(\beta)\propto h(\beta)\exp\left\{\left[\beta,-\alpha_g(\beta)\right]^T\gamma\right\}, \quad (3.60)$$

where

$$\gamma=\left[\gamma_1+\sum_{j=1}^{n}T(z_j,x_j),\gamma_2+n\right]^T. \quad (3.61)$$

Therefore, the global variational update is

$$\gamma=\left[\gamma_1+\sum_{j=1}^{n}E_{\varphi_j}\left[T(z_j,x_j)\right],\gamma_2+n\right]^T. \quad (3.62)$$

Now we calculate $ELBO(q(\beta),q(z))$. Extending equation (3.7) to the conditionally conjugate models, we obtain

$$ELBO(q(\beta),q(z))=E\left[\log p(\beta,z,x)\right]-E\left[\log q(\beta,z)\right]. \quad (3.63)$$

Using equations (3.54) and (3.55), we obtain

$$E\left[\log p(\beta,z,x)\right]=\left(\gamma_1+\sum_{j=1}^{n}E_{\varphi_j}\left[T(z_j,x_j)\right]\right)^T E_\gamma(\beta)-(\gamma_2+n)E_\gamma\left[\alpha_g(\beta)\right]. \quad (3.64)$$

Using equations (3.57) and (3.58), we obtain

$$E\left[\log q(z)\right]=\sum_{j=1}^{n}\left\{\varphi_j^T E_{\varphi_j}\left[t(z_j)\right]-\alpha_l(\varphi_j)\right\}. \quad (3.65)$$

Similarly, we have

$$E\left[\log q(\beta)\right]=\gamma^T E_\gamma\left[T(\beta)\right]-\alpha_g(\gamma). \quad (3.66)$$

Combining equations (3.65) and (3.66), we obtain

$$E\left[\log q(\beta,z)\right]=E\left[\log q(\beta)\right]+E\left[\log q(z)\right]$$
$$=\gamma^T E_\gamma\left[T(\beta)\right]-\alpha_g(\gamma)+\sum_{j=1}^{n}\left\{\varphi_j^T E_{\varphi_j}\left[t(z_j)\right]-\alpha_l(\varphi_j)\right\}. \quad (3.67)$$

Using equations (3.64) and (3.67), we can calculate $ELBO(q(\beta),q(z))$.

We derive all formulas for local and global variational updates. CAVI algorithms for the generic exponential family are summarized as follows.

CAVI algorithms for the Generic Exponential Family.

Data and Parameter Input: $n \times D$ dimensional data matrix and exponential family.

Output: Variational posterior approximation on the global and local parameters $q(\beta, z)$.

Step 1: Initialization. Set β^0, $\varphi_j^0 = E\left[\eta\left(\beta^0, x_j\right)\right]$, $j = 1, \ldots, n$, $\gamma = \left[\alpha_1 + \sum_{j=1}^{n} E_{\varphi_j^0}\left[T\left(z_j, x_j\right)\right], \alpha_2 + n\right]^T$

Set $t = 0$.

Step 2: Set $t = t + 1$.

Step 3: For $j \in \{1, \ldots, n\}$ do

Set

$$\varphi_j^t = E\left[\eta\left(\beta^t, x_j\right)\right].$$

End

Step 4:

Set

$$\gamma^t = \left[\gamma_1^t, \gamma_2^t\right]^T = \left[\gamma_1^{t-1} + \sum_{j=1}^{n} E_{\varphi_j^{t-1}}\left[T\left(z_j, x_j\right)\right], \gamma_2^t + n\right]^T.$$

end

Step 5: Compute ELBO

Compute

$$ELBO\left(q^t(\beta), q^t(z)\right) = E\left[\log p^t\left(\beta^t, z, x\right)\right] - E\left[\log q^t\left(\beta^t, z\right)\right],$$

$$E\left[\log p^t\left(\beta^t, z, x\right)\right] = \left(\gamma_1^t + \sum_{j=1}^{n} E_{\varphi_j^t}\left[\left[T\left(z_j, x_j\right)\right]\right]\right)^T \times E_{\gamma^t}\left(\beta^t\right) - \left(\gamma_2^t + n\right) E_{\gamma^t}\left[\alpha\left(\beta^t\right)\right],$$

$$E\left[\log q^t\left(\beta^t, z\right)\right] = \left(\gamma^t\right)^T E_{\gamma^t}\left[T\left(\beta^t\right)\right] - \alpha\left(\gamma^t\right) + \sum_{j=1}^{n} \left(\varphi_j^t\right)^T E_{\varphi_j^T}\left[z_j\right] - \alpha\left(\varphi_j^t\right).$$

Step 6: Check convergence

If $\left|ELBO^t - ELBO^{t-1}\right| \le \varepsilon$ then stop and output

$$q^t\left(\beta^t, z\right);$$

else go to Step 2.

3.1.5 Stochastic Variational Inference

In the previous section, we showed VI formulated the Bayesian inference as an optimization problem. In general, we assume that sampling is independent; the variation objective can be decomposed into sum over contributions from all individual data points. As a consequence, such optimization problems can be efficiently solved using stochastic optimization (Hoffman et al. 2013; Zhang et al. 2019). The widely used stochastic optimization methods are gradient methods. However, the traditional gradient methods implicitly depend on the Euclidean distance metric. VI optimizes the parameters in the distributions. The Euclidean distance metric for the parameters cannot fully capture the major properties such as the shape of the distribution. Natural gradient is an extension of gradient from Euclidean space to distributional space to overcome this limitation. In this section, we mainly introduce the natural gradient-based stochastic optimization methods for VI.

3.1.5.1 Natural Gradient Decent

Let $L(\theta)$ be an objective function with parameters θ. The Newton method for optimizing $l(\theta)$ with respect to θ is derived as follows. The necessary condition for optimizing $l(\theta)$ is

$$\frac{\partial l(\theta)}{\partial \theta} = 0. \tag{3.68}$$

Taylor expansion of $\frac{\partial l(\theta + d)}{\partial \theta}$ is

$$\frac{\partial l(\theta + d)}{\partial \theta} = \frac{\partial l(\theta)}{\partial \theta} + \frac{\partial^2 l(\theta)}{\partial \theta \partial \theta^T} d \approx 0,$$

which implies

$$d = -(H)^{-1} \frac{\partial l(\theta)}{\partial \theta}. \tag{3.69}$$

where $H = \frac{\partial^2 l(\theta)}{\partial \theta \partial \theta^T}$ is the Hessian matrix of the objective function $l(\theta)$ with respect to θ.

KL distance is a metric to measure difference between two distributions. To derive the formula similar to equation (3.69), we first derive the gradient of KL distance $KL\left(p(x|\theta)\|\, p(x|\theta')\right)$ (Exercise 3.4):

$$\frac{\partial KL\left(p(x|\theta)\|\, p(x|\theta')\right)}{\partial \theta'} = -\int p(x|\theta)\frac{\partial \log p(x|\theta')}{\partial \theta'}dx. \tag{3.70}$$

The first order Taylor series expansion of $\frac{\partial KL\left(p(x|\theta)\|p(x|\theta')\right)}{\partial \theta'}$ is

$$\frac{\partial KL\left(p(x|\theta)\|\, p(x|\theta+d)\right)}{\partial \theta'} = \frac{\partial KL\left(p(x|\theta)\|\, p(x|\theta')\right)}{\partial \theta'}\Big|_{\theta'=\theta} + H_{KL\left(p(x|\theta)\|p(x|\theta')\right)}\Big|_{\theta} d, \tag{3.71}$$

where $\theta' = \theta + d$ and H is the Hessian matrix of KL distance:

$$H_{KL\left(p(x|\theta)\|p(x|\theta')\right)} = \frac{\partial^2 KL\left(p(x|\theta)\|\, p(x|\theta')\right)}{\partial \theta' (\partial \theta')^T}.$$

We can show that the Hessian matrix of the KL distance is equal to the Fisher matric F defined as (Exercise 3.5):

$$F = -\int p(x|\theta)\frac{\partial^2 \log p(x|\theta)}{\partial \theta \partial \theta^T}dx = H_{KL\left(p(x|\theta)\|p(x|\theta')\right)}. \tag{3.72}$$

Substituting equation (3.72) into equation (3.71) yields

$$\frac{\partial KL\left(p(x|\theta)\|\, p(x|\theta+d)\right)}{\partial \theta'} = \frac{\partial KL\left(p(x|\theta)\|\, p(x|\theta')\right)}{\partial \theta'}\Big|_{\theta'=\theta} + Fd. \tag{3.73}$$

Using equation (3.70), we obtain

$$\begin{aligned}\frac{\partial KL\left(p(x|\theta)\|\, p(x|\theta')\right)}{\partial \theta'}\Big|_{\theta'=\theta} &= -\int p(x|\theta)\frac{\partial \log p(x|\theta')}{\partial \theta'}dx \\ &= -\int p(x|\theta)\frac{1}{p(x|\theta)}\frac{\partial p(x|\theta)}{\partial \theta}dx \\ &= -\int \frac{\partial p(x|\theta)}{\partial \theta}dx = -\frac{\partial}{\partial \theta}\int p(x|\theta)dx = -\frac{\partial}{\partial \theta}1 = 0.\end{aligned} \tag{3.74}$$

Substituting equation (3.74) into equation (3.73), we obtain

$$\frac{\partial KL\left(p(x|\theta)\|\, p(x|\theta+d)\right)}{\partial \theta'} = \frac{\partial KL\left(p(x|\theta)\|\, p(x|\theta+d)\right)}{\partial d} = Fd. \tag{3.75}$$

Using equations (3.72) and (3.74), we obtain the second order Taylor expansion of KL distance $KL\left(p(x|\theta)\|\, p(x|\theta')\right)$:

$$KL\left(p(x|\theta)\|\, p(x|\theta+d)\right) \approx \frac{1}{2}d^T Fd. \tag{3.76}$$

Now it is ready to derive the deepest decent direction to minimize the loss function or the objective function $l(\theta)$ of VI in distribution space. In other words, we want to find out in which direction we can decrease the KL distance the most, i.e., we attempt to solve the following optimization problem:

$$\begin{aligned}&\min_d l(\theta+d) \\ &s.t.\ KL\left(p(x|\theta)\|p(x|\theta+d)\right) = c.\end{aligned} \tag{3.77}$$

Using the Lagrangian method, the constraint optimization problem (3.77) can be reformulated as the following unconstrained optimization problem:

$$\min_d L(d) = l(\theta+d) + \lambda\left(KL\left(p(x|\theta)\|p(x|\theta+d)\right) - c\right). \tag{3.78}$$

The necessary optimum condition of $L(d)$ is

$$\frac{\partial L(d)}{\partial d} = 0, \tag{3.79}$$

where

$$\frac{\partial L(d)}{\partial d} = \frac{\partial l(\theta)}{\partial \theta} + \lambda\frac{\partial}{\partial d}KL\left(p(x|\theta)\|p(x|\theta+d)\right). \tag{3.80}$$

Substituting equation (3.75) into equation (3.80), we obtain

$$\frac{\partial L(d)}{\partial d} = \frac{\partial l(\theta)}{\partial \theta} + \lambda Fd. \tag{3.81}$$

Combining equations (3.79) and (3.81) yields

$$\frac{\partial l(\theta)}{\partial \theta} + \lambda F d = 0. \tag{3.82}$$

Solving equation (3.82) for d, we obtain the steepest direction in the distribution space:

$$d = -\frac{1}{\lambda} F^{-1} \frac{\partial l(\theta)}{\partial \theta}, \tag{3.83}$$

which implies the following natural gradient definition.

Definition 3.2: Natural Gradient

Natural gradient of the function $l(\theta)$ is defined as

$$\tilde{\nabla}_\theta l(\theta) = F^{-1} \frac{\partial l(\theta)}{\partial \theta}. \tag{3.84}$$

Natural gradient adjusts the direction of the traditional gradient using Fisher information matrix.

3.1.5.2 Revisit Variational Distribution for Exponential Family

Instead of using equation (3.16) or (3.17) to define variational distribution $q(z)$ and equation (3.59) or (3.60) to define variational distribution $q(\beta)$, we define variational distribution $q(z,\beta)$ as follows (Hoffman et al. 2013).

$$q(z,\beta) = q(\beta \mid \lambda) \prod_{j=1}^{n} \prod_{m=1}^{M} q(z_{jm} \mid \varphi_{jm}), \tag{3.85}$$

where λ represents a vector of global parameters for β and $\varphi_j = [\varphi_{j1},\ldots,\varphi_{jM}]^T$ are a vector of local parameters for z_j. The variational distributions $q(\beta \mid \lambda)$ and $q(z_{jm} \mid \varphi_{jm})$ are assumed to be in exponential families and defined as

$$q(\beta|\lambda) = h(\beta)\exp\{\lambda^T T(\beta) - a_g(\lambda)\}, \tag{3.86}$$

$$q(z_{jm}|\varphi_{jm}) = h(z_{jm})\exp\{\varphi_{jm}^T T(z_{jm}) - a_l(\varphi_{jm})\}. \tag{3.87}$$

Recall that if we introduce a vector of fixed parameters α and write a distribution of global latent variables as $p(\beta \mid \alpha)$ and then joint distribution factorization equation (3.50) can be written as

$$p(x,z,\beta|\alpha) = p(\beta \mid \alpha) \prod_{j=1}^{n} p(x_j, y_j \mid \beta). \tag{3.88}$$

A complete conditional distribution of global variables β in equation (3.52) that is the conditional probability of global variables β, given the other latent variables z and observations x can be written as

$$\begin{aligned} &p(\beta|x,z,\alpha) \\ &= h(\beta)\exp\{\eta_g(x,z,\alpha)^T T(\beta) - a_g(\eta_g(x,z,\alpha))\}. \end{aligned} \tag{3.89}$$

Equation (3.51) can be written as

$$p(x_j, z_j|\beta) = h(x_j, z_j)\exp\{\beta^T T(x_j, z_j) - a_l(\beta)\}. \tag{3.90}$$

The prior distribution $p(\beta \mid \alpha)$ that is also in an exponential family use given by

$$p(\beta|\alpha) = h(\beta)\exp\{\alpha^T T(\beta) - a_g(\alpha)\}. \tag{3.91}$$

Now we calculate ELBO as a function of λ. Similar to equation (3.63), we have

$$\begin{aligned} l(\lambda) = ELBO(\lambda) &= E_{q(z,\beta)}[\log p(x,z,\beta \mid \alpha)] \\ &\quad - E_{q(z,\beta)}[\log q(z,\beta)] \\ &= E_{q(z)}[\log p(x,z)] + E_{q(z,\beta)} \log p(\beta \mid x,z,\alpha)] \\ &\quad - E_{q(z,\beta)}[\log q(z,\beta)] \\ &= E_{q(z,\beta)} \log p(\beta \mid x,z,\alpha)] \\ &\quad - E_{q(z,\beta)}[\log q(z,\beta)] + const \\ &= \eta_g(x,z,\alpha)^T E_{q(\beta)}[T(\beta)] - E_{q(z,\beta)}[\log q(\beta \mid \lambda)] \\ &\quad - \sum_{j=1}^{n} \sum_{m=1}^{M} E_{q(z,\beta)}[\log q(z_{jm} \mid \varphi_{jm})] + const \\ &= \eta_g(x,z,\alpha)^T E_{q(\beta)}[T(\beta)] \\ &\quad - \lambda^T E_{q(\beta)}[T(\beta)] + a_g(\lambda) + const. \end{aligned} \tag{3.92}$$

It is well known that for the exponential family the expectation of the sufficient statistics is equal to the gradient of the log normalizer, i.e.,

$$E_{q(\beta)}\left[T(\beta)\right]=\frac{\partial a_g(\lambda)}{\partial \lambda}. \quad (3.93)$$

Substituting equation (3.93) into equation (3.92), we obtain

$$l(\lambda)=\eta_g(x,z,\alpha)^T\frac{\partial a_g(\lambda)}{\partial \lambda}-\lambda^T\frac{\partial a_g(\lambda)}{\partial \lambda} + a_g(\lambda)+const. \quad (3.94)$$

Taking gradient on both sides of equation (3.94), we obtain

$$\begin{aligned}\frac{\partial l(\lambda)}{\partial \lambda}&=\frac{\partial^2 a_g(\lambda)}{\partial\lambda\partial\lambda^T}\eta_g(x,z,\alpha)-\frac{\partial^2(\lambda)}{\partial\lambda\partial\lambda^T}\lambda -\frac{\partial a_g(\lambda)}{\partial \lambda}+\frac{\partial a_g(\lambda)}{\partial \lambda}\\ &=\frac{\partial^2 a_g(\lambda)}{\partial\lambda\partial\lambda^T}\left(\eta_g(x,z,\alpha)-\lambda\right). \end{aligned} \quad (3.95)$$

Recall that for the exponential family Fisher information matrix is equal to

$$F_\beta=\frac{\partial^2 a_g(\lambda)}{\partial\lambda\partial\lambda^T}.$$

Multiplying both sizes of equation (3.95) by F_β^{-1}, we obtain the natural gradient of ELBO with respect to λ

$$\tilde{\nabla}_\lambda l(\lambda)=F_\beta^{-1}\frac{\partial l(\lambda)}{\partial\lambda}=\eta_g(x,z,\alpha)-\lambda. \quad (3.96)$$

Next we calculate $ELBO(\varphi)=l(\varphi)$. Again, similarly, we have

$$\begin{aligned} l(\varphi)=ELBO(\varphi)&=E_{q(z,\beta)}\left[\log p(x,z,\beta\,|\,\alpha)\right] - E_{q(z,\beta)}\left[\log q(z,\beta)\right]\\ &=E_{q(z,\beta)}\left[\log p(\beta)\right]+E_{q(z,\beta)}\left[\log p(x,z\,|\,\beta)\right] - E_{q(z,\beta)}\left[\log q(z,\beta)\right]\\ &=E_{q(z,\beta)}\left[\log p(x,z\,|\,\beta)\right]-E_{q(z,\beta)}\left[\log q(z,\beta)\right]+const\\ &=\sum_{j=1}^{n}\sum_{m=1}^{M}E_{q(z,\beta)}\left[\log p\left(z_{jm}\,|\,x_j,\,z_{j,-m},\,\beta\right)\right] - E_{q(z,\beta)}\left[\log q(z,\beta)\right]+const\\ &=\sum_{j=1}^{n}\sum_{m=1}^{M}E_{q(z,\beta)}\left[\eta_l\left(x_j,z_{j,-m,}\beta\right)\right]^T E_{q(z)}\left[T\left(z_{j,m}\right)\right] - E_{q(z,\beta)}\left[\log q(z,\beta)\right]+const. \end{aligned} \quad (3.97)$$

Recall that

$$E_{q(z,\beta)}\left[\log q(z,\beta)\right] =\sum_{j=1}^{n}\sum_{m=1}^{M}\left\{\phi_{j,m}^T E_{q(z)}\left[T\left(z_{j,m}\right)\right]-a_l\left(\varphi_{j,m}\right)\right\}, \quad (3.98)$$

$$E_{q(z)}\left[T\left(z_{j,m}\right)\right]=\frac{\partial a_l\left(\varphi_{j,m}\right)}{\partial\varphi_{j,m}}. \quad (3.99)$$

Substituting equations (3.98) and (3.99) into equation (3.97) yields

$$l(\varphi)=\sum_{j=1}^{n}\sum_{m=1}^{M}\eta_l\left(x_j,z_{j,-m,}\beta\right)^T\frac{\partial a_l\left(\varphi_{j,m}\right)}{\partial\varphi_{j,m}} -\sum_{j=1}^{n}\sum_{m=1}^{M}\left\{\phi_{j,m}^T\frac{\partial a_l\left(\varphi_{j,m}\right)}{\partial\varphi_{j,m}}-a_l\left(\varphi_{j,m}\right)\right\}+const. \quad (3.100)$$

Taking partial derivative with respect to φ_{jm} on both sides of equation (3.100), we obtain

$$\begin{aligned}\frac{\partial l(\varphi)}{\partial\varphi_{jm}}&=\frac{\partial^2 a_l\left(\varphi_{j,m}\right)}{\partial{\varphi_{j,m}}^2}E_{q(z,\beta)}\left[\eta_l\left(x_j,z_{j,-m,}\beta\right]\right. -\frac{\partial^2 a_l\left(\varphi_{j,m}\right)}{\partial{\varphi_{j,m}}^2}\varphi_{j,m}+\frac{\partial a_l\left(\varphi_{j,m}\right)}{\partial\varphi_{j,m}}-\frac{\partial a_l\left(\varphi_{j,m}\right)}{\partial\varphi_{j,m}}\\ &=\frac{\partial^2 a_l\left(\varphi_{j,m}\right)}{\partial{\varphi_{j,m}}^2}\left(E_{q(z,\beta)}\left[\eta_l\left(x_j,z_{j,-m,}\beta\right]-\varphi_{j,m}\right)\right.\\ &=F_\varphi\left(E_{q(z,\beta)}\left[\eta_l\left(x_j,z_{j,-m,}\beta\right]-\varphi_{j,m}\right)\right.. \end{aligned} \quad (3.101)$$

Multiplying inverse of Fisher information matrix on both sides of equation (3.101), we obtain the following natural gradient:

$$\tilde{\nabla}_{\varphi_{jm}}=E_{\lambda,\varphi_{j,m}}\left[\eta_l\left(x_j,z_{j,-m,}\beta\right]-\varphi_{j,m}\right).\right. \quad (3.102)$$

3.2 VARIATIONAL AUTOENCODER

Variational autoencoders (VAEs) are the types of generative models. The generative models attempt to find interesting patterns, cluster, statistical correlations, and causal structures of the data. Suppose that data points X are distributed over some potentially high-dimensional space. The VAE attempts to generate model of distributions $P(X)$ over all the variables X. In other words, the generative model learns how the data are generated. The generative model has two remarkable features (Kingma and Welling 2019). First, the generative model formulates the major part in physical laws and constraints into the data generative process and takes the unknown minor or unimportant part as noise. The theory or hypothesis can be tested by the generative model and data. Second, the generative model describes causal relations of the world. Causal relations allow better generalization than mere association.

In the past several years, the great progress in the development and application of VAE has been made. In this section, we will introduce the VAE for cross-sectional data following the tutorial of Kingma and Welling (2019) and the extension of the VAE to temporal data based on paper *"Dynamical Variational Autoencoders: A Comprehensive Review"* (Girin et al. 2020).

3.2.1 Autoencoder

The architecture of VAE is two coupled, but independently parameterized encoder or recognition model, and the decoder or generative model. Before introducing VAE, we first introduce classical autoencoder.

Autoencoders consist of two connected networks, an encoder and a decoder (Figure 3.1), for learning latent representations. The encoder receives an input data point X and converts it to a representation in a lower dimensional hidden space Z. The decoder reconstructs the original input X as close as possible from the low-dimensional representation Z. The reconstruction accuracy is determined by the quality of low-dimensional representation Z. Both encoder and decoder are implemented by neural networks. The aim of the autoencoder is to compress the high-dimensional original data to the low-dimensional representation. Because the neural networks can learn nonlinear relationships, the autoencoder is a nonlinear extension of the principal component analysis (PCA).

The accuracy of reconstruction can be measured by loss function. A widely used loss function is defined as mean square error between the original data X and reconstructed $\hat{X}$:

$$L(X,\hat{X}) = \|X - \hat{X}\|^2.$$

3.2.2 Deep Latent Variable Models and Intractability of Likelihood Function

Deep latent variable models whose distribution is parameterized by neural networks are a powerful tool for data modeling (Kingma and Welling 2019). Latent variables are unobserved. They are part of the model, but are not the part of the dataset. Let z be a set of latent variables, $p_\theta(x)$ be a marginal distribution of the observed variables x and $p_\theta(x,z)$ be a joint distribution of the observed variables and latent variables z.

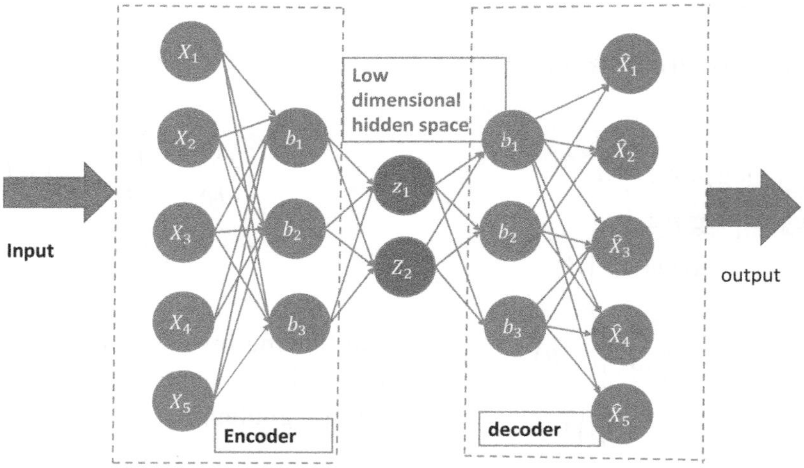

FIGURE 3.1 Outline of autoencoder.

The marginal distribution of the observed variables, or marginal likelihood, $p_\theta(x)$ is given by

$$p_\theta(x)=\int p_\theta(x,z)dz. \tag{3.103}$$

The deep latent variable models can also be conditional on some context, which is denoted by

$$p_\theta(x,z\,|\,y).$$

Since the joint distributions $p_\theta(x,z)$ are parameterized by neural networks, a remarkable feature of the deep latent variables is that, although each factor in the directed model may be simple, the marginal distribution $p_\theta(x)$ derived from the joint distribution $p_\theta(x,z)$ can be very complex. Therefore, the complicated marginal distribution can be well approximated by the deep latent variable models.

Example 3.4: Decoder for Gaussian Data

Consider a simple deep latent variable models for continuous data (Kingma and Welling 2014). Assume that a Gaussian latent space with $p(z)=N(0,I)$. Given the latent variables z, the logarithm of conditional distribution of output of decoder is given by

$$\log p(x|z)=\log N(x,\mu,\sigma^2 I).$$

The parameters μ, σ^2 in the conditional distribution $p(x\,|\,z)$ can be modeled by two layers of multilayered perceptrons (MLPs):

$$\mu=W_4\tanh(W_3 z+b_3)+b_4,$$

$$\log\sigma^2=W_5\tanh(W_3 z+b_3)+b_5.$$

Computing the marginal distribution using equation (3.103) involves integration. Maximum likelihood estimation of the parameters in integral $\int p_\theta(x,z)dz$ does not, in general, have analytic solution or even efficiently differentiate the integral with respect to the parameters θ. The maximum likelihood estimation of the parameters in the marginal distribution $p_\theta(x)$ is intractable. The intractability of the marginal distribution $p_\theta(x)$, in turn, affects the intractability of the posterior distribution $p_\theta(z\,|\,x)$. In fact, the posterior distribution $p_\theta(z\,|\,x)$ is defined as

$$p_\theta(z|x)=\frac{p_\theta(x,z)}{p_\theta(x)}. \tag{3.104}$$

Although, the joint distribution $p_\theta(x,z)$ can be efficiently computed, $p_\theta(x)$ is still involved in equation (3.104). Since MLE of $p_\theta(x)$ is intractable, the MLE of $p_\theta(z\,|\,x)$ must be intractable.

To overcome intractability problem in the computation of both $p_\theta(x)$ and $p_\theta(z\,|\,x)$, approximate inference techniques should be developed to allow us to approximate the marginal distribution $p_\theta(x)$ and posterior distribution $p_\theta(z\,|\,x)$ in the deep latent variable models.

3.2.3 Approximate Techniques and Recognition Model

An essential issue in machine learning is to discover a complex probability distribution $p_\theta(x)$ with only a limited set of high-dimensional data points x sampled from this distribution. Directly deriving the probability distribution $p_\theta(x)$ is intractable. Approximation methods are sought. Two distribution approximation methods exist. One is a Markov Chain Monte Carlo sampling method. However, the Markov Chain Monte Carlo method is computationally expensive and is difficult to apply to large data sets. The second method is the so-called VI, a deterministic approximation technique. The VI forms the core in VAE.

The key idea in applying the approximation techniques to the deep latent variable models is to introduce an encoder or a recognition model $q_\varnothing(z\,|\,x)$, which approximates the posterior distribution $p_\theta(z\,|\,x)$. The parameters $\varnothing$ are called the variational parameters. The variational parameters $\varnothing$ are chosen to make $q_\varnothing(z\,|\,x)$ to approach $p_\theta(z\,|\,x)$ as accurate as possible.

The recognition models have two remarkable features (Kingma and Welling 2019). First, similar to the deep latent variable models, the recognition models can be any directed graphical models and can be factorized as

$$q_\varnothing(z|x)=q_\varnothing(z_1,\ldots,z_K|x)=\prod_{j=1}^{K} q_\varnothing(z_j\,|\,P_a(z_j),x),$$

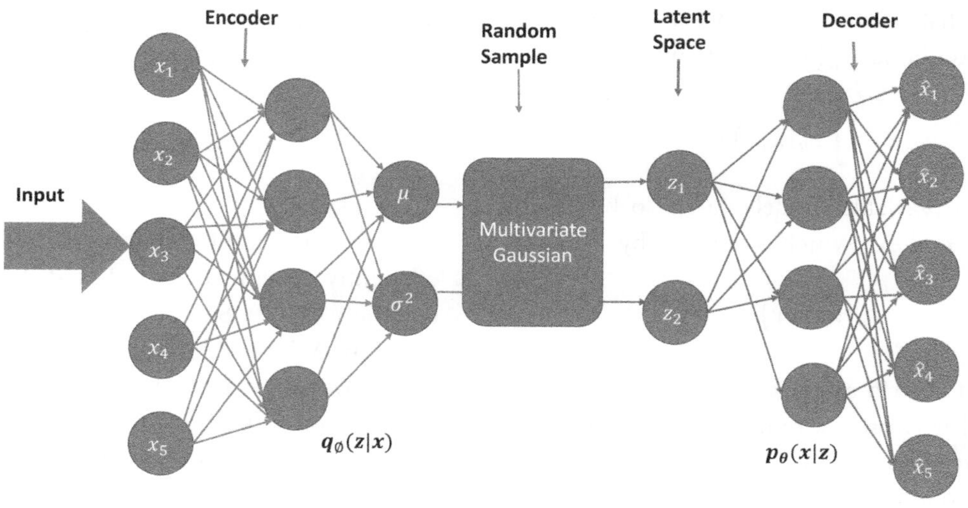

FIGURE 3.2 Variational autoencoder.

where K is the dimension of the latent space, $P_a(z_j)$ denotes the set of parent variables of the variable z_j in the directed graph.

Second, the neural networks are used to parameterize the distribution $q_\varnothing(z|x)$, where the variational parameters $\varnothing$ includes the parameters of the neural networks. We often assume that $q_\varnothing(z|x)$ follows a multivariate normal distribution with a mean vector μ and a diagonal covariance matrix diag (σ):

$$q_\varnothing(z|x) = N(\mu, \text{diag}(\sigma)), \tag{3.105}$$

where the parameters μ and σ are modeled by neural networks:

$$\mu = \text{NeuralNet}_{\varnothing_1}(x), \tag{3.106}$$

$$\log \sigma = \text{NeuralNet}_{\varnothing_2}(x), \tag{3.107}$$

where variation parameters are shared across the data points.

3.2.4 Framework of VAE

The serious limitation of autoencoder is lack of ability to generate similar, but diversified data. The autoencoder could only generate data that are similar to the original inputs. To overcome this limitation, VAEs have been developed (Kingma and Welling 2014; Rezende et al. 2014). The remarkable property that distinguishes the VAEs from AEs is that the latent spaces of VAEs are, by design, continuous, and can be easy random sampled and interpolated. The architecture of VAE is shown in Figure 3.2.

The VAE consists of encoder and decoder. The encoder is a neural network and is also called a reference network or the recognition model $q(z|x)$. It takes data x as input whose distribution is typically complicated, and generates latent representation z (code) whose prior distribution can be simple such as Gaussian distribution. The encoder learns a mapping between an observed x-space and a latent z-space. The decoder is also a neural network and is called a generator. Given a code z, it generates a distribution over the possible corresponding values of the original data x.

Now we derive variational lower bound or ELBO in estimation of log-likelihood function $\log p_\theta(x)$ (Kingma and Welling 2019). Using the Bayesian rule, the log-likelihood can be changed to

$$\begin{aligned} \log p_\theta(x) &= \log \frac{p_\theta(x,z)}{p_\theta(z|x)} \\ &= \log \frac{p_\theta(x,z)}{q_\varnothing(z|x)} \frac{q_\varnothing(z|x)}{p_\theta(z|x)} \\ &= \log \frac{p_\theta(x,z)}{q_\varnothing(z|x)} + \log \frac{q_\varnothing(z|x)}{p_\theta(z|x)} \\ &= \log p_\theta(x,z) - \log q_\varnothing(z|x) + \log \frac{q_\varnothing(z|x)}{p_\theta(z|x)}. \end{aligned} \tag{3.108}$$

Taking expectation with respect to $q_\varnothing(z|x)$ on both sites of equation (3.108), we obtain

$$\log p_\theta(x) = E_{q_\varnothing(z|x)}[\log p_\theta(x,z)] - E_{q_\varnothing(z|x)}[\log q_\varnothing(z|x)] + E_{q_\varnothing(z|x)}\left[\log \frac{q_\varnothing(z|x)}{p_\theta(z|x)}\right]. \quad (3.109)$$

Recall that the last term in equation (3.109) is a the KL distance between the distribution $q_\varnothing(z|x)$ and distribution $p_\theta(z|x)$:

$$KL(q_\varnothing(z|x)p_\theta(z|x)) = E_{q_\varnothing(z|x)}\left[\log \frac{q_\varnothing(z|x)}{p_\theta(z|x)}\right]. \quad (3.110)$$

Substituting equation (3.110) into equation (3.109) yields

$$\log p_\theta(x) = E_{q_\varnothing(z|x)}[\log p_\theta(x,z)] - E_{q_\varnothing(z|x)}[\log q_\varnothing(z|x)] + KL(q_\varnothing(z|x)p_\theta(z|x)). \quad (3.111)$$

Recall that KL distance is always non-negative, i.e.,

$$KL(q_\varnothing(z|x)p_\theta(z|x)) \geq 0, \text{ which implies}$$
$$\log p_\theta(x) \geq E_{q_\varnothing(z|x)}[\log p_\theta(x,z)] - E_{q_\varnothing(z|x)}[\log q_\varnothing(z|x)]. \quad (3.112)$$

$$\mathcal{L}_{\theta,\varnothing}(x) = E_{q_\varnothing(z|x)}[\log p_\theta(x,z)] - E_{q_\varnothing(z|x)}[\log q_\varnothing(z|x)] \quad (3.113)$$

is called ELBO or variational lower bound.

The ELBO can also be expressed as (Exercise 3.6)

$$\mathcal{L}_{\theta,\varnothing}(x) = E_{q_\varnothing(z|x)}[\log p_\theta(x|z)] - KL(q_\varnothing(z|x)||p(z)).$$

Therefore, the log-likelihood $\log p_\theta(x)$ is equal to

$$\log p_\theta(x) = \mathcal{L}_{\theta,\varnothing}(x) + KL(q_\varnothing(z|x)p_\theta(z|x)). \quad (3.114)$$

The gap between the log-likelihood $\log p_\theta(x)$ and ELBO $\mathcal{L}_{\theta,\varnothing}(x)$ is the KL distance $KL(q_\varnothing(z|x)p_\theta(z|x))$. The more accurate $q_\varnothing(|x)$ approximates the posterior distribution $p_\theta(z|x)$, the smaller the gap.

3.2.5 Optimization of the ELBO and Stochastic Gradient Method

The maximum likelihood estimation of parameters is transformed to maximize the ELBO:

$$\max_{\theta,\varnothing} \mathcal{L}_{\theta,\varnothing}(x) = \max_{\theta,\varnothing} \sum_{i=1}^{n} \mathcal{L}_{\theta,\varnothing}(x_i), \quad (3.115)$$

where x_i are the independently and identically sampled data points.

Widely used optimization method is stochastic gradient method. A general procedure of stochastic gradient method is given by

$$\begin{bmatrix} \theta^{k+1} \\ \varnothing^{k+1} \end{bmatrix} = \begin{bmatrix} \theta^{k} \\ \varnothing^{k} \end{bmatrix} + \eta \begin{bmatrix} \nabla_\theta \mathcal{L}_{\theta,\varnothing}(x) \\ \nabla_\varnothing \mathcal{L}_{\theta,\varnothing}(x) \end{bmatrix}_{\theta^k,\varnothing^k}. \quad (3.116)$$

A key step for iterative procedure (3.116) is to calculate gradients. It follows from equation (3.113) that

$$\begin{aligned}\nabla_\theta \mathcal{L}_{\theta,\varnothing}(x) &= \nabla_\theta E_{q_\varnothing(z|x)}[\log p_\theta(x,z) - \log q_\varnothing(z|x)] \\ &= E_{q_\varnothing(z|x)}[\nabla_\theta(\log p_\theta(x,z) - \log q_\varnothing(z|x))] \\ &= E_{q_\varnothing(z|x)}[\nabla_\theta \log p_\theta(x,z)] \\ &\approx \nabla_\theta \log p_\theta(x,z). \end{aligned} \quad (3.117)$$

However, it is difficult to compute $\nabla_\varnothing \mathcal{L}_{\theta,\varnothing}(x)$ as the following formula shown:

$$\begin{aligned}\nabla_\varnothing \mathcal{L}_{\theta,\varnothing}(x) &= \nabla_\varnothing E_{q_\varnothing(z|x)}[\log p_\theta(x,z) - \log q_\varnothing(z|x)] \\ &\neq E_{q_\varnothing(z|x)}[\nabla_\varnothing(\log p_\theta(x,z) - \log q_\varnothing(z|x))].\end{aligned}$$

To overcome this difficulty, a reparameterization trick for computing $\nabla_\varnothing \mathcal{L}_{\theta,\varnothing}(x)$ has been developed (Kingma and Welling 2014, 2019; Rezende et al. 2014).

3.2.6 Reparameterization Trick

A key idea of reparameterization trick is to make a change of variables, which lead to simplified computation of expectation with respect to a set of new changed variables. We assume that the set of latent variables z is transformed from a set of random variables with simple distributions:

$$z = g(\varepsilon, \varnothing, x), \quad (3.118)$$

where random variables ε are independent of x or $\varnothing$.

Then, using statistical theory of changes of variables, we obtain

$$q_\varnothing(z|x)=\frac{p(\varepsilon)}{d(\varnothing,x)}, \tag{3.119}$$

and

$$\log q_\varnothing(z|x)=\log p(\varepsilon)-\log d(\varnothing,x),$$

where $q_\varnothing(z|x)$ is the conditional probability density function of the variable z, $d(\varnothing,x)$ is the absolute value of the determinant of Jacobian matrix:

$$d(\varnothing,x)=|\det(J)|=\left|\det\begin{bmatrix}\frac{\partial z_1}{\partial \varepsilon_1} & \cdots & \frac{\partial z_1}{\partial \varepsilon_m}\\ \vdots & \vdots & \vdots \\ \frac{\partial z_m}{\partial \varepsilon_1} & \cdots & \frac{\partial z_m}{\partial \varepsilon_m}\end{bmatrix}\right|. \tag{3.120}$$

Let $f(z)$ be a nonlinear function of z. Then, we have

$$\begin{aligned}E_{q_\varnothing}[f(z)]&=\int q_\varnothing(z|x)f(z)dz\\ &=\int\frac{p(\varepsilon)}{d(\varnothing,x)}f(g(\varepsilon,\varnothing,x))d(\varnothing,x)d\varepsilon\\ &=\int p(\varepsilon)f(g(\varepsilon,\varnothing,x))d\varepsilon\\ &=E_{p(\varepsilon)}\left[f(g(\varepsilon,\varnothing,x))\right].\end{aligned}$$

Therefore, the ELBO can be calculated as follows:

$$\begin{aligned}\mathcal{L}_{\theta,\varnothing}(x)&=E_{q_\varnothing(z|x)}\left[(\log p_\theta(x,z)]-\log q_\varnothing(z|x))\right]\\ &=E_{p(\varepsilon)}\left[\log p_\theta(x,g(\varepsilon,\varnothing,x))-\log\frac{p(\epsilon)}{d(\varnothing,x)}\right]\\ &=E_{p(\varepsilon)}\left[\log p_\theta(x,g(\varepsilon,\varnothing,x))-\log p(\varepsilon)\right.\\ &\qquad\left.+\log d(\varnothing,x))\right]\\ &=\log d(\varnothing,x))+E_{p(\varepsilon)}\left[\log p_\theta(x,g(\varepsilon,\varnothing,x))\right.\\ &\qquad\left.-\log p(\varepsilon)\right].\end{aligned}$$

The above results can be summarized to

Lemma 3.1

Define

$z=g(\varepsilon,\varnothing,x)$, where random variables ε are independent of x or $\varnothing$ and $f(z)$ be a nonlinear function. Then, we have

$$E_{q_\varnothing(z|x)}[f(z)]=E_{p(\varepsilon)}\left[f(g(\varepsilon,\varnothing,x))\right], \tag{3.121}$$

$\mathcal{L}_{\theta,\varnothing}(x)=\log d(\varnothing,x))+E_{p(\varepsilon)}\left[\log p_\theta(x,g(\varepsilon,\varnothing,x))-\log p(\varepsilon)\right]$, and a simple Monte Carlo estimator $\tilde{\mathcal{L}}_{\theta,\varnothing}(x)$ of the individual-data point ELBO is

$$\tilde{\mathcal{L}}_{\theta,\varnothing}(x)=\log p_\theta(x,z)-\log q_\varnothing(z|x). \tag{3.122}$$

Lemma 3.1 provides a simple formula for computing $E_{q_\varnothing(z|x)}[f(z)]$ and ELBO.

3.2.7 Gradient of Expectation and Gradient of ELBO

In our setting, the expectation and gradient become commutative. Therefore, it follows from equation (3.121) and simple Monte Carlo estimation that

$$\begin{aligned}\nabla_\varnothing E_{q_\varnothing(z|x)}[f(z)]&=\nabla_\varnothing E_{p(\varepsilon)}\left[f(g(\varepsilon,\varnothing,x))\right]\\ &=E_{p(\varepsilon)}\left[\nabla_\varnothing f(g(\varepsilon,\varnothing,x))\right]\\ &\approx\nabla_\varnothing f(g(\varepsilon,\varnothing,x)).\end{aligned} \tag{3.123}$$

Figure 3.3 illustrates the procedures for gradient of expectation under the original form and reparametrized trick form. We can see that under the reparametrized form, the gradient of expectation of function is reduced to gradient of function, which is much easier to compute.

Using equations (3.122) and (3.123), we obtain

$$\nabla_{\theta,\varnothing}E_{q_\varnothing(z|x)}\left[\mathcal{L}_{\theta,\varnothing}(x)\right]\approx\nabla_{\theta,\varnothing}\mathcal{L}_{\theta,\varnothing}(x). \tag{3.124}$$

3.2.8 Bernoulli Generative Model

Consider a Gaussian latent space with Gaussian prior:

$$p(z)=N(z;0,I).$$

The probability for Bernoulli distribution is defined by a neural network:

$$p=\text{DecoderNeuralNet}_\theta(z).$$

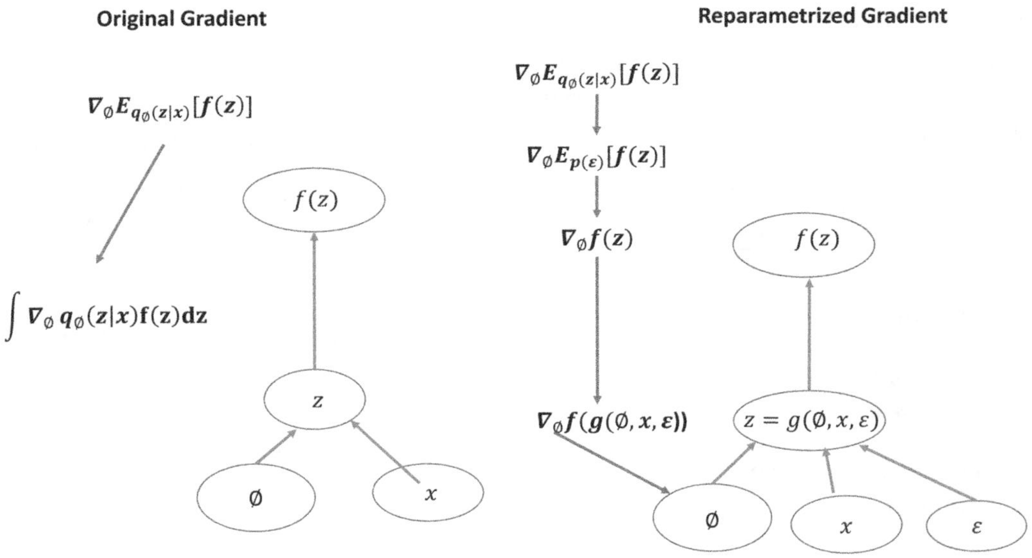

FIGURE 3.3 Illustration of gradient of expectation.

The generative model with Bernoulli distribution is

$$\log p_\theta(x|z) = \sum_{j=1}^{d} \log p(x_j | z)$$
$$= \sum_{j=1}^{d} \left(x_j \log p_j + (1 - x_j)\log(1 - p_j)\right). \quad (3.125)$$

Therefore, we have

$$\log p_\theta(x, z) = \log p_\theta(z) + \log p_\theta(x | z)$$
$$= -\frac{1}{2}\sum_{i=1}^{m}\left(\log(2\pi) + z_i^2\right)$$
$$+ \sum_{j=1}^{d}\left(x_j \log p_j + (1 - x_j)\log(1 - p_j)\right). \quad (3.126)$$

3.2.9 Factorized Gaussian Encoder

Consider m-dimensional latent space with Gaussian latent variable $q_\emptyset(z|x) = N(z;\mu, \text{diag}(\sigma^2))$ (Kingma and Welling 2019). The posterior distribution $q_\emptyset(z|x)$ can be factorized as

$$q_\emptyset(z|x) = \prod_{i=1}^{m} q_\emptyset(z_i | x) = \prod_{i=1}^{m} N(z_i; \mu_i, \sigma_i^2). \quad (3.127)$$

We further assume that the parameters μ and σ^2 are defined by neural networks:

$$(\mu, \log\sigma^2) = \text{EncoderNueralNet}_\emptyset(x), \quad (3.128)$$

where $\mu = [\mu_1, \ldots, \mu_m]^T$ and $\sigma = [\sigma_1, \ldots, \sigma_m]^T$.

The transformation $z = g(\varepsilon, \emptyset, x)$ can be defined as

$$z = \mu + \sigma \odot \varepsilon, \quad (3.129)$$

where $\odot$ is the element-wise product and

$$\varepsilon \sim N(0, I). \quad (3.130)$$

The Jacobean matrix of the transformation is

$$J = \frac{\partial z}{\partial \varepsilon} = \begin{bmatrix} \sigma_1 & \cdots & 0 \\ \vdots & \vdots & \vdots \\ 0 & 0 & \sigma_m \end{bmatrix}. \quad (3.131)$$

Combining equations (3.120) and (3.131), we obtain

$$d(\emptyset, x) = \prod_{i=1}^{m} \sigma_i. \quad (3.132)$$

Let $q_\emptyset(z|x) = p(z)$. It follows from equations (3.119) that

$$\log q_\emptyset(z|x) = \log p(\varepsilon) - \log d(\emptyset, x)$$
$$= \sum_{i=1}^{m}\left(\log N(\varepsilon_i; 0, 1) - \log\sigma_i\right)$$
$$= -\frac{1}{2}\sum_{i=1}^{m}\left(\log(2\pi) + \varepsilon_i^2 + \log\sigma_i\right). \quad (3.133)$$

3.2.10 Full Gaussian Encoder

Now we introduce latent variance space with full covariance Gaussian. The full covariance Gaussian posterior distribution is given by

$$q_{\varnothing}(z|x) = N\left(z;\mu,\sum\right).$$

Using Cholesky decomposition, we obtain

$$\sum = LL^T,$$

where L is a lower triangular matrix.

Consider the transformation:

$$z = \mu + L\varepsilon, \varepsilon \sim N(0,I).$$

The Jacobian matrix of the above transformation is

$$J = \frac{\partial z}{\partial \varepsilon^T} = L, \text{ which implies}$$

$$\log d(\varnothing,x) = \sum_{i=1}^{m} \log L_{ii}. \quad (3.134)$$

Using equation (3.119), we obtain

$$\log q_{\varnothing}(z|x) = \log p(\varepsilon) - \sum_{i=1}^{m} \log|L_{ii}|$$

$$= -\frac{1}{2}\sum_{i=1}^{m}\left(\log(2\pi) + \varepsilon_i^2 + \log|L_{ii}|\right). \quad (3.135)$$

Next we discuss how to define the lower triangular matrix L. First, define

$$L_{mask} = \begin{bmatrix} 0 & 0 & \cdots & 0 \\ 1 & 0 & \cdots & 0 \\ \vdots & \vdots & \vdots & \vdots \\ 1 & 1 & \cdots & 0 \end{bmatrix},$$

where the elements on and above diagonal are zeros, and elements below diagonal are ones. Then, use neural networks to approximate the parameters in the distributions:

$$(\mu, \log\sigma, L_1) \leftarrow \text{EncodeNeuralNet}_{\varnothing}(x), \quad (3.136)$$

$$L \leftarrow L_{mask} \odot L_1 + \text{diag}(\sigma). \quad (3.137)$$

It is clear that

$$L_{ii} = \sigma_i. \quad (3.138)$$

Combining equations (3.134) and (3.138), we obtain

$$\log d(\varnothing,x) = \sum_{i=1}^{m} \log L_{ii} = \sum_{i=1}^{m} \log \sigma_i. \quad (3.139)$$

Substituting equation (3.139) into equation (3.135) yields

$$\log q_{\varnothing}(z|x) = -\frac{1}{2}\sum_{i=1}^{m}\left(\log(2\pi) + \varepsilon_i^2 + \log\sigma_i\right). \quad (3.140)$$

3.2.11 Algorithms for Computing ELBO

Combining equations (3.120), (3.126), and (3.128) yields the ELBO with the Bernoulli generative model. The ELBO is summarized as Lemma 3.2.

Lemma 3.2

Assume the Bernoulli generative model. The approximate ELBO is given by

$$\tilde{L}_{\theta,\varnothing}(x) = -\frac{1}{2}\sum_{i=1}^{m}\left(\log(2\pi) + z_i^2\right)$$
$$+ \sum_{j=1}^{d}\left(x_j \log p_j + (1-x_j)\log(1-p_j)\right)$$
$$+ \frac{1}{2}\sum_{i=1}^{m}\left(\log(2\pi) + \varepsilon_i^2 + \log\sigma_i\right). \quad (3.141)$$

The procedures for computation of estimate of single data point ELBO are summarized in Algorithm 3.1, assuming a full-covariance Gaussian inference model and a factorized Bernoulli generative model (Kingma and Welling 2019).

Algorithm 3.1: Computation of Unbiased Estimate of Single Point ELBO

Input Data

x: An observed data point

ε: A sample from multivariate standard normal distribution $N(0, \boldsymbol{I})$

θ: Parameters in generative model

$\varnothing$: Parameters in recognition model

$q_\varnothing(z|x)$: Conditional distribution of latent variables, given the observed data

$p_\theta(x,z)$: Joint distribution of observed data and latent variables

Output Result

$\tilde{\mathcal{L}}_{\theta,\varnothing}(x)$: Unbiased estimate of the single data point ELBO $\mathcal{L}_{\theta,\varnothing}(x)$

Step 1: Encoder

$$(\mu, \log\sigma, L_1) \leftarrow EncoderNeuralNet_\varnothing(x)$$

$$L \leftarrow L_{mask} \odot L_1 + \text{diag}(\sigma)$$

$$\varepsilon \sim N(0,I)$$

$$z \leftarrow L\varepsilon + \mu$$

Step 2: Calculate ELBO due to encoder

$$\tilde{\mathcal{L}}_{\log q_\varnothing(z|x)} = -\frac{1}{2}\sum_{i=1}^{m}\left(\varepsilon_i^2 + \log(2\pi) + \log\sigma_i\right)$$

Step 3: Calculate ELBO due to decoder

$$\text{p} \leftarrow \text{DecoderNeuralNet}_\theta(z)$$

$$\tilde{\mathcal{L}}_{\log p_\theta(z)} \leftarrow -\frac{1}{2}\sum_{i=1}^{m}\left(z_i^2 + \log(2\pi)\right)$$

$$\tilde{\mathcal{L}}_{\log p_\theta(x|z)} \leftarrow \sum_{i=1}^{d}\left(x_i \log p_i + (1-x_i)\log(1-p_i)\right)$$

Step 4: Calculate total ELBO

$$\tilde{\mathcal{L}}_{\theta,\varnothing}(x) = \tilde{\mathcal{L}}_{\log q_\varnothing(z|x)} + \tilde{\mathcal{L}}_{\log p_\theta(z)} + \tilde{\mathcal{L}}_{\log p_\theta(x|z)}.$$

3.2.12 Improve the Lower Bound

There are two ways to improve the lower bound on the estimator of log-likelihood $p_\theta(x)$. The first approach is importance sampling (Burda et al. 2015). The second approach is to increase the flexibility of the generative model (Kingma and Welling 2019).

3.2.12.1 Importance Weighted Autoencoder

It is clear that

$$\log p_\theta(x) = \log\int p_\theta(x,z)dz$$

$$= \log\int q_\varnothing(z|x)\frac{p_\theta(x,z)}{q_\varnothing(z|x)}dz$$

$$= \log E_{q_\varnothing(z|x)}\left[\frac{p_\theta(x,z)}{q_\varnothing(z|x)}\right].$$

Assume that $\{z_l, l = 1,2,\ldots,k\}$ are randomly sampled from the distribution $q_\varnothing(z|x)$. A Monte Carlo estimator of the expectation gives

$$\log p_\theta(x) = \log\frac{1}{k}\sum_{l=1}^{k}\frac{p_\theta(x,z^l)}{q_\varnothing(z^l|x)}.$$

Let $L_k = \log\frac{1}{k}\sum_{l=1}^{k}\frac{p_\theta(x,z^l)}{q_\varnothing(z^l|x)}$. Then, we have $L_{k+1} \geq L_k$ (Exercise 3.7). When k increases, the lower bound L_k increases. When k goes to infinite, L_k converges to the log-likelihood $\log p_\theta(x)$.

When $k = 1$, L_1 equals the ELBO estimator of the standard VAE.

3.2.12.2 Connection between ELBO and KL Distance

We further investigate connection between ELBO and KL distance. Consider i.i.d dataset $D = \{x^{(1)},\ldots,x^{(n_D)}\}$. Let $q_D(x)$ be empirical data distribution and can be expressed as

$$q_D(x) = \frac{1}{n_D}\sum_{i=1}^{n_D} q_D^{(i)}(x), \quad (3.142)$$

where $q_D^{(i)}(x)$ is a Dirac delta distribution centered at value $x^{(i)}$ such that

$$\int q_D^{(i)}(x)f(x)dx = f(x^{(i)}). \quad (3.143)$$

Assume that data are independently and identically distributed, log-likelihood function is

$$\log p_\theta(D) = \frac{1}{n_D}\sum_{i=1}^{n_D}\log p_\theta(x^{(i)}) = \frac{1}{n_D}\sum_{x\in D}\log p_\theta(x). \quad (3.144)$$

Next we show that

$$E_{q_D(x)}\left[\log p_\theta(x)\right] = \log p_\theta(D). \quad (3.145)$$

In fact, using definition of expectation, and equations (3.142) and (3.143), we obtain

$$\begin{aligned} E_{q_D(x)}\left[\log p_\theta(x)\right] &= \int q_D(x)\log p_\theta(x)dx \\ &= \int \frac{1}{n_D}\sum_{i=1}^{n_D} q_D^{(i)}(x)\log p_\theta(x)dx \\ &= \frac{1}{n_D}\sum_{i=1}^{n_D}\int q_D^{(i)}(x)\log p_\theta(x)dx \\ &= \frac{1}{n_D}\sum_{i=1}^{n_D}\log p_\theta\left(x^{(i)}\right) \\ &= \log p_\theta(D). \end{aligned}$$

The KL distance between data distribution $q_D(x)$ and model distribution $p_\theta(x)$ can be written as

$$\begin{aligned} KL\left(q_D(x)\|p_\theta(x)\right) = & E_{q_D(x)}\left[\log q_D(x)\right] \\ & - E_{q_D(x)}\left[\log p_\theta(x)\right]. \end{aligned} \quad (3.146)$$

Substituting equation (3.145) into equation (3.146) yields

$$\begin{aligned} KL\left(q_D(x)\|p_\theta(x)\right) = & E_{q_D(x)}\left[\log q_D(x)\right] \\ & -\log p_\theta(D), \text{ which implies} \end{aligned}$$

$$\log p_\theta(D) = -KL\left(q_D(x)\|p_\theta(x)\right) - \mathcal{H}\left(q_D(x)\right), \quad (3.147)$$

where $\mathcal{H}\left(q_D(x)\right)$ is an entropy of $q_D(x)$.

Equation (3.147) shows Lemma 3.3.

Lemma 3.3

Maximization of the data log-likelihood $\log p_\theta(D)$ is equivalent to minimization of KL distance $KL\left(q_D(x)p_\theta(x)\right)$ between the data distribution $q_D(x)$ and model distribution $p_\theta(x)$.

Next we study the connection between the ELBO and KL distance. Let $q_{D,\varnothing}(x,z)$ be a joint distribution over data x and latent variables z. The joint distribution $q_{D,\varnothing}(x,z)$ can be factorized to

$$q_{D,\varnothing}(x,z) = q_D(x)q_\varnothing(z\mid x). \quad (3.148)$$

KL distance $KL\left(q_{D,\varnothing}(x,z)\|\, p_\theta(x,z)\right)$ can be reduced to

$$\begin{aligned} & KL\left(q_{D,\varnothing}(x,z)\|\, p_\theta(x,z)\right) \\ & = E_{q_{D,\varnothing}(x,z)}\left[\log\frac{q_{D,\varnothing}(x,z)}{p_\theta(x,z)}\right]. \end{aligned} \quad (3.149)$$

Substituting equation (3.148) into equation (3.149) yields

$$\begin{aligned} KL\left(q_{D,\varnothing}(x,z)\|\, p_\theta(x,z)\right) &= E_{q_{D,\varnothing}(x,z)}\left[\log\frac{q_D(x)q_\varnothing(z\mid x)}{p_\theta(x,z)}\right] \\ &= E_{q_D(x)}\left[E_{q_\varnothing(z|x)}\left[\log\frac{q_D(x)q_\varnothing(z|x)}{p_\theta(x,z)}\right]\right] \\ &= E_{q_D(x)}\left[E_{q_\varnothing(z|x)}\left[-\log p_\theta(x,z) + \log q_D(x) + \log q_\varnothing(z|x)\right]\right]. \end{aligned} \quad (3.150)$$

Substituting equation (3.113) into equation (3.150) leads to

$$\begin{aligned} & KL\left(q_{D,\varnothing}(x,z)\|\, p_\theta(x,z)\right) \\ & = -E_{q_D(x)}\left[\mathcal{L}_{\theta,\varnothing}(x) - \log q_D(x)\right] \\ & = -\mathcal{L}_{\theta,\varnothing}(D) - \mathcal{H}\left(q_D(x)\right). \end{aligned} \quad (3.151)$$

Results in equation (3.151) can be summarized in Lemma 3.4.

Lemma 3.4

Maximization of the ELBO is equivalent to the minimization of the KL distance $KL\left(q_{D,\varnothing}(x,z)\|\, p_\theta(x,z)\right)$.

3.3 OTHER TYPES OF VARIATIONAL AUTOENCODER

3.3.1 Convolutional Variational Autoencoder

Convolutional VAE (CVAE) can be used to extract the important features of two- or three-dimensional data such as the captured videos or images, which reduce the dimension of the datasets while keeping the most of the valuable information (Nugroho et al. 2020).

3.3.1.1 Encoder

The structure of CVAE is shown in Figure 3.4. It consists of three main parts: the encoder, the latent space, and the decoder. Assume the CVAE is fully convolutional VAE. All layers in the network architecture are convolutional layers. A typical CVAE assumes that the

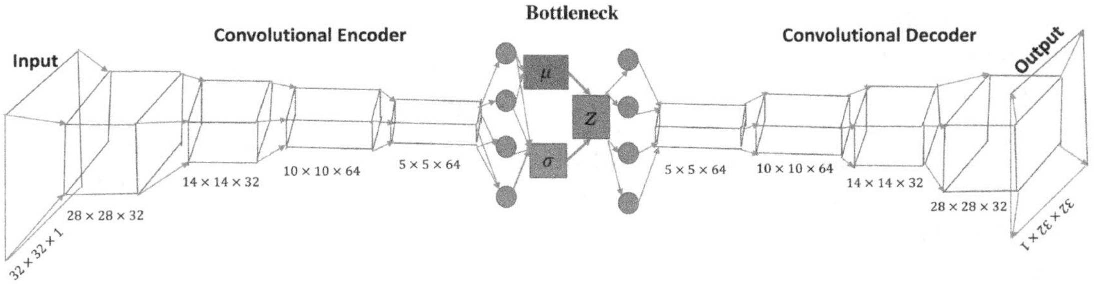

FIGURE 3.4 Scheme of convolutional variational autoencoder.

input is two- or three-dimensional images. In the standard VAE, the encoder is implemented by the multilayer feedforward neural networks. In the CVAE, the multilayer feedforward neural networks are replaced by convolutional neural networks (CNNs). The CNN consists of convolutional layers with ReLU activation function, pooling layer, flattened layer, and fully connected neural networks. Let n_w^l and n_h^l be the width and height dimension of the convolutional layer, respectively. Let p be the padding size, f be the kernel size, and s be the stride size. The width and height dimensions of the convolutional layer are, respectively, determined by

$$n_w^l = \frac{n_w^{l-1} + 2p - f}{s} + 1$$

$$n_h^l = \frac{n_h^{l-1} + 2p - f}{s} + 1.$$

Detailed descriptions of the CNN are in Section 1.1.2.

3.3.1.2 Bottleneck

The block following encoder is the bottleneck that generates latent variables and latent space. The encoder outputs two vectors: mean vector and covariance matrix. Define the mean vector and variance matrix as

$$\mu = \begin{bmatrix} \mu_1 \\ \vdots \\ \mu_k \end{bmatrix} \text{and covariance matrix } \Sigma = \begin{bmatrix} \sigma_{11} & \cdots & \sigma_{1k} \\ \vdots & \vdots & \vdots \\ \sigma_{k1} & \cdots & \sigma_{kk} \end{bmatrix}.$$

Let $\Sigma = L_1 L_1^T$.

The output of encoder is given by

$$(\mu, \log\sigma, L_1) \leftarrow EncoderConvolutionalNet_{\varnothing}(x)$$

The latent variables are generated by the following procedures:

$$\begin{aligned} L &\leftarrow L_{mask} \odot L_1 + \text{diag}(\sigma) \\ \varepsilon &\sim N(0, I_k) \\ z &= \mu + L\varepsilon. \end{aligned} \quad (3.152)$$

If we assume the diagonal variance matrix

$$\Sigma = \begin{bmatrix} \sigma_{11} & \cdots & 0 \\ \vdots & \vdots & \vdots \\ 0 & \cdots & \sigma_{kk} \end{bmatrix},$$

Then the latent variables are generated by

$$\varepsilon_i \sim N(0,1),$$

$$z_i = \mu_i + \sqrt{\sigma_{ii}}\,\varepsilon_i,\ i = 1,\ldots,k. \quad (3.153)$$

The dimension k of mean vector and covariance matrix specifies the dimension of the features extracted from the input data (image).

3.3.1.3 Decoder

The decoder generates the output from the sampled latent variables z to reconstruct the input. The decoder is also implemented by CNN, but in a reverse version of CNN in the encoder.

3.3.2 Graphic Convolutional Variational Autoencoder

Graphic data analysis has broad applications including node clustering and classification (Kipf and Welling 2016; Zhang et al. 2019, 2020), link prediction (Pan et al. 2018), and molecular design (Liu et al. 2018). The graph neural networks were initially investigated for recurrent graph neural networks

(Gori et al. 2005; Scarselli et al. 2009; Gallicchio and Micheli 2010). Convolutional graph neural networks that redefine the "convolution" for graphs have been developed following the recurrent graph neural networks. The convolutional graph neural networks are divided into the spectral-based approaches and the spatial-based approaches (Micheli 2009; Bruna et al. 2013; Henaff et al. 2015; Defferrard et al. 2016; Niepert et al. 2016; Gilmer et al. 2017; Kipf and Welling 2017). In addition, many alternative methods for graph neural networks such as graph autoencoders (GAEs) and spatial-temporal graph neural networks have been developed in the past few years (Wu et al. 2019). In general, graph neural networks can be classified into recurrent graph neural networks, convolutional graph neural network, GAEs, and spatial-temporal graph neural networks (Wu et al. 2019; Zhang et al. 2020). In this section, we focus on graph autoencoders, specifically, on graphic convolutional VAEs.

3.3.2.1 Notation and Basic Concepts for Graph Autoencoder

Graph autoencoders (GAEs) (Kipf and Welling 2016; Zhang et al. 2020) encode the node features and graph structure into the latent representation and decode the latent representation back into graphs. Consider an undirected graph $G=\{V, E, X\}$, where $V=\{v_1, v_2,\ldots, v_n\}$ denotes a set of nodes, E denotes a set of edges with $e_{ij}=(v_i, v_j)\in E$ denoting the edge between the node v_i and v_j, $x_i \in R^m$ denotes a m-dimensional vector of features for the ith node, and $X=[x_1,x_2,\ldots,x_n]^T$ be an $n\times m$-dimensional matrix of features. The neighborhood of a node v is defined as a set of nodes $N(v)=\{u\in V|(v,u)\in E\}$.

Define an adjacency matrix:

$$A=\begin{bmatrix} a_{11} & \cdots & a_{1n} \\ \vdots & \vdots & \vdots \\ a_{n1} & \cdots & a_{nn} \end{bmatrix},$$

where

$$a_{ij}=\begin{cases} 1 & x_i \text{ is connected to } x_j \\ 0 & \text{otherwise} \end{cases}.$$

The number $|V|$ of the nodes is n. The weight associated with an edge is a real number and measures the degree of connection.

Example 3.5: Adjacency Matrix

An adjacency matrix is Shown in Figure 3.5. The graph G is mapped to latent space. Assume that a k-dimensional vector z_i in the latent space is the latent representation of the node v_i with the m-dimensional feature vector x_i. Define the latent representation matrix:

$$Z=\begin{bmatrix} z_1^T \\ \vdots \\ z_n^T \end{bmatrix}=\begin{bmatrix} z_{11} & \cdots & z_{1k} \\ \vdots & \vdots & \vdots \\ z_{n1} & \cdots & z_{nk} \end{bmatrix}.$$

The latent representation matrix Z contains the topological structure and node features information of the graph. Let $f:(A,x)\rightarrow Z$ be a map from the graph to the latent space.

3.3.2.2 Spectral-Based Convolutional Graph Neural Networks

Convolution is a key operation in CNNs. The CNNs are designed to analyze images. Convolution operation uses a filter to convolve the pixels in the images for extraction of features. The pixels in the images are regularly arranged into a matrix. In other words, the images have Euclidean structure. However, graphs have non-Euclidean structure or topology structure. Convolution operation cannot be directly applied to graphs without any changes. Therefore, CNNs cannot be directed extended to graphs. To extend CNN to graph data, we first need to extend convolution mathematic operation to graphs using spectral graph theory with eigenvalue analysis of matrices associated with the graph, such as its adjacency matrix or Laplacian matrix.

3.3.2.2.1 Laplacian Matrix There are three types of Laplacian matrices:

1. Combinatorial Laplacian that is defined as

$$L=D-A,$$

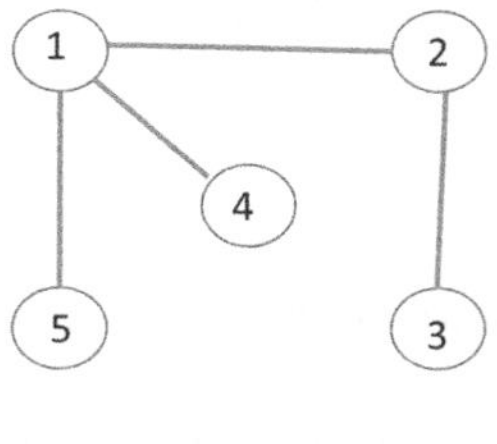

$$A=\begin{bmatrix} 0 & 1 & 0 & 1 & 1 \\ 1 & 0 & 1 & 0 & 0 \\ 0 & 1 & 0 & 0 & 0 \\ 1 & 0 & 0 & 0 & 0 \\ 1 & 0 & 0 & 0 & 0 \end{bmatrix}$$

FIGURE 3.5 Example 3.5.

where A is an adjacency matrix, D is a diagonal matrix with $D_{ii} = \sum_j A_{ij}$.

2. Symmetric normalized Laplacian that is defined as

$$L_{sys} = D^{-1/2} L D^{-1/2}.$$

3. Random walk normalized Laplacian that is defined as

$$L_{rw} = D^{-1} L.$$

Example 3.6: Laplacian Matrix (Example 3.5 Continue)

Combinatorial Laplacian:

$$L = \begin{bmatrix} 3 & -1 & 0 & -1 & -1 \\ -1 & 2 & -1 & 0 & 0 \\ 0 & -1 & 1 & 0 & 0 \\ -1 & 0 & 0 & 1 & 0 \\ -1 & 0 & 0 & 0 & 1 \end{bmatrix},$$

Symmetric normalized Laplacian:

$$L_{sys} = \begin{bmatrix} 1 & -\frac{1}{\sqrt{6}} & 0 & 0 & 0 \\ -\frac{1}{\sqrt{6}} & 1 & -\frac{1}{\sqrt{2}} & 0 & 0 \\ 0 & -\frac{1}{\sqrt{2}} & 1 & 0 & 0 \\ -\frac{1}{\sqrt{3}} & 0 & 0 & 1 & 0 \\ -\frac{1}{\sqrt{3}} & 0 & 0 & 0 & 1 \end{bmatrix}, \text{ and}$$

Random walk normalized Laplacian:

$$L_{rw} = \begin{bmatrix} 0 & \frac{1}{3} & 0 & \frac{1}{3} & \frac{1}{3} \\ \frac{1}{2} & 0 & \frac{1}{2} & 0 & 0 \\ 0 & 1 & 0 & 0 & 0 \\ 1 & 0 & 0 & 0 & 0 \\ 1 & 0 & 0 & 0 & 0 \end{bmatrix}.$$

3.3.2.2.2 Graph Fourier Transform

3.3.2.2.2.1 General Formula for Graph Fourier Transform The inner product is defined as

$$\langle f, g \rangle = \int f(x)\overline{g(x)}dx,$$

where $\overline{g(x)}$ is the conjugate function of the complex function $g(x)$.

The classical Fourier transform of the function $f(x)$ (Exercise 3.8) is

$$\hat{f}(\omega) = \mathcal{F}(f(x)) = \langle f, e^{i\omega x} \rangle = \int f(x) e^{-i\omega x} dx. \quad (3.154)$$

If the Fourier transform $\hat{f}(\omega)$ is known, the function $f(x)$ can be obtained by inverse Fourier transform of $\hat{f}(\omega)$:

$$\mathcal{F}^{-1}(\hat{f}(\omega)) = \int \hat{f}(\omega) e^{i\omega x} d\omega. \quad (3.155)$$

Let $\Delta = \frac{\partial^2}{\partial x^2}$ be the one-dimensional Laplace operator. Then, we have

$$\Delta(e^{-i\omega x}) = -\omega^2 e^{-i\omega x}. \quad (3.156)$$

Equation (3.156) clearly shows that function $e^{-i\omega x}$ is an egenfunction of the Laplace operator. Convolution theorem states that the Fourier transform of a convolution of two signals is equal to the product of their Fourier transforms (https://en.wikipedia.org/wiki/Convolution_theorem).

Now we study how to apply Fourier transform theory to the convolution of graph. Suppose that the Laplacian matrix can be decomposed to

$$L = U \Lambda U^T, \quad (3.157)$$

where $U = [u_1, \ldots, u_n]$ is an orthonormal matrix of eigenvectors ordered by eigenvalues, i.e., $UU^T = I$, Λ is a diagonal matrix of eigenvalues (spectrum):

$$U = \begin{bmatrix} u_{11} & \cdots & u_{1n} \\ \vdots & \vdots & \vdots \\ u_{n1} & \cdots & u_{nn} \end{bmatrix}, \Lambda = \begin{bmatrix} \lambda_1 & \cdots & 0 \\ \vdots & \vdots & \vdots \\ 0 & \cdots & \lambda_n \end{bmatrix}. \quad (3.158)$$

Now we define the Fourier transform of the graph. Comparing Laplacian operator and Laplacian matrix, we can think that an eigenvector of the Laplacian matrix corresponds to an eigenfunction of the Laplacian operator.

Fourier transform in equation (3.154) can be discretized to

$$\hat{f}(\omega)=\sum_{k=1}^{n} f(x_k)e^{-i\omega x_k}=f^T u_\omega, \tag{3.159}$$

where

$$f=\begin{bmatrix} f(x_1) \\ \vdots \\ f(x_n) \end{bmatrix} \text{ and } u_\omega=\begin{bmatrix} e^{-i\omega x_1} \\ \vdots \\ e^{-i\omega x_n} \end{bmatrix}.$$

Similar to equation (3.159), for the Fourier transform of graph, the feature vector of the nodes in the graph $f=[f_1,\ldots,f_n]^T$ can be taken as $f=[f(x_1),\ldots,f(x_n)]^T$ and eigenvector $u_l=[u_{1l},\ldots,u_{nl}]^T$ of the Laplacian matrix can be taken as u_ω. The graph Fourier transform for one particular eigenvalue λ_l is then defined as

$$f(\lambda_l)=\langle f,u_l\rangle=f^T u_l. \tag{3.160}$$

The whole graph Fourier transform is then defined as

$$\begin{bmatrix} \hat{f}(\lambda_1) \\ \hat{f}(\lambda_2) \\ \vdots \\ \hat{f}(\lambda_n) \end{bmatrix}=\begin{bmatrix} u_{11} & \cdots & u_{l1} & \cdots & u_{n1} \\ u_{12} & \cdots & u_{l2} & \cdots & u_{n2} \\ \vdots & \vdots & \vdots & \vdots & \vdots \\ u_{1n} & \cdots & u_{ln} & \cdots & u_{nn} \end{bmatrix}\begin{bmatrix} f_1 \\ f_2 \\ \vdots \\ f_n \end{bmatrix}. \tag{3.161}$$

Let $\hat{f}(\lambda)=\left[\hat{f}(\lambda_1),\ldots,\hat{f}(\lambda_n)\right]^T$. Then, equation (3.161) can be written in the following matrix form:

$$\hat{f}(\lambda)=U^T f. \tag{3.162}$$

The inverse graph Fourier transform for specific feature f_i is given by

$$f_i=\sum_{l=1}^{n} u_{il}\hat{f}(\lambda_l), \tag{3.163}$$

and the inverse graph Fourier transform is given by

$$f=U\hat{f}(\lambda), \tag{3.164}$$

or

$$\begin{bmatrix} f_1 \\ \vdots \\ f_n \end{bmatrix}=\begin{bmatrix} u_{11} & \cdots & u_{1n} \\ \vdots & \vdots & \vdots \\ u_{n1} & \cdots & u_{nn} \end{bmatrix}\begin{bmatrix} \hat{f}(\lambda_1) \\ \vdots \\ \hat{f}(\lambda_n) \end{bmatrix}.$$

3.3.2.2.2.2 Graph Convolution Now we introduce the formula for graph convolution. Let h be a kernel and f be a feature vector of a graph. A convolution of the kernel h with the graph feature vector f is denoted by $f*h$. Then, the Fourier transform of convolution $f*h$ is equal to the product of the Fourier transform of the feature vector f and the Fourier transform of the kernel h:

$$\mathcal{F}(f*h)(\lambda_l)=\mathcal{F}(f)(\lambda_l)\mathcal{F}(h)(\lambda_l). \tag{3.165}$$

Using equation (3.162) we obtain

$$\mathcal{F}(f)=\hat{f}=U^T f, \tag{3.166}$$

$$\mathcal{F}(h)=\hat{h}=U^T h, \tag{3.167}$$

where

$$f=\begin{bmatrix} f_1 \\ \vdots \\ f_n \end{bmatrix}, \hat{f}=\begin{bmatrix} \hat{f}(\lambda_1) \\ \vdots \\ \hat{f}(\lambda_n) \end{bmatrix}, h=\begin{bmatrix} h_1 \\ \vdots \\ h_n \end{bmatrix} \text{and } \hat{h}=\begin{bmatrix} \hat{h}(\lambda_1) \\ \vdots \\ \hat{h}(\lambda_n) \end{bmatrix}.$$

Thus, we obtain

$$\mathcal{F}(f*h)(\lambda_l)=\hat{f}(\lambda_l)\hat{h}(\lambda_l), \tag{3.168}$$

$$\mathcal{F}(f*h)=\begin{bmatrix} \hat{f}(\lambda_1)\hat{h}(\lambda_1) \\ \vdots \\ \hat{f}(\lambda_n)\hat{h}(\lambda_n) \end{bmatrix}. \tag{3.169}$$

It follows from equation (3.169) that

$$\mathcal{F}(f*h)=\hat{h}\odot\hat{f}=(U^T h)\odot(U^T f), \tag{3.170}$$

where Hadamard product $\odot$ denotes element-wise multiplication.

Let

$$\hat{h}_D = \begin{bmatrix} \hat{h}(\lambda_1) & \cdots & 0 \\ \vdots & \vdots & \vdots \\ 0 & \cdots & \hat{h}(\lambda_n) \end{bmatrix}.$$

Then,

$$\hat{h}_D \hat{f} = \begin{bmatrix} \hat{h}(\lambda_1) & \cdots & 0 \\ \vdots & \vdots & \vdots \\ 0 & \cdots & \hat{h}(\lambda_n) \end{bmatrix}$$

$$\hat{f} = \begin{bmatrix} \hat{f}(\lambda_1)\hat{h}(\lambda_1) \\ \vdots \\ \hat{f}(\lambda_n)\hat{h}(\lambda_n) \end{bmatrix} = \hat{h} \odot \hat{f}. \quad (3.171)$$

Using equations (3.164), (3.170), and (3.171), we obtain the convolution of graphic kernel with the graph feature vector:

$$f * h = U\left(\left(U^T h\right) \odot \left(U^T f\right)\right) = U\left(\hat{h}_D U^T f\right). \quad (3.172)$$

Lemma 3.5

The spectral graph convolution of the graph feature vector f with a kernel h is

$$f * h = U\left(\left(U^T h\right) \odot \left(U^T f\right)\right) = U\left(\hat{h}_D U^T f\right).$$

For the convenience of computation, the Fourier-domain kernel matrix $\hat{h}_D$ can be reduced to a simple learnable kernel matrix

$$g_\theta(\Lambda) = \begin{bmatrix} \theta_1 & \cdots & 0 \\ \vdots & \vdots & \vdots \\ 0 & \cdots & \theta_n \end{bmatrix}, \quad (3.173)$$

where $\theta_1, \ldots, \theta_n$ are learnable parameters.

In this case, equation (3.172) is simplified to

$$f * g_\theta = U\left(g_\theta(\Lambda) U^T f\right. \quad (3.174)$$

Equation (3.174) has two serious limitations. Firstly, each time forward update requires multiplications of three matrices. Secondly, complete specification of a kernel needs n parameters. To overcome these limitations, $\hat{h}_D(\lambda_l)$ is defined as

$$\hat{h}_D(\lambda_l) = \sum_{j=0}^{K} \alpha_j \lambda_l^j,$$

which implies

$$g_\theta(\Lambda) = \sum_{j=0}^{K} \alpha_j \Lambda^j, \quad (3.175)$$

where

$$\Lambda = \begin{bmatrix} \lambda_1 & \cdots & 0 \\ \vdots & \vdots & \vdots \\ 0 & \cdots & \lambda_n \end{bmatrix}.$$

Recall that

$$L = U \Lambda U^T, \; U^T U = I,$$

$$L^j = U \Lambda U^T U \Lambda U^T \ldots U \Lambda U^T U \Lambda U^T = U \Lambda^j U^T,$$

which implies that

$$\Lambda^j = U^T L^j U. \quad (3.176)$$

Combining equations (3.174)–(3.176), we obtain

$$f * g_\theta = \sum_{j=0}^{K} \alpha_j L^j f. \quad (3.177)$$

The kernel $g_\theta(\Lambda)$ defined in equation (3.175) has three remarkable features:

1. The number of parameters K is, in general, much less than the number of nodes n in the graph.
2. Computation of graph convolution does not require eigenvector decomposition of the Laplacian matrix of the graph.
3. This type of graph kernel has good spatial localization. The number K is the receptive field. Each computation of graph convolution summarizes the features of K-hop neighbors of the central node with the weights α_j.

3.3.2.2.2.3 Chebyshev Polynomials Approximation Recall that $g_\theta(\Lambda)$ in equation (3.174) is a function of the eigenvalues of the normalized graph Laplacian L. It can be well approximated by Chebyshev polynomials (Hammond et al. 2011; Kipf and Welling 2017):

$$g_\theta(\Lambda)=\sum_{j=0}^{K}\alpha_j T_j(\tilde{\Lambda}), \tag{3.178}$$

where $T_j(x)$ is the jth order Chebyshev polynomials and $\tilde{\Lambda}=\frac{2}{\lambda_{max}}\Lambda-I$.

The Chebyshev polynomials can be recursively defined by

$$T_0(x)=1$$

$$T_1(x)=x$$

$$T_n(x)=2xT_{n-1}(x)-T_{n-2}(x), \tag{3.179}$$

where $-1\le x\le 1$.

We can show (Exercise 3.11) that

$$-1\le\tilde{\Lambda}=\frac{2}{\lambda_{max}}\Lambda-I\le 1.$$

Therefore, $T_j(\tilde{\Lambda})$ is well defined.

A convolution of a feature vector of the graph with the kernel $g_\theta(\Lambda)$ is given by

$$g_\theta(\Lambda)*f=U\sum_{j=0}^{K}\alpha_j T_j(\tilde{\Lambda})U^T f. \tag{3.180}$$

Since $T_j(\tilde{\Lambda})$ is a diagonal matrix, we have

$$U\alpha_j T_j(\tilde{\Lambda})U^T=\alpha_j T_j(U\tilde{\Lambda}U^T)=\alpha_j T_j(\tilde{L}), \tag{3.181}$$

where

$$\tilde{\text{L}}=\frac{2}{\lambda_{max}}\text{L}-I. \tag{3.182}$$

Equation (3.181) implies

$$g_\theta(\Lambda)*f=\sum_{j=0}^{K}\alpha_j T_j(\tilde{\text{L}})f=\sum_{j=0}^{K-1}\theta_j T_j(\tilde{\text{L}})f.$$

This shows that convolution with a spectral graph filter can be performed using Chebyshev polynomials in Laplacian without Fourier transform, and hence no eigen-decomposition is required.

Lemma 3.6

By approximating the kernel with an expansion of kth-order Chebyshev polynomials, the convolution of graph kernel with the node feature vector is given by

$$g_\theta(\Lambda)*f=\sum_{j=0}^{K}\theta_j T_j(\tilde{\text{L}})f. \tag{3.183}$$

3.3.2.2.2.4 Linear Model for Graphic Convolution To simplify the computation, we can limit the order of Chebyshev polynomials $K=1$ and consider a linear model for graphic convolution with $\lambda_{max}\approx 2$ and $\tilde{\text{L}}=L-I$. The graph convolution equation (3.183) is reduced to

$$g_\theta(\Lambda)*f=\theta_0 f+\theta_1(L-I)f. \tag{3.184}$$

Recall that the Laplacian matrix L is defined as

$$L=I-D^{-\frac{1}{2}}AD^{-\frac{1}{2}}. \tag{3.185}$$

Substituting equation (3.185) into equation (3.184), we obtain

$$g_\theta(\Lambda)*f=\theta_0 f-\theta_1 D^{-\frac{1}{2}}AD^{-\frac{1}{2}}f, \tag{3.186}$$

with two parameters θ_0 and θ_1.

To avoid overfitting and minimizing the number of operations, we further reduce two parameters θ_0 and θ_1 into one parameter θ. Let $\theta_0=\theta$ and $\theta_1=-\theta$. Then, equation (3.186) is reduced to

$$g_\theta(\Lambda)*f=\theta\left(I+D^{-\frac{1}{2}}AD^{-\frac{1}{2}}\right)f. \tag{3.187}$$

Since the eigenvalues of the matrix $\left(I+D^{-\frac{1}{2}}AD^{-\frac{1}{2}}\right)$ is in the ranger [0, 2], its gradient may be exploding (vanishing) when the largest eigenvalue is larger than 1 (is close to zero). To overcome this limitation, we change equation (3.187) to

$$g_\theta(\Lambda)*f=\theta\tilde{D}^{-\frac{1}{2}}\tilde{A}\tilde{D}^{-\frac{1}{2}}, \tag{3.188}$$

where $\tilde{A}=A+I$ and $\tilde{D}_{ii}=\sum_j\tilde{A}_{ij}$.

Consider c features (c channels) for each node and m kernels (feature maps), equation (3.188) can be generalized to

$$Z=\tilde{D}^{-\frac{1}{2}}\tilde{A}\tilde{D}^{-\frac{1}{2}}XW,$$

where $X \in R^{n\times c}$ is a signal matrix, $W \in R^{c\times m}$ is a parameter matrix and $Z \in R^{n\times m}$ is the convolved signal matrix (Kipf and Welling 2017).

Lemma 3.7: Convolution of a General Graph

Consider c features (c channels) for each node and m kernels (feature maps). The convolution of m kernels with the n nodes and c channels are given by

$$Z=\tilde{D}^{-\frac{1}{2}}\tilde{A}\tilde{D}^{-\frac{1}{2}}XW. \tag{3.189}$$

3.3.2.2.3 Graph Convolutional Networks Consider the lth layer with c_l channels and m_l kernels. Let $Z^l \in R^{m_l\times c_l}$ be input matrix for convolution in the lth layer, $A \in R^{n\times n}$ be an adjacency matrix and $W^l \in R^{c_l\times m_l}$ be the learnable parameter (weight) matrix, $Z^{l+1} \in R^{m_{l+1}\times c_{l+1}}$ be the output matrix after convolution, and σ be a element-wise nonlinear activation function. Let $Z^0 = X$ and $\hat{A}=\tilde{D}^{-\frac{1}{2}}\tilde{A}\tilde{D}^{-\frac{1}{2}}$. Then, a layer-wise transformation by a spectral graph convolution function $\sigma(Z^l, A|W^l)$ is given by

$$Z^{l+1}=\sigma(Z^l, A|W^l) \tag{3.190}$$

$$=\sigma(\hat{A}Z^lW^{l+1}), \tag{3.191}$$

where $\hat{A}=\tilde{D}^{-\frac{1}{2}}\tilde{A}\tilde{D}^{-\frac{1}{2}}$.

A two-layer graph convolutional network (GCN) for supervised or semi-supervised node classification is defined by

$$Z=\text{softmax}\left(\hat{A}\text{ReLU}\left(\hat{A}XW^0\right)W^1\right), \tag{3.192}$$

where softmax is defined as

$$\sigma(Z)_i=\frac{e^{Z_i}}{\sum_{j=1}^{q}e^{Z_j}}.$$

Let Ω_q be the set of node indices that have labels. For multi-class classification, the cross-entropy error over all labeled examples can be used as objective function for classification (Kipf and Welling 2017):

$$\mathcal{L}=-\sum_{q\in\Omega_q}\sum_{f=1}^{F}Y_{qf}\log Z_{qf}. \tag{3.193}$$

3.3.2.3 Graph Convolutional Encoder

The graph convolutional encoder (GCE) has L_e transformation layers. The input data are an adjacency matrix $A \in R^{n\times n}$ and a feature matrix $X \in R^{n\times c}$. The GCE consists of GCNs. The GCNs learn a layer-wise transformation by a spectral graph convolution mapping defined in equations (3.190) and (3.191) (Kipf and Welling 2016; Zhang et al. 2020).

Similar to equations (3.128) and (3.129), the latent variables can be generated by

$$q_\varnothing(Z|X,A)=\prod_{j=1}^{K}q_\varnothing(Z_j|X,A)$$

$$=\prod_{j=1}^{K}N(Z_j;\mu_j,\sigma_j^2), \tag{3.194}$$

where

$$(\mu,\log\sigma^2)=\text{EncoderNueralNet}_\varnothing(x)$$

$$=GCN_\varnothing(X,A). \tag{3.195}$$

Specifically, the inference model is defined as

$$Z^{l+1}=\sigma\left(\hat{A}Z^lW^{l+1}\right), l=1,\ldots,L_e-2, \tag{3.196}$$

$$\mu=\sigma\left(\hat{A}Z^{L_e-1}W_\mu^{L_e}\right), \tag{3.197}$$

$$\log\sigma^2=\sigma\left(\hat{A}Z^{L_e-1}W_\sigma^{L_e}\right), \tag{3.198}$$

or

$$(\mu,\log\sigma^2)=\sigma\left(\hat{A}Z^{L_e-1}W^{L_e}\right), \tag{3.199}$$

where $\hat{A}=\tilde{D}^{-\frac{1}{2}}\tilde{A}\tilde{D}^{-\frac{1}{2}}$, $\tilde{A}=A+I$, $Z^1=X$, and $\tilde{D}_{ii}=\sum_j\tilde{A}_{ij}$.

Example 3.7: Three-Layer GCN Inference Model

The three-layer GCN inference model with ReLU as the activation function σ is defined as

$$\mu = \hat{A}\mathrm{ReLU}\left(\hat{A}\mathrm{ReLU}\left(\hat{A}XW^1\right)W^2\right)W_\mu^3,$$

$$\log\sigma^2 = \hat{A}\mathrm{ReLU}\left(\hat{A}\mathrm{ReLU}\left(\hat{A}XW^1\right)W^2\right)W_\sigma^3.$$

If we combine two of two-layer GCNs together, we get

$$\left(\mu, \log\sigma^2\right) = \hat{A}\mathrm{ReLU}\left(\hat{A}\mathrm{ReLU}\left(\hat{A}XW^1\right)W^2\right)W^3.$$

Parameterization trick is used to calculate the latent variables Z:

$$Z = \mu + \sigma * \varepsilon,$$

where $\varepsilon \sim N(0, I)$.

3.3.2.4 Graph Convolutional Decoder

The output of decoder or generative model consists of two parts: a reconstructed graph structure and node features (Zhang et al. 2020). The node features decoder is the mirror process of the encode. Let $Z^{l_e} = Z$. Similar to equation (3.196), we have

$$Z^{l+1} = \sigma\left(\hat{A}Z^l W^{l+1}\right), l = l_e, \ldots, l_d - 1, \quad (3.200)$$

where $\hat{X} = Z^{l_d}$ is the reconstructed node features.

Let A^* be reconstructed adjacency matrix. It is generated by an inner product between latent variables (Kipf and Welling 2016):

$$P\left(A^* | Z\right) = \prod_{i=1}^{n}\prod_{j=1}^{n} P\left(A_{ij}^* \mid Z_i, Z_j\right), \quad (3.201)$$

Or

simply by

$$A^* = \sigma\left(ZZ^T\right), \quad (3.202)$$

where

$$P\left(A_{ij}^* = 1 | Z_i, Z_j\right) = \sigma\left(Z_i^T Z_j\right), \quad (3.203)$$

A_{ij}^* are the elements of the reconstructed adjacency matrix A^* and σ is the logistic sigmoid function.

3.3.2.5 Loss Function

The loss function for learning graph autoencoder consists of four parts (Zhang et al. 2020). The first part is the reconstruction loss of node feature. The reconstruction loss of node feature is calculated as

$$\mathcal{L}_x = \left\|X - \hat{X}\right\|_F^2, \quad (3.204)$$

where $\hat{X} = Z^{l_d}$ and $\|\cdot\|_F$ is the Frobenius Norm.

The second part is the reconstruction loss of the graph structure. It is calculated as

$$\mathcal{L}_A = \left\|A - A^*\right\|_F^2, \quad (3.205)$$

where $A^* = \sigma\left(ZZ^T\right)$.

The third part is the loss of connection between nodes in the latent space. It is defined as

$$\mathcal{L}_L = \sum_{i=1}^{n}\sum_{j=1}^{n}\left(\left\|Z_i - Z_j\right\|_2^2 a_{ij} + \gamma a_{ij}^2\right)$$

$$= \mathrm{Tr}\left(ZLZ^T\right) + \gamma\left\|A\right\|_F^2, \quad (3.206)$$

$$\text{s.t. } a_i^T 1 = 1, 0 \le \boldsymbol{a}_i \le 1,$$

where $a_i \in R^n$ is the ith column vector of adjacency matrix A, the Laplacian matrix L is defined as

$$L = \frac{1}{2}(D - A), D_{ii} = \sum_{j=1}^{n}\frac{a_{ij} + a_{ji}}{2}.$$

Finally, the fourth part is the regularizer term $\mathcal{L}_R$ to prevent overfitting. It is defined as

$$\mathcal{L}_R = \frac{1}{2}\sum_{i=1}^{l_d}\left\|W^i\right\|_F^2.$$

The total loss function is then given by

$$\mathcal{L}_{GVAE} = \mathcal{L}_X + \mathcal{L}_A + \alpha\mathcal{L}_L + \lambda\mathcal{L}_R. \quad (3.207)$$

3.3.2.6 A Typical Approach to Variational Graph Autoencoders

This section is an extension from Kingma and Welling (2014). There are two types of graph data. The first type of graph data has a unique feature vector of nodes $X=[x_1,\ldots,x_n]^T$ and an adjacency matrix A. The second type of graph data has m feature vectors of nodes $X=[X^1,\ldots,X^m]$, $X^j=[x_1^j,\ldots,x_n^j]^T$ and an adjacency matrix A. The first type of graph data is a special case of the second type of graph data when $m=1$. In the following discussion, we only consider the second type of graph data. Let $Z\in R^{n\times d}$ be a latent space with each node $z_i\in R^d$.

The marginal likelihood can be written as $\log p_\theta(X^1,\ldots,X^m,A)=\sum_{j=1}^m p_\theta(X^j,A)$ if the data are independently sampled. Since stochastic gradient method is used for optimization, we can consider marginal likelihood of one data point (Appendix 3B):

$$\begin{aligned}\log p_\theta(X^i,A) &= \int q_\varnothing(Z\mid X^i,A)\log p_\theta(X^i,A)dZ\\ &= KL\left(q_\varnothing(Z\mid X^i,A)\,||\,p_\theta(Z\mid X^i,A)\right)+\mathcal{L}(\theta,\varnothing,X^i,A)\\ &\geq \mathcal{L}(\theta,\varnothing,X^i,A),\end{aligned} \tag{3.208}$$

where

$$\begin{aligned}\mathcal{L}(\theta,\varnothing,X^i,A) &= E_{q_\varnothing(Z|X^i,A)}\left[\log p_\theta(X^i,A\mid Z)\right]\\ &\quad - KL\left(q_\varnothing(Z|X^i,A)\,||\,p_\theta(Z)\right).\end{aligned} \tag{3.209}$$

Our goal is to use lower bound $\mathcal{L}(\theta,\varnothing,X^i,A)$ to estimate both variation parameters $\varnothing$ and generate parameters θ using gradient methods. If we assume that both $q_\varnothing(Z|X^i,A)$ and $p_\theta(Z)$ are normally distributed, the KL distance between these two distributions can be analytically calculated.

Define the following variable change:

$$Z=g_\varnothing(\varepsilon,X,A),\ \varepsilon\sim p(\varepsilon), \tag{3.210}$$

where $g_\varnothing(\varepsilon,X,A)$ can be a vector of linear or nonlinear functions.

Monte Carlo estimates of expectations of $E_{q_\varnothing(Z|X^i,A)}\left[\log p_\theta(X^i,A\mid Z)\right]$, the first term in equation (3.209) is given by

$$\begin{aligned}&E_{q_\varnothing(Z|X^i,A)}\left[\log p_\theta(X^i,A\mid Z)\right]\\ &\approx\frac{1}{L}\sum_{l=1}^{L}\log p_\theta(X^i,A\mid Z^{i,l}),\end{aligned} \tag{3.211}$$

where $Z^{i,l}=g_\varnothing(\varepsilon^l,X^i,A)$, $\varepsilon^l\sim p(\varepsilon)$.

Now we want to calculate the KL distance between the prior distribution and posterior distribution. We first assume that the prior over latent matrix Z has standard matrix normal $MN_{nd}(0,I,I)$. Its density function is given by

$$\log p(Z)=-\frac{nd}{2}\log(2\pi)-\frac{1}{2}\mathrm{Tr}\left(vec(Z)(vec(Z))^T\right),$$

where Tr denotes the trace of the matrix: $\mathrm{Tr}(A)=\sum_{i=1}^n A_{ii}$.

It can be shown that expectation of $\log p(Z)$ is given by (Appendix 3B)

$$\begin{aligned}E_{q_\varnothing(Z|X,A)}\left[\log p(Z)\right] &= -\frac{nd}{2}\log(2\pi)\\ &\quad-\frac{1}{2}\sum_{i=1}^{n}\sum_{j=1}^{d}\left(\Omega_{ii}M_{jj}+\mu_{ij}^2\right)\end{aligned} \tag{3.212}$$

where $\Sigma_z=\Omega\otimes M$.

Next we assume that the variational approximate posterior $q_\varnothing(Z|X,A)$ has the following matrix normal $MN_{nd}(\mu_z,\Omega_z,V_z)$ (Appendix 3C):

$$\begin{aligned}&\log q_\varnothing(Z|X,A)\\ &=-\frac{nd}{2}\log(2\pi)-\frac{1}{2}\log\left|\sum\nolimits_z\right|\\ &\quad-\frac{1}{2}\mathrm{Tr}\left(\sum\nolimits_z^{-1}\left(vec(Z-\mu)(vec(Z-\mu))^T\right)\right).\end{aligned} \tag{3.213}$$

Its expectation is given by

$$\begin{aligned}&E_{q_\varnothing(Z|X,A)}\left[\log q_\varnothing(Z|X,A)\right]\\ &=-\frac{nd}{2}\log(2\pi)-\frac{1}{2}\log\left|\sum\nolimits_z\right|-\frac{nd}{2}.\end{aligned} \tag{3.214}$$

Combining equations (3.212) and (3.214), we obtain the KL distance between the prior and posterior distributions:

$$\begin{aligned}&-KL\left(q_\varnothing(Z|X,A)\,||\,p(z)\right)\\ &=\frac{1}{2}\left\{\log\left|\sum\nolimits_z\right|+nd-\sum_{i=1}^{n}\sum_{j=1}^{d}\left(\Omega_{ii}M_{jj}+\mu_{ij}^2\right)\right\}.\end{aligned} \tag{3.215}$$

If we assume that

$$\sum_z = \text{diag}\left(S_{11},\ldots,S_{n1},S_{12},\ldots,S_{n2},\ldots,S_{1d,\ldots},S_{nd}\right)$$
$$= S_z, S_{ij} = \Omega_{ii} M_{jj},$$

then equation (3.215) is reduced to

$$-KL\left(q_{\varnothing}(Z|X,A)||p(z)\right)$$
$$= \frac{1}{2}\left(\sum_{i=1}^{n}\sum_{j=1}^{d}\left(\log S_{ij} + 1 - S_{ij} - \mu_{ij}^2\right)\right). \quad (3.216)$$

If we assume that $d = 1$ and n samples are independent, then equation (3.216) is reduced to

$$-KL(q_{\varnothing}(Z|X)||\, p(z)) = \frac{1}{2}\sum_{i=1}^{n}(\log\sigma_i^2 + 1 - \sigma_i^2 - \mu_i^2). \quad (3.217)$$

The three-layer GCN inference model with ReLU as the activation function is defined as

$$(\mu_Z, \log S_Z) = \hat{A}\text{ReLU}\left(\hat{A}\text{ReLU}\left(\hat{A}XW^1\right)W^2\right)W^3, \quad (3.218)$$

where $\hat{A} = \tilde{D}^{-\frac{1}{2}}\tilde{A}\tilde{D}^{-\frac{1}{2}}$, $\tilde{A} = A + I$, $\tilde{D}_{ii} = \Sigma_j \tilde{A}_{ij}$, and W^1, W^2, W^3 are weight matrices in the GCN.

The generative model is defined as

$$p\left(X,A|Z\right) = p_{\theta_1}(A\,|\,Z)p_{\theta_2}(X\,|\,Z), \quad (3.219)$$

where we assume that graph structure generating process and node feature generating process are independent.

The elements of the adjacent matrix A are generated by an inner product between latent variables:

$$p_{\theta_1}\left(A\,|\,Z\right) = \prod_{i=1}^{n}\prod_{j=1}^{n} p\left(A_{ij}\,|\,Z_i, Z_j\right), \quad (3.220)$$

where $p\left(A_{ij} = 1\,|\,Z_i, Z_j\right) = f\left(Z_i^T Z_j\right)$, A_{ij} are the elements of adjacent matrix A and f is the logistic sigmoid function.

The node features decoder is the mirror process of the encoder. We assume that distribution $p_{\theta_2}\left(X\,|\,Z\right)$ is matrix normal distribution. The latent variable Z can be sampled by

$$Z^l = \mu_Z + S_Z \odot \varepsilon^l, \varepsilon^l \sim MN_{nd}(0, I, I), l = 1, 2, \ldots, L. \quad (3.221)$$

Define the three-layer GCN decoder model for continuous feature sector of node with ReLU as the activation function:

$$[\mu_x^l, \log S_x^l] = \hat{A}\text{ReLU}\left(\hat{A}\text{ReLU}\left(\hat{A}Z^l W_x^1\right)W_x^2\right)W_x^3, \quad (3.222)$$

where $\hat{A} = \tilde{D}^{-\frac{1}{2}}\tilde{A}\tilde{D}^{-\frac{1}{2}}$, $\tilde{A} = A + I$, $\tilde{D}_{ii} = \Sigma_j \tilde{A}_{ij}$, and W_x^1, W_x^2, W_x^3 are weight matrices in the GCN.

Then, we can obtain the following expectation:

$$E_{q_{\varnothing}(Z|X,A)}\left[\log p_{\theta_2}(X\,|\,Z)\right]$$
$$\approx -\frac{1}{L}\sum_{l=1}^{L}\left(\frac{nd}{2}\log(2\pi) + \frac{1}{2}\left(\sum_{i=1}^{n}\sum_{j=1}^{d}\left(\log\left(S_x\right)_{ij}^l + \frac{\left(X_{ij} - \left(\mu_x^l\right)_{ij}\right)^2}{\left(S_x\right)_{ij}^l}\right)\right)\right).$$

The Monte Carlo approximation of the ELBO:

$$\mathcal{L}(\theta,\varnothing,X,A) \approx \frac{1}{2}\left(\sum_{i=1}^{n}\sum_{j=1}^{d}\left(\log S_{ij} + 1 - S_{ij} - \mu_{ij}^2\right)\right)$$
$$+\frac{1}{L}\sum_{l=1}^{L}\sum_{i=1}^{n}\sum_{j=1}^{n}\log f\left(\left(Z_i^l\right)^T Z_j^l\right)$$
$$-\frac{1}{L}\sum_{l=1}^{L}\left(\frac{nd}{2}\log(2\pi) + \frac{1}{2}\left(\sum_{i=1}^{n}\sum_{j=1}^{d}\left(\log\left(S_x\right)_{ij}^l + \frac{\left(X_{ij} - \left(\mu_x^l\right)_{ij}\right)^2}{\left(S_x\right)_{ij}^l}\right)\right)\right). \quad (3.223)$$

3.3.2.7 Directed Graph Variational Autoencoder

Now we extend undirected graph VAE to directed VAE. Unlike undirected graph where the adjacency matrix is symmetric, the directed graph must have asymmetric measures (Gasulla 2015; Salha et al. 2019). Essential issue is how to effectively map directed graph to latent space using the graph VAE and to reconstruct asymmetric relationships from latent variables.

3.3.2.7.1 Directed Graph Representation in Latent Space

In this section, we introduce an asymmetric

graph decoding scheme for coding directed graph in latent space, which is inspired by Newton's theory of universal gravitation (Salha et al. 2019). We first define the adjacency matrix A as the transition matrix: $A_{ij} = p_{ij}$, where p_{ij} represents the probability of the node i is directed to the node j. Typically, we define

$$p_{ij} = \begin{cases} 1 \text{ if node } i \text{ is directed to node } j \\ 0 \text{ otherwise} \end{cases} \quad (3.224)$$

We assume that a graph encoder model assigns a d-dimensional latent vector Z_i to each node i of the graph. In order to represent directed graph in the latent space, we add a mass parameter $m_i \in R^+$ as one additional component of the latent vector Z_i for each node to determine the direction of the edge. Consider a pair of nodes i and j with the feature vectors Z_i and Z_j. Define the measure:

$$a_{i \to j} = \frac{m_j}{r_{ij}^2},$$

where $r_{ij}^2 = Z_i - Z_j{}^2$. If $a_{j \to i} > a_{i \to j}$ then the node j is directed to i. According to universal gravitation model, if we view m_i and m_j as the mass of nodes i and j, respectively, $a_{j \to i} > a_{i \to j}$ implies that $m_i > m_j$ then the node j will move toward the node i.

Define the elements of the adjacency matrix in the latent space as

$$\tilde{A}_{ij} = f\left(\log a_{i \to j}\right) = f\left(\log\left(Gm_j\right) - \log\left\|Z_i - Z_j\right\|^2\right). \quad (3.225)$$

3.3.2.7.2 Encoder To encode direction, we need to augment matrix $Z \in R^{n \times d}$ to the matrix $\tilde{Z} \in R^{n \times (d+1)}$, where last column denotes the model's estimate of $\tilde{M} = \left[\tilde{m}_1, \ldots, \tilde{m}_n\right]^T$, where $\tilde{m}_i = \log Gm_i$. The normalized adjacency matrix in the classical graph VAE is replaced by $D_{out}^{-1}(A+I)$, where A is the adjacency matrix, defined in equation (3.224), D_{out} represents the diagonal out-degree matrix of $A+I$, where $D_{out}(i,i) = \sum_j A_{ij} + 1$ denotes the number of edges going out of node i, plus one.

The three-layer GCN inference model with ReLU as the activation function is defined as

$$\left(\tilde{\mu}_Z, \log \tilde{S}_Z\right) = \hat{A}\mathrm{ReLU}\left(\hat{A}\mathrm{ReLU}\left(\hat{A}XW^1\right)W^2\right)W^3, \quad (3.226)$$

where $\tilde{\mu}_Z \in R^{n \times (d+1)}$, the number of diagonal elements in the matrix $\tilde{S}_z$ is $n(d+1)$, $\hat{A} = \tilde{D}^{-\frac{1}{2}}\tilde{A}\tilde{D}^{-\frac{1}{2}}$, $\tilde{A} = A + I$, $\tilde{D}_{ii} = \sum_j \tilde{A}_{ij}$, and W^1, W^2, W^3 are weight matrices in the GCN.

3.3.2.7.3 Decoder The generative model is defined as

$$p\left(X, A|\tilde{Z}\right) = p_{\theta_1}\left(A \mid \tilde{Z}\right)p_{\theta_2}\left(X \mid \tilde{Z}\right), \quad (3.227)$$

where $\tilde{Z} = \left[Z, \tilde{M}\right]$, we assume that graph structure generating process and node feature generating process are independent.

The elements of the adjacent matrix A are generated by an inner product between latent variables:

$$p_{\theta_1}\left(A|Z, \tilde{M}\right) = \prod_{i=1}^{n}\prod_{j=1}^{n} p\left(A_{ij} \mid Z_i, Z_j, \tilde{m}_j\right), \quad (3.228)$$

where

$$p\left(A_{ij} = 1 \mid Z_i, Z_j, \tilde{m}\right) = f\left(\log m_j - \log\left\|Z_i - Z_j\right\|^2\right), \quad (3.229)$$

or

$$p\left(A_{ij} = 1|Z_i, Z_j, \tilde{m}\right) = f\left(\log m_j - \lambda \log\left\|Z_i - Z_j\right\|^2\right), \quad (3.230)$$

where A_{ij} are the elements of adjacent matrix A and f is the logistic sigmoid function.

Sampling latent variable Z:

$$\tilde{Z}^l = \tilde{\mu}_Z + \tilde{S}_Z \odot \varepsilon^l, \ \varepsilon^l \sim MN_{nd}(0, I, I), \ l = 1, 2, \ldots, L. \quad (3.231)$$

where $\tilde{Z}^l = \left[Z^l, \tilde{m}^l\right]$

The Monte Carlo estimation of $E_{q\varnothing(\tilde{Z}|X,A)}\left[\log p_{\theta_1}\left(A \mid \tilde{Z}\right)\right]$:

$$E_{q\varnothing(\tilde{Z}|X,A)}\left[\log p_{\theta_1}\left(A \mid \tilde{Z}\right)\right] \approx \frac{1}{L}\sum_{l=1}^{L}\sum_{i=1}^{n}\sum_{j=1}^{n} \log f\left(\tilde{m}_j - \log\left\|Z_i^l - Z_j^l\right\|^2\right). \quad (3.232)$$

Define the three-layer GCN decoder model for continuous feature sector of node with ReLU as the activation function:

$$\left[\mu_x^l, \log S_x^l\right] = \hat{A}ReLU\left(\hat{A}ReLU\left(\hat{A}\tilde{Z}^l W_x^1\right)W_x^2\right)W_x^3, \quad (3.233)$$

where $\hat{A}=\tilde{D}^{-\frac{1}{2}}\tilde{A}\tilde{D}^{-\frac{1}{2}}$, $\tilde{A}=A+I$, $\tilde{D}_{ii}=\Sigma_j \tilde{A}_{ij}$, and W_x^1, W_x^2, W_x^3 are weight matrices in the GCN.

The Monte Carlo estimation of the ELBO is given by

$$\mathcal{L}(\theta,\varnothing,X,A) \approx \frac{1}{2}\left(\sum_{i=1}^{n}\sum_{j=1}^{d+1}\left(\log\left(S_x\right)_{ij}+1-\left(S_x\right)_{ij}-\left(\mu_x\right)_{ij}^2\right)\right) + \frac{1}{L}\sum_{l=1}^{L}\sum_{i=1}^{n}\sum_{j=1}^{n}\log f\left(\tilde{m}_j-\log\left\|Z_i^l-Z_j^l\right\|^2\right) - \frac{1}{L}\sum_{l=1}^{L}\left(\frac{nd}{2}\log(2\pi)+\frac{1}{2}\left(\sum_{i=1}^{n}\sum_{j=1}^{d}\left(\log\left(S_x\right)_{ij}^l+\frac{\left(x_{ij}-\left(\mu_x^l\right)_{ij}\right)^2}{\left(S_x\right)_{ij}^l}\right)\right)\right). \tag{3.234}$$

Next we consider discrete categorical variables X. Categorical variable can be coded by a one-hot vector, where the "1" indicates the value of the corresponding variable, others takes the "0" values. Let P_x be probability matrix for X. The probability matrix P_x is generated by

$$P_x^l = softmax\left(\hat{A}\mathrm{ReLU}\left(\hat{A}\mathrm{ReLU}\left(\hat{A}\tilde{Z}^l W_x^1\right)W_x^2\right)W_x^3\right). \tag{3.235}$$

The probability distribution $p_{\theta_2}\left(X\,|\,\tilde{Z}\right)$ for the categorical variables is given by

$$p_{\theta_2}\left(X|\tilde{Z}^l\right)=\sum_{i=1}^{n}\sum_{j=1}^{d}X_{ij}\log\left(P_x^l\right)_{ij}. \tag{3.236}$$

Since we assume that the prior and posterior distributions are not changes, which in turn implies that KL distance between them is also unchanged. Therefore, the evidence lover bound for categorical variables is given by

$$\mathcal{L}(\theta,\varnothing,X,A) \approx \frac{1}{2}\left(\sum_{i=1}^{n}\sum_{j=1}^{d+1}\left(\log\left(S_x\right)_{ij}+1-\left(S_x\right)_{ij}-\left(\mu_x\right)_{ij}^2\right)\right) + \frac{1}{L}\sum_{l=1}^{L}\sum_{i=1}^{n}\sum_{j=1}^{n}\log f\left(\tilde{m}_j-\log\left\|Z_i-Z_j\right\|^2\right) + \frac{1}{L}\sum_{l=1}^{L}\sum_{i=1}^{n}\sum_{j=1}^{d}X_{ij}\log\left(P_x^l\right)_{ij}. \tag{3.237}$$

3.3.2.8 Graph VAE for Clustering

Clustering aims to group individuals into several classes without label information (Yang et al. 2019). There are model-based and similarity-based approaches for clustering. Due to space limitation, in this section, we only introduce spectrum approach for graph clustering (Wang et al. 2017). The spectral graph clustering offers the estimation of the number of optimal clusters and similarity measures. The procedures of the spectral graph clustering are given below (Wang et al. 2017).

Step 1: Let $Z_0=Z^L$, where

$$Z^L=\mu_x^L+S_x^L\odot\varepsilon^l, \text{ the final output of Graph VAE} \tag{3.238}$$

Step 2: Construct similarity matrix:

$$Z_1=Z_0Z_0^T.$$

Step 3: Construct normalized Laplacian Matrix:

$$Z_2=\frac{1}{2}\left(\left|Z_1\right|+\left|Z_1^T\right|\right).$$

Step 4: Eigenvalue decomposition:

$$Z_2=U\Lambda U^T, \tag{3.239}$$

where

$$U=\begin{bmatrix} u_{11} & \cdots & u_{1n} \\ \vdots & \vdots & \vdots \\ u_{k1} & \cdots & u_{kn} \\ \vdots & \vdots & \vdots \\ u_{n1} & \cdots & u_{nn} \end{bmatrix} \text{ and } \Lambda=\begin{bmatrix} \lambda_1 & \cdots & 0 \\ \vdots & \ddots & \vdots \\ 0 & \cdots & \lambda_n \end{bmatrix}, \lambda_1\ge\lambda_2\ge\ldots\ge\lambda_k\ge\ldots\ge\lambda_n.$$

Step 5: Cluster analysis:

Take k eigenvectors corresponding to the k largest eigenvalues as points in R^k space:

$$Y=\begin{bmatrix} u_{11} & \cdots & u_{k1} \\ \vdots & \vdots & \vdots \\ u_{n1} & \cdots & u_{nk} \end{bmatrix}.$$

Step 6: Renormalize the rows of Y to have unit length.

$$\tilde{Y}_{ij} = \frac{Y_{ij}}{\sqrt{\sum_j Y_{ij}^2}}.$$

Let $\tilde{Y} = \left(\tilde{Y}_{ij}\right)_{n\times k}$.

Use clustering algorithms, for example K-means algorithm to cluster k vectors in the matrix $\tilde{Y}$.

SOFTWARE PACKAGE

Code for VAE for directed graph can be downloaded from:

https://github.com/muhanzhang/D-VAE

Code for "Gravity-Inspired Graph Autoencoders for Directed Link Prediction" is posted on the website:

https://github.com/deezer/gravity_graph_autoencoders

Code for the paper "SPECTRALNET: spectral clustering using deep neural networks" is published on the website: https://github.com/KlugerLab/SpectralNet

Software for VAE for regression is posted on the website:

https://github.com/QingyuZhao/VAE-for-Regression

Code for "Deep Clustering by Gaussian Mixture Variational Autoencoders with Graph Embedding" can be downloaded from:

https://github.com/dodoyang0929/DGG.git

Code for the paper "DAG-GNN: DAG Structure Learning with Graph Neural Networks" is posted on the website:

https://github.com/fishmoon1234/DAG-GNN

APPENDIX 3A

Define

$$\mathrm{F}\left(q\left(z_j\right)\right) = \int q_j\left(z_j\right) E_{q(z_{-j})} \log p\left(x, z_j | z_{-j}\right) dz_j - \int q\left(z_j\right) \log q\left(z_j\right) dz_j. \quad \text{(3A1)}$$

Now we use calculus of variation to find minimizer of $\mathrm{F}\left(q\left(z_j\right)\right)$. The necessary condition for the optimum of functional $\mathrm{F}\left(q\left(z_j\right)\right)$ is the first variation of a functional $\mathrm{F}\left(q\left(z_j\right)\right)$ must be equal to zero. The first variation of a functional $\mathrm{F}\left(q\left(z_j\right)\right)$ is given by

$$\delta \mathrm{F}\left(q\left(z_j\right), h\right) = \frac{\partial F\left(q\left(z_j\right) + \varepsilon h\right)}{\partial \varepsilon}. \quad \text{(3A2)}$$

Note that

$$F\left(q_j\left(z_j\right) + \varepsilon h\right) = \int \left(q_j\left(z_j\right) + \varepsilon h\right) E_{q(z_{-j})} \log p\left(x, z_j | z_{-j}\right) dz_j. - \int \left(q_j\left(z_j\right) + \varepsilon h\right) \log\left(q_j\left(z_j\right) + \varepsilon h\right) dz_j. \quad \text{(3A3)}$$

Substituting equation (3A3) into equation (3A2), we obtain

$$\frac{\partial F\left(q\left(z_j\right) + \varepsilon h\right)}{\partial \varepsilon} = \int h E_{q(z_{-j})} \log p\left(x, z_j | z_{-j}\right) dz_j - \int \left[h \log q_j\left(z_j\right) + h\right] dz_j$$
$$= \int \left\{E_{q(z_{-j})} \log p\left(x, z_j | z_{-j}\right) - \left[\log q_j\left(z_j\right) + 1\right]\right\} h dz_j. \quad \text{(3A4)}$$

It follows from equation (3A4) that $\frac{\partial F\left(q\left(z_j\right) + \varepsilon h\right)}{\partial \varepsilon} = 0$ requies

$$\log q_j^*\left(z_j\right) = E_{q(z_{-j})} \log p\left(x, z_j | z_{-j}\right) - 1, \quad \text{(3A5)}$$

or

$$q_j^*\left(z_j\right) \propto \exp\left\{E_{q(z_{-j})} \log p\left(x, z_j | z_{-j}\right)\right\}. \quad \text{(3A6)}$$

APPENDIX 3B: DERIVATION OF ALGORITHMS FOR VARIATIONAL GRAPH AUTOENCODERS

3B1 Evidence of Lower Bound

This appendix is mainly from Kingma and Welling (2014). There are two types of graph data. The first type of graph data has a unique feature vector of n nodes $X = \left[x_1, \ldots, x_n\right]^T$ and an adjacency matrix A. The second type of graph data has m feature vectors of nodes $X = \left[X^1, \ldots, X^m\right]$, $X^j = \left[x_1^j, \ldots, x_n^j\right]^T$ and an adjacency matrix A. The first type of graph data is a special case of the second type of graph data when

$m=1$. In the following discussion, we only consider the second type of graph data. Let $Z \in R^{n\times d}$ be a latent space with each node $z_i \in R^d$.

The marginal likelihood can be written as $\log p_\theta(X^1,\ldots,X^m,A)=\sum_{j=1}^{m} p_\theta(X^j,A)$, if the data are independently sampled. Since stochastic gradient method is used for optimization, we can consider marginal likelihood of one data point:

$$\begin{aligned}\log p_\theta(X^i,A) &= \int q_\varnothing(Z|X^i,A)\log p_\theta(X^i,A)dZ\\ &= \int q_\varnothing(Z|X^i,A)\log\frac{p_\theta(X^i,A,Z)}{p_\theta(Z|X^i,A)}dZ\\ &= \int q_\varnothing(Z|X^i,A)\log\frac{q_\varnothing(Z|X^i,A)}{q_\varnothing(Z|X^i,A)}\frac{p_\theta(X^i,A,Z)}{p_\theta(Z|X^i,A)}dZ\\ &= -\int q_\varnothing(Z|X^i,A)\log q_\varnothing(Z|X^i,A)dZ\\ &\quad + KL\left(q_\varnothing(Z|X^i,A)\,||\,p_\theta(Z|X^i,A)\right)\\ &\quad + \int q_\varnothing(Z|X^i,A)\log p_\theta(X^i,A,Z)dZ\\ &= KL\left(q_\varnothing(Z|X^i,A)\,||\,p_\theta(Z|X^i,A)\right)+\mathcal{L}(\theta,\varnothing,X^i,A)\\ &\ge \mathcal{L}(\theta,\varnothing,X^i,A),\end{aligned} \tag{3B1}$$

where

$$\mathcal{L}(\theta,\varnothing,X^i,A) = E_{q_\varnothing(Z|X^i,A)}\left[-\log q_\varnothing(Z|X^i,A)+\log p_\theta(X^i,A,Z)\right]. \tag{3B2}$$

Note that

$$\log p_\theta(X^i,A,Z)=\log p_\theta(Z)+\log p_\theta(X^i,A|Z) \tag{3B3}$$

Substituting equation (3B3) into equation (3B2) yields

$$\mathcal{L}(\theta,\varnothing,X^i,A)=E_{q_\varnothing(Z|X^i,A)}\left[\log p_\theta(X^i,A|Z)\right] - KL\left(q_\varnothing(Z|X^i,A)\,||\,p_\theta(Z)\right). \tag{3B4}$$

Our goal is to use lower bound $\mathcal{L}(\theta,\varnothing,X^i,A)$ to estimate both variation parameters $\varnothing$ and generate parameters θ using gradient methods. If we assume that both $q_\varnothing(Z|X^i,A)$ and $p_\theta(Z)$ are normally distributed, the KL distance between these two distributions can be analytically calculated. The remaining term in the lower bound involves to derive gradient

$$\begin{aligned}&\nabla_\varnothing E_{q_\varnothing(Z|X,A)}\left[f(Z)\right]\\ &= E_{q_\varnothing(Z|X,A)}\left[f(Z)\nabla_{q_\varnothing(Z|X,A)}\log q_\varnothing(Z|X,A)\right]\\ &\approx \frac{1}{L}\sum_{l=1}^{L} f(Z^l)\nabla_{q_\varnothing(Z^l|X,A)}\log q_\varnothing(Z^l|X,A),\end{aligned} \tag{3B5}$$

where $Z^l \sim q_\varnothing(Z^{l-1}|X,A)$.

Calculation of equation (3B5) requires iteration, and hence is intractable.

3B2 The Reparameterization Trick

A key challenge in computation of $\nabla_\varnothing E_{q_\varnothing(Z|X,A)}\left[f(Z)\right]$ is that expectation involves variational parameters $\varnothing$. If we can transfer $\varnothing$ from the distribution of $q_\varnothing(Z|X,A)$ to the function $f(Z)$, then the problem of estimation of the variation parameters $\varnothing$ is mitigated. Changing variable technique in the distribution theory can achieve this. Define the following variable change:

$$Z=g_\varnothing(\varepsilon,X,A),\ \varepsilon \sim p(\varepsilon), \tag{3B6}$$

where $g_\varnothing(\varepsilon,X,A)$ can be a vector of linear or nonlinear functions.

Let $\frac{\partial g_\varnothing(\varepsilon,X,A)}{\partial \varepsilon^T}$ be the Jacobian matrix of the transformation $g_\varnothing(\varepsilon,X,A)$. The distribution $p(Z|X,A)$ is given by

$$p(Z|X,A)=\frac{p(\varepsilon)}{\left|\det\left(\frac{\partial g_\varnothing(\varepsilon,X,A)}{\partial \varepsilon^T}\right)\right|}, \tag{3B7}$$

where det (.) denotes determinant of a matrix. The distribution $p(\varepsilon)$ involves no variational parameters $\varnothing$. The distribution $p(Z|X,A)$ can be used to approximate $q_\varnothing(Z|X,A)$. Therefore,

$$q_\varnothing(Z|X,A)dZ = p(Z|X,A)dz. \tag{3B8}$$

Substituting equation (3B7) into equation (3B8) yields

$$q_\varnothing(Z|X,A)dZ = \frac{p(\varepsilon)}{\left|\det\left(\frac{\partial g_\varnothing(\varepsilon,X,A)}{\partial \varepsilon^T}\right)\right|}dZ. \tag{3B9}$$

It follows from equation (3B6) that

$$dZ = \left| \det\left(\frac{\partial g_{\varnothing}(\varepsilon, X, A)}{\partial \varepsilon^T} \right) \right| d\varepsilon. \quad (3B10)$$

Substituting equation (3B10) into equation (3B9), we obtain

$$q_{\varnothing}(Z|X,A)dZ = p(\varepsilon)d\varepsilon, \quad (3B11)$$

which implies that conditional probability $q_{\varnothing}(Z \mid X, A)dZ$ can be calculated by $p(\varepsilon)d\varepsilon$. Therefore,

$$E_{q_{\varnothing}(Z|X,A)}[f(Z)] = \int f(Z) q_{\varnothing}(Z \mid X,A) dZ$$
$$= \int f(g_{\varnothing}(\varepsilon, X,A)) p(\varepsilon) d\varepsilon. \quad (3B12)$$

Now quantity $\int f(g_{\varnothing}(\varepsilon, X,A)) p(\varepsilon) d\varepsilon$ involves only random variable ε.

3B3 Stochastic Gradient Variational Bayes (SGVB) Estimator

Now Monte Carlo estimates of expectations of some function $f(Z)$ can be calculated by

$$E_{q_{\varnothing}(Z|X,A)}[f(Z)] \approx \frac{1}{L} \sum_{l=1}^{L} f(g_{\varnothing}(\varepsilon^l, X,A)), \ \varepsilon \sim p(\varepsilon). \quad (3B13)$$

We first apply equation (3B13) to equation (3B2). Let

$$f(Z) = -\log q_{\varnothing}(Z|X^i,A) + \log p_{\theta}(X^i,A,Z).$$

Then, it follows from equations (3B2) and (3B13), we obtain the estimator of the first version of lower bound:

$$\hat{\mathcal{L}}^A(\theta,\varnothing,X^i,A) \approx \frac{1}{L} \sum_{l=1}^{L} \left[\log p_{\theta}(X^i, A, Z^{i,l}) - \log q_{\varnothing}(Z^{i,l}|X^i,A) \right], \quad (3B14)$$

where $Z^{i,l} = g_{\varnothing}(\varepsilon^l, X^i, A)$, $\varepsilon^l \sim p(\varepsilon)$.

Now we consider lower bound in equation (3B4). Both prior distribution $p_{\theta}(Z)$ and approximate posterior distribution $q_{\varnothing}(Z|X^i,A)$ are often assumed normal distribution. The second term KL distance can be calculated analytically and can be interpreted as regularizing variation parameters $\varnothing$ to enforce the approximate posterior $q_{\varnothing}(Z|X^i,A)$ to be close to the prior $p_{\theta}(Z)$. We only need to use Stochastic Gradient Variational Bayes to estimate the first term, reconstruction error $E_{q_{\varnothing}(Z|X^i,A)}[\log p_{\theta}(X^i, A \mid Z)]$, in equation (3B4):

$$E_{q_{\varnothing}(Z|X^i,A)}\left[\log p_{\theta}(X^i, A \mid Z)\right] \approx \frac{1}{L} \sum_{l=1}^{L} \log p_{\theta}(X^i, A \mid Z^{i,l}), \quad (3B15)$$

where $Z^{i,l} = g_{\varnothing}(\varepsilon^l, X^i, A)$, $\varepsilon^l \sim p(\varepsilon)$.

Substituting equation (3B15) into equation (3B4), we obtain the estimator of the second version of the lower bound:

$$\hat{\mathcal{L}}^B(\theta,\varnothing,X^i,A) = \frac{1}{L} \sum_{l=1}^{L} \log p_{\theta}(X^i, A \mid Z^{i,l}) - KL(q_{\varnothing}(Z \mid X^i,A) \| p_{\theta}(Z)). \quad (3B16)$$

Suppose that N data points $\{X^i, i = 1, \ldots, N, A\}$ are sampled. Let M be size of minibatch. Using minibatch, the marginal likelihood can be estimated by

$$\mathcal{L}(\theta, \varnothing, X, A) \approx \hat{\mathcal{L}}^M(\theta, \varnothing, X^M, A) = \frac{N}{M} \sum_{i=1}^{M} \hat{\mathcal{L}}(\theta,\varnothing,X^i, A), \quad (3B17)$$

where $X^M = \{X^i, i = 1, \ldots, M\}$ are randomly sampled M data points from the entire dataset X with sample size N. The number of sample L is often set to 1 when the minibatch size M is larger than 100.

There are three main choices of selecting a differentiable transformation function $g_{\varnothing}(\varepsilon,X,A)$ (Kingma and Welling 2014).

1. Tractable inverse cumulative distribution function (CDF) of $q_{\varnothing}(Z \mid X,A)$. Let $y = F(z) = P(Z \le z)$ be the CDF. The inverse of CDF is $F^{-1}(y)$. For example, let $F(z) = 1 - e^{-z}$ be a CDF of exponential distribution. Then, its inverse CDF is $-\log(1-y)$. Inverse CDF of Exponential, Cauchy, Logistic, Rayleigh, Pareto, Weibull, Reciprocal, Gompertz, Gumbel, and Erlang distributions are typical examples for transformation functions.

2. Location-scale family distribution. Let μ denote location and σ denote scale. The standard location-scale distribution with $\mu=0$ (or matrix μ) and $\sigma=1$ (or covariance matrix $\Sigma=I$) can be taken as the auxiliary variable ε. Define the transformation function $g(Z|X,A)=\mu+\Sigma\odot\varepsilon$, $\varepsilon\sim MN(0, I, I)$. Location-scale family distribution includes Laplace, Elliptical, Student's t, Logistic, Uniform, Triangular, and Gaussian distributions when scalar variable is considered.
3. Composition: Transformation function can be composite function of tractable transformation function. Typical examples include Log-Normal, Gamma, Dirichlet, Beta, Chi-Squared, and F distributions.

3B4 Neural Network Implementation

The inference (recognition) model and generative model can be implemented by feedforward neural networks, CNNs, and recurrent neural networks.

3B4.1 Encoder (Inference Model)

Recall that $Z\in R^{n\times d}$ is a latent space with each node $z_i\in R^d$. Assume that the prior over latent matrix Z has standard matrix normal $MN_{nd}(0, I, I)$. Its density function is given by

$$p(Z)=(2\pi)^{-\frac{nd}{2}}exp\left\{-\frac{1}{2}(vec(Z))^T vec(Z)\right\}$$

$$=(2\pi)^{-\frac{nd}{2}}exp\left\{-\frac{1}{2}\text{Tr}\left((vec(Z))^T vec(Z)\right)\right\}$$

$$=(2\pi)^{-\frac{nd}{2}}exp\left\{-\frac{1}{2}\text{Tr}(vec(Z)(vec(Z))^T\right\}, \text{ which implies}$$

$$\log p(Z)=-\frac{nd}{2}\log(2\pi)-\frac{1}{2}\text{Tr}(vec(Z)(vec(Z))^T. \tag{3B18}$$

Next we assume that the variational approximate posterior $q_\varnothing(Z|X)$ has the following matrix normal $MN_{nd}(\mu_z, \Omega_z, V_z)$ (Appendix 3C):

$$q_\varnothing(Z|X,A)=(2\pi)^{-\frac{nd}{2}}\left|\sum_z\right|^{-\frac{1}{2}} exp\left\{-\frac{1}{2}\text{Tr}\left(\sum_z^{-1}(vec(Z-\mu)(vec(Z-\mu))^T\right)\right\} \text{ or}$$

$$\log q_\varnothing(Z|X,A)=-\frac{nd}{2}\log(2\pi)-\frac{1}{2}\log\left|\sum_z\right| -\frac{1}{2}\text{Tr}\left(\sum_z^{-1}(vec(Z-\mu)(vec(Z-\mu))^T\right), \tag{3B19}$$

where $\Sigma_z=\Omega_z\odot V_z$.

$$E_{q\varnothing(Z|X,A)}[\log p(Z)]=-\frac{nd}{2}\log(2\pi)-\frac{1}{2}\text{Tr} \times\left(E_{q\varnothing(Z|X,A)}\left[vec(Z)(vec(Z))^T\right]\right). \tag{3B20}$$

Note that

$$E_{q\varnothing(Z|X,A)}\left[vec(Z)(vec(Z))^T\right] =\sum_z+vec(\mu)(vec(\mu))^T, \text{ which implies}$$

$$\text{Tr}\left(E_{q\varnothing(Z|X,A)}\left[vec(Z)(vec(Z))^T\right]\right) =\sum_{i=1}^n\sum_{j=1}^d\left(\Omega_{ii}M_{jj}+\mu_{ij}^2\right), \tag{3B21}$$

where $\Sigma_z=\Omega\otimes M$.

Substituting equation (3B21) into equation (3B20) yields

$$E_{q\varnothing(Z|X,A)}[\log p(Z)] =-\frac{nd}{2}\log(2\pi)-\frac{1}{2}\sum_{i=1}^n\sum_{j=1}^d\left(\Omega_{ii}M_{jj}+\mu_{ij}^2\right). \tag{3B22}$$

Next it follows from equation (3B19) that

$$E_{q\varnothing(Z|X,A)}\left[\log q_\varnothing(Z|X,A)\right] =-\frac{nd}{2}\log(2\pi)-\frac{1}{2}\log\left|\sum_z\right| -\frac{1}{2}\text{Tr}\left(\sum_z^{-1}E\left[vec(Z-\mu)(vec(Z-\mu))^T\right]\right)$$

$$=-\frac{nd}{2}\log(2\pi)-\frac{1}{2}\log\left|\sum_z\right|-\frac{1}{2}\text{Tr}\left(\sum_z^{-1}\sum_z\right)$$

$$=-\frac{nd}{2}\log(2\pi)-\frac{1}{2}\log\left|\sum_z\right|-\frac{nd}{2}. \tag{3B23}$$

Combining equations (3B22) and (3B23), we obtain

$$-KL\left(q_\varnothing(Z|X,A)||p(z)\right) =\frac{1}{2}\left\{\log\left|\sum_z\right|+nd-\sum_{i=1}^n\sum_{j=1}^d\left(\Omega_{ii}M_{jj}+\mu_{ij}^2\right)\right\}. \tag{3B24}$$

If we assume that

$$\sum_z = \text{diag}\left(S_{11}, \ldots, S_{n1}, S_{12}, \ldots, S_{n2}, \ldots, S_{1d,\ldots,} S_{nd}\right)$$
$$= S_z, S_{ij} = \Omega_{ii} M_{jj}, \tag{3B25}$$

then equation (3B24) will be reduced to

$$-KL\left(q_\varnothing(Z|X,A)||p(z)\right) = \frac{1}{2}\left(\sum_{i=1}^{n}\sum_{j=1}^{n}\left(\log S_{ij} + 1 - S_{ij} - \mu_{ij}^2\right)\right). \tag{3B26}$$

If we assume that $d=1$ and n samples are independent, then equation (3B24) is reduced to

$$-KL\left(q_\varnothing(Z|X)||\,p(z)\right) = \frac{1}{2}\sum_{j=1}^{n}\left(\log\sigma_j^2 + 1 - \sigma_j^2 - \mu_j^2\right). \tag{3B27}$$

The three-layer GCN inference model with ReLU as the activation function is defined as

$$\left(\mu_Z, \log S_Z\right) = \hat{A}\text{ReLU}\left(\hat{A}\text{ReLU}\left(\hat{A}XW^1\right)W^2\right)W^3, \tag{3B28}$$

where $\hat{A} = \tilde{D}^{-\frac{1}{2}}\tilde{A}\tilde{D}^{-\frac{1}{2}}$, $\tilde{A} = A + I, \tilde{D}_{ii} = \Sigma_j \tilde{A}_{ij}$, and W^1, W^2, W^3 are weight matrices in the GCN.

3B4.2 Decoder (Generative Model)

The generative model is defined as

$$p(X,A|Z) = p_{\theta_1}(A\,|\,Z)\,p_{\theta_2}(X\,|\,Z), \tag{3B29}$$

where we assume that graph structure generating process and node feature generating process are independent.

The elements of the adjacent matrix A are generated by an inner product between latent variables:

$$p_{\theta_1}(A|Z) = \prod_{i=1}^{n}\prod_{j=1}^{d} p\left(A_{ij}\,|\,Z_i, Z_j\right), \tag{3B30}$$

where $p\left(A_{ij}=1\,|\,Z_i, Z_j\right) = f\left(Z_i^T Z_j\right)$, A_{ij} are the elements of adjacent matrix A and f is the logistic sigmoid function.

The node features decoder is the mirror process of the encoder. We assume that distribution $p_{\theta_2}(X\,|\,Z)$ is matrix normal distribution:

$$p_{\theta_2}(X|Z) = (2\pi)^{-\frac{nd}{2}}|S_x|^{-\frac{1}{2}} \exp\left\{-\frac{1}{2}\left(vec(X-\mu)\right)^T S_x^{-1} vec(X-\mu)\right\}, \tag{3B31}$$

where $S_x = \text{diag}\left((S_x)_{11}, \ldots, (S_x)_{n1}, (S_x)_{12}, \ldots, (S_x)_{1n}, (S_x)_{nn}\right)$.

Thus, $\log p_{\theta_2}(X|Z)$ is given by

$$\log p_{\theta_2}(X|Z) = -\frac{nd}{2}\log(2\pi) - \frac{1}{2}\log|S_x| - \frac{1}{2}\left(vec(X-\mu)\right)^T S_x^{-1} vec(X-\mu)$$
$$= -\frac{nd}{2}\log(2\pi) - \frac{1}{2}\left(\sum_{i=1}^{n}\sum_{j=1}^{d}\left(\log(S_x)_{ij} + \frac{\left(X_{ij} - (\mu_x)_{ij}\right)^2}{(S_x)_{ij}}\right)\right). \tag{3B32}$$

Sampling latent variable Z:

$$Z^l = \mu_Z + S_Z \odot \varepsilon^l, \varepsilon^l \sim MN_{nd}(0, I, I), l = 1, 2, \ldots, L. \tag{3B33}$$

Combining equations (3B30) and (3B33), we obtain the Monte Carlo estimation of $E_{q\varnothing(Z|X,A)}\left[\log p_{\theta_1}(A\,|\,Z)\right]$:

$$E_{q\varnothing(Z|X,A)}\left[\log p_{\theta_1}(A\,|\,Z)\right] \approx \frac{1}{L}\sum_{l=1}^{L}\sum_{i=1}^{n}\sum_{j=1}^{n}\log f\left(\left(Z_i^l\right)^T Z_j^l\right). \tag{3B34}$$

Define the three-layer GCN decoder model for continuous feature sector of node with ReLU as the activation function:

$$\left[\mu_x^l, \log S_x^l\right] = \hat{A}\text{ReLU}\left(\hat{A}\text{ReLU}\left(\hat{A}Z^l W_x^1\right)W_x^2\right)W_x^3, \tag{3B35}$$

where $\hat{A} = \tilde{D}^{-\frac{1}{2}}\tilde{A}\tilde{D}^{-\frac{1}{2}}$, $\tilde{A} = A + I, \tilde{D}_{ii} = \Sigma_j \tilde{A}_{ij}$, and W_x^1, W_x^2, W_x^3 are weight matrices in the GCN.

The Monte Carlo estimation of $E_{q\varnothing(Z|X,A)}\left[\log p_{\theta_2}(X\,|\,Z)\right]$ is given by

$$E_{q\varnothing(Z|X,A)}\left[\log p_{\theta_2}(X\,|\,Z)\right] \approx \frac{1}{L}\sum_{l=1}^{L}\log p_{\theta_2}(X\,|\,Z^l)$$

$$= -\frac{1}{L}\sum_{l=1}^{L}\left(\frac{nd}{2}\log(2\pi)+\frac{1}{2}\left(\sum_{i=1}^{n}\sum_{j=1}^{d}\left(\log\left(S_x\right)_{ij}^{l}+\frac{\left(X_{ij}-\left(\mu_x\right)_{ij}\right)^2}{\left(S_x\right)_{ij}}\right)\right)\right). \quad (3B36)$$

Therefore, using equations (3B16), (3B26), (3B34), and (3B36), we obtain the Monte Carlo approximation of the ELBO:

$$\mathcal{L}(\theta,\varnothing,X,A)\approx\frac{1}{2}\left(\sum_{i=1}^{n}\sum_{j=1}^{n}\left(\log S_{ij}+1-S_{ij}-\mu_{ij}^2\right)\right)+\frac{1}{L}\sum_{l=1}^{L}\sum_{i=1}^{n}\sum_{j=1}^{n}\log f\left(\left(Z_i^l\right)^T Z_j^l\right)$$
$$-\frac{1}{L}\sum_{l=1}^{L}\left(\frac{nd}{2}\log(2\pi)+\frac{1}{2}\left(\sum_{i=1}^{n}\sum_{j=1}^{d}\left(\log\left(S_x\right)_{ij}^{l}+\frac{\left(X_{ij}-\left(\mu_x\right)_{ij}\right)^2}{\left(S_x\right)_{ij}}\right)\right)\right). \quad (3B37)$$

Next we consider discrete categorical variables. Categorical variable can be coded by a one-hot vector, where the "1" indicates the value of the corresponding variable, others takes the "0" values. For example, suppose that we have four discrete variables X_1, X_2, X_3, and X_4. One-hot vector representation can be as follows:

X_1	1	0	0	0
X_2	0	1	0	0
X_3	0	0	1	0
X_4	0	0	0	1

Assumptions about the latent variables and latent space are not changed. We still assume that the prior and the posterior distributions are matrix normal distributions. Let $p(X|Z)$ be conditional distribution of X, given latent variables Z, and P_x be its probability matrix where each row denotes a vector of probability for the corresponding variable. The probability matrix P_x is generated by

$$P_x^l = softmax\left(\hat{A}\text{ReLU}\left(\hat{A}\text{ReLU}\left(\hat{A}Z^l W_x^1\right)W_x^2\right)W_x^3\right). \quad (3B38)$$

The probability distribution $p_{\theta_2}(X|Z)$ for the categorical variables is given by

$$p_{\theta_2}\left(X|Z^l\right)=\sum_{i=1}^{n}\sum_{j=1}^{d}X_{ij}\log\left(P_x^l\right)_{ij}. \quad (3B39)$$

Since we assume that the prior and posterior distributions are not changes, which in turn imply that KL distance between them is also unchanged. Therefore, the evidence lover bound for categorical variables is given by

$$\mathcal{L}(\theta,\varnothing,X,A)\approx\frac{1}{2}\left(\sum_{i=1}^{n}\sum_{j=1}^{n}\left(\log S_{ij}+1-S_{ij}-\mu_{ij}^2\right)\right)+\frac{1}{L}\sum_{l=1}^{L}\sum_{i=1}^{n}\sum_{j=1}^{n}\log f\left(\left(Z_i^l\right)^T Z_j^l\right)+\frac{1}{L}\sum_{l=1}^{L}\sum_{i=1}^{n}\sum_{j=1}^{d}X_{ij}\log\left(P_x^l\right)_{ij}. \quad (3B40)$$

Now we study $E_{q_\phi(Z|X,A)}\left[\log p(X,A|Z)\right]$. We assume that the variables in the latent space are independent and prior over each latent variable is the centered isotropic multivariate Gaussian $p_\theta(z_i)=N(z_i;0,I_d)$. The generative model will produce the feature vector.

APPENDIX 3C: MATRIX NORMAL DISTRIBUTION

In this appendix, we introduce basic properties of matrix normal distribution, mainly based on Ding et al. (2014).

3C1 Notations and Definitions

Consider a random matrix $X=\left(X_{ij}\right)_{n_r \times n_c}$. The mean and covariance matrix of X are, respectively, defined as

$$\mu=E[X]=\left(\mu_{ij}\right)_{n_r\times n_c} \text{ and }$$
$$\sum=E\left[vec\,(X-\mu)vec(X-\mu)^T\right], \quad (3C1)$$

where

$$X=\begin{bmatrix} X_{11} & \cdots & X_{1n_c} \\ \vdots & \cdots & \vdots \\ X_{n_r1} & \cdots & X_{n_rn_c}\end{bmatrix}=\begin{bmatrix} X_{.1} & \cdots & X_{.n_c}\end{bmatrix},$$

$$\mu = \begin{bmatrix} \mu_{11} & \cdots & \mu_{1n_c} \\ \vdots & \vdots & \vdots \\ \mu_{n_1 1} & \cdots & \mu_{n_r n_c} \end{bmatrix} = \begin{bmatrix} \mu_{.1} & \cdots & \mu_{.n_r} \end{bmatrix},$$

$vec\ (X-\mu) \in R^{n_r n_l}$ is the vectorization of the matrix $X-\mu$:

$$vec\ (X-\mu) = \begin{bmatrix} X_{.1}-\mu_{.1} \\ \vdots \\ X_{.n_c}-\mu_{.n_c} \end{bmatrix}.$$

Before we introduce density function for random normal matrix, we show an equality:

$$\begin{aligned} &\text{Tr}(AB) = \left(vec\left(A^T\right)\right)^T vec(B), \\ &\text{or } Tr\left(A^T B\right) = \left(vec(A)\right)^T vec(B). \end{aligned} \quad (3C2)$$

Proof.

Let

$$vec\left(A^T\right) = \begin{bmatrix} A_{1.}^T \\ \vdots \\ A_{n.}^T \end{bmatrix} \text{ and } vec(B) = \begin{bmatrix} B_{.1} \\ \vdots \\ B_{.n} \end{bmatrix}.$$

Then,

$$\text{Tr}(AB) = A_{1.}B_{1.} + \cdots + A_{n.}B_{.n} = \left(vec\left(A^T\right)\right)^T vec(B).$$

We assume that $vec(X)$ follows a multivariate normal distribution with density function:

$$\begin{aligned} f_X(x) = f_{vec(X)}\left(vec(X)\right) = (2\pi)^{-\frac{n_r n_c}{2}} \left|\sum\right|^{-\frac{1}{2}} \\ \times \exp\left\{-\frac{1}{2}\left(vec(X-\mu)\right)^T \sum{}^{-1} vec(X-\mu)\right\}. \end{aligned} \quad (3C3)$$

Dimension of the covariance matrix $\Sigma = E \times \left[vec(X-\mu)\left(vec(X-\mu)\right)^T\right]$ is $(n_r n_c)^2$ is often very high. Computation of the covariance matrix Σ may be very heavy. To reduce the computational time of the covariance matrix Σ we want to decompose the covariance matrix Σ into the Kronecker product of two positive definite matrices Ω with dimension n_r and M with dimension n_c i.e., $\Sigma = \Omega \otimes \text{M}$. Recall that $\Sigma^{-1} = \Omega^{-1} \otimes M^{-1}$ and $|\Sigma| = |\Omega|^{n_r}|M|^{n_c}$. Then, density function of the random matrix X in equation (3C3) can be reduced to

$$\begin{aligned} &f_{vec(X)}\left(vec(X)\right) = (2\pi)^{-\frac{n_r n_c}{2}} |\Omega|^{-\frac{n_r}{2}} |M|^{-\frac{n_c}{2}} \\ &\times \exp\left\{-\frac{1}{2}\left(vec(X-\mu)\right)^T \Omega^{-1} \otimes M^{-1} vec(X-\mu)\right\}. \end{aligned} \quad (3C4)$$

Recall that

$$\begin{aligned} &(\Omega^{-1} \otimes M^{-1}) vec(X-M) \\ &= vec\left(M^{-1}(X-M)\Omega^{-1}\right), \text{ which implies} \\ &\left(vec(X-\mu)\right)^T \Omega^{-1} \otimes M^{-1} vec(X-\mu) \\ &= \left(vec(X-\mu)\right)^T vec\left(M^{-1}(X-M)\Omega^{-1}\right). \end{aligned} \quad (3C5)$$

Using equation (3C2), we obtain

$$\begin{aligned} &\left(vec(X-\mu)\right)^T vec\left(M^{-1}(X-M)\Omega^{-1}\right) \\ &= \text{Tr}\left((X-\mu)^T M^{-1}(X-M)\Omega^{-1}\right) \\ &= \text{Tr}\left(\Omega^{-1}(X-\mu)^T M^{-1}(X-M)\right). \end{aligned} \quad (3C6)$$

Substituting equation (3C6) into equation (3C4), we obtain

$$\begin{aligned} &f_{vec(X)}\left(vec(X)\right) = (2\pi)^{-\frac{n_r n_c}{2}} |\Omega|^{-\frac{n_r}{2}} |M|^{-\frac{n_c}{2}} \\ &\times \exp\left\{-\frac{1}{2}\text{Tr}\left(\Omega^{-1}(X-\mu)^T M^{-1}(X-M)\right)\right\}. \end{aligned} \quad (3C7)$$

Now we can define the following matrix normal distribution.

3C1.1 Definition: Matrix Normal Distribution

If the covariance matrix of the random matrix X can be decomposed as the Kronecker product of two positive definite matrices Ω and M, and $vec\ (X)$ follows a multivariate normal distribution with mean $vec(\mu)$ and covariance matrix $\Sigma = \Omega \otimes M$, then the random matrix X has a matrix normal distribution and denoted as $N_{n_r \times n_c}(\mu, \Omega, M)$. The density function of matrix normal distribution is given by

$$\begin{aligned} &f_{vec(X)}\left(vec(X)\right) = (2\pi)^{-\frac{n_r n_c}{2}} |\Omega|^{-\frac{n_r}{2}} |M|^{-\frac{n_c}{2}} \\ &\times \exp\left\{-\frac{1}{2}\left(vec(X-\mu)\right)^T \Omega^{-1} \otimes M^{-1} vec(X-\mu)\right\} \end{aligned} \quad (3C8)$$

or

$$f_{vec(X)}(vec(X))=(2\pi)^{-\frac{n_r n_c}{2}}|\Omega|^{-\frac{n_r}{2}}|M|^{-\frac{n_c}{2}}$$
$$\times\exp\left\{-\frac{1}{2}\mathrm{Tr}\left(\Omega^{-1}(X-\mu)^T M^{-1}(X-M)\right)\right\}. \quad (3C9)$$

3C2 Properties of Matrix Normal Distribution

Lemma 3C.1

The second moments of X are

$$E\left[(X-\mu)(X-\mu)^T\right]=M\mathrm{Tr}(\Omega), \quad (3C10)$$

$$E\left[(X-\mu)^T(X-\mu)\right]=\Omega\mathrm{Tr}(M). \quad (3C11)$$

We intuitively show equations (3C10) and (3C11). Let

$$X-\mu=\begin{bmatrix} X_{.1}-\mu_{.1} & \cdots & X_{.n_c}-\mu_{.n_c}\end{bmatrix} \text{ and}$$

$$(X-\mu)^T=\begin{bmatrix}(X_{.1}-\mu_{.1})^T \\ \vdots \\ (X_{.n_c}-\mu_{.n_c})^T\end{bmatrix}.$$

Then, we have the following second moment:

$$E\left[(X-\mu)(X-\mu)^T\right]$$
$$=E\left\{\begin{bmatrix} X_{.1}-\mu_{.1} & \cdots & X_{.n_c}-\mu_{.n_c}\end{bmatrix}\begin{bmatrix}(X_{.1}-\mu_{.1})^T \\ \vdots \\ (X_{.n_c}-\mu_{.n_c})^T\end{bmatrix}\right\}$$
$$=\sum_{j=1}^{n_c}cov(X_{.j},X_{.j}), \quad (3C12)$$

where $cov(X_{.j},X_{.j})=E\left\{(X_{.j}-\mu_{.j})(X_{.j}-\mu_{.j})^T\right\}$.

Recall that

$$\sum=E\left[vec(X-\mu)(vec(X-\mu))^T\right]$$
$$=E\left\{\begin{bmatrix}X_{.1}-\mu_{.1} \\ \vdots \\ X_{.n_c}-\mu_{.n_c}\end{bmatrix}\begin{bmatrix}(X_{.1}-\mu_{.1})^T & \cdots & (X_{.n_c}-\mu_{.n_c})^T\end{bmatrix}\right\}$$
$$=\begin{bmatrix}cov(X_{.1},X_{.1}) & \cdots & cov(X_{.1},X_{.n_c}) \\ \vdots & \vdots & \vdots \\ Cov(X_{.n_c},X_{.1}) & \cdots & cov(X_{.n_c},X_{.n_c})\end{bmatrix}$$
$$=\begin{bmatrix}\Omega_{11}M & \cdots & \Omega_{1n_r}M \\ \vdots & \vdots & \vdots \\ \Omega_{n_r1}M & \cdots & \Omega_{n_rn_r}M\end{bmatrix}, \quad (3C13)$$

where

$$cov(X_{.j},X_{.j})=\Omega_{jj}M. \quad (3C14)$$

Combining equations (3C12) and (3C14), we obtain

$$E\left[(X-\mu)(X-\mu)^T\right]=\sum_{j=1}^{n_c}cov(X_{.j},X_{.j})$$
$$=\sum_{j=1}^{n_c}\Omega_{jj}M=\left(\sum_{j=1}^{n_c}\Omega_{jj}\right)M=\mathrm{Tr}(\Omega)M.$$

This shows equation (3C10). Similarly, we can show equation (3C11).

Next we derive the distribution of the transpose of matrix.

Lemma 3C.2: Transpose

Distribution of the transpose of matrix with matrix normal distribution is also matrix normal distribution and denoted as $X^T \sim MN_{n_c \times n_r}(\mu^T, M, \Omega)$.

Proof

For simplicity, we consider 2×2 dimensional matrix

$$X=\begin{bmatrix}X_{11} & X_{12} \\ X_{21} & X_{22}\end{bmatrix}.$$

Recall that

$$E\left[vec(X)(vex(X))^T\right]$$
$$=\begin{bmatrix}cov(X_{11},X_{11}) & cov(X_{11},X_{21}) & cov(X_{11},X_{12}) & cov(X_{11},X_{22}) \\ cov(X_{21},X_{11}) & cov(X_{21},X_{21}) & cov(X_{21},X_{12}) & cov(X_{21},X_{22}) \\ cov(X_{12},X_{11}) & cov(X_{12},X_{21}) & cov(X_{12},X_{12}) & cov(X_{12},X_{22}) \\ cov(X_{22},X_{11}) & cov(X_{22},X_{21}) & cov(X_{22},X_{12}) & cov(X_{22},X_{22})\end{bmatrix}$$

$$= \Omega \times M = \begin{bmatrix} \Omega_{11}M_{11} & \Omega_{11}M_{12} & \Omega_{12}M_{11} & \Omega_{12}M_{12} \\ \Omega_{11}M_{21} & \Omega_{11}M_{22} & \Omega_{12}M_{21} & \Omega_{12}M_{22} \\ \Omega_{21}M_{11} & \Omega_{21}M_{12} & \Omega_{22}M_{11} & \Omega_{22}M_{12} \\ \Omega_{21}M_{21} & \Omega_{21}M_{22} & \Omega_{22}M_{21} & \Omega_{22}M_{22} \end{bmatrix}. \quad (3C15)$$

$$E\left[vec(X^T)(vec(X^T)^T\right]$$

$$= \begin{bmatrix} cov(X_{11}, X_{11}) & cov(X_{11}, X_{12}) & cov(X_{11}, X_{21}) & cov(X_{11}, X_{22}) \\ cov(X_{12}, X_{11}) & cov(X_{12}, X_{12}) & cov(X_{12}, X_{21}) & cov(X_{12}, X_{22}) \\ cov(X_{21}, X_{11}) & cov(X_{21}, X_{12}) & cov(X_{21}, X_{21}) & cov(X_{21}, X_{22}) \\ cov(X_{22}, X_{11}) & cov(X_{22}, X_{12}) & cov(X_{22}, X_{21}) & cov(X_{22}, X_{22}) \end{bmatrix}. \quad (3C16)$$

Substituting $cov(X_{ij}, X_{kl})$ in equation (3C15) into equation (3C16), we obtain

$$E\left[vec(X^T)(vec(X^T)^T\right]$$

$$= \begin{bmatrix} \Omega_{11}M_{11} & \Omega_{12}M_{11} & \Omega_{11}M_{12} & \Omega_{12}M_{12} \\ \Omega_{21}M_{11} & \Omega_{22}M_{11} & \Omega_{21}M_{12} & \Omega_{22}M_{12} \\ \Omega_{11}M_{21} & \Omega_{12}M_{21} & \Omega_{11}M_{22} & \Omega_{12}M_{22} \\ \Omega_{21}M_{21} & \Omega_{22}M_{21} & \Omega_{21}M_{22} & \Omega_{22}M_{22} \end{bmatrix} = M \otimes \Omega. \quad (3C17)$$

This shows that $X^T \sim MN_{n_c \times n_r}(\mu^T, M, \Omega)$.

Lemma 3C.3: Transformation

Let $D \in R^{r \times n_r}, C \in R^{n_c \times s}$. Then, the random matrix DXC has a matrix normal distribution $DXC \sim MN_{rs}(D\mu C, D\Omega D^T, C^T MC)$ (Exercise 3.12).

Lemma 3C.4: Expectation of Quadratic Forms

Let $A \in R^{n_c \times n_c}, B \in R^{n_r \times n_r}, C \in R^{n_c \times n_r}, X \in R^{n_r \times n_c}$. Then, we have the following formula for computing expectation of quadratic forms:

$$E[XAX^T] = M\mathrm{Tr}(A\Omega) + \mu A \mu^T, \quad (3C18)$$

$$E[X^T BX] = \Omega \mathrm{Tr}(MB^T) + \mu^T B\mu, \quad (3C19)$$

$$E[XCX] = \Omega C^T M + \mu C \mu. \quad (3C20)$$

Proof

Let

$$X = \begin{bmatrix} X_{1.} \\ \vdots \\ X_{n_r.} \end{bmatrix}.$$

$$\text{Then, } [XAX^T] = E\left[(X-\mu+\mu)A(X-\mu+\mu)^T\right]$$
$$= E\left[(X-\mu)A(X-\mu)^T\right] + \mu A \mu^T. \quad (3C21)$$

Note that

$$E\left[(X-\mu)A(X-\mu)^T\right] = \begin{bmatrix} E[X_{1.}AX_{1.}^T] & E[X_{2.}^T AX_{1.}] \\ E[X_{2.}AX_{1.}^T] & E[X_{2.}AX_{2.}^T] \end{bmatrix}.$$

Using theorem of quadratic form of the vector, we obtain

$$E[X_{1.}AX_{1.}^T] = E\left[\mathrm{Tr}(X_{1.}AX_{1.}^T)\right] = \mathrm{Tr}\left(\mathrm{AE}[X_{1.}^T X_{1.}]\right). \quad (3C22)$$

Recall that

$$E\left[vec(X^T)(vec(X^T))^T\right] = E\left\{\begin{bmatrix} X_{1.}^T X_{1.} & X_{1.}^T X_{2.} \\ X_{2.}^T X_{1.} & X_{2.}X_{2.}^T \end{bmatrix}\right\}. \quad (3C23)$$

Substituting equation (3C17) into equation (3C23), we obtain

$$E\left[vec(X^T)(vec(X^T))^T\right] = M \otimes \Omega = \begin{bmatrix} M_{11}\Omega & M_{12}\Omega \\ M_{21}\Omega & M_{22}\Omega \end{bmatrix}. \quad (3C24)$$

Combining equations (3C22) and (3C24), we obtain

$$E[X_{1.}^T X_{1.}] = M_{11}\Omega,$$

which implies that

$$E[X_{1.}AX_{1.}^T] = \mathrm{Tr}(M_{11}A\Omega) = M_{11}\mathrm{Tr}(A\Omega). \quad (3C25)$$

Using the similar arguments, we obtain

$$E[X_{2.}^T AX_{1.}] = M_{12}\mathrm{Tr}(A\Omega), \quad (3C26)$$

$$E[X_{2.}AX_{1.}^T] = M_{21}\text{Tr}(A\Omega), \qquad (3C27)$$

$$E[X_{2.}AX_{2.}^T] = M_{22}\text{Tr}(A\Omega). \qquad (3C28)$$

Combining equations (3C21), (3C22), (3C25–3C28), we obtain

$E[XAX^T] = M\text{Tr}(A\Omega) + \mu A\mu^T$, which proves equation (3C18). By the similar argument, we can prove equations (3C19) and (3C20).

EXERCISES

EXERCISE 3.1

Write CAVI for a Bayesian mixture of Gaussians, Assuming $k = 2$ and $n = 2$.

EXERCISE 3.2

Derive priors for the multinomial distribution and Poisson distribution.

EXERCISE 3.3

Write CAVI for Bernoulli distribution.

EXERCISE 3.4

Show that

$$\frac{\partial KL(p(x|\theta)||p(x|\theta'))}{\partial\theta'} = -\int p(x|\theta)\frac{\partial \log p(x|\theta')}{\partial\theta'}.$$

EXERCISE 3.5

Show that

$$H_{KL(p(x|\theta)||p(x|\theta'))} = \frac{\partial^2 KL(p(x|\theta)||p(x|\theta'))}{\partial\theta'(\partial\theta')^T}$$

$$= F = -\int p(x|\theta)\frac{\partial^2 \log p(x|\theta)}{\partial\theta\partial\theta^T}dx.$$

EXERCISE 3.6

Show that

$$\mathcal{L}_{\theta,\varnothing}(x) = E_{q_\varnothing(z|x)}[\log p_\theta(x|z)] - KL(q_\varnothing(z|x)||p(z)).$$

EXERCISE 3.7

Please show

Let $L_k = \log\frac{1}{k}\sum_{l=1}^{k}\frac{p_\theta(x,z^l)}{q_\varnothing(z^l|x)}$. Show $L_{k+1} \geq L_k$.

EXERCISE 3.8

Consider a Fourier expansion:

$$f(x) = \sum_{n=-\infty}^{\infty} c_n e^{2\pi i\frac{n}{T}x}.$$

Show that

$$c_n = \frac{1}{T}\int_{-\frac{T}{2}}^{\frac{T}{2}} f(x)e^{-2\pi i\left(\frac{n}{T}\right)x}dx,$$

$$\hat{f}\left(\frac{n}{T}\right) = \int_{-\frac{T}{2}}^{\frac{T}{2}} f(x)e^{-2\pi i\left(\frac{n}{T}\right)x}dx \to \int_{-\infty}^{\infty} f(x)e^{-i\omega x}dx$$

$= \hat{f}(\omega)$ when $T \to \infty$, and

$$f(x) = \int_{-\infty}^{\infty}\hat{f}(\omega)e^{i\omega x}d\omega.$$

EXERCISE 3.9

Let $f = [1, 2, 0\ 4, 3, 1]$ and

$$L = \begin{bmatrix} 3 & -1 & 0 & -1 & -1 \\ -1 & 2 & -1 & 0 & 0 \\ 0 & -1 & 1 & 0 & 0 \\ -1 & 0 & 0 & 1 & 0 \\ -1 & 0 & 0 & 0 & 1 \end{bmatrix}.$$

Calculate its graph Fourier transform.

EXERCISE 3.10

The *Chebyshev polynomials of the first kind* (T_n) are defined as

$$T_n(\cos(\theta)) = \cos(n\theta).$$

Let $x = \cos(\theta)$. Show that

$$T_0(x) = 1$$

$$T_1(x) = x$$

$$T_2(x) = 2xT_1(x) - 1.$$

EXERCISE 3.11

Assume that λ_{max} represents the largest eigenvalue of the normalized graph Laplacian matrix L

Show that

$$-1 \le \tilde{\Lambda} = \frac{2}{\lambda_{max}}\Lambda - I \le 1.$$

EXERCISES 3.12

Let $D \in R^{r \times n_r}, C \in R^{n_c \times s}$. Then, the random matrix DXC has a matrix normal distribution $DXC\tilde{\ } MN_{rs}\left(D\mu C, D\Omega D^T, C^T MC\right)$.

CHAPTER 4

Generative Adversarial Networks

4.1 INTRODUCTION

In the past decade, great progress in artificial intelligence (AI) has been made. Generative adversarial networks (GANs) that can produce large number of samples of data distribution without precise modeling and learn deep representations (Salehi et al. 2020) have become one of the most promising advances in AI in the past decade (Goodfellow et al. 2014). The GANs can now (1) generate text, images, image-to-image translation, clinical data, and even music that is indistinguishable as fake to human generated (Beaulieu-Jones et al. 2017; Brock et al. 2018; Emami et al. 2018; Galbusera et al. 2018), (2) translate from voice to face (Wen et al. 2019), (3) be applied to survival analysis (Chapfuwa et al. 2018), (4) regression (Aggarwal et al. 2020), (5) time series generation (Yoon et al. 2019; Smith and Smith 2020), (6) signal processing and speech recognition (Kong et al. 2020), (7) hypothesis testing (Bellot and van der Schaar 2019), and (8) provide tools for causal inference (Athey and Imbens 2016; Hartford et al. 2016; Lopez-Paz and Oquab 2016; Goudet et al. 2017; Hartford et al. 2017; Kocaoglu et al. 2017; Goudet et al. 2018; Kalainathan et al. 2018; Li et al. 2018; Ramachandra 2018; Bica et al. 2020).

Although they are powerful generative models (Miyato et al. 2018), the GANs suffer from training instability and fail to converge in some cases (Qin et al. 2020). To improve the training stability and sample quality, several techniques, including Wasserstein GAN, ordinary differential equations approach, and style-based generator architecture, and three new architectures: convolutional, conditional, and autoencoder for GANs have been developed (Mirza and Osindero 2014; Radford et al. 2015; Makhzani et al. 2016; Arjovsky et al. 2017; Arjovsky and Bottou 2017; Gulrajani et al. 2017; Fathoby and Goela 2018; Wei et al. 2018; Karras et al. 2019; Bica et al. 2020; Chrysos et al. 2020; Li et al. 2020; Pidhorskyi et al. 2020; Salehi et al. 2020).

In this section, we will introduce the main concepts and the theoretic models of GANs, their computational algorithms, the evaluation metrics, and widely used three types of GANs: convolutional, conditional, and autoencoder. The application of the GAN to regression, hypothesis testing, and time series will be discussed. Using GANs for causal inference will be discussed in Chapters 5, 6, and 7.

4.2 GENERATIVE ADVERSARIAL NETWORKS

4.2.1 Framework and Architecture of GAN

The GANs learn the models that capture the statistical distribution of training data via adversarial process, allowing us to synthesize samples from the learned models which are very similar to the real data distribution (Goodfellow et al. 2014; Salehi et al. 2020). The GAN consists of two major components: a generator and a discriminator (Figure 4.1). The generator tries to generate fake data that is indistinguishable from real-world data by learning the distribution of the trained real data. On the other hand, discriminator tries to discriminate whether a given data is the synthesized fake data or the data from real world. Both generator and discriminator are adversarial networks. They are in constant conflict, simultaneously optimize themselves and compete against each other to improve their capabilities during training process, finally reaching a stationary state where neither the generator nor the discriminator can improve its benefit. The

DOI: 10.1201/9781003028543-4

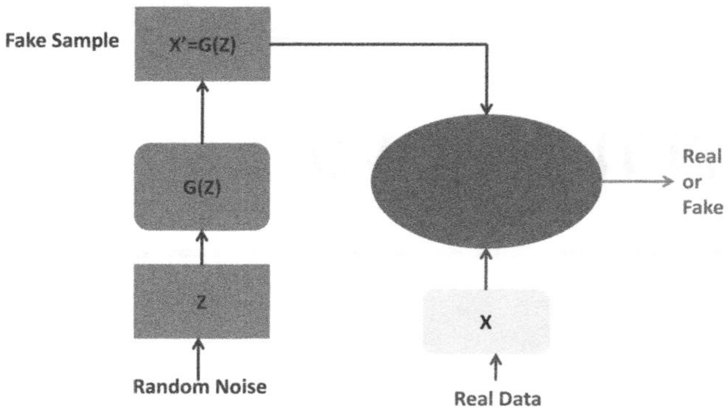

FIGURE 4.1 Scheme of generative adversarial networks.

generator and discriminator can be represented by neural networks (either multilayer feedforward neural networks or convolution neural networks) (Salehi et al. 2020).

Random noise vector $z \in R^d$ with distribution $p_z(z)$, e.g., a uniform distribution or a Gaussian distribution, are inputted to the generator that transforms z to $G(z;\theta_g)$, where G is a differentiable nonlinear function and represented by a neural network with parameters θ_g. The generator outputs (synthesizes) the faked sample $x' = G(z;\theta_g)$. The discriminator $D(x;\theta_d)$ transforms input x to a single scalar that represents the probability of x being from real data, denoted by x_{data}, rather than from faked data.

The generator network can be considered as mapping from some representation space, called a latent space z, to the space of the data. Let $x' = G(z)$. The density function $p_g(x')$ of the faked data x' can be derived as by changes of variables theorem (https://mathworld.wolfram.com/ChangeofVariablesTheorem.html):

$$p_g(x') = \frac{p_z(z)}{\left|\frac{\partial G(z)}{\partial z}\right|}, \tag{4.1}$$

where $p_z(z)$ is a density function of the random vector z and $\left|\frac{\partial G(z)}{\partial z}\right|$ is the absolute value of the Jacobian matrix of the transformation $G(z)$.

The discriminator function $D(x)$ is defined as

$$D(x) = \begin{cases} 1 & x \text{ is from real data} \\ 0 & x \text{ is from the generator} \end{cases}. \tag{4.2}$$

Equation (4.2) shows that the discriminator is a binary classification and discriminate the real x_{data} sample from fake $G(Z)$ sample. When a generator $G(z)$ is fixed, the discriminator, D, is trained to classify the input data as either being from the real data (close to one) or from a fixed generator (close to zero). The GAN plays game of two players. The generator and discriminator are in competition with each other.

4.2.2 Loss Function

The objective of the generator is to produce samples that are as close to real data samples as possible; while the discriminator is to identify the fake samples produced by the generator and distinguish between real data samples and the fake samples produced by the generator.

In summary, the goal of training D is to maximize the probability of correctly classify the real data samples and fake samples. The discriminator learns to recognize the generator's fake samples, while the generator learns to fool the discriminator.

Recall that $D(x)$ is the probability of the discriminator receiving the data from real samples and $1-D(x)$ is the probability of the discriminator receiving the data from the samples produced by the generator. We assume that $D=1$ indicates that the data are from the real sample and $D=0$ from fake sample. The discriminator classifies the data from x with probability $D(x)$ and classifies the data from the fake sample with the probability $1-D(G(z))$.

The likelihood and log likelihood of observing the output of the discriminator are

$$L_1 = D(x)(1-D(G(z)), \tag{4.3}$$

$$\log L_1 = \log D(x) + \log(1-D(G(z)), \tag{4.4}$$

respectively.

If it is known that the data inputing to discriminator are from the generator, its loglikelihood is

$$L_2 = \log(D(G(z)). \tag{4.5}$$

Taking expectations of loglikelihood, we obtain

$$l_1 = E_{x\sim p_{data}(x)}[\log D(x)] + E_{z\sim p_z(z)}[\log(1-D(G(z))], \tag{4.6}$$

$$l_2 = E_{z\sim p_z(z)}[\log(D(G(z))], \tag{4.7}$$

where $p_z(z)$ is a density function of the latent variable z and $p_{data}(x)$ is a density function of the real data x_{data}.

The maximum likelihood estimators of $D(x)$ and $G(z)$ can be obtained by

$$max_D l_1 = E_{x\sim p_{data}(x)}\left[\log D(x)\right] + E_{z\sim p_{z}(z)}\left[\log\left(1-D\left(G(z)\right)\right)\right], \tag{4.8}$$

$$max_G l_2 = E_{z\sim p_z(z)}\left[\log D\left(G(z)\right)\right]. \tag{4.9}$$

Optimization problem (4.9) can be transferred to

$$min_G E_{z\sim p_{z}(z)}\left[\log\left(1-D\left(G(z)\right)\right)\right]. \tag{4.10}$$

Combining optimization problems (4.8) and (4.10) together gives

$$min_G max_D l(D,G) = E_{x\sim p_{data}(x)}\left[\log D(x)\right] + E_{z\sim p_z(z)}\left[\log\left(1-D\left(G(z)\right)\right)\right]. \tag{4.11}$$

4.2.3 Optimal Solutions

Now we first solve optimization problem (4.8). The optimal solution is summarized in Proposition 4.1 (Goodfellow et al. 2014).

Proposition 4.1: Given generator $G(z)$, the optimal discriminator $D(x)$, denoted by $\hat{D}_G(x)$, is

$$\hat{D}_G(x) = \frac{p_{data}(x)}{p_{data}(x)+p_g(x)}, \tag{4.12}$$

where $p_g(x)$ is the density function of the faked data.

Proof

We first transform the loglikelihood function l_1 as follows:

$$l_1 = E_{x\sim p_{data}(x)}\left[\log D(x)\right] + E_{z\sim p_{z}(z)}\left[\log\left(1-D\left(G(z)\right)\right)\right]$$
$$= \int p_{data}(x)\log D(x)dx + \int p_z(z)\log\left(1-D\left(G(z)\right)\right)dz. \tag{4.13}$$

Let $x = G(z)$. Making change in variable z and using equation (4.1), we obtain

$$l_1 = \int p_{data}(x)\log D(x)dx + \int |G'(z)| p_g(x)\log\left(1-D\left(G(z)\right)\right)\frac{1}{|G'(z)|}dx$$

$$= \int p_{data}(x)\log D(x)dx + \int p_g(x)\log(1-D(x)dx. \tag{4.14}$$

Since $D(x)$ is a function, the traditional calculus is difficult to apply for solving optimization problem (4.14). We employ variation of calculus to find optimal discriminator $D(x)$ (Xiong 2018a).

Assume that function $D(x)$ moves to $D(x)+\varepsilon\delta(x)$, where $\delta(x)$ is a differential. Define

$$F(\varepsilon) = \int p_{data}(x)\log\left(D(x)+\varepsilon\delta(x)\right)dx + \int p_g(x)\log\left(1-D(x)-\varepsilon\delta(x)\right)dx. \tag{4.15}$$

Taking derivative of $F(\varepsilon)$ with respect to ε, we obtain

$$\frac{dF(\varepsilon)}{d\varepsilon}|_{\varepsilon=0} = \int \frac{p_{data}(x)}{D(x)}\delta(x)dx - \frac{p_g(x)}{1-D(x)}\delta(x)dx. \tag{4.16}$$

Let

$$\delta(x) = \frac{p_{data(x)}}{D(x)} - \frac{p_{g(x)}}{1-D(x)}. \tag{4.17}$$

Substituting equation (4.17) into equation (4.16) yields

$$\frac{dF(\varepsilon)}{d\varepsilon}|_{\varepsilon=0} = \int\left[\frac{p_{data}(x)}{D(x)} - \frac{p_g(x)}{1-D(x)}\right]^2 dx. \tag{4.18}$$

Optimality of l_1 requires

$$\frac{dF(\varepsilon)}{d\varepsilon}|_{\varepsilon=0} = 0, \tag{4.19}$$

which implies

$$\frac{p_{data}(x)}{D(x)} - \frac{p_g(x)}{1-D(x)} = 0 \text{ or}$$
$$\hat{D}(x) = \frac{p_{data}(x)}{p_{data}(x)+p_g(x)}. \tag{4.20}$$

Substituting equation (4.12) into equations (4.11) and (4.14), we obtain

$$min_G max_D l(D,G) = min_G E_{x\sim p_{data}(x)}\left[\log \hat{D}(x)\right] + E_{x\sim p_g}\left[\log\left(1-\hat{D}(x)\right]\right.$$

$$= min_G \int p_{data}(x)\log\frac{p_{data}(x)}{p_{data}(x)+p_g(x)}dx$$
$$+\int p_g(x)\log\left(1-\frac{p_{data}(x)}{p_{data}(x)+p_g(x)}\right)dx$$
$$= min_G \int p_{data}(x)\log\frac{p_{data}(x)}{p_{data}(x)+p_g(x)}dx$$
$$+\int p_g(x)\log\left(\frac{p_g(x)}{p_{data}(x)+p_g(x)}\right)dx. \quad (4.21)$$

Let

$$p_m(x)=\frac{p_{data}(x)+p_g(x)}{2}. \quad (4.22)$$

Then, the optimization problem can be transformed to

$$min_G\begin{bmatrix} KL\left(p_{data}(x)\,||\,p_m(x)\right) \\ +KL\left(p_g(x)\,||\,p_m(x)\right)-2\log 2\end{bmatrix}, \quad (4.23)$$

where $KL(p_r \,||\, p_\theta)$ denotes the Kullback-Leibler (KL) distance between two distributions p_r and p_θ (Xiong 2018b).

Define

$$\hat{p}_g(x)=p_{data}(x). \quad (4.24)$$

Then, we have

$$\begin{bmatrix} KL\left(p_{data}(x)\,||\,p_m(x)\right) \\ +KL\left(\hat{p}_g(x)\,||\,p_m(x)\right)-2\log 2\end{bmatrix}=-2\log 2. \quad (4.25)$$

Since

$$\begin{bmatrix} KL\left(p_{data}(x)\,||\,p_m(x)\right) \\ +KL\left(p_g(x)\,||\,p_m(x)\right)-2\log 2\end{bmatrix}\gg -2\log 2$$

then, $\hat{p}_g(x)=p_{data}(x)$ is the optimal solution to the problem (4.21).

In summary, the solution to GAN is summarized in Theorem 4.1.

Theorem 4.1

The optimal solution to the GAN problem (4.11) is

$$\hat{p}_g(x)=p_{data}(x) \text{ and } \hat{D}(x)=\frac{1}{2}. \quad (4.26)$$

The optimal value of the objective function is

$$min_G max_D l(D,G)=-2\log 2. \quad (4.27)$$

4.2.4 Algorithm

According to equations (4.20) and (4.24), ideally, the discriminator is trained until optimal with respect to the current generator; then the generator is again updated. The discriminator and generator are represented by neural networks. The update of the discriminator and generator in the iterations is implemented by updating their weights via backpropagation. The algorithm is summarized as follows (Goodfellow et al. 2014).

Algorithm 4.1

Let k be the number of steps which updating discriminator takes before updating the generator and n be the number of samples in the real dataset.

For number of training iterations **do**

For k steps **do**

Sample minibatch of n noise samples $\{z_{(1)},\ldots,z_{(n)}\}$ from distribution $p_z(z)$.

Input minibatch of n real data samples $\{x_{(1)},\ldots,x_{(n)}\}$ that have distribution p_{data}(x).

Search optimal solution for discriminator, given the generator by updating weights of the discriminator via ascending its stochastic gradient:

$$\nabla_{\theta_d}\frac{1}{n}\sum_{i=1}^{n}\left[\log D\left(x_{(i)},\theta_d\right)+\log\left(1-D\left(G\left(z_{(i)},\theta_g\right),\theta_d\right)\right)\right].$$

end for

Sample minibatch of n noise samples $\{z_{(1)},\ldots,z_{(n)}\}$ from distribution $p_z(z)$.

Update the weights of the generator by descending its stochastic gradient to search the optimal solution to the generator:

$$\nabla_g\frac{1}{n}\sum_{i=1}^{n}\left[\log\left(1-D\left(G\left(z_{(i)},\theta_g\right)\right)\right)\right].$$

end for

Despite the theoretical guarantee of unique solutions, GAN training is challenging for several reasons.

First, the training is often unstable. Second, the generative model may collapse, which generates very similar samples for different inputs. Third, the gradient may quickly vanish. To overcome these limitations, Arjovsky et al. (2017) developed a new version of GAN, Wasserstein GAN that is based on the Wasserstein distance.

4.2.5 Wasserstein GAN

4.2.5.1 Different Distances

There are two approaches to learning the generative models. One approach is to directly learn the probability density function p_θ via maximum likelihood function. Second approach is to learn a function that transforms the existing distribution z into the density function p_θ via $p_\theta = G_\theta(z)$. We first study the maximum likelihood estimator.

Assume that real data samples $\{x_1,\ldots,x_n\}$ with density function $p_r(x)$ being given. We define a parametric family of density $p_\theta(x)$. The maximum likelihood estimator is to maximize the likelihood of the data:

$$max_\theta \frac{1}{n}\sum_{i=1}^{n}\log p_\theta(x_i). \quad (4.28)$$

By law of large numbers, we have

$$\frac{1}{n}\sum_{i=1}^{n}\log p_\theta(x_i) \to E_{x\sim p_r}\left[\log p_\theta(x)\right]. \quad (4.29)$$

Combining equations (4.28) and (4.29), we obtain

$$\lim_{n\to\infty}\max_\theta \frac{1}{n}\sum_{i=1}^{n}\log p_\theta(x_i) = \max_\theta \int p_r(x)\log p_\theta(x)dx. \quad (4.30)$$

Recall that the KL distance between two distributions $p_r(x)$ and $p_\theta(x)$ is defined as

$$KL(p_r\,||p_\theta) = \int p_r(x)\log\frac{p_r(x)}{p_\theta(x)}dx. \quad (4.31)$$

Using equation (4.31), the maximization problem (4.30) can be transformed to the following optimization problem:

$$\begin{aligned}\lim_{n\to\infty}\max_\theta \frac{1}{n}\sum_{i=1}^{n}\log p_\theta(x_i) &= \max_\theta \int p_r(x)\log p_\theta(x)dx\\ &= -\min_\theta \int p_r(x)\log p_\theta(x)dx\\ &= \min_\theta \int p_r(x)\log p_r(x)dx - \int p_r(x)\log p_\theta(x)dx\\ &= \min_\theta \int p_r(x)\log\frac{p_r(x)}{p_\theta(x)}dx\\ &= \min_\theta KL(p_r\,||p_\theta). \end{aligned} \quad (4.32)$$

Equation (4.32) implies that maximum likelihood estimation is equivalent to minimization of KL distance. In other words, we want to find the density function p_θ of the generative model that makes the KL distance as small as possible.

Major limitation of the maximum likelihood estimation of the generative models is that when two distributions p_θ and p_r are not completely overlapped, for example, $p_\theta(x_0)=0$, where $p_r(x_0)>0$, then KL distance goes to $+\infty$. This will happen when the distribution p_θ has low dimensional support, because p_r may be high dimensional and it is unlikely that all of p_r lies in the low dimensional support of p_θ.

The solution is to learn a nonlinear function map $g_\theta : Z \to \chi$ (typically a neural network) that to transform a random variable z with a known distribution $p(z)$ (typically, uniform or normal distribution) to a certain distribution p_θ. By varying θ, we can change distribution p_θ, making it close to the real data distribution p_r. To measure the degree of closeness, we need to develop distance or divergence between two distributions p_θ and p_r.

4.2.5.1.1 Total Variation (TV) Distance Assume that p_r and p_θ are two probability distributions defined on the space Ω. Then, TV distance between two distributions is defined as

$$\delta(P_r, P_\theta) = \sup_{A\in F}\left|p_r(A) - p_\theta(A)\right|, \quad (4.33)$$

where $F \in \Omega$ is a subset of Ω.

Example 4.1

Consider probability distributions defined over probability space R^2. Let $Z \sim U[0,1]$ be the uniform distribution on the unit interval. Let p_0 be the distribution defined on $(0,Z)$ and p_θ be the distribution defined on (θ,Z), where θ is a single real parameter. Define

$$A=\{(0,z),\ z\in[0,1].$$

Then, $p_0(A)=z$ and $p_\theta(A)=0$ when $\theta\neq 0$. Thus, when $\theta\neq 0$, we have
$|p_r(A)-p_\theta(A)|=z$, which implies

$$\sup_{A\in\mathcal{F}}|p_r(A)-p_\theta(A)|=1.$$

When, $\theta=0$, we obtain $p_0(A)=p_\theta(A)$. Therefore,

$$\delta(p_0,p_\theta)=\begin{cases}1 & \theta\neq 0\\ 0 & \theta=0\end{cases}.$$

4.2.5.1.2 The Kullback-Leibler (KL) Divergence The KL divergence between two distributions is defined as

$$KL(p_r\,||p_\theta)=\int p_r\log\frac{p_r}{p_\theta}dx. \tag{4.34}$$

Example 4.1 (continue)

$$KL(p_r\,||p_\theta)=\begin{cases}+\infty & \theta\neq 0\\ 0 & \theta=0\end{cases}.$$

4.2.5.1.3 The Jenson-Shannon (JS) Divergence Let $p_m=\frac{p_r+p_\theta}{2}$. The JS divergence is defined as

$$JS(p_r\,||p_\theta)=KL(p_r\,||p_m)+KL(p_\theta\,||\,p_m). \tag{4.35}$$

Example 4.1 (continue)

When $\theta\neq 0$, where $p_0\neq 0$, $p_\theta=0$, then we have $p_m=\frac{p_r}{2}$. Similarly, when $\theta=0$ where $p_\theta=p_0=1$, we have $p_m=\frac{p_\theta+p_r}{2}=1$. Thus, we obtain

$$JS(p_0,p_\theta)=\begin{cases}\log 2 & \theta\neq 0\\ 0 & \theta=0\end{cases}.$$

4.2.5.1.4 Earth Mover (EM) or Wasserstein Distance When we consider discrete probability distributions, the Wasserstein distance is also descriptively referred to as the earth mover's distance (EMD). Consider two discrete distributions p_r and p_θ, each will possible l states x or y. The distribution $p_r(x)$ is defined as the probability mass of each point. The distribution $p_\theta(y)$ can be similarly defined. Our goal is to move mass around to change the distribution $p_r(x)$ into $p_\theta(y)$. Let $r(x,y)$ be a transportation plan that moves $r(x,y)$ mass from x to y. The distance from x to y is denoted as $\|x-y\|$. The cost moving $r(x,y)$ mass from x to y is equal to $r(x,y)\|x-y\|$. The moving plan $r(x,y)$ can be viewed as a joint distribution. It should satisfy the following constraints:

$$p_r(x)=\sum_y r(x,y) \text{ and } p_\theta(y)=\sum_x r(x,y).$$

Let $\Pi(p_r,p_\theta)$ be the set of all possible joint distributions with marginal distributions $p_r(x)$ and $p_\theta(y)$. The EMD is then defined as the minimum cost required to move $r(x,y)$ mass from x to y:

$$\begin{aligned}W(p_r,p_\theta)&=\inf_{r\in\Pi(p_r,p_\theta)}\sum_x\sum_y r(x,y)\|x-y\|\\&=\inf_{r\in\Pi(p_r,p_\theta)}E_{(x,y)\sim r}\left[\|x-y\|\right].\end{aligned} \tag{4.36}$$

If the variables are continuous, equation (4.36) can be extended to

$$W(p_r,p_\theta)=\inf_{r\in\Pi(p_r,p_\theta)}\int_x\int_y r(x,y)\|x-y\|dxdy. \tag{4.37}$$

Example 4.1 (continue)

Consider two distributions $p_0(z)$ and $p_\theta(z)$, where $Z\sim U(0,1)$. Let $x=[0,z]$ and $y=[\theta,z]$. It is clear that $\|x-y\|=|\theta|$.

$$W(p_r,p_\theta)=\inf_{r\in\Pi(p_r,p_\theta)}\int_x\int_y r(x,y)\|x-y\|dxdy$$

$$=|\theta|\inf_{r\in\Pi(p_r,p_\theta)}\int_x\int_y r(x,y)dxdy=|\theta|.$$

4.2.5.2 The Kantorovich-Rubinstein Duality

Now we briefly introduce the Kantorovich-Rubinstein duality theory (Herrmann 2021). We first formulate

Wasserstein distance in terms of linear programming (LP). Define the matrices

$$\Gamma=\left(r(x,y)\right)=\begin{bmatrix} r(x_1,y_1) & \cdots & r(x_1,y_l) \\ \vdots & \vdots & \vdots \\ r(x_l,y_1) & \cdots & r(x_l,y_l) \end{bmatrix} \text{ and}$$

$$D=\left(\|x-y\|\right)=\begin{bmatrix} \|x_1-y_1\| & \cdots & \|x_1-y_l\| \\ \vdots & \vdots & \vdots \\ \|x_l-y_1\| & \cdots & \|x_l-y_l\| \end{bmatrix}.$$

We can write Wasserstein distance as

$$W\left(p_r,p_\theta\right)=\inf_{r\in\Pi(p_r,\,p_\theta)} <D,\Gamma>_F, \tag{4.38}$$

where $<,>_F$ is the Frobenius inner product (sum of all the element-wise products).

Let

$$x=vec\left(\Gamma^T\right),\ c=vec(D) \text{ and}$$

$$b=\begin{bmatrix} p_r \\ p_\theta \end{bmatrix}.$$

To pose constraints: $p_r(x)=\Sigma_y r(x,y)$ and $p_\theta(y)=\Sigma_x r(x,y)$, we define matrix A such that

$$Ax=b. \tag{4.39}$$

To make presentation easy, we consider $l=2$. Let

$$A=\begin{bmatrix} 1 & 1 & 0 & 0 \\ 0 & 0 & 1 & 1 \\ 1 & 0 & 1 & 0 \\ 0 & 1 & 0 & 1 \end{bmatrix},\ x=\begin{bmatrix} r(x_1,y_1) \\ r(x_1,y_2) \\ r(x_2,y_1) \\ r(x_2,y_2) \end{bmatrix}.$$

Then, we have

$$Ax=\begin{bmatrix} p_r(x_1) \\ p_r(x_2) \\ p_\theta(y_1) \\ p_\theta(y_2) \end{bmatrix}=b.$$

In general, we define

$$A=\begin{bmatrix} 1 & 1 & \cdots & 0 & 0 & \cdots & \cdots & 0 & 0 & \cdots \\ 0 & 0 & \cdots & 1 & 1 & \cdots & \cdots & 0 & 0 & \cdots \\ \cdots & \cdots & \cdots & \cdots & \cdots & \cdots & \cdots & \cdots & \cdots & \cdots \\ 0 & 0 & \cdots & 0 & 0 & \cdots & \cdots & 1 & 1 & \cdots \\ 1 & 0 & \cdots & 1 & 0 & \cdots & \cdots & 1 & 0 & \cdots \\ 0 & 1 & \cdots & 0 & 1 & \cdots & \cdots & 0 & 1 & \cdots \\ \cdots & \cdots & \cdots & \cdots & \cdots & \cdots & \cdots & \cdots & \cdots & \cdots \\ 0 & \cdots & 1 & 0 & \cdots & 1 & \cdots & 0 & \cdots & 1 \end{bmatrix}.$$

Then, we have

$$Ax=b.$$

It is clear that $<D,\Gamma>_F=c^Tx$. Then, the problem of calculating the Wasserstein distance can be formulated as the following primal LP problem:

$$\begin{aligned} \min_x \quad & z=c^Tx \\ s.t. \quad & Ax=b \\ & x\geq 0. \end{aligned} \tag{4.40}$$

However, solving the primal LP problem of the Wasserstein distance is impractical for two major reasons. The first, in general the dimension is high. The number of variables in x exponentially increases as the dimension of input variable increases. The second, the distribution p_θ appears in the right side of the constraint. It is not easy to calculate gradient $\nabla_{p_\theta}W\left(p_r,p_\theta\right)$. Overcome these limitations, we consider its dual form:

$$\begin{aligned} \max_y \quad & \tilde{z}=b^T\,y \\ s.t. \quad & A^Ty\leq c. \end{aligned} \tag{4.41}$$

It is clear that

$$z=c^Tx\geq y^TAx=y^Tb=\tilde{z}, \tag{4.42}$$

which implies that the optimal solution $\tilde{z}$ of dual LP is a lower bound of the optimal solution of the primal LP. In fact, we can show that $z=\tilde{z}$.

Let $y=\begin{bmatrix} f \\ g \end{bmatrix}$. Then, $\tilde{z}=f^Tp_r+g^Tp_\theta$. Consider $l=2$. Its dual form is

$$\begin{aligned} \max_y \quad & \tilde{z}=f(x_1)p_r(x_1)+f(x_2)p_r(x_2) \\ & +g(x_1)p_\theta(x_1)+g(x_2)p_\theta(x_2) \end{aligned}$$

$$\begin{bmatrix} 1 & 0 & 1 & 0 \\ 1 & 0 & 0 & 1 \\ 0 & 1 & 1 & 0 \\ 0 & 1 & 0 & 1 \end{bmatrix} \begin{bmatrix} f(x_1) \\ f(x_2) \\ g(x_1) \\ g(x_2) \end{bmatrix} \le \begin{bmatrix} D_{11} \\ D_{21} \\ D_{12} \\ D_{22} \end{bmatrix}$$

or

$$f(x_1)+g(x_1)\le D_{11}$$
$$f(x_1)+g(x_2)\le D_{21}$$
$$f(x_2)+g(x_1)\le D_{12}$$
$$f(x_2)+g(x_2)\le D_{22}.$$

In general, we have

$$f(x_i)+g(x_j)\le D_{ji},\ i=1,\dots,l,\ j=1,\dots,l. \quad (4.43)$$

When $i=j$, then $f(x_i)+g(x_i)\le 0$. It is known from optimization theory (Kuhn-Tucker conditions) that the optimal solutions are in boundary. Therefore, we have

$$g(x_i)=-f(x_i). \quad (4.44)$$

Substituting equation (4.44) into equation (4.43), we obtain

$$f(x_i)-f(x_j)\le D_{ji} \text{ and}$$
$$f(x_j)-f(x_i)\le D_{ij}.$$
$$\text{But, } D_{ij}=D_{ji}=\left\|x_i-x_j\right\|.$$

Combining the above two inequality, we obtain Lipschitz condition with Lipschitz constant 1:

$$\left|f(x_i)-f(x_j)\right|\le\left\|x_i-x_j\right\|. \quad (4.45)$$

The objective function can be written as

$$\sum_i f(x_i)p_r(x_i)+\sum_i g(x_i)p_\theta(x_i)$$
$$=\sum_i f(x_i)p_r(x_i)-\sum_i f(x_i)p_\theta(x_i)$$
$$=E_{x\sim p_r}f(x)-E_{x\sim p_\theta}f(x). \quad (4.46)$$

Combining equations (4.45) and (4.46), dual problem can be reduced to

$$W(p_r,p_\theta)=\sup_{\|f\|_L\le 1}E_{x\sim p_r}f(x)-E_{x\sim p_\theta}f(x). \quad (4.47)$$

4.2.5.3 Wasserstein GAN

Now we introduce the Wasserstein GAN (Arjovsky et al. 2017; Gulrajani et al. 2017; Wei et al. 2018). Optimization problem (4.47) is a constrained optimization. Solving optimization problem under Lipschitz constraints with constant 1 is intractable. To alleviate the problem, we release the Lipschitz constraints with constant 1 to Lipschitz constraints with constant K, i.e.,

$$\left\|f(x_i)-f(x_j)\right\|\le K\left\|x_i-x_j\right\|.$$

Let $g(x)=\frac{f(x)}{K}$. Then, we have $\left\|g(x_i)-g(x_j)\right\|\le\left\|x_i-x_j\right\|$, i.e., $g(x)$ is a Lipschitz function with constant 1.

The supremum over K-Lipschitz constraints is still hard to implement, but it allows easier to approximate. Consider a parametrized function family $\{f_w\}_{w\in W}$, where W is the set of all possible weights. These functions are also K-Lipschitz functions. Then, we have

$$\sup_{\|f\|_L\le K}E_{x\sim p_r}f(x)-E_{x\sim p_\theta}f(x)=K\sup_{\|g\|_L\le 1}E_{x\sim p_r}g(x)$$
$$-E_{x\sim p_\theta}g(x)=KW(p_r,p_\theta),\text{ which implies}$$
$$\max_{w\in W}E_{x\sim p_r}f_w(x)-E_{x\sim p_\theta}f_w(g_\theta(z))\le\sup_{\|f\|_L\le K}E_{x\sim p_r}f(x)$$
$$-E_{x\sim p_\theta}f(x)=KW(p_r,p_\theta).$$

Therefore, the optimization problem (4.47) can be approximated by

$$\max_{w\in W}E_{x\sim p_r}f_w(x)-E_{x\sim p_\theta}f_w(x). \quad (4.48)$$

Let

$$x=g_\theta(z). \quad (4.49)$$

Substituting equation (4.49) into equation (4.48), we obtain

$$W(p_r,p_\theta)\ge\max_{w\in W}E_{x\sim p_r}f_w(x)-E_{z\sim p(z)}f_w(g_\theta(z)). \quad (4.50)$$

Both functions $f_w(x)$ and $g_\theta(z)$ can be modeled by neural networks and parameters w and θ are the weights in the neural networks. Our goal is to find the generator $g_\theta(z)$, making the Wasserstein distance between the distribution p_r of the real data and distribution p_θ of the generator as small as possible, i.e.,

$$\min_\theta W(p_r, p_\theta). \tag{4.51}$$

The Wasserstein distance is calculated by its lower bound in equation (4.50). In summary, to find the generator $g_\theta(z)$, we solve the following optimization problem:

$$\min_\theta \max_{w \in W} E_{x \sim p_r} f_w(x) - E_{z \sim p(z)} f_w(g_\theta(z)). \tag{4.52}$$

Numerically, the optimization problem (4.52) can be solved by Algorithm 4.2 (Arjovsky et al. 2017).

Algorithm 4.2: WGAN

Input: α, the learning rate; c, the clipping parameter; m, the batch size; n_d, the number of iterations of discriminator neural network update per generator iteration, w_0, initial weight parameters of the network for function f_w; θ_0, initial generator's parameters.

While θ has not converged do

for $t = 0, \ldots, n_d$ do

Sample $\{x^{(i)}\}_{i=1}^m \sim p_r$ a batch from the real data.

Sample $\{z^{(i)}\}_{i=1}^m \sim p(z)$, a batch from the noises.

$$g_w \leftarrow \nabla_w \left[\frac{1}{m} \sum_{i=1}^m f_w(x^{(i)}) - \frac{1}{m} \sum_{i=1}^m f_w(g_\theta(z^{(i)})) \right]$$

$w \leftarrow w + \alpha RMSProp(w, g_w)$

$w \leftarrow clip(w, -c, c)$

end for

Sample $\{z^{(i)}\}_{i=1}^m \sim p(z)$, a batch from the noises.

$$g_\theta \leftarrow -\nabla_\theta \frac{1}{m} \sum_{i=1}^m f_w(g_\theta(z^{(i)}))$$

$$\theta \leftarrow \theta - \alpha RMSPtop(\theta, g_\theta)$$

end while

4.3 TYPES OF GAN MODELS

4.3.1 Conditional GAN

4.3.1.1 Classical CGAN

Conditional GAN (CGAN) is a type of important GAN model (Mirza and Osindero 2014). Unlike GAN where the noises or latent variables are mapped to a target space, the CGAN not only map the noises or latent variables, but also simultaneously map the observed variables, e.g., labels, styles, the predictors, to the target space (Radford et al. 2015; Karras et al. 2019). The conditional information is feed into both generator and discriminator as an additional input layer.

Let $x \in R^d$ be conditional information, $z \in R^l$ be a vector of noises or latent variables with known distribution, e.g., uniform distribution, or Gaussian distribution, and $y \in R^m$ be real data (target signals). Let G denote the mapping, which outputs the data produced by the generator. Then, the output of the generator is given by

$$\tilde{G} = G(x, z). \tag{4.53}$$

Let $p_g(\tilde{G})$ be the distribution of the output $\tilde{G}$ of the generator. The goal of the GAN attempts to make $p_g(\tilde{G}) = p_{data}(y)$, where $p_{data}(y)$ is the distribution of real data y. Thus, mathematically, the CGAN attempts to learn

$$y = G(x, z). \tag{4.54}$$

The optimization problem for learning the CGAN is

$$\begin{aligned} min_G max_D l(D,G) = {} & E_{y,x \sim p_{data}(y,x)}\left[\log D(y,x)\right] \\ & + E_{x \sim p_{data}(x),\, z \sim p_z(z)} \\ & \left[\log\left(1 - D(G(x,z),x)\right)\right]. \end{aligned} \tag{4.55}$$

Equation (4.54) shows that the CGAN is similar to nonlinear regression which is a function of missing or noisy variables z and predictors x.

Similar to Proposition 4.1 and Theorem 4.1 in the GAN, we can derive Proposition 4.2 and Theorem 4.2 to summarize the properties of CGAN. Again, the density function of $G(x,z)$ is given by

$$p_g(y \mid x) = \frac{p_a(z)}{\left| \frac{\partial G(x,z)}{\partial z^T} \right|}. \tag{4.56}$$

Using equation (4.56), the loss function $l(D,G)$ can be expressed as

$$l(D,G)=\int p_{data}(y,x)\log D(y,x)dydx + \int p_{data}(x)p_z(z)\log(1-D(G(x,z),x))dxdz$$

$$= \int p_{data}(y,x)\log D(y,x)dydx + \int p_{data}(x)p_g(y|x)\left|\frac{\partial G(x,z)}{\partial z^T}\right|\log(1-D(y,x)) dx\frac{dy}{\left|\frac{\partial G(x,z)}{\partial z^T}\right|},$$

$$= \int p_{data}(y,x)\log D(y,x)dydx + \int p_{data}(x)p_g(y|x)\log(1-D(y,x)) dydx. \quad (4.57)$$

Using calculus of variation theory (Sagan 1969), we obtain

$$\left.\frac{\partial l(D+\varepsilon\delta,G)}{\partial\varepsilon}\right|_{\varepsilon=0} \quad (4.58)$$

$$= \int\left[\frac{p_{data}(y,x)\delta}{D(y,x)}-\frac{p_g(y,x)\delta}{1-D(y,x)}\right] = dydx.$$

Let

$$\delta = \frac{p_{data}(y,x)}{D(y,x)}-\frac{p_g(y,x)}{1-D(y,x)}. \quad (4.59)$$

Substituting equation (4.59) into equation (4.58), we obtain

$$\left.\frac{\partial l(D+\varepsilon\delta,G)}{\partial\varepsilon}\right|_{\varepsilon=0} = \int\left(\frac{p_{data}(y,x)}{D(y,x)}-\frac{p_g(y,x)}{1-D(y,x)}\right)^2 dydx. \quad (4.60)$$

Setting $\left.\frac{\partial l(D+\varepsilon\delta,G)}{\partial\varepsilon}\right|_{\varepsilon=0} = 0$, we obtain

$\frac{p_{data}(y,x)}{D(y,x)}-\frac{p_g(y,x)}{1-D(y,x)} = 0$, which implies

$$D(y,x)=\frac{p_{data}(y,x)}{p_{data}(y,x)+p_g(y,x)}. \quad (4.61)$$

Equation (4.61) shows the Proposition 4.2.

Proposition 4.2

Given generator $G(x,z)$, the optimal discriminator $D(y,x)$, denoted by $\hat{D}_G(y,x)$, is

$$\hat{D}_G(y,x)=\frac{p_{data}(y,x)}{p_{data}(y,x)+p_g(y,x)}.$$

Again, we can approve that the solution to CGAN can be summarized in Theorem 4.2 (Exercise 4.1).

Theorem 4.2

The optimal solution to the CGAN problem (4.55) is

$$\hat{p}_g(y,x)=p_{data}(y,x) \text{ and } \hat{D}(y,x)=\frac{1}{2}. \quad (4.62)$$

The optimal value of the objective function is

$$min_G max_D l(D,G)=-2\log 2.$$

4.3.1.2 Robust CGAN

To improve robustness to noise in the generator, the authors (Chrysos et al. 2020) developed a robust CGAN (RoCGAN) with a generator that included two pathways (Figure 4.2(b)). One pathway is the original pathway from the input x to the output $G(x)$ or $G_d(G_e(x))$ of the generator. The authors called this pathway a reg pathway. The second additional pathway is from the input y to the output $G_{AE}(y)=AE_d(AE_e(y))$ of an autoencoder that reconstructs the real data y. The second pathway is called AE pathway. Similar encoder-decoder structures are used for the structures of two pathways. The weights of the decoders in two pathways are shared. The discriminator in the CGAN is not changed in the RoCGAN. To mitigate the impact of the noise on the generated data y, we require that the latent spaces of the two pathways are the same. To achieve this, in addition to the loss function $l(D,G)$ in CGAN, we need to introduce other loss functions to implement these requirements.

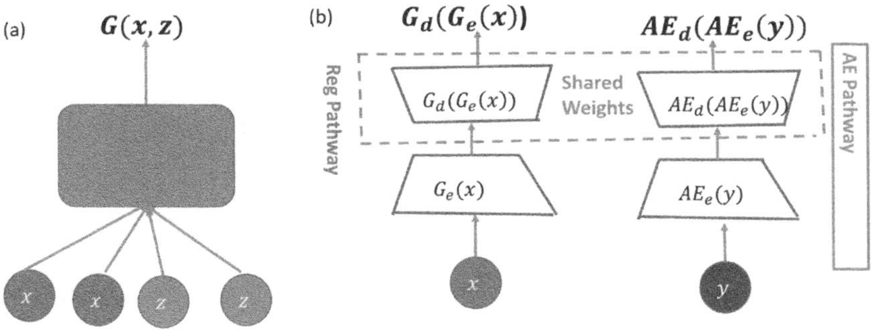

FIGURE 4.2 (a) Schematics of the generators of cGAN and (b) RoCGAN.

The first additional loss function is defined as

$$l_{AE}(y) = f_{AE}(y, G_{AE}(y)), \tag{4.63}$$

where f_{AE} is a nonlinear function to measure the distance between the output $G_{AE}(y)$ of the second pathway and real data y.

The second additional loss function intends to reduce the distance between two latent representations in the autoencoders. By reducing the distance between the outputs of two pathways, we enforce the outputs of two pathways as close as possible. The second additional loss function is defined as

$$l_{lat} = E_{y,x \sim p_{data}(y,x)}\left[f_{lat}\left(G(x), G_{AE}(y)\right)\right], \tag{4.64}$$

where f_{lat} can be any divergence function.

Finally, the total loss function of the RoCGAN is defined as summation of the original loss function l_{CGAN} for the CGAN and two additional loss functions defined in equations (4.63) and (4.64):

$$l_{RoCGAN} = l_{CGAN} + \lambda_{AE} l_{AE} + \lambda_{lat} l_{lat}, \tag{4.65}$$

where λ_{AE} and λ_{lat} are penalty parameters.

4.3.2 Adversarial Autoencoder and Bidirectional GAN

4.3.2.1 Adversarial Autoencoder (AAE)

One of the limitations of the autoencoder network is that the latent variables generated by the encoder are not distributed smoothly over the latent space, resulting in non-smooth distributions (Nauata et al. 2020; Salehi et al. 2020). To overcome this limitation, the AAE which combines the adversarial network with autoencoder has been developed (Makhzani et al. 2016). As it is shown in Figure 4.3, through the similar mechanism of the discriminator in the GAN, the AAE imposes the arbitrary prior distribution $p(z)$ on the distribution of the latent variables generated by the encoder to ensure no presence of gap in the latent space. Assume that the prior distribution $p(z)$ is smooth. The discriminator computes the probability $D(z)$ that point z is sampled from the prior distribution $p(z)$, rather than from the latent space. Through training process, we finally reach the optimal solution of the GAN: $q(z) = p(z)$, which leads to the smooth latent distribution $q(z)$. Therefore, upon completion of the training process, the decoder can reconstruct the meaningful distribution.

4.3.2.2 Bidirectional GAN

The architecture of bidirectional GAN (BiGAN) (Donahue et al. 2017; Salehi et al. 2020) is shown in Figure 4.4. In BiGAN, we consider two maps: (1) map noises or latent variables z to the target space x and (2) map the target variable x back to the latent space z and still use adversarial principle to train the two mapping. The first mapping is still implemented by

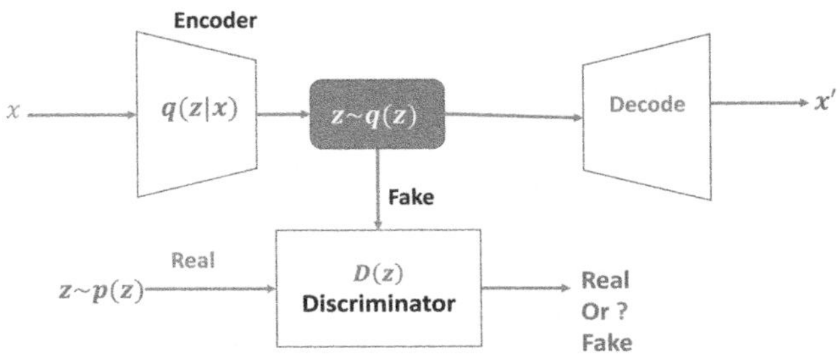

FIGURE 4.3 Architecture of an adversarial autoencoder.

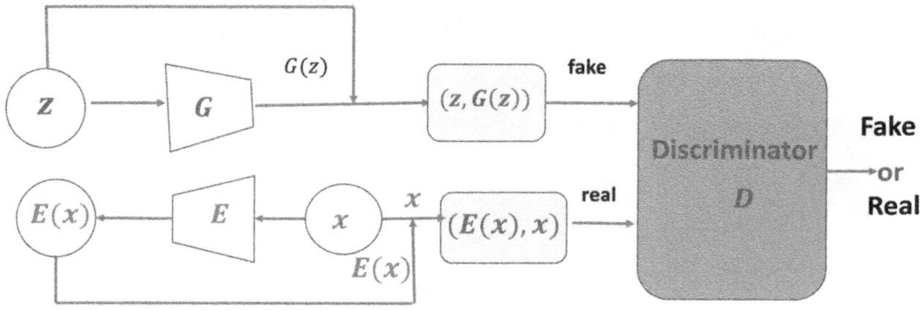

FIGURE 4.4 Architecture of BiGAN.

generator as done in the original GAN and is denoted by $G(z)$. We add an encoder for an inverse mapping which maps the target variable x back to the latent space. The second mapping is denoted by $E(x)$. The encoder extracts features from the data, which are inputted to the discriminator. The data features contained in the data are the output of the encoder. The BiGAN adds an encoder to the original GAN, and hence has three basic components: generator, discriminator, and encoder.

The first discriminator input for the generated data is a tuple $(z,G(z))$ of the latent variables z and the output $G(z)$ of its corresponding generator. The second discriminator input for the inverse mapping (real samples) is a tuple $(E(x),x)$ of the output of the encoder $E(x)$ and real data x. The discriminator computes the probability $D(x,z)$ that the discriminator distinguishes the pair $((E(x),x)$ from the pair $(z,G(z))$. The training process is implemented by (Mattia et al. 2019).

$$\min_{G,E}\max_{D} V(D,G,E)=E_{x\sim p_{data}(x)}\left[E_{z\sim p_E(z|x)}[\log D(x,z)]\right]$$
$$+E_{z\sim p_z(z)}\left[E_{x\sim p_G(x|z)}[\log 1-D(x,z)]\right]. \quad (4.66)$$

We can show that optimal solution to the optimization problem (4.66) (Exercise 4.2)

$$p_{data}(x)p_E(z|x)=p_z(z)p_G(x\,|\,z). \quad (4.67)$$

After the training process is completed, the following equalities hold:

$$x=G(z) \text{ and } z=E(x). \quad (4.68)$$

4.3.2.3 Anomaly Detection by BiGAN

BiGAN can be used for anomaly detection (Zenati et al. 2018; Mattia et al. 2019). Trained only on positive samples, BiGAN learns mappings $\hat{x}=G(z)$ from the latent space representation z to the real date x and invers mapping $\hat{z}=E(x)$ from the real data x to the latent representation z. Then, the BiGAN can use the trained inverse mapping to map the new, unseen, samples $\tilde{x}$ back to the latent space $\tilde{z}=E(\tilde{x})$.

After training a BiGAN on normal samples is completed, a new sample x_{new} is presented and mapped to the latent space representation z_{new}, again z_{new} is mapped to the real data space through $\hat{x}_{new}=G(z_{new})$. The whole test process is summarized as follows:

$$x_{new}\to E(x_{new})\to z_{new}\to G(z_{new})\to \hat{x}_{new}. \quad (4.69)$$

Then, the dissimilarity between the query sample x_{new} and the generated sample $G(z_{new})$ is used to measure the degree of anomaly:

$$\mathcal{L}_G(x)=\|x_{new}-\hat{x}_{new}\|_1=\|x_{new}-G(z_{new})\|_1, \quad (4.70)$$

where $\|.\|_1$ is a L_1 norm.

Next we use discriminator-based loss $\mathcal{L}_D(x)$ to measure the anomaly level. There are two ways to define the discriminator-based loss (Zenati et al. 2018). The discriminator is a binary classifier. It classifies the real sample as true (class 1) and the generated sample as false (class 0). Therefore, one way of defining $\mathcal{L}_D(x)$ is to use the cross-entropy loss $\sigma(D(x,E(x)),1)$ from the discriminator of the input sample being the real data sample (class 1), i.e., $\mathcal{L}_D(x)=\sigma(D(x,E(x)),1)$. Alternatively, we can use feature matching loss to define $\mathcal{L}_D(x)$. Let f_D be extracting features from a discriminator layer f. Then, the discriminator-based loss $\mathcal{L}_D(x)$ is defined as

$$\mathcal{L}_D(x)=\left\|f_D(x,E(x))-f_D(G(E(x)),E(x))\right\|_1, \quad (4.71)$$

which measures the similarity between the features of the generated data and the features of the real data.

Finally, anomaly measure $A(x)$ is defined as a combination of $\mathcal{L}_G(x)$ and $\mathcal{L}_D(x)$:

$$A(x) = \lambda \mathcal{L}_G(x) + (1-\lambda)\mathcal{L}_D(x). \tag{4.72}$$

The more large anomaly measure $A(x)$ is, the more likely to be anomalous, the new sample is.

4.3.3 Graph Representation in GAN

Analyzing graph data has a wide range of application, including link prediction (Rossi et al. 2020), node clustering and classification (Salha et al. 2020), and graph classification and clustering (Li et al. 2020). The graph data have complex topology and high dimensions. The graph representation or network embedding is to map the graph into a low-dimension vector space, while keeping information about the original graph topology and node contents (Cao et al. 2016). The methods for simple sequences or classical grids design are not suitable for graph representation and network analysis. To identify "good" latent representations for graphs is an essential issue for graph representation and learning. High nonlinearity, complex structures, property preserving, and sparsity are four major challenges. Autoencoder is emerging as a powerful tool for graph representation (Zheng et al. 2020). The adjacency matrix and the node feature characterize the graph and learning. However, the most of recent autoencoder-based graph representation methods reconstruct either the adjacency matrix or node features, rather both (Cao et al. 2016; Kipf and Welling 2016; Wang et al. 2016; Wang et al. 2017; Wang et al. 2018; Li et al. 2019; Park et al. 2019). To improve graph representation, the methods that simultaneously preserve both the topology and node features in the graph representation have been developed. In this section, we will introduce adversarially regularized graph autoencoder (ARGA) (Pan et al. 2018), graph variational GAN (Yang et al. 2019), and BiGAN (Zheng et al. 2020) for graph representation. All these three methods encode the topology and node content in the graph.

4.3.3.1 Adversarially Regularized Graph Autoencoder

Graph representation maps graph data into a low dimensional, compact, and continuous feature space (Pan et al. 2018). After the graph representations are completed, the ordinary algorithms for clustering and classification can be applied to the low dimensional representations of the graphs for node and graph clustering and classification. The essential issue for graph representation is how to preserve the topology and node feature information.

The most methods for graph representation are not regularized (Pan et al. 2018). The learned latent code space by unregularized graph representation approaches is often free of structure, which leads to the poor representation of the graphs in the real world. To overcome this problem, ARGA and adversarially regularized variational graph autoencoder (ARVGA) for graph representation learning have been developed (Pan et al. 2018). Unlike the classical approaches where graph representation learning algorithms attempt only minimize the reconstruction errors of graphs, the learning objective functions of the ARGA and ARVGA algorithms consist of two terms: (1) minimization of the reconstruction errors of graphs and (2) enforcing the latent codes to match a prior distribution.

4.3.3.1.1 Mathematic Formulation of the Graph Representation Problem

We first introduce mathematic formulation of the graph representation problem (Pan et al. 2018). Assume that a graph has n nodes. Denote the graph as $G = (V, E, X)$, where V denotes a set of nodes $V = \{v_i, i = 1, \ldots, n\}$, v_i represents the ith node, e_{ij} denotes the edge between the nodes v_i and v_j, $E = \{e_{ij}\}$ denotes a set of edges, $x_i \in R^m$ denotes an m-dimensional vector of features of the ith node v_i, and $X \in R^{n \times d}$ denotes the matrix of the features. Define an adjacency matrix A for the topology structure of the graph: $A = (a_{ij})_{n \times n}$, where $a_{ij} = 1$ if $e_{ij} \in E$, otherwise, $a_{ij} = 0$.

A graph is modeled by the adjacency matrix A and the matrix of the node features X. Therefore, a graph can be mathematically represented by $G = (A, X)$. Let f be an n-dimensional vector of nonlinear functions. The graph representation attempts to map the graph $G = (A, X)$ to the low dimensional latent space Z. The nonlinear mapping is denoted by

$$Z = f(A, X), \tag{4.73}$$

where

$$Z = \begin{bmatrix} z_1 \\ \vdots \\ z_n \end{bmatrix} = \begin{bmatrix} z_{11} & \cdots & z_{1d} \\ \vdots & \vdots & \vdots \\ z_{n1} & \cdots & z_{nd} \end{bmatrix},$$

which is called a graph representation matrix and preserves the structure A and node features X.

The ARGA consists of two major components: graph autoencoder (graph convolutional autoencoder or variational graph autoencoder) and adversarial network (Pan et al. 2018).

The graph encoder of the ARGA converts the graph $G=(A,X)$ into its latent representation Z which contains the graph structure information A and the node feature information X. The decoder of the ARGA reconstructs the graph adjacency matrix A from the latent graph representation matrix Z. The architecture of the adversarial network component of the ARGA has only discriminator. It takes the encoder of the ARGA as a generator and the latent graph representation code z as fake data and the samples generated from a prior distribution $p(z)$ as real data. The discriminator enforces the distribution of the latent graph representation to match a prior distribution by distinguishing whether a sample is from latent space or from a prior distribution.

4.3.3.1.2 Graph Convolutional Encoder and Variational Graph Encoder In Section 3.3.2.2.3, we introduced graph convolutional network (GCN) and in Section 3.3.2.6, we introduced variational graph autoencoder. Both graph GCN and variational graph autoencoder can be applied to ARGA. We first study GCN for learning graph representation. A spectral convolution is a powerful tool to map both graph topology A and node features X to the latent space. Recall that the number of nodes is n, the number of features of each node is m, and the dimension of each latent variable is d. The feature matrix of the node is denoted by $X \in R^{n\times m}$, and the latent graph representation matrix is denoted by $Z \in R^{n\times d}$. Define $\tilde{A}=A+I$ and $\tilde{D}_{ii}=\Sigma_j \tilde{A}_{ij}$. Let $W^{(0)} \in R^{m\times d}$ be the weight matrix of the filter in the first layer of the network. The convolution of m kernels (features) with the n nodes and d dimensions in the latent space are given by

$$Z^{(1)}=\sigma\left(\tilde{D}^{-\frac{1}{2}}\tilde{A}\tilde{D}^{-\frac{1}{2}}XW^{(0)}\right). \tag{4.74}$$

Equation (4.74) can be extended to the higher layer of the graph encoder. Consider the lth layer with d dimensional latent space Let $Z^l \in R^{n\times d}$ be input matrix for convolution in the lth layer, $A \in R^{n\times n}$ be an adjacency matrix and $W^l \in R^{d\times d}$ be the learnable parameter (weight) matrix, $Z^{l+1} \in R^{n\times d}$ be the output matrix after convolution, and σ be a element-wise nonlinear activation function. Let $Z^0=X$ and $\hat{A}=\tilde{D}^{-\frac{1}{2}}\tilde{A}\tilde{D}^{-\frac{1}{2}}$. Then, a layer-wise transformation by a spectral graph convolution function $\sigma(Z^l, A\,|\,W^l)$ is given by

$$Z^{l(+1)}=\sigma\left(Z^{(l)}, A\,|\,W^l\right) \tag{4.75}$$

$$=\sigma\left(\hat{A}Z^{(l)}\left(W^{l+1}\right)\right), \tag{4.76}$$

where $\hat{A}=\tilde{D}^{-\frac{1}{2}}\tilde{A}\tilde{D}^{-\frac{1}{2}}$.

A two-layer graph encoder $q(Z\,|\,A,X)$ for encoding both graph topology and node features is defined by

$$Z=q(Z|A,X)=Z^{(2)}=\text{ReLu}\left(\hat{A}\text{ReLU}\left(\hat{A}XW^0\right)W^1\right), \tag{4.77}$$

where $q(Z\,|\,A,X)$ is a deterministic function.

For a variational graph encoder, $q(Z\,|\,A,X)$ represents a distribution (Pan et al. 2018) (Section 3.3.2.6). We assume that the variational approximate posterior $q_{\varnothing}(Z\,|\,X,A)$ has the following matrix normal $MN_{nd}\left(\mu_z, \Omega_z, V_z\right)$:

$$\log q_{\varnothing}(Z|X,A)=-\frac{nd}{2}\log(2\pi)-\frac{1}{2}\log\left|\sum_z\right| -\frac{1}{2}\text{Tr}\left(\sum_z^{-1}\left(vec(Z-\mu)\right)\left(vec(Z-\mu)\right)^T\right). \tag{4.78}$$

If we assume that

$$\sum_z=\text{diag}\left(S_{11},\ldots,S_{n1},S_{12},\ldots,S_{n2},\ldots,S_{1d},\ldots,S_{nd}\right)$$
$$=S_z,\ S_{ij}=\Omega_{ii}M_{jj}.$$

A three-layer variational graph encoder with ReLU as the activation function is defined as

$$(\mu_Z,\log S_Z)=\hat{A}\text{ReLU}\left(\hat{A}\text{ReLU}\left(\hat{A}XW^1\right)W^2\right)W^3, \tag{4.79}$$

where $\hat{A}=\tilde{D}^{-\frac{1}{2}}\tilde{A}\tilde{D}^{-\frac{1}{2}}$, $\tilde{A}=A+I$, $\tilde{D}_{ii}=\Sigma_j \tilde{A}_{ij}$, and W^1, W^2, W^3 are weight matrices in the variational graph encoder and are similarly defined as before. The distribution $q_{\varnothing}(Z|X,A)$ contains the graph topology A and node features X. The latent variable Z can be sampled by

$$Z^l = \mu_Z + S_Z \odot \varepsilon^l,\ \varepsilon^l \sim MN_{nd}(0, I, I),\ l = 1, 2, \ldots, L. \quad (4.80)$$

4.3.3.1.3 Graph Convolutional Decoder In Section 3.3.2.4, we introduced graph convolutional decoder. The decoder consists of two parts: a reconstructed graph structure A and node features X. The node features decoder is the mirror process of the encode. Let $Z^{l_e} = Z$, where l_e is the index of the layer in the graph encoder-decoder, i.e., the output layer of the graph encoder or the latent variables in the latent space. Then, the output Z^{l+1} of the $(l+1)^{th}$ layer in the encoder-decoder is given by

$$Z^{l+1} = \sigma\left(\hat{A}Z^l W^{l+1}\right),\ l = l_e, \ldots, l_d - 1, \quad (4.81)$$

where $\hat{A}$ and W^{l+1} are defined as before.

The reconstructed node features is defined as $\hat{X} = Z^{l_d}$.

Let A^* be reconstructed adjacency matrix. It is generated by an inner product between latent variables (Kipf and Welling 2016):

$$P(A^*|Z) = \prod_{i=1}^{n}\prod_{j=1}^{n} P\left(A^*_{ij} \mid Z_i, Z_j\right), \quad (4.82)$$

Or

simply by

$$A^* = \sigma\left(ZZ^T\right), \quad (4.83)$$

where σ is an activation function, and

$$P\left(A^*_{ij} = 1|Z_i, Z_j\right) = \sigma\left(Z_i^T Z_j\right), \quad (4.84)$$

A^*_{ij} are the elements of the reconstructed adjacency matrix A^*.

4.3.3.1.4 Variational Graph Decoder Recall that the variational graph decoder which is introduced in Section 3.3.2.6 (equation 3.219) is defined as

$$p(X, A|Z) = p_{\theta_1}(A \mid Z)\, p_{\theta_2}(X \mid Z), \quad (4.85)$$

where we assume that the node feature variables X and graph structure A are conditionally independent, given the graph representation Z.

The elements of the adjacent matrix A are generated by an inner product between latent variables:

$$p_{\theta_1}(A|Z) = \prod_{i=1}^{n}\prod_{j=1}^{n} p\left(A_{ij} \mid Z_i, Z_j\right), \quad (4.86)$$

where $p\left(A_{ij} = 1|Z_i, Z_j\right) = \sigma\left(Z_i^T Z_j\right)$, A_{ij} are the elements of adjacent matrix A and σ is the logistic sigmoid function.

The node feature decoder is the mirror process of the encoder. We assume that distribution $p_{\theta_2}(X \mid Z)$ is matrix normal distribution. The latent variable Z can be sampled by

$$Z^l = \mu_Z + S_Z \odot \varepsilon^l,\ \varepsilon^l \sim MN_{nd}(0, I, I),\ l = 1, 2, \ldots, L, \quad (4.87)$$

where $\odot$ denotes the element-wise multiplication.

Define the three-layer variational graph decoder model for continuous feature sector of node with ReLU as the activation function:

$$\left[\mu_x^l, \log S_x^l\right] = \hat{A}\,\mathrm{ReLU}\left(\hat{A}\,\mathrm{ReLU}\left(\hat{A}Z^l W_x^1\right)W_x^2\right)W_x^3, \quad (4.88)$$

where $\hat{A} = \tilde{D}^{-\frac{1}{2}}\tilde{A}\tilde{D}^{-\frac{1}{2}}$, $\tilde{A} = A + I$, $\tilde{D}_{ii} = \sum_j \tilde{A}_{ij}$, and W_x^1, W_x^2, W_x^3 are weight matrices in the variational graph decoder.

4.3.3.1.5 Loss Function

4.3.3.1.5.1 Graph Autoencoder The loss function for learning graph autoencoder consists of four parts (Zhang et al. 2020). The first part is the reconstruction loss of node feature. The reconstruction loss of node feature is calculated as

$$\mathcal{L}_x = \left\|X - \hat{X}\right\|_F^2, \quad (4.89)$$

where $\hat{X} = Z^{l_d}$ and $\|.\|_F$ are the Frobenius Norms.

The second part is the reconstruction loss of the graph structure. It is calculated as

$$\mathcal{L}_A = \left\|A - A^*\right\|_F^2, \quad (4.90)$$

where $A^* = \sigma\left(ZZ^T\right)$.

The third part is the loss of connection between nodes in the latent space. It is defined as

$$\mathcal{L}_L = \sum_{i=1}^{n}\sum_{j=1}^{n}\left(\left\|Z_i - Z_j\right\|_2^2 a_{ij} + \gamma a_{ij}^2\right)$$

$$= \mathrm{Tr}\left(ZLZ^T\right) + \gamma\left\|A\right\|_F^2, \tag{4.91}$$

$$\text{s.t. } a_i^T\mathbf{1} = 1, \mathbf{0} \le \boldsymbol{a}_i \le \mathbf{1},$$

where $a_i \in R^n$ is the ith column vector of adjacency matrix A, the Laplacian matrix L is defined as

$$L = \frac{1}{2}\left(D - A\right), D_{ii} = \sum_{j=1}^{n}\frac{a_{ij} + a_{ji}}{2}.$$

Finally, the fourth part is the regularizer term $\mathcal{L}_R$ to prevent overfitting. It is defined as

$$\mathcal{L}_R = \frac{1}{2}\sum_{i=1}^{l_d}\left\|W^i\right\|_F^2.$$

The total loss function is then given by

$$\mathcal{L}_{GVAE}\left(X, A, Z, \theta_g\right) = \mathcal{L}_X + \mathcal{L}_A + \alpha\mathcal{L}_L + \lambda\mathcal{L}_R. \tag{4.92}$$

4.3.3.1.5.2 Variational Graph Autoencoder The Monte Carlo approximation of the evidence lower bound:

$$\mathcal{L}(\theta, \varnothing, X, A, Z^l) \approx \frac{1}{2}\left(\sum_{i=1}^{n}\sum_{j=1}^{d}\left(\log S_{ij} + 1 - S_{ij} - \mu_{ij}^2\right)\right) + \frac{1}{L}\sum_{l=1}^{L}\sum_{i=1}^{n}\sum_{j=1}^{n}\log f\left(\left(Z_i^l\right)^T Z_j^l\right)$$

$$-\frac{1}{L}\sum_{l=1}^{L}\left(\frac{nd}{2}\log(2\pi) + \frac{1}{2}\left(\sum_{i=1}^{n}\sum_{j=1}^{d}\left(\log\left(S_x\right)_{ij}^l + \frac{\left(X_{ij} - \left(\mu_x^l\right)_{ij}\right)^2}{\left(S_x\right)_{ij}^l}\right)\right)\right), \tag{4.93}$$

where

$$\left(\mu_Z, \log S_Z\right) = \hat{A}\mathrm{ReLU}\left(\hat{A}\mathrm{ReLU}\left(\hat{A}XW^1\right)W^2\right)W^3, \tag{4.94}$$

$\hat{A} = \tilde{D}^{-\frac{1}{2}}\tilde{A}\tilde{D}^{-\frac{1}{2}}, \tilde{A} = A + I, \tilde{D}_{ii} = \sum_j \tilde{A}_{ij}$, and W^1, W^2, W^3 are weight matrices in the variational graph encoder, $\Sigma_z = \mathrm{diag}\left(S_{11}, \ldots, S_{n1}, S_{12}, \ldots, S_{n2}, \ldots, S_{1d}, \ldots, S_{nd}\right) = S_z$,

$$\left[\mu_x^l, \log S_x^l\right] = \hat{A}\mathrm{ReLU}\left(\hat{A}\mathrm{ReLU}\left(\hat{A}Z^lW_x^1\right)W_x^2\right)W_x^3, \tag{4.95}$$

where $\hat{A} = \tilde{D}^{-\frac{1}{2}}\tilde{A}\tilde{D}^{-\frac{1}{2}}$, $\tilde{A} = A + I$, $\tilde{D}_{ii} = \sum_j \tilde{A}_{ij}$, and W_x^1, W_x^2, W_x^3 are weight matrices in the variational graph decoder.

4.3.3.1.6 Adversarial Discriminator Model The goal of both ARGA and ARGVA models is to match the distribution of the latent graph representation Z with a prior distribution (Pan et al. 2018). It follows from equations (4.74)–(4.77) that the latent representation can be expressed as a nonlinear function of the node features X and adjacency matrix A:

$$Z = G(X, A), \tag{4.96}$$

where G is a nonlinear function.

The discriminator in the GAN takes samples from the prior distribution as real data input and latent graph representation Z as fake data input. Recall that loss function of the GAN is defined as

$$l_1 = E_{x \sim p_{data}(x)}\left[\log D(x)\right] + E_{z \sim p_z(z)}\left[\log(1 - D(G(z)))\right]. \tag{4.97}$$

In the ARGA and ARGVA models, the variables Z with prior distribution $p(z)$ correspond to the real data X with the distribution $p_{data}(x)$. Therefore, the first term in equation (4.97) will be replaced by

$$E_{z \sim p_z(z)}\left[\log D(z)\right]. \tag{4.98}$$

The generator in the GAN is replaced by the encoder of the ARGA and ARGVA. The node feature variables $X \sim p_X(x)$ correspond to the noises $z \sim p_z(z)$ in the second term in equation (4.97) and function $G(X, A)$ corresponds to the function $G(z)$. Therefore, the second term in equation (4.97) will be replaced by

$$E_{X \sim P_X(x)}\left[\log(1 - D(G(X, A)))\right]. \tag{4.99}$$

Combining equations (4.98) and (4.99), we obtain the loss function for the ARGA and ARGVA:

$$V(G,D)=E_{z\sim p_z(z)}\left[\log D(z)\right] + E_{X\sim P_X(x)}\left[\log\left(1-D\left(G(X,A)\right)\right)\right]. \quad (4.100)$$

The graph representation of the ARGA and ARGVA can be obtained by solving the following optimization problem:

$$\min_G \max_D \mathrm{V}(G,D)=E_{z\sim p_z(z)}\left[\log D(z)\right] + E_{X\sim P_X(x)}\left[\log(1-D\left(G(X,A)\right))\right]. \quad (4.101)$$

Let θ_d be the all parameters in the discriminator (the neural networks implement the discrimination) and θ_g be the all parameters in the graph encoder (the neural networks implementing the graph encoder). The loss function $V(G,D)$ can be rewritten as

$$V(G,D)=E_{z\sim p_z(z)}\left[\log D(z,\theta_d)\right] + E_{X\sim P_X(x)}\left[\log(1-D\left(G\left(X,A,\theta_g\right),\theta_d\right)\right], \quad (4.102)$$

And the optimization problem (4.101) can also be rewritten as

$$\min_{\theta_g}\max_{\theta_d} V(G,D)=E_{z\sim p_z(z)}\left[\log D(z,\theta_d)\right]+E_{X\sim P_X(x)} \left[\log(1-D\left(G\left(X,A,\theta_g\right),\theta_d\right)\right]. \quad (4.103)$$

Sampling approximation $\hat{V}(G,D)$ is given by

$$\hat{V}(G,D)=\frac{1}{m}\sum_{i=1}^{m}\begin{bmatrix}\log D(z_i,\theta_d)\\ +\log\left(1-D\left(G^i\left(X,A,\theta_g\right),\theta_d\right)\right)\end{bmatrix}, \quad (4.104)$$

where $G^i\left(X,A,\theta_g\right)=q\left(Z^i|A,X\right)$, which is the ith sample from $Z^{(2)}$ in equation (4.77).

Procedures for solutions to optimization problem (4.103) is summarized in Algorithm 4.3 (Pan et al. 2018).

Algorithm 4.3

Step 1: Input and Initialization.
Assume that the graph is given. Input the adjacency matrix $A\in R^{n\times n}$ and node feature matrix $X\in R^{n\times d}$.
Define

T: the number of iterations;

K: the number of steps for iterating discriminator;

$Z\in R^{n\times d}$: Latent Matrix.

Randomly assign the weights to the graph encoder and discriminator.

Step 2: For Number of Training Iterations T do
Step (a):
Generate latent variables matrix $Z=Z^{(2)}=\mathrm{ReLu}\left(\hat{A}\mathrm{ReLU}\left(\hat{A}XW^0\right)W^1\right)$ (equation (4.77))

Step (b): Update the parameters in the discriminator.
For K steps **do**

Sample minibatch of m samples $\left\{G^1,\ldots,G^m\right\}$ from latent variable matrix.

Input minibatch of m real data samples $\left\{z^1,\ldots,z^m\right\}$ from the prior distribution $p_z(z)$.

Search optimal solution for discriminator, given the graph encoder by updating weights of the discriminator via ascending its stochastic gradient:

$$\nabla_{\theta_d}\frac{1}{m}\sum_{i=1}^{m}\left[\log D(z^i,\theta_d)+\log\left(1-D\left(G^i,\theta_d\right)\right)\right]. \quad (4.105)$$

end for

Step (c): Update the parameters in the graph autoencoder.

Sample minibatch of m samples $\left\{G^1,\ldots G^m\right\}$ from latent variable matrix Z.

Update the weights of the generator by descending its stochastic gradient to search the optimal solution to the graph encoder:

1. ARGA

$$\nabla_g\frac{1}{m}\sum_{l=1}^{m}\left[\mathcal{L}_{GVAE}\left(X,A,G^l,\theta_g\right)\right]. \quad (4.106)$$

2. ARGVA

$$\nabla_g \frac{1}{m}\sum_{l=1}^{m}\mathcal{L}\left(\theta_g, \varnothing_l, X, A, Z^l\right). \tag{4.107}$$

end for

Step 3:
Return latent variables matrix $Z \in R^{n\times d}$ generated in **Step 2.**

We first randomly assign the weights to the neural networks in graph encoder and discriminator. Then, we carry out T iterations. Each iteration consists of three steps: step (a) generate latent variables matrix Z from equation (4.77), step (b) update the parameters in the discriminator K times via ascending its stochastic gradient, and step (c) update the parameters in the graph autoencoder via descending its stochastic gradient to search the optimal solution to the graph encoder. We use different objective functions to update the parameters in the graph encoder in the ARGA and variational graph encoder in the ARGVA. Finally, when iteration stops, we take the latent variables matrix $Z \in R^{n\times d}$ generated in Step 2 as the final latent variant matrix.

4.3.3.2 Cycle-Consistent Adversarial Networks

4.3.3.2.1 Architecture Cycle-consistent adversarial networks (CycleGANs) are a powerful tool for learning mapping from a source domain X to a target domain Y in the absence of paired examples (Zhu et al. 2017). The CycleGAN was originally developed for unpaired image-to-image translation. The architecture of the CycleGAN is shown in Figure 4.5.

Consider two domains X and Y and two training datasets $\{x_i\}_{i=1}^n$, where $x_i \in X$ and $\{y_i\}_{i=1}^n$, where $y_i \in Y$. Let $p_{data}(x)$ be the distribution of x and $p_{data}(y)$ be the distribution of y. Consider two generators $y' = g(x)$ and $x' = F(y)$. The CycleGAN includes two discriminators D_x and D_y. The inputs to the discriminator D_x are the generated data y' and real data y. The discriminator D_x intends to distinguish real image data x from generated image $F(y)$. The inputs to the discriminator D_y are the generated images $G(x)$ and the real images y. The discriminator D_y is to distinguish the generated images $G(x)$ form the real images y.

4.3.3.2.2 Loss Function The loss function of the CycleGAN includes two types of loss: (1) adversarial loss and (2) cycle consistency loss. The adversarial loss measures the errors of matching the distribution of generated images to the real data distribution in the target domain. The adversarial loss for the GAN ($G: X \to Y$ and discriminator D_y) is defined as

$$\mathcal{L}_{GAN}\left(G, D_y, X, Y\right) = E_{y\sim p_{data}(y)}\left[\log D_y(y)\right] + E_{x\sim p_{data}(x)}\left[\log\left(1 - D_y\left(G(x)\right)\right)\right]. \tag{4.108}$$

The discriminator function D_y tries to maximize its ability to distinguish between the generated images $G(x)$ and real sample data y. The generator $G(x)$ tries to confuse the discriminator D_y and minimize the loss. Therefore, the goal of the GAN (: $X \to Y$ and discriminator D_y) is

$$\min_G \max_{D_y} \mathcal{L}_{GAN}\left(G, D_y, X, Y\right). \tag{4.109}$$

Similarly, we can define the GAN (F, D_x,Y,X):

$$\min_F \max_{D_x} \mathcal{L}_{GAN}(F, D_x, Y, X), \tag{4.110}$$

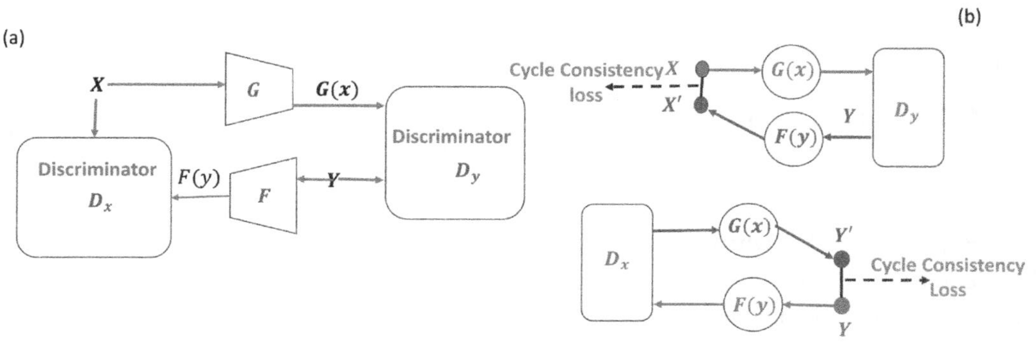

FIGURE 4.5 (a) Architecture of CycleGAN, (b) two cycle consistency losses.

where

$$\mathcal{L}_{GAN}\left(F, D_x, Y, X\right) = E_{x \sim p_{data}(x)}\left[\log D_x(x)\right] + E_{y \sim p_{data}(y)}\left[\log\left(1 - D_x\left(F(y)\right)\right)\right]. \quad (4.111)$$

The optimal solution to the GAN loss function can only match the distribution of the output of the generator to the target distribution. However, there is an infinite number of the generator outputs satisfies this requirement. The solutions are not unique. To overcome this limitation, we introduce second type of loss, i.e., cycle consistency loss (Figure 4.5(b)). The learned generators should be cycle consistent.

Architecture of the CycleGAN shows that its learning process forms a cycle. Starting a specific sample x from domain X, we generate $G(x)$ that matches with target value y from domain Y, than take y as an input to another generator F, which in turn generates $\hat{x} = F(G(x))$. Finally, we bring x back to $\hat{x}$ and complete a cycle, which is called forward cycle consistency. Similarly, we can complete another cycle: $y \to F(y) \to G(F(y)) \to \hat{y} = G(F(y))$, which is called backward cycle consistency. Putting all together, we can define the total of cycle consistency loss:

$$\mathcal{L}_{cyc}(G,F) = E_{x \sim p_{data}(x)}\left[\left\|F(G(x)) - x\right\|_1\right] + E_{y \sim p_{data}(y)}\left[\left\|G(F(y)) - y\right\|_1\right]. \quad (4.112)$$

Combining equations (4.108), (4.111), and (4.112), we can define the loss function of the CycleGAN as

$$\mathcal{L}\left(G,F,D_x, D_y\right) = \mathcal{L}_{GAN}\left(G, D_y, X, Y\right) + \mathcal{L}_{GAN}\left(F, D_x, Y, X\right) + \lambda \mathcal{L}_{cyc}(G,F), \quad (4.113)$$

where λ is a penalty parameter which balances the relative importance of the GAN loss and cycle consistency loss. Two generators G and F, and two discriminators D_x and D_y can be determined by solving the following optimization problem:

$$\min_{G,F} \max_{D_x, D_y} \mathcal{L}\left(G,F,D_x, D_y\right). \quad (4.114)$$

As Zhu et al. (2017) pointed out, the Cycle GAN can be viewed as two autoencoders: $F(G(x)): X \to X$ and $G(F(y)): Y \to Y$. We use an adversarial loss to train the bottleneck layer of an autoencoder to match the output of the generator (latent representation) with an arbitrary target distribution (an input to the another generator). Therefore, the CycleGAN can also be viewed as AAEs.

4.3.3.3 *Conditional Variational Autoencoder and Conditional Generative Adversarial Networks*

Recently there have been many interesting works on combining conditional variational autoencoder (CVAE) and conditional generative adversarial network (CGAN) which is abbreviated as CVAE-GAN (Bao et al. 2017; Lim et al. 2018; Faez et al. 2020; Wang et al. 2020). The CVAE-GANs have been successfully applied to image generation, anomaly detection, molecular design, and graph classification. In this section, we will introduce the architecture of the CVAE-GAN and its loss functions.

4.3.3.3.1 Architecture of CVAE-GAN The architecture of CVAE-GAN is shown in Figure 4.6. It consists of four subnetworks: (1) the encoder network E, (2) the generative (decoder) network G, (3) the discriminative network D, and (4) classification network (C). If the problem is not a classification problem, then the classification network can be removed.

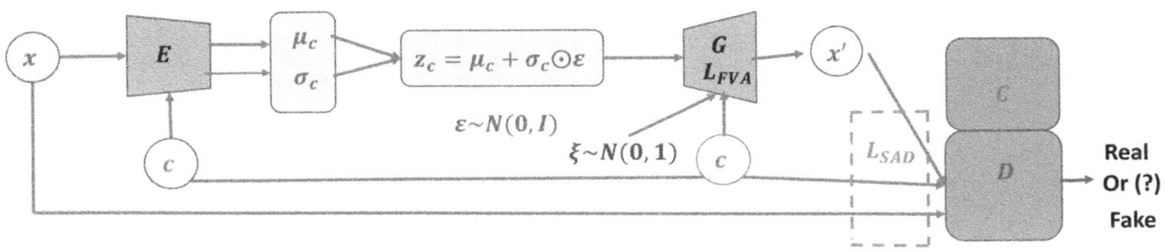

FIGURE 4.6 Architecture of CVAE-GAN.

Assume that N subjects are sampled. Consider L categories $c=\{c_1,\ldots,c_L\}$ and real sample dataset $\{x_1,\ldots,x_N\}$, $x_i \in R^d$. The categorical variable c is associated with each sample x_i. Let $z \in R^m$ be a latent variable in an m-dimensional latent space.

The encoder E maps the data sample x and its associated category c to a latent space z in the latent space with a conditional distribution $p(z\mid x,c)$. The generative (decoder) network decodes the latent variable z to reconstruct the original sample data x' by sampling from z and c with distribution $p(x\mid z,c)$. The CVAE and CGAN share the generator network G, which connects the CVAE and CGAN. The discriminator network D tries to discriminate "real" data x from "fake" (reconstructed) data x'. The classification network attempts to classify the data x into L categories with the posterior probability $p(c\mid x)$.

4.3.3.3.2 Encoder Encoder maps images (data) x and their associated categories c into the latent variable z. The variational approximate posterior $q_\varnothing(z\mid x,c)$ can be factorized as products of independent normal distributions:

$$q_\varnothing(z|x,c)=\prod_{i=1}^{m} q_\varnothing(z_i|x,c)=\prod_{i=1}^{m} N\left(\mu_i(x,c),\sigma_i^2(x,c)\right). \tag{4.115}$$

Assume that a prior distribution is given by

$$p(z)=\prod_{i=1}^{m} N(0,1).$$

Then, the KL distance between the variational approximate posterior $q_\varnothing(z\mid x,c)$ and the prior distribution $p(z)$ is

$$L_{KL}=E_{q_\varnothing(z|x,c)}\log\frac{q_\varnothing(z|x,c)}{p(z)}$$

$$=-\frac{1}{2}\sum_{i=1}^{m}\left(\log\sigma^2(x_i,c_i)+1-\sigma^2(x_i,c_i)-\mu^2(x_i,c_i)\right), \tag{4.116}$$

where

$$\left[\mu(x,c)\,|\log\sigma^2(x,c)\right]=MLP(x,c,\varnothing), \tag{4.117}$$

$$\mu(x,c)=\left[\mu_1(x,c),\ldots,\mu_m(x,c)\right]^T,\ \log\sigma^2(x,c)$$
$$=\left[\log\sigma_1^2(x,c),\ldots,\log\sigma_m^2(x,c)\right]^2,$$

MLP denotes multilayered perceptrons.

4.3.3.3.3 Decoder The network G in Figure 4.6 has two functions. Its first function is decoder that reconstructs the original data x from sampled latent variable z. Its second function is to generate fake sample for the network D of CGAN. Now we study the first function decoder of the network G.

Recall that the reconstruction loss is

$$L_{\text{Recost}}=E_{q_\varnothing(z|x,c)}\left[\log p_\theta(x\mid c,z)\right]. \tag{4.118}$$

If the data x are continuous, then

$$\log p_\theta(x|c,z)=\log N\left(\mu_x(z,c),\sigma_x^2(z,c)I\right).$$

If the data x are discrete, then

$$\log p_\theta(x\mid c,z)=x\log y_\theta(z,c)+(1-x)\log\left(1-y_\theta(z,c)\right).$$

4.3.3.3.4 Discriminator and Generator The GAN consists of two subnetworks: the generator and the discriminator. The vanilla CGAN tries to minimize the following two loss functions:

$$\begin{aligned}L_D&=-E_{x,c\sim p_{data}(x,c)}\left[\log D(x,c)\right]\\&\quad-E_{z,c\sim p(z,c)}\left[\log\left(1-D\left(G(z,c),c\right)\right)\right],\end{aligned} \tag{4.119}$$

$$L_G=-E_{z,c\sim p(z,c)}\left[\log D\left(G(z,c),c\right)\right]. \tag{4.120}$$

The goal of the CGAN is to close the distance between the real distribution $p_{data}(x,c)$ and the generated fake distribution $p_g(x',c)$. However, the supports of the distributions $p_{data}(x,c)$ and $p_g(x',c)$ may not be completely overlapped. There is an empty intersection between two distributions (Wang et al. 2020). For example, when the discriminator is trained, we have $D(x,c)\to 1$ and $D\left(G(z,c),c\right)\to 0$. Thus, $\frac{\partial L_G}{\partial D(G(z,c),c)}=-E\left[\frac{1}{D(G(z,c),c)}\right]\to-\infty$, which in turn causes the gradient-vanishing and algorithm unstable.

To overcome this limitation, Wang et al. (2020) proposed to replace the original two-player game loss function with least-squares loss function. The least-squares loss function can effectively move the fake samples toward the right decision boundary and lead to the stable algorithm. The loss functions for the D and G can be modified as follows (Wamg et al. 2020):

$$\begin{aligned} L_D = & -\frac{1}{2}E_{x,c\sim p_{data}(x,c)}\left[(D(x,c)-1)^2\right] \\ & -\frac{1}{2}E_{z,c\sim(z,c)}\left[(D(G(z,c),c)-0)^2\right], \end{aligned} \tag{4.121}$$

$$L_G = -\frac{1}{2}E_{z,c\sim p(z,c)}\left[(D(G(z,c),c)-1)^2\right], \tag{4.122}$$

where 1 and 0 in equation (4.121), respectively, denote the real data and fake data, and 1 in equation (4.122) denotes the decision value of the discriminator which classifies the data as fake data.

In the regular GAN, the normal distributed noise $\xi \sim N(0,1)$ is taken as an input to the generator to generate the fake data $x' = G(z,c,\xi)$. Therefore, the loss functions L_D and L_G should be modified as

$$\begin{aligned} L_D = & -\frac{1}{2}E_{x,c\sim p_{data}(x,c)}\left[(D(x,c)-1)^2\right] \\ & -\frac{1}{2}E_{z,c\sim p(z,c)}\left[(D(G(z,c),c)-0)^2\right] \\ & -\frac{1}{2}E_{\xi\sim N(0,1),\, c\sim p(c)}\left[(D(\xi,c)-0)^2\right], \end{aligned} \tag{4.123}$$

$$\begin{aligned} L_G = & -\frac{1}{2}E_{z,c\sim p(z,c)}\left[(D(G(z,c),c)-1)^2\right] \\ & -\frac{1}{2}E_{\xi\sim N(0,1),\, c\sim p(c)}\left[(D(\xi,c)-1)^2\right]. \end{aligned} \tag{4.124}$$

4.3.3.3.5 Loss Function Before defining the total loss function, we introduce two additional loss functions. One loss function tries to further measure the difference between the generated samples and real data samples. This loss function is defined as follows (Wang et al. 2020):

$$L_{SAD} = \frac{1}{b_s}\sum_{(x,c)\sim p_{g,\xi}(x,c)}\left[\begin{array}{c}\left(\dfrac{G(z,c)^T x}{\|G(z,c)\|_2\|x\|_2}+1\right) \\ +\left(\dfrac{G(\xi,c)^T x}{\|G(\xi,c)\|_2\|x\|_2}+1\right)\end{array}\right], \tag{4.125}$$

where b_s is the batch size. L_{SAD} includes two terms. The first term is the normalized inner product between the sample generated from the latent variable and real data sample. The second term is the normalized inner product between the sample generated from the noise variable ξ and the real data sample. The smaller the loss function L_{SAD}, the more similar the synthesized sample and the real sample, and the more reliable the generator will be.

To further reduce the likelihood of mode collapse, we want the feature space of the generative samples to cover the training data distribution as more as possible. To achieve this, similar to L_{SAD}, we introduce the second loss function to measure the correlation between the pairwise feature of an intermediate layer. We take the penultimate layer of the generator network as the intermediate layer. Let $F(.)$ denote the features of the penultimate layer of the generator network. We consider all possible feature pairs of the penultimate layer and define the second loss function as follows (Wang et al. 2020):

$$L_{FVA} = \frac{1}{L}\sum_{l=1}^{L}\frac{1}{m^l(m^l-1)}\sum_{k=1}^{m^l}\sum_{n\neq k}^{m^l}\left(\frac{F(z^k|c_l)^T F(z^n|c_l)}{\|F(z^k|c_l)\|_2\|F(z^n|c_l)\|_2}+1\right), \tag{4.126}$$

where the normalized inner product of the features of the penultimate layer is computed among the samples which belong to the same category l. The loss function L_{FVA} is the average of the normalized inner products of all possible pair-wise features of the penultimate layer and measure the diversity of the generated samples. To increase the diversity of the generated samples, we need to minimize the loss function L_{FVA}.

The network D can be modified as the classifier C where the categorical vector in the original network D is replaced by a zero vector, and the last layer is replaced by a softmax layer to obtain the posterior probability $p(x|c)$. Then, the loss function for the network C is defined as

$$L_c = -E_{x\sim p_{data}(x)}\left[\log p(c|x)\right]. \tag{4.127}$$

The loss function of CVAE-GAN includes three parts: (1) loss from the CVAE, (2) loss from the CGAN, and

(3) additional loss functions L_{SAD}, L_{FVA}, and L_c. The loss from the CVAE consists of L_{KL} and L_{Recost}, and the loss from the CGAN includes L_D and L_G. The total loss function of the CVAE-GAN is defined as follows:

$$L = L_D + L_G + L_c + L_{KL} + \lambda_1 L_{FVA} + \lambda_2 L_{SAD} + \lambda_3 L_{Recost}. \tag{4.128}$$

The computational procedures for optimizing loss function L are summarized in Algorithm 4.4 (Wang et al. 2020).

Algorithm 4.4

Step 1: Input and Initialization.
Input: Training dataset $T_r = \{(x_1, c_1), \ldots, (x_N, c_N)\}$, $x_i \in R^d$, $c_i = (1, \ldots, L)$, test dataset $T_s = \{(x_1, c_1), \ldots, (x_s, c_s)\}$.
Initialize: Network: Encode E, generator G, and discriminator D. Batch size b_s, $\lambda_1 = 0.3$, $\lambda_2 = 0.6$, $\lambda_3 = 1$.

Step 2: Training.
For number of training iterations **do**

Sample $(x_i, c_i, i = 1, \ldots, b_s)$ from training dataset T_r.

Calculate

$$[\mu(x_i, c_i) | \log \sigma^2(x_i, c_i)] = MLP(x_i, c_i, \varnothing),$$

$$L_{KL} \leftarrow -\frac{1}{2}\sum_{i=1}^{b_s}\left(\log\sigma^2(x_i, c_i) + 1 - \sigma^2(x_i, c_i) - \mu^2(x_i, c_i)\right).$$

Define the transformation:

$$z_i^k = \mu(x_i, c_i) + \sigma(x_i, c_i) \odot \varepsilon^{k,}, \ k = 1, \ldots, b_s, \ \varepsilon \sim N(0,1).$$

$$L_{\text{Recost}} \leftarrow \frac{1}{b_s}\sum_{k=1}^{b_s} \log p_\theta(x_i | c_i, z_i^k).$$

If the data x are continuous, then

$$\log p_\theta(x_i | c_i, z_i^k) = \log N\left(\mu_{x_i}(z_i^k, c_i), \sigma_{x_i}^2(z_i^k, c_i) I\right).$$

If the data x are discrete, then

$$\log p_\theta(x_i | c_i, z_i^k) = x_i \log G(z_i^k, c_i) + (1 - x_i)\log\left(1 - G_\theta(z_i^k, c_i)\right).$$

Input x, c, z and ξ to the generator network G. Let $p_{g,\xi}(x,c)$ be the distribution of the output of the generator G. Sample $(x,c) \sim p_{g,\xi}(x,c)$ and then calculate the loss function L_{SAD}:

$$L_{SAD} \leftarrow \frac{1}{b_s}\sum_{(x,c)\sim p_{g,\xi}(x,c)}\left[\left(\frac{G(z,c)^T x}{\|G(z,c)\|_2 \|x\|_2} + 1\right) + \left(\frac{G(\xi,c)^T x}{\|G(\xi,c)\|_2 \|x\|_2} + 1\right)\right].$$

$$L_{FVA} \leftarrow \frac{1}{L}\sum_{l=1}^{L}\frac{1}{m^l(m^l - 1)}\sum_{k=1}^{m^l}\sum_{n \neq k}^{m^l}\left(\frac{F(z^k | c_l)^T F(z^n | c_l)}{\|F(z^k | c_l)\|_2 \|F(z^n | c_l)\|_2} + 1\right).$$

When $c = 1$, sample $(x,c) \sim p_{data}(x,c)$, and when $c = 0$, sample $(x,c) \sim p_{g,\xi}(x,c)$, $(x,y) \sim p_{g.z}(x,y)$.

$$L_D \leftarrow -\frac{1}{2}E_{(x,c)\sim p_{data}(x,c)}\left[(D(x,c) - 1)^2\right] - \frac{1}{2}E_{(x,c)\sim p_{g,z}(x,c)}\left[(D(x,c))^2 - \frac{1}{2}E_{(x,c)\sim p_{g,\xi}(x,c)}\left[(D(x,c))^2\right]\right.$$

$$L_G \leftarrow -\frac{1}{2}E_{(x,c)\sim p_{g,z}(x,c)}\left[(D(x,c) - 1)^2\right] - \frac{1}{2}E_{(x,c)}\left[(D(x,c) - 1)^2\right].$$

$$L_c \leftarrow -\frac{1}{b_s}\sum_{i=1}^{b_s}\sum_{j=1}^{L}\log p(c_j | x_i).$$

Step 3: Calculate Gradients.

Let θ_D be the parameters in the discriminate network D. Calculate the gradients:

$$\nabla_{\theta_D}\left(\frac{1}{b_s}L_D\right)$$

Calculate the cost of classification based on the samples: $(x,c) \sim p_{data}(x,c)$,

$$(x,c) \sim p_{g,z}(x,c), (x,c) \sim p_{g,\xi}(x,c).$$

Let $\theta_{Dsoftmax}$ be the parameters in the softmax layer of the discriminator D. Calculate their gradient.

$$\nabla_{\theta_{Dsoftmax}} L_{Ds} = \left[\frac{1}{b_s} \left(\begin{array}{l} \sum_{(x,c)\sim p_{g,z}(x,c)} \left[-\log p(c \mid x) \right] \\ + \sum_{(x,c)\sim p_{data}(x,c)} \left[-\log p(c \mid x) \right] \end{array} \right) \right].$$

Let θ_g be the parameters in the generator network G and θ_e be the parameters in the encoder network E. Calculate gradients:

$$\nabla_{\theta_g} L_{qG} = \left[\frac{1}{b_s} \left(L_G + \frac{\lambda_1}{b_s - 1} L_{FVA} + \lambda_2 L_{SAD} + \lambda_3 L_{Recost} \right) \right],$$

$$\nabla_{\theta_e} L_{qE} = \left[\frac{1}{b_s} \left(\begin{array}{l} L_G + \frac{\lambda_1}{b_s - 1} L_{FVA} + \lambda_2 L_{SAD} \\ + \lambda_3 L_{Recost} + L_{KL} \end{array} \right) \right].$$

Step 4: Update Parameters.

$$\theta_D \leftarrow \theta_D - \eta \nabla_{\theta_D} \left(\frac{1}{b_s} L_D \right),$$

$$\theta_{Dsoftmax} \leftarrow \theta_{Dsoftmax} - \eta \nabla_{\theta_{Dsoftmax}} L_{Ds},$$

$$\theta_g \leftarrow \theta_g - \eta \nabla_{\theta_g} L_{qG},$$

$$\theta_e \leftarrow \theta_e - \eta \nabla_{\theta_e} L_{qE}.$$

End for

4.3.3.4 Integrated Conditional Graph Variational Adversarial Networks

In Section 4.3.3.3, we introduced the integrated conditional VAE and conditional GAN analysis. However, conditional VAE and conditional GAN can only generate and analyze unstructured data. In this section, we will extend the integrated conditional VAE and conditional GAN (ICVAE-GAN) to the structured data which can generate graphs (Yang et al. 2019).

4.3.3.4.1 Models Consider two sets of data: (1) a set of graphs $= \{G_1, \ldots, G_n\}$, where $G_i = \{V_i, E_i\}$, V_i denotes a set of nodes and E_i denotes a set of edges, and (2) a set of conditions $C = \{C_1, \ldots, C_n\}$, where C_i is a condition vector of characterizing some typical graph contexts, and is associated with the graph G_i. Let $\mathcal{H} = \{G_i, C_i\}_{i=1}^n$.

To generate the conditional structures of the graphs, we require two assumptions. The first assumption is that given a new condition $C \in \mathcal{H}$, generate new graphs that mimicking the structures of the graphs in the training set $\mathcal{H}$. The second assumption is that given a new condition $\notin \mathcal{H}$, we can generate novel graphs that are similar to the graphs in the training set with similar context, while getting insight into the unobserved world (Yang et al. 2019).

Based on all context-structure pairs in $\mathcal{H}$, a single model should jointly capture both the contexts and the structures of the graphs. The structure of a graph is often represented by an adjacency matrix A. However, the adjacency matrix$A$ depends on the order of the nodes. The permutations can change the order of the nodes, which in turn change the adjacency matrix A, but cannot change the underlying structure of the graph. To generate a unique underlying graph, the graph generative models should be permutation-invariant.

4.3.3.4.2 Encoder The encoder of the ICVAE-GAN should include the following features (Yang et al. 2019). Firstly, the encoder should effectively convert node-level encoding into permutation-invariant graph-level encoding. Secondly, arbitrary numbers of graphs can be learned and graphs with variable sizes can be generated. Semantic conditions such as attributes and labels which can be taken as features are included in node contents.

A given graph $G = \{V, E\}$ can be taken as a plain network with the adjacency matrix A and node features $X = X(A)$, which can be represented as the standard k-dim vector. The graph convolution network (GCN) discussed in Chapter 3 is used as an encoder (Figure 4.7). The encoder maps the adjacency matrix A and node feature vector to the latent variables z in the latent space. The encoder is defined as the posterior distribution $q(Z|X, A)$ of the latent variables Z, given the observed adjacency matrix A and node feature vector X. The posterior distribution $q(Z|X, A)$ can be further factorized as

$$q(Z|X, A) = \prod_{i=1}^{n} q(z_i \mid X, A), \tag{4.129}$$

where $z_i \in Z$ can be viewed as the node embedding of $v_i \in V$.

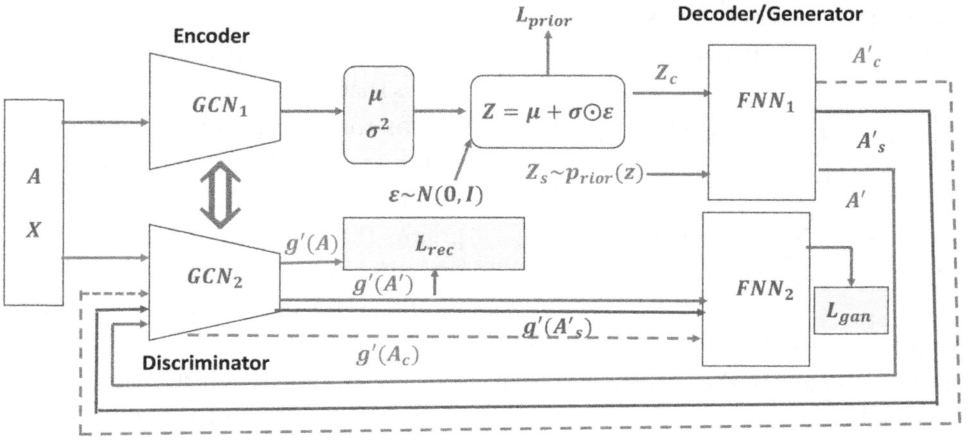

FIGURE 4.7 Architecture of ICVAE-GAN.

Similar to equation (4.77), we can define the matrix:

$$g(X,A)=\left[g_\mu(X,A)g_\sigma(X,A)\right]=\hat{A}\text{ReLU}\left(\hat{A}XW_0\right)W_1, \tag{4.130}$$

where $\tilde{A}=A+I$, $\tilde{D}_{ii}=\Sigma_j\tilde{A}_{ij}$, and $\hat{A}=\tilde{D}^{-\frac{1}{2}}\tilde{A}\tilde{D}^{-\frac{1}{2}}$, W_0 and W_1 are weight matrices of the MLP.

Assume that z_i are independent and identically distributed as a normal distribution:

$$q(z_i|X,A)=N\left(\bar{z};\bar{\mu},diag\left(\bar{\sigma}^2\right)\right), \tag{4.131}$$

where $\bar{\mu}=\frac{1}{n}\Sigma_{i=1}^n g_\mu(X,A)_i$ and $\bar{\sigma}^2=\frac{1}{n^2}\Sigma_{i=1}^n g_\sigma(X,A)_i^2$, $g(X,A)_i$ denotes the ith row of the matrix $g(X,A)$. The mean $\bar{z}$ of the latent variables is the mean of latent variables z_i over graph G and can be viewed as the graph embedding of G. The KL distance between the posterior distribution $q(Z|X,A)$ and prior distribution $p(Z)$ is (Exercise 4.3)

$$\begin{aligned} L_{KL} &= -KL\left(q(Z|X,A)||p(Z)\right) \\ &= \frac{n}{2}\left(1+\log\bar{\sigma}^2-\bar{\mu}^2-\bar{\sigma}^2\right). \end{aligned} \tag{4.132}$$

To show that encoding is permutation-invariant, we first introduce a concept of permutation matrix. An $n\times n$ dimensional permutation matrix P is defined as a representation of permutation of n elements. A permutation matrix is a square matrix with exactly one entry of 1 in each row and each column and 0 elsewhere. Consider a permutation π:

$$\begin{pmatrix} 1 & 2 & \cdots & n \\ \pi(1) & \pi(2) & \cdots & \pi(n) \end{pmatrix}.$$

For example, $\pi=\begin{pmatrix} 1 & 2 & 3 & 4 \\ 3 & 4 & 1 & 2 \end{pmatrix}$.

A permutation matrix has two representations: column representation and row representation. Here, we introduce column representation. Define a unit vector e with only one entry of 1 and 0 elsewhere. Define $e_{\pi(i)}$ as

$$e_{\pi(i)}=\begin{bmatrix} 0 \\ \vdots \\ \pi(i) \\ \vdots \\ 0 \end{bmatrix}.$$

Then, a permutation matrix can be defined as

$$P_\pi=\begin{bmatrix} e_{\pi(1)} \\ e_{\pi(2)} \\ \vdots \\ e_{\pi(n)} \end{bmatrix}.$$

The permutation $\pi=\begin{pmatrix} 1 & 2 & 3 & 4 \\ 3 & 4 & 1 & 2 \end{pmatrix}$ corresponds to the following permutation matrix:

$$P=\begin{bmatrix} 0 & 0 & 1 & 0 \\ 0 & 0 & 0 & 1 \\ 1 & 0 & 0 & 0 \\ 0 & 1 & 0 & 0 \end{bmatrix}.$$

We can show that $g(X,A)$ is permutation-invariant (Exercise 4.2). In other words, we can show that

$$g(PX, PAP^T)=g(X,A),$$

which implies that $g(PX, PAP^T)_i = g(X, A)_i$ and $\Sigma_{i=1}^n g(PX, PAP^T)_i = \Sigma_{i=1}^n g(X, A)_i$. This shows that $\bar{z}$, $\bar{\mu}$, and $\bar{\sigma}$ are also permutation-invariant. Permutation of the node order will not change the mean $\bar{z}$ of the latent variables, i.e., encoding is permutation-invariant.

4.3.3.4.3 Decoder Decoder consists of two layers of fully connected feedforward neural networks, denoted as f (Yang et al. 2019). We sample from the posterior z_i^k, where

$$z_i^k = \bar{\mu} + \bar{\sigma} \odot \varepsilon^k, \ \varepsilon^k \sim N(0,1), i = 1, \ldots, n, \ k = 1, \ldots, K. \tag{4.133}$$

The decoder for reconstruction of the adjacency matrix A is defined as

$$p(A|Z) = \prod_{i=1}^{n}\prod_{j=1}^{n} p(A_{ij} \mid z_i, z_j), \tag{4.134}$$

where

$$p(A_{ij}|z_i, z_j) = \sigma\left(f(z_i)^T f(z_j)\right), \tag{4.135}$$

σ is an activation function. The adjacency matrix that is generated by equation (4.135) is denoted by $\mathcal{G}(Z)$.

Link reconstruction loss function L_{rec} is defined as

$$\begin{aligned} L_{rec} &= E_{q(Z|X,A)}\left[\log p(A \mid Z)\right] \\ &\approx \frac{1}{K}\sum_{k=1}^{K}\sum_{i=1}^{n}\sum_{j=1}^{n} \log\sigma\left(f(z_i^k)^T f(z_j^k)\right). \end{aligned} \tag{4.136}$$

Combining equations (4.132) and (4.136), we obtain the loss function L_{VAE} for the VAE that consists of encoder and decoder:

$$L_{VAE} = L_{rec} + L_{KL}. \tag{4.137}$$

In the encoder section, we showed that the encoding process is permutation-invariant. However, the decoding process is not permutation-invariant. Since the node orders of generated adjacency matrix $A' = \mathcal{G}(Z)$ and the original adjacency matrix A may be different, the computation of the permutation-variant L_{rec} may not be accurate. To overcome this limitation, the reconstruction loss function L_{rec} needs to be redefined, which will be discussed in the next section.

4.3.3.4.4 Discriminator in the GAN A discriminator D consists of a two-layer GCN followed by a two-layer FNN, where GCN and FNN are constructed as before. As shown before, the GCN is permutation-invariant, and hence such constructed discriminator D is also permutation-invariant.

To jointly train the discriminator D together with the encoder $\mathcal{E}$ and decoder/generator $\mathcal{G}$, we optimize the following GAN loss function (Yang et al. 2019):

$$L_{gan} = \log D(A) + \log\left(1 - D(A')\right), \tag{4.138}$$

where

$D(A) = f'\left(g'(X(A), A)\right)$, X, g', f' are the input spectral embedding, GCN, and FNN, respectively.

The discriminator has three inputs: the original spectral embedding, the adjacency matrix A, and the generated adjacency matrix A'. Similar to g, the output g' of the GCN is also permutation-invariant, i.e., $g'\left(X(PAP^T), PAP^T\right) = g'\left(X(A), A\right)$. Recall that $A' = \mathcal{G}(Z)$, i.e., $A' = PAP^T$, the adjacency matrix A' can be obtained by permutation of A. Therefore, $g'\left(X(A'), A'\right) = g'\left(X(PAP^T), PAP^T\right) = g'\left(X(A), A\right)$ does not depend on the order of the adjacency matrix A. Now we can redefine the reconstruction loss function L_{rec} as follows:

$$L_{rec} = \left\|g'(A) - g'(A')\right\|_2^2. \tag{4.139}$$

Thus, redefined loss L_{rec} does not depend on the order of the matrices A and A', and hence is permutation-invariant.

Similar to AAE in Section 4.3.2, we consider two sources of generating adjacency matrix A': (1) from the latent variables $Z_c = \mathcal{E}(A)$ and (2) sampled graph encoding Z_s from the prior distribution. Then, the GAN loss function L_{gan} in equation (4.138) should be modified as follows:

$$\begin{aligned} L_{gan} = &\log\left(D(A)\right) + \log\left(1 - D\left(\mathcal{G}(Z_s)\right)\right) \\ &+ \log\left(1 - D\left(\mathcal{G}(Z_c)\right)\right). \end{aligned} \tag{4.140}$$

Finally, we assume that the parameters in two GCN (GCN module g in the encoding and GCN module g' in the discriminator) are shared. Similar to CycleGAN discussed in Section 4.4.3.2, mapping consistency between the context and structure spaces should be enforced. In other words, we require that the reconstructed adjacency matrix A' can be mapped back to the latent space of graph encoding with contexts $Z \odot C\left(g(A') \rightarrow Z \odot C\right)$, which can also be mapped from the original graph A by the same graph encoding g $\left(g(A) \rightarrow Z \odot C\right)$.

4.3.3.4.5 Total Loss Function Combining equations (4.132), (4.139), and (4.140), we can define the total loss function for the ICVAE-GAN as follows (Yang et al. 2019):

$$L_{ICVAE-GAN} = L_{rec} + \lambda_1 L_{KL} + \lambda_2 L_{gan}, \qquad (4.141)$$

where λ_1 and λ_2 are penalty parameters to make trade-off between the reconstruction loss L_{rec}, and two other losses: L_{KL} and L_{gan}.

Solving the optimization problem (4.141), we jointly train the encoder $\mathcal{E}$, decoder/generator $\mathcal{G}$, and discriminator D. Let θ_E be the parameters in the encoder, θ_G be the parameters in the decoder/generator, and θ_D be the parameters in the discriminator. Thus, parameters θ_E are only involved in the loss functions L_{rec} and L_{KL}, the parameters θ_G are only involved in the loss functions L_{rec} and L_{gan}, and parameters θ_D are only involved in the loss function L_{gan}. Let

$$\theta = \begin{bmatrix} \theta_E \\ \theta_G \\ \theta_D \end{bmatrix}.$$

Then,

$$\nabla_\theta L_{ICVAE-GAN} = \begin{bmatrix} \nabla_{\theta_E}\left(L_{rec} + \lambda_1 L_{KL}\right) \\ \nabla_{\theta_G}\left(L_{rec} + \lambda_2 L_{gan}\right) \\ \nabla_{\theta_D}\left(\lambda_2 L_{gan}\right) \end{bmatrix}.$$

Therefore, the parameter updating rules in each training batch are given by

$$\begin{bmatrix} \theta_E^{t+1} = \theta_E^t - \eta \nabla_{\theta_E}\left(L_{rec} + \lambda_1 L_{KL}\right) \\ \theta_G^{t+1} = \theta_G^t - \eta \nabla_{\theta_G}\left(L_{rec} + \lambda_2 L_{gan}\right) \\ \theta_D^{t+1} = \theta_D^t - \eta \nabla_{\theta_D}\left(\lambda_2 L_{gan}\right) \end{bmatrix}. \qquad (4.142)$$

4.3.4 Deep Convolutional Generative Adversarial Network

Deep convolutional generative adversarial network (DCGAN) is a powerful tool for generation of imaging (Salehi et al. 2020). DCGAN has been successfully applied to data augmentation of images, image classification, object detection, image segmentation (Venu and Ravula 2021) and fault detection (Viola et al. 2021). The DCGAN can produce meaningful, sufficient and realistic samples, and generate the new samples for improving the performance of target networks and enhancing generalization (Shamsolmoali et al. 2020).

4.3.4.1 Architecture of DCGAN

The DCGAN is an extension of the GAN where the multilevel feedforward neural networks of the discriminator and the generator are replaced by a convolutional neural network (CNN) and convolutional-transpose network (CTN), respectively (Venu and Ravula 2021). The architecture of the DCGAN is shown in Figure 4.8. The CNN and CTN networks in the DCGAN should have some specific features (Viola et al. 2021). The first feature is that except for the output layer of the generator and input layer of the discriminator, the DCGAN adds the batch normalization layers to all the hidden layers. The second feature is that the

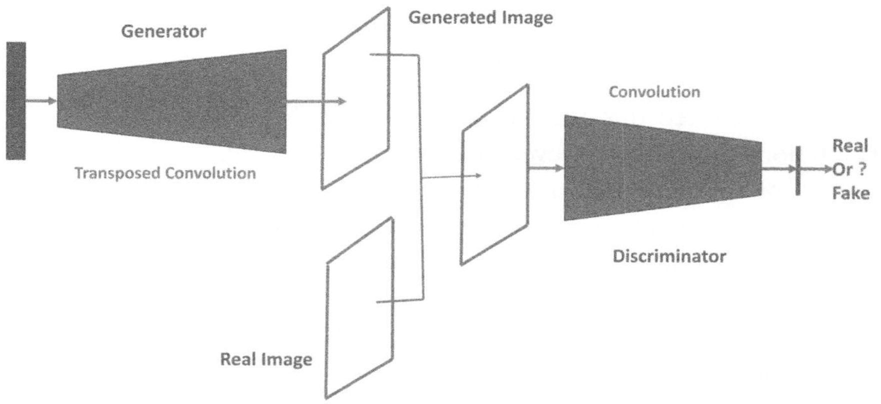

FIGURE 4.8 Architecture of deep convolutional generative adversarial network.

pooling layers are replaced with strided convolutions for the discriminator, and fractional-strided convolutions for the generator. The third feature is to remove full layers connections in the hidden layers of both generator and discriminator, except for the output layer of the discriminator which will still be fully connected. The fourth feature is to take the ReLU as activation function for the input and hidden layers of the generator, and Tanh as activation function for the output of the generator. The fifth feature also includes to take LeakyReLU as activation function for all the layers of discriminator.

4.3.4.2 Generator

Now we introduce the generator (Viola et al. 2021). The generator network mainly consists of transpose-convolution layers. The input to the generator is a random noise vector with uniform or normal distributions, followed is a dense layer. The dense layer with ReLU activation and batch normalization reshapes the random vector to a three-dimensional representation. A series of convolution-transpose layers follow the output of the dense layer to upsample the representation. Each convolution-transpose layer consists of convolution transpose, ReLU, and batch normalization operations. The output layer uses the regular convolution operation and tanh activation. Batch normalization is carried out for all layers except for output layer. After batch normalization, each variable has zero mean and unit variance, leading to stable learning process.

The key operation is the transposed convolution. The transposed convolution with learnable parameters attempts to upsample the input feature map to a desired output feature map. Figure 4.9 shows the basic procedures of a transposed convolution operation which can be summarized as follows:

Step 1: Consider a 2×2 encoded input feature map, which needs to be upsampled to 3×3 output feature map via a 2×2 kernel.

Step 2: Take the upper left element of the input feature map and multiply it with every element of the kernel with unit stride and zero padding.

Step 3: Repeat process for all the remaining elements of the input feature map.

Step 4: Summarize all elements of the overlapping positions, resulting the final 3×3 dimensional output feature map.

Now we briefly introduce the mathematical derivation of the transposed convolution. Consider a 3×3 dimensional input feature matrix and a kernel matrix K:

$$Input = \begin{bmatrix} x_1 & x_2 & x_3 \\ x_4 & x_5 & x_6 \\ x_7 & x_8 & x_9 \end{bmatrix}, K = \begin{bmatrix} w_{11} & w_{12} \\ w_{21} & w_{22} \end{bmatrix}.$$

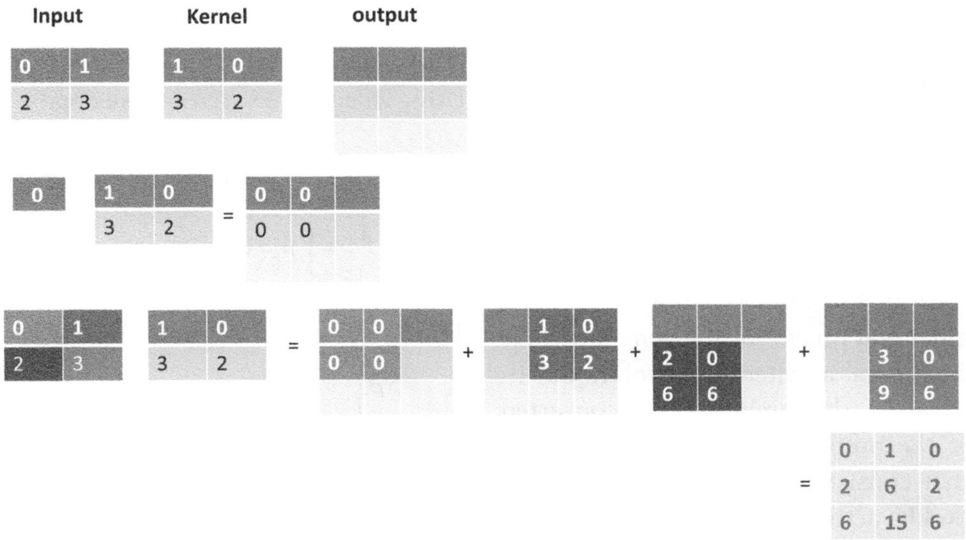

FIGURE 4.9 Illustration of convolution transpose operation.

Assume that stride = 1 and padding = 0, i.e., $i=3, k=2, s=1, p=0$. Then, the size of the output matrix is $o=\frac{i+2p-k}{s}+1=2$. The output feature matrix is denoted as

$$output=\begin{bmatrix} y_1 & y_2 \\ y_3 & y_4 \end{bmatrix}.$$

Let

$$X=[x_1, x_2, x_3, x_4, x_5, x_6, x_7, x_8, x_9]^T,$$

$$Y=[y_1, y_2, y_3, y_4]^T, \text{ and}$$

$$C=\begin{bmatrix} w_{11} & w_{12} & 0 & w_{21} & w_{22} & 0 & 0 & 0 & 0 \\ 0 & w_{11} & w_{12} & 0 & w_{21} & w_{22} & 0 & 0 & 0 \\ 0 & 0 & 0 & w_{11} & w_{12} & 0 & w_{21} & w_{22} & 0 \\ 0 & 0 & 0 & 0 & w_{11} & w_{12} & 0 & w_{21} & w_{22} \end{bmatrix}.$$

Then, we have

$$Y=CX. \tag{4.143}$$

Its inverse operation gives

$$X=C^TY. \tag{4.144}$$

Equation (4.144) shows that the inverse of convolution is multiplication of the transpose of the matrix C by the output vector Y corresponding to the output matrix. Their product corresponds to the input feature matrix X. We use the transposed convolution that performs the inverse of the convolution operation can convert the output feature matrix back to the input feature matrix.

4.3.4.3 Discriminator Network

The discriminate network tries to distinguish the real image from the fake image (Viola et al. 2021). The input images of the discriminator come from either the original real images or the generated fake images from the generator. After its convolution with LeakyReLu activation, the input image is converted to a signal for the output by a series of convolutional layers that consists of convolution operation with a LeakyReLU activation function and Batch Normalization operation. The output layer uses a sigmoid activation function to assess whether the converted image is real or fake.

4.3.5 Multi-Agent GAN

The GAN can be generalized to multi-agent GAN to solve the problem of mode collapse (Ghosh et al. 2018).

Let

$$x_j=G_j\left(z,\theta_g^j\right),\ j=1,\ldots,k \text{ and } x=\sum_{j=1}^{k}x_j.\ \text{Define } p_{g_j}=P\left(x_j\right)=P\left(G_j\left(z,\theta_g^j\right)\right) \text{ and } p_g(x)=P\left(\sum_{j}^{k}G_j\left(z,\theta_g^j\right)\right).$$

Define objective function

$$\begin{aligned} F &= E_{x\sim p_d}\log D_{k+1}(x;\theta_d) + E_{z\sim p(z)}\log\left(1-D_{k+1}\left(\sum_{j=1}^{k}G_j\left(z,\theta_g^j\right),\theta_d\right)\right) \\ &= E_{x\sim p_d}\log D_{k+1}(x;\theta_d)+E_{x\sim p_g}\log\left(1-D_{k+1}(x,\theta_d)\right). \end{aligned} \tag{4.145}$$

Our goal is

$$\min_{G_1,\ldots,G_k}\max_{D_{k+1}} F=E_{x\sim p_d}\log D_{k+1}(x;\theta_d) + E_{z\sim p(z)}\log\left(1-D_{k+1}\left(\sum_{j=1}^{k}G_j\left(z,\theta_g^j\right),\theta_d\right)\right). \tag{4.146}$$

Using calculus of variation, we obtain

$$F(\varepsilon)=\int\begin{bmatrix} p_d(x)\log\left(D_{k+1}(x)+\varepsilon\Delta D_{k+1}\right) \\ +\int p_g(x)\log\left(1-D_{k+1}(x,\theta_d)-\varepsilon\Delta D_{k+1}\right)\end{bmatrix}dx$$

Taking

$$\frac{\partial F}{\partial\varepsilon}\Big|_{\varepsilon=0}=\int p_d(x)\frac{1}{D_{k+1}(x)}\Delta D_{k+1}dx - \int\sum_{j=1}^{k}p_{g_j}\frac{1}{1-D_{k+1}(x,\theta_d)}\Delta D_{k+1}\,dx$$

$$= \int \begin{bmatrix} p_d(x)\frac{1}{D_{k+1}(x)}dx \\ -\sum_{j=1}^{k} p_{g_j} \frac{1}{1-D_{k+1}(x,\theta_d)} \end{bmatrix} \Delta D_{k+1}dx$$

and setting it to zero, yields

$$\int \begin{bmatrix} p_d(x)\frac{1}{D_{k+1}(x)} \\ -\sum_{j=1}^{k} p_{g_j} \frac{1}{1-D_{k+1}(x,\theta_d)} \end{bmatrix} \Delta D_{k+1}dx = 0. \quad (4.147)$$

Taking

$$\Delta D_{k+1} = p_d(x)\frac{1}{D_{k+1}(x)} - \sum_{j=1}^{k} \frac{p_{g_j}}{1-D_{k+1}(x,\theta_d)}$$

and substituting it to equation (4.147), we obtain

$$p_d(x)\frac{1}{D_{k+1}(x)} - \sum_{j=1}^{k} p_{g_j} \frac{1}{1-D_{k+1}(x,\theta_d)} = 0. \quad (4.148)$$

Solving equation (4.148) with respect to $D_{k+1}(x)$ yields

$$D_{k+1}(x) = \frac{p_d(x)}{p_d(x) + \sum_{j=1}^{k} p_{g_j}(x)}. \quad (4.149)$$

Let

$$p_g = \frac{\sum_{j=1}^{k} p_{g_j}(x)}{k} \text{ and } p_{avg} = \frac{p_d(x) + \sum_{j=1}^{k} p_{g_j}(x)}{k+1}.$$

Substituting equation (4.149) into equation (4.145), we obtain

$$\begin{aligned} F &= -(k+1)\log(k+1) + k\log k \\ &+ KL(p_d \| p_{avg}) + kKL(p_g \| p_{avg}). \end{aligned} \quad (4.150)$$

When

$$p_g = p_{avg}, \quad (4.151)$$

both KL distance terms are equal to zero, equation (4.150) is reduced to

$$F = -(k+1)\log(k+1) + k\log k. \quad (4.152)$$

Equation (4.151) also implies that

$$p_g = p_d. \quad (4.153)$$

Define a vector of indicator variables:

$$\delta = \begin{bmatrix} \delta_1 \\ \vdots \\ \delta_k \end{bmatrix}, \text{ where}$$

$$\delta_j = \begin{bmatrix} 1 \text{ if the sample belongs to the j-th generator} \\ 0 \quad \text{otherwise} \end{bmatrix}.$$

Then, we have

$$E[\delta] = \begin{bmatrix} p_{g_1} \\ \vdots \\ p_{g_k} \end{bmatrix}.$$

Define

$D_j = 1$, if the sample belongs to the jth generator.

When a classification task has more than two classes, machine learning often uses a softmax output layer. The softmax function predicts discrete probability distribution over the classes (generators). Let $s_j = h_j^T \theta_g^j$ be the input to the $j$the output unit. Then, the softmax activation of the jth output unit is

$$p_{g_j} = \frac{e^{s_j}}{\sum_{m=1}^{k} e^{s_m}}, \ j = 1,\ldots,k. \quad (4.154)$$

The cross entropy error function for multiclass generator output is defined as

$$E_\delta \left[\sum_{j=1}^{k} \delta_j \log D_j \right] = \sum_{j=1}^{k} p_{g_j} \log D_j. \quad (4.155)$$

To find the optimal D_j, we attend to solve the following constrained optimization problem:

$$\max_{D_1,\ldots,D_k, \sum_{j=1}^{k} D_j = 1} \sum_{j=1}^{k} p_{g_j} \log D_j. \quad (4.156)$$

Solving optimization problem, we obtain

$$D_j = \frac{p_{g_j}}{\sum_{m=1}^{k} p_{g_m}}, \ j = 1,\ldots,k. \quad (4.157)$$

We now summarize the procedures for multiple GANs as Algorithm 4.5.

Algorithm 4.5

Let m be the number of steps which updating discriminator takes before updating the generator, k be the number of generators and n be the number of samples in the real dataset.

For number of training iterations **do**

For m steps **do**

Sample minibatch of n noise samples $\{z_{(1)},\ldots,z_{(n)}\}$ from distribution $p_z(z)$.

Input minibatch of n real data samples $\{x_{(1)},\ldots,x_{(n)}\}$ that have distribution $p_{data}(x)$.

Search optimal solution for discriminator given generators by updating weights of the discriminator via ascending its stochastic gradient:

$$\nabla_{\theta_d}\frac{1}{n}\sum_{i=1}^{n}\left[\begin{array}{l}\log D_{k+1}\left(x_{(i)}\right)\\ +\log\left(1-D_{k+1}\left(\sum_{j=1}^{k}G_j\left(z_{(i)},\theta_g^j\right),\theta_d\right)\right)\end{array}\right]$$

end for

Sample minibatch of n noise samples $\{z_{(1)},\ldots,z_{(n)}\}$ from distribution $p_z(z)$.

Update the weights of the generators by descending its stochastic gradient to search the optimal solution to the generator:

$$\nabla_{\theta_g^j}\frac{1}{n}\sum_{i=1}^{n}\left[\log\left(1-D_{k+1}\left(\sum_{j=1}^{k}G_j\left(z_{(i)},\theta_g^j\right),\theta_d\right)\right)\right],$$

$j=1,\ldots,k.$

end for

Compute Softmax score

$$D_j\left(x_{(i)}\right)=\frac{e^{G_j\left(z_{(i)},\theta_g^j\right)}}{\sum_{l=1}^{k}e^{G_l\left(z_{(i)},\theta_g^j\right)}},\ i=1,\ldots,n,\ j=1,\ldots,k.$$

To identify the generator (or subpopulation):

$$j_{(i)}^{*}=\underset{j=1,\ldots,k}{\operatorname{argmax}}\,D_j\left(x_{(i)}\right)$$

4.4 GENERATIVE IMPLICIT NETWORKS FOR CAUSAL INFERENCE WITH MEASURED AND UNMEASURED CONFOUNDERS

Unmeasured confounding is a key issue that hampers application of causal inference. Causal inference in observational studies often assumes that there are no unmeasured confounders. When the unmeasured confounders are present, they often lead to the biased causal effect estimators. The focus of this section is to develop generative implicit networks from causal inference with both measured and unmeasured confounders to mitigate confounding bias (Mohamed and Lakshminarayanan 2017).

4.4.1 Generative Implicit Models

Let Y be an outcome, t be a treatment or cause, W be a vector of observed confounding variables, Z be a vector of unobserved confounding variables, and X be a vector of features that can be used as proxy variables of the unobserved confounders Z. Causal graph among these variables is shown in Figure 4.10. Generative implicit models use two hidden random variables Z and N and transform them using nonlinear deterministic functions that are realized by neural networks. The generative implicit models for Figure 4.11 are given by

$$W=g_{\theta_w}(N), \tag{4.158}$$

$$X=g_{\theta_x}(Z), \tag{4.159}$$

$$t=g_{\theta_t}(Z,W), \tag{4.160}$$

$$Y=g_{\theta_Y}(Z,W,t). \tag{4.161}$$

The observed data are X, t, W, and Y. Let $p^*(x,t,w,y)$ be the density function of the true data X, t, W and

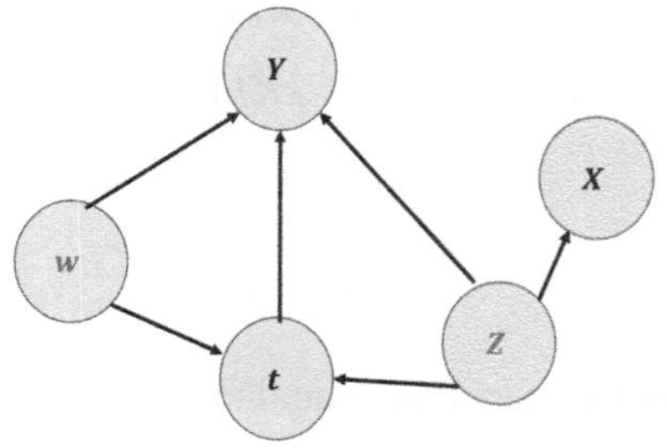

FIGURE 4.10 Causal models with observed and unobserved confounders.

Y, $q_\theta(w,x,t,y)$ be the density function of the model that generates the data X, t, W, and Y, and $q(n,z)$ be a density over hidden variables N and Z.

4.4.2 Loss Function

4.4.2.1 *Bernoulli Loss*

It is often difficult to explicitly specify likelihood functions. There are increasingly interests in developing likelihood free methods for inference. Widely used likelihood free methods are based on generative explicit models. Using the models, we can generate samples. Learning rules can be derived by comparing samples from the true data distribution with samples from the model distribution. Therefore, we first discuss the loss function that is based on density estimation-by-comparison. Estimation-by-comparison procedure consists of two steps: comparison and estimation. The comparison is to test the hypothesis that the true data distribution $p^*(w,x,t,y)$ and model distribution $q_\theta(w,x,t,y)$ are equal. Define the density difference:

$$d(w,x,t,y)=p^*(w,x,t,y)-q_\theta(w,x,t,y), \quad (4.162)$$

and the density ratio:

$$r(w,x,t,y)=\frac{p^*(w,x,t,y)}{q_\theta(w,x,t,y)}. \quad (4.163)$$

The comparator that measures $d(w,x,t,y)$ or $r(w,x,t,y)$ provides information about the departure of the distribution of the generative model from the true data distribution. Then, we use this information to learn the parameters of the generative model.

The density ration can be estimated using a classifier that discriminates the observed data from the data generated by the model. Specifically, let

$$u=\begin{bmatrix} w \\ x \\ t \\ y \end{bmatrix},$$

the data matrix of the observed m samples

$$U_p=\left\{u_1^{(p)},\ldots,u_m^{(p)}\right\},$$

and the data matrix of m' samples generated by the model

$$U_q=\left\{u_1^{(q)},\ldots,u_{m'}^{(q)}\right\}.$$

Define a new dataset:

$$D=\begin{cases} \left[v=1, U_p\right] \\ \left[v=0, U_q\right] \end{cases}.$$

Therefore, the true data distribution $p^*(w,x,t,y)$ and the model distribution $q_\theta(w,x,t,y)$ can be represented by $p^*(w,x,t,y)=p(w,x,t,y\,|\,v=1)$ and

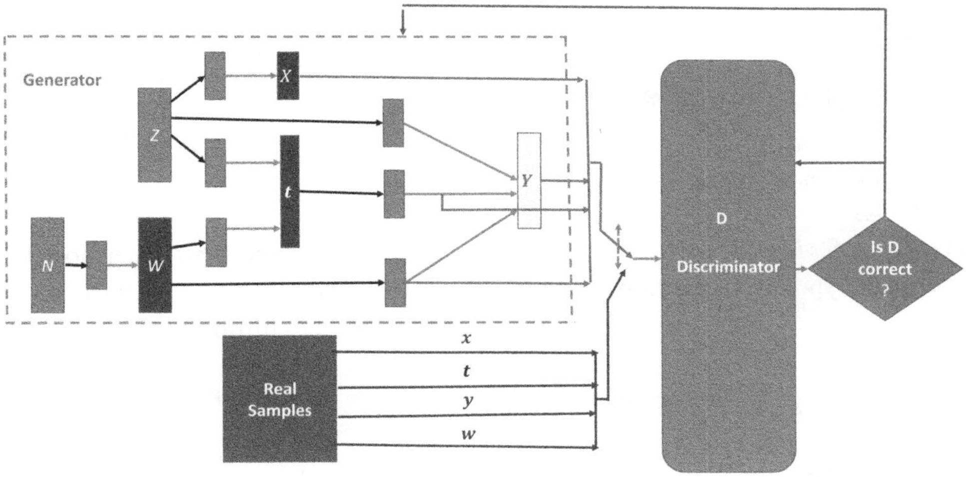

FIGURE 4.11 Concept of GAN for causal inference with confounders.

$q_\theta(w,x,t,y) = p(w,x,t,y \mid v=0)$, respectively. Let $\pi = P(v=1)$. Then, the density ratio can be calculated by

$$r(w,x,t,y) = \frac{p^*(w,x,t,y)}{q_\theta(w,x,t,y)} = \frac{p(w,x,t,y \mid v=1)}{p(w,x,t,y \mid v=0)}$$

$$= \frac{\dfrac{p(w,x,t,y,v=1)}{p(v=1)}}{\dfrac{p(w,x,t,y,v=0)}{p(v=0)}} \tag{4.164}$$

$$= \frac{\dfrac{p(w,x,t,y)p(v=1 \mid w,x,t,y)}{p(v=1)}}{\dfrac{p(w,x,t,y)p(v=0 \mid w,x,t,y)}{p(v=0)}}$$

$$= \frac{p(v=1 \mid w,x,t,y)}{p(v=0 \mid w,x,t,y)} \frac{1-\pi}{\pi}.$$

If we assume $m = m'$, then $\pi = 0.5$ and equation (4.164) is reduced to

$$r(w,x,t,y) = \frac{p(v=1 \mid w,x,t,y)}{p(v=0 \mid w,x,t,y)}. \tag{4.165}$$

Let $D(w,x,t,y,\varphi) = p(v=1 \mid w,x,t,y)$. We can view the classifier that discriminates the true real data from model generated data as a Bernoulli process with the probability $D(w,x,t,y,\varphi)$ of being class $v=1$. The log-likelihood function for the Bernoulli process is

$$L(\varphi,\theta) = v\log D(w,x,t,y,\varphi) + (1-v)\log(1-D(w,x,t,y,\varphi)). \tag{4.166}$$

4.4.2.2 Loss Function for the Generative Implicit Models

Taking expectation of $L(\varphi,\theta)$ with respect to $p(w,x,t,y,v) = p(w,x,t,y|v)p(v)$, equation (4.166) is reduced to

$$\mathcal{L}(\varphi,\theta) = E_{p(w,x,t,y|v)p(v)}\begin{bmatrix} v\log D(w,x,t,y,\varphi)+(1-v) \\ \log(1-D(w,x,t,y,\varphi)) \end{bmatrix} \tag{4.167}$$

$$= \pi E_{p^*(w,x,t,y)}\begin{bmatrix} \log D(w,x,t,y,\varphi) \\ +(1-\pi)E_{q_\theta(w,x,t,y)} \\ \left[\log(1-D(w,x,t,y,\varphi))\right] \end{bmatrix}.$$

It is known that the variables w,x,t,y are generated by the model:

$$\begin{bmatrix} w \\ x \\ t \\ y \end{bmatrix} = G_\theta(n,z), \tag{4.168}$$

where $q(n,z)$ is a density over hidden variables N and Z. Substituting equation (4.168) into equation (4.167), we obtain

$$\mathcal{L}(\varphi,\theta) = \pi E_{p^*(w,x,t,y)}\begin{bmatrix} \log D(w,x,t,y,\varphi) \\ +(1-\pi)E_{q_\theta(n,z)} \\ \left[\log(1-D(G_\theta(n,z),\varphi))\right] \end{bmatrix}. \tag{4.169}$$

The loss function (4.169) is exactly the objective function in GANs (Goodfellow et al. 2014). The parameter in the Bernoulli process is the probability function $D(w,x,t,y,\varphi)$. The maximum likelihood estimator of the parameter is obtained by

$$\max_\varphi \mathcal{L}(\varphi,\theta) = \pi E_{p^*(w,x,t,y)}\begin{bmatrix} \log D(w,x,t,y,\varphi) \\ +(1-\pi)E_{q_\theta(n,z)} \\ \left[\log(1-D(G_\theta(n,z),\varphi))\right] \end{bmatrix}. \tag{4.170}$$

The Generator needs to learn how to generate data that are as close to the true data as possible. To achieve this, we need to minimize (Figure 4.11)

$$\min_\theta \max_\varphi \mathcal{L}(\varphi,\theta). \tag{4.171}$$

The global optimum is achieved if and only if $p^*(w,x,t,y) = q_\theta(w,x,t,y)$. Recall that KL distance is defined as $KL(p\|q)=E_p\log\frac{p}{q}$. Define the Jensen-Shannon divergence $JSD(p\|q)$ as

$$JSD = KL\left(p\left\|\frac{p+q}{2}\right.\right)+KL\left(q\left\|\frac{p+q}{2}\right.\right).$$

Then, we have

$$\max_{\varphi}\mathcal{L}(\varphi,\theta)=-\log 4$$
$$+2JSD\left(p^*(w,x,t,y)\|q_\theta(w,x,t,y)\right).$$

When $p^*(w,x,t,y)=q_\theta(w,x,t,y)$ we obtain $\min_\theta\max_\varphi\mathcal{L}(\varphi,\theta)=-\log(4)$. The goal of the GAN is to minimize the Jensen-Shannon divergence.

4.4.3 Divergence Minimization

In this section, we will introduce f-divergence measure that can be used as an objective function to learn the generative model (Nowozin et al. 2016). Given two distributions P and Q with two density functions p and q, the f-divergence between two distributions is defined as

$$D_f(P\|Q)=\int_\chi q(x)f\left(\frac{p(x)}{q(x)}\right)dx, \tag{4.172}$$

where $f:R_+\to R$ is a convex generator function with $f(1)=0$.

We note that

1. When $P=Q$ then $D_f(P\|Q)=\int_\chi q(x)f(1)dx=0$;
2. For any P and Q, we have

$$D_f(P\|Q)=\int_\chi q(x)f\left(\frac{p(x)}{q(x)}\right)dx\ge f\left(\int_\chi q(x)\frac{p(x)}{q(x)}dx\right)$$
$$=f(1)=0.$$

Example 4.1: KL Divergence

Let $f=x\log x$. Then,

$$D_f(P\|Q)=\int_\chi q(x)\frac{p(x)}{q(x)}\log\frac{p(x)}{q(x)}dx$$
$$=\int_\chi p(x)\log\frac{p(x)}{q(x)}dx.$$

Example 4.2: Reverse KL Divergence

Let $f=-\log x$. Then,

$$D_f(P\|Q)=-\int_\chi q(x)\log\frac{p(x)}{q(x)}dx$$
$$=\int_\chi q(x)\log\frac{q(x)}{p(x)}dx.$$

Example 4.3: Pearson χ^2

Let $f=(u-1)^2$. Then,

$$D_f(P\|Q)=\int_\chi q(x)\left(\frac{p(x)}{q(x)}-1\right)^2dx$$
$$=\int_\chi\frac{(p(x)-q(x))^2}{q(x)}dx.$$

Example 4.4: Squared Hellinger

Let $f=\left(\sqrt{u}-1\right)^2$. Then,

$$D_f(P\|Q)=\int_\chi q(x)\left(\sqrt{\frac{p(x)}{q(x)}}-1\right)^2dx$$
$$=\int_\chi\left(\sqrt{p(x)}-\sqrt{q(x)}\right)^2dx.$$

Example 4.5: Jensen-Shannon

Let $f=-(u+1)\log\frac{1+u}{2}+u\log u$. Then,

$$D_f(P\|Q)=\int_\chi q(x)\left[\begin{array}{l}-\left(\frac{p(x)}{q(x)}+1\right)\log\frac{1+\frac{p(x)}{q(x)}}{2}\\+\frac{p(x)}{q(x)}\log\frac{p(x)}{q(x)}dx\end{array}\right]$$

$$=\int_\chi\left[\begin{array}{l}(-p(x)-q(x))\log\frac{1+\frac{p(x)}{q(x)}}{2}\\+p(x)\log\frac{p(x)}{q(x)}\end{array}\right]dx$$

$$= \int_{\chi} \begin{bmatrix} p(x)\log\frac{2p(x)}{p(x)+q(x)} \\ +q(x)\log\frac{2q(x)}{p(x)+q(x)} \end{bmatrix} dx.$$

Example 4.6: GAN

Let $f = u\log u-(u+1)\log(u+1)$. Then, we have

$$D_f(P\|Q) = \int_{\chi} q(x) \begin{bmatrix} \frac{p(x)}{q(x)}\log\frac{p(x)}{q(x)} \\ -\left(\frac{p(x)}{q(x)}+1\right)\log\left(\frac{p(x)}{q(x)}+1\right) \end{bmatrix} dx$$

$$= \int_{\chi} \begin{bmatrix} p(x)\log\frac{p(x)}{q(x)} - (p(x)+q(x)) \\ \log\frac{p(x)+q(x)}{q(x)} \end{bmatrix} dx$$

$$= \int_{\chi} p(x)\log\frac{p(x)}{p(x)+q(x)}dx + \int_{\chi} q(x)\log\frac{q(x)}{p(x)+q(x)}dx$$

$$= \int_{\chi} p(x)\log\frac{2p(x)}{2(p(x)+q(x))}dx + \int_{\chi} q(x)\log\frac{2q(x)}{2(p(x)+q(x))}dx$$

$$= -\log 4 + \int_{\chi} p(x)\log\frac{2p(x)}{p(x)+q(x)}dx + \int_{\chi} q(x)\log\frac{2q(x)}{p(x)+q(x)}dx.$$

Fenchel Conjugate

The Fenchel conjugate of the convex function f is defined as

$$f^*(t) = \sup_{x\in dom_f}\{tx - f(x)\}. \qquad (4.173)$$

The function f^* is a convex function. Indeed, consider a point $\lambda t_1 + (1-\lambda)t_2$. We have the following inequality:

$$\begin{aligned} f^*(\lambda t_1 + (1-\lambda)t_2) &\geq (\lambda t_1 + (1-\lambda)t_2)x - f(x) \\ &= \lambda t_1 x - \lambda f(x) + (1-\lambda)t_2 x \\ &\quad -(1-\lambda)f(x) \\ &\geq \lambda \sup_{x\in dom_f}\{t_1 x - f(x)\} \\ &\quad +(1-\lambda)\sup_{x\in dom_f}\{t_2 x - f(x)\} \\ &= \lambda f^*(t_1) + (1-\lambda)f^*(t_2). \end{aligned}$$

Fenchel's inequality: Definition (4.173) directly implies

$$f^*(t) + f(x) \geq tx \text{ for all } t, x. \qquad (4.174)$$

Example 4.7: Quadratic Function

Let $f(x) = \frac{1}{2}x^T Ax + b^T x + c$. Assume that A^{-1} exists. Then, by definition (4.173) we have

$$t - Ax - b = 0.$$

Solving the above equation for x, we obtain

$$x = A^{-1}(t-b),$$

which implies

$$\begin{aligned} f^*(t) &= t^T A^{-1}(t-b) - \frac{1}{2}(t-b)^T \\ &\quad A^{-1}(t-b) - b^T A^{-1}(t-b) - c \\ &= \frac{1}{2}(t-b)^T A^{-1}(t-b) - c. \end{aligned} \qquad (4.175)$$

Example 4.8: Negative Entropy

Negative entropy is defined as

$$f(x) = \sum_{i=1}^{n} x_i \log x_i.$$

Let $(x) = \sum_{i=1}^{n} t_i x_i - \sum_{i=1}^{n} x_i \log x_i$. Then, we have

$$\frac{\partial g}{\partial x_i} = t_i - \log x_i - 1 = 0.$$

Solving the above equation, we obtain

$$x_i = e^{t_i - 1}.$$

Therefore,

$$f^*(t) = \sum_{i=1}^{n} t_i e^{t_i-1} - \sum_{i=1}^{n} e^{t_i-1}(t_i - 1) = \sum_{i=1}^{n} e^{t_i-1}. \tag{4.176}$$

KL Divergence

The generator function for KL divergence is $f(x) = x\log x$, which is a special case of negative entropy. Therefore, its corresponding conjugate function is

$$f^*(t) = e^{t-1}, \tag{4.177}$$

Example 4.9: Negative Logarithm

Logarithm is defined as

$$f(x) = -\sum_{i=1}^{n} \log x_i.$$

By the similar arguments, we obtain

$$f^*(t) = -\sum_{i=1}^{n} \log(-t_i) - n. \tag{4.178}$$

Example 4.10: Matrix Logarithm

Matrix is defined as

$$f(X) = \log|X|.$$

Define

$$g(X) = Tr(YX) + \log|X|. \tag{4.179}$$

Then,

$$\frac{\partial g(X)}{\partial X} = Y^T + X^{-T} = 0, \text{ which implies}$$

$$X = -Y^{-1}. \tag{4.180}$$

Substituting equation (4.180) into equation (4.179) yields

$$f^*(Y) = -n - \log(-Y). \tag{4.181}$$

Example 4.11: Reverse KL

The generator function for reverse KL is defined as

$$f(x) = -\log x.$$

Let

$$g(x) = tx + \log x. \tag{4.182}$$

Then, to find conjugate function, setting $g'(x) = t + \frac{1}{x} = 0$, which implies

$$x = -\frac{1}{t}. \tag{4.183}$$

Substituting equation (4.183) into equation leads to

$$f^*(t) = -1 - \log(-t). \tag{4.184}$$

Example 4.12: Pearson χ^2

The generator function of Pearson χ^2 is

$$f(x) = (x-1)^2.$$

Let

$$g(x) = tx - (x-1)^2. \tag{4.185}$$

Then, setting $g'(x) = t - 2(x-1) = 0$ leads to

$$x = 1 + \frac{t}{2}. \tag{4.186}$$

Substituting equation (4.186) to equation (4.185), we obtain the conjugate function:

$$f^*(t) = \frac{t^2}{4} + t. \tag{4.187}$$

Example 4.13: Squared Hellinger

The generator function for the squared Hellinger divergence is

$$f(x) = \left(\sqrt{x} - 1\right)^2.$$

Let

$$g(x) = tx - \left(\sqrt{x} - 1\right)^2. \tag{4.188}$$

Setting

$$g'(x)=t-\left(1-\frac{1}{\sqrt{x}}\right)=0,$$

We obtain

$$x=\frac{1}{(1-t)^2}. \quad (4.189)$$

Substituting equation (4.189) into equation (4.188) yields the conjugate function of the generator function for the squared Hellinger divergence:

$$f^*(t)=\frac{t}{1-t}. \quad (4.190)$$

Example 4.14: Jensen-Shannon Divergence

The generator function of Jensen-Shannon divergence is

$$f(x)=x\log x-(x+1)\log\frac{1+x}{2}.$$

Let

$$g(x)=tx-x\log x+(x+1)\log\frac{1+x}{2}. \quad (4.191)$$

Then, setting $g'(x)=t+\log\frac{1+x}{2x}=0$, we obtain

$$x=\frac{1}{2e^{-t}-1}. \quad (4.192)$$

Substituting equation (4.192) into equation (4.191), we obtain the conjugate function of the Jensen-Shannon divergence:

$$f^*(t)=-\log\left(2-e^t\right). \quad (4.193)$$

Example 4.15: GAN

The generator function for the GAN is

$$f(x)=x\log x-(x+1)\log(x+1).$$

Let

$$g(x)=tx-x\log x+(x+1)\log(x+1). \quad (4.194)$$

Setting
$g'(x)=t-\log x+\log(x+1)=0$ leads to

$$x=\frac{e^t}{1-e^t}. \quad (4.195)$$

Substituting equation (4.195) into equation (4.194), we obtain the conjugate function of the GAN:

$$f^*(t)=-\log\left(1-e^t\right). \quad (4.196)$$

Example 4.16: Indicator Function and Norm

Indicator of a convex set C is defined as

$$\delta_C(x)=\begin{cases}0 & x\in C\\ \infty & otherwise\end{cases}.$$

The conjugate of the indicator is

$$\delta_C^*(t)=\sup_{x\in C} t^T x. \quad (4.197)$$

Let $f(x)=\|x\|$.

The dual of a norm is defined as

$$\|t\|_* = \sup_{\|x\|\ll 1} t^T x.$$

If $\|t\|_* \le 1$ then we have
$t^T\frac{x}{\|x\|}\le 1$, which implies

$$t^T x\le\|x\|. \quad (4.198)$$

When $x=0$ then $t^T x=\|x\|$. By definition of the conjugate function, we obtain

$$f^*(t)=\sup_x t^T x-\|x\|=0 \quad (4.199)$$

when $\|t\|_* \le 1$.

Now we assume $\|t\|_* > 1$. Then, by definition of dual norm we can always find a x_0 such that

$$t^T x_0-\|x_0\|>0. \quad (4.200)$$

Therefore, when $\lambda\to\infty$

$$\begin{aligned}f^*(t)=\sup_x t^T x-\|x\|&\ge t^T(\lambda x_0)\\ -\|\lambda x_0\|&=\lambda\left(t^T x_0-\|x_0\|\right)\to\infty.\end{aligned} \quad (4.201)$$

Combining equations (4.199) and (4.201), we obtain

$$f^*(t)=\begin{cases} 0 & \|t\|_* \le 1 \\ \infty & \|t\|_* > 1 \end{cases}. \tag{4.202}$$

We can show that

$$f^{**}=f. \tag{4.203}$$

4.4.4 Lower Bound of the f-Divergence

By definition, we obtain

$$D_f(P\|Q)=\int_\chi q(x)f\left(\frac{p(x)}{q(x)}\right)dx. \tag{4.204}$$

Substituting equation (4.203) into equation (4.204), we obtain

$$\begin{aligned} D_f(P\|Q) &= \int_\chi q(x)f^{**}\left(\frac{p(x)}{q(x)}\right)dx \\ &= \int_\chi q(x)\sup_{u\in dom_{f^*}}\left\{u\frac{p(x)}{q(x)}-f^*(u)\right\}dx \\ &\ge \sup_{U\in\mathcal{H}}\left(\int_\chi\left(p(x)U(x)-q(x)f^*(U(x))\right)dx\right) \\ &= \sup_{U\in\mathcal{H}}\left(E_{x\sim P}[U(x)]-E_{x\sim Q}[f^*(U(x))]\right), \end{aligned} \tag{4.205}$$

where $\mathcal{H}$ is a set of functions $\mathcal{H}: \chi\to R$.

Lemma 4.1: Conjugates and Subgradients

If $f(x)$ is closed and convex, then

$$y\in\partial f(x)\leftrightarrow x\in\partial f^*(y)\leftrightarrow f^*(y)=x^Ty-f(x). \tag{4.206}$$

If f is closed and strong convex then

$$\nabla f^*=\underset{x}{argmax}\left(y^Tx-f(x)\right). \tag{4.207}$$

Proof

By definition of conjugate function, we have

$$f^*(y)=\sup_u\{u^Ty-f(u)\}.$$

If $y\in\partial f(x)$ then x satisfies the necessary condition for optimal solution. Thus,

$$f^*(y)=\sup_u\{u^Ty-f(u)\}=x^Ty-f(x). \tag{4.208}$$

Note that

$$\begin{aligned} f^*(y+\Delta y) &= \sup_u\{u^T(y+\Delta y)-f(u)\} \\ &\ge x^T(y+\Delta y)-f(x) \\ &= x^Ty-f(x)+x^T\Delta y \\ &= f^*(y)+x^T\Delta, \end{aligned}$$

which implies $x\in\partial f^*(y)$.

Now assume $x\in\partial f^*(y)$. Then, from above formula, we obtain $y\in\partial f^{**}(x)$. But, $f^{**}(x)=f(x)$. Therefore, we have $y\in\partial f(x)$.

Now we show that if f is closed and strong convex then $\nabla f^*=argmax\left(y^Tx-f(x)\right)$.

Indeed, if $\overset{x}{f}$ is closed and strong convex then f^* is defined for all y and differentiable everywhere. It is also known that $y^Tu-f(u)$ has a unique maximizer x for every y, i.e.,

$$f^*(y)=y^Tx-f(x).$$

Using equation (4.206), we have $x\in\partial f^*(y)=\nabla f^*(y)$. In other words, we have

$$\nabla f^*(y)=\underset{x}{argmax}\left(y^Tx-f(x)\right). \tag{4.209}$$

4.4.4.1 Tighten Lower Bound of the f-Divergence

Now we find solution $U(x)$ to $\sup_{U\in\mathcal{H}}\left(E_{x\sim P}[U(x)]-E_{x\sim Q}[f^*(U(x))]\right)$.

$$F(\varepsilon)=E_{x\sim P}[U(x)+\varepsilon\Delta U]-E_{x\sim Q}[f^*(U(x)+\varepsilon\Delta U)].$$

By variation of calculus, we have

$$\delta F=\frac{\partial F}{\partial\varepsilon}=E_{x\sim P}[\Delta U]-E_{x\sim Q}\left[\frac{\partial f^*}{\partial U}\Delta U\right]. \tag{4.210}$$

From equation (4.209) it follows that

$$\frac{\partial f^*}{\partial U} = \underset{t}{argmax}\left(y^T t - f(t)\right) = t^*, \text{ or}$$

$$f'(t^*) = U^*(x). \tag{4.211}$$

Equation (4.210) can be rewritten as

$$\delta F = \int_{\chi} \left[p(x) - q(x)t^*\right] \Delta U dx. \tag{4.212}$$

Taking $\Delta U = p(x) - q(x)t^*$, $\delta F = 0$ implies

$$p(x) - q(x)t^* = 0 \text{ or}$$

$$t^* = \frac{p(x)}{q(x)}. \tag{4.213}$$

Substituting equation (4.213) into equation (4.211) yields

$$U^*(x) = f'\left(\frac{p(x)}{q(x)}\right). \tag{4.214}$$

Therefore, the lower bound of $D_f(P\|Q)$ is given by

$$D_f(P\|Q) \geq E_{x\sim P}\left[U^*(x)\right] - E_{x\sim Q}\left[f^*\left(U^*(x)\right),\right. \tag{4.215}$$

where $U^*(x) = f'\left(\frac{p(x)}{q(x)}\right)$.

4.4.5 Representation for the Variational Function

Let the lower bound on the f-divergence $D_f(P\|Q)$ be

$$F(\theta,\varphi) = E_{x\sim P}\left[U_\varphi(x)\right] - E_{x\sim Q\theta}\left[f^*\left(U_\varphi(x)\right)\right]. \tag{4.216}$$

Before we optimize the lower bound, we need to restrict $U_\varphi(x)$ to the domain of the conjugate functions f^*. To satisfy this condition, we define $U_\varphi(x) = g_f\left(V_\varphi(x)\right)$, where $V_\varphi(x): \chi \to R$ without any range constraints on the output, and $g_f : R \to dom_{f^*}$ is an output activation function. Then the objective function in equation (4.216) is reduced to the f-divergence.

$$F(\theta,\varphi) = E_{x\sim P}\left[g_f\left(V_\varphi(x)\right] - E_{x\sim Q\theta}\left[f^*\left(g_f(V_\varphi(x)\right)\right)\right]. \tag{4.217}$$

Example 4.17: Output Activation Function

KL divergence. Its domain of f^* is R. The function v can be taken as its output activation function $g_f(v) = v$.

Reverse KL divergence. Its domain of f^* is R_-. Then, the output activation function is $g_f(v) = -e^v$.

Pearson χ^2 divergence. Its domain of f^* is R. The function v can be taken as its output activation function $g_f(v) = v$.

GAN. Its domain of f^* is R_-. The output activation function is $g_f(v) = -\log(1+e^{-v})$.

Example 4.18: Objective Function

KL divergence. Note that conjugate function for the KL divergence is $f^*(t) = e^{t-1}$. The objective function for the KL divergence is

$$F(\theta,\varphi) = E_{x\sim P}\left[V_\varphi(x)\right] - E_{x\sim Q\theta}\left[e^{V_\varphi(x)-1}\right]. \tag{4.218}$$

Reverse KL divergence. The conjugate function of the KL divergence is $f^*(t) = -1 - \log(-t)$. The objective function for the reverse KL divergence is

$$F(\theta,\varphi) = -E_{x\sim P}\left[e^{-V_\varphi(x)}\right] + E_{x\sim Q\theta}\left[1 - V_\varphi(x)\right]. \tag{4.219}$$

Pearson χ^2 divergence. The conjugate function of the Pearson χ^2 divergence is $f^*(t) = \frac{1}{4}t^2 + t$. The objective function for the Pearson χ^2 divergence is

$$F(\theta,\varphi) = E_{x\sim P}\left[V_\varphi(x)\right] - E_{x\sim Q\theta}\left[\frac{1}{4}V_\varphi^2(x) + V_\varphi(x)\right]. \tag{4.220}$$

GAN. The output activation function and conjugate function for GAN are $g_f(v) = -\log(1+e^{-v})$ and $f^*(t) = -\log(1-e^t)$, respectively. Let $D_\varphi(x) = \frac{1}{1+e^{-V_\varphi(x)}}$. Then, the objective function for GAN is

$$F(\theta,\varphi) = E_{x\sim P}\left[\log D_\varphi(x)\right] + E_{x\sim Q\theta}\left[\log\left(1 - D_\varphi(x)\right)\right]. \tag{4.221}$$

4.4.6 Single-Step Gradient Method for Variational Divergence Minimization (VDM)

Since $Q_\theta(x)$ is unknown and the goal of learning is to find a distribution $Q_\theta(x)$ to approximate the distribution $P(x)$ as close as possible, we can use a neural network and noise to generate $Q_\theta(x)$. Define transformation that is realized by a neural network as follows: $x = G_\theta(z)$. Then, the objective function in equation (4.217) can be rewritten as

$$F(\theta,\varphi) = E_{x\sim P}\left[g_f\left(V_\varphi(x)\right]\right. - E_{z\sim P_z(z)}\left[f^*\left(g_f\left(V_\varphi\left(G(z)\right)\right)\right)\right]. \tag{4.222}$$

The expectation in equation (4.222) can be calculated by sampling mean. The optimization consists of two tasks: (1) tightening the lower bound by maximizing the objective function with respect to $V_\varphi(x)$ and (2) generating data with distribution $Q_\theta(x)$ approximating distribution $P(x)$ as close as possible by minimizing the objective function with respect to $Q_\theta(x)$ or $G(z)$. Assume that both $V_\varphi(x)$ and $G(z)$ are implemented by neural networks. Therefore, our goal is

$$\min_{G_\theta(z)}\max_{V_\varphi(x)} F\left(V_\varphi(x), G_{\theta(z)}\right) = \min_\theta \max_\varphi F(\theta,\varphi). \tag{4.223}$$

A general algorithm for the single-step gradient method is given below.

Algorithm 4.6

Step 1: Sampling minibatch of m noise samples $\{z_1,\ldots,z_m\}$ from noise distribution $P_z(z)$. Sampling minibatch of m real data samples $\{x_1,\ldots,x_m\}$. Calculating sample mean of $F(\theta,\varphi)$:

$$\hat{F}(\theta,\varphi) = \frac{1}{m}\sum_{i=1}^{m}\left[g_f\left(V_\varphi(x_i) - f^*\left(g_f\left(V_\varphi\left(G(z_i)\right)\right)\right)\right]. \tag{4.224}$$

Step 2: Repeat until convergence

$$\varphi^{k+1} = \varphi^k + \alpha\nabla_\varphi \hat{F}(\theta^k,\varphi^k),$$
$$\theta^{k+1} = \theta^k - \alpha\nabla_\theta \hat{F}(\theta^k,\varphi^k).$$

end

4.4.7 Random Vector Functional Link Network for Pearson χ^2 Divergence

GAN models are trained by using gradient-based backpropagation methods. The problems with backpropagation methods are slow convergence processes. This limitation can be overcome by randomization-based methods. Here, we introduce random vector functional link network (RVFL) for f-GAN (Suganthan 2018).

RVFL is a single layer feed-forward neural network. Weights and biases of the hidden neurons in the RVFL are randomly generated. Unlike the standard neural networks where the inputs of output neurons consist of only nonlinearly transformed signals from hidden layer neurons, the inputs of output neurons in the RVFL consist of nonlinearly transformed signals from hidden layer neurons and original input data directly from input layer.

We begin with description of RVFL for the generation of $G(Z)$. Assume that the number of samples is m. The input random vector is denoted by $Z = \left[\, z^1 \cdots z^m \,\right]$ and $z^j = \begin{bmatrix} z_1^j \\ \vdots \\ z_k^j \end{bmatrix}$, where k is the dimension of the input noise vector and z^j is the vector of the noises corresponding to the jth sample. Assume that the number of hidden layer neurons in RVFL for the generator is d_z. Let a vector of output of hidden layer neurons in RVFL for the generator be $H_z = \left[H_z^1,\ldots,H_z^m\right]$, $H_z^j = \begin{bmatrix} H_{z_1}^j \\ \vdots \\ H_{z_{d_z}}^j \end{bmatrix}$, where H_{z_j} is the vector of output of all hidden layer neurons for the jth sample. Define the weight matrix W_{HZ} that connects the output neurons to the hidden layer neurons:

$(W_{HZ})^T = \begin{bmatrix} (w_{H_z}^1)^T \\ \vdots \\ (w_{H_z}^k)^T \end{bmatrix}$ and $w_{H_z}^j = \begin{bmatrix} w_{H_{z1}}^j \\ \vdots \\ w_{H_{zd_z}}^j \end{bmatrix}$, where $w_{H_z}^j$ is a vector of weights that connect the jth neuron in the output layer to all hidden layer neurons. Define the weight matrix W_z that connects the output neurons to the input noise neurons:

$(W_z)^T = \begin{bmatrix} (w_z^1)^T \\ \vdots \\ (w_z^k)^T \end{bmatrix}$ and $w_z^j = \begin{bmatrix} w_{z_1}^j \\ \vdots \\ w_{z_k}^j \end{bmatrix}$, where w_z^j is a vector of weights that connect the jth neuron in the output layer to all input neurons. Define the output vector $G(Z)$ of the RVFL for the generator as

$$G(z)=\begin{bmatrix} G(z^1) & \cdots & G(z^m) \end{bmatrix} \text{ and } G(z^j)=\begin{bmatrix} G_1(z^j) \\ \vdots \\ G_k(z^j) \end{bmatrix}$$

Then, the output vector can be expressed as a linear function of the output of hidden layer neurons and input neurons as:

$$G(z)=(W_{HZ})^T H_z+(W_z)^T Z. \qquad (4.225)$$

Next we study the RVFL for the discriminator. The input data vector is denoted as

$X=\begin{bmatrix} x^1 & \cdots & x^m \end{bmatrix}$ and $x^j=\begin{bmatrix} x_1^j \\ \vdots \\ x_k^j \end{bmatrix}$, where k is the dimension of the input data vector and x^j is the vector of the data corresponding to the jth sample. Assume that the number of hidden layer neurons in RVFL for the discriminator is d. Let a vector of output of hidden layer neurons in RVFL for the discriminator be $H=[\, H^1 \; \cdots \; H^m \,], H^j=\begin{bmatrix} H_1^j \\ \vdots \\ H_d^j \end{bmatrix}$, where H^j is the output of the hidden layer neurons for the jth sample. Define the weight vector W_H that connects the output neuron to the hidden layer neurons: $W_H=\begin{bmatrix} w_{H_1} \\ \vdots \\ w_{H_d} \end{bmatrix}$. Define the weight vector W_x that connects the output neurons to the input neurons: $W_x=\begin{bmatrix} w_{x_1} \\ \vdots \\ w_{x_k} \end{bmatrix}$. Define the vector of outputs of the RVFL for the discriminator as $V_\varphi(X)=\begin{bmatrix} V_\varphi(x_1) & \cdots & V_\varphi(x_m) \end{bmatrix}$.

Then, the total output $V_\varphi(x)$ of the RVFL for the discriminator is

$$V_\varphi(X)=W_H^T H+W_x^T X. \qquad (4.226)$$

The sampling formula for $F(\theta,\varphi)$ in equation (4.220) is

$$F(\theta,\varphi)=\frac{1}{m}\sum_{j=1}^m V_\varphi(x^j)-\frac{1}{m}\sum_{j=1}^m \begin{bmatrix} \frac{1}{4}V_\varphi^2\left(G_\theta\left(z_j\right)\right) \\ +V_\varphi\left(G_\theta\left(z_j\right)\right) \end{bmatrix} \qquad (4.227)$$

$$=\frac{1}{m}\sum_{j=1}^m V_{(W_H,W_x)}(x^j)$$

$$-\frac{1}{m}\sum_{j=1}^m \begin{bmatrix} \frac{1}{4}V_{(W_H,W_x)}^2\left(G_{(W_{H_z},W_z)}\left(z_j\right)\right) \\ +V_{(W_H,W_x)}\left(G_{(W_{H_z},W_z)}\left(z_j\right)\right) \end{bmatrix}$$

It follows from equations (4.225) and (4.226) that

$$V_{(W_H,W_x)}(G(Z))=W_H^T H+W_x^T\left((W_{HZ})^T H_z+(W_z)^T Z\right). \qquad (4.228)$$

Then, equation (4.227) can be reduced to

$$\begin{aligned} F(\theta,\varphi)=&\frac{1}{m}V_{(W_H,W_x)}(X)\mathbf{1} \\ &-\frac{1}{4m}V_{(W_H,W_x)}(G(Z))\big(V_{(W_H,W_x)} \\ &(G(Z))\big)^T-\frac{1}{m}V_{(W_H,W_x)}(G(Z))\mathbf{1}. \end{aligned} \qquad (4.229)$$

Using matrix calculus, we obtain the saddle point for $\max_{V(W_H,W_x)} F(W_H,W_x,W_{HZ},W_z)$:

$$\frac{\partial F}{\partial W_H}=\frac{1}{m}\left[H\mathbf{1}-\frac{1}{2}V_{(W_H,W_x)}(G(Z))H-H\mathbf{1}\right], \qquad (4.230)$$

$$\frac{\partial F}{\partial W_x}=\frac{1}{m}\begin{bmatrix} X\mathbf{1}-\frac{1}{2}V_{(W_H,W_x)}(G(Z)) \\ \left((W_{HZ})^T H_z+(W_z)^T Z\right) \\ -\left((W_{HZ})^T H_z+(W_z)^T Z\right)\mathbf{1} \end{bmatrix}, \qquad (4.231)$$

Let

$$\begin{aligned} C(W_{HZ},W_z)=&-\frac{1}{4m}V_{(W_H,W_x)}(G(Z))\big(V_{(W_H,W_x)} \\ &(G(Z))\big)^T-\frac{1}{m}V_{(W_H,W_x)}(G(Z))\mathbf{1} \end{aligned}$$

Similarly, using matric calculus, we obtain the saddle point for $\min_{W_{HZ},W_z} C(W_{HZ},W_z)$:

$$\frac{\partial C}{\partial W_{HZ}}=-\frac{1}{m}\left(\frac{1}{2}\begin{bmatrix} W_x W_H^T H H_z^T \\ +H_z H_z^T W_{H_z} W_x W_x^T \\ +H_z Z^T W_z W_x^T \end{bmatrix}+H_z\mathbf{1}W_x^T\right), \qquad (4.232)$$

$$\frac{\partial F}{\partial W_z} = -\frac{1}{m}\left[\frac{1}{2}\left(ZH^T W_H + ZH_z^T W_{H_z} W_x + ZZ^T W_z + Z1\right)\right]. \quad (4.233)$$

Setting equations (4.230) and (4.231) equal to zero and solving it, we obtain

$$\begin{bmatrix} W_x \\ W_H \end{bmatrix} = \begin{pmatrix} G(z)G^T(z) & G(z)H^T \\ HG^T(z) & HH^T \end{pmatrix}^{-1} \begin{bmatrix} 2(X - G(z))\mathbf{1} \\ 0 \end{bmatrix}. \quad (4.234)$$

Setting equations (4.232) and (4.233) equal to zero, we obtain

$$W_z = -(ZZ^T)^{-1}\left(2Z\mathbf{1} + ZH^T W_{H_z} W_x\right), \quad (4.235)$$

$$W_{H_z} W_x W_x^T = \left(H_z H_z^T - H_z Z^T (ZZ^T)^{-1} ZH_z^T\right)^{-1} \begin{bmatrix} H_z Z^T (ZZ^T)^{-1}\left(2Z\mathbf{1} + ZH^T W_H\right) \\ W_x^T - 2H_z \mathbf{1} W_x^T - W_x W_H^T HH_z^T \end{bmatrix}. \quad (4.236)$$

Then, solving the matrix equation (4.236).

SOFTWARE PACKAGE

All code and hyperparameters for Generative Adversarial Networks are available at http://www.github.com/goodfeli/adversarial

The code of the paper "A Style-Based Generator Architecture for Generative Adversarial Networks" is available at https://github.com/NVlabs/stylegan

The code for the paper "Image-to-Image Translation with Conditional Adversarial Networks" is posted on the website: https://github.com/phillipi/pix2pix

The code for the paper "Graphic Generative Adversarial Networks" is available at https://github.com/zhenxuan00/graphical-gan

The software for implementing "Efficient-GAN-Anomaly-Detection" is posted on the website (https://github.com/houssamzenati/Efficient-GAN-Anomaly-Detection)

The source code for "Distribution-induced Bidirectional Generative Adversarial Network for Graph Representation Learning" is released in https://github.com/SsGood/DBGAN

All code and some data sources in the paper "Conditional structure generation through graph variational generative adversarial nets" are made available on GitHub: https://github.com/KelestZ/CondGen

All software implementations of these GAN methods and datasets have been collected and made available in one place https://github.com/pshams55/GAN-Case-Study

EXERCISES

EXERCISE 4.1
Show that

The optimal solution to the CGAN problem (4.55) is

$$\hat{p}_g(y, x) = p_{data}(y, x) \text{ and } \hat{D}(y, x) = \frac{1}{2}.$$

The optimal value of the objective function is

$$min_G max_D l(D, G) = -2\log 2.$$

EXERCISE 4.2
Show that $g(X, A)$ is permutation-invariant, i.e., for any permutation matrix P, we have $g(PX, PAP^T) = g(X, A)$.

EXERCISE 4.3
Show that

$$L_{KL} = KL\left(q(Z|X, A) \| p(Z)\right) = \frac{n}{2}\left(1 + \log\bar{\sigma}^2 - \bar{\mu}^2 - \bar{\sigma}^2\right).$$

CHAPTER 5

Deep Learning for Causal Inference

Causal inference is a fundamental problem in science, engineering, and philosophy (Spirtes et al. 2000; Pearl 2009, Peters et al. 2017; Xiong 2018b). Although randomized experiments are gold standard for causal discovery, they are often expansive, time consuming, infeasible, and even unethical. Fortunately, recent advances in causal research allow to identify causal relationships from observational data under some assumptions (Mooij et al. 2016; Marx and Vreeken 2017; Yu et al. 2019). The problems for causal discovery can be divided into two classes: bivariate causal discovery and causal network discovery (Blöbaum et al. 2018; Li et al. 2020). There are many methods for inferring causal relationships. A popular method that can be used for both bivariate causal discovery and causal network discovery is properly constrained functional models, including linear and nonlinear functions (Zhang et al. 2015; Kalainathan et al. 2018). The functional causal models can be used to model the observational variables and latent variables (Bühlmann et al. 2014; Cai et al. 2019). Neural network is a universal approximation to any function (Lu and Lu 2020). In this chapter, we will introduce deep learning as a general framework for causal inference. Two types of generative models: generative adversarial networks (GANs) and variational autoencoder (VAE) will be explored as major tools for both bivariate causal discovery and causal network reconstruction with discrete and continuous variables.

5.1 FUNCTIONAL ADDITIVE MODELS FOR CAUSAL INFERENCE

5.1.1 Correlation, Causation, and *Do*-Calculus

Correlation between two variables measures dependence between them and can be precisely defined. However, causation lacks consensus definition. There are several approaches to causation analysis. First approach is potential outcome framework developed by statistician Neyman (1923/1990). Potential outcomes include observed factual and unobserved counterfactuals. If potential outcomes are defined for individual, then counterfactual cannot be observed. However, if the potential outcomes are defined for all individuals in the population, then all potential outcomes can be observed in the population (Rubin 1974). Second approach to causal inference is artificial intelligence (AI), which uses directed acyclic graphs (DAGs) to infer causal relations and can trace back to the path analysis of Wright (1921; Pearl 1995; Dablander 2020). The major component of the AI is structural causal models (SCMs), which incorporate interventions into the model.

The key component of the SCM is intervention and intervention calculus (*do*-calculus), which was proposed in 1995 to identify causal effects in nonparametric models (Pearl 1995). There are two types of data: observational data (“seeing”) and interventional data (“doing”). An intervention is defined as forcing variables to take fixed values, which simulates physical interventions (Pearl 2012). The purpose of intervention calculus is to describe the mathematical conditions under which we can make causal inference from observational data.

We define an operator *do*() as making an action on a variable or a set of variables. We consider two observed variables X and Y. A causal model can be defined by intervention (action) as follows. If we $do\,(X = x)$ (forcing the random variable X to take a specified value x), then Y will be affected. This intervention can be

DOI: 10.1201/9781003028543-5

mathematically denoted as $(Y \mid do(X=x))$. Causation analysis investigates prediction of the effects of action that perturbs the observed system (Mooij et al. 2016). A variable X causing a variable Y can be mathematically defined as

$$P(Y \mid do(X_1)) \neq P(Y \mid do(X_2)) \quad \text{for some } X_1, X_2, X_1 \neq X_2. \tag{5.1}$$

If X causes $Y(X \rightarrow Y)$, then in general, we have

$$P(Y|X) = P(Y \mid do(X)) \neq P(Y), \tag{5.2}$$

where in the causal direction, the conditional distribution $P(Y \mid X)$ is equal to the distribution $P(Y \mid do(X))$ of Yconditioned on taking action on X (this can be statistically approved in Section 5.3 using the Pearl's second rule (Pearl 2012).

However, in the anticausal direction, since changing the effect value Y will not cause any change of X, or distribution $P(X)$, we infer that $P(X|do\,(Y)) = P(X) \neq P(X \mid Y)$, i.e., the probability $P(X \mid do\,(Y))$ is not equal to the conditional probability $P(X \mid Y)$.

Using the similar arguments, if Y causes X $(Y \rightarrow X)$ then

$$P(X \mid do(Y)) = P(X) \neq P(X \mid Y).$$

Although in statistics, the joint probability can be factorized in terms of marginal distribution and conditional distribution as

$$P(XY) = P(X)P(Y|X) = P(Y)P(X \mid Y),$$

which is symmetric. If X causes Y $(X \rightarrow Y)$, we have the factorization: $P(XY) = P(X)P(Y \mid do(X))$, but in this case $(X \rightarrow Y)$, we do not have $P(XY) = P(Y)$ $P(X \mid do(Y))$, i.e., $P(XY) \neq P(Y)P(X \mid do(Y))$, the joint probability of X and Y cannot be factorized in terms of marginal distribution $P(Y)$ and interventional probability distribution $P(X \mid do(Y)$. For the *do*-calculus, symmetric does not hold. In summary, under association assumption, the joint distribution $P(X,Y)$ can be symmetrically factorized into the product of the marginal distribution and conditional distribution in two directions. However, under causal assumption, the joint distribution $P(X,Y)$ can be factorized into the product of the marginal distribution and interventional distribution in the cause-effect direction. It cannot be factorized into the product of the marginal distribution and interventional distribution in the anticausal direction.

Do-calculus can also be defined as $E[Y \mid do(X)]$. If the effect variable Y is a binary variable, then we have

$$E[Y|do(X)] = P(Y=1 \mid do(X)). \tag{5.3}$$

5.1.2 The Rules of *Do*-Calculus

Before introducing the rules of *do*-calculus, we first introduce the d-separation. The d-separation is a graphical interpretation of conditional independence. Consider three disjoint sets of nodes: A, B, and C. The d-separation is a tool for assessing whether a set of variables A is independent of another set of variables B, given a third set of variables C and is denoted by $A \perp\!\!\!\perp B \mid C$.

Definition 5.1: d-separation

The sets A and B are d-separated by the set C if and only if for all directed paths P from A to B, at least one of three conditions holds:

1. Path P contains a "chain" "$A \rightarrow C \rightarrow B$" with an observed middle node C;
2. Path P contains a fork "$A \leftarrow C \rightarrow B$" with an observed parent node C;
3. Path P contain a collider "$A \rightarrow Z \leftarrow B$", and the collider Z is not a member of the conditioning set C or does not have a descendent in C.

The *do*-calculus has three inference rules that provide tools to map interventional distributions and observational distributions under some conditions in the causal diagram (Pearl 2012). Now we introduce some graph notations necessary for describing three inference rules. Consider a graph G (Figure 5.1(a)) with the disjoint subsets of variables: W, X, Y, Z. Let $G_{\overline{X}}$ (Figure 5.1(c)) denotes the perturbed graph that removes all edges pointing to X, $G_{\underline{x}}$ (Figure 5.1(b)) denotes the perturbed graph that removes all edges pointing from X, $Z(W)$ denote the set of nodes in Z, which do not include the ancestors of W and $G_{\overline{X}\underline{Z}}$ (Figure 5.1(e)) denote the perturbed graph that removes both incoming to X and outgoing from Z edges.

Now we introduce three inference rules of *do*-calculus (Pearl 2012).

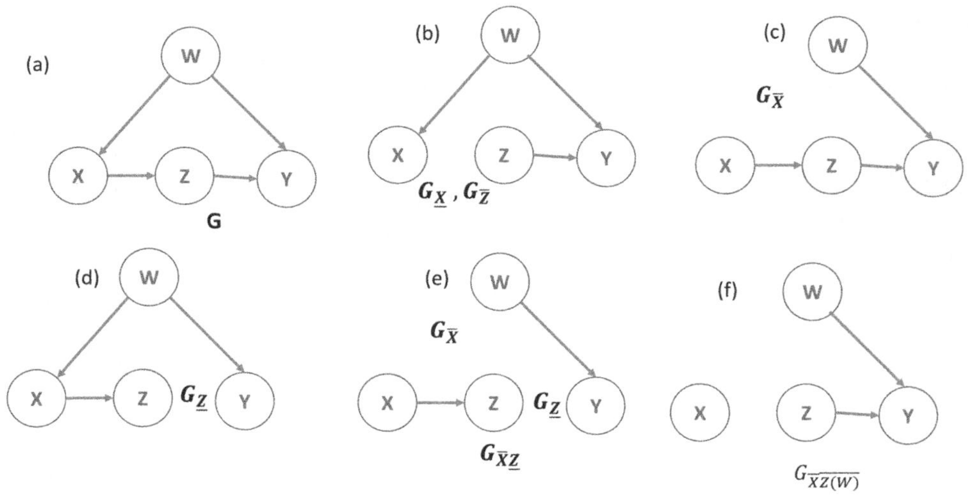

FIGURE 5.1 Graph notations for rules of *do*-calculus.

Theorem 5.1 The Pearl's Rules of *Do*-Calculus

Rule 1: Ignoring Insertion/Deletion of Observations

If the four sets of variables X, Y, Z, and W in the graph $G_{\underline{x}}$ (Figure 5.2(b)) satisfies the following independent assumption:

$$Y \perp\!\!\!\perp Z \mid X, W, \tag{5.4}$$

then we have

$$P(Y \mid do(X), Z, W) = P(Y \mid do(X), W), \tag{5.5}$$

where Z can be ignored.

Rule 2: Backdoor Criterion (Action/Observation Exchange)

If the four sets of variables X, Y, Z, and W in the graph $G_{\overline{X}\underline{Z}}$ (Figure 5.1(e)) satisfies the following independent assumption:

$$Y \perp\!\!\!\perp Z \mid X, W,$$

then, we have

$$P(Y|Do(X), Do(Z), W) = P(Y|Do(X), Z, W). \tag{5.6}$$

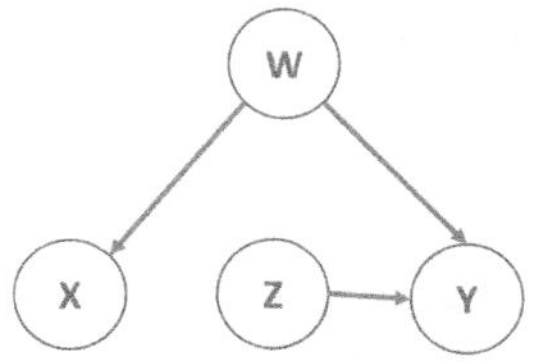

FIGURE 5.2 Illustration of Example 5.3.

Rule 3: Ignoring Actions/Interventions

If the four sets of variables X, Y, Z, and W in the graph $G_{\overline{XZ}}$ (Figure 5.2(f)) satisfies the following independent assumption:

$$Y \perp\!\!\!\perp Z \mid X, W,$$

then, we have

$$P(Y|Do(X), Do(Z), W) = P(Y \mid Do(X), W). \tag{5.7}$$

Three rules can be used to prove the identifiability of causal models. Repeatedly applying the rules of *do*-calculus to the causal model until the final expression of the causal models no longer contains a do-operator, we then claim that the causal model is identifiable.

Example 5.1 Illustration of the Pearl's First Rule

Consider several scenarios for the Pearl's first rule.

Case 1: Independence in statistics. We assume that there are no X and W.

In this case, equation (5.4) becomes $Y \perp\!\!\!\perp Z$ and equation (5.5) is simplified to

$$P(Y|Z) = P(Y). \tag{5.8}$$

It is clear that if $Y \perp\!\!\!\perp Z$ then equation (5.8) holds.

Case 2: We observe Z, but do not observe X.

In this case, equation (5.4) is reduced to

$$\left(Y \perp\!\!\!\perp Z \mid W\right)_G, \tag{5.9}$$

equation (5.5) is reduced to

$$P(Y|Z,W)=P(Y\,|\,W). \tag{5.10}$$

Since equation (5.9) holds, then Y and Z are d-separated, which implies that conditioning on W, Y, and Z are independent, and hence equation (5.10) holds.

Case (5.3): We observe X, but do not observe W. In this case, equation (5.4) is reduced to

$$\left(Y \perp\!\!\!\perp Z\,|\,X\right)_{G_{\overline{X}}}, \tag{5.11}$$

and equation (5.5) is reduced to

$$P(Y|do(X),Z)=P(Y\,|\,do(X)) \tag{5.12}$$

Equation (5.11) implies that $(X \to Y\ Z)$. Therefore, intervening on Y will not have effect on Z, and hence equation (5.12) holds.

Example 5.2 Compute $P(Z\,|\,do(X))$

Goal of computing $P(Z\,|\,do(X))$ is to exchange action with observation, i.e., converting $P(Z\,|\,do(X))$ to $P(Z\,|\,X)$. To achieve this, only Pearl's second rule can be used. Set $Y=Z, X=\varnothing, Z=X, W=\varnothing$. Then, condition $(Y \perp\!\!\!\perp Z\,|\,X,W)_{G_{\overline{X}\underline{Z}}}$ is changed to $(Z \perp\!\!\!\perp X)_{G_{\underline{X}}}$. It is clear from Figure 5.2 that since the path from X to Z is blocked by collider Y, X is independent of Z. Therefore, condition $(Z \perp\!\!\!\perp X)_{G_{\underline{X}}}$ holds, which in turn implies that the condition $(Y \perp\!\!\!\perp Z\,|\,X,W)_{G_{\overline{X}\underline{Z}}}$ for the Pearl's second rule holds. Thus, if the graph is Figure 5.2, then we obtain

$$P(Z|do(X))=P(Z\,|\,X). \tag{5.13}$$

Example 5.3 Compute P(Y| *do* (X))

We first use total probability to transform computation of $P(Y\,|\,do(X))$ to G

$$P(Y|do(X))=\sum_{Z}P(Y,Z\,|\,do(X))$$

$$=\sum_{Z}P(Y|do(X),Z)P(Z\,|\,do(X). \tag{5.14}$$

Now our goal is to reduce $P(Z\,|\,do(X))$. To use the second rule of *do*-calculus, we set $Y=Z, Z=X, X=\varnothing, W=\varnothing$. Then, the graph $G_{\overline{X}\underline{Z}}$ should be changed to $G_{\underline{X}}$ (Figure 5.2). Since the only one path from X to Z contains a fork $X \leftarrow W \rightarrow Y$, the path from X to Z is blocked. Therefore, we have $X \perp\!\!\!\perp Z$ in $G_{\underline{X}}$, which satisfies the condition of the second rule of *do*-calculus. Using the second rule, we obtain

$$P(Z|do(X))=P(Z\,|\,X). \tag{5.15}$$

Substituting equation (5.15) into equation (5.14), we obtain

$$P(Y|do(X))=\sum_{Z}P(Y|do(X),Z)P(Z\,|\,X). \tag{5.16}$$

Next we show that

$$P(Y|do(Z),do(X))=P(Y\,|\,Z,do(X)). \tag{5.17}$$

To use the second rule of the *do*-calculus, we need to check the condition $Y \perp\!\!\!\perp Z\,|\,X$ in the graph $G_{\overline{X}\underline{Z}}$ (Figure 5.2(e)). Since the path from Z to Y via X is blocked, then the condition $Y \perp\!\!\!\perp Z\,|\,X$ holds, which implies equation (5.17) holds. Substituting equation (5.17) into equation (5.16), we obtain

$$P(Y|do(X))=\sum_{Z}P(Y|do(X),do(Z))P(Z\,|\,X). \tag{5.18}$$

Next we show $P(Y|do(X),do(Z))=P(Y\,|\,do(Z))$ using the third rule of *do*-calculus. Set $W=\varnothing$. Check condition $Y \perp\!\!\!\perp Z\,|\,X$ in the graph $G_{\overline{X}\overline{Z}}$ (Figure 5.2(f)). It is clear that any path from X to Z is blocked. Therefore, the condition $Y \perp\!\!\!\perp Z\,|\,X$ in the graph $G_{\overline{X}\overline{Z}}$ holds. We obtain

$$P(Y|do(X),do(Z))=P(Y|do(Z)). \tag{5.19}$$

Substituting equation (5.19) into equation (5.18), we obtain

$$P(Y|do(X))=\sum_{Z}P(Y|do(Z))(Z\,|\,X). \tag{5.20}$$

Again, using the total probability rule, we obtain

$$P(Y|do(Z))=\sum_{x}P(Y|X,do(Z))P(X\,|\,do(Z)). \tag{5.21}$$

To apply the second rule of the *do*-calculus to $P(Y|X, do(Z))$, we need to check conditions: $Y \perp\!\!\!\perp Z$ in $G_{\underline{Z}}$.

Set $W = X$, $X = \varnothing$. In $G_{\underline{Z}}$, the path $Z \rightarrow Y$ is blocked by X. Therefore, we have $Y \perp\!\!\!\perp Z$ in $G_{\underline{Z}}$. Using the second rule, we obtain

$$P(Y|X, do(Z)) = P(Y \mid X, Z). \quad (5.22)$$

Using the third rule of the *do*-calculus, we can show (Exercise 5.1)

$$P(X|do(Z)) = P(X). \quad (5.23)$$

Substituting equations (5.22) and (5.23) into equation (5.21), we obtain

$$P(Y|do(Z)) = \sum_x P(Y \mid X, Z) P(X). \quad (5.24)$$

Combining equations (5.20) and (5.24), we obtain

$$P(Y|do(X)) = \sum_Z \sum_X P(Y|X,Z) P(Z|X) P(X). \quad (5.25)$$

Consider a simple case $X \rightarrow Y \leftarrow e$. To use the second rule of *do*-calculus, we set $Z = X$, $X = \varnothing$, $W = e$. The graph $G_{\overline{X}\underline{Z}}$ is reduced to $G_{\underline{Z}}$ or $G_{\underline{X}}$ in this case. Therefore, the condition $Y \perp\!\!\!\perp X \mid e$ holds, which implies that

$$P(Y|do\,(X)) = P(Y \mid X). \quad (5.26)$$

Consider $X \rightarrow Y \leftarrow e$ and $Y \leftarrow W$. Then, using the similar argument, we can prove (Exercise 5.2)

$$P(Y|do\,(X), W) = P(Y \mid X, W). \quad (5.27)$$

5.1.3 Structural Equation Models and Additive Noise Models for Two or Two Sets of Variables

Equation (5.3) can be extended to any real values. Substituting expectations for probabilities facilitate to use functional models to assess causal information (Pearl et al. 2016). The structural equation models (SEMs) were originally developed for causal inference with more than three variables. Here, we modified the original multivariate SEM to the bivariate SEM. In the language of expectation, equation (5.26) can be transformed to

$$E[Y|do(X)) = E[Y \mid X]. \quad (5.28)$$

Consider a nonlinear regression model:

$$Y = f(x) + \varepsilon, \quad (5.29)$$

where we assume $E[\varepsilon] = 0$.

Taking conditional expectation on both sides of equation (5.29), we obtain

$$E[Y \mid X] = f(X) + E[\varepsilon \mid X]. \quad (5.30)$$

If we assume $X \perp\!\!\!\perp \varepsilon$, then equation (5.30) is reduced to

$$E[Y|X] = f(x),$$

which implies

$$E[Y|do(X)] = f(X). \quad (5.31)$$

If we jointly consider cause X and covariate W and extend equation (5.29) to

$$Y = f(X, W) + \varepsilon, \quad (5.32)$$

where $X \perp\!\!\!\perp \varepsilon$, $W \perp\!\!\!\perp \varepsilon$.

Taking conditional expectation on both sides of equation (5.32), we obtain

$$E[Y|X,W] = f(X,W). \quad (5.33)$$

Replacing the probability by expectation, equation (5.27) can be extended to

$$E[Y|do(x), W] = E[Y \mid X,W]. \quad (5.34)$$

Combining equations (5.33) and (5.34), we obtain

$$E[Y|do(x), W] = f(X,W). \quad (5.35)$$

This motives us to define the additive noise model (ANM). The ANMs are widely used methods for bivariate causal discovery (Mooij et al. 2016). We assume that there is no confounding, no selection bias, and no feedback between the cause and effects (Mooij et al. 2016). We define the ANMs as follows.

Definition 5.2: ANMs with Continuous Variables

Consider continuous variables X and Y. An ANM from X to Y is defined as

$$Y = f_Y(X) + N_Y,\ X \perp\!\!\!\perp N_Y, \text{ or} \quad (5.36)$$

$$Y = f_Y(X, W) + N_Y,\ X \perp\!\!\!\perp N_Y,\ W \perp\!\!\!\perp N_Y, \quad (5.37)$$

where Y is a potential effect, X is a potential cause, W is a covariate, N_Y is a noise, f_Y is a nonlinear function, and X and N_Y, $W$and N_Y are independent.

An ANM is called reversible if there is also an ANM:

$$X = f_X(Y) + N_X,\ Y \perp\!\!\!\perp N_X, \quad (5.38)$$

where f_X is a nonlinear deterministic function, N_X is a noise, and Y and N_X are independent.

Now we define the ANMs for discrete variables.

Definition 5.3: ANMs with Integer Variables

Let m be an integer. Assume that $Z = km + r, r = 0, 1, \ldots, m-1$. Z is called an m-cyclic random variable, if Z takes the remainder r as its value. Now we define discrete ANMs for genetic causation analysis.

An ANM from X to Y is defined as (Peters et al. 2011)

$$Y = f_y(X) + N_y,\ X \perp\!\!\!\perp N_y, \quad (5.39)$$

where f_y is an integer function, N_y is a cyclic noise variable, and X and N_y are independent.

An ANM is called reversible if there is also an ANM:

$$X = f_x(Y) + N_x,\ Y \perp\!\!\!\perp N_x, \quad (5.40)$$

where $f_x(Y)$ is a nonlinear integer function, N_x is a cyclic noise variable, and Y and N_x are independent.

Example 5.4

Let Y and X be continuous variables. The model

$$Y = 2X^3 + e^X + \varepsilon,\ \varepsilon \perp\!\!\!\perp X$$

is an ANM from X to Y.

Example 5.5

Define an indicator variable for the genotype at a locus:

$$X = \begin{cases} 0 & aa \\ 1 & Aa \\ 2 & AA \end{cases},$$

where A is a disease allele. Let Y be a binary variable to indicate disease status: $Y = 1$, presence of disease and $Y = 0$, normal. Let m be an integer. Assume that $Z = km + r, r = 0, 1, \ldots, m-1$. Z is called an m-cyclic random variable, if Z takes the remainder r as its value. Thus, X and Y be 3- and 2-cyclic random variables, respectively.

The model (Peters et al. 2011)

$$Y = f_y(X) + N_y,\ X \perp\!\!\!\perp N_y, \quad (5.41)$$

where f_y is an integer function, N_y is a 2-cyclic noise variable, and X and N_y are independent, is an ANM from X to Y.

If there is

$$X = f_x(Y) + N_x,\ Y \perp\!\!\!\perp N_x, \quad (5.42)$$

where $f_x(Y)$ is a nonlinear integer function, N_x is a 3-cyclic noise variable, and Y and N_x are independent, then the model (5.42) is a reversible ANM.

Next we study causal relations between two sets of variables. For example, we consider causal relationships between two biochemical pathways. The set of effect variables are often called responses and the set of causal variables are often called predictors. We can consider four cases: (1) one response variable and one predictor, (2) one response variable and one set of predictor variables, (3) one set of response variables and one predictor variables, and (4) one set of response variables and one set of predictor variables. Four cases can be unified into one vector ANM.

Definition 5.4: Vector ANMs

Consider vectors $X = [X_1, \ldots, X_m]^T$ and $Y = [Y_1, \ldots, Y_n]^T$. A vector ANM from vector X to Y is defined as (Figure 5.3)

$$Y = f_Y(X) + N_Y,\ X \perp\!\!\!\perp N_Y,\ N_{Yk} \perp\!\!\!\perp N_{Yl},\ \forall k,l, \quad (5.43)$$

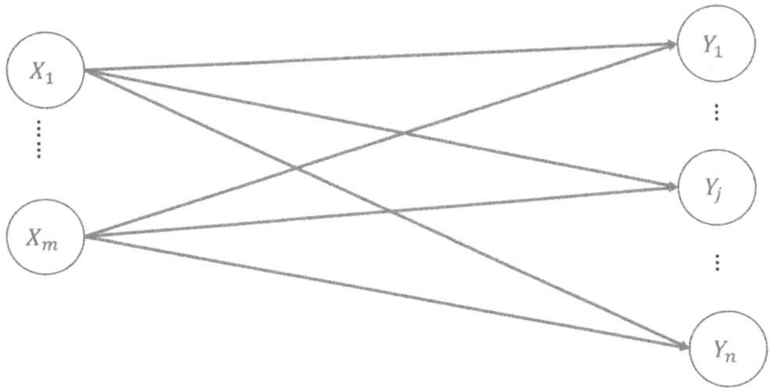

FIGURE 5.3 Scheme of vector X causing vector Y.

where

$$f_Y(X)=\begin{bmatrix} f_{Y1}(X_1,\ldots,X_m) \\ \vdots \\ f_{Yn}(X_1,\ldots,X_m)\end{bmatrix}, N_Y=\begin{bmatrix} N_{Y1} \\ \vdots \\ N_{Yn}\end{bmatrix},$$

$$X_i \perp\!\!\!\perp N_{Yj}, N_{Yk} \perp\!\!\!\perp N_{Yl}, \forall i,j,k,l.$$

If there is

$$X = f_X(Y)+N_X, Y \perp\!\!\!\perp N_X, N_{xk} \perp\!\!\!\perp N_{xl}, \forall k,l, \quad (5.44)$$

where

$$f_X(Y)=\begin{bmatrix} f_{X1}(Y_1,\ldots,Y_n) \\ \vdots \\ f_{X_m}(Y_1,\ldots,Y_n)\end{bmatrix}, N_X=\begin{bmatrix} N_{X1} \\ \vdots \\ N_{Xm}\end{bmatrix},$$

$$Y_i \perp\!\!\!\perp N_{Xj}, N_{xk} \perp\!\!\!\perp N_{xl}, \forall k,l, \forall \mathrm{i,j,k,l},$$

then the model (5.44) is a reversible vector ANM from Y to X.

5.1.4 VAE and ANMs for Causal Analysis

Several methods for bivariate causal discovery have been developed. These methods include the linear non-Gaussian acyclic model (LiNGAM) (Shimizu et al. 2006), the ANMs, the information-geometric approach for causal inference (IGCI) (Janzing et al. 2012), unsupervised inverse regression (CURE) (Sgouritsa et al. 2015), the SLOPE algorithm using regression to estimate the Kolmogorov complexities (Marx and Vreeken 2017), Regression Error based Causal Inference (RECI) (Bloebaum et al. 2018). This book will focus on application of AI to causal inference. Therefore, in this section we will introduce application of VAE to the ANMs for causal inference.

The ANMs require fitting a nonlinear regression and testing independence of the cause and regression residuals. It is well known that the VAE allows us to efficiently learn deep latent-variable models and to fit the data very well (Khemakhem et al. 2019). Now we study how to use VAE to fit the functional model and evaluate the causal relationships between the cause and effect (Cai et al. 2019).

5.1.4.1 Evidence Lower Bound (ELBO) for ANM

The ANMs consist of three parts: (1) the selection of nonlinear function, (2) estimation of parameters, and (3) testing independence between the cause and residue. The widely used methods for parameter estimation in statistics are maximum likelihood estimation. A key issue for maximum likelihood estimation is to compute likelihood for the ANMs. In this section, we will derive the likelihood function for the ANMs and use variational principle to approximate the likelihood function, which leads to the evidence lower bound (ELBO) for the ANMs (Kingma and Welling 2013; Rezende et al. 2014; Cai et al. 2019).

The ANMs have three quantities: the observed potential cause X, the effect Y, and unobserved noise N_Y, which can be taken as a latent variable. The likelihood function for the ANMs is denoted by $P_\theta(X,Y)$. The data are generated by the ANMs, involving an unobserved continuous random variable N_Y. For the convenience of presentation, N_Y is denoted by Z. The unknown distribution of Z is denoted by $P_\theta(Z)$. We assume that the latent variable Z is generated from prior distribution $P_\theta(Z)$ and the observed data (X,Y) are generated from conditional distribution $P_\theta(X,Y \mid Z)$. Thus, the joint distribution of the observed variables X,Y and latent variable Z is given by

$$P_\theta(X,Y,Z)=P_\theta(X,Y \mid Z)P_\theta(Z), \quad (5.45)$$

where the conditional distribution $P_\theta(X,Y \mid Z)$ is often implemented by neural networks (NNs).

Since the distribution of the latent variable is unknown, the marginal likelihood

$$P_\theta(X,Y)=\int P_\theta(X,Y \mid Z)P_\theta(Z)dZ \quad (5.46)$$

can model a rich class of data distribution $P_\theta(X,Y)$, but is intractable. Since the posterior distribution

$P_\theta(Z|X,Y\} = \frac{P_\theta(Z)P_\theta(X,Y|Z)}{P_\theta(X,Y)}$ involves the marginal distribution $P_\theta(X,Y)$, the posterior distribution $P_\theta(Z|X,Y\}$ is also intractable.

Recall that in Chapter 3, the latent variable is interpreted as code, the VAE consists of encoder that maps the observed variables X and Y to the code Z in the latent space and decoder that maps the latent variable Z back to the original observed variables X and Y.

Assume that data $D = \{X^{(1)}, Y^{(1)}, \ldots, X^{(n)}, Y^{(n)}\}$ are given. The log-likelihood of marginal distribution is

$$\log P_\theta\left(X^{(1)}, Y^{(1)}, \ldots, X^{(n)}, Y^{(n)}\right) = \sum_{i=1}^{n} \log P_\theta\left(X^{(i)}, Y^{(i)}\right). \tag{5.47}$$

It can be shown (Appendix 5A) that

$$\log P_\theta\left(X^{(i)}, Y^{(i)}\right) \geq \tilde{\mathcal{L}}\left(\theta, \varnothing, X^{(i)}, Y^{(i)}\right), \tag{5.48}$$

where

$$\tilde{\mathcal{L}}\left(\theta, \varnothing, X^{(i)}, Y^{(i)}\right) = E_{q_\varnothing(Z|X^{(i)}, Y^{(i)})}\left[\log \frac{P_\theta\left(X^{(i)}, Y^{(i)}, Z\right)}{q_\varnothing\left(Z|X^{(i)}, Y^{(i)}\right)}\right]. \tag{5.49}$$

Equation (5.49) can be reduced to

$$\tilde{\mathcal{L}}\left(\theta, \varnothing, X^{(i)}, Y^{(i)}\right) = E_{q_\varnothing(Z|X^{(i)}, Y^{(i)})}\left[\log P_\theta\left(X^{(i)}, Y^{(i)} \mid Z\right)\right] - KL\left(q_\varnothing\left(Z|X^{(i)}, Y^{(i)}\right) \| P_\theta(Z)\right). \tag{5.50}$$

Equation (5.50) can be further reduced to (Appendix 5A)

$$\tilde{\mathcal{L}}\left(\theta, \varnothing, X^{(i)}, Y^{(i)}\right) = \log P_\theta\left(X^{(i)}\right) + E_{q_\varnothing(Z|X^{(i)}, Y^{(i)})}\left[\log P_\theta\left(Y^{(i)} = f\left(X^{(i)}, Z\right)\right)\right] - KL\left(q_\varnothing\left(Z \mid X^{(i)}, Y^{(i)}\right) \| P_\theta(Z)\right). \tag{5.51}$$

Since $P_\theta\left(X^{(i)}\right)$ does not involve the latent variable Z, it can be ignored in optimization of $\tilde{\mathcal{L}}\left(\theta, \varnothing, X^{(i)}, Y^{(i)}\right)$. Equation (5.51) can be simplified to

$$\tilde{\mathcal{L}}\left(\theta, \varnothing, X^{(i)}, Y^{(i)}\right) = E_{q_\varnothing(Z|X^{(i)}, Y^{(i)})}\left[\log P_\theta\left(Y^{(i)} = f\left(X^{(i)}, Z\right)\right)\right] - KL\left(q_\varnothing\left(Z|X^{(i)}, Y^{(i)}\right) \| P_\theta(Z)\right). \tag{5.52}$$

Thus, the ELBO is

$$\mathcal{L}(\theta, \varnothing, X, Y) = \frac{n}{M} \sum_{i=1}^{M} \tilde{\mathcal{L}}\left(\theta, \varnothing, X^{(i)}, Y^{(i)}\right), \tag{5.53}$$

where M is the sample size in the minibatch.

5.1.4.2 Computation of the ELBO

The Monte Carlo estimator of the ELBO for single point is (Appendix 5B)

$$\tilde{\mathcal{L}}\left(\theta, \varnothing, X^{(i)}, Y^{(i)}\right) = \frac{1}{L} \sum_{l=1}^{L} \log P_\theta\left(Y^{(i)} = f\left(X^{(i)}, Z^{(i,l)}\right)\right) - KL\left(q_\varnothing\left(Z|X^{(i)}, Y^{(i)}\right) \| P_\theta(Z)\right), \tag{5.54}$$

where the first term

$Z^{(i,l)} = g_\varnothing\left(\varepsilon^l, X^i, Y^{(i)}\right)$, $\varepsilon^{(l)} \sim P(\varepsilon)$, $i = 1, \ldots, n$, $l = 1, \ldots, L$, and $g_\varnothing(\varepsilon, X, Y)$ be a differentiable transformation function.

The ELBO for the entire dataset is

$$\mathcal{L}(\theta, \varnothing, X, Y) = \frac{n}{M} \sum_{i=1}^{M} \tilde{\mathcal{L}}\left(\theta, \varnothing, X^{(i)}, Y^{(i)}\right). \tag{5.55}$$

Using equation (5.54) to computer the ELBO, we need to define the transformation function $g_\varnothing\left(\varepsilon^l, X^i, Y^{(i)}\right)$, which is referred to as encoder (Figure 5.4). The encoder maps the observed data (X,Y) to the latent space. For any "location-scale" family of distributions, a simple way to define the transformation function is to set

$$Z^{(i,l)} = \mu^{(i)} + \sigma^{(i)} \varepsilon^{(l)}, \quad \varepsilon^{(l)} \sim N(0,1). \tag{5.56}$$

The distribution of the variable $Z^{(i,l)}$ approximates the posterior distribution $q_\varnothing\left(Z^{(l)} \mid X^{(i)}, Y^{(i)}\right)$. In this setting,

$$q_\varnothing\left(Z^{(l)} \mid X^{(i)}, Y^{(i)}\right) = N\left(Z^{(l)}; \mu^{(i)}, \left(\sigma^{(i)}\right)^2\right), \tag{5.57}$$

where $\mu^{(i)}, \left(\sigma^{(i)}\right)^2$ are nonlinear functions of $X^{(i)}, Y^{(i)}$, which are implemented by NNs. Multilayered perceptrons (MLPs) are typical simple NNs to implement nonlinear transformation.

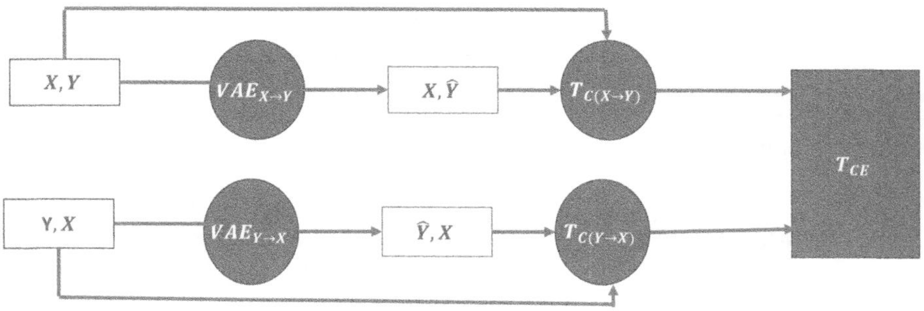

FIGURE 5.4 ANM variational autoencoder.

Example 5.6 Two Layer MLP for Encoder (Figure 5.4)

A typical two layer MLP for encoder is given below.

$$\log q_{\varnothing}\left(Z^{(l)} \mid X^{(i)}, Y^{(i)}\right)=\log N\left(Z^{(l)}, \mu^{(i)},\left(\sigma^{(i)}\right)^{2}\right) \tag{5.58}$$

$$\mu^{(i)}=W_{\mu} h^{i}+\sigma_{\mu} \tag{5.59}$$

$$\log\left(\sigma^{(i)}\right)^{2}=W_{\sigma} h^{i}+b_{\sigma} \tag{5.60}$$

$$h^{i}=\tanh\left(W_{h}\left(\begin{array}{c} X^{(i)} \\ Y^{(i)} \end{array}\right)+\left(\begin{array}{c} b_{x} \\ b_{y} \end{array}\right)\right) \tag{5.61}$$

$$Z^{(i,l)}=\mu^{(i)}+\sigma^{(i)} \varepsilon^{(l)}, \varepsilon^{(l)} \sim N(0,1). \tag{5.62}$$

Next we introduce decoder which reconstructs the causal relationships using the latent variables. To achieve this, we first define $P_{\theta}(Y \mid X, Z)$ to be a Gaussian for real-valued data or Bernoulli for binary data. The distribution parameters are computed from the observed cause X and latent variable Z with a neural network. Here is a typical 2-year MLP to implement decoder.

Example 5.7 Two Layer MLP for Decoder with Continuous Data (Figure 5.4)

The equations for implementing MLP for decoder with continuous variable Y are given below.

$$\begin{aligned} &\log P_{\theta}\left(Y^{(i,l)} | X^{(i)}, Z^{(l)}\right) \\ &=\log N\left(Y^{(i,l)}, \mu\left(X^{(i)}, Z^{(l)}\right), \sigma^{2}\left(X^{(i)}, Z^{(l)}\right)\right) \end{aligned} \tag{5.63}$$

$$\mu\left(X^{(i)}, Z^{(l)}\right)=W_{1} h_{1}^{(i,l)}+b_{1} \tag{5.64}$$

$$\log \sigma^{2}\left(X^{(i)}, Z^{(l)}\right)=W_{2} h^{(i,l)}+b_{2} \tag{5.65}$$

$$h^{(i,l)}=\tanh\left(W_{3}\left[\begin{array}{c} X^{(i)} \\ Z^{(l)} \end{array}\right]+b_{3}\right). \tag{5.66}$$

Example 5.8 Two Layer MLP for Decoder with Binary Data (Figure 5.4)

For the binary data, $P_{\theta}\left(Y^{(i)} | X^{(i)}, Z^{(l)}\right)$ follows Bernoulli distribution. The equations for implementing MLP for decoder with binary data are defined as follows.

$$\begin{aligned} &\log P_{\theta}\left(Y^{(i)} | X^{(i)}, Z^{(l)}\right) \\ &=Y^{(i)} \log g^{(i)}+\left(1-Y^{(i)}\right) \log\left(1-g^{(i)}\right) \end{aligned} \tag{5.67}$$

$$g^{(i)}=f_{\sigma}\left(W_{5} \tanh\left(W_{4}\left[\begin{array}{c} X^{(i)} \\ Z^{(l)} \end{array}\right]+b_{4}\right)+b_{5}\right). \tag{5.68}$$

5.1.4.3 Computation of the KL Distance Now we calculate $KL\left(q_{\varnothing}\left(Z | X^{(i)}, Y^{(i)}\right) \| P_{\theta}(Z)\right)$ in equation (5.54). Under some assumptions, the KL distance can often be integrated analytically. We assume that $P_{\theta}(Z)=N(0,1)$. Recall equation (5.53):

$$q_{\varnothing}\left(Z^{(l)} \mid X^{(i)}, Y^{(i)}\right)=N\left(Z^{(l)} ; \mu^{(i)},\left(\sigma^{(i)}\right)^{2}\right).$$

In Appendix 5C, we show

$$\begin{aligned} &KL\left(q_{\varnothing}\left(Z | X^{(i)}, Y^{(i)}\right) \| P_{\theta}(Z)\right) \\ &=-\frac{1}{2}\left[1+\log\left(\sigma^{(i)}\right)^{2}-\left(\mu^{(i)}\right)^{2}-\left(\sigma^{(i)}\right)^{2}\right]. \end{aligned} \tag{5.69}$$

Substituting equation (5.69) into equation (5.54), we obtain

$$\tilde{\mathcal{L}}\left(\theta, \varnothing, X^{(i)}, Y^{(i)}\right)=\frac{1}{L}\sum_{l=1}^{L}\log P_\theta\left(Y^{(i)}=f\left(X^{(i)},Z^{(i,l)}\right)\right.$$

$$+\frac{1}{2}\begin{bmatrix}1+\log\left(\left(\sigma^{(i)}\right)^2\right)\\ -\left(\mu^{(i)}\right)^2-\left(\sigma^{(i)}\right)^2\end{bmatrix}, \quad (5.70)$$

and

$$\mathcal{L}(\theta, \varnothing, X, Y)=\frac{n}{M}\sum_{i=1}^{M}\tilde{\mathcal{L}}\left(\theta, \varnothing, X^{(i)}, Y^{(i)}\right), \quad (5.71)$$

where if the decoder is two layer MLP with continuous effect, then the first term in equation (5.70) $\log P_\theta\left(Y^{(i)}=f\left(X^{(i)}+Z^{(i,l)}\right)\right.$ is calculated by equations (5.63)– (5.66), and if the decoder is two layer MLP with binary effect, then the first term in equation (5.70) $\log P_\theta(Y^{(i)}=f\left(X^{(i)}+Z^{(i,l)}\right)$ is calculated by equations (5.67) and (5.68).

Now it is easy to calculate the gradient at one data point as follows:

$$\nabla_\theta\tilde{\mathcal{L}}\left(\theta, \varnothing, X^{(i)}, Y^{(i)}\right)=\frac{1}{2}\sum_{l=1}^{L}\frac{\nabla_\theta P_\theta\left(Y^{(i)}=f\left(X^{(i)}, Z^{(i,l)}\right)\right.}{P_\theta\left(Y^{(i)}=f\left(X^{(i)}, Z^{(i,l)}\right)\right)}, \quad (5.72)$$

$$\nabla_\varnothing\tilde{\mathcal{L}}\left(\theta, \varnothing, X^{(i)}, Y^{(i)}\right)=\begin{bmatrix}-\mu^{(i)}\\ \dfrac{1}{\sigma^{(i)}}-\sigma^{(i)}\end{bmatrix}. \quad (5.73)$$

Iteration of the parameters using stochastic gradient is given by

$$\begin{bmatrix}\theta^{(k+1)}\\ \varnothing^{(k+1)}\end{bmatrix}=\begin{bmatrix}\theta^{(k)}\\ \varnothing^{(k)}\end{bmatrix}+\gamma\begin{bmatrix}\nabla_\theta\tilde{\mathcal{L}}\left(\theta, \varnothing, X^{(i)}, Y^{(i)}\right)\\ \nabla_\varnothing\tilde{\mathcal{L}}\left(\theta, \varnothing, X^{(i)}, Y^{(i)}\right)\end{bmatrix}, \quad (5.74)$$

where γ is a learning rate.

5.1.5 Classifier Two-Sample Test for Causation

Two-sample tests that include the t-test, the Wilcoxon-Mann-Whitney test, the Kolmogorov-Smirnov tests, kernel methods, Maximum Mean Discrepancy (MMD test (Gretton et al. 2012), the Mean Embedding test (Jitkrittum et al. 2016), and classifier two-sample test (Lopez-Paz and Oquab 2017) can be used to assess causal direction (Lopez-Paz and Oquab 2017). Below we adapt the classifier two-sample test that has already been applied to the bivariate causal discovery using conditional generative adversarial network (CGAN) and VAE (Lopez-Paz and Oquab 2017).

If $X \to Y$, then the output $\hat{Y}$ fits the model $\hat{Y}=f_Y\left(X, \hat{N}_Y\right)$ very well. We are unable to distinguish the fitted dataset $D_{X\to Y}=\left\{X^{(i)}, f_Y\left(X^{(i)}, \left(N_Y\right)^{(i)}\right), i=1,\ldots,n\right\}$ and the original dataset $D_t=\left\{X^{(i)}, Y^{(i)}, i=1,\ldots,n\right\}$. Let the distribution of $D_{X\to Y}$ be P and the distribution of D_t be Q. The null hypothesis of $X \to Y$ is then transferred to $P=Q$. Since $X^{(i)}, i=1,\ldots,n$ in two datasets are the same, we remove them from two datasets. For simplicity, let $u_i=\hat{Y}^{(i)}$ and $v_i=Y^{(i)}$. We also assign label to the data. Define

$$l=\begin{cases}1 & u_i\\ 0 & v_i\end{cases}.$$

Define the dataset

$$D=\left\{\left[\left(u_1,1\right),\ldots,\left(u_n,1\right)\right]\cup\left[\left(v_1,0\right),\ldots,\left(v_n,0\right)\right]\right\}$$
$$=\left\{\left(h_1,l_1\right),\ldots,\left(h_{2n},l_{2n}\right)\right\}.$$

The data points in the set D are randomly shuffled. The shuffled data are split it into the disjoint training and testing subsets $D=D_{tr}\cup D_{te}$. Train a binary classy classifier that can be implemented by a neural network or any classical classifier on the training dataset D_{tr}. Let $n_{te}=\left|D_{te}\right|$. Then, the trained classifier is applied to the test dataset D_{te}. Let $P\left(l_i=1\,|\,h_i\right)$ be the probability of classifying the sample that is from the fitted data as $l=1$ and $f\left(h_i\right)$ be its estimator. If $f\left(h_i\right)>\frac{1}{2}$, then $l_i=1$ is assigned.

Define the test statistic as the classification accuracy that is the proportion of correctly classified samples:

$$T_{C(X\to Y)}=\frac{1}{n_{te}}\sum_{(h_i,l_i)\in D_{te}} w_i, \quad (5.75)$$

where $w_i=I\left[I\left(f\left(h_i\right)>\frac{1}{2}\right)=l_i\right]$ and I is an indicator function.

The random variable w_i follows an independent Bernoulli (P_i) distribution. Under the null hypothesis

$$H_0 : P = Q,$$

the probability of classifying correctly the example h_i should be equal to $\frac{1}{2}\left(P_i = \frac{1}{2}\right)$. Using the central limit theorem, under the null hypothesis, the classifier two-sample test statistic $T_{c(X\to Y)}$ asymptotically follows a normal distribution $N\left(\frac{1}{2}, \frac{1}{4n_{te}}\right)$. Intuitively, the closer to 0.5 the test statistic $T_{c(X\to Y)}$, the less accurate the binary classifier, i.e., the closer two distributions. Therefore, the closer to 0.5 the test statistic $T_{c(X\to Y)}$, the stronger evidence that $X \to Y$.

Now we consider two causal directions: $X \to Y$ and $Y \to X$, let $u_i = \hat{X}^{(i)}$ and $v_i = X^{(i)}$. Define

$$T_{C(Y\to X)} = \frac{1}{n_{te}} \sum_{(h_i, l_i)\in D_{te}} g_i, \tag{5.76}$$

where $g_i = I\left[I\left(f(h_i) > \frac{1}{2}\right) = l_i\right]$.

There are four cases to consider for testing causation: (1) $T_{C(X\to Y)}$ is small, but $T_{C(Y\to X)}$ is large (Only $X \to Y$ is possible), (2) $T_{C(Y\to X)}$ is small, but $T_{C(X\to Y)}$ is large (Only $Y \to X$) is possible, (3) both $T_{C(X\to Y)}$ and $T_{C(Y\to X)}$ are small (both $X \to Y$ and $Y \to X$ are possible, it is inclusive), and (4) both $T_{C(X\to Y)}$ and $T_{C(Y\to X)}$ are large (No causation). Only cases (1) and (2) are true causal. This implies that $|T_{C(X\to Y)} - T_{C(Y\to X)}|$ should be large. Now we develop statistics to test for it.

The null hypothesis is that the following conditions hold.

$$H_0 :$$

1. The distributions of $\hat{Y}$ and Y are indistinguishable, and
2. The distributions of $\hat{X}$ and X are indistinguishable.

We can show that under the null hypothesis H_0, the variance $\sigma^2 = var\left(T_{c(X\to Y)} - T_{C(Y\to X)}\right)$ can be estimated by

$$\sigma^2 = 2\left(\frac{1}{4n_{te}} - \frac{\sum_{i=1}^{n_{te}}\left(w_i - T_{c(X\to y)}\right)\left(g_i - T_{c(Y\to X)}\right)}{n_{te} - 1}\right). \tag{5.77}$$

The statistic for testing the causation between X and Y is defined as

$$T_{CE} = \frac{\left(T_{C(X\to Y)} - T_{C(Y\to X)}\right)^2}{\sigma^2}. \tag{5.78}$$

Under the null hypothesis, T_{CE} is asymptotically distributed as a central $\chi^2_{(1)}$ distribution.

Now we summarize the procedures for testing the causation between two variables, which is referred to as VAE Classifier Two Sample Test (VCTST).

5.1.5.1 *Procedures of the VCTEST (Figure 5.5)*

1. Use VAE to fit the ANMs $X \to Y$ ($\hat{Y} = f_Y(X, N_Y)$) and generate the dataset $D_{X\to Y} = \left\{\left(X_i, \hat{Y}_i\right)\right\}_{i=1}^{n}$.

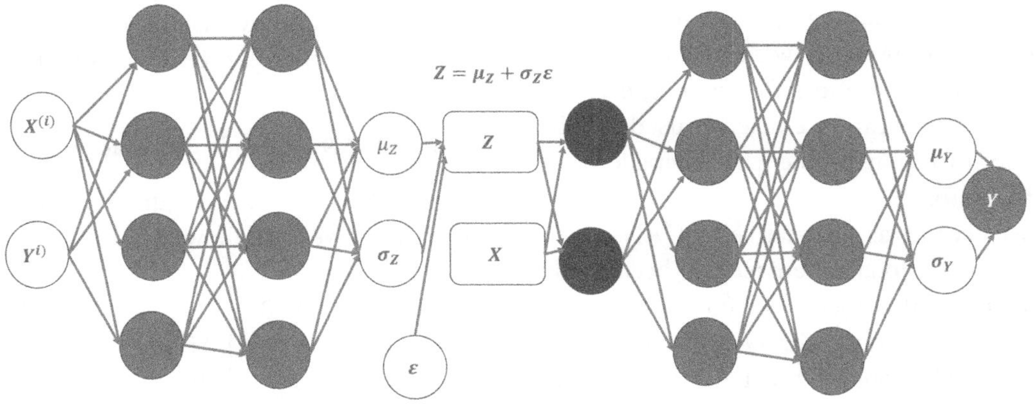

FIGURE 5.5 Scheme of classifier two-sample test for causation using VAE.

2. Split the data into the training dataset and testing dataset: $D_{tr(X \to Y)}$ and $D_{te(X \to Y)}$. Use $D_{tr(X \to Y)}$ to train the classifier and apply the trained classifier to the $D_{te(X \to Y)}$ to calculate $T_{C(X \to Y)} = \frac{1}{n_{te}} \sum_{(h_i, l_i) \in D_{te}} w_i$.
3. Use VAE to fit the ANMs $Y \to X$ ($\hat{X} = f_X(Y, N_X)$ and generate the dataset $D_{Y \to X} = \left\{ (\widehat{X}_i, Y_i) \right\}_{i=1}^{n}$.
4. Split the data into the training dataset and testing dataset: $D_{tr(Y \to X)}$ and $D_{te(Y \to X)}$. Use $D_{tr(Y \to X)}$ to train the classifier, and apply the trained classifier to the $D_{te(Y \to X)}$ and calculate $T_{C(Y \to X)} = \frac{1}{n_{te}} \sum_{(h_i, l_i) \in D_{te}} g_i$.
5. Calculate $T_{CE} = \frac{\left(T_{C(X \to Y)} - T_{C(Y \to X)}\right)^2}{\sigma^2}$ and test for the causation.
6. If test shows the significant causation, then determine the cause direction. If $T_{C(X \to Y)} < T_{C(Y \to X)}$ then $X \to Y$. Otherwise, $Y \to X$.

5.2 LEARNING STRUCTURAL CAUSAL MODELS WITH GRAPH NEURAL NETWORKS

There are two basic approaches to learning SCM: constrain-based approaches and score-based approaches (Yu et al. 2019). Constraint-based methods learn causal structure via testing for conditional independencies and for d-separations. Score-based methods formulate a causal network learning problem as a combinatorial optimization problem (Bühlmann et al. 2014; Manzour et al. 2019; Hu et al. 2020). The intractable search space seriously limits the size of causal graphs, which we attempt to reconstruct. To overcome combinatorial optimization limitation and to make the causal graph search tractable, Zheng et al. (2018) formulate causal graph learning as a continuous optimization problem with linear structural equations and least square loss function. The widely used gradient methods can be efficiently used to solve continuous optimization problem (Lachapelle et al. 2019. Yu et al. 2019).

In the past decade, as alternative to GANs, great progress in the variational inference has been made. The graph variational autoencoder (GVAE) has two remarkable features. Firstly, the GVAE can capture the complex distributions of the DAG in the data. Secondly, the GVAE can incorporate the model into their mathematic formulation (Yu et al. 2019). We observe the successful applications of the GVAE to causal inference (Zhang et al. 2020). In this section, we will introduce two GVAE-based approaches to the causal graph learning. The first approach of the GVAE is to take the graph structure as input (Lachapelle et al. 2019; Yu et al. 2019). The second approach of the GVAE is to explicitly map causal graph structure to the latent space, where latent variables form causal graphs (Leeb et al. 2020; Shen et al. 2020; Xie et al. 2020; Yang et al. 2020).

5.2.1 A General Framework for Formulation of Causal Inference into Continuous Optimization

The classical methods for learning DAGs are often formulated as a discrete optimization problem, where the search space of DAGs is combinatorial, and hence the number of search increases superexponentially with the number of nodes (Zheng et al. 2018). Therefore, inferring DAGs is an NP hard problem (Chickering et al. 2004). The problem comes from the acyclic constraint. Zheng et al. (2018) formulate acyclic constraint in term of smooth continuous function, avoiding combinatorial formulization.

5.2.1.1 Score Function and New Acyclic Constraint

The score-based methods for learning DAGs consists of two parts: score (or objective function) and constraint. The likelihood functions or some loss functions are often used as a score. Consider a DAG (V, E) with $M = |V|$ nodes. A d-dimensional feature vector $Y_i \in R^d$ is associated with each node and $Y \in R^{M \times d}$ be a feature matrix of M nodes. If there is an edge from node i to node j, then weight is denoted as a_{ij}. Let $A = \left(a_{ij}\right)_{M \times M}$ be the weighted adjacency matrix. For an unweighted adjacency matrix, if an edge is directed from node i to node j, then $a_{ij} = 1$, otherwise, $a_{ij} = 0$.

Example 5.9 Example of Adjacency Matrix

The DAG is shown in Figure 5.6. The adjacency matrix is given as follows:

$$A = \begin{bmatrix} 0 & a_{12} & 0 & 0 \\ 0 & 0 & 0 & 0 \\ a_{31} & 0 & 0 & a_{34} \\ a_{41} & 0 & 0 & 0 \end{bmatrix} \text{ and } A^T = \begin{bmatrix} 0 & 0 & a_{31} & a_{41} \\ a_{12} & 0 & 0 & 0 \\ 0 & 0 & 0 & 0 \\ 0 & 0 & a_{34} & 0 \end{bmatrix}.$$

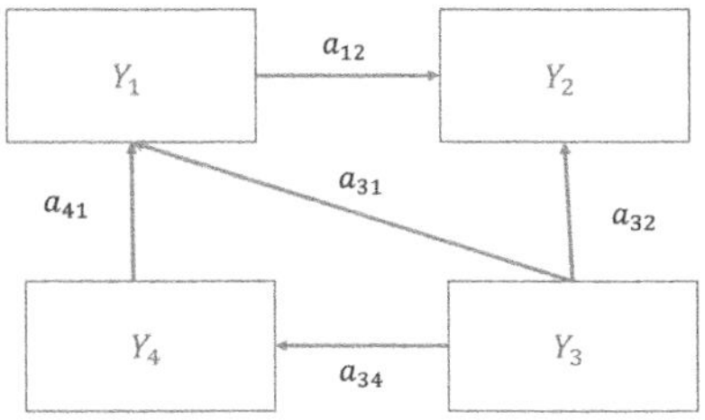

FIGURE 5.6 Example of DAG.

Equation for Figure 5.6 is

$$Y_1 = a_{31}Y_3 + a_{41}Y_4$$
$$Y_2 = a_{12}Y_1$$

$Y_4 = a_{34}Y_3$, which can be written in a matrix form:

$$Y = \begin{bmatrix} Y_1 \\ Y_2 \\ Y_3 \\ Y_4 \end{bmatrix} = A^T Y = \begin{bmatrix} 0 & 0 & a_{31} & a_{41} \\ a_{12} & 0 & 0 & 0 \\ 0 & 0 & 0 & 0 \\ 0 & 0 & a_{34} & 0 \end{bmatrix} \begin{bmatrix} Y_1 \\ Y_2 \\ Y_3 \\ Y_4 \end{bmatrix}.$$

A linear structural equation for modeling a DAG can be written as

$$Y = A^T Y + Z, \quad (5.79)$$

where Z is a vector of noises and are uncorrelated with Y.

Linear structural equation (5.79) contains only node variables which are often called endogenous variables. In practice, we may also need exogenous variables. The SEM can also be extended to general SEM, including exogenous variables. Variables in a system can be classified into two basic types of variables: observed variables that can be measured and the residual error variables that cannot be measured and represent all other unmodeled causes of the variables. Most observed variables are random. Some observed variables may be nonrandom or control variables (e.g., drug dosage) whose values remain the same in repeated random sampling or might be manipulated by the experimenter. The observed variables will be further classified into exogenous variables, which lie outside the model, and endogenous variables, whose values are determined through joint interaction with other variables within the system. All nonrandom can be viewed as exogenous variables. The terms exogenous and endogenous are model specific. It may be that an exogenous variable in one model is endogenous in another.

Let $Y = [Y_1, \ldots, Y_M]^T$ be a vector of the M endogenous variables and $X = [X_1, \ldots, X_K]^T$ be a vector of q exogenous variables. Occasionally, one or more of the X's are nonrandom. We denote the errors by e. We assume that $E[e] = 0$ and that e is uncorrelated with the exogenous variables in X. We also assume that e_i is homoscedastic and nonautocorrelated (BOLLEN 1989; Xiong 2018b). Assume that the sample size is n, the number of endogenous variables is M, and the number of exogenous variables is k. Then, the structural equations for modeling a system are given by

$$Y = Y\Gamma + XB + E, \quad (5.80)$$

where

$$Y = \begin{bmatrix} y_{11} & \cdots & Y_{1M} \\ \vdots & \vdots & \vdots \\ y_{n1} & \cdots & Y_{nM} \end{bmatrix} = [Y_1, \ldots, Y_M]$$

$$X = \begin{bmatrix} X_{11} & \cdots & X_{1K} \\ \vdots & \vdots & \vdots \\ X_{n1} & \cdots & X_{nK} \end{bmatrix} = [X_1, \ldots, X_K]$$

$$\Gamma = \begin{bmatrix} \gamma_{11} & \cdots & \gamma_{1M} \\ \vdots & \vdots & \vdots \\ \gamma_{M1} & \cdots & \gamma_{MM} \end{bmatrix} = [\Gamma_1, \ldots, \Gamma_M],$$

$$B = \begin{bmatrix} \beta_{11} & \cdots & \beta_{1M} \\ \vdots & \vdots & \vdots \\ \beta_{K1} & \cdots & \beta_{KM} \end{bmatrix} = [B_1, \ldots, B_M], \text{ and}$$

$$E = \begin{bmatrix} e_{11} & \cdots & e_{1M} \\ \vdots & \vdots & \vdots \\ e_{n1} & \cdots & e_{nM} \end{bmatrix} = [e_1, \ldots, e_M].$$

Example 5.10 Yeast Cell Cycle

In Figure 5.7, we assume that the expression levels of the genes Cdc 28, Clb1, and CLb3, denoted

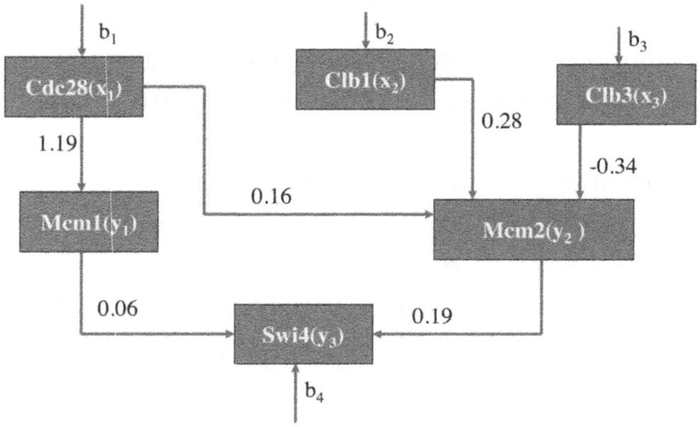

FIGURE 5.7 Small genetic network of yeast cell cycle.

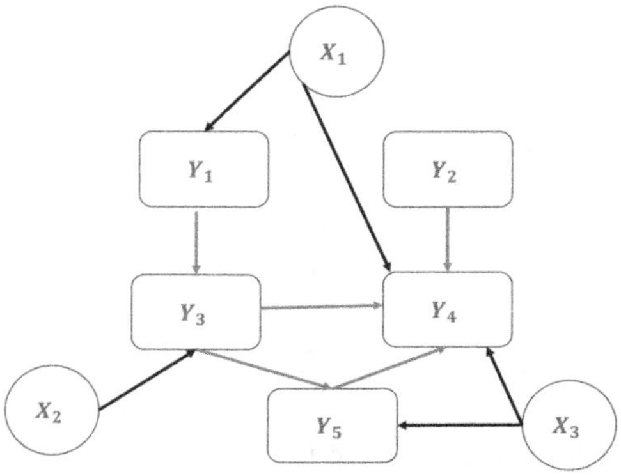

FIGURE 5.8 Graph representation of SEM.

by x_1, x_2, and x_3, respectively, are exogenous variables and the expression levels of the genes Mcm1, Mcm2, and Swi4 are denoted by y_1, y_2, and $y_3$respectively, are endogenous variables. The structural equations for the small genetic network of yeast cell cycle are written as

$$y_1 = 1.19x_1 + e_1$$

$$y_2 = 0.16x_1 + 0.28x_2 - 0.34x_3 + e_2$$

$$y_3 = 0.06y_1 + 0.19y_2 + e_3.$$

Assume that the sample size is n. Let

$$Y = \begin{bmatrix} y_{11} & y_{12} & y_{13} \\ \vdots & \vdots & \vdots \\ y_{n1} & y_{n2} & y_{n3} \end{bmatrix}, X = \begin{bmatrix} x_{11} & x_{12} & x_{13} \\ \vdots & \vdots & \vdots \\ x_{n1} & x_{n2} & x_{n3} \end{bmatrix},$$

$$E = \begin{bmatrix} e_{11} & e_{12} & e_{13} \\ \vdots & \vdots & \vdots \\ e_{n1} & e_{n2} & e_{n3} \end{bmatrix}, \Gamma = \begin{bmatrix} 0 & 0 & 0.06 \\ 0 & 0 & 0.19 \\ 0 & 0 & 0 \end{bmatrix},$$

$$B = \begin{bmatrix} 1.19 & 0.16 & 0 \\ 0 & 0.28 & 0 \\ 0 & -0.34 & 0 \end{bmatrix}.$$

SEM can be written in a matrix form (Figure 5.8)

$$Y = Y\Gamma + XB + E.$$

5.2.2 Parameter Estimation and Optimization

After matrix exponential is introduced as a constraint in Section 5.2.4, we can form the following optimization problem (5.81) for learning DAGs using the SEMs:

$$\begin{aligned} &\min_{\Gamma \in R^{M\times M},\, B \in R^{K\times M}} l(\Gamma, B, \lambda; X, Y) \\ &= \frac{1}{2n}\|Y - Y\Gamma - XB\|_F^2 + \lambda_1\|\Gamma\|_1 + \lambda_2\|B\|_1, \end{aligned} \tag{5.81}$$

Subject to $h(\Gamma) = \text{Tr}(e^{\Gamma \circ \Gamma}) - M = 0$,

where $e^{\Gamma \circ \Gamma}$ is a matrix exponential and 0 denotes the Hadamard product of matrices.

Now the combinatorial optimization problem needs to be transformed to equality constrained continuous, but nonsmooth optimization problem. Solutions consist of two major steps: (1) using Lagrange multiplier method to transform the equality constrained optimization problem into a sequence of non-constrained optimization problem and (2) solving non-smooth optimization problem.

5.2.2.1 *Transform the Equality Constrained Optimization Problem into Unconstrained Optimization Problem*

The classical augmented Lagrange multiple method (Bertsekas 1996) is used to transform the constrained optimization (5.81) into the unconstrained optimization problem:

The primal problem:

$$\begin{aligned} D(\alpha) &= \min_W \mathcal{L}^\rho(\Gamma, B, \lambda, \rho, \alpha; X, Y) \\ &= l(\Gamma, B, \lambda; X, Y) + \frac{\rho}{2}|h(\Gamma)|^2 + \alpha h(\Gamma), \end{aligned} \tag{5.82a}$$

where $l(\Gamma, B, \lambda; X,Y) = \frac{1}{2n}\|Y - Y\Gamma - XB\|_F^2 + \lambda_1\|\Gamma\|_1 + \lambda_2\|B\|_1$, $\frac{\rho}{2}|h(\Gamma)|^2$ is an augmentation term and $\alpha h(\Gamma)$ is a Lagrange term.

The dual problem:

$$\max_{\alpha} D(\alpha). \tag{5.82b}$$

The optimization problem (5.82a) can be separated into smooth optimization and nonsmooth optimization problem:

$$\mathcal{L}^{\rho}(\Gamma, B, \lambda, \rho, \alpha; X,Y) = f(\Gamma, B, \rho, \alpha; X,Y) + \lambda_1\|\Gamma\|_1 + \lambda_2\|B\|_1, \tag{5.83}$$

where

$$f(\Gamma, B, \rho, \alpha; X,Y) = \frac{1}{2n}\|Y - Y\Gamma - XB\|_F^2 + \frac{\rho}{2}|h(\Gamma)|^2 + \alpha h(\Gamma)$$

is the smooth part of the objective function. For simplicity of notation, we define $\|\Gamma\|_1 = \|vec(\Gamma)\|_1$ and $\|B_1\| = \|vec(B)\|_1$. Let $W = \begin{bmatrix}\Gamma \\ B\end{bmatrix} = \begin{bmatrix}W_1 \\ W_2\end{bmatrix}$ and $\omega = vec(W)$. Then, equation (5.83) can be rewritten as

$$\mathcal{L}^{\rho}(W, \lambda, \rho, \alpha; X,Y) = f(W, \rho, \alpha; X,Y) + \lambda\|W_1\|, \tag{5.84a}$$

where

$$f(W, \rho, \alpha; X,Y) = \frac{1}{2n}Y - ZW_F^2 + \frac{\rho}{2}|h(\Gamma)|^2 + \alpha h(\Gamma), Z = [Y, X]. \tag{5.84b}$$

The classic Newton's method is an efficient method for solving unconstrained smooth optimization problem, but is not appropriate for solving nonsmooth optimization problem. We will introduce proximal methods which can be viewed as an extension of Newton's method from solving smooth optimization problems to nonsmooth optimization problems (Parikh and Boyd 2013). In general, the optimization problem (5.83) can be solved by the proximal method (Bach et al. 2012; Parikh and Boyd 2013). In other words, to solve the optimization problem (5.83), at each iteration we often expand the function $f(\Gamma, B, \rho, \alpha; X,Y)$ in a neighborhood of the current iterate (Γ_i, B_i) by a Taylor expansion:

$$f(\omega, \rho, \alpha; X,Y) \approx f(\omega_i, \rho, \alpha; X,Y) + (\nabla_w^T f)(\omega_i - \omega) + \frac{1}{2}(\omega_i - \omega)^T H_i(\omega_i - \omega), \tag{5.85}$$

where $f(\omega_i, \rho, \alpha; X,Y) = \frac{1}{2n}\|vec(Y) - (I \otimes Z)vec(W)\|_2^2 + \frac{\rho}{2}|h(\Gamma)|^2 + \alpha h(\Gamma)$, H_i is the Hessian matrix and ω is viewed as $vec(W)$.

The Hessian matrix H_i can be approximated by BFGS correction (Byrd et al. 1987). We can show that BFGS correction E_i is given by (Appendix 5D),

$$E_{i+1} = E_i + \frac{g_i g_i^T}{g_i^T s_i} - \frac{E_i s_i s_i^T E_i}{s_i^T E_i s_i}, \tag{5.86}$$

where

$$g_i = \nabla_{\omega} f(\omega_{i+1}) - \nabla_{\omega} f(\omega_i), \tag{5.87}$$

$$s_i = \omega_{i+1} - \omega_i. \tag{5.88}$$

Thus, equation (5.85) can be reduced to

$$f(\omega, \rho, \alpha; X,Y) \approx f(\omega_i, \rho, \alpha; X,Y) + q_i^T s_i + \frac{1}{2}s_i^T E_i s_i, \tag{5.89}$$

where $q_i = \nabla_{\omega} f(\omega_i)$.

Combining equations (5.83) and (5.89), optimization problem (5.83) at iteration i can be reduced to

$$\mathcal{L}(\omega, \lambda, d) = q_i^T d + \frac{1}{2}d^T E_i d + \lambda|\omega_i + d|_1. \tag{5.90}$$

Coordinate descent algorithms, which solve optimization problems by successively perform optimization along coordinate directions and often have a closed form solution, can be used to solve nonsmooth optimization problem (5.90). Let e_j be a unit vector with the jth component being 1 and all other components being zero. For simplicity, the changes of the jth component of $\omega_i + d$ is denoted by $\omega_j + d_j + ze_i$, where z is a search variable. For the jth component, equation (5.90) is reduced to

$$\min_z \frac{1}{2}E_{jj}z^2 + (q_j + (Ed)_j)z + \lambda\|\omega_j + d_j + z\|_1. \tag{5.91a}$$

The solution is (Appendix 5E)

$$d_j^{i+1} = d_j^i + z^*, \quad (5.91b)$$

where

$$z^* = -\omega_j - d_j + \text{sign}\left(\omega_j + d_j - \frac{q_j + (Ed)_j}{E_{jj}}\right)$$
$$\left(\left|\omega_j + d_j - \frac{q_j + (Ed)_j}{E_{jj}}\right| - \frac{\lambda}{E_{jj}}\right)_+, \quad (5.92)$$
$$(a)_+ = \max(0, a).$$

5.2.2.2 *Compact Representation for the Hessian Approximation E_k and Limited-Memory-BFGS*

To improve computation and save memory, in Appendix 5D, we introduce the compact representation for the Hessian approximation E_k:

$$E_k = E_0 - \begin{bmatrix} E_0 S_k & Y_k \end{bmatrix} \begin{bmatrix} S_k^T E_0 S_k & L_k \\ L_k^T & -D_k \end{bmatrix}^{-1} \begin{bmatrix} S_k^T E_0 \\ Y_k^T \end{bmatrix}, \quad (5.93)$$

where L_k is defined as

$$(L_k)_{ij} = \begin{cases} s_{i-1}^T g_{j-1} & i > j \\ 0 \text{ otherwise} \end{cases}, \quad (5.94)$$

$$g_k = \nabla_\omega f(\omega_{k+1}) - \nabla_\omega f(\omega_k), \; s_k = \omega_{k+1} - \omega_k,$$
$$S_k = [s_0, \ldots, s_{k-1}], \; Y_k = [g_0, \ldots, g_{k-1}],$$
$$D_k = \text{diag}(s_0^T g_0, \ldots, s_{k-1}^T g_{k-1}),$$

E_0 is often set as $E_0 = \gamma_k I$ and $\gamma_k = \frac{g_{k-1}^T s_{k-1}}{s_{k-1}^T s_{k-1}}$.
Let

$$Q = \begin{bmatrix} E_0 S_k & Y_k \end{bmatrix}, \; R = \begin{bmatrix} S_k^T E_0 S_k & L_k \\ L_k^T & -D_k \end{bmatrix}^{-1}, \; \tilde{Q} = RQ^T.$$

Then, equation (5.93) can be rewritten as

$$E_k = E_0 - QRQ^T = E_0 - Q\tilde{Q}. \quad (5.95)$$

Equations (5.93) or (5.95) shows that k BFGS updates for the Hessian approximation E_k can be written in the compact forms, BFGS updates have limited memory implementation (Byrd et al. 1994). The data for E_k corrections only require pairs (s_i, g_i). We can keep the m most recent pairs and use equations (5.93) or (5.95) to implement iterations. At $k+1$ iteration, we remove the most old pair (s_0, g_0) and add the a newly generated pair (s_k, g_k) in to form a new matrices $S_{k+1} = [s_{k-m+1}, \ldots, s_k]$, $Y_{k+1} = [g_{k-m+1}, \ldots, g_k]$ so that the matrices Q and $\tilde{Q}$ have the fixed dimensions $d \times 2m$ and $2m \times d$, respectively. Therefore, $Q(\tilde{Q})$) is updated just on the columns (rows).

Algorithm 5.1 Zheng et al. 2018

Step 1: Input initial values (W_0, α_0), learning rate $c \in (0,1)$, tolerance error $\varepsilon > 0$, and threshold $\omega_h > 0$.

Step 2: For $i = 0, 1, 2, \ldots$ **Do**

1. Solve the primal problem:
   $W_{i+1} \leftarrow \underset{W}{\text{argmin}}\, \mathcal{L}^\rho(W, \alpha_i)$, defined in equations (5.84a) and (5.84b), with ρ such that $h(W_{i+1}) < ch(W_i)$. The primal problem is equivalent to solving the non-smooth optimization problem:

$$\mathcal{L}^\rho(W\; \lambda, \rho, \alpha; X, Y) = f(W, \rho, \alpha; X, Y) + \lambda \|W\|_1,$$

   $f(W, \rho, \alpha; X, Y) = \frac{1}{2n}\|Y - ZW\|_F^2 + \frac{\rho}{2}|h(\Gamma)|^2 + \alpha h(\Gamma)$, $Z = [Y, X]$, which can be solved by coordinate decent method (Wright 2015). Performed algorithm 2 for Proximal Quasi-Newton to solve the unconstrained optimization problem.

Algorithm 5.2

1. Input initial values ω_0 and gradient $\frac{\partial f(\omega_0)}{\partial \omega}$ and active set $S = [p]$.
2. For $k = 0, 1, 2, \ldots$
   a. Remove components with $\omega_j = 0$ or small subgradient $|\partial_j \mathcal{L}^\rho(\omega)|$ from the activist S and shrink S
   b. Check shrinking stopping criterion. If it is satisfied then
      i. Rest $S = [p]$ and update approximation Hessian matrix by Limited-memory BFGS (Section 5.2.3.2) to obtain E_k.
      ii. Using equations (5.87)–(5.89) to calculate $\omega_j$and $|\partial_j \mathcal{L}^\rho(\omega)|$. Update shrinking stopping criteria and continue.

c. Solve the optimization problem (5.90) for decent direction

$$d_k = \underset{d}{\operatorname{argmin}}\, \mathcal{L}(\omega, \lambda, d)$$

and using solution (5.92) to update $d \leftarrow d + z^* e_j$ on activing set S.

d. Using the classical line search for step size η until Armijo rule is satisfied:

$$f(\omega_k + \eta d_k) \le f(\omega_k) + \eta c_1 \left(\lambda \|\omega_k + d_k\|_1 - \lambda \|\omega\|_k + q_k^T d_k \right),$$

where c_1 is a prespecified small constant, usually set to 10^{-3} or 10^{-4}.

e. Update weights $\omega_{k+1} \leftarrow \omega_k + \eta d_k$.

f. Check convergence. If $\|\omega_{k+1} - \omega_k\| \le \varepsilon$, then stop, otherwise, Update parameters $q, s, g, Q, \tilde{Q}$ restricted to S, go to (a).

3. Solve the dual problem.
   Dual update: $\alpha_{i+1} \leftarrow \alpha_i + \rho h(W_{i+1})$.
4. If $h(W_{i+1}) < \varepsilon$, then set $\tilde{W}_s = W_{i+1}$, and break.
   Return

Step 3: Return the threshold matrix $\hat{W} = \tilde{W}_s \circ I\left(|\tilde{W}| > \omega_h\right)$.

In Algorithm 5.2, we need to use the following formula for calculation of gradients.

$$\frac{\partial f}{\partial \omega} = \begin{bmatrix} \dfrac{\partial f}{\partial w_1} \\ \dfrac{\partial f}{\partial w_2} \end{bmatrix},$$

$$\frac{\partial f}{\partial w_1} = -\frac{1}{n}\left[I_{(2M)\times(2M)} 0 \right](I \otimes Z^T)(vec(Y)) - (I \otimes Z)\omega) + \rho vec\left(h(\Gamma)(e^{\Gamma \circ \Gamma})^T \circ (2\Gamma)\right) + \alpha vec\left((e^{\Gamma \circ \Gamma})^T \circ (2\Gamma)\right).$$

$$\frac{\partial f}{\partial w_2} = \frac{1}{2}\left[0 \;\; I_{(KM)\times(KM)} \right]\left[I_{(2M)\times(2M)} \;\; 0 \right] \times \left[(I \otimes Z^T)(vec(Y) - (I \otimes Z)\omega) \right].$$

5.2.3 VAE for Learning Structural Models and DAG among Observed Variables

VAE can be used to learn causal relations between observed variables (Yu et al. 2019) or latent variables (Xie et al. 2020). In this section, we introduce VAE for learning DAG structure models with observed variables (Yu et al. 2019).

5.2.3.1 Linear Structure Equation Model and Graph Neural Network Model

VAE generalizes linear SEMs and learns the weighted adjacency matrix of a DAG. Mathematical representation of a DAG and linear SEMs are defined as that in Section 5.2.1.1. A linear SEMs with d dimensional feature vector is given by

$$Y = A^T Y + BX + Z, \tag{5.96}$$

where $Y \in R^{m \times d}$, $A \in R^{m \times m}$ is a weighted adjacency matrix, $B \in R^{m \times m_1}$, $X \in R^{m_1 \times d}$, $Z \in R^{m \times d}$ and elements in X can take either continuous or categorical values. If $I - A^T$ is non-singular, then equation (5.96) can be rewritten as

$$Y = (I - A^T)^{-1}(BX + Z). \tag{5.97}$$

Equation (5.96) can be generalized to a semi-nonlinear SEMs:

$$f_1^{-1}(Y) = A^T f_1^{-1}(Y) + Bf_2(X) + f_3(Z)), \tag{5.98}$$

where f_1, f_2, and f_3 are nonlinear functions and can be implemented by NNs. Similar to equation (5.97), we can obtain from equation (5.98) that

$$Y = f_1\left(\left((I - A^T)^{-1}(Bf_2(X) + f_3(Z))\right)\right) \tag{5.99}$$

and

$$Z = f_3^{-1}\left((I - A^T) f_1^{-1}(Y) - Bf_2(X)\right). \tag{5.100}$$

Since equation (5.99) is complicated, it is not easy to directly write likelihood function of the variables Y. We will use VAE to approximate likelihood function $P_\theta(Y)$ and estimate the structures and parameters.

5.2.3.2 ELBO for Learning the Generative Model

Assume that n data point $(Y^i \in R^{m \times d}, X^i \in R^{m \times d})$, $i = 1, \ldots, n$ are sampled. Let $Y = [Y^1, Y^2, \ldots, Y^n]$ and $X = [X^1, X^2, \ldots, X^n]$. The marginal likelihood can be written as $\log p_\theta(Y^1, X^1, \ldots, Y^n, X^n) = \sum_{i=1}^n \log p_\theta(Y^i, X^i)$, if the data are independently sampled. We can show (Appendix 5F) that the

ELBO with one observed data point (Y^i, X^i) for learning the SEMs is given by

$$\log p_\theta(Y^i, X^i) \geq \mathcal{L}(\theta, \varnothing, Y^i, X^i), \quad (5.101)$$

where

$$\mathcal{L}(\theta, \varnothing, Y^i, X^i) = E_{q_\varnothing(Z|Y^i, X^i)}\left[\log p_\theta(Y^i, X^i \mid Z)\right] - KL\left(q_\varnothing(Z|Y^i, X^i) \| p_\theta(Z)\right), \quad (5.102)$$

$q_\varnothing(Z \mid Y^i, X^i)$ is a variational posterior which approximate the true posterior $p_\theta(Z \mid Y^i, X^i)$, $p_\theta(Z)$ is a prior distribution of latent variables Z and $p_\theta(Y^i, X^i \mid Z)$ is a conditional distribution of observed variable (Y^i, X^i), given Z. Quantity $\mathcal{L}(\theta, \varnothing, Y^i, X^i)$ is referred to as the ELBO.

The ELBO provides a general framework for VAE. The VAE consists of encoder and decoder. The posterior $p_\theta(Z \mid Y^i, X^i)$ represents to encode the observed endogenous variables Y and exogenous variables X into latent variables Z and the conditional distribution $p_\theta(Y, X \mid Z)$ represents to decode the latent variables Z back to the original variables Y and X.

5.2.3.3 Computation of ELBO

5.2.3.3.1 Encoder The architecture of computing ELBO is shown in Figure 5.9. We first study encoder and discuss how to calculate the second term KL distance in equation (5.102). Recall that $Z \in R^{m \times d}$. Instead of assuming classical normal distribution, we assume that the prior $p_\theta(Z)$ follows a matrix normal distribution $MN_{md}(0, I, I)$(Appendix 3C). Its density function is given by

$$p_\theta(Z) = (2\pi)^{-\frac{md}{2}} exp\left\{-\frac{1}{2}\mathrm{Tr}\left(vec(Z)\left(vec(Z)\right)^T\right)\right\}, \text{ or} \quad (5.103)$$

$$\log p_\theta(Z) = -\frac{md}{2}\log(2\pi) - \frac{1}{2}\mathrm{Tr}(vec(Z)(vec(Z))^T. \quad (5.104)$$

Next we assume that the variational approximate posterior $q_\varnothing(Z \mid Y, X)$ has the following matrix normal $MN_{md}(\mu_z, \Omega_z, V_z)$ (Appendix 3C):

$$q_\varnothing(Z|Y, X) = (2\pi)^{-\frac{md}{2}} \left|\Sigma_z\right|^{-\frac{1}{2}} exp\left\{-\frac{1}{2}\mathrm{Tr}\left(\Sigma_z^{-1}(vec(Z-\mu)\left(vec(Z-\mu)\right)^T\right)\right\}, \quad (5.105)$$

where $\Sigma_z = \Omega_z \otimes V_z$.

If we assume that

$$\Sigma_z = \mathrm{diag}\left(S_{11}, \ldots, S_{m1}, S_{12}, \ldots, S_{m2}, \ldots, S_{1d,\ldots,} S_{md}\right) = S_z, \; S_{ij} = \Omega_{ii} V_{jj}.$$

We show in Appendix 5F that

$$-KL\left(q_\varnothing(Z^n|Y^n, X^n) \| p(Z^n)\right) = \frac{1}{2}\left(\sum_{i=1}^{m}\sum_{j=1}^{d}\left(\log(S_Z^n)_{ij} + 1 - (S_Z^n)_{ij} - (\mu_Z^n)_{ij}^2\right)\right). \quad (5.106)$$

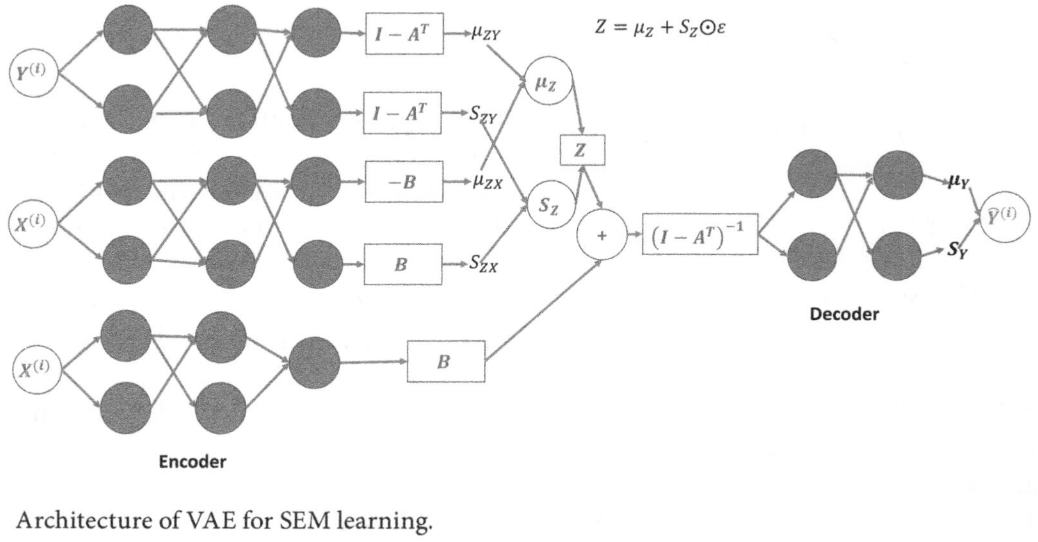

FIGURE 5.9 Architecture of VAE for SEM learning.

If we assume that $d=1$, then equation (5.106) is reduced to

$$-KL(q_{\varnothing}(Z^n|Y^n,X^n)\|p(Z^n))$$
$$=\frac{1}{2}\sum_{j=1}^{m}\left(\log(\sigma^n)_j^2+1-(\sigma^n)_j^2-(\mu^n)_j^2\right). \quad (5.107)$$

Define the transformation function (Appendix 5F):

$$Z^n=g_{\varnothing}(Z^n|Y^n,X^n)=\mu_Z^n+\Sigma_Z^n\odot\varepsilon,\ \varepsilon\sim MN(0,I,I). \quad (5.108)$$

Assume that the distribution $p_{\theta}(Z^n|Y^n,X^n)=g_{\varnothing}(Z^n|Y^n,X^n)$. Recall equation (5.100):

$$Z^n=f_3^{-1}\left((I-A^T)f_1^{-1}(Y^n)-Bf_2(X^n)\right).$$

The two-layer NNs with Matrix normal output, and the ReLU and tanh activation function as the encoder is defined as (Figure 5.9)

$$\left[(\mu^n)_Y\mid S_Y^n\right]=(I-A^T)\text{ReLu}(Y^nW^1)W^2,$$
$$\left[(\mu^n)_X\mid S_X^n\right]=B\tanh(X^nW^3)W^4,$$
$$\mu_Z^n=\mu_Y^X-\mu_X^n,$$
$$S_Z^n=S_Y^n+S_X^n,$$

where W^1, W^2, W^3 and W^4 are weight matrices in the NNs.

5.2.3.3.2 Decoder Next we study decoder and discuss how to calculate the first term in equation (5.102). The decoder is defined as

$$p_{\theta}(Y^n,X^n|Z^n)=p_{\theta_1}(Y^n|X^n,Z^n)p_{\theta_2}(X^n|Z^n)$$
$$=p_{\theta_1}(Y^n|X^n,Z^n)p_{\theta_2}(X^n), \quad (5.109)$$

where we assume that X and $Z$are independent.

We assume that distribution $p_{\theta_1}(Y^n\mid X^n,Z^n)$ is matrix normal distribution:

$$p_{\theta_1}(Y^n|X^n,Z^n)=(2\pi)^{-\frac{md}{2}}\left|S_Y^n\right|^{-\frac{1}{2}}$$
$$\exp\left\{-\frac{1}{2}\left(vec(Y^n-\mu_Y^n)\right)^T S_Y^{-1}vec(Y^n-\mu_Y^n)\right\}, \quad (5.110)$$

where $S_Y^n=\text{diag}\left((S_Y^n)_{11},\ldots,(S_Y^n)_{m1},(S_Y^n)_{12},\ldots,(S_Y^n)_{1d},(S_Y^n)_{md}\right)$.

Thus, $\log p_{\theta_1}(Y^n|X^n,Z^n)$ is given by

$$\log p_{\theta_1}(Y^n|X^n,Z^n)$$
$$=-\frac{md}{2}\log(2\pi)-\frac{1}{2}\log\left|S_Y^n\right|-\frac{1}{2}\left(vec(Y^n-\mu_Y^n)\right)^T$$
$$S_Y^{-1}vec(Y^n-\mu_Y^n)=-\frac{md}{2}\log(2\pi)$$
$$-\frac{1}{2}\left(\sum_{i=1}^{m}\sum_{j=1}^{d}\left(\log(S_Y^n)_{ij}+\frac{\left(Y_{ij}^n-(\mu_Y^n)_{ij}\right)^2}{(S_Y^n)_{ij}}\right)\right). \quad (5.111)$$

Sampling latent variable Z:

$$Z_n^l=\mu_Z^n+S_Z^n\odot\varepsilon^l,\ \varepsilon^l\sim MN_{md}(0,I,I),\ l=1,2,\ldots,L. \quad (5.112)$$

The Monte Carlo estimation of $E_{q_{\varnothing}(Z^n|Y^n,X^n)}\left[\log p_{\theta_1}(Y^n|X^n,Z^n)\right]$ is given by

$$E_{q_{\varnothing}(Z^n|Y^n,X^n)}\left[\log p_{\theta_1}(Y^n|X^n,Z^n)\right]$$
$$\approx\frac{1}{L}\sum_{l=1}^{L}\log p_{\theta_1}(Y^n\mid X^n,Z_n^l)$$
$$=-\frac{1}{L}\sum_{l=1}^{L}\left(\frac{md}{2}\log(2\pi)\right.$$
$$\left.+\frac{1}{2}\left(\sum_{i=1}^{m}\sum_{j=1}^{d}\left(\log(S_Y^n)_{ij}^l+\frac{\left(Y_{ij}^n-(\mu_Y^n)_{ij}^l\right)^2}{(S_y^n)_{ij}^l}\right)\right)\right). \quad (5.113)$$

Recall that

$$Y=f_1\left(\left((I-A^T)^{-1}(Bf_2(X)+f_3(Z))\right)\right).$$

We assume that f_3 is an identity mapping, f_1 and f_2 are implemented by two-layer NNs. Then, the two-layer NNs for generating μ_Y and $S_Y$in equation (5.113) with ReLU and tanh as the activation function are given by (Figure 5.9),

$$\left[\mu_Y^n|S_Y^n\right]$$
$$=\text{ReLu}\left((I-A^T)^{-1}\left(B\tanh(X^nW^5)W^6)+Z^n\right)W^7\right)W^8\Big) \quad (5.114)$$

where W^5, W^6, W^7, and W^8 are weight matrices. Finally, combining equations (5.102), (5.106), and (5.113), we obtain the estimation of ELBO:

$$\mathcal{L}(\theta,\varnothing,Y^n,X^n)=\frac{1}{L}\sum_{l=1}^{L}\log p_\theta\left(Y^n,X^n\mid Z^{n,l}\right)$$

$$-KL\left(q_\varnothing\left(Z|Y^n,X^n\right)||p_\theta(Z)\right)=-\frac{1}{L}\sum_{l=1}^{L}$$

$$\left(\frac{md}{2}\log(2\pi)+\frac{1}{2}\left(\sum_{i=1}^{m}\sum_{j=1}^{d}\left(\begin{array}{c}\log\left(S_Y^n\right)_{ij}^l\\ +\dfrac{\left(Y_{ij}^n-\left(\mu_Y^n\right)_{ij}^l\right)^2}{\left(S_Y^n\right)_{ij}^l}\end{array}\right)\right)\right)$$

$$+\frac{1}{2}\left(\sum_{i=1}^{m}\sum_{j=1}^{d}\left(\log\left(S_Z^n\right)_{ij}+1-\left(S_Z^n\right)_{ij}-\left(\mu_Z^n\right)_{ij}^2\right)\right), \quad (5.115)$$

where

$$\left[\mu_Y^n\mid S_Y\right]=\left(I-A^T\right)\mathrm{ReLu}\left(Y^nW^1\right)W^2,$$

$$\left[\mu_X^n\mid S_X^n\right]=B\tanh\left(X^nW^3\right)W^4,$$

$$\mu_Z^n=\mu_Y^n-\mu_X^n,$$

$$S_Z^n=S_Y^n+S_X^n,$$

$$Z_n^l=\mu_Z^n+S_Z^n\odot\varepsilon^l,\ \varepsilon^l\sim MN_{md}(0,I,I),\ l=1,2,\ldots,L,$$

$$\left[\mu_{\hat{Y}}^{n,l}|S_{\hat{Y}}^{n,l}\right]=\mathrm{ReLu}((I-A^T)^{-1}(B\tanh(X^nW^5)W^6)+Z_n^l)W^7)W^8).$$

Next we consider categorical endogenous variables Y. The first term in equation (5.115) can be viewed as the reconstruction error and the second term can be viewed as a regularization term.

In Appendix 5F, we show that the ELBO for categorical variables is given by

$$\mathcal{L}(\theta,\varnothing,Y^n,X^n)\approx\frac{1}{L}\sum_{l=1}^{L}\sum_{i=1}^{m}\sum_{j=1}^{d}Y_{ij}^n\log\left(\left(p_Y^n\right)^l\right)_{ij}$$

$$+\frac{1}{2}\left(\sum_{i=1}^{m}\sum_{j=1}^{d}\left(\log\left(S_Z^n\right)_{ij}+1-\left(S_Z^n\right)_{ij}-\left(\mu_Z^n\right)_{ij}^2\right)\right), \quad (5.116)$$

where the probability matrix p_Y^n is generated

$$p_Y^{n,l}=softmax(\mathrm{ReLu}((I-A^T)^{-1}(B\tanh(X^nW^5)W^6)+Z_n^l)W^7)W^8)),$$

where W^5, W^6, W^7, and W^8 are weight matrices.

The ELBO for entire dataset is

$$\mathcal{L}(\theta,\varnothing,Y,X)=\frac{1}{N}\sum_{n=1}^{N}\mathcal{L}(\theta,\varnothing,Y^n,X^n). \quad (5.117)$$

5.2.3.4 Optimization Formulation for Learning DAG

In this section, we show that the ELBO is a nonlinear function of the adjacency matrix A and coefficient matrix B. Maximization of ELBO attempts to fit the data for making the estimators of the matrices A and B as close the true matrices A and B as possible. However, the estimated adjacency matrix A cannot ensure that the estimated directed graph is acyclic. We need to impose constraint to guarantee the acyclicity of the estimated graph. In the following, we will illustrate that the ELBO has close connection with the loss function of SEMs (Yu et al. 2019) and introduce more convenient constraint for acyclicity than the proposed by Zeng et al. (2018).

5.2.3.4.1 Connection of the ELBO with the Loss Function of the SEM

In order to establish the connection of the ELBO with the loss function of the SEM, we first simplify the formulation of the ELBO. Define $\hat{Y}_{ij}=\left(\mu_Y\right)_{ij}$ and assume that $S_Y=1$. Then, the reconstruction error (the first term) in equation (5.115) is reduced to

$$-\left(\frac{md}{2}\log(2\pi)+\frac{1}{2L}\sum_{i=1}^{m}\sum_{j=1}^{d}\left(Y_{ij}-\hat{Y}_{ij}\right)^2\right).$$

$$=-\left(\frac{md}{2}\log(2\pi)+\frac{1}{2L}\left\|Y-\hat{Y}\right\|_F^2\right). \quad (5.118)$$

If we define $Z_{ij}=\left(\mu_Z\right)_{ij}$ and assume that $S_Z=1$, then the second term in equation (5.115) is reduced to

$$-\frac{1}{2}\sum_{i=1}^{m}\sum_{j=1}^{d}Z_{ij}^2=-\|Z\|_F^2. \quad (5.119)$$

If we assume that $f_1 = f_2 = f_3 = 1$, then equation (5.100) is reduced to

$$Z = (I - A^T)Y - BX. \tag{5.120}$$

Substituting equation (5.120) into equation (5.119), we obtain

$$-\|Y - A^T Y - BX\|_F^2. \tag{5.121}$$

which approximates $l(\Gamma, B; X, Y)$.

Combining equations (5.118) and (5.121), we obtain

$$\begin{aligned} -ELBO &\approx \|Y - A^T Y - BX\|_F^2 \\ &+ \left(\frac{md}{2} \log(2\pi) + \frac{1}{2L} \|Y - \hat{Y}\|_F^2 \right) \\ &\approx l(\Gamma, B; X, Y) + \left(\frac{md}{2} \log(2\pi) + \frac{1}{2L} \|Y - \hat{Y}\|_F^2 \right). \end{aligned} \tag{5.122}$$

Comparing with loss function (5.81), the first term $l(\Gamma, B; X, Y)$ in equation (5.122) is the OLS loss function in equation (5.81). The second term in equation (5.122) also plays similar regularization role as that the second term in equation (5.81) plays.

5.2.3.4.2 Acyclicity Constraint The VAE intends to fit the model to data, but cannot ensure the corresponding graph of the resulting estimated adjacency matrix is acyclic. To overcome this limitation, in Section 5.2.2 we introduced the equality constraint to ensure the acyclicity of the inferred graph (Zheng et al. 2018). In this section, we will introduce a modified equality constraint to ensure the acyclicity of the resulting graph which was proposed by Yu et al. (2019).

We can show that if all elements in the adjacency matrix A are positive, then $\mathrm{Tr}(A^k)$ counts the number of the length-k closed path. The existence of a cycle indicates that the number of the length-m closed path is not equal to zero. Therefore, the corresponding directed graph of the adjacency matrix A is acyclic if and only if $\mathrm{Tr}(A^m) = 0$. To ensure nonnegativity of the elements of the matrix, we can define $C = A \circ A$, where $\circ$ indicates the element-wise multiplication. Yu et al. (2019) introduced the following constraint to characterize the acyclicity of the graph:

$$h(A) = \mathrm{Tr}\left[(I + \alpha A \circ A)^m\right] - m = 0. \tag{5.123}$$

5.2.3.4.3 Optimization Formulation for Learning a DAG Learning a DAG defined by SEMs can be formulated as the following equality constrained optimization problem (Yu et al. 2019):

$$\min_{A,B,\theta,\varnothing} F(A,B,\theta,\varnothing) = -ELBO = -\mathcal{L}(\theta,\varnothing,Y,X) \tag{5.124}$$

Subject to $h(A) = \mathrm{Tr}\left[(I + \alpha A \circ A)^m\right] - m = 0$,

where the unknown parameters include the matrices A, B and all parameters θ of the VAE. All techniques for solving optimization problem can be applied here.

To solve the equality constrained optimization problem (5.124), we use augmented Lagrange method and duality theory in nonlinear programming (Geoffrion 1971). The optimization problem (5.124) is called the primal problem. Using Lagrange multipliers to add constraint $h(A)$ to the object function, we define the augmented Lagrangian:

$$\begin{aligned} \mathcal{L}^\rho(A, B, \theta, \varnothing, \rho, \lambda) &= F(A, B, \theta, \varnothing) \\ &+ \frac{\rho}{2}(h(A))^2 + \lambda h(A), \end{aligned} \tag{5.125}$$

where ρ is a penalty parameter and λ is a Lagrange multiplier. Increasing ρ will gradually enforce constraint $h(A)$ or acyclicity of the graph. When $\rho = +\infty$ the minimizer will enforce $h(A) = 0$. Then, the objective function $\mathcal{L}^\rho(A,B,\theta,\varnothing,\rho,\lambda)$ is reduced to $-\mathcal{L}(\theta,\varnothing,Y,X)$. The constrained optimization problem (5.124) is solved using duality theory and augmented Lagrangian method (Geoffrion 1971).

Define Lagrange dual function:

$$D(\lambda) = \min_{A,B,\theta,\varnothing} \mathcal{L}^\rho(A,B,\theta,\varnothing,\rho,\lambda). \tag{5.126}$$

Define

$$\max_\lambda D((\lambda)) = \max_\lambda \min_{A,B,\theta.\varnothing} \mathcal{L}^\rho(A,B,\theta,\varnothing,\rho,\lambda).$$

Let F^* be the optimal value of the primal problem:

$$\min_{A,B,\theta,\varnothing} F(A,B,\theta,\varnothing) = -ELBO = -\mathcal{L}(\theta,\varnothing,Y,X) \tag{5.127}$$

Subject to $h(A) = \mathrm{Tr}\left[(I + \alpha A \circ A)^m\right] - m = 0$.

Then, under some conditions, we have

$$\max_{\lambda} D((\lambda)) = F^*. \quad (5.128)$$

Solution procedures are iterated between the primal problem and dual problem. For fixed Lagrangian λ_k, we solve the primal problem using stochastic gradient method:

$$[A_k^*, B_k^*, \theta_k^*, \varnothing_k^*] = \underset{A,B,\theta,\varnothing}{\operatorname{argmin}} \mathcal{L}^{\rho}(A,B,\theta,\varnothing, \rho, \lambda). \quad (5.129)$$

Then, for fixed $A_k^*, B_k^*, \theta_k^*, \varnothing_k^*$, we solve the dual problem using gradient ascent method. The gradient is given by

$$\frac{\partial \mathcal{L}^{\rho}(A_k^*, B_k^*, \theta_k^*, \varnothing_k^*, \rho_k^*, \lambda)}{\partial \lambda} = h(A_k^*). \quad (5.130)$$

The Lagrangian λ_k is updated by ascend:

$$\lambda_{k+1} = \lambda_k + \gamma_k h(A_k^*), \quad (5.131)$$

where γ_k is a tune learning rate.

The penalty parameter ρ_k is updated by (Yu et al. 2019)

$$\rho_{k+1} = \begin{cases} \eta\rho_k & if\ |h(A_k)| > \xi |h(A_{k-1})| \\ \rho_k & otherwise \end{cases}, \quad (5.132)$$

where $\eta > 1$ and $\xi < 1$.

Algorithm for solving primal-dual is summarized as Algorithm 5.3.

Algorithm 5.3

Step 1: Initialization. Input initial values $(A_0, B_0, \theta_0, \phi_0)$ and tolerance $\varepsilon > 0$.

Step 2: For $k = 0, 1, 2, \ldots$:

1. Solving the primal $[A_k^*, B_k^*, \theta_k^*, \varnothing_k^*] = \underset{A,B,\theta,\varnothing}{\operatorname{argmin}} \mathcal{L}^{\rho}(A,B,\theta,\varnothing, \rho, \lambda)$.
2. Solving the dual by ascent

$$\lambda_{k+1} = \lambda_k + \gamma_k h(A_k^*),$$

$$\rho_{k+1} = \begin{cases} \eta\rho_k & if\ |h(A_k)| > \xi |h(A_{k-1})| \\ \rho_k & otherwise \end{cases},$$

where $\eta > 1$ and $\xi < 1$.

3. If $h(A_{k+1}) < \varepsilon$, then set

$$A^* = A_{k+1}, B^* = B_{k+1}, \theta^* = \theta_{k+1}, \text{and } \phi^* = \phi_{k+1}.$$

Step 3: Return the network. Set $\hat{A} = A^* \circ I[|A^*| > \omega_A]$ and $\hat{B} = B^* \circ I[|B^*| > \omega_b]$.

5.2.4 Loss Function and Acyclicity Constraint

If we assume that the error E and Y are uncorrelated, and E and X are uncorrelated, then the ordinary least-squares (OLS) methods can be used to estimate the parameters in the SEM. However, when the residuals E and Y are correlated, the OLS for parameter estimation, in general, will be inconsistent. Other methods, including two-stage least square methods and instrument variable methods should be used (Xiong 2018b; Bun and Harrison 2019). This section will focus on the OLS loss function combined with acyclicity constraints and penalty function for sparse SEM.

5.2.4.1 OLS Loss Function

The OLS loss function for linear SME is defined as

$$l(\Gamma, B; X, Y) = \frac{1}{2n} \|Y - Y\Gamma - XB\|_F^2, \quad (5.133)$$

where n is the number of sampled data points, $\|.\|_F$ is the Frobenius norm of the matrix. It can be proved that under some conditions, the minimizer of the OLS loss can infer a true DAG and is consistent for both Gaussian and non-Gaussian SEM (Loh and Buhlmann 2014; Aragam et al. 2016; Zheng et al. 2018).

In practice, DAGs, in general, are sparse, which implies that the corresponding SEMs are sparse. To develop sparse SEMs, we should incorporate the penalty terms into the loss function. Define L_1 norm of the matrix W as $\|W\|_1 = \sum_{i=1}^{p} \sum_{j=1}^{q} |W_{ij}|$. To ensure that DAGs are sparse, we add L_1-penalization term to the loss function $l(\Gamma, B; X, Y)$, leading to (Zheng et al. 2018)

$$l(\Gamma, B, \lambda; X, Y) = \frac{1}{2n} \|Y - Y\Gamma - XB\|_F^2 + \lambda_1 \|\Gamma\|_1 + \lambda_2 \|B\|_1, \quad (5.134)$$

where λ_1, λ_2 are penalty parameters.

Only minimization of the loss function $l(\Gamma, B, \lambda; X, Y)$ cannot ensure that the inferred graphs are acyclic. To make the inferred graphs acyclic, we need to impose acyclicity constraints. Let $G(\Gamma)$ be inferred directed graphs and DA be a set of DAGs. Imposing acyclicity constraints leads to the following optimization problem for inferring DAGs:

$$\min_{\Gamma\in R^{M\times M},\, B\in R^{K\times M}} l(\Gamma, B, \lambda; X, Y) \tag{5.135}$$

$$\text{Subject to } G(\Gamma)\in DA.$$

The acyclicity constraints $G(\Gamma, B)$ are often specified by combinatorial methods. The optimization problem (5.135) is, in nature, a combinatorial optimization problem, and hence is an NP hard problem (Chickering et al. 2004). The elegant exact algorithms for solving the combinatorial optimization problem includes dynamic programming (Eaton and Murphy 2007, Koivisto 2012), A* algorithm (Yuan and Malone 2013), and integer programming algorithm (Bartlett and Cussens 2017; Hu et al. 2020; Manzour et al. 2020). To meet this computational challenge, Zheng et al. (2018) proposed to transform these combinatorial optimization problems into continuous optimization problem using a new smooth acyclicity constraint.

5.2.4.2 A New Characterization of Acyclicity

To transform a combinatorial optimization problem for inferring DAGs using SEMs, Zheng et al. (2018) proposed to replace a combinatorial constraint $G(\Gamma)\in DA$ with a single smooth equality constraint $h(\Gamma)=0$. They specified that the smooth function $h(\Gamma)\in R^{M\times M}$ should satisfy the following four conditions:

1. $h(\Gamma)=0$ if and only if Γ is acyclic;
2. The value of $h(\Gamma)$ measures how far the graph is away from DAG;
3. $h(\Gamma)$ is a smooth function;
4. Computation of $h(\Gamma)$ and its derivative is simple.

Under new constraints, the optimization problem (5.135) will be transformed to

$$\min_{\Gamma\in R^{M\times M},\, B\in R^{K\times M}} l(\Gamma, B, \lambda; X, Y) \tag{5.136}$$

Subject to $h(\Gamma)=0$.

Now we introduce smooth constraint function $h(\Gamma)$ proposed by Zheng et al. (2018). We begin with binary adjacent matrices.

Example 5.11 Trace of the Adjacency Matrix for a DGA

The adjacency matrix A corresponding to Γ matrix in Figure 5.8 and its power are given by

$$A=\begin{bmatrix}0&0&1&0&0\\0&0&0&1&0\\0&0&0&1&1\\0&0&0&0&0\\0&0&0&1&0\end{bmatrix},\ A^2=\begin{bmatrix}0&0&0&1&1\\0&0&0&0&0\\0&0&0&1&0\\0&0&0&0&0\\0&0&0&0&0\end{bmatrix},$$

$$A^3=\begin{bmatrix}0&0&0&0&1\\0&0&0&0&0\\0&0&0&0&0\\0&0&0&0&0\\0&0&0&0&0\end{bmatrix},\text{ and } A^4=\begin{bmatrix}0&0&0&0\\0&0&0&0\\0&0&0&0\\0&0&0&0\\0&0&0&0\end{bmatrix}.$$

It is clear that $\mathrm{Tr}(A)=\mathrm{Tr}(A^2)=\mathrm{Tr}(A^3)=\mathrm{Tr}(A^4)=0$.

Example 5.11 can be extended a generic case. Intuitively, the elements in the diagonal of the matrix $A^2$is calculated as

$$\tilde{a}_{ii}^{(2)}=\sum_{j=1}^{M} a_{ij}a_{ji}.$$

Since there is no cycle, $a_{ij}=1$ imlies $a_{ji}=0$. Therefore, all $\tilde{a}_{ii}^{(2)}=0, i=1,\ldots,M$. Thus, $\mathrm{Tr}(A)=\mathrm{Tr}(A^2)=0$. Consider off diagonal of the matrix A^2. Then, we have $\tilde{a}_{ij}^{(2)}=\sum_{l=1}^{M} a_{il}a_{lj}$, which indicates length-2 path $(i\to j)$. In general, $\tilde{a}_{ii}^{(k)}=\sum_{l_1=1}^{M}\cdots\sum_{l_k=1}^{M} a_{il_1}\ldots a_{l_k i}$. The diagonal element $\tilde{a}_{ii}^{(k)}$ of the matrix A^k indicates the length-k closed path $(i\to j_1\ldots\to j_{k-1}\to i)$ in a directed graph. Thus, $\mathrm{Tr}(A^k)$ counts the number of the length-k closed path $(i\to j_1\ldots\to j_{k-1}\to i)$. Since there is no cycle, we conclude $\mathrm{Tr}(A^k)=0$. Using similar argument, we can show that an acyclic graph will have

$$\mathrm{Tr}(A^k)=0,\ \forall k=1,2,.,\infty. \tag{5.137}$$

Lemma 5.1: One Simple Characterization of Acyclicity (Zheng et al. 2018)

Assume an $M \times M$ dimensional adjacency matrix A with binary elements and its spectral radius $r(A) < 1$. The adjacency matrix A is a DAG if and only if the following equality holds:

$$\mathrm{Tr}(I - A)^{-1} = M. \tag{5.138}$$

Proof

Following Zheng et al. (2018) approach, we can easily approve it. Using Taylor expansion, when $r(A) < 1$, we have

$$(I - A)^{-1} = I + \sum_{k=1}^{\infty} A^k, \tag{5.139}$$

which implies

$$\mathrm{Tr}\left((I - A)^{-1} = M + \sum_{k=1}^{\infty} \mathrm{Tr}(A^k). \tag{5.140}$$

Substituting equation (5.137) into equation (5.140) yields

$$\mathrm{Tr}(I - A)^{-1} = M.$$

Although equation (5.138) can characterize acyclicity, it has several limitations. First, it requires $r(A) < 1$. Second, its numeral computation is unstable. The elements of A^k are small and often exceed machine precision, which leads to unstable evaluations of A^k and its derivatives. The following lemma for characterization of acyclicity using matrix exponential is practically more useful.

Let A be an $M \times M$ matrix. Similar to the exponential function, the exponential of matrix A, denoted by e^A or $\exp(A)$ is defined as

$$e^A = I + A + \frac{A^2}{2!} + \cdots + \frac{A^k}{k!} + \cdots = \sum_{k=0}^{\infty} \frac{A^k}{k!}. \tag{5.141}$$

It is easy to see that

$$e^0 = I,$$

$$e^A e^B = e^{A+B} \text{ if } AB = BA,$$

$$e^A e^{-A} = I.$$

Example 5.12

Let

$$A = \begin{bmatrix} 0 & X & 0 \\ 0 & 0 & Y \\ 0 & 0 & 0 \end{bmatrix}.$$

Then,

$$A^2 = \begin{bmatrix} 0 & 0 & XY \\ 0 & 0 & 0 \\ 0 & 0 & 0 \end{bmatrix}, e^A = \begin{bmatrix} 1 & X & \frac{1}{2}XY \\ 0 & 1 & Y \\ 0 & 0 & 1 \end{bmatrix}.$$

Lemma 5.2: Matrix Exponential Characterization of Acyclicity

An $M \times M$ binary adjacency matrix A represents a DAG if and only if

$$\mathrm{Tr}(e^A) = M. \tag{5.142}$$

Proof

Combining equations (5.137) and (5.141), we obtain

$$\mathrm{Tr}(e^A) = M + \sum_{k=1}^{\infty} \mathrm{Tr}(A^k) = M.$$

It is clear that if the binary adjacency matrix A is replaced by nonnegative weighted matrix, then equations (5.137), and hence (5.142) still hold. Now if some elements of the weighted matrix W are negative, then both equations (5.137) and (5.142) do not hold. To solve this problem, we define a new weight matrix as the Hadamard product $W \circ W$, i.e., all multiplications are element-wise multiplications. It is clear that all elements of $W \circ W$ are nonnegative. Therefore, equations (5.137) and (5.142) hold for $W \circ W$.

Theorem 5.2: Characterization of Acyclicity for the General Weighted Adjacency Matrices (Zheng et al. 2018)

A weight matrix W is a DAG if and only if

$$h(W) = \mathrm{Tr}\left(e^{W \circ W}\right) - M = 0. \tag{5.143}$$

Partial derivative of the constraint function $h(W)$ with respect to matrix W is given by

$$\frac{\partial h(W)}{\partial W} = \left(e^{W \circ W}\right)^T \circ 2W, \quad (5.144)$$

where

$$\frac{\partial h(W)}{\partial W} = \left(\frac{\partial h(W)}{\partial W_{ij}}\right)_{M \times M}.$$

Proof

In the end of Lemma 5.2, we have shown the equation (5.143). We can show equation (5.144) (Exercise 5.4). These show that constraint function $h(W)$ satisfies the following conditions (1), (3), and (4). Now we intuitively show that $h(W)$ can measure the degree of acyclicity. It is clear that

1. $\text{Tr}\left(A + A^2 + \cdots + A^k + \cdots\right)$ counts the number of closed paths in A,
2. These counts are reweighted by matrix exponential e^W,
3. Since the weight of each edge (i, j) in $W \circ W$ is w_{ij}^2, replacing A with $W \circ W$ is to count the weighted closed paths, and
4. Inequality $h(W_1) > h(W_2)$ indicates that either W_1 has more cycles than W_2 or the cycles in W_1 have more weights than in W_2.

5.3 LATENT CAUSAL STRUCTURE

In Section 5.2, we discussed to infer causal networks among observational data. However, in many scenarios, the observed variables may not have direct causal structure. They are generated by latent causal variables or causally related confounders (O'Shaughnessy et al. 2020; Xie et al. 2020; Yang et al. 2020). In Section 5.2.4, we assume that latent variables in the VAE are independent. To model the causal structure between latent variables, in this section, we assume that the latent variables that generates the observed variables form causal structure. We will focus on introducing VAE with latent causal networks (Yang et al. 2020).

5.3.1 Latent Space and Latent Representation

Classical VAE learns mutually independent latent factors. Its general framework takes the standard multivariate normal distribution as a prior of the latent variables and approximates the unknown true posterior $p(Z|Y)$ by a variational posterior $q(Z|Y)$. However, unsupervised VAE does not perform well when factors form complex causal networks. Now we will introduce VAE with latent causal representation.

Consider two sets of observed variables: $Y = (Y_1, \ldots, Y_d)$ and $X = (X_1, \ldots, X_k)$. Assume that X is low dimensional $(k \le d)$. Assume that N subjects are sampled. The observed dataset is denoted by $D = \{(Y^1, X^1), \ldots, (Y^N, X^N,)\}$, where $Y^i = [Y_1^i, \ldots, Y_d^i]$, and $X^i = [X_1^i, \ldots, X_k^i]$. The data X can be class label associated with the data Y, time index or previous data points in a time series, or another concurrently observed variable, for example, Y are observed gene expressions and X are observed genotypes (Khemakhem et al. 2019). The data are mapped to a latent space. The latent space contains two sets of latent variables: $Z = (Z_1, \ldots, Z_m)$ and $\varepsilon = (\varepsilon_1, \ldots, \varepsilon_m)$.

Unlike the classical VAE, where the latent variables are independent, here we assume that the latent variables are connected by a DAG with an adjacency matrix A (Figure 5.10). For simplicity, we assume that the DAG is modeled by linear SEMs:

$$Z = A^T Z + \varepsilon = \left(I - A^T\right)^{-1} \varepsilon, \quad (5.145)$$

where ε are independent exogenous factors that follow Multivariate Gaussian $\varepsilon \tilde{} N(0, I)$ and Z are causal representation of latent endogenous factors. The model with unsupervised learning is unidentifiable (Louizos et al. 2017; Yang et al. 2020). To make the model identifiable, we need to provide additional information.

Equation (5.145) introduces a causal layer to learn latent causal representation (Figure 5.10). The inverse matrix $\left(I - A^T\right)^{-1}$ is parameters in the latent causal layer. A key feature of latent causal representation is that once the latent causal graph and SEMs are learned, we can intervene the latent codes to generate artificial data which are not present in the training data.

5.3.2 Mapping Observed Variables to the Latent Space

VAE is unsupervised learning. However, Locatello et al. (2018) showed that the inferred latent representation using unsupervised learning is not identifiable. To ensure that the latent causal model is identifiable, we must provide the additional supervising information associated with the true causal concepts. The

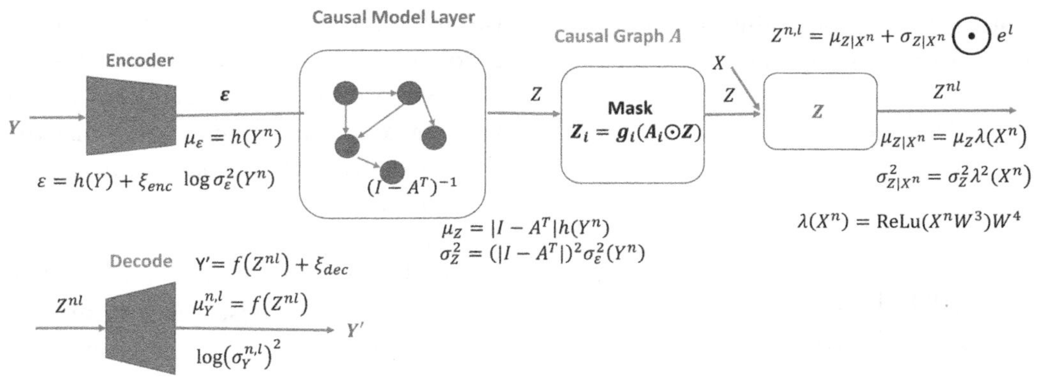

FIGURE 5.10 VAE with latent causal graph.

supervising information includes the label, pixel level observation, and genetic variants in the gene regulatory networks. Therefore, the observed variables include the concept variables Y and supervising information variables X. The latent variables are Z and ε. The additional information X serves two purposes (Yang et al. 2020). Firstly, a conditional prior $p(Z|X)$ can regularize the learned posterior of the latent variables Z and ensure the identifiability of the learned models. Secondly, the additional information X can be used to learn the causal model of the latent variables Z. The latent substantive variables Z are dependent, but the latent variables ε are independent. We assume that Z and ε are independent. The variables Y are mapped to the latent variables Z and ε, and the variables X are only mapped to the latent variables Z.

5.3.2.1 *Mask Layer*

To further improve the initially learned causal structure, a mask layer is introduced (Yang et al. 2020). The mask layer generates the data from the parental variables to their children. Recall that the adjacency matrix for the inferred causal network is defined as $A = [A_1 | \ldots | A_m]$, where A_i is an edge weight vector and A_{ji} denotes the weight from node Z_j to node Z_i. Let Z_i be the ith component of the vector Z. Define

$$A_i \odot Z = \begin{bmatrix} A_{1i} Z_1 \\ \vdots \\ A_{ji} Z_j \\ \vdots \\ A_{mi} Z_m \end{bmatrix},$$

where $\odot$ denotes the element-wise multiplication and $A_{ji} Z_j$ represents the contribution of the parent node Z_j to the node Z_i $(Z_j \to Z_i)$. Therefore, $A_i \odot Z$ contains only the parent information of the node Z_i and masks information of its all non-parent nodes. Define a mask:

$$Z_i = g_i(A_i \odot Z) + e_i,\ i = 1, \ldots, m,$$

where g_i is a nonlinear and invertible function.

By minimizing the reconstruction error, the mask layer can improve the estimation of the adjacency matrix A.

5.3.2.2 *Encoder and Decoder for Latent Causal Graph*

Recall that $\varepsilon \in R^m$ is the latent exogenous independent variables and $Z \in R^m$ is the latent endogenous variables with the linear SEM:

$$Z = A^T Z + \varepsilon = C\varepsilon,$$

where $C = (I - A^T)^{-1}$.

Consider encoding function $h(Y)$ and decoding function $f(Z)$. Similar to the VAE discussed in Section 5.2, the observed variables Y must be mapped to independent latent variables ε. Therefore, the model for the encoder is given by

$$\varepsilon = h(Y) + \xi_{enc}, \tag{5.146}$$

where ξ_{enc} are the vectors of independent noise with density function $p_{\xi_{enc}}$.

Decoding equation is defined as

$$Y = f(Z) + \xi_{dec}, \tag{5.147}$$

where ξ_{dec} are the vectors of independent noise with density function $p_{\xi_{dec}}$ and $f(Z) = [f_1(Z), \ldots, f_m(Z)]^T$.

Let $\xi = (\xi_{enc}, \xi_{dec})$ and p_ξ be density function. When ξ is small, the encoder and decoder distributions can be taken as deterministic ones.

Given supervising information X, the conditional distribution for generating the variables Y, Z, and ε can be factorized to

$$p_\theta(Y,Z,\varepsilon \mid X) = p_\theta(Z,\varepsilon \mid X) p_\theta(Y \mid Z,\varepsilon,X). \quad (5.148)$$

Define the **generating process:**

$$p_\theta(Y|Z,\varepsilon,X) = p_\theta(Y|Z) = p_{\xi_{Dec}}(Y|Z) = p_{\xi_{Dec}}(Y - f(Z), \quad (5.149)$$

and

the **inference process:**

$$q_\phi(\varepsilon \mid Y,X) = q_\phi(\varepsilon \mid Y) = p_{\xi_{Enc}}(\varepsilon - h(Y)). \quad (5.150)$$

Instead of the prior $p_\theta(Z)$ in the classic VAE, we define the joint prior $p_\theta(\varepsilon, Z \mid X)$ for the latent variables ε and Z as

$$p_\theta(\varepsilon, Z \mid X) = p_\varepsilon(\varepsilon) p_\theta(Z \mid X), \quad (5.151)$$

where $p_\varepsilon(\varepsilon) = N(0, I)$.

We assume that the prior of latent substantive variables $p_\theta(Z \mid X)$ is a factorized Gaussian distribution (Yang et al. 2020):

$$p_\theta(Z \mid X) = \prod_{i=1}^{m} N(Z; \lambda_1(X_i), \lambda_2^2(X_i), \quad (5.152)$$

where λ_1 and λ_2 are nonlinear functions and approximated by NNs. Define

$$\begin{aligned} T(Z) &= [\mu(Z), \sigma(Z)] \\ &= [\mu_1(Z_1), \sigma_1(Z_1), \ldots, \mu_m(Z_m), \sigma_m(Z_m)] \\ &= T[T_{1,1}(Z_1), T_{1,2}(Z_1), \ldots, T_{m,1}(Z_m), T_{m,2}(Z_m)]. \end{aligned} \quad (5.153)$$

Let $\theta = (f, h, C, T, \lambda)$ be the parameters in the model.

5.3.3 ELBO for the Log-Likelihood $\log p_\theta(Y \mid X)$

Variational Bayes $q_\phi(\varepsilon, Z \mid Y, X)$ will be used to approximate the true posterior $p_\theta(\varepsilon, Z \mid Y, X)$. Using the similar techniques of the Variational Bayes discussed in Section 5.2, we obtain the following ELBO for the log-likelihood $\log p_\theta(Y \mid X)$:

$$\begin{aligned} E_{q_D(Y^n, X^n)}[\log p_\theta(Y^n \mid X^n)] &\ge ELBO(Y^n, X^n) \\ &= \mathcal{L}(\theta, \varnothing, Y^n, X^n), \end{aligned}$$

where

$$\begin{aligned} &\mathcal{L}(\theta, \varnothing, Y^n, X^n) \\ &= E_{q_D(Y^n, X^n)}\Big\{E_{q_\varnothing(Z, \varepsilon|Y^n, X^n)}[\log p_\theta(Y^n \mid Z, \varepsilon, X^n)] \\ &\quad - KL(q_\varnothing(Z, \varepsilon|Y^n, X^n) \| p_\theta \|(Z, \varepsilon \mid X^n))\Big\}. \end{aligned} \quad (5.154)$$

The ELBO for entire dataset is

$$\mathcal{L}(\theta, \varnothing, Y, X) = \frac{1}{N}\sum_{n=1}^{N} \mathcal{L}(\theta, \varnothing, Y^n, X^n). \quad (5.155)$$

To compute the ELBO, we need to specify the distributions $p_\theta(Y^n \mid Z, \varepsilon, X^n)$, $q_\varnothing(Z, \varepsilon|Y^n, X^n)$, $p_\theta(Z, \varepsilon \mid X^n)$ and define transformation for Z, ε. Using equation (5.145), we can define

$$\begin{aligned} q_\varnothing(Z, \varepsilon|Y^n, X^n) &= q_\varnothing(\varepsilon \mid Y^n, X^n) I_{Z=C\varepsilon}(Z) \\ &= q_\varnothing(Z \mid Y^n, X^n) I_{\varepsilon=C^{-1}Z}(\varepsilon). \end{aligned} \quad (5.156)$$

Before calculating ELBO, we further transform equation (5.154). Using equations (5.149) and (5.156), the first term in equation (5.154) can be reduced to

$$E_{q_D(Y^n, X^n)}\Big\{E_{q_\varnothing(Z|Y^n, X^n)}[\log p_\theta(Y^n \mid Z)]\Big\}. \quad (5.157)$$

Using equations (5.151) and (5.156), the second term in equation (5.154) can be reduced to

$$\begin{aligned} &\int q_\varnothing(Z, \varepsilon|Y^n, X^n) \\ &\times \log \frac{\left(q_\varnothing(\varepsilon \mid Y^n, X^n) I_{Z=C\varepsilon}(Z) q_\varnothing(Z \mid Y^n, X^n) I_{\varepsilon=C^{-1}Z}(\varepsilon)\right)^{\frac{1}{2}}}{p_\varepsilon(\varepsilon) p_\theta(Z \mid X^n)} \\ &\times d\varepsilon dZ \\ &= \frac{1}{2}\left[E_{q_\varnothing(\varepsilon|Y^n, X^n)} \log \frac{q_\varnothing(\varepsilon \mid Y^n, X^n)}{p_\varepsilon(\varepsilon)} \right. \\ &\quad \left. + E_{q_\varnothing(Z|Y^n, X^n)} \log \frac{q_\varnothing(Z \mid Y^n, X^n)}{p_\theta(Z \mid X^n)}\right] \\ &= \frac{1}{2}\Big[KL(q_\varnothing(\varepsilon \mid Y^n, X^n) \| p_\varepsilon(\varepsilon)) \\ &\quad + KL(q_\varnothing(Z \mid Y^n, X^n) \| p_\theta(Z \mid X^n))\Big]. \end{aligned} \quad (5.158)$$

Combining equations (5.157) and (5.158), we obtain

$$\mathcal{L}(\theta,\varnothing,Y^n,X^n)$$
$$=E_{q_D(Y^n,X^n)}\left\{E_{q_\varnothing(Z|Y^n,X^n)}\left[\log p_\theta(Y^n|Z)\right]\right.$$
$$-\frac{1}{2}\left[KL\left(q_\varnothing(\varepsilon|Y^n,X^n)\|p_\varepsilon(\varepsilon)\right)\right.$$
$$\left.\left.+KL\left(q_\varnothing(Z|Y^n,X^n)\|p_\theta(Z|X^n)\right)\right]\right\}. \quad (5.159)$$

5.3.4 Computation of ELBO

5.3.4.1 *Encoder*

We first model autoencoder and calculate KL distance between the variational approximate posterior and prior over the latent variables. Recall from equation (5.150) that

$$q_\phi(\varepsilon|Y^n,X^n)=q_\phi(\varepsilon|Y^n)=p_{\xi_{Enc}}(\varepsilon-h(Y^n)).$$

We assume that $q_\phi(\varepsilon|Y,X)$ follows a multivariate Gaussian with a diagonal covariance structure:

$$\log q_\phi(\varepsilon|Y^n,X^n)=\log N(\varepsilon;h(Y^n),\sigma_\varepsilon^2(Y^n)),$$

where mean $h(Y^n)$ and variance $\sigma_\varepsilon^2(Y^n)$ are output of nonlinear encoding functions of data point Y^n and the variational parameters $\varnothing$, defined as

$$[h(Y^n)|\log\sigma_\varepsilon^2(Y^n)]=\mathrm{Relu}(Y^nW^1)W^2 \quad (5.160)$$

We assume that $p_\varepsilon(\varepsilon)$ is the centered isotropic multivariate Gaussian $p_\varepsilon(\varepsilon)=N(0,I_m)$. By the same argument as before, we can show that (Exercise 5.5)

$$-KL\left(q_\varnothing(\varepsilon|Y^n,X^n)\|p_\varepsilon(\varepsilon)\right)$$
$$=\frac{1}{2}\sum_{j=1}^m\left[1+\log(\sigma_\varepsilon)_j^2-(h_j(Y^n))^2-(\sigma_\varepsilon)_j^2\right]. \quad (5.161)$$

Next we calculate $KL\left(q_\varnothing(Z|Y^n,X^n)\|p_\theta(Z|X^n)\right)\big]$. It follows from equation (5.156) that (Exercise 5.6)

$$q_\varnothing(Z|Y^n,X^n)=\frac{q_\phi(\varepsilon|Y^n,X^n))}{|C|}, \quad (5.162)$$

which implies that (Exercise 5.6)

$$\mu_Z(Y^n,X^n)=|I-A^T|h(Y^n) \text{ and}$$
$$\sigma_Z^2(Y^n,X^n)=(|I-A^T|)^2\sigma_\varepsilon^2(Y^n). \quad (5.163)$$

It follows from equation (5.152) that (Exercise 5.7)

$$\log p_\theta(Z|X^n)$$
$$=\sum_{i=1}^m\log N\left(\mu_{Z|X^n}^i(Z_i)\lambda(X_i),\ \sigma_{Z|X^n}^i(Z_i)\lambda(X_i)\right), \quad (5.164)$$

where

$$\mu_{Z|X^n}=\mu_Z(Y^n,X^n)\lambda(X^n), \quad (5.165)$$

$$\sigma_{Z|X^n}^2=\sigma_Z^2(Y^n,X^n)\lambda^2(X^n), \quad (5.166)$$

$$\lambda(X^n)=\mathrm{ReLu}(X^nW^3)W^4 \quad (5.167)$$

We can show that (Exercise 5.8) that

$$-KL\left(q_\varnothing(Z|Y^n,X^n)\|p_\theta(Z|X^n)\right)\big]$$
$$=\frac{1}{2}\sum_{j=1}^m\left(1+\log\left(\frac{\sigma_Z^j}{\sigma_{Z|X^n}^j}\right)^2-\left(\frac{\sigma_Z^j}{\sigma_{Z|X^n}^j}\right)^2-\left(\frac{\mu_z^j-\mu_{Z|X^n}^j}{\sigma_{Z|X^n}^j}\right)^2\right) \quad (5.168)$$

Combining equations (5.161) and (5.168), we obtain

$$-\frac{1}{2}\left\{\left[KL\left(q_\varnothing(\varepsilon|Y^n,X^n)\|p_\varepsilon(\varepsilon)\right)\right.\right.$$
$$\left.\left.+KL\left(q_\varnothing(Z|Y^n,X^n)\|p_\theta(Z|X^n)\right)\right]\right\}$$
$$=\frac{1}{4}\sum_{j=1}^m\left(2+\log(\sigma_\varepsilon^j)^2-(h_j(Y^n))^2-(\sigma_\varepsilon^j)^2\right.$$
$$\left.+\log\left(\frac{\sigma_Z^j}{\sigma_{Z|X^n}^j}\right)^2-\left(\frac{\sigma_Z^j}{\sigma_{Z|X^n}^j}\right)^2-\left(\frac{\mu_z^j-\mu_{Z|X^n}^j}{\sigma_{Z|X^n}^j}\right)^2\right) \quad (5.169)$$

5.3.4.2 *Decoder*

The decoder is defined as

$$Y=f(Z)+\xi_{Dec}.$$

We assume that ξ_{Dec} follows a multivariate Gaussian $N(0,\sigma^2I)$. Thus, we obtain

$$p_\theta(Y^n|Z)=\frac{1}{(2\pi)^{1/2}}\frac{1}{\prod_{j=1}^m\sigma_Y^j}\exp\left\{-\frac{1}{2}\frac{(Y_j^n-f_j(Z))^2}{(\sigma_{Yj}^n)^2}\right\}, \quad (5.170)$$

where

$$\left[\mu_Y^{n,l} = f(Z^{n,l}) \,|\, \log\left(\sigma_Y^{n,l}\right)^2\right] = \text{Relu}\left(Z^{n,l}W^5\right)W^6, \tag{5.171}$$

$$Z^{n,l} = \mu_Z\left(Y^n, X^n\right) + \sigma_Z\left(Y^n, X^n\right) \odot e^l, \, e^l \sim N(0, I). \tag{5.172}$$

The Monte Carlo estimator of the first term $E_{q_\varnothing(Z|Y^n, X^n)}\left[\log p_\theta\left(Y^n \,|\, Z\right)\right]$ in equation (5.159) is given by

$$E_{q_\varnothing(Z|Y^n, X^n)}\left[\log p_\theta\left(Y^n \,|\, Z\right)\right] \approx \frac{1}{L}\sum_{l=1}^{L} \log p_\theta\left(Y^n \,|\, Z^{n,l}\right)$$

$$= -\frac{1}{2L}\sum_{l=1}^{L}\left[\log(2\pi) + \sum_{j=1}^{m}\left(\log\left(\sigma_{Yj}^{n,l}\right)^2 + \frac{\left(Y_j^n - \mu_{Y,j}^{n,l}\right)^2}{\left(\sigma_{Y,j}^{n,l}\right)^2}\right].\tag{5.173}$$

Combining equations (5.167) and (5.173), we obtain

$$\mathcal{L}(\theta, \varnothing, Y^n, X^n) = -\frac{1}{2L}\sum_{l=1}^{L}\left[\log(2\pi) + \sum_{j=1}^{m}\left(\log\left(\sigma_{Yj}^{n,l}\right)^2 + \frac{\left(Y_j^n - \mu_{Y,j}^{n,l}\right)^2}{\left(\sigma_{Y,j}^{n,l}\right)^2}\right]$$
$$+\frac{1}{4}\sum_{j=1}^{m}\left(2 + \log\left(\sigma_\varepsilon^j\right)^2 - \left(h_j(Y^n)\right)^2 - \left(\sigma_\varepsilon^j\right)^2 + \log\left(\frac{\sigma_Z^j}{\sigma_{Z|X^n}^j}\right)^2 - \left(\frac{\sigma_Z^j}{\sigma_{Z|X^n}^j}\right)^2 - \left(\frac{\mu_z^j - \mu_{Z|X^n}^j}{\sigma_{Z|X^n}^j}\right)^2\right). \tag{5.174}$$

5.3.4.3 Learning Latent Causal Graph

Recall that the mask layer is defined as

$$Z_i = g_i\left(A_i \odot Z, \eta_i\right) + e_i, \, i = 1, \ldots, m, \tag{5.175}$$

where g_i is a nonlinear and invertible function, η_i are parameters.

To learn the mask and refine the estimated adjacency matrix A, we impose the constraint:

$$l_m\left(Y^n, X^n\right) = E_{Z \sim q_\varnothing(Z|Y^n, X^n)}\sum_{i=1}^{m}\left\|Z_i - g_i\left(A_i \odot Z, \eta_i\right)\right\|^2 \le e_M$$

or its sampling approximation

$$\mathcal{L}_m\left(Y^n, X^n\right) = \frac{1}{L}\sum_{l=1}^{m}\sum_{i=1}^{m}\left\|Z_i^{n,l} - g_i\left(A_i \odot Z^{n,l}, \eta_i\right)\right\|^2 \le e_M.$$

The ELBO for entire dataset is

$$\mathcal{L}(\theta, \varnothing, Y, X) = \frac{1}{N}\sum_{n=1}^{N}\mathcal{L}(\theta, \varnothing, Y^n, X^n). \tag{5.176}$$

5.3.5 Optimization for Learning the Latent DAG

Similar to Section 5.2.3.4, to ensure acyclicity of the learned network, we can also introduce constraint:

$$h(A) = \text{Tr}\left[\left(I + \alpha A \circ A\right)^m\right] - m = 0. \tag{5.177}$$

Optimization formulation of learning latent DAG can be formulated as

$$\min_{A,B,\theta,\varnothing} F(A, B, \theta, \varnothing) = -ELBO = -\mathcal{L}(\theta, \varnothing, Y, X) \tag{5.178}$$

Subject to

$$h(A) = \text{Tr}\left[\left(I + \alpha A \circ A\right)^m\right] - m = 0,$$

$$\frac{1}{N}\sum_{n=1}^{N}\mathcal{L}_m\left(Y^n, X^n\right) \le e_M.$$

Again, duality theory and Lagrange multiplier methods can be used to solve the equality constrained optimization problem (5.178). For details, please read Section 5.2.3.4.

5.4 CAUSAL MEDIATION ANALYSIS

Mediation analysis is a powerful tool for understanding mechanism of intervention and exposures (Cai et al. 2013; Cashin et al. 2020). Mediation analysis attempts to decipher complex causal mechanisms, where the total effect is decomposed into causal direct and indirect effects (Rijnhart et al. 2020). Mediation analysis can be classified into classical mediation analysis and modern causal mediation analysis, or single variate mediation and multivariate mediation analysis, cross sectional and longitudinal mediation analysis (Zhao and Luo 2019; Blum et al. 2020). Mediational analysis has rich literature. In this section, we will focus on application of VAE to cascade unobserved mediator models (Cai et al. 2019).

5.4.1 Basics of Mediation Analysis

Mediation analysis is to study how independent variables (causes, interventions) influence the outcomes (effects, outcomes) through intermediate variables (mediators) (Gunzler et al. 2013). By mediation analysis, we can identify the pathways from causes (interventions) to the effects (outcomes). Mediation analysis allows us to gain insight and discover deep mechanism of actions.

5.4.1.1 Univariate Mediation Model

A typical simple framework for mediation analysis is shown in Figure 5.11(a). Mediation analysis consists of three variables: intervention variable (X), outcome variable (Y), and mediator variable (Z). Outcome and mediator variables are often called endogenous variables and intervention variable is called exogenous variable. The SEMs are often used for mediation analysis. The SEM for Figure 5.11(a) is given by

$$X = \varepsilon_X, \tag{5.179}$$

$$Z = \beta_{XZ} X + \varepsilon_Z, \tag{5.180}$$

$$Y = \gamma_{XY} X + \gamma_{ZY} Z + \varepsilon_Y. \tag{5.181}$$

The adjacency matrix A is given by

$$A = \begin{bmatrix} 0 & \beta_{XZ} & \gamma_{XY} \\ 0 & 0 & \gamma_{ZY} \\ 0 & 0 & 0 \end{bmatrix}.$$

Let $W = [X, Y, Z]^T$ and $\varepsilon = [\varepsilon_X, \varepsilon_Y, \varepsilon_Z]^T$ Then, equations (5.179)– (5.181) can be written in a matrix form:

$$W = A^T W + \varepsilon. \tag{5.182}$$

5.4.1.2 Multivariate Mediation Analysis

A graphical representation of a multiple mediation model is shown in Figure 5.11(b). There are k potential mediators $M_1, \ldots, M_k$ between the intervention variable X and outcome variable Y.

Let $W = [X, M_1, \ldots, M_k, Y]^T$ and $\varepsilon = [\varepsilon_X, \varepsilon_{M_1}, \ldots, \varepsilon_{M_k}, \varepsilon_Y]^T$. The adjacency matrix A is given by

$$A = \begin{bmatrix} 0 & \alpha_1 & \cdots & \alpha_k & 0 \\ 0 & 0 & \cdots & 0 & \beta_1 \\ \vdots & \vdots & \vdots & \vdots & \vdots \\ 0 & 0 & \cdots & 0 & \beta_k \\ 0 & 0 & \cdots & 0 & 0 \end{bmatrix}.$$

The SEM is

$$\begin{bmatrix} X \\ M_1 \\ \vdots \\ M_k \\ Y \end{bmatrix} = \begin{bmatrix} 0 & 0 & \cdots & 0 & 0 \\ \alpha_1 & 0 & \cdots & 0 & 0 \\ \vdots & \vdots & \vdots & \vdots & \vdots \\ \alpha_k & 0 & \cdots & 0 & 0 \\ 0 & \beta_1 & \cdots & \beta_k & 0 \end{bmatrix} \begin{bmatrix} X \\ M_1 \\ \vdots \\ M_k \\ Y \end{bmatrix} + \begin{bmatrix} \varepsilon_X \\ \varepsilon_{M_1} \\ \vdots \\ \varepsilon_{M_k} \\ \varepsilon_Y \end{bmatrix}, \tag{5.183}$$

or

$$W = A^T W + \varepsilon. \tag{5.184}$$

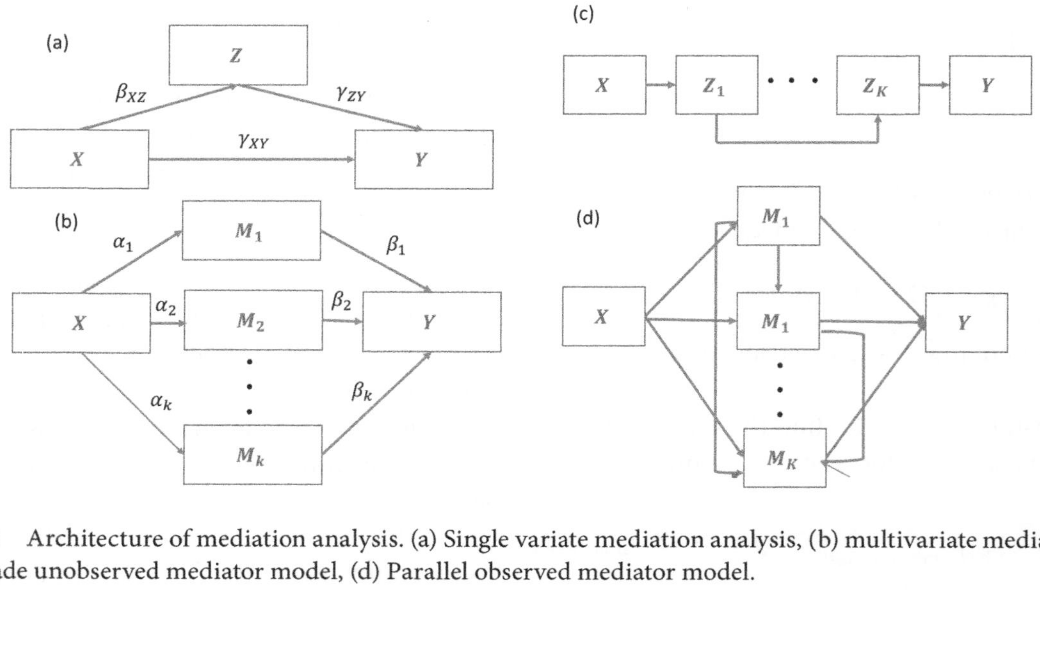

FIGURE 5.11 Architecture of mediation analysis. (a) Single variate mediation analysis, (b) multivariate mediation analysis, (c) cascade unobserved mediator model, (d) Parallel observed mediator model.

5.4.1.3 Cascade Unobserved Mediator Model

All previous mediation models assume that the mediators are observed. However, in many cases, the mediators are unobserved. In this section, we introduce cascade unobserved additive mediation models proposed by Cai et al. (2019). The cascade additive mediation model is shown in Figure 5.11(c). Let X be a cause and Y be its effect. There are k unobserved mediators $(Z_1, \ldots, Z_k)$ between them. Define the cascade ANM:

$$M_1 = f_1(X) + Z_1, \tag{5.185}$$

$$M_k = f_k\left(M_{pa(k)}\right) + Z_k, \tag{5.186}$$

$$Y = f_{K+1}\left(M_{pa(Y)}\right) + \varepsilon, \tag{5.187}$$

where $X, Z_k\,(k = 1, \ldots, K)$, and ε are mutually independent, $M_{pa(k)}$ and $M_{pa(Y)}$ denote parents of M_k and Y, respectively, and the causal relations among M_k are recursive.

5.4.1.4 Unobserved Multivariate Mediation Model

Again, let X be a cause and Y be its effect. There are K parallel unobserved mediators $M_1, \ldots, M_K$ between them as shown in Figure 11(d). Define the model:

$$M = A^T M + f(X) + Z, \tag{5.188}$$

$$Y = g(M_1, \ldots, M_K) + \varepsilon, \tag{5.189}$$

where $f(X) = [f_1(x), \ldots, f_K(X)]^T$, $M = [M_1, \ldots, M_K]^T$, $Z = [Z_1, \ldots, Z_K]^T$, $Z_1, \ldots, Z_K$, and ε are mutually independent, $f_1, \ldots, f_K$ and g are nonlinear functions, A is an adjacency matrix.

5.4.2 VAE for Cascade Unobserved Mediator Model

Due to space limitation, we will focus on application of VAE to cascade unobserved mediator model which was developed by Cai et al. (2019).

5.4.2.1 ELBO for Cascade Mediator Model

Consider a dataset $D = \{X^n, Y^n\}_{n=1}^N$. We can view the unobserved mediators as the latent variables. Using the model equations (5.185)–(5.187), the marginal density function $p_\theta(X^n, Y^n)$ can be calculated by integral over the latent variables M_k:

$$p_\theta(X^n, Y^n) = \int p_\theta(X^n, Y^n, M)dM$$

$$= \int p_\theta(X^n) p_\theta \left(Y^n \mid M_{pa(Y)}\right) \prod_{k=2}^{K} p_\theta\left(M_k | M_{pa(k)}\right) p_\theta(M_1 | X^n) dM. \tag{5.190}$$

However, the latent variables $M_k, k = 1, \ldots, K$ can be mapped to noise variables $Z_k, k = 1, \ldots, K$. Recall that by assumption of the additive model that Z_k is independent of $M_{pa(k)}$. Thus, we have

$$p_\theta\left(M_k | M_{pa(k)}\right) = p_\theta\left(Z_k = M_k - f_k\left(M_{pa(k)}\right) | M_{pa(k)}\right)$$

$$= p_\theta(Z_k), \text{ which implies} \prod_{k=2}^{K} p_\theta \left(M_k | M_{pa(k)}\right) p_\theta(M_1 | X^n) = \prod_{k=1}^{K} p_\theta(Z_k), \tag{5.191}$$

Similarly, we obtain

$$p_\theta\left(\left(Y^n | M_{pa(Y)}\right) = p_\theta(\varepsilon^n = Y^n - f_{K+1}\left(M_{pa(Y)}\right)\right)$$

$$= p_\theta\left(\varepsilon^n = Y^n - f_{K+1}\left(\ldots f_2\left(f_1(X) + Z_1\right)\right)\right)$$

$$= p_\theta\left(\varepsilon^n = Y^n - f(X, Z)\right). \tag{5.192}$$

Substituting equations (5.191) and (5.192) into equation (5.190) yields

$$p_\theta(X^n, Y^n)$$

$$= \int p_\theta(X^n) p_\theta\left(\left(\varepsilon^n = Y^n - f(X, Z)\right)\right) \prod_{k=1}^{K} p_\theta(Z_k) \mathrm{dz}$$

$$= \int p_\theta(X^n, \varepsilon^n, Z) dZ, \tag{5.193}$$

where

$$p_\theta(X^n, \varepsilon^n, Z)$$

$$= p_\theta(X^n) p_\theta\left(\left(\varepsilon^n = Y^n - f(X, Z)\right)\right) \prod_{k=1}^{K} p_\theta(Z_k). \tag{5.194}$$

Similar to Appendix 5F, we can show that the ELBO is given by

$$\log p_\theta\left(X^n, \varepsilon^n, Z\right) \geq \mathcal{L}\left(\theta, \varnothing, Y^n, X^n\right), \quad (5.195)$$

where

$$\mathcal{L}\left(\theta, \varnothing, Y^n, X^n\right) = \log p\left(X^n\right) + E_{q_\varnothing(Z|Y^n, X^n)}\left[\log p\left(\varepsilon^n = Y^n - f\left(X^n, Z, \theta\right)\right)\right] - KL\left(q_\varnothing\left(Z|Y^n, X^n\right) \| p_\theta(Z)\right). \quad (5.196)$$

5.4.2.2 Encoder and Decoder

5.4.2.2.1 Encoder Encoder is to find approximate posterior $q_\varnothing\left(Z|Y^n, X^n\right)$ using NNs. The cause X^n and effect Y^n are mapped to latent space Z. The representation trick (Kingma and Welling 2013; Cai et al. 2019) is used to make approximate posterior $q_\varnothing\left(Z|Y^n, X^n\right)$. Define the following differentiable transformation:

$$Z = g_\varnothing(\tau, Y, X), \tau \sim p(\tau). \quad (5.197)$$

Assume that $q_\varnothing\left(Z|Y^n, X^n\right)$ follows a Gaussian distribution

$$q_\varnothing\left(Z|Y^n, X^n\right) = N\left(Z; \mu_\varnothing\left(Y^n, X^n\right), \sigma_\varnothing\left(Y^n, X^n\right)\right), \quad (5.198)$$

where

$$\left[\mu_\varnothing\left(Y^n, X^n\right) | \log \sigma_\varnothing\left(Y^n, X^n\right)\right] = \mathrm{Relu}\left(W^1\left(Y^n, X^n\right)^T\right) W^2, \quad (5.199)$$

and

Z also follows a Gaussian distribution: $N(0, I)$. We can show that

$$-KL\left(q_\varnothing\left(Z|Y^n, X^n\right) p_\theta(Z)\right) = \frac{1}{2}\left\{\sum_{k=1}^{K}\left(1 + \log\left(\sigma_\varnothing^k\left(Y^n, X^n\right)\right)^2 - \left(\mu_\varnothing^k\left(Y^n, X^n\right)\right)^2 - \left(\sigma_\varnothing^k\left(Y^n, X^n\right)\right)^2\right\}. \quad (5.200)$$

5.4.2.2.2 Decoder The decoder is designed as $\log p\left(\varepsilon^n = Y^n - f\left(X^n, Z, \theta\right)\right)$. We assume that ε follows a standard normal distribution $N(0,1)$. Then, we obtain

$$\log p\left(\varepsilon^n = Y^n - f\left(X^n, Z, \theta\right)\right) = -\frac{1}{2}\left(\log(2\pi) + \left(Y^n - f\left(X^n, Z, \theta\right)\right)^2\right), \quad (5.201)$$

where $f\left(X^n, Z, \theta\right)$ is implemented by NNs, for example,

$$f\left(X^n, Z, \theta\right) = \mathrm{Relu}\left(W^3\begin{bmatrix} X^n \\ Z \end{bmatrix}\right) W^4. \quad (5.202)$$

The latent variable $Z^{n,l}$ is defined as

$$Z^{n,l} = \mu_\varnothing\left(Y^n, X^n\right) + \sigma_\varnothing\left(Y^n, X^n\right) \odot u^l, u^l \sim N(0,1), \quad l = 1, \ldots, L, \quad (5.203)$$

where $\mu_\varnothing\left(Y^n, X^n\right)$ and $\sigma_\varnothing\left(Y^n, X^n\right)$ are defined in equation (5.199).

The Monte Carlo estimation of $E_{q_\varnothing(Z|Y^n, X^n)}\left[\log p\left(\varepsilon^n = Y^n - f\left(X^n, Z, \theta\right)\right)\right]$ is given by

$$E_{q_\varnothing(Z|Y^n, X^n)}\left[\log p\left(\varepsilon^n = Y^n - f\left(X^n, Z, \theta\right)\right)\right] \approx -\frac{1}{2L}\sum_{l=1}^{L}\left(\log(2\pi) + \left(Y^n - f\left(X^n, Z^{n,l}, \theta\right)\right)^2\right), \quad (5.204)$$

where $f\left(X^n, Z^{n,l}, \theta\right)$ is defined in equation (5.202).

The ELBO for entire dataset is

$$\mathcal{L}(\theta, \varnothing, Y, X) = \frac{1}{N}\sum_{n=1}^{N}\mathcal{L}\left(\theta, \varnothing, Y^n, X^n\right), \quad (5.205)$$

where

$$\mathcal{L}\left(\theta, \varnothing, Y^n, X^n\right) = -\frac{1}{2L}\sum_{l=1}^{L}\left(\log(2\pi) + \left(Y^n - f\left(X^n, Z^{n,l}, \theta\right)\right)^2\right) + \frac{1}{2}\left\{\sum_{k=1}^{K}\left(1 + \log\left(\sigma_\varnothing^k\left(Y^n, X^n\right)\right)^2 - \left(\mu_\varnothing^k\left(Y^n, X^n\right)\right)^2 - \left(\sigma_\varnothing^k\left(Y^n, X^n\right)\right)^2\right\}. \quad (5.206)$$

Architecture of VAE for cascade unobserved mediator model is similar to Figure 5.5.

5.4.2.3 Test Statistics

Classifier two-sample test (Section 5.1.5) can be used to test causation and detect cause direction. Using equation (5.75), we define the test statistics $T_{C(X \to Y)}$, $T_{C(Y \to X)}$, and

$$T_{CE} = \frac{\left(T_{C(X \to Y)} - T_{C(Y \to X)}\right)^2}{\sigma^2}.$$

Under the null hypothesis of no causation, T_{CE} is distributed as a central $\chi^2_{(1)}$ distribution. If test shows the significant causation, then determine the cause direction. If $T_{C(X \to Y)} < T_{C(Y \to X)}$ or $(\mathcal{L}_{X \to Y} > \mathcal{L}_{Y \to X})$ then $X \to Y$. Otherwise, $Y \to X$.

5.5 CONFOUNDING

Confounders that affect both an intervention and its outcome obscure the real effect of the cases. Confounders will cause bias. It is a major obstacle to making valid causal inferences from observational data (Suttorp et al. 2015). Widely used classical methods for causal analysis with unobserved confounders are instrumental variable (IV) methods, proxy variable methods (Kallus et al. 2018; Veitch et al. 2019), regression discontinuity design and identification by enumeration of mechanism (front door methods) (Guo et al. 2020). In the recent years, deep learning and latent variable models for causal inference with unobserved confounders have been developed (Louizos et al. 2017; Dinga et al. 2020). Since VAE make very weak assumption about the data generating processes and causal structure of the latent confounders, it is a powerful tool for causal inference with unobserved confounders. This section, we will mainly introduce latent variable models and IV methods for causal inference with unobserved confounders.

5.5.1 Deep Latent Variable Models for Causal Inference under Unobserved Confounders

A key issue for causal analysis from observational data is confounding. The recently developed methods for causal inference with confounders include VAE and representation learning (Louizos et al. 2017; Kallus et al. 2018; Miao et al. 2018), Identifiability motivated representation learning (Khemakhem et al. 2019; Wu and Fukumizu 2020; Yang et al. 2020), and GANs (Yoon et al. 2018; Ge et al. 2020). In this section, we will mainly introduce VAE and latent variable models for causal inference with unobserved confounders (Louizos et al. 2017).

Consider a simple model for causal inference with unobserved confounder. Let $Y \in R^d$ be an effect, $T \in R^d$ be causal or intervention variable, $Z \in R^d$ be an unobserved confounder and $X \in R^d$ be a proxy variable. Their relationships are shown in Figure 5.12(a).

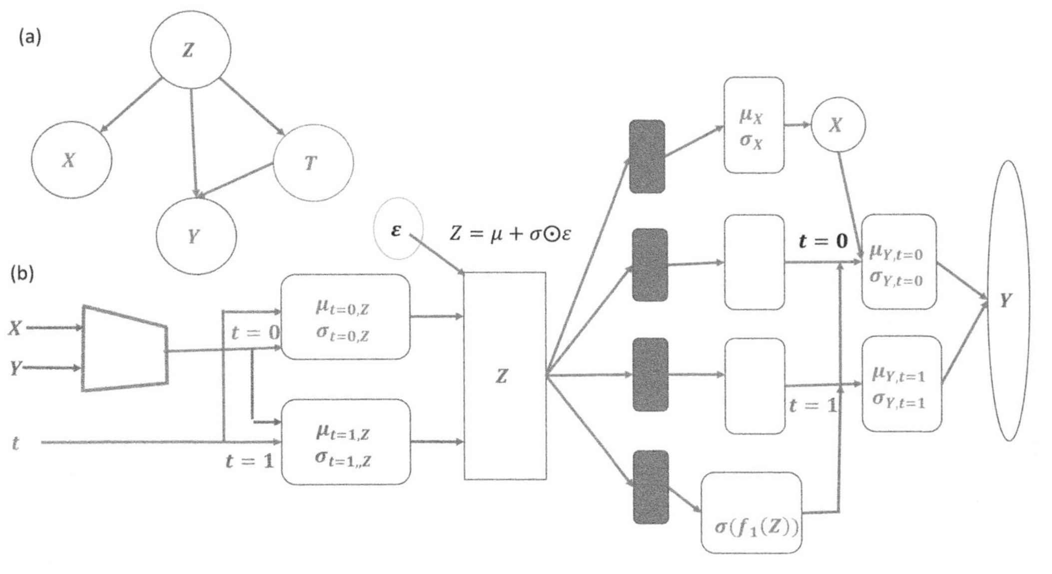

FIGURE 5.12 (a) Unobserved confound model, (b) Architecture of causal VAE.

The four variables form a DAG. The adjacent matrix A is given by

$$A=\begin{bmatrix} 0 & 0 & 0 \\ 0 & 0 & \alpha_{TY} \\ 0 & 0 & 0 \end{bmatrix}.$$

The linear structural equations for Figure 5.12(a) are

$$X=Z_x, \quad (5.207)$$

$$T=Z_T, \quad (5.208)$$

$$Y=\alpha_{TY}T+Z_Y. \quad (5.209)$$

Let

$$W=\begin{bmatrix} X \\ T \\ Y \end{bmatrix} \text{ and } Z=\begin{bmatrix} Z_X \\ Z_T \\ Z_Y \end{bmatrix}.$$

Then, equations (5.207)–(5.209) can be written in a matrix form:

$$W=A^TW+Z. \quad (5.210)$$

Similar to Yu et al. (2019), equation (5.210) can be transformed to

$$f_2^{-1}(W)=A^Tf_2^{-1}(W)+f_1(Z), \quad (5.211)$$

where f_1, f_2 are nonlinear functions, and f_2^{-1} is inverse of function f_2.

It follows from equation (5.211) that

$$W=f_2\left(\left(I-A^T\right)^{-1}f_1(Z)\right). \quad (5.212)$$

Consider a special case: $f_2=1$ and

$$f_1(Z)=\begin{bmatrix} \beta_XZ_0 \\ \beta_TZ_0 \\ \beta_YZ_0 \end{bmatrix}.$$

Then, we obtain

$$X=\beta_XZ_0,$$

$$T=\beta_TZ_0,$$

$$Y=(\alpha_{TY}\beta_T+\beta_Y)Z_0.$$

It follows from equation (5.211) that

$$Z=f_1^{-1}\left(\left(I-A^T\right)f_2^{-1}(W)\right)$$
$$=f_4\left(\left(I-A^T\right)f_3(W)\right). \quad (5.213)$$

Define encoder:

$$Z=f_4\left(\left(I-A^T\right)f_3(W)\right) \text{ and}$$

Decoder:

$$W=f_2\left(\left(I-A^T\right)^{-1}f_1(Z)\right),$$

where we assume that $Z_X=Z_T=Z_Y=Z$. Using techniques in Yu et al. (2019) or discussed in Section (5.2.4), we can estimate the causal effect and remove the bias from the unobserved confounder.

5.5.2 Treatment Effect Formulation for Causal Inference with Unobserved Confounder

Next we formulate the causal inference with unobserved confounder in terms of treatment effect estimation developed by Louizos et al. (2017). We assume that X and t, and X and Y are independent, given Z (Figure 5.12), i.e.,

$$p(t|X,Z)=p(t\,|\,Z), \quad (5.214)$$

$$p(Y|t,Z,X)=p(Y\,|\,t,Z). \quad (5.215)$$

This is a typical application of VAE. We first introduce decoder (Louizos et al. 2017).

5.5.2.1 Decoder

The latent variable Z is viewed as the unobserved confounder. Let X_i be the input feature vector, t_i be the treatment assignment variable (cause variable), and Y_i be the outcome (effect) of the particular treatment of the ith sample. The decoder is defined as $p_\theta(Y,t,X\,|\,Z)$ (Louizos et al. 2017). Using equations (5.214) and (5.215), we can factorize the encoding probability as

$$p_\theta(Y,t,X|Z)=p_\theta(Y\,|\,t,Z)p_\theta(t\,|\,Z)p_\theta(X\,|\,Z). \quad (5.216)$$

We can consider multiple confounders and multiple features. Therefore, we assume that Z and X are vectors. Here, we only consider a binary treatment assignment. Let Z_{ij}, $j=1,\ldots,k_z$ be the component of the vector Z, where k_z is the dimension of the vector Z.

The prior distributions of Z_{ij} are assumed a Gaussian distribution and given by

$$p(Z_i) = \prod_{j=1}^{k_Z} N(Z_{ij} \mid 0,1). \quad (5.217)$$

If we assume that X are continuous variables, then the log-conditional distributions of X, given Z, is

$$\log p(X_i|Z_i) = -\frac{1}{2}\sum_{j=1}^{k_Z}\left(\log(2\pi) + \log\sigma_{xij}^2 + \frac{(X_{ij} - \mu_{xij})^2}{\sigma_{xij}^2}\right), \quad (5.218)$$

where

$$[\mu_{xi} \mid \sigma_{xi}^2] = \text{Relu}(W_x^1 Z_i) W_x^2.$$

If the variables X are discrete, then Bernoulli distribution or other discrete distribution are used for $p(X_i|Z_i)$. Let Bern denote Bernoulli distribution and σ be the logistic function. The log-conditional distribution of t, given Z is

$$\begin{aligned}\log p(t_i|Z_i) &= t_i \log\sigma(f_1(Z_i)) \\ &+ (1-t_i)\log(1-\sigma(f_1(Z_i))),\end{aligned} \quad (5.219)$$

where $\sigma(f_1(Z_i)) = \text{logit}(W_t^2 \tanh(W_t^1 Z_i + b_z^1) + b_t^2)$.

Consider both continuous and discrete outcomes. The distribution of continuous outcome is assumed to follow a Gaussian distribution. The conditional distribution of Y_i, given t_i, Z_i, is

$$p(Y_i|t_i, Z_i) = N(Y_i; \hat{\mu}_i, \hat{V}), \hat{\mu}_i = t_i f_2(Z_i) + (1-t_i) f_3(Z_i),$$

$$\text{or } \log p(Y_i|t_i, Z_i) = -\frac{1}{2}\left(\log(2\pi + \hat{V}) + \frac{(Y_i - \mu_{Yi})^2}{\hat{V}}\right), \quad (5.220)$$

where

$$\mu_{Yi} = t_i \text{Relu}(W_Y^2 Z_i) W_Y^1 + (1-t_i)\text{Relu}(W_Y^4 Z_i) W_Y^4 \text{ and}$$
$$\hat{V} = \text{Relu}(W_Y^6 Z) W_Y^5, \ Z = [Z_1, \ldots, Z_n]^T.$$

For the binary outcome, the conditional distribution of Y_i, given t_i, Z_i, is

$$\begin{aligned} p(Y_i|t_i, Z_i) &= Bern(\pi = \hat{\pi}_i), \hat{\pi}_i \\ &= \sigma((t_i f_2(Z_i) + (1-t_i) f_3(Z_i)), \\ \text{or } \log p(Y_i|t_i, Z_i) &= Y_i \log\hat{\pi}_i + (1-Y_i)\log(1-\hat{\pi}_i), \end{aligned} \quad (5.221)$$

where

$$\begin{aligned}\hat{\pi}_i = \text{sigm}\big(&t_i(W_Y^2 \tanh(W_Y^1 Z_i + b_Y^1) + b_Y^2) \\ &+ (1-t_i)(W_Y^4 \tanh(W_Y^3 Z_i + b_Y^3) + b_Y^4\big).\end{aligned}$$

These nonlinear functions f_j are implemented by NNs with parameters θ_j.

5.5.2.2 Encoder

The encoder is defined as the approximate estimator of the posterior distribution of Z_i, given the observed data Y_i, t_i, X_i (Louizos et al. 2017)

$$q_\varnothing(Z_i|Y_i, t_i, X_i) = N(Z_i; \mu_{ij} = \bar{\mu}_{ij}, \sigma_{ij}^2 = \bar{\sigma}_{ij}^2), \quad (5.222)$$

where

$$\bar{\mu}_i = t_i \mu_{t=1,i} + (1-t_i)\mu_{t=0,i}, \bar{\sigma}_i^2 = t_i \sigma_{t=1,i}^2 + (1-t_i)\sigma_{t=0,i}^2, \quad (5.223)$$

$$\begin{aligned}[\mu_{t=1,i}|\sigma_{t=1,i}^2] &= g_2 \odot g_1(X_i, Y_i), [\mu_{t=0,i}|\sigma_{t=0,i}^2] \\ &= g_3 \odot g_1(X_i, Y_i),\end{aligned} \quad (5.224)$$

where g_1, g_2, g_3 are vectors of functions and implemented by NNs with parameters $\varnothing_k$, and $\odot$ denotes element-wise multiplications.

5.5.3 ELBO

Log-likelihood function $\log p_\theta(Y_i, t_i, X_i)$ for the DAG in Figure 5.12(a) is bounded by (Kingma and Welling 2013; Louizos et al. 2017)

$$\log p_\theta(Y_i, t_i, X_i) \geq \mathcal{L}(\theta, \varnothing, Y_i, t_i, X_i), \quad (5.225)$$

where

$$\begin{aligned}&\mathcal{L}(\theta, \varnothing, Y_i, t_i, X_i) \\ &= E_{q_\varnothing(Z_i|Y_i, t_i, X_i)}\big[\log p_\theta(Y_i, t_i, X_i|Z_i) \\ &\quad - KL(q_\varnothing(Z_i|Y_i, t_i, X_i) \| p(Z_i))\big],\end{aligned} \quad (5.226)$$

$\mathcal{L}(\theta, \varnothing, Y_i, t_i, X_i)$ is the ELBO for data point (Y_i, t_i, X_i).

Note that

$$\begin{aligned}&\log p_\theta(Y_i, t_i, X_i|Z_i) \\ &= \log p_\theta(Y_i \mid t_i, Z_i) + \log p_\theta(X_i, t_i \mid Z_i).\end{aligned} \quad (5.227)$$

Substituting equation (5.227) into equation (5.226), we obtain

$$\mathcal{L}(\theta, \varnothing, Y_i, t_i, X_i)$$
$$= E_{q_{\varnothing}(Z_i|Y_i, t_i, X_i)}\left[\log p_\theta(Y_i \mid t_i, Z_i) + \log p_\theta(X_i, t_i \mid Z_i)\right.$$
$$\left. - KL\left(q_{\varnothing}(Z_i|Y_i, t_i, X_i) \| p(Z_i)\right)\right]. \tag{5.228}$$

We can show (Exercise 5.10) that

$$-KL\left(q_{\varnothing}(Z_i|Y_i, t_i, X_i) \| p(Z_i)\right)\}$$
$$= \frac{1}{2}\sum_{j=1}^{k_Z}\left(1 + \log \bar{\sigma}_{ij}^2 - \bar{\sigma}_{ij}^2 - \bar{\mu}_{ij}^2\right). \tag{5.229}$$

Next we calculate $\log p_\theta(Y_i, t_i, X_i \mid Z_i^l)$. First, define transformation:

$$Z_i^l = \bar{\mu}_i + \bar{\sigma}_i \odot \varepsilon^l, \varepsilon^l \sim N(0, I), l = 1, 2, \ldots, L,$$
$$\bar{\mu}_i = t_i \mu_{t=1,i} + (1 - t_i)\mu_{t=0,i}, \bar{\sigma}_i^2 = t_i \sigma_{t=1,i}^2 + (1 - t_i)\sigma_{t=0,i}^2,$$
$$\left[\mu_{t=1,i} | \sigma_{t=1,i}^2\right] = g_2 \odot g_1(X_i, Y_i), \left[\mu_{t=0,i} | \sigma_{t=0,i}^2\right]$$
$$= g_3 \odot g_1(X_i, Y_i),$$

where g_1, g_2, g_3 are vectors of functions and implemented by NNs with parameters $\varnothing_k$, and $\odot$ denotes element-wise multiplications.

1. **Continuous Outcome**

   Combining equations (5.218)–(5.220), we obtain

$$\log p_\theta(Y_i, t_i, X_i \mid Z_i^l)$$
$$= -\frac{1}{2}\left[\sum_{j=1}^{k_x}\left(\log(2\pi) + \log\left(\sigma_{xij}^l\right)^2 + \frac{\left(X_{ij} - \mu_{xij}^l\right)^2}{\left(\sigma_{xij}^l\right)^2}\right)\right.$$
$$\left. + \log(2\pi) + \log \hat{V}^l \frac{\left(Y_i - \mu_{Y,i}^l\right)^2}{\hat{V}^l}\right]$$
$$+ t_i \log \sigma\left(f_1(Z_i^l)\right) + (1 - t_i)\log\left(1 - \sigma\left(f_1(Z_i^l)\right)\right), \tag{5.230}$$

   where

$$\left[\mu_{xi} \mid \sigma_{xi}^2\right] = \text{Relu}\left(W_x^1 Z_i\right) W_x^2, \tag{5.231}$$

$$\mu_{Y,i}^l = t_i \text{Relu}\left(W_Y^2 Z_i^l\right) W_Y^1 + (1 - t_i)\text{Relu}\left(W_Y^4 Z_i^l\right) W_Y^4 \text{ and} \tag{5.232}$$

$$\hat{V}^l = \text{Relu}\left(W_Y^6 Z^l\right) W_Y^5, \tag{5.233}$$

$$\sigma\left(f_1(Z_i^l)\right) = \text{logit}\left(W_t^2 \tanh\left(W_t^1 Z_i + b_z^1\right) + b_t^2\right), \tag{5.234}$$

2. **Binary Outcome**

$$\log p_\theta(Y_i, t_i, X_i \mid Z_i^l)$$
$$= -\frac{1}{2}\left[\sum_{j=1}^{k_x}\left(\log(2\pi) + \log\left(\sigma_{xij}^l\right)^2 + \frac{\left(X_{ij} - \mu_{xij}^l\right)^2}{\left(\sigma_{xij}^l\right)^2}\right)\right]$$
$$+ Y_i \log \hat{\pi}_i^l + (1 - Y_i)\log\left(1 - \hat{\pi}_i^l\right) + t_i \log \sigma\left(f_1(Z_i^l)\right)$$
$$+ (1 - t_i)\log\left(1 - \sigma\left(f_1(Z_i^l)\right)\right), \tag{5.235}$$

where

$$\hat{\pi}_i^l = \text{sigm}(t_i\left(W_Y^2 \tanh\left(W_Y^1 Z_i^l + b_Y^1\right) + b_Y^2\right)$$
$$+ (1 - t_i)\left(W_Y^4 \tanh\left(W_Y^3 Z_i^l + b_Y^3\right) + b_Y^4\right). \tag{5.236}$$

Therefore, the ELBO for data point (Y_i, t_i, X_i) is given by (Exercise 5.10)

$$\mathcal{L}(\theta, \varnothing, Y_i, t_i, X_i) = \frac{1}{L}\sum_{l=1}^{L}\log p_\theta(Y_i, t_i, X_i \mid Z_i^l)$$
$$+ \frac{1}{2}\sum_{j=1}^{k_Z}\left(1 + \log \bar{\sigma}_{ij}^2 - \bar{\sigma}_{ij}^2 - \bar{\mu}_{ij}^2\right). \tag{5.237}$$

The ELBO for the entire data is

$$\mathcal{L}(\theta, \varnothing, Y, t, X) = \frac{1}{N}\sum_{i=1}^{N}\mathcal{L}(\theta, \varnothing, Y_i, t_i, X_i). \tag{5.238}$$

Finally, we study that when a new patient comes, only data X are available. How we can estimate the treatment effect and select the optimal treatment. To solve this problem, we first need to estimate the treatment assignment t and outcome Y. Louizos et al. (2017) gave the following prediction distribution. Define

$$q(t_i|X_i) = \text{Bern}(\pi = \sigma\left(g_4(X_i)\right). \tag{5.239}$$

For continuous outcome, we define the conditional distribution of outcome Y, given t and X:

$$q(Y_i|X_i,t_i)=N(\mu=\bar{\mu}_i,\sigma^2=\bar{v}), \quad (5.240)$$

where

$$\bar{\mu}_i=t_i(g_6\circ g_5(X_i))+(1-t_i)(g_7\circ g_5(X_i)).$$

For the binary outcome, we define

$$q(Y_i|X_i,t_i)=Bern(\pi=\bar{\pi}_i), \quad (5.241)$$

where

$$\bar{\pi}_i=t_i(g_6\circ g_5(X_i))+(1-t_i)(g_7\circ g_5(X_i). \quad (5.242)$$

Adding these two terms to the ELBO, we obtain

$$\tilde{\mathcal{L}}(\theta,\varnothing,Y,t,X)=\mathcal{L}(\theta,\varnothing,Y,t,X)$$
$$+\sum_{i=1}^{N}(\log q(t_i=t_i^*|X_i^*)+\log(Y_i=Y_i^*|X_i^*,t_i^*)), \quad (5.243)$$

where Y_i^*, t_i^*, X_i^* are the observed values for the outcome, treatment, and input covariate variables in the training set.

5.6 INSTRUMENTAL VARIABLE MODELS

Another general approach to causal inference with unobserved confounding is instrumental variable (IV) models. The widely used IV model is IV regression (Hartford et al. 2017; Bennett et al. 2019; Xu et al. 2020). The IV regression attempts to regress the treatment (intervention or cause) on the instrument variable. Classical IV regression consists of two-stage least square (2SLS) estimation (Xu et al. 2020). In the first stage, the treatment is regressed on the IV. In the second stage, the outcome is regressed on the conditional expectation of the treatment (cause), given the IV. The classical IV regression only considers linear regression, but can be extended to nonlinear regression (Hartford et al. 2017; Xu et al. 2020). In this section, we will focus on linear IV regression. Deep learning inspired nonlinear IV regression will be introduced in Chapter 7.

5.6.1 Simple Linear IV Regression and Mendelian Randomization

Linear IV regression for causal inference has been well developed (Baiocchi et al. 2014; Vanderweele et al. 2014; DiPrete et al. 2019). Consider causal model in Figure 5.13:

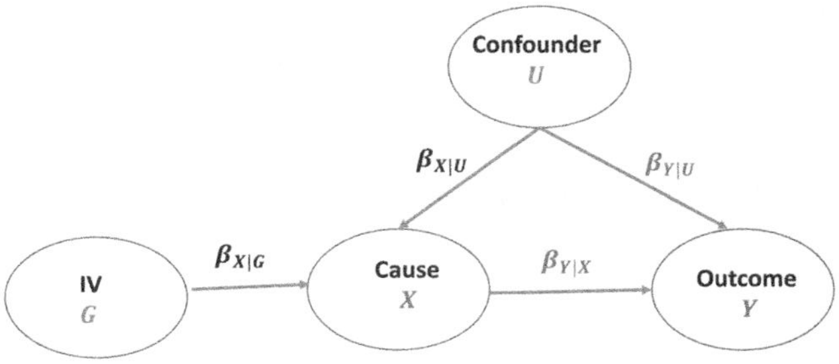

FIGURE 5.13 Illustration of simple linear IV regression.

$$X=\beta_{X|G}G+\beta_{X|U}U \quad (5.244)$$

$$Y=\beta_{Y|X}X+\beta_{Y|U}U, \quad (5.245)$$

where X represents an exposure (treatment, or cause), Y outcomes (effect or response), U an unmeasured confounding variable, and G (IV) (e.g., genotype). We assume that the instrument variable G is independent of the confounding variable U and is associated with the exposure X. Our goal is to estimate a causal effect of the exposure X on the outcome Y. Three methods can be employed to estimate the causal effect of the exposure X on outcome Y.

1. **Covariance Analysis**

Without loss of generality, we can assume that $Var(G)=1$ and $Var(X)=1$. By the assumption that $cov(G,U)=0$ (Figure 5.13).

Taking covariance with G on both sides of equation (5.244) yields

$$cov(X,G)=\beta_{X|G}. \quad (5.246)$$

Similarly, taking covariance with G on both sides of equation (5.245), we obtain $cov(Y,G)=\beta_{Y|X}cov(X,G)$, which implies that

$$\beta_{Y|X}=\frac{cov(Y,G)}{cov(X,G)}. \quad (5.247)$$

Define the regression of Y on G:

$$Y=\beta_{Y|G}G. \quad (5.248)$$

Taking covariance with G, we obtain

$$cov(Y,G) = \beta_{Y|G}. \tag{5.249}$$

Substituting equations (5.246) and (5.249) into equation (5.247) leads to

$$\beta_{Y|X} = \frac{\beta_{Y|G}}{\beta_{X|G}}. \tag{5.250}$$

The causal effect of the exposure X on outcome Y is estimated by $\hat{\beta}_{Y|X} = \frac{\beta_{Y|G}}{\beta_{X|G}}$.

1. **Generalized Least Square Estimator**

Structural equations for modeling linear IV regression in equations (5.244) and (5.245) are

$$-Y + \beta_{Y|X}X + \beta_{Y|U}U + \varepsilon_1 = 0, \tag{5.251}$$

$$-X + \beta_{X|G}G + \beta_{X|U}U + \varepsilon_2 = 0, \tag{5.252}$$

where $\varepsilon_1, \varepsilon_2$ are uncorrelated with U.

Let

$$\beta_{Y|U}U + \varepsilon_1 = e_1, \tag{5.253}$$

$$\beta_{X|U}U + \varepsilon_2 = e_2. \tag{5.254}$$

Substituting equations (5.253) and (5.254) into equations (5.251) and (5.252), we obtain

$$-Y + \beta_{Y|X}X + e_1 = 0, \tag{5.255}$$

$$-X + \beta_{X|G}G + e_2 = 0, \tag{5.256}$$

here Y, X are endogenous variables and denoted by $W = [Y \ X]$, G is exogenous variable and e_1, e_2 are uncorrelated with G.

Let

$$\Gamma = \begin{bmatrix} -1 & 0 \\ \beta_{Y|X} & -1 \end{bmatrix}, B = \begin{bmatrix} 0 & \beta_{X|G} \end{bmatrix}, E = \begin{bmatrix} e_1 & e_2 \end{bmatrix}.$$

Then, $\Gamma^{-1} = \begin{bmatrix} -1 & -\beta_{Y|X} \\ 0 & -1 \end{bmatrix}$.

Then, equations (5.255) and (5.256) can be rewritten in a matrix form as (Xiong 2018b)

$$W\,\Gamma + GB + E = 0. \tag{5.257}$$

Multiplying by Γ^{-1} on both sides of equation (5.257), we obtain

$W + GB\Gamma^{-1} + E\Gamma^{-1} = 0$, which can be reduced to

$$\begin{aligned} W &= -GB\Gamma^{-1} - E\Gamma^{-1} \\ &= G\Pi + V, \end{aligned} \tag{5.258}$$

where

$$\Pi = -B\Gamma^{-1}, \Pi = \begin{bmatrix} 0 & \beta_{X|G} \end{bmatrix}, \text{and} = -E\Gamma^{-1}.$$

Using least square methods to solve optimization problem (5.258), we obtain

$$\Pi = (G^T G)^{-1} G^T W. \tag{5.259}$$

Using the relation $\Pi\Gamma = -B$, we obtain

$(G^T G)^{-1} G^T W\Gamma = -B$, which implies

$$G^T W\Gamma = -G^T GB, \text{ or}$$

$$G^T \begin{bmatrix} Y & X \end{bmatrix} \begin{bmatrix} -1 & 0 \\ \beta_{Y|X} & -1 \end{bmatrix} = -G^T \begin{bmatrix} 0 & G\beta_{X|G} \end{bmatrix}. \tag{5.260}$$

Expanding equation (5.260), we obtain

$$G^T \begin{bmatrix} -Y + X\beta_{Y|X} & -X \end{bmatrix} = G^T \begin{bmatrix} 0 & G\beta_{X|G} \end{bmatrix}.$$

Therefore, we have

$$G^T Y = G^T \ X\beta_{Y|X}, \tag{5.261}$$

$$G^T \ X = G^T \ G\beta_{X|G}. \tag{5.262}$$

Equations (5.261) and (5.262) can be rewritten as

$$G^T \begin{bmatrix} Y \\ X \end{bmatrix} = G^T \begin{bmatrix} X & 0 \\ 0 & G \end{bmatrix} \begin{bmatrix} \beta_{Y|X} \\ \beta_{X|G} \end{bmatrix}. \tag{5.263}$$

Equations (5.255) and (5.256) can be rewritten as

$$\begin{aligned} \begin{bmatrix} Y \\ X \end{bmatrix} &= \begin{bmatrix} X & 0 \\ 0 & G \end{bmatrix} \begin{bmatrix} \beta_{Y|X} \\ \beta_{X|G} \end{bmatrix} + \begin{bmatrix} e_1 \\ e_2 \end{bmatrix} \\ &= Z\delta + e, \end{aligned} \tag{5.264}$$

where

$$Z = \begin{bmatrix} X & 0 \\ 0 & G \end{bmatrix}, \delta = \begin{bmatrix} \beta_{Y|X} \\ \beta_{X|G} \end{bmatrix} \text{ and } e = \begin{bmatrix} e_1 \\ e_2 \end{bmatrix}.$$

Multiplying both sizes of equation (5.264) by G^T, we obtain

$$G^T W^T = G^T Z\delta + G^T e. \tag{5.265}$$

Note that

$$\Sigma = cov\left(G^T e, G^T e\right) = G^T \Lambda G, \tag{5.266}$$

where

$$\Lambda = \begin{bmatrix} \left(\beta_{Y|U}\right)^2 \sigma_{\tilde{U}}^2 + \sigma_{\varepsilon_1}^2 & \beta_{Y|U}\beta_{X|U}\sigma_{\tilde{U}}^2 \\ \beta_{Y|U}\beta_{X|U}\sigma_{\tilde{U}}^2 & \left(\beta_{X|U}\right)^2 \sigma_{\tilde{U}}^2 + \sigma_{\varepsilon_2}^2 \end{bmatrix}$$

Using weighted least square methods, we obtain the generalized least square estimator:

$$\hat{\delta} = \left[Z^T G(G^T \ \Lambda G)^{-1} G^T Z\right]^{-1} Z^T G(G^T \ \Lambda G)^{-1} G^T W^T. \tag{5.267}$$

If we assume $\Lambda = \sigma^2 I$, then equation (5.267) is reduced to

$$\hat{\delta} = \left[Z^T G(G^T G)^{-1} G^T Z\right]^{-1} Z^T G(G^T G)^{-1} G^T W^T. \tag{5.268}$$

5.6.1.1 Two-Stage Least Square Method

Assume that the outcome Y are continuous and the cause-effect relationships between X and Y are linear. A 2-stage least squares (2SLS) regression can be used to estimate the causal effect of the exposure X on outcome Y.

Stage 1:

In the first stage, the treatment variable X is regressed on the IV G:

$$X = \alpha_{X|G} G + \varepsilon_X.$$

Then, the estimator $\hat{\alpha}_{X|G}$ is given by

$$\hat{\alpha}_{X|G} = Cov(X, G). \tag{5.269}$$

The estimator $\hat{X}$ is

$$\hat{X} = \hat{\alpha}_{X|G} G,$$

Stage 2:

In the second stage, the outcome is regressed on the conditional expectation of the treatment, given IV:

$$Y = \alpha_{Y|X} E\{X|G\} + \varepsilon_Y \text{ or}$$

$$Y = \alpha_{Y|X} \hat{\alpha}_{X|G} G + \varepsilon_Y. \tag{5.270}$$

Taking covariance with G on both sides of equations (5.270), we obtain

$$\alpha_{Y|X} \hat{\alpha}_{X|G} = cov\ (Y, G),$$

which implies that

$$\hat{\alpha}_{Y|X} = \frac{cov\ (Y, G)}{\hat{\alpha}_{X|G}}. \tag{5.271}$$

Substituting equation (5.269) into equation (5.271), we obtain

$$\hat{\alpha}_{Y|X} = \frac{cov\ (Y, G)}{Cov(X, G)}. \tag{5.272}$$

Substituting equation (5.247) into equation (5.272), we obtain

$$\hat{\alpha}_{Y|X} = \hat{\beta}_{Y|X},$$

which shows that the estimator of the regression coefficient of the regression of outcome on the estimator of the conditional expectation of the treatment, given IV is the estimator of the causal effect of the exposure X on the outcome Y.

Now we study the relationship between the generalized least square estimator and two-stage least square estimator.

Let

$$\hat{Z} = \begin{bmatrix} \hat{X} & 0 \\ 0 & G \end{bmatrix}, \tag{5.273}$$

where

$$\hat{X} = G(G^T G)^{-1} G^T X.$$

Equation (5.273) can be reduced to

$$\hat{Z} = \begin{bmatrix} G(G^T G)^{-1} G^T X & 0 \\ 0 & GG^T G)^{-1} G^T G \end{bmatrix} = G(G^T G)^{-1} G^T Z. \tag{5.274}$$

Consider two-stage least square estimator:

Stage 1:

Solving the regression which regress the treatment on the IV:

$$X = G\beta_{X|G} + e_X, \tag{5.275}$$

we obtain

$$\hat{\beta}_{X|G} = (G^TG)^{-1}G^TX \text{ and}$$
$$\hat{X} = G(G^TG)^{-1}G^TX. \tag{5.276}$$

Stage 2:

Solving the regression

$$W^T = \hat{Z}\delta + e_w,$$

we obtain

$$\hat{\delta} = \left(\hat{Z}^T\hat{Z}\right)^{-1}\hat{Z}^TW^T. \tag{5.277}$$

Substituting equation (5.274) into equation (5.277), we obtain

$$\begin{aligned}\hat{\delta} &= \left[Z^TG(G^TG)^{-1}G^TG(G^TG)^{-1}G^TZ\right]^{-1}Z^TG(G^TG)^{-1}G^TW^T \\ &= \left[Z^TG(G^TG)^{-1}G^TZ\right]^{-1}Z^TG(G^TG)^{-1}G^TW^T,\end{aligned} \tag{5.278}$$

which is the same as equation (5.268).

5.6.1.2 Assumptions of IV

Instruments were originally taken as exogenous variables in SEMs (Lousdal 2018). IV concept can be parallel to randomization. IV is defined as a variable that is associated with the exposure (cause) but is not associated with the outcome if the exposure (cause) is given.

There are three core assumptions for IV:

1. The IV is associated with the exposure (potential cause) and has to predict the exposure.
   The association of the IV with the exposure is either due to direct causation of the IV to the exposure or due to a shared common cause of the IV and exposure.
2. The IV affects the outcome only through the exposure. The IV cannot have any direct associations with the outcome.
3. The IV does not share common cause with the outcome. The IV is independent of factors (measured and unmeasured) that confound both exposure and outcome.

It is well known that the randomly inherited genetic factors are not changed by traits and over time. They are independent of socioeconomic and many environmental factors (Teumer 2018). Therefore, genetic factors known as Mendelian randomization are popular IV in epidemiology studies (Lousdal 2018). Other IVs include geographic variation variable, physical distance to a facility, and treatment preference. However, identifying a valid IV is a challenge problem.

5.6.2 IV and Deep Latent Variable Models

Comparing Figure 5.13 with Figure 5.12(a), we found that the IV regression and unobserved confound model are quite similar. In this section, we apply the VAE to IV regression. For the easy of presentation, the architecture of IV for causal inference with VAE is shown in Figure 5.14. We first define decoder.

5.6.2.1 Decoder

Again, the latent variable Z is viewed as the unobserved confounder. Let X_i be the IV, T_i be the treatment variable (cause variable), Y_i be the outcome (effect) of the particular treatment (cause) of the ith sample. The decoder is defined as $p_\theta(Y,T,X|Z)$. The decoding probability $p_\theta(Y,T,X|Z)$ can be factorized as

$$p_\theta(Y,T,X|Z) = p_\theta(Y|T,X,Z)p_\theta(T|X,Z)p_\theta(X|Z). \tag{5.279}$$

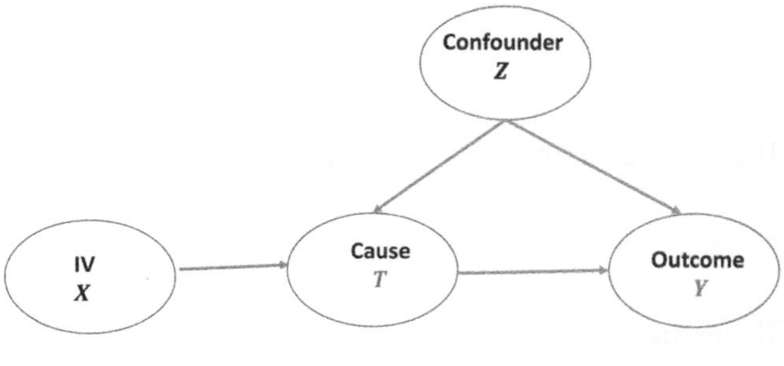

FIGURE 5.14 Model of the IV for causal inference with VAE.

By assumption three of the IV, the IV X is independent of the confounder Z, i.e.,

$$p_\theta(X|Z) = p_\theta(X). \tag{5.280}$$

Substituting equation (5.280) into equation (5.279), we obtain

$$p_\theta(Y,T,X|Z) = p_\theta(Y|T,X,Z)\,p_\theta(T|X,Z)\,p_\theta(X). \tag{5.281}$$

The prior distributions of Z are assumed to follow a Gaussian distribution:

$$p(Z_i) = N(Z_i|0,1). \tag{5.282}$$

If we assume that X is continuous normal variable, then the log distributions of X is

$$\log p_\theta(X_i) = -\frac{1}{2}\left(\log(2\pi) + \log\sigma_X^2 + \frac{(X_i - \mu_X^2)^2}{\sigma_X^2}\right). \tag{5.283}$$

If X is a discrete variable, then Bernoulli distribution or other discrete distribution are used for $p_\theta(X)$.

If the exposure T is a continuous variable, then the log-conditional distribution of T, given X and Z is

$$\log p_\theta(T_i|X_i,Z_i) = -\frac{1}{2}\left(\log(2\pi) + \log\sigma_T^2 + \frac{(T_i - \mu_T)^2}{\sigma_T^2}\right), \tag{5.284}$$

where

$$[\mu_T|\sigma_T^2] = \text{Relu}\left(W_T^1\begin{bmatrix} X \\ Z \end{bmatrix}\right)W_T^2. \tag{5.285}$$

Let Bern denote Bernoulli distribution and σ be the logistic function. If the exposure T is a discrete variable, then the log-conditional distribution of T, given Z is

$$\begin{aligned} \log p(T_i|X_i,Z_i) &= T_i \log\sigma(f_1(X_i, Z_i)) \\ &+ (1-T_i)\log(1-\sigma(f_1(X_i, Z_i))), \end{aligned} \tag{5.286}$$

where

$$\sigma(f_1(X_i, Z_i)) = \text{logit}\left(W_T^2 \tanh\left(W_t^1\begin{bmatrix} X_i \\ Z_i \end{bmatrix} + b_z^1\right) + b_T^2\right). \tag{5.287}$$

Consider both continuous and discrete outcomes. The distribution of continuous outcome is assumed to follow a Gaussian distribution. The conditional distribution of Y_i, given T_i, Z_i, is

1. **Binary Exposure**

$$\begin{aligned} p(Y_i|T_i, Z_i) &= N(Y_i; \hat{\mu}_i, \hat{V}), \\ \hat{\mu}_i &= T_i f_2(Z_i) + (1-T_i) f_3(Z_i), \end{aligned} \tag{5.288}$$

where

$$f_2(Z_i) = \text{Relu}(W_Y^6 Z_i)W_Y^5, f_3(Z_i) = \text{Relu}(W_Y^8 Z_i)W_Y^7;$$

2. **Continuous Exposure**

$$\log p(Y_i|T_i, Z_i) = -\frac{1}{2}\left(\log(2\pi) + \log\hat{V} + \frac{(Y_i - \mu_{Yi})^2}{\hat{V}}\right), \tag{5.289}$$

where

$$\mu_{Yi} = \text{Relu}\left(W_Y^2\begin{bmatrix} T_i \\ Z_i \end{bmatrix}\right)W_Y^1 \text{ and}$$

$$\hat{V} = \text{Relu}\left(W_Y^6\begin{bmatrix} T \\ Z \end{bmatrix}\right)W_Y^5,$$

$$T = [T_1, \ldots, T_n]^T, Z = [Z_1, \ldots, Z_n]^T.$$

For the binary outcome, the conditional distribution of Y_i, given T_i, Z_i, is

$$\begin{aligned} p(Y_i|T, Z_i) &= Bern(\pi = \hat{\pi}_i), \\ \hat{\pi}_i &= \sigma((T_i f_2(Z_i) + (1-T_i) f_3(Z_i)), \text{ or} \\ \log p(Y_i|T_i, Z_i) &= Y_i \log\hat{\pi}_i + (1-Y_i)\log(1-\hat{\pi}_i), \end{aligned} \tag{5.290}$$

where

$$\hat{\pi}_i = \text{sigm}\left(T_i\left(W_Y^2 \tanh\left(W_Y^1 Z_i + b_Y^1\right) + b_Y^2\right)\right.$$
$$\left. + (1 - T_i)\left(W_Y^4 \tanh\left(W_Y^3 Z_i + b_Y^3\right) + b_Y^4\right)\right).$$

These nonlinear functions f_j are implemented by NNs with parameters θ_j.

5.6.2.2 Encoder

The encoder is defined as the approximate estimator of the posterior distribution of Z_i, given the observed data Y_i, T, X_i (Louizos et al. 2017)

$$q_{\varnothing}(Z_i|Y_i, T_i, X_i) = N(Z_i; \mu_i, \sigma_i^2), \quad (5.291)$$

where

1. **Binary Exposure**

$$\mu_i = T_i \mu_{t=1,i} + (1 - T_i)\mu_{t=0,i}, \sigma_i^2 = T_i \sigma_{t=1,i}^2 + (1 - T_i)\sigma_{t=0,i}^2, \quad (5.292)$$

$$\left[\mu_{T=1,i}|\sigma_{t=1,i}^2\right] = g_2 \odot g_1(X_i, Y_i), \left[\mu_{T=0,i}|\sigma_{T=0,i}^2\right] = g_3 \odot g_1(X_i, Y_i), \quad (5.293)$$

where g_1, g_2, g_3 are nonlinear functions and implemented by NNs with parameters $\varnothing$, and $\odot$ denotes element-wise multiplications.

2. **Continuous Exposure**

$$\mu_i = g_4(Y_i, T_i, X_i), \sigma_i^2 = g_5(Y_i, T_i, X_i), \quad (5.294)$$

where g_4, g_5 are nonlinear functions and implemented by NNs.

5.6.2.3 ELBO

Log-likelihood function $\log p_\theta(Y_i, T_i, X_i)$ for the DAG in Figure 5.12(a) is bounded by (Kingma and Welling 2013; Louizos et al. 2017)

$$\log p_\theta(Y_i, T_i, X_i) \geq \mathcal{L}(\theta, \varnothing, Y_i, T_i, X_i), \quad (5.295)$$

where

$$\mathcal{L}(\theta, \varnothing, Y_i, T, X_i) = E_{q_{\varnothing}(Z_i|Y_i, T_i, X_i)}\left[\log p_\theta(Y_i, T_i, X_i|Z_i) - KL\left(q_{\varnothing}(Z_i|Y_i, T_i, X_i)||p(Z_i)\right)\right], \quad (5.296)$$

$\mathcal{L}(\theta, \varnothing, Y_i, T_i, X_i)$ is the ELBO for data point (Y_i, T_i, X_i).

Note that

$$\log p_\theta(Y_i, T_i, X_i|Z_i) = \log p_\theta(Y_i|T_i, Z_i) + \log p_\theta(T_i|X_i, Z_i) + \log p_\theta(X). \quad (5.297)$$

Substituting equation (5.297) into equation (5.296), we obtain

$$\mathcal{L}(\theta, \varnothing, Y_i, T_i, X_i) = E_{q_{\varnothing}(Z_i|Y_i, T_i, X_i)} \times\left[\log p_\theta(Y_i|T_i, Z_i) + \log p_\theta(T_i|X_i, Z_i) + \log p_\theta(X) - KL\left(q_{\varnothing}(Z_i|Y_i, T_i, X_i)||p(Z_i)\right)\right]. \quad (5.298)$$

We can show (Exercise 5.11) that

$$-KL\left(q_{\varnothing}(Z_i|Y_i, T_i, X_i)||p(Z_i)\right)] = \frac{1}{2}\left(1 + \log \bar{\sigma}_i^2 - \bar{\sigma}_i^2 - \bar{\mu}_i^2\right), \quad (5.299)$$

where

1. **Binary Exposure**

$$\bar{\mu}_i = t_i \mu_{t=1,i} + (1 - t_i)\mu_{t=0,i}, \bar{\sigma}_i^2 = t_i \sigma_{t=1,i}^2 + (1 - t_i)\sigma_{t=0,i}^2, \quad (5.300)$$

$$\left[\mu_{t=1,i}|\sigma_{t=1,i}^2\right] = g_2 \odot g_1(X_i, Y_i), \left[\mu_{t=0,i}|\sigma_{t=0,i}^2\right] = g_3 \odot g_1(X_i, Y_i), \quad (5.301)$$

2. **Continuous Exposure**

$$\bar{\mu}_i = g_4(Y_i, T_i, X_i), \bar{\sigma}_i^2 = g_5(Y_i, T_i, X_i), \quad (5.302)$$

where g_4, g_5 are nonlinear functions and implemented by NNs.

Next we calculate $\log p_\theta(Y_i, T, X_i \mid Z_i^l)$. First, define transformation:

$$Z_i^l = \bar{\mu}_i + \bar{\sigma}_i \odot \varepsilon^l, \varepsilon^l \sim N(0, I), l = 1, 2, \ldots, L. \quad (5.303)$$

To calculate $\log p_\theta(Y_i, T, X_i \mid Z_i^l)$, each of the IV, exposure, and outcome can be either binary or continuous variables. There are four combinations to be discussed. We only consider three combinations: (1) binary exposure and IV, and continuous outcome, (2) all exposure, IV, and outcome are binary, and (3) all exposure, IV, and outcomes are continuous. The fourth combination is left for exercises.

1. **Continuous Outcome, Binary Exposure, and Binary IV**

Combining equations (5.283)–(5.290), we obtain

$$\log p_\theta\left(Y_i, T_i, X_i \mid Z_i^l\right) = X_i \log p_X + \left(1-X_i\right)\log\left(1-p_x\right)$$
$$+ T_i \log\sigma\left(f_1\left(X_i, Z_i^l\right)\right) + \left(1-T_i\right)\log\left(1-\sigma\left(f_1\left(X_i, Z_i^l\right)\right)\right)$$
$$-\frac{1}{2}\left(\log(2\pi) + \log\hat{V}^l + \frac{\left(Y_i-\mu_{Yi}^l\right)^2}{\widehat{\hat{V}^l}}\right), \tag{5.304}$$

where

$$\sigma\left(f_1\left(X_i, Z_i^l\right)\right) = \text{logit}\left(W_T^2 \tanh\left(W_t^1\begin{bmatrix} X_i \\ Z_i^l\end{bmatrix} + b_z^1\right) + b_T^2\right), \tag{5.305}$$

$$\mu_{Y,i}^l = T_i \text{Relu}\left(W_Y^2 Z_i^l\right)W_Y^1 + \left(1-T_i\right)\text{Relu}\left(W_Y^4 Z_i^l\right)W_Y^4 \text{ and} \tag{5.306}$$

$$\hat{V}^l = \text{Relu}\left(W_Y^6 Z^l\right)W_Y^5. \tag{5.307}$$

2. **Binary Outcome, Binary Exposure, and Binary IV**

In this case, we have

$$\log p_\theta\left(Y_i, T_i, X_i \mid Z_i^l\right) = X_i \log p_X + \left(1-X_i\right)\log\left(1-p_x\right)$$
$$+ T_i \log\sigma\left(f_1\left(X_i, Z_i^l\right)\right) + \left(1-T_i\right)\log\left(1-\sigma\left(f_1\left(X_i, Z_i^l\right)\right)\right)$$
$$-Y_i \log\hat{\pi}_i^l - \left(1-Y_i\right)\log\left(1-\hat{\pi}_i^l\right). \tag{5.308}$$

where

$$\sigma\left(f_1\left(X_i, Z_i^l\right)\right) = \text{logit}\left(W_T^2 \tanh\left(W_t^1\begin{bmatrix} X_i \\ Z_i^l\end{bmatrix} + b_z^1\right) + b_T^2\right), \tag{5.309}$$

$$\hat{\pi}_i^l = \text{sigm}\Big(T_i\left(W_Y^2 \tanh\left(W_Y^1 Z_i^l + b_Y^1\right) + b_Y^2\right)$$
$$+\left(1-T_i\right)\left(W_Y^4 \tanh\left(W_Y^3 Z_i^l + b_Y^3\right) + b_Y^4\right). \tag{5.310}$$

3. **Continuous Outcome, Continuous Exposure, and Continuous IV**

$$\log p_\theta\left(Y_i, T_i, X_i \mid Z_i^l\right) = -\frac{1}{2}\left(\log(2\pi) + \log\sigma_X^2 + \frac{\left(X_i-\mu_X^2\right)^2}{\sigma_X^2}\right)$$
$$-\frac{1}{2}\left(\log(2\pi) + \log\left(\sigma_T^l\right)^2 + \frac{\left(T_i-\mu_T^l\right)^2}{\left(\sigma_T^l\right)^2}\right) \tag{5.311}$$
$$-\frac{1}{2}\left(\log(2\pi) + \log\hat{V}^l + \frac{\left(Y_i-\mu_{Yi}^l\right)^2}{\widehat{\hat{V}^l}}\right).$$

where

$$\left[\mu_T^l \middle| \left(\sigma_T^l\right)^2\right] = \text{Relu}\left(W_T^1\begin{bmatrix} X \\ Z_i^l\end{bmatrix}\right)W_T^2, \tag{5.312}$$

$$\mu_{Yi}^l = \text{Relu}\left(W_Y^2\begin{bmatrix} T_i \\ Z_i^l\end{bmatrix}\right)W_Y^1, \tag{5.313}$$

$$\widehat{\hat{V}^l} = \text{Relu}\left(W_Y^6\begin{bmatrix} T \\ Z^l\end{bmatrix}\right)W_Y^5,\ T = \left[T_1, \ldots, T_n\right]^T,$$
$$Z = \left[Z_1, \ldots, Z_n\right]^T. \tag{5.314}$$

Therefore, the ELBO $\mathcal{L}(\theta, \varnothing, Y_i, T, X_i)$ for data point (Y_i, T, X_i) is approximated by

$$\mathcal{L}(\theta, \varnothing, Y_i, T, X_i) = \frac{1}{L}\sum_{l=1}^{L}\log p_\theta\left(Y_i, T_i, X_i | Z_i^l\right)$$
$$+\frac{1}{2}\left(1+\log\bar{\sigma}_i^2 - \bar{\sigma}_i^2 - \bar{\mu}_i^2\right). \tag{5.315}$$

The ELBO for the entire data is

$$\mathcal{L}(\theta, \varnothing, Y, t, X) = \frac{1}{N}\sum_{i=1}^{N}\mathcal{L}\left(\theta, \varnothing, Y_i, T_i, X_i\right). \tag{5.316}$$

SOFTWARE PACKAGE

Software for implementing Causal Discovery with Cascade Nonlinear Additive Noise Models can be downloaded from https://github.com/DMIRLAB-Group/CANM

Software for implementing the classifier two sample tests can be downloaded from https://github.com/lopezpaz/classifier_tests

DAGs with NO TEARS: Continuous Optimization for Structure Learning are publicly available at https://github.com/xunzheng/notears

Using Embeddings to Correct for Unobserved Confounding in Networks Code is available at github.com/vveitch/causal-network-embeddings

The code for "Disentangled Generative Causal Representation Learning" is available at https://github.com/xwshen51/DEAR

Code for "Deep generalized method of moments for instrumental variable analysis" can be downloaded from https://github.com/CausalML/DeepGMM

APPENDIX 5A: DERIVE EVIDENCE LOWER BOUND (ELBO) FOR ANM

Since $P_\theta\left(X^{(i)}, Y^{(i)}\right)$ does not explicitly contain latent variable Z, the following equality holds:

$$\log P_\theta\left(X^{(i)}, Y^{(i)}\right) = \int q_\varnothing\left(Z \mid X^{(i)}, Y^{(i)}\right)\log P_\theta\left(X^{(i)}, Y^{(i)}\right)dZ. \quad (5A1)$$

Note that

$$P_\theta\left(X^{(i)}, Y^{(i)}\right) = \frac{P_\theta\left(X^{(i)}, Y^{(i)}, Z\right)}{P_\theta\left(Z \mid X^{(i)}, Y^{(i)}\right)}. \quad (5A2)$$

Substituting equation (5A2) into equation (5A1) yields

$$\begin{aligned}\log P_\theta\left(X^{(i)}, Y^{(i)}\right) &= \int q_\varnothing\left(Z \mid X^{(i)}, Y^{(i)}\right)\log\frac{P_\theta\left(X^{(i)}, Y^{(i)}, Z\right)}{P_\theta\left(Z \mid X^{(i)}, Y^{(i)}\right)}dZ \\ &= \int q_\varnothing\left(Z|X^{(i)}, Y^{(i)}\right)\log P_\theta\left(X^{(i)}, Y^{(i)}, Z\right)dZ \\ &\quad - \int q_\varnothing\left(Z|X^{(i)}, Y^{(i)}\right)\log P_\theta\left(Z|X^{(i)}, Y^{(i)}\right)dZ \\ &\quad - \int q_\varnothing\left(Z|X^{(i)}, Y^{(i)}\right)\log q_\varnothing\left(Z|X^{(i)}, Y^{(i)}\right)dZ \\ &\quad + \int q_\varnothing\left(Z|X^{(i)}, Y^{(i)}\right)\log q_\varnothing\left(Z|X^{(i)}, Y^{(i)}\right)dZ.\end{aligned} \quad (5A3)$$

Combining the first term and third term in equation (5A3), we obtain

$$\begin{aligned}&\int q_\varnothing\left(Z|X^{(i)}, Y^{(i)}\right)\log P_\theta\left(X^{(i)}, Y^{(i)}, Z\right)dZ \\ &- \int q_\varnothing\left(Z|X^{(i)}, Y^{(i)}\right)\log q_\varnothing\left(Z|X^{(i)}, Y^{(i)}\right)dZ \\ &= \int q_\varnothing\left(Z|X^{(i)}, Y^{(i)}\right)\log\frac{P_\theta\left(X^{(i)}, Y^{(i)}, Z\right)}{q_\varnothing\left(Z|X^{(i)}, Y^{(i)}\right)}dZ.\end{aligned} \quad (5A4)$$

Combining the second term and the fourth term in equation (5A3), we obtain

$$\begin{aligned}&\int q_\varnothing\left(Z|X^{(i)}, Y^{(i)}\right)\log q_\varnothing\left(Z|X^{(i)}, Y^{(i)}\right)dZ \\ &- \int q_\varnothing\left(Z|X^{(i)}, Y^{(i)}\right)\log P_\theta\left(Z|X^{(i)}, Y^{(i)}\right)dZ \\ &= KL\left(q_\varnothing\left(Z|X^{(i)}, Y^{(i)}\right) \| P_\theta\left(Z|X^{(i)}, Y^{(i)}\right)\right).\end{aligned} \quad (5A5)$$

Substituting equations (5A4) and (5A5) into equation (5A3) yields

$$\begin{aligned}\log P_\theta\left(X^{(i)}, Y^{(i)}\right) &= E_{q_\varnothing\left(Z|X^{(i)}, Y^{(i)}\right)}\left[\log\frac{P_\theta\left(X^{(i)}, Y^{(i)}, Z\right)}{q_\varnothing\left(Z|X^{(i)}, Y^{(i)}\right)}\right] \\ &\quad + KL\left(q_\varnothing\left(Z|X^{(i)}, Y^{(i)}\right) \| P_\theta\left(Z|X^{(i)}, Y^{(i)}\right)\right).\end{aligned} \quad (5A6)$$

Sine KL distance $KL\left(q_\varnothing\left(Z|X^{(i)}, Y^{(i)}\right) \| P_\theta\left(Z|X^{(i)}, Y^{(i)}\right)\right) \geq 0$, equation (5A6) provides the lower bound:

$$\log P_\theta\left(X^{(i)}, Y^{(i)}\right) \geq E_{q_\varnothing\left(Z|X^{(i)}, Y^{(i)}\right)}\left[\log\frac{P_\theta\left(X^{(i)}, Y^{(i)}, Z\right)}{q_\varnothing\left(Z|X^{(i)}, Y^{(i)}\right)}\right]. \quad (5A7)$$

Define the ELBO as

$$\mathcal{L}\left(\theta, \varnothing, X^{(i)}, Y^{(i)}\right) = E_{q_\varnothing\left(Z|X^{(i)}, Y^{(i)}\right)}\left[\log\frac{P_\theta\left(X^{(i)}, Y^{(i)}, Z\right)}{q_\varnothing\left(Z|X^{(i)}, Y^{(i)}\right)}\right]. \quad (5A8)$$

Note that

$$\log P_\theta\left(X^{(i)}, Y^{(i)}, Z\right) = \log P_\theta(Z) + \log P_\theta\left(X^{(i)}, Y^{(i)} \mid Z\right). \quad (5A9)$$

Substituting equation (5A9) into equation (5A8) leads to

$$\begin{aligned}\mathcal{L}\left(\theta, \varnothing, X^{(i)}, Y^{(i)}\right) &= E_{q_\varnothing\left(Z|X^{(i)}, Y^{(i)}\right)}\left[\log P_\theta\left(X^{(i)}, Y^{(i)} \mid Z\right)\right] \\ &\quad - KL\left(q_\varnothing\left(Z|X^{(i)}, Y^{(i)}\right) \| P_\theta(Z)\right).\end{aligned} \quad (5A10)$$

Using conditional probability concept, we obtain

$$P_\theta\left(X^{(i)}, Y^{(i)}|Z\right)=P_\theta\left(X^{(i)}\,|\,Z\right)P_\theta\left(Y^{(i)}\,|\,X^{(i)},Z\right). \quad (5A11)$$

The ANM model (5.36) gives

$$P_\theta\left(Y^{(i)}|X^{(i)},Z\right)=P_\theta\left(N_Y^{(i)}=Y^{(i)}-f_Y\left(X^{(i)}\right)\right). \quad (5A12)$$

Substituting equation (5A12) into equation (5A11), we obtain

$$\begin{aligned}P_\theta\left(X^{(i)}, Y^{(i)}|Z\right)&=P_\theta\left(X^{(i)}\,|\,Z\right)P_\theta\left(N_Y^{(i)}=Y^{(i)}-f_Y\left(X^{(i)}\right)\right)\\&=P_\theta\left(X^{(i)}\right)P_\theta\left(N_Y^{(i)}=Y^{(i)}-f_Y\left(X^{(i)}\right)\right). \quad (5A13)\end{aligned}$$

Substituting equation (5A13) into equation (5A10), we have

$$\begin{aligned}\mathcal{L}\left(\theta,\varnothing, X^{(i)}, Y^{(i)}\right)&=\log P_\theta\left(X^{(i)}\right)\\&+E_{q_\varnothing\left(Z|X^{(i)},Y^{(i)}\right)}\left[\log P_\theta\left(N_Y^{(i)}=Y^{(i)}-f_Y\left(X^{(i)}\right)\right)\right]\\&-KL\left(q_\varnothing\left(Z|X^{(i)}, Y^{(i)}\right)\|P_\theta(Z)\right). \quad (5A14)\end{aligned}$$

APPENDIX 5B: APPROXIMATION OF EVIDENCE LOWER BOUND (ELBO) FOR ANM

A key to the computation of the ELBO is how to approximate the posterior distribution $q_\varnothing\left(Z|X^{(i)},Y^{(i)}\right)$. Kingma and Welling (2013) proposed to use the reparameterization trick and variational Bayesian method to approximate the posterior distribution $q_\varnothing\left(Z|X^{(i)},Y^{(i)}\right)$. Let $g_\varnothing(\varepsilon, X,Y)$ be a differentiable function. Define transformation of the latent variable Z to a new variable $\tilde{Z}$ as

$$\tilde{Z}=g_\varnothing(\varepsilon, X,Y), \quad (5B1)$$

where $\varepsilon \sim P(\varepsilon)$ is a random variable. By appropriate selection of the transformation function and random variable ε, we make the distribution of $\tilde{Z}$ to approximate the distribution $q_\varnothing(Z\,|\,X,Y)$. Using the variable change theorem in statistics, we can establish the relationships between the distribution of the transformed variable Z and the distribution of the random variable ε:

$$P\left(\tilde{Z}\right)=\frac{P(\varepsilon)}{\left|\frac{\partial g_\varnothing(\varepsilon, X,Y)}{\partial\varepsilon}\right|}. \quad (5B2)$$

Our goal is to make

$$P\left(\tilde{Z}\right)=\frac{P(\varepsilon)}{\left|\frac{\partial g_\varnothing(\varepsilon, X,Y)}{\partial\varepsilon}\right|}=q_\varnothing(Z\,|\,X,Y). \quad (5B3)$$

Using equation (5B3) we can calculate expectation $E_{q_\varnothing(Z|X,Y)}\left[f(Z)\right]$ as follows:

$$\begin{aligned}E_{q_\varnothing(Z|X,Y)}\left[f(Z)\right]&=\int q_\varnothing(Z|X,Y)f(Z)dZ\\&=\int\frac{P(\varepsilon)}{\left|\frac{\partial g_\varnothing(\varepsilon, X,Y)}{\partial\varepsilon}\right|}f\left(g_\varnothing(\varepsilon, X,Y)\right)\\&\quad\times\left|\frac{\partial g_\varnothing(\varepsilon, X,Y)}{\partial\varepsilon}\right|d\varepsilon\\&=\int P(\varepsilon)f\left(g_\varnothing(\varepsilon, X,Y)\right)d\varepsilon\\&=E_\varepsilon\left[f\left(g_\varnothing(\varepsilon, X,Y)\right)\right]. \quad (5B4)\end{aligned}$$

Sampling $\varepsilon^{(l)} \sim P(\varepsilon)$, $l=1, 2, \ldots, L$ and using equation (5B1), we obtain the sampled latent variable

$$Z^{(i,l)}=g_\varnothing\left(\varepsilon^l, X^i,Y^{(i)}\right), i=1,\ldots,n, l=1,\ldots,L. \quad (5B5)$$

Using this sampling scheme, the Monte Carlo estimator of the ELBO is

$$\begin{aligned}\tilde{\mathcal{L}}\left(\theta,\varnothing, X^{(i)}, Y^{(i)}\right)&=\frac{1}{L}\sum_{l=1}^{L}\log P_\theta(Y^{(i)}=f\left(X^{(i)},Z^{(i,l)}\right)\\&-KL\left(q_\varnothing\left(Z|X^{(i)},Y^{(i)}\right)\|P_\theta(Z)\right). \quad (5B6)\end{aligned}$$

Assume that n data points are sampled. Using minibatch principle, the ELBO for the entire dataset can be approximated by

$$\mathcal{L}(\theta,\varnothing, X, Y)=\frac{n}{M}\sum_{i=1}^{M}\tilde{\mathcal{L}}\left(\theta,\varnothing, X^{(i)}, Y^{(i)}\right), \quad (5B7)$$

where M is the sample size in the minibatch.

APPENDIX 5C: COMPUTATION OF KL DISTANCE

Since we assume $P_\theta(Z)=N(0,1)$, we have (Kingma and Welling 2013)

$$\log P_\theta(Z)=-\frac{1}{2}\log(2\pi)-\frac{1}{2}Z^2. \quad (5C1)$$

Taking expectation on both sides of equation (5C1) yields

$$E_{q_\varnothing(Z^{(l)}|X^{(i)},\,Y^{(i)})}\left[\log P_\theta(Z)\right] = -\frac{1}{2}\log(2\pi) - \frac{1}{2}\left(\left(\mu^{(i)}\right)^2 + \left(\sigma^{(i)}\right)^2\right). \tag{5C2}$$

Recall that

$$\log q_\varnothing\left(Z^{(l)}|X^{(i)},\,Y^{(i)}\right) = -\frac{1}{2}\log(2\pi) - \frac{1}{2}\log\left(\left(\sigma^{(i)}\right)^2\right) - \frac{1}{2}\frac{\left(Z^{(l)}-\mu^{(i)}\right)^2}{\left(\sigma^{(i)}\right)^2}. \tag{5C3}$$

Again, taking expectation on both sides of equation (5C3), we obtain

$$E_{q_\varnothing(Z^{(l)}|X^{(i)},\,Y^{(i)})}\left[\log q_\varnothing\left(Z^{(l)}|X^{(i)},\,Y^{(i)}\right)\right] = -\frac{1}{2}\log(2\pi) - \frac{1}{2}\log\left(\left(\sigma^{(i)}\right)^2\right) - \frac{1}{2}. \tag{5C4}$$

Combining equations (5C2) and (5C4), we obtain

$$KL\left(q_\varnothing\left(Z|X^{(i)},Y^{(i)}\right)\|P_\theta(Z)\right) = -\frac{1}{2}\left[1 + \log\left(\sigma^{(i)}\right)^2 - \left(\mu^{(i)}\right)^2 - \left(\sigma^{(i)}\right)^2\right]. \tag{5C5}$$

APPENDIX 5D: BFGS AND LIMITED BFGS UPDATING ALGORITHM

The Broyden-Fletcher-Goldfarb-Shanno (BFGS) (Liu 2014) updating techniques that is an optimization algorithm in the family of quasi-Newton methods can be used to approximate the Hessian matrix. Consider a second order Taylor expansion of the function $f(W)$ at the current iterate W_{i+1}:

$$f(W) \approx f(W_{i+1}) + \nabla^T f(W_{i+1})(W - W_{i+1}) + \frac{1}{2}\Delta^T W H(W_{i+1})\Delta W, \tag{5D1}$$

where H is a Hessian matrix.

Taking gradient at W_i on both sides of equation (5D1) yields

$$\nabla f(W_i) = \nabla f(W_{i+1}) + H(W_{i+1})(W_i - W_{i+1}). \tag{5D2}$$

Let $E_{i+1} = H(W_{i+1})$. Then, it follows from equation (5D2) that

$$E_{i+1}(W_{i+1} - W_i) = \nabla f(W_{i+1}) - \nabla f(W_i). \tag{5D3}$$

Let

$$g_i = \nabla f(W_{i+1}) - \nabla f(W_i), \tag{5D4}$$

$$s_i = W_{i+1} - W_i. \tag{5D5}$$

Substituting equations (5D4) and (5D5) into equation (5D3) leads to

$$E_{i+1}s_i = g_i. \tag{5D6}$$

Let

$$E_{i+1} = E_i + F_i = E_i + \alpha u_i u_i^T + \beta v_i v_i^T, \tag{5D7}$$

Substituting equations (5D7) into equation (5D6), we obtain

$$\left(\alpha u_i^T s_i\right)u_i + \left(\beta v_i^T s_i\right)v_i = g_i - E_i s_i. \tag{5D8}$$

Define

$$u_i = E_i s_i \text{ and } v_i = g_i. \tag{5D9}$$

Substituting equation (5D9) into equation (5D8), we obtain

$$\left(1 + \alpha u_i^T s_i\right)E_i s_i + \left(\beta v_i^T s_i - 1\right)g_i = 0. \tag{5D10}$$

To make equation (5.10) holds for any values $E_i s_i$ and g_i, it must be that

$$1 + \alpha u_i^T s_i = 0, \beta v_i^T s_i - 1 = 0, \tag{5D11}$$

which implies

$$\alpha = -\frac{1}{u_i^T s_i} = -\frac{1}{s_i^T E_i s_i}, \tag{5D12}$$

$$\beta = \frac{1}{v_i^T s_i} = \frac{1}{g_i^T s_i}. \tag{5D13}$$

Substituting equations (5D9), (5D12), and (5D13) into equation (5D7), we obtain the BFGS update:

$$E_{i+1} = E_i + \frac{g_i g_i^T}{g_i^T s_i} - \frac{E_i s_i s_i^T E_i}{s_i^T E_i s_i}.$$

Before derivation, we introduce Lemma 5D.1.

Lemma 5D.1: Sherman-Morrison-Woodbury Formula

Let A be an $n\times n$ dimensional matrix, u and v be $n\times k$ dimensional matrix, and W be a $k\times k$ dimensional matrix. Then,

$$\left(A+UWV^T\right)^{-1}=A^{-1}-A^{-1}UW\left(I+V^TA^{-1}UW\right)^{-1}V^TA^{-1}. \tag{5D14}$$

Now we derive recursive formula for approximation of the inverse of Hessian matrix. Let $G_i=E_i^{-1}$. Then,

$$G_{i+1}=E_{i+1}^{-1}=\left(E_i+\frac{g_ig_i^T}{g_i^Ts_i}-\frac{E_is_is_i^TE_i}{s_i^TE_is_i}\right)^{-1}. \tag{5D15}$$

Let

$$A=E_i+\frac{g_ig_i^T}{g_i^Ts_i},\ u=-E_is_i,\ v^T=s_i^TE_i,\ W=\frac{1}{s_i^TE_is_i}.$$

Then, using Sherman-Morrison-Woodbury Formula, we obtain

$$G_{i+1}=\left(E_i+\frac{g_ig_i^T}{g_i^Ts_i}\right)^{-1}+\left(E_i+\frac{g_ig_i^T}{g_i^Ts_i}\right)^{-1}\times\frac{E_is_is_i^TE_i}{s_i^TE_is_i-s_i^TE_i\left(E_i+\frac{g_ig_i^T}{g_i^Ts_i}\right)^{-1}E_is_i}\left(E_i+\frac{g_ig_i^T}{g_i^Ts_i}\right)^{-1}. \tag{5D16}$$

Again, applying Sherman-Morrison-Woodbury Formula to $\left(E_i+\frac{g_ig_i^T}{g_i^Ts_i}\right)^{-1}$, we obtain

$$\left(E_i+\frac{g_ig_i^T}{g_i^Ts_i}\right)^{-1}=E_i^{-1}-\frac{E_i^{-1}g_ig_i^TE_i^{-1}}{g_i^Ts_i+g_i^TE_i^{-1}g_i}, \tag{5D17}$$

which implies

$$\begin{aligned}&s_i^TE_is_i-s_i^TE_i\left(E_i+\frac{g_ig_i^T}{g_i^Ts_i}\right)^{-1}E_is_i\\&=s_i^TE_is_i-s_i^TE_i\left(E_i^{-1}-\frac{E_i^{-1}g_ig_i^TE_i^{-1}}{g_i^Ts_i+g_i^TE_i^{-1}g_i}\right)E_is_i\\&=s_i^TE_is_i-\left(s_i^TE_is_i-\frac{s_i^Tg_ig_i^Ts_i}{g_i^Ts_i+g_i^TE_i^{-1}g_i}\right)\\&=\frac{s_i^Tg_ig_i^Ts_i}{g_i^Ts_i+g_i^TE_i^{-1}g_i}.\end{aligned} \tag{5D18}$$

Substituting equations (5D17) and (5D18) into equation (5D16) yields

$$G_{i+1}=E_i^{-1}-\frac{E_i^{-1}g_ig_i^TE_i^{-1}}{g_i^Ts_i+g_i^TE_i^{-1}g_i}+\left(E_i^{-1}-\frac{E_i^{-1}g_ig_i^TE_i^{-1}}{g_i^Ts_i+g_i^TE_i^{-1}g_i}\right)\times\frac{E_is_is_i^TE_i}{\frac{s_i^Tg_ig_i^Ts_i}{g_i^Ts_i+g_i^TE_i^{-1}g_i}}\left(E_i^{-1}-\frac{E_i^{-1}g_ig_i^TE_i^{-1}}{g_i^Ts_i+g_i^TE_i^{-1}g_i}\right). \tag{5D19}$$

Note that

$$\begin{aligned}&\left(E_i^{-1}-\frac{E_i^{-1}g_ig_i^TE_i^{-1}}{g_i^Ts_i+g_i^TE_i^{-1}g_i}\right)E_is_is_i^TE_i\left(E_i^{-1}-\frac{E_i^{-1}g_ig_i^TE_i^{-1}}{g_i^Ts_i+g_i^TE_i^{-1}g_i}\right)\\&=\left(s_i-\frac{E_i^{-1}g_ig_i^Ts_i}{g_i^Ts_i+g_i^TE_i^{-1}g_i}\right)\left(s_i^T-\frac{s_i^Tg_ig_i^TE_i^{-1}}{g_i^Ts_i+g_i^TE_i^{-1}g_i}\right)\\&=s_is_i^T-\frac{s_is_i^Tg_ig_i^TE_i^{-1}}{g_i^Ts_i+g_i^TE_i^{-1}g_i}-\frac{E_i^{-1}g_ig_i^Ts_is_i^T}{g_i^Ts_i+g_i^TE_i^{-1}g_i}\\&+\frac{E_i^{-1}g_ig_i^TE_i^{-1}}{g_i^Ts_i+g_i^TE_i^{-1}g_i}\frac{s_i^Tg_ig_i^TE_i^{-1}}{g_i^Ts_i+g_i^TE_i^{-1}g_i}.\end{aligned} \tag{5D20}$$

Combining the third term in equation (5D19) and equation (5D20), we obtain

$$\begin{aligned}&\left(E_i^{-1}-\frac{E_i^{-1}g_ig_i^TE_i^{-1}}{g_i^Ts_i+g_i^TE_i^{-1}g_i}\right)\frac{E_is_is_i^TE_i}{\frac{s_i^Tg_ig_i^Ts_i}{g_i^Ts_i+g_i^TE_i^{-1}g_i}}\\&\times\left(E_i^{-1}-\frac{E_i^{-1}g_ig_i^TE_i^{-1}}{g_i^Ts_i+g_i^TE_i^{-1}g_i}\right)=\frac{s_is_i^T\left(g_i^Ts_i+g_i^TE_i^{-1}g_i\right)}{s_i^Tg_ig_i^Ts_i}\\&-\frac{s_is_i^Tg_ig_i^TE_i^{-1}}{s_i^Tg_ig_i^Ts_i}-\frac{E_i^{-1}g_ig_i^Ts_is_i^T}{s_i^Tg_ig_i^Ts_i}\\&+\frac{E_i^{-1}g_ig_i^TE_i^{-1}s_i^Tg_ig_i^TE_i^{-1}}{s_i^Tg_ig_i^Ts_i\left(g_i^Ts_i+g_i^TE_i^{-1}g_i\right)}\\&=\frac{s_is_i^T\left(g_i^Ts_i+g_i^TE_i^{-1}g_i\right)}{\left(s_i^Tg_i\right)^2}-\frac{s_ig_i^TE_i^{-1}}{s_i^Tg_i}\\&-\frac{E_i^{-1}g_is_i^T}{s_i^Tg_i}+\frac{E_i^{-1}g_ig_i^TE_i^{-1}}{g_i^Ts_i+g_i^TE_i^{-1}g_i}.\end{aligned} \tag{5D21}$$

Substituting equation (5D21) into equation (5D19), we obtain

$$G_{i+1} = E_i^{-1} + \frac{s_i s_i^T}{s_i^T g_i} + \frac{s_i s_i^T g_i^T E_i^{-1} g_i}{\left(s_i^T g_i\right)^2} - \frac{s_i g_i^T E_i^{-1}}{s_i^T g_i} - \frac{E_i^{-1} g_i s_i^T}{s_i^T g_i}$$

$$= E_i^{-1} - \frac{E_i^{-1} g_i s_i^T}{s_i^T g_i} - \frac{s_i g_i^T E_i^{-1}}{s_i^T g_i} + \frac{s_i g_i^T E_i^{-1} g_i s_i^T}{\left(s_i^T g_i\right)^2} + \frac{s_i s_i^T}{s_i^T g_i}$$

$$= E_i^{-1}\left(I - \frac{g_i s_i^T}{s_i^T g_i}\right) - \frac{s_i g_i^T E_i^{-1}}{s_i^T g_i}\left(I - \frac{g_i s_i^T}{s_i^T g_i}\right) + \frac{s_i s_i^T}{s_i^T g_i}$$

$$= \left(E_i^{-1} - \frac{s_i g_i^T E_i^{-1}}{s_i^T g_i}\right)\left(I - \frac{g_i s_i^T}{s_i^T g_i}\right) + \frac{s_i s_i^T}{s_i^T g_i}$$

$$= \left(I - \frac{s_i g_i^T}{s_i^T g_i}\right) E_i^{-1}\left(I - \frac{g_i s_i^T}{s_i^T g_i}\right) + \frac{s_i s_i^T}{s_i^T g_i},$$

$$= \left(I - \frac{s_i g_i^T}{s_i^T g_i}\right) G_i\left(I - \frac{g_i s_i^T}{s_i^T g_i}\right) + \frac{s_i s_i^T}{s_i^T g_i}. \tag{5D22}$$

Let $\rho_i = \frac{1}{s_i^T g_i}$ and

$$V_i = I - \rho_i g_i s_i^T. \tag{5D23}$$

Then, equation (5D22) can be rewritten as

$$G_{i+1} = V_i^T G_i V_i + \rho_i s_i s_i^T. \tag{5D24}$$

A key observation from equation (5D24) is that the inverse of Hessian matrix G_{i+1} is obtained by updating a pair (s_i, g_i). The following Lemma 5D.2 (Byrd et al. 1994) is useful for computation of the Hessian matric and its inverse via a sequence of pairs (s_i, g_i).

Lemma 5D.2: Compact Representation of BFGS Matrices

Let n be the number of variables. Define the $n \times k$ matrices S_k and Y_k as

$$S_k = [s_0, \ldots, s_{k-1}] \text{ and } Y_k = [g_0, \ldots, g_{k-1}]. \tag{5D25}$$

A sequence of matrices V_i in equation (5D23) satisfies

$$V_0 \ldots V_{k-1} = I - Y_k R_k^{-1} S_k^T, \tag{5D26}$$

where

$$\left(R_k\right)_{ij} = \begin{cases} s_{i-1}^T g_{j-1} & i \le j \\ 0 \text{ otherwise} \end{cases}. \tag{5D27}$$

Proof

Proof is by induction.

It follows from equation (5D23) that

$$V_0 = I - \frac{g_0 S_0^T}{S_0^T g_0}. \tag{5D28}$$

We first show that equation (5D26) holds for $k = 1$. In fact, when $k = 1$ equation (5D27) is reduced to

$$R_0 = S_0^T g_0 \text{ and} \tag{5D29}$$

$$S_0 = s_0,\ Y_0 = g_0. \tag{5D30}$$

Equation (5D26) gives solution:

$$V_0 = I - Y_0 R_0^{-1} S_0^T. \tag{5D31}$$

Substituting equations (5D29) and (5D30) into equation (5D31), we obtain

$$V_0 = I - \frac{g_0 s_0^T}{S_0^T g_0}, \tag{5D32}$$

which coincides with equation (5D28). This proves that Lemma 5D.2 holds for $k = 1$.

Now we assume that equation (5D26) holds for k and proves that it will also hold for $k+1$. It follows from equation (5D27) that

$$R_{k+1} = \begin{bmatrix} R_k & S_k^T g_k \\ 0 & \dfrac{1}{\rho_k} \end{bmatrix}.$$

Its inverse matrix is given by

$$R_{k+1}^{-1} = \begin{bmatrix} R_k^{-1} & -\rho_k R_k^{-1} S_k^T g_k \\ 0 & \rho_k \end{bmatrix}. \tag{5D33}$$

Combining equations (5D25) and (5D33), we obtain

$$I - Y_{k+1} R_{k+1}^{-1} S_{k+1}^T$$

$$= I - \begin{bmatrix} Y_k & g_k \end{bmatrix} \begin{bmatrix} R_k^{-1} & -\rho_k R_k^{-1} S_k^T g_k \\ 0 & \rho_k \end{bmatrix} \begin{bmatrix} S_k^T \\ s_k^T \end{bmatrix}$$

$$= I - \begin{bmatrix} Y_k & g_k \end{bmatrix} \begin{bmatrix} R_k^{-1} S_K^T - \rho_k R_k^{-1} S_k^T g_k s_k^T \\ \rho_k s_k^T \end{bmatrix}$$

$$= I - Y_k R_k^{-1} S_K^T + \rho_k g_k R_k^{-1} S_k^T g_k s_k^T - \rho_k g_k s_k^T$$

$$= \left(I - Y_k R_k^{-1} S_K^T\right)\left(I - \rho_k g_k s_k^T\right). \tag{5D34}$$

Since by the hypothesis that equation (5D26) holds for k, then we have

$$V_0 \ldots V_{k-1} = I - Y_k R_k^{-1} S_k^T,$$

which implies that

$$V_0 \ldots V_{k-1} V_k = \left(I - Y_k R_k^{-1} S_k^T\right)\left(I - \rho_k g_k s_k^T\right). \tag{5D35}$$

Substituting equation (5D34) into equation (5D35) yields

$$V_0 \ldots V_{k-1} V_k = I - Y_{k+1} R_{k+1}^{-1} S_{k+1}^T. \tag{5D36}$$

This shows that equation (5D26) also holds for $k+1$, and hence we prove the Lemma 5D.2.

Now we are ready to introduce a compact representation of the approximation of the inverse Hessian matrix G_{k+1} in equation (5D24) (Byrd et al. 1994).

Theorem 5D.1

Let G_0 be symmetric and positive definite. Assume that the k pairs $\left(s_i, g_i, i = 1, \ldots, k\right)$ satisfies $s_i^T g_i > 0$. Also assume that G_k are updated by pairs $\left(s_i, g_i, i = 0, \ldots, k-1\right)$ and equation (5D24). Then,

$$G_k = G_0 + \begin{bmatrix} S_k & G_0 Y_k \end{bmatrix} \begin{bmatrix} R_k^{-1}\left(D_k + Y_k^T G_0 Y_k\right) R_k^T & -R_k^{-T} \\ -R_k^{-1} & 0 \end{bmatrix} \times \begin{bmatrix} S_k^T \\ Y_k^T G_0 \end{bmatrix}, \tag{5D37}$$

where

$$D_k = \operatorname{diag}\left(s_0^T g_0, \ldots, s_{k-1}^T g_{k-1}\right). \tag{5D38}$$

Proof (Byrd et al. 1994)

Equation (5D24) can be rewritten as

$$G_k = M_k + N_k, \tag{5D39}$$

where M_k and N_k are recursively defined as

$$M_0 = G_0,\; M_{k+1} = V_k^T M_k V_k, \tag{5D40}$$

$$N_1 = \rho_0 s_0 s_{0,}^T\; N_{k+1} = V_k^T N_k V_k + \rho_k s_k s_k^T. \tag{5D41}$$

Now we derive the explicit expression of the matrix M_k. By definition of M_k (equation (5D40)), we obtain

$$M_k = V_{k-1}^T M_{k-1} V_{k-1}$$
$$= V_{k-1}^T V_{k-2}^T M_{k-2} V_{k-2} V_{k-1}$$
$$= \ldots$$
$$= V_{k-1}^T V_{k-2}^T \ldots V_0^T G_0 V_0 \ldots V_{k-2} V_{k-1}. \tag{5D42}$$

Substituting equation (5D26) into equation (5D42) yields

$$M_k = \left(I - S_k R_k^{-1} Y_k^T\right) G_0 \left(I - Y_k R_k^{-1} S_k^T\right). \tag{5D43}$$

Using induction, next we will show that

$$N_k = S_k R_k^{-T} D_k R_k^{-1} S_k^T. \tag{5D44}$$

First, we check $k = 1$. It follows from equations (5D25), (5D27), and (5D38) that

$$S_1 = s_0,\; R_1 = s_0^T g_0,\; D_0 = s_0^T g_0. \tag{5D45}$$

Substituting equation (5D45) into equation (5D44), we obtain

$$N_1 = s_0 \left(s_0^T g_0\right)^{-T} s_0^T g_0 \left(s_0^T g_0\right)^{-1} s_0^T = \rho_0 s_0 s_0^T. \tag{5D46}$$

This coincides with equation (5D41).

We assume equation (5D44) holds for k. Then, we show that equation (5D44) also holds for $k+1$. Combining equations (5D41) and (5D44), we obtain

$$N_{k+1} = V_k^T S_k R_k^{-T} D_k R_k^{-1} S_k^T V_k + \rho_k s_k s_k^T. \tag{5D47}$$

Now we further simplify equation (5D47). It follows from equation (5D23) that

$$S_k^T V_k = S_k^T \left(I - \rho_k g_k s_k^T\right) = \begin{bmatrix} I & -\rho_k S_k^T g_k \end{bmatrix} \begin{bmatrix} S_k^T \\ s_k^T \end{bmatrix}. \tag{5D48}$$

It follows from equation (5D25) that

$$S_{k+1}^T = \begin{bmatrix} S_k^T \\ s_k^T \end{bmatrix}. \tag{5D49}$$

Substituting equation (5D49) into equation (5D48), we obtain $S_k^T V_k = \begin{bmatrix} I & -\rho_k S_k^T g_k \end{bmatrix} S_{k+1}^T$, which implies that

$$R_k^{-1} S_k^T V_k = \begin{bmatrix} R_k^{-1} & -\rho_k R_k^{-1} S_k^T g_k \end{bmatrix} S_{k+1}^T \tag{5D50}$$

It follows from equation (5D33) that

$$\begin{bmatrix} R_k^{-1} & -\rho_k R_k^{-1} S_k^T g_k \end{bmatrix} = \begin{bmatrix} I & 0 \end{bmatrix} R_{k+1}^{-1}. \tag{5D51}$$

Substituting equation (5D51) into equation (5D50), we obtain

$$R_k^{-1} S_k^T V_k = \begin{bmatrix} I & 0 \end{bmatrix} R_{k+1}^{-1} S_{k+1}^T. \tag{5D52}$$

It follows from equation (5D33) that

$$R_{k+1}^{-T} e_{k+1} = \begin{bmatrix} R_k^{-1} & 0 \\ -\rho_k R_k^{-1} S_k^T g_k & \rho_k \end{bmatrix} \begin{bmatrix} 0 \\ 1 \end{bmatrix} = \begin{bmatrix} 0 \\ \rho_k \end{bmatrix}. \tag{5D53}$$

Thus,

$$S_{k+1} R_{k+1}^{-T} e_{k+1} = \begin{bmatrix} S_k & s_k \end{bmatrix} \begin{bmatrix} 0 \\ \rho_k \end{bmatrix} = \rho_k s_k, \text{ which implies that}$$

$$s_k = S_{k+1} R_{k+1}^{-T} e_{k+1} \frac{1}{\rho_k}. \tag{5D54}$$

Substituting equations (5D52) and (5D54) into equation (5D47), we obtain

$$N_{k+1} = S_{k+1} R_{k+1}^{-1} \begin{bmatrix} I \\ 0 \end{bmatrix} D_k \begin{bmatrix} I & 0 \end{bmatrix} R_{k+1}^{-1} S_{k+1}^T + S_{k+1} R_{k+1}^{-T}$$

$$\times \begin{bmatrix} 0 & & & \\ & \ddots & & \\ & & 0 & \\ & & & \frac{1}{\rho_k} \end{bmatrix} R_{k+1}^{-T} S_{k+1}^T$$

$$= S_{k+1} R_{k+1}^{-1} \left(\begin{bmatrix} D_k & 0 \\ 0 & 0 \end{bmatrix} + \begin{bmatrix} 0 & 0 \\ 0 & \frac{1}{\rho_k} \end{bmatrix} \right) R_{k+1}^{-T} S_{k+1}^T$$

$$= S_{k+1} R_{k+1}^{-1} \begin{bmatrix} D_k & 0 \\ 0 & \frac{1}{\rho_k} \end{bmatrix} R_{k+1}^{-T} S_{k+1}^T$$

$$= S_{k+1} R_{k+1}^{-1} D_{k+1} R_{k+1}^{-T} S_{k+1}^T. \tag{5D55}$$

This proves equation (5D44) for $k+1$.

Combining equations (5D43) and (5D44), we obtain

$$\begin{aligned} M_k + N_k &= \left(I - S_k R_k^{-1} Y_k^T\right) G_0 \left(I - Y_k R_k^{-1} S_k^T\right) \\ &\quad + S_k R_k^{-T} D_k R_k^{-1} S_k^T \\ &= G_0 - S_k R_k^{-1} Y_k^T G_0 - G_0 Y_k R_k^{-1} S_k^T \\ &\quad + S_k R_k^{-1} Y_k^T G_0 Y_k R_k^{-1} S_k^T + S_k R_k^{-T} D_k R_k^{-1} S_k^T \\ &= G_0 - S_k R_k^{-1} Y_k^T G_0 - G_0 Y_k R_k^{-1} S_k^T \\ &\quad + S_k R_k^{-1} \left(D_k + Y_k^T G_0 Y_k\right) R_k^{-1} S_k^T. \end{aligned} \tag{5D56}$$

This shows that

$$M_k + N_k = G_0 + \begin{bmatrix} S_k & G_0 Y_k \end{bmatrix} \times \begin{bmatrix} R_k^{-1}\left(D_k + Y_k^T G_0 Y_k\right) R_k^T & -R_k^{-T} \\ -R_k^{-1} & 0 \end{bmatrix} \begin{bmatrix} S_k^T \\ Y_k^T G_0 \end{bmatrix}. \tag{5D57}$$

Combining equations (5D39) and (5D57), we obtain

$$G_k = M_k + N_k = G_0 + \begin{bmatrix} S_k & G_0 Y_k \end{bmatrix} \times \begin{bmatrix} R_k^{-1}\left(D_k + Y_k^T G_0 Y_k\right) R_k^T & -R_k^{-T} \\ -R_k^{-1} & 0 \end{bmatrix} \begin{bmatrix} S_k^T \\ Y_k^T G_0 \end{bmatrix}, \tag{5D58}$$

which proves the equation (5D37).

Recall that the BFGS update for Hessian matrix is given by

$$E_{k+1} = E_k + \frac{g_k g_k^T}{g_k^T s_k} - \frac{E_k s_k s_k^T k}{s_k^T E_k s_k}. \tag{5D59}$$

Similar to Theorem 5D.1 for the inverse of Hessian matrix, we have Theorem 5D.2 for the Hessian Matrix (Byrd et al. 1994).

Theorem 5D.2

Let E_0 be symmetric and positive definite. Assume that k pairs $\left(s_i, g_i, i = 1, \ldots, k-1\right)$ satisfy $s_i^T g_i > 0$. Furthermore, assume that the matrix E_k is obtained by updating E_0, k times using equation (5D59). Then, we have

$$E_k = E_0 - \begin{bmatrix} E_0 S_k & Y_k \end{bmatrix} \begin{bmatrix} S_k^T E_0 S_k & L_k \\ L_k^T & -D_k \end{bmatrix}^{-1} \begin{bmatrix} S_k^T E_0 \\ Y_k^T \end{bmatrix}, \quad (5D60)$$

where L_k is defined as

$$(L_k)_{ij} = \begin{cases} s_{i-1}^T g_{j-1} & i > j \\ 0 & \text{otherwise} \end{cases}. \quad (5D61)$$

Proof (Byrd et al. 1994)

Let

$$U_k = \begin{bmatrix} S_k & G_0 Y_k \end{bmatrix},$$

$$C_k = \begin{bmatrix} R_k^{-T}(D_k + Y_k^T G_0 Y_k) R_k^{-1} & -R_k^{-T} \\ -R_k^{-1} & 0 \end{bmatrix}. \quad (5D62)$$

Then, equation (5D58) can be rewritten as

$$G_k = G_0 + U_k C_k U_k^T. \quad (5D63)$$

Let

$$C = \begin{bmatrix} A & B^T \\ B & 0 \end{bmatrix}. \quad (5D64)$$

Using transformation method, we can find the inverse of the matrix C:

$$C^{-1} = \begin{bmatrix} 0 & B^{-1} \\ B^{-T} & -B^T A B^{-1} \end{bmatrix}. \quad (5D65)$$

Let $A = R_k^{-1}(D_k + Y_k^T G_0 Y_k) R_k^T$, $B = -R_k^{-1}$. Then, using equation (5D65), we can find

$$C^{-1} = \begin{bmatrix} 0 & -R_k \\ -R_k^T & -(D_k + Y_k^T G_0 Y_k) \end{bmatrix}. \quad (5D66)$$

Using Lemma 5D.1 (Sherman-Morrison-Woodbury Formula) and equation (5D63), we obtain

$$E_k = E_0 - E_0 U_k C_k \left(I + U_k^T E_0 U_k C_k\right)^{-1} U_k^T E_0$$
$$= E_0 - E_0 U_k \left(C_k^{-1} + U_k^T E_0 U_k\right)^{-1} U_k^T E_0. \quad (5D67)$$

Note that

$$B_0 G_0 = I \text{ and}$$

$$U_k^T E_0 U_k = \begin{bmatrix} S_k^T \\ Y_k^T G_0^T \end{bmatrix} B_0 \begin{bmatrix} S_k & G_0 Y_k \end{bmatrix}$$
$$= \begin{bmatrix} S_k^T B_0 S_k & S_k^T Y_k \\ Y_k^T S_k & Y_k^T G_0 Y_k \end{bmatrix}. \quad (5D68)$$

Combining equations (5D66) and (5D68), we obtain

$$C_k^{-1} + U_k^T E_0 U_k = \begin{bmatrix} S_k^T B_0 S_k & S_k^T Y_k - R_k \\ Y_k^T S_k - R_k^T & -D_k \end{bmatrix}. \quad (5D69)$$

Recall that

$$S_k = [s_0, \ldots, s_{k-1}] \text{ and } Y_k = [g_0, \ldots, g_{k-1}]. \quad (5D70)$$

It follows from equation (5D70) that

$$S_k^T Y_k = \begin{bmatrix} s_0^T g_0 & \cdots & s_0^T g_{k-1} \\ \vdots & \vdots & \vdots \\ s_{k-1}^T & \cdots & s_{k-1}^T g_{k-1} \end{bmatrix}. \quad (5D71)$$

It follows from equations (5D27) and (5D61) that

$$R_k = \begin{bmatrix} s_0^T g_0 & \cdots & s_0^T g_{k-1} \\ \vdots & \vdots & \vdots \\ 0 & \cdots & s_{k-1}^T g_{k-1} \end{bmatrix} \text{ and } L_k = \begin{bmatrix} 0 & \cdots & 0 \\ \vdots & \vdots & \vdots \\ s_{k-1}^T g_0 & \cdots & 0 \end{bmatrix}. \quad (5D72)$$

Combining equations (5D71) and (5D72), we obtain

$$L_k = S_k^T Y_k - R_k. \quad (5D73)$$

It follows from equations (5D72) and (5D73) that

$$(L_k)_{ij} = \begin{cases} s_{i-1}^T g_{j-1} & i > j \\ 0 & \text{otherwise} \end{cases} \quad (5D74)$$

APPENDIX 5E: NONSMOOTH OPTIMIZATION ANALYSIS

Let

$$a = E_{jj}, b = q_j + (Ed)_j, c = w_j + d_j, u = c + z. \quad (5E1)$$

Then, objective in equation (5.91a) can be simplified to

$$f(u)=\frac{1}{2}a(u-c)^2+b(u-c)+\lambda|u|. \quad (5E2)$$

Setting $\frac{\partial f(u)}{\partial u}=0$, we obtain

$$\frac{\partial f(u)}{\partial u}=a(u-c)+b+\lambda\,\partial|u|=0, \quad (5E3)$$

where $\partial|u|$ is subdifferential and given by

$$\partial|u|=\begin{cases}1 & u>0\\ [-1,1] & u=0\\ -1 & u<0\end{cases}. \quad (5E4)$$

Dividing by a on both sides of equation (5E4) and solving equation (5E3) for u, we obtain

$$u=c-\frac{b}{a}-\frac{\lambda}{a}\partial|u|. \quad (5E5)$$

Next we show that u and $c-\frac{b}{a}$ have the same sign. In fact, if we assume $c-\frac{b}{a}>0$, then $u>0$. Otherwise, if $u<0$, then from equation (5E4) it follows that $\partial|u|=-1$, which leads to $c-\frac{b}{a}<0$ (using equation (5E5)). This contradicts the hypothesis $c-\frac{b}{a}>0$. Thus, it must be $u>0$. Using the similar argument, we can prove that $c-\frac{b}{a}<0$ implies $u<0$. Therefore, the solution to equation (5E5) is

$$u=\text{sign}\left(c-\frac{b}{a}\right)\left(\left|c-\frac{b}{a}\right|-\frac{\lambda}{a}\right)_+. \quad (5E6)$$

Substituting equation (5E1) into equation (5E6), we obtain

$$z^*=-w_j-d_j+\text{sign}\left(w_j+d_j-\frac{q_j+(Ed)_j}{E_{jj}}\right)\times\left(\left|w_j+d_j-\frac{q_j+(Ed)_j}{E_{jj}}\right|-\frac{\lambda}{E_{jj}}\right)_+. \quad (5E7)$$

APPENDIX 5F: COMPUTATION OF ELBO FOR LEARNING SEMS

5F1 ELBO for SEMs

Assume that n data point ($Y^i\in R^{m\times d}$, $X^i\in R^{m\times d}$), $i=1,\ldots,n$ are sampled. Let $Y=[Y^1,Y^2,\ldots,Y^n]$ and $X=[X^1,X^2,\ldots,X^n]$. The marginal likelihood can be written as $\log p_\theta(Y^1,X^1,\ldots,Y^n,X^n)=\sum_{i=1}^n\log p_\theta(Y^i,X^i)$ if the data are independently sampled. Since stochastic gradient method is used for optimization, we can consider marginal likelihood of one data point (Yu et al. 2019):

$$\begin{aligned}\log p_\theta(Y^i,X^i)&=\int q_\varnothing(Z|Y^i,X^i)\log p_\theta(Y^i,X^i)dZ\\&=\int q_\varnothing(Z|Y^i,X^i)\log\frac{p_\theta(Y^i,X^i,Z)}{p_\theta(Z|Y^i,X^i)}dZ\\&=\int q_\varnothing(Z|Y^i,X^i)\log\frac{q_\varnothing(Z|Y^i,X^i)}{q_\varnothing(Z|Y^i,X^i)}\frac{p_\theta(Y^i,X^i,Z)}{p_\theta(Z|Y^i,X^i)}dZ\\&=-\int\begin{bmatrix}q_\varnothing(Z|Y^i,X^i)\log q_\varnothing(Z|Y^i,X^i)dZ\\+KL\left(q_\varnothing(Z|Y^i,X^i)\|p_\theta(Z|Y^i,X^i)\right)\end{bmatrix}dz\\&\quad+\int q_\varnothing(Z|Y^i,X^i)\log p_\theta(Y^i,X^i,Z)dZ\\&=KL\left(q_\varnothing(Z|Y^i,X^i)\|p_\theta(Z|Y^i,X^i)\right)+\mathcal{L}(\theta,\varnothing,Y^i,X^i)\\&\geq\mathcal{L}(\theta,\varnothing,Y^i,X^i),\end{aligned} \quad (5F1)$$

where

$$\mathcal{L}(\theta,\varnothing,Y^i,X^i)=E_{q_\varnothing(Z|Y^i,X^i)}\left[-\log q_\varnothing(Z|Y^i,X^i)+\log p_\theta(Y^i,X^i,Z)\right], \quad (5F2)$$

$q_\varnothing(Z|Y^i,X^i)$ is a variational posterior, which approximates the true posterior $p_\theta(Z|Y^i,X^i)$.

Note that

$$\log p_\theta(Y^i,X^i,Z)=\log p_\theta(Z)+\log p_\theta(Y^i,X^i|Z). \quad (5F3)$$

Substituting equation (5F3) into equation (5F2) yields

$$\mathcal{L}(\theta,\varnothing,Y^i,X^i)=E_{q_\varnothing(Z|Y^i,X^i)}\left[\log p_\theta(Y^i,X^i|Z)\right]-KL\left(q_\varnothing(Z|Y^i,X^i)\|p_\theta(Z)\right). \quad (5F4)$$

Our goal is to use lower bound $\mathcal{L}(\theta,\varnothing,Y^i,X^i)$ to estimate both variation parameters $\varnothing$ and generate parameters θ using gradient methods. If we assume that both $q_\varnothing(Z|Y^i,X^i)$ and $p_\theta(Z)$ are normally distributed, the KL distance between these two distributions can be analytically calculated. The remaining term in the lower bound involves to derive gradient

$$\nabla_{\varnothing} E_{q_{\varnothing}(Z|Y,X)}\left[f(Z)\right] = E_{q_{\varnothing}(Z|Y,X)}$$

$$\times\left[f(Z)\nabla_{q_{\varnothing}(Z|Y,X)}\log q_{\varnothing}(Z|Y,X)\right]$$

$$\approx \frac{1}{L}\sum_{l=1}^{L} f(Z^l)\nabla_{q_{\varnothing}(Z^l|Y,X)}\log q_{\varnothing}(Z^l|Y,X), \quad (5F5)$$

where $Z^l \sim q_{\varnothing}(Z|Y,X)$.

Calculation of equation (5F5) requires iteration, and hence is intractable.

5F2 The Reparameterization Trick

A key challenge in computation of $\nabla_{\varnothing} E_{q_{\varnothing}(Z|Y,X)}\left[f(Z)\right]$ is that expectation involves variational parameters $\varnothing$. If we can transfer $\varnothing$ from the distribution of $q_{\varnothing}(Z|Y,X)$ to the function $f(Z)$, then the problem of estimation of the variation parameters $\varnothing$ is mitigated. Changing variable technique in the distribution theory can achieve this. Define the following variable change:

$$Z = g_{\varnothing}(\varepsilon, Y, X), \varepsilon \sim p(\varepsilon), \quad (5F6)$$

where $g_{\varnothing}(\varepsilon, Y, X)$ can be a vector of linear or nonlinear functions.

Let $\frac{\partial g_{\varnothing}(\varepsilon, Y, X)}{\partial \varepsilon^T}$ be the Jacobian matrix of the transformation $g_{\varnothing}(\varepsilon, Y, X)$. The distribution $p(Z|Y,X)$ is given by

$$p(Z|Y,X) = \frac{p(\varepsilon)}{\left|\det\left(\frac{\partial g_{\varnothing}(\varepsilon, Y, X)}{\partial \varepsilon^T}\right)\right|}, \quad (5F7)$$

where det (.) denotes determinant of a matrix. The distribution $p(\varepsilon)$ involves no variational parameters $\varnothing$. The distribution $p(Z|Y,X)$ can be used to approximate $q_{\varnothing}(Z|Y,X)$. Therefore,

$$q_{\varnothing}(Z|Y,X)dZ = p(Z|Y,X)dz. \quad (5F8)$$

Substituting equation (5F7) into equation (5F8) yields

$$q_{\varnothing}(Z|Y,X)dZ = \frac{p(\varepsilon)}{\left|\det\left(\frac{\partial g_{\varnothing}(\varepsilon, Y, X)}{\partial \varepsilon^T}\right)\right|} dZ. \quad (5F9)$$

It follows from equation (5F6) that

$$dZ = \left|\det\left(\frac{\partial g_{\varnothing}(\varepsilon, Y, X)}{\partial \varepsilon^T}\right)\right| d\varepsilon. \quad (5F10)$$

Substituting equation (5F10) into equation (5F9), we obtain

$$q_{\varnothing}(Z|Y,X)dZ = p(\varepsilon)d\varepsilon, \quad (5F11)$$

which implies that conditional probability $q_{\varnothing}(Z|Y,X)dZ$ can be calculated by $p(\varepsilon)d\varepsilon$. Therefore,

$$E_{q_{\varnothing}(Z|Y,X)}\left[f(Z)\right] = \int f(Z) q_{\varnothing}(Z|Y,X)dZ$$

$$= \int f\left(g_{\varnothing}(\varepsilon, Y, X)\right) p(\varepsilon)d\varepsilon. \quad (5F12)$$

Now quantity $\int f\left(g_{\varnothing}(\varepsilon, Y, X)\right) p(\varepsilon)d\varepsilon$ involves only random variable ε.

5F3 Stochastic Gradient Variational Bayes (SGVB) Estimator

Now Monte Carlo estimates of expectations of some function $f(Z)$ can be calculated by

$$E_{q_{\varnothing}(Z|Y,X)}\left[f(Z)\right] \approx \frac{1}{L}\sum_{l=1}^{L} f\left(g_{\varnothing}(\varepsilon^l, Y, X)\right), \varepsilon \sim p(\varepsilon). \quad (5F13)$$

We first apply equation (5F13) to equation (5F2). Let

$$f(Z) = -\log q_{\varnothing}(Z|Y^i, X^i) + \log p_{\theta}(Y^i, X^i, Z)$$

Then, it follows from equations (5F2) and (5F13), we obtain the estimator of the first version of lower bound:

$$\hat{\mathcal{L}}^A(\theta, \varnothing, Y^i, X^i) \approx \frac{1}{L}\sum_{l=1}^{L}\left[\log p_{\theta}(Y^i, X^i, Z^{i,l})\right.$$

$$\left. - \log q_{\varnothing}(Z^{i,l}|Y^i, X^i)\right], \quad (5F14)$$

where $Z^{i,l} = g_{\varnothing}(\varepsilon^l, Y^i, X^i)$, $\varepsilon^l \sim p(\varepsilon)$.

Now we consider lower bound in equation (5F4). Both prior distribution $p_{\theta}(Z)$ and approximate posterior distribution $q_{\varnothing}(Z|Y^i, X^i)$ are often assumed normal distribution. The second term KL distance can be calculated analytically and can be interpreted as regularizing variation parameters $\varnothing$ to enforce the approximate posterior $q_{\varnothing}(Z|Y^i, X^i)$ to be close to the prior $p_\theta(Z)$. We only need to use Stochastic Gradient Variational Bayes to estimate the first term, reconstruction error $E_{q_{\varnothing}(Z|Y^i,X^i)}\left[\log p_\theta\left(Y^i, X^i \mid Z\right)\right]$, in equation (5F4):

$$E_{q_{\varnothing}(Z|Y^i,X^i)}\left[\log p_\theta\left(Y^i,X^i \mid Z\right)\right] \approx \frac{1}{L}\sum_{l=1}^{L}\log p_\theta\left(Y^i,X^i \mid Z^{i,l}\right), \tag{5F15}$$

where $Z^{i,l} = g_{\varnothing}\left(\varepsilon^l, Y^i, X^i\right)$, $\varepsilon^l \sim p(\varepsilon)$.

Substituting equation (5F15) into equation (5F4), we obtain the estimator of the second version of the lower bound:

$$\hat{\mathcal{L}}^B\left(\theta,\varnothing,Y^i,X^i\right) = \frac{1}{L}\sum_{l=1}^{L}\log p_\theta\left(Y^i,X^i \mid Z^{i,l}\right) - KL\left(q_{\varnothing}\left(Z|Y^i,X^i\right)||p_\theta(Z)\right). \tag{5F16}$$

Suppose that N data points $\{(Y^i, X^i), i = 1, \ldots, N\}$ are sampled. Let K be size of minibatch. Using minibatch, the marginal likelihood can be estimated by

$$\mathcal{L}(\theta, \varnothing, Y, X) \approx \hat{L}^K(\theta, \varnothing, Y^K, X^K) = \frac{N}{K}\sum_{i=1}^{K}\hat{\mathcal{L}}(\theta,\varnothing,Y^i, X^i), \tag{5F17}$$

where $Y^K = \{Y^i, i = 1, \ldots, K\}$, $X^K = \{X^i, i = 1, \ldots, K\}$ are randomly sampled K data points from the entire dataset (X, Y) with sample size N. The number of sample L is often set to 1 when the minibatch size K is larger than 100.

There are three main choices of selecting a differentiable transformation function $g_{\varnothing}(\varepsilon, Y, X)$ (Kingma and Welling 2013).

1. Tractable inverse cumulative distribution function (CDF) of $q_{\varnothing}(Z \mid Y, X)$. Let $u = F(z) = P(Z \le z)$ be the CDF. The inverse of CDF is F^{-1}(u). From example, let $F(z) = 1 - e^{-z}$ be a CDF of exponential distribution. Then, its inverse CDF is $-\log(1-u)$. Inverse CDF of Exponential, Cauchy, Logistic, Rayleigh, Pareto, Weibull, Reciprocal, Gompertz, Gumbel, and Erlang distributions are typical examples for transformation functions.
2. Location-scale family distribution. Let μ denote location and σ denote scale. The standard location-scale distribution with $\mu = 0$ (or matrix μ) and $\sigma = 1$ (or covariance matrix $\Sigma = I$) can be taken as the auxiliary variable ε. Define the transformation function $g_{\varnothing}(Z|Y,X) = \mu + \Sigma \odot \varepsilon, \varepsilon \sim MN(0, I, I)$. Location-scale family distribution includes Laplace, Elliptical, Student's distribution, Logistic, Uniform, Triangular, and Gaussian distributions when scalar variable is considered.
3. Composition. Transformation function can be composite function of tractable transformation function. Typical examples include Log-Normal, Gamma, Dirichlet, Beta, Chi-Squared, and F-distributions.

3F4 Neural Network Implementation

The inference (recognition) model and generative model can be implemented by feedforward NNs, convolutional NNs, and recurrent NNs.

3F4.1 Encoder (Inference Model)

Recall that $Z \in R^{m \times d}$ is a latent space with each node $z_i \in R^d$. Assume that the prior over latent matrix Z has standard matrix normal $MN_{md}(0, I, I)$ (Appendix 3C). Its density function is given by

$$\begin{aligned} p(Z) &= (2\pi)^{-\frac{md}{2}} exp\left\{-\frac{1}{2}\left(vec(Z)\right)^T vec(Z)\right\} \\ &= (2\pi)^{-\frac{md}{2}} exp\left\{-\frac{1}{2}\mathrm{Tr}\left(\left(vec(Z)\right)^T vec(Z)\right)\right\} \\ &= (2\pi)^{-\frac{md}{2}} exp\left\{-\frac{1}{2}\mathrm{Tr}(vec(Z)\left(vec(Z)\right)^T\right\}, \end{aligned}$$

which implies

$$\log p(Z) = -\frac{md}{2}\log(2\pi) - \frac{1}{2}\mathrm{Tr}\left(vec(Z)\left(vec(Z)\right)^T\right.. \tag{5F18}$$

Next we assume that the variational approximate posterior $q_{\varnothing}(Z \mid Y, X)$ has the following matrix normal $MN_{md}\left(\mu_{z,}\Omega_z, V_z\right)$ (Appendix 3C):

$$q_{\varnothing}\left(Z|Y, X\right)=\left(2\pi\right)^{-\frac{md}{2}}\left|\Sigma_z\right|^{-\frac{1}{2}}$$

$$exp\left\{-\frac{1}{2}\mathrm{Tr}\left(\sum_z^{-1}\left(vec\left(Z-\mu\right)\left(vec\left(Z-\mu\right)\right)^T\right)\right\} \text{ or}$$

$$\log q_{\varnothing}\left(Z|Y, X\right)=-\frac{md}{2}\log\left(2\pi\right)-\frac{1}{2}\log\left|\sum_z\right| -\frac{1}{2}\mathrm{Tr}\left(\sum_z^{-1}\left(vec\left(Z-\mu\right)\left(vec\left(Z-\mu\right)\right)^T\right),\quad (5F19)$$

where $\Sigma_z=\Omega_z\otimes V_z$.

$$E_{q\varnothing(Z|Y, X)}\left[\log p\left(Z\right)\right]=-\frac{md}{2}\log\left(2\pi\right) -\frac{1}{2}\mathrm{Tr}\left(E_{q\varnothing(Z|Y,X)}\left[vec\left(Z\right)\left(vec\left(Z\right)\right)^T\right]\right).\quad (5F20)$$

Note that

$$E_{q\varnothing(Z|Y, X)}\left[vec\left(Z\right)\left(vec\left(Z\right)\right)^T\right] =\sum_z+vec\left(\mu\right)\left(vec\left(\mu\right)\right)^T, \text{ which implies}$$

$$\mathrm{Tr}\left(E_{q\varnothing(Z|Y,X)}\left[vec\left(Z\right)\left(vec\left(Z\right)\right)^T\right]\right) =\sum_{i=1}^m\sum_{j=1}^d\left(\Omega_{ii}M_{jj}+\mu_{ij}^2\right),\quad (5F21)$$

where $\Sigma_z=\Omega\otimes M$.

Substituting equation (5F21) into equation (5F20) yields

$$E_{q\varnothing(Z|Y,X)}\left[\log p\left(Z\right)\right]=-\frac{md}{2}\log\left(2\pi\right) -\frac{1}{2}\sum_{i=1}^m\sum_{j=1}^d\left(\Omega_{ii}M_{jj}+\mu_{ij}^2\right).\quad (5F22)$$

Next it follows from equation (5F19) that

$$E_{q\varnothing(Z|Y,X)}\left[\log q_{\varnothing}\left(Z|Y,X\right)\right]$$

$$=-\frac{md}{2}\log(2\pi)-\frac{1}{2}\log\left|\sum_z\right| -\frac{1}{2}\mathrm{Tr}\left(\sum_z^{-1}E\left[\left(vec\left(Z-\mu\right)\left(vec\left(Z-\mu\right)\right)^T\right]\right)$$

$$=-\frac{md}{2}\log(2\pi)-\frac{1}{2}\log\left|\sum_z\right|-\frac{1}{2}\mathrm{Tr}\left(\sum_z^{-1}\sum_Z\right)$$

$$=-\frac{md}{2}\log(2\pi)-\frac{1}{2}\log\left|\sum_z\right|-\frac{md}{2}.\quad (5F23)$$

Combining equations (5F22) and (5F23), we obtain

$$-KL\left(q_{\varnothing}\left(Z|Y,X\right)||p(z)\right)=\frac{1}{2}\left\{\log\left|\sum_z\right|+md -\sum_{i=1}^m\sum_{j=1}^d\left(\Omega_{ii}M_{jj}+\mu_{ij}^2\right)\right\}.\quad (5F24)$$

If we assume that

$$\sum_z=\mathrm{diag}\left(S_{11},\ldots,S_{m1},S_{12},\ldots,S_{m2},\ldots,S_{1d,\ldots,}S_{md}\right) =S_z,\ S_{ij}=\Omega_{ii}M_{jj},\quad (5F25)$$

where when $d=1$, we have $M_{11}=1$.

Then equation (5F24) will be reduced to

$$-KL\left(q_{\varnothing}\left(Z|Y,X\right)||p(z)\right) =\frac{1}{2}\left(\sum_{i=1}^m\sum_{j=1}^d\left(\log\left(S_Z\right)_{ij}+1-\left(S_Z\right)_{ij}-\left(\mu_Z\right)_{ij}^2\right)\right).\quad (5F26)$$

If we assume that $d=1$, then equation (5F26) is reduced to

$$-KL(q_{\varnothing}\left(Z|Y,X\right)||\ p(z)) =\frac{1}{2}\sum_{j=1}^m\left(\log\sigma_j^2+1-\sigma_j^2-\mu_j^2\right).\quad (5F27)$$

The two-layer NNs for inference model with ReLU and tanh as the activation function is defined as

$$[\mu_Y \mid S_Y] = (I - A^T)\text{ReLu}(YW^1)W^2, \quad (5F28)$$

$$[\mu_X \mid S_X] = B\tanh(XW^3)W^4, \quad (5F29)$$

$$\mu_Z = \mu_Y - \mu_X, \quad (5F30)$$

$$S_Z = S_Y + S_X, \quad (5F31)$$

where W^1, W^2, W^3, and W^4 are weight matrices in the NNs.

3F4.2 Decoder (Generative Model)

The generative model is defined as

$$\begin{aligned} p_\theta(Y, X|Z) &= p_{\theta_1}(Y|X, Z)\, p_{\theta_2}(X|Z) \\ &= p_{\theta_1}(Y|X, Z)\, p_{\theta_2}(X), \end{aligned} \quad (5F32)$$

where we assume that X and Z are independent.

The endogenous variable Y decoder is the mirror process of the encoder. We assume that distribution $p_{\theta_1}(Y \mid X, Z)$ is matrix normal distribution:

$$p_{\theta_2}(Y|X, Z) = (2\pi)^{-\frac{md}{2}} |S_Y|^{-\frac{1}{2}} \times \exp\left\{-\frac{1}{2}\left(vec(Y - \mu_Y)\right)^T S_Y^{-1} vec(Y - \mu_Y)\right\}, \quad (5F33)$$

where

$S_Y = \text{diag}\left((S_Y)_{11}, \ldots, (S_Y)_{m1}, (S_Y)_{12}, \ldots, (S_Y)_{1d}, (S_Y)_{md}\right)$.

Thus, $\log p_{\theta_1}(Y \mid X, Z)$ is given by

$$\begin{aligned} \log p_{\theta_1}(Y \mid X, Z) &= -\frac{md}{2}\log(2\pi) - \frac{1}{2}\log|S_Y| \\ &\quad - \frac{1}{2}\left(vec(Y - \mu_Y)\right)^T S_Y^{-1} vec(Y - \mu_Y) \\ &= -\frac{md}{2}\log(2\pi) - \frac{1}{2}\left(\sum_{i=1}^{m}\sum_{j=1}^{d}\left(\log(S_Y)_{ij} + \frac{\left(Y_{ij} - (\mu_Y)_{ij}\right)^2}{(S_Y)_{ij}}\right)\right). \end{aligned} \quad (5F34)$$

Sampling latent variable Z:

$$Z^l = \mu_Z + S_Z \odot \varepsilon^l, \ \varepsilon^l \sim MN_{md}(0, I, I), \ l = 1, 2, \ldots, L. \quad (5F35)$$

The Monte Carlo estimation of $E_{q\varnothing(Z|Y,X)}\left[\log p_{\theta_1}(Y \mid X, Z)\right]$ is given by

$$\begin{aligned} E_{q\varnothing(Z|Y,X)}\left[\log p_{\theta_2}(Y \mid X, Z)\right] &\approx \frac{1}{L}\sum_{l=1}^{L}\log p_{\theta_1}(Y \mid X, Z^l) \\ &= -\frac{1}{L}\sum_{l=1}^{L}\left(\frac{md}{2}\log(2\pi) + \frac{1}{2}\right. \\ &\quad \left. + \frac{1}{2}\left(\sum_{i=1}^{m}\sum_{j=1}^{d}\left(\log(S_Y)_{ij}^l + \frac{\left(Y_{ij} - (\mu_Y)_{ij}^l\right)^2}{(S_Y)_{ij}^l}\right)\right)\right). \end{aligned} \quad (5F36)$$

The two-layer NNs for generating μ_Y and S_Y in equation (5F36) with ReLU and tanh as the activation function are given by (Figure 5.9)

$$[\mu_{\hat{Y}} | S_{\hat{Y}}] = \text{ReLu}\left(\left(I - A^T\right)^{-1}\left(B\tanh(XW^5)W^6\right) + Z)W^7\right)W^8\right), \quad (5F37)$$

where W^5, W^6, W^7, and W^8 are weight matrices. Finally, combining equations (5F16), (5F26), (5F36) and (5F37), we obtain the estimation of ELBO:

$$\begin{aligned} \hat{\mathcal{L}}^B(\theta, \varnothing, Y^i, X^i) &= \frac{1}{L}\sum_{l=1}^{L}\log p_\theta(Y^i, X^i \mid Z^{i,l}) \\ &\quad - KL\left(q_\varnothing(Z|Y^i, X^i) || p_\theta(Z)\right) \\ &= -\frac{1}{L}\sum_{l=1}^{L}\left(\frac{md}{2}\log(2\pi) + \frac{1}{2}\right. \\ &\quad \left. \times\left(\sum_{i=1}^{m}\sum_{j=1}^{d}\left(\log(S_Y)_{ij}^l + \frac{\left(Y_{ij} - (\mu_Y)_{ij}^l\right)^2}{(S_Y)_{ij}^l}\right)\right)\right) \\ &\quad + \frac{1}{2}\left(\sum_{i=1}^{m}\sum_{j=1}^{d}\left(\log(S_Z)_{ij} + 1 - (S_Z)_{ij} - (\mu_Z)_{ij}^2\right)\right), \end{aligned} \quad (5F38)$$

where

$$[\mu_Y \mid S_Y] = (I - A^T)\text{ReLu}(YW^1)W^2,$$

$$[\mu_X \mid S_X] = B\tanh(XW^3)W^4,$$

$$\mu_Z = \mu_Y - \mu_X,$$

$$S_Z = S_Y + S_X,$$

$$Z^l = \mu_Z + S_Z \odot \varepsilon^l,\ \varepsilon^l \sim MN_{md}(0, I, I), l = 1, 2, \ldots, L,$$

$$\left[\mu_{\hat{Y}} | S_{\hat{Y}}\right] = \text{ReLu}\left(\left(I - A^T\right)^{-1}\left(B\text{ReLu}\left(XW^5\right)W^6\right)\right.$$

$$\left.+Z\right)W^7\Big)W^8\Big).$$

Next we consider categorical endogenous variables Y. Categorical variable can be coded by a one-hot vector, where the "1" indicates the value of the corresponding variable, others take the "0" values. For example, suppose that we have four discrete variables Y_1, X_2, Y_3, and Y_4. One-hot vector representation can be given as follows:

Y_1	1	0	0	0
Y_2	0	1	0	0
Y_3	0	0	1	0
Y_4	0	0	0	1

Assumptions about the latent variables and latent space are not changed. We still assume that the prior and the posterior distributions are matrix normal distributions. Let $p\left((Y|X,Z)\right)$ be conditional distribution of Y, given X and latent variables Z, and p_Y be its probability matrix where each row denotes a vector of probability for the corresponding variable. The probability matrix P_Y is generated by

$$p_Y = softmax\Big(\text{ReLu}\Big(\left(I - A^T\right)^{-1}\left(B\tanh\left(XW^5\right)W^6\right)$$

$$+Z\big)W^7\big)W^8\big)\Big), \tag{5F39}$$

where W^5, W^6, W^7, and W^8 are weight matrices.

The probability distribution $p_{\theta_2}\left(Y|X,Z\right)$ for the categorical variables is given by

$$p_{\theta_1}\left(Y|X, Z^l\right) = \sum_{i=1}^{m}\sum_{j=1}^{d} Y_{ij} \log\left(p_Y^l\right)_{ij}. \tag{5F40}$$

Since we assume that the prior and posterior distributions are not changed, which in turn imply that KL distance between them is also unchanged. Therefore, the ELBO for categorical variables is given by

$$\mathcal{L}(\theta, \varnothing, Y, X) \approx \frac{1}{L}\sum_{l=1}^{L}\sum_{i=1}^{m}\sum_{j=1}^{d} Y_{ij}\log\left(p_Y^l\right)_{ij}$$

$$+\frac{1}{2}\left(\sum_{i=1}^{m}\sum_{j=1}^{d}\left(\log\left(S_Z\right)_{ij} + 1 - \left(S_Z\right)_{ij} - \left(\mu_Z\right)_{ij}^2\right)\right). \tag{5F41}$$

EXERCISES

EXERCISE 5.1

Show equation (5.23)

$$P\left(X|do(Z)\right) = P\left(Y|X\right).$$

EXERCISE 5.2

Consider $X \to Y \leftarrow e$ and $Y \leftarrow W$. Prove

$$P\left(Y|do\,(X), W\right) = P\left(Y|X, W\right).$$

EXERCISE 5.3

Show that under the null hypothesis H_0, the variance $\sigma^2 = var\left(T_{c(X\to Y)} - T_{C(Y\to X)}\right)$ can be estimated by

$$\sigma^2 = 2\left(\frac{1}{4n_{te}} - \frac{\sum_{i=1}^{n_{te}}\left(w_i - T_{c(X\to y)}\right)\left(g_i - T_{c(Y\to X)}\right)}{n_{te}-1}\right).$$

EXERCISE 5.4

Show

$$\frac{\partial h(W)}{\partial W} = \left(e^{W \circ W}\right)^T \circ 2W,$$

where

$$\text{h}(\text{W}) = \text{Tr}\left(e^{W \circ W}\right) - M.$$

EXERCISE 5.5

Show

$$-KL\left(q_{\varnothing}\left(\varepsilon|Y^n, X^n\right) \| p_\varepsilon(\varepsilon)\right)$$

$$= \frac{1}{2}\sum_{j=1}^{m}\left[1 + \log\left(\sigma_\varepsilon\right)_j^2 - \left(h_j(Y^n)\right)^2 - \left(\sigma_\varepsilon\right)_j^2\right].$$

EXERCISE 5.6

Show

$$q_{\varnothing}(Z|Y^n, X^n) = \frac{q_{\phi}(\varepsilon|Y^n, X^n)}{|C|}.$$

EXERCISE 5.7

Show that

$$\log p_{\theta}(Z|X^n) = \sum_{i=1}^{m} \log N \times \left(N\left(\mu_{Z|Z}^{i}(Z_i)\lambda(X_i), \sigma_{Z|Z}^{i}(Z_i)\lambda(X_i)\right)\right).$$

EXERCISE 5.8

Show that

$$-KL\left(q_{\varnothing}(Z \mid Y^n, X^n) \| p_{\theta}(Z \mid X^n)\right)\Big]$$

$$= \frac{1}{2}\sum_{j=1}^{m}\left(1 + \log\left(\frac{\sigma_Z^j}{\sigma_{Z|X^n}^j}\right)^2 - \left(\frac{\sigma_Z^j}{\sigma_{Z|X^n}^j}\right)^2 - \left(\frac{\mu_z^j - \mu_{Z|X^n}^j}{\sigma_{Z|X^n}^j}\right)^2\right).$$

EXERCISE 5.9

Show that

$$- KL\left(q_{\varnothing}(Z|Y^n, X^n) \| p_{\theta}(Z)\right)$$

$$= \frac{1}{2}\left\{\sum_{k=1}^{K}\left(1 + \log\left(\sigma_{\varnothing}^{k}(Y^n, X^n)\right)\right)^2 - \left(\mu_{\varnothing}^{k}(Y^n, X^n)\right)^2 - \left(\sigma_{\varnothing}^{k}(Y^n, X^n)\right)^2\right\}.$$

EXERCISE 5.10

Show that

$$\mathcal{L}(\theta, \varnothing, Y_i, t_i, X_i) = \frac{1}{L}\sum_{l=1}^{L} \log p_{\theta}(Y_i, t_i, X_i \mid Z_i^l) + \frac{1}{2}\sum_{j=1}^{k_Z}\left(1 + \log \bar{\sigma}_{ij}^2 - \bar{\sigma}_{ij}^2 - \bar{\mu}_{ij}^2\right).$$

EXERCISE 5.11

Show that

$$> -KL\left(q_{\varnothing}(Z_i|Y_i, T_i, X_i) \| p(Z_i)\right)\Big]$$

$$= \frac{1}{2}\left(1 + \log \bar{\sigma}_i^2 - \bar{\sigma}_i^2 - \bar{\mu}_i^2\right)$$

CHAPTER 6

Causal Inference in Time Series

6.1 INTRODUCTION

Causal inference in time series is a fundamental problem in science and engineering (Eichler 2013; Glymour et al. 2019). An essential difference between time series and cross-sectional data is that the time series data have temporal order, but cross-sectional data do not have any order. As a consequence, the causal inference methods for cross-sectional data which were discussed in the previous chapters cannot be directly applied to time series data. In this chapter we introduce causal inference methods in time series. First, we introduce four basic concepts of causality for multiple time series: intervention, structural, Granger and Sims causality. Then, we focus on investigation of two major causal graphical models: Granger graphical and dynamic direct acyclic graphical (DAG) models (Khanna and Tan 2019; Pamfil et al. 2020). Finally, applications of the dynamic causal graphical models to public health are investigated.

6.2 FOUR CONCEPTS OF CAUSALITY FOR MULTIPLE TIME SERIES

6.2.1 Granger Causality

Earliest and widely used concept of causality for time series data is Granger causality (Granger 1969; Eichler 2013). Underlying Granger causality is the following two principles:

1. Effect does not precede the cause in time;
2. The effect series contains unique cause series information, which is not present elsewhere.

The first principle is intuitively clear and widely accepted. However, the second principle requires further elaboration. Causal series information needs specific definition and be separated from other possible information (Eichler 2013). Two types of information can be considered. The first type of information includes the set of all possible information up to time t and is denoted by I_t. The second type of information consists of remaining information which is left from the first type of information after removing information contained in the cause series X, and is denoted by $I_{-X,t}$.

Consider three time series: X_t, Y_t and Z_t Let $X^t = \{X_s, s \le t\}$, $Y^t = \{Y_s, s \le t\}$ and $Z^t = \{Z_s, s \le t\}$. Let $X \perp\!\!\!\perp Y \mid Z$ denote the independence between two time series X and Y, give time series Z. Now I give the formal definition of Granger causality.

Definition 6.1: Granger Causality

Time series X does not Granger-cause time series Y if $Y_{t+1} \perp\!\!\!\perp X^t \mid I_{-X,t}$ for all time t. Otherwise, time series X Granger-causes time series Y.

Definition 6.1 states that given information $I_{-X,t}$, time series Y_{t+1} contains no information of all-time series data X_t before time t if time series X does not Granger-cause time series Y. A big problem in Definition 6.1 is that information $I_{-X,t}$ is not exactly defined. In practice, only background information is available. Suppose that all information is included in the data sets $\{X, Y, Z\}$. Let $I_t = \{X_t, Y_t, Z_t\}$ and $I_{-X,t} = \{Y_t, Z_t\}$, then Definition 6.1 can be modified to Definition 6.2 (Eichler 2013).

DOI: 10.1201/9781003028543-6

Definition 6.2: Granger Causality

Time series X is not Granger-cause of time series Y with respect to background information $V=(X,Y,Z)$ if $Y_{t+1} \perp\!\!\!\perp X^t \mid Y_t, Z_t$ for all time t. Otherwise, the time series X is Granger-causes the time series Y with respect to background information V.

Granger causality is a basic concept for discovering the causal structures among multiple time series. Definition 6.2 implies that the basic information should be as large as possible. We need to collect as many time series as we can. All collected time series should be included in the dataset. All-time series except for target time series X_t and Y_t are included in the dataset Z_t. Given the set of time series $V=(X,Y,Z)$, if we have $Y_{t+1} \perp\!\!\!\perp I_t \mid V^t$, then all direct structural causes of Y_t are included in the set of time series $V=(X,Y,Z)$.

In general, Granger causality is defined in terms of a fixed time delay. In other words, every time point of effect time series Y_t is affected by the time series X_t with a fixed time delay. However, in practice, the assumption of the fixed time delay often does not hold. To overcome this limitation, Granger causality with a fixed time delay can be generalized to Granger causality with arbitrary time delays (Amornbunchornvej et al. 2019).

Definition 6.3: Variable-Lag Granger Causality

Time series X is not variable-lag Granger-cause of time series Y with respect to background information $V=(X,Y,Z)$ if $Y_{t+\tau_t} \perp\!\!\!\perp X^t \mid Y_t, Z_t$ for all time t, where τ_t is a variable time lag. Otherwise, the time series X is Granger-causes the time series Y with variable-lag time under background information V.

6.2.2 Sims Causality

In the previous section, we pointed out that Granger causality only consider direct causation. Now I introduce Sims causality that considers both direct and indirect causations (Sims 1972; Eichler 2013). To include both direct and indirect causation, we need to changes conditional independent statement $Y_{t+1} \perp\!\!\!\perp X^t \mid Y_t, Z_t$ to $Y_\tau \perp\!\!\!\perp X_t \mid X^{t-1}, Y_t, Z_t$ for all and $\tau > t$. Then, we can consider indirect causation of X^t through other paths to reach Y_τ, $\tau > t$.

Definition 6.4: Sims Causality

Time series X is not Sims causal for time series Y with respect to $V=(X,Y,Z)$ if $Y_\tau\ (\tau > t) \perp\!\!\!\perp X_t \mid X^{t-1}, Y_t, Z_t$ for all time t. Otherwise, time series X is called Sims causal for time series Y with respect to V.

If the set of time series Z includes all confounders that cause both time series X and time series Y, Sims causality implies the presence of a causal effect of time series X on time series Y at least one lag.

6.2.3 Intervention Causality

The gold standard for discovering and assessing causal effects is experiments. Experiments allow us to intervene the system and assess the response of the system to intervention. However, experiments are expensive, time consuming and sometimes are infeasible. Similar to intervention in experiments, I introduce intervention causality. Intervention causality is the most rigorously defined causality concept and is widely used (Hagmayer et al. 2007; Eichler 2013). Taking actions and examining their response are crucial for inferring cause-effect relations. Consider two events A and B. If the event A is enforced to change, which leads to the change of the event B, then the change of the event B is caused by the event A. Otherwise, the event A and the event B do not have causal relationship.

An intervention can be either point-wise or stochastic process. Here, I focus on the point-wise intervention. Suppose that an intervention acts on the variable X. For the point-wise intervention, we assume that time series is stationary. Let σ_t be an intervention indicator at the time t and takes values in $(\varnothing, \delta \in R)$. We consider four types of point-wise interventions:

1. Without intervention. In this case, $\sigma_t = \varnothing$. The time series X is naturally generated without intervention.

2. Current point-wise intervention. Let $\sigma_t = x^*$. Intervention forces X_t to take value $X_t = x^*$ at time t.

3. Lagged intervention. Let $\sigma_t = g(X_{t-1})$, where g is a function. The intervention enforces the time series X_t to take value $X_t = g(X_{t-1})$ at the time t, which depends on the observation at the previous time $t-1$.

4. Random intervention. Let $\sigma_t = Z$ whose distribution is D. Random intervention enforces the time series X_t to randomly take value $X_t = Z$ with the distribution D.

Definition 6.5: Intervention Causal Effect

Intervention causal effect is measured by any functional of the post-intervention distribution and denoted by $P(Y_\tau, \tau > t \mid do\left(X_t = x^*\right)$.

The most common intervention causal effect measure is the average causal effect defined in Definition 6.6.

Definition 6.6: Average Intervention Causal Effect

Average intervention causal effect is defined as the average increase or decrease in value caused by the intervention. Without loss of generality, assume that $E\left[Y_t = 0\right]$. Mathematically, the average intervention causal effect is defined as

$$ACE_{\sigma_t} = E\left[Y_\tau \middle| do\left(X_t = \sigma_t\right)\right] - E\left[Y_\tau\right] \text{ for all } \tau > t. \quad (6.1)$$

In equation (6.1), the expectation E is taken with respect to distribution of the original system without intervention.

Intervention causal effect can also be defined as difference in variance between intervention and without intervention:

$$Var\left[Y_\tau \middle| do\left(X_t = \sigma_t\right)\right] - Var\left[Y_\tau\right] \text{ for all } \tau > t. \quad (6.2)$$

In the previous discussion, we did not consider causal structures. In other words, we did not consider specific structures that generate the data. Now we study causal-effect relationships among the stochastic processes with recursive dynamic structures (White and Lu 2010; Eichler 2013; Aragam 2020).

6.2.4 Structural Causality

Let time series X be a potential causal and time series Y be a response of time series X. Let Z be the set of all relevant observed time series. Suppose that two time series X and Y are recursively generated by

$$X_t = f_t^x\left(X^{t-1}, Y^{t-1}, Z^{t-1}, U_t^x\right), \quad (6.3)$$

$$Y_t = f_t^y\left(X^{t-1}, Y^{t-1}, Z^{t-1}, U_t^y\right), \quad (6.4)$$

where f_t^x and f_t^y are unknown functions, U_t^x and U_t^y are unobserved stochastic processes. The structural relations may be linear, nonlinear, and non-monotonic. The observed and unobserved processes may be non-separable. Equations (6.1) and (6.2) may generate either stationary or nonstationary or both stationary and nonstationary. Now we are ready to define structural causality in terms of structural equations (Eichler 2013).

Definition 6.7: Structural Causality

If function $f_t^y\left(x^{t-1}, y^{t-1}, z^{t-1}, u_t^y\right)$ is constant in x^{t-1} for all admissible values for $x^{t-1}, y^{t-1}, z^{t-1}, u_t^y$, then time series X does not directly structurally cause time series Y. Otherwise, time series X is called to directly structurally cause time series.

6.3 STATISTICAL METHODS FOR GRANGER CAUSALITY INFERENCE IN TIME SERIES

Granger causality means that given two time series X and Y, if the past information on X helps in forecasting Y better than using only the past information on Y, then X Granger-causes Y. Granger causality focuses on time series. Defining causality in time series is much easier than defining general causality in general data. Granger causality depends on the concept of "predictive causality". The many Granger causality tests are developed within this framework (Mazzarisi et al. 2020).

6.3.1 Bivariate Granger Causality Test

In this section, we introduce the linear and nonlinear Granger causality tests to identify the causality relationships between two time series or two stochastic processes.

6.3.1.1 Bivariate Linear Granger Causality Test

Consider two single variable time series X and Y. Suppose that X is a potential cause and Y is an effect or response. We assume that both time series X and Y are stationary. Define two linear time series models. The first model is the full model that includes all the past information of both cause X and response Y:

$$Y_t = \alpha_0 + \sum_{i=1}^{p} \alpha_i Y_{t-i} + \sum_{j=1}^{p} \beta_j X_{t-j} + \varepsilon_t, t = 1, \ldots, T, \quad (6.5)$$

where p is the optimal lag and ε_t is a time dependent noise.

The second model is the restricted model that includes only the past data of the response Y:

$$Y_t = \gamma_0 + \sum_{i=1}^{p} \gamma_i Y_{t-i} + e_t, t = 1, \ldots, T, \quad (6.6)$$

where p is the optimal lag and e_t is a time dependent noise.

The parameters in the full model and restricted models are estimated by least square method. The sum of square errors in the full and restricted models are, respectively, calculated by

$$SSE_{full} = \sum_{t=1}^{T} \left(Y_t - \hat{\alpha}_0 - \sum_{i=1}^{p} \hat{\alpha}_i Y_{t-i} - \sum_{j=1}^{p} \hat{\beta}_j X_{t-j} \right)^2, \quad (6.7)$$

and

$$SSE_R = \sum_{t=1}^{T} \left(Y_t - \hat{\gamma}_0 - \sum_{i=1}^{p} \hat{\gamma}_i Y_{t-i} \right)^2. \quad (6.8)$$

The null hypothesis for testing the Granger causality of time series X to the time series Y is

$$H_0 : \beta_1 = \cdots = \beta_p = 0.$$

An F statistic for testing Granger causality is given by

$$F_{GC} = \frac{\left(SSE_R - SSE_{full}\right) / p}{SSE_{full} / (T - 2p - 1)} \sim F_{p, T-2p-1}. \quad (6.9)$$

6.3.1.2 Bivariate Nonlinear Causality Test

Before introducing the definition of nonlinear Granger causality, I first define two notations. For any stationary and weekly dependent time series Y_t, the m-length lead vector and l_y- length lag vector of Y_t are, respectively, defined by

$$Y_t^m = \left(Y_t, Y_{t+1}, \ldots, Y_{t+m-1}\right), m = 1, 2, \ldots, t = 1, 2, \ldots \text{ and}$$

$$Y_{t-l_y}^{l_y} = \left(Y_{t-l_y}, \ldots, Y_{t-1}\right), l_y = 1, 2, \ldots, t = l_y + 1, l_y + 2, \ldots.$$

The m-length lead vector and l_x- length lag vector of X_t can be similarly defined. Now we define nonlinear Granger causality (Bai et al. 2010).

Definition 6.8: Nonlinear Granger Causality

Time series X_t does not strictly Granger cause another time series Y_t nonlinearly if and only if

$$P\left(\left\|Y_t^m - Y_s^m\right\| < e \left\| \left\|X_{t-l_x}^{l_x} - X_{s-l_x}^{l_x}\right\| < e, \left\|Y_{t-l_y}^{l_y} - Y_{s-l_y}^{l_y}\right\| < e\right)$$

$$= P\left(\left\|Y_t^m - Y_s^m\right\| < e \,\middle|\, \left\|Y_{t-l_y}^{l_y} - Y_{s-l_y}^{l_y}\right\| < e\right), \quad (6.10)$$

where $\|\cdot\|$ denotes the maximal norm defined as $X - Y = \max\left(x_1 - y_1, \ldots, x_n - y_n\right)$ for any two vectors $X = \left(x_1, \ldots, x_n\right)$ and $Y = \left(y_1, \ldots, y_n\right)$, $e > 0$ is a pre-specified constant.

Recall that

$$X_{t-l_x}^{m+l_x} = \left(X_{t-l_x}, \ldots, X_{t-1}, X_t, \ldots, X_{t+m-1}\right) \text{ and}$$

$$Y_{t-l_y}^{m+l_y} = \left(Y_{t-l_y}, \ldots, Y_{t-1}, Y_t, \ldots, Y_{t+m-1}\right).$$

Note that

$$X_{t-l_x}^{m+l_x} = \left(x_{t-l_x}^{l_x}, x_t^m\right) \text{ and}$$

$$X_{t-l_x}^{m+l_x} - X_{s-l_x}^{m+l_x} = \left(X_{t-l_x}^{l_x} - X_{s-l_x}^{l_x}, X_t^m - X_s^m\right).$$

It is easy to see that

$$\left\|X_{t-l_x}^{m+l_x} - X_{s-l_x}^{m+l_x}\right\| = \max\left(\left\|X_{t-l_x}^{l_x} - X_{s-l_x}^{l_x}\right\|, \left\|X_t^m - X_s^m\right\|\right). \quad (6.11)$$

Thus,

$$\left\|X_{t-l_x}^{m+l_x} - X_{s-l_x}^{m+l_x}\right\| < e \text{ implies}$$

$$\left\|X_{t-l_x}^{l_x} - X_{s-l_x}^{l_x}\right\| < e \text{ and } \left\|X_t^m - X_s^m\right\| < e,$$

and

$$\left\|X_{t-l_x}^{l_x} - X_{s-l_x}^{l_x}\right\| < e, \left\|X_t^m - X_s^m\right\| < e \text{ implies}$$

$$\left\|X_{t-l_x}^{m+l_x} - X_{s-l_x}^{m+l_x}\right\| < e.$$

Therefore, we obtain

$$P\left(\left\|Y_t^m - Y_s^m\right\| < e \,\middle|\, \left\|X_{t-l_x}^{l_x} - X_{s-l_x}^{l_x}\right\| < e, \left\|Y_{t-l_y}^{l_y} - Y_{s-l_y}^{l_y}\right\| < e\right)$$

$$= \frac{P\left(\left\|Y_t^m - Y_s^m\right\| < e, \left\|Y_{t-l_y}^{l_y} - Y_{s-l_y}^{l_y}\right\| < e, \left\|X_{t-l_x}^{l_x} - X_{s-l_x}^{l_x}\right\| < e\right)}{P\left(\left\|X_{t-l_x}^{l_x} - X_{s-l_x}^{l_x}\right\| < e, \left\|Y_{t-l_y}^{l_y} - Y_{s-l_y}^{l_y}\right\| < e\right)}$$

$$= \frac{P\left(\left\|Y_{t-l_y}^{m+l_y} - Y_{s-l_y}^{m+l_y}\right\| < e, \left\|X_{t-l_x}^{l_x} - X_{s-l_x}^{l_x}\right\| < e\right)}{P\left(\left\|X_{t-l_x}^{l_x} - X_{s-l_x}^{l_x}\right\| < e, \left\|Y_{t-l_y}^{l_y} - Y_{s-l_y}^{l_y}\right\| < e\right)}. \quad (6.12)$$

Similarly, we have

$$P\left(\left\|Y_t^m - Y_s^m\right\| < e \,\middle|\, \left\|Y_{t-l_y}^{l_y} - Y_{s-l_y}^{l_y}\right\| < e\right) = \frac{P\left(\left\|Y_t^m - Y_s^m\right\| < e, \left\|Y_{t-l_y}^{l_y} - Y_{s-l_y}^{l_y}\right\| < e\right)}{P\left(\left\|Y_{t-l_y}^{l_y} - Y_{s-l_y}^{l_y}\right\| < e\right)}$$

$$= \frac{P\left(\left\|Y_{t-l_y}^{m+l_y} - Y_{s-l_y}^{m+l_y}\right\| < e\right)}{P\left(\left\|Y_{t-l_y}^{l_y} - Y_{s-l_y}^{l_y}\right\| < e\right)}. \quad (6.13)$$

Let

$$C1\left(m+l_y, l_x, e\right) = P\left(\left\|Y_{t-l_y}^{m+l_y} - Y_{s-l_y}^{m+l_y}\right\| < e, \left\|X_{t-l_x}^{l_x} - X_{s-l_x}^{l_x}\right\| < e\right), \quad (6.14)$$

$$C2\left(l_y, l_x, e\right) = P\left(\left\|X_{t-l_x}^{l_x} - X_{s-l_x}^{l_x}\right\| < e, \left\|Y_{t-l_y}^{l_y} - Y_{s-l_y}^{l_y}\right\| < e\right), \quad (6.15)$$

$$C3\left(m+l_y, e\right) = P\left(\left\|Y_{t-l_y}^{m+l_y} - Y_{s-l_y}^{m+l_y}\right\| < e\right), \quad (6.16)$$

$$C4\left(l_y, e\right) = P\left(\left\|Y_{t-l_y}^{l_y} - Y_{s-l_y}^{l_y}\right\| < e\right). \quad (6.17)$$

Substituting equations (6.14) and (6.15) into equation (6.12) yields

$$P\left(\left\|Y_t^m - Y_s^m\right\| < e \,\middle|\, \left\|X_{t-l_x}^{l_x} - X_{s-l_x}^{l_x}\right\| < e, \left\|Y_{t-l_y}^{l_y} - Y_{s-l_y}^{l_y}\right\| < e\right) = \frac{C1\left(m+l_y, l_x, e\right)}{C2\left(l_y, l_x, e\right)}. \quad (6.18)$$

Similarly, substituting equations (6.16) and (6.17) into equation (6.13) yields

$$P\left(\left\|Y_t^m - Y_s^m\right\| < e \,\middle|\, \left\|Y_{t-l_y}^{l_y} - Y_{s-l_y}^{l_y}\right\| < e\right) = \frac{C3\left(m+l_y, e\right)}{C4\left(l_y, e\right)}. \quad (6.19)$$

Substituting equations (6.18) and (6.19) into equation (6.10) leads to

$$\frac{C1\left(m+l_y, l_x, e\right)}{C2\left(l_y, l_x, e\right)} = \frac{C3\left(m+l_y, e\right)}{C4\left(l_y, e\right)}. \quad (6.20)$$

Define an indicator function:

$$I\left(\left\|Z_1 - Z_2\right\| < e\right) = \begin{cases} 1 & \left\|Z_1 - Z_2\right\| < e \\ 0 & otherwise \end{cases}.$$

Let $n = T + 1 - m - \max\left(l_x, l_y\right)$. Using definition of probability as a frequency of event, we have the following estimations:

$$C1\left(m+l_y, l_x, e\right) = \frac{2}{n(n-2)} \sum\sum_{t<s} I\left(\left\|Y_{t-l_y}^{m+l_y} - Y_{s-l_y}^{m+l_y}\right\| < e\right) \times I\left(\left\|X_{t-l_x}^{l_x} - X_{s-l_x}^{l_x}\right\| < e\right), \quad (6.21)$$

$$C2\left(l_y, l_x, e\right) = \frac{2}{n(n-2)} \sum\sum_{t<s} I\left(\left\|X_{t-l_x}^{l_x} - X_{s-l_x}^{l_x}\right\| < e\right) \times I\left(\left\|Y_{t-l_y}^{l_y} - Y_{s-l_y}^{l_y}\right\| < e\right), \quad (6.22)$$

$$C3\left(m+l_y, e\right) = \frac{2}{n(n-2)} \sum\sum_{t<s} I\left(\left\|Y_{t-l_y}^{m+l_y} - Y_{s-l_y}^{m+l_y}\right\| < e\right), \quad (6.23)$$

$$C4\left(l_y, e\right) = \frac{2}{n(n-2)} \sum\sum_{t<s} I\left(\left\|Y_{t-l_y}^{l_y} - Y_{s-l_y}^{l_y}\right\| < e\right). \quad (6.24)$$

The null hypothesis for nonlinear Granger test is

$$H_0: X \text{ does not strictly Granger cause } Y.$$

The nonlinear Granger causality test statistic is defined as

$$T_{nlG} = \sqrt{n}\left(\frac{C1\left(m+l_y, l_x, e\right)}{C2\left(l_y, l_x, e\right)} - \frac{C3\left(m+l_y, e\right)}{C4\left(l_y, e\right)}\right). \quad (6.25)$$

Theorem 6.1

Assume that time series X and Y are strictly stationary and weakly dependent. Under the null hypothesis H_0, the test statistic T_{nlG} is asymptotically distributed as normal distribution $N(0, \sigma^2(m, l_x, l_y, e))$, where variance $\sigma^2(m, l_x, l_y, e)$ is given in Appendix 6A.

6.3.2 Multivariate Granger Causality Test

In this section, we will extent bivariate linear and nonlinear Granger causality tests to multivariate linear and nonlinear Granger causality tests (Bai et al. 2010). We begin with vector autoregressive regression (VAR) and multivariate linear Granger causality test.

6.3.2.1 Multivariate Linear Granger Causality Test

6.3.2.1.1 VAR Models Define a VAR model:

$$Y_t = A_0 + A(L)Y_{t-1} + e_t, \quad (6.26)$$

where

$$Y_t = \begin{bmatrix} Y_{1t} \\ \vdots \\ Y_{nt} \end{bmatrix}, A_0 = \begin{bmatrix} A_{10} \\ \vdots \\ A_{n0} \end{bmatrix}, A(L) = \begin{bmatrix} A_{11}(L) & \cdots & A_{1n}(L) \\ \vdots & \vdots & \vdots \\ A_{n1}(L) & \cdots & A_{nn}(L) \end{bmatrix},$$

$$Y_{t-1} = \begin{bmatrix} Y_{1,t-1} \\ \vdots \\ Y_{n,t-1} \end{bmatrix}, e_t = \begin{bmatrix} e_{1t} \\ \vdots \\ e_{nt} \end{bmatrix},$$

L is the backward operation, where $LY_t = Y_{t-1}$, $L^p Y_t = Y_{t-p}$, $A_{ij}(L) = a_{ij}(1)L + a_{ij}(2)L^2 + \cdots a_{ij}(p)L^p$, residuals e_t are distributed as a normal $N(0, \Sigma)$. In practice, a common order p will be selected for all the lag polynomials $A_{ij}(L)$. If the Gauss-Markov assumptions for the residuals are made, the ordinary least square estimation (OLSE) methods can be used to estimate the parameters in the VAR model.

6.3.2.1.2 Likelihood Ratio Test Suppose that we test the linear Granger causality relationship between two vectors of time series:

$$X_t = \begin{bmatrix} X_{1t} \\ \vdots \\ X_{n_1 t} \end{bmatrix} \text{ and } Y_t = \begin{bmatrix} Y_{1t} \\ \vdots \\ Y_{n_2 t} \end{bmatrix}, \text{ where } n = n_1 + n_2.$$

Consider the following VAR model:

$$\begin{bmatrix} X_t \\ Y_t \end{bmatrix} = \begin{bmatrix} A_x \\ A_y \end{bmatrix} + \begin{bmatrix} A_{xx}(L) & A_{xy}(L) \\ A_{yx}(L) & A_{yy}(L) \end{bmatrix} \begin{bmatrix} X_{t-1} \\ Y_{t-1} \end{bmatrix} + \begin{bmatrix} e_x \\ e_y \end{bmatrix}, \quad (6.27)$$

where A_x and A_y are n_1-dimensional and n_2-dimensional vector of intercept terms, $A_{xx}(L)_{n_1 \times n_1}$, $A_{xy}(L)_{n_1 \times n_2}$, $A_{yx}(L)_{n_2 \times n_1}$, and $A_{yy}(L)_{n_2 \times n_2}$ are matrices of lag polynomials.

There are four different cases of causal relationships between two vectors of time series X_t and Y_t (Bai et al. 2010):

1. If $A_{xy}(L)$ is significantly different from the zero, while $A_{yx}(L)$ shows no significantly different from zero, then there exists a unidirectional Ganger causality from time series Y_t to X_t;
2. If $A_{yx}(L)$ is significantly different from zero, while $A_{xy}(L)$ shows no significantly difference from zero, then there exists a unidirectional Ganger causality from X_t to Y_t;
3. If both coefficients $A_{xy}(L)$ and $A_{yx}(L)$ are significantly different from zero, then there exists bidirectional Granger causality between X_t and Y_t;
4. If both coefficients $A_{xy}(L)$ and $A_{yx}(L)$ are not significantly different from zero, then X_t and Y_t are not rejected to be independent.

The four statements imply that Ganger causal relationships between X_t and Y_t depend on the coefficients $A_{xy}(L)$ and $A_{yx}(L)$. Therefore, the null hypotheses for testing the Ganger causality between X_t and Y_t are

1. $H_0^1 : A_{xy}(L) = 0$,
2. $H_0^2 : A_{yx}(L) = 0$, and
3. Both H_0^1 and $H_0^2 : A_{xy}(L) = 0$ and $A_{yx}(L) = 0$.

Define the residual covariance matrix of the full VAR model in equation (6.27) as

$$\Sigma = \sum_{2n\times 2n} = \begin{bmatrix} cov(e_x, e_x) & cov(e_x, e_y) \\ cov(e_y, e_x) & cov(e_y, e_y) \end{bmatrix}.$$

The likelihood function and log-likelihood function for the full VAR model with T periods of time are, respectively, given by

$$L = \frac{1}{(2\pi)^{nT}\left|\Sigma\right|^{\frac{T}{2}}}\exp\left(-\frac{T}{2}\begin{pmatrix} e_x^T & e_y^T \end{pmatrix}\Sigma^{-1}\begin{pmatrix} e_x \\ e_y \end{pmatrix}\right), \quad (6.28)$$

and

$$l = \log L = -nT\log(2\pi) - \frac{T}{2}\log\left|\Sigma\right| - \frac{T}{2}\begin{pmatrix} e_x^T & e_y^T \end{pmatrix}\Sigma^{-1}\begin{pmatrix} e_x \\ e_y \end{pmatrix}. \quad (6.29)$$

Similarly, the likelihood and log-likelihood function for the reduced model under the null hypothesis are, respectively, given by

$$L_0 = \frac{1}{(2\pi)^{nT}\left|\Sigma_0\right|^{\frac{T}{2}}}\exp\left(-\frac{T}{2}\begin{pmatrix} e_{x0}^T & e_{y0}^T \end{pmatrix}\Sigma^{-1}\begin{pmatrix} e_{x0} \\ e_{y0} \end{pmatrix}\right), \quad (6.30)$$

and

$$l_0 = \log L_0 = -nT\log(2\pi) - \frac{T}{2}\log\left|\Sigma_0\right| - \frac{T}{2}\begin{pmatrix} e_{x0}^T & e_{y0}^T \end{pmatrix}\Sigma^{-1}\begin{pmatrix} e_{x0} \\ e_{y0} \end{pmatrix}. \quad (6.31)$$

We can show that (Exercise 6.1)

$$\hat{\Sigma} = \begin{pmatrix} e_x \\ e_y \end{pmatrix}\begin{pmatrix} e_x^T & e_y^T \end{pmatrix}, \quad (6.32)$$

and

$$\hat{\Sigma}_0 = \begin{pmatrix} e_{x0} \\ e_{y0} \end{pmatrix}\begin{pmatrix} e_{x0}^T & e_{y0}^T \end{pmatrix}. \quad (6.33)$$

We can also show that

$$\begin{pmatrix} e_x^T & e_y^T \end{pmatrix}\hat{\Sigma}^{-1}\begin{pmatrix} e_x \\ e_y \end{pmatrix} = n \text{ and } \begin{pmatrix} e_{x0}^T & e_{y0}^T \end{pmatrix}\hat{\Sigma}_0^{-1}\begin{pmatrix} e_x \\ e_y \end{pmatrix} = n. \quad (6.34)$$

Substituting equations (6.32) and (6.34) into equation (6.30), and equations (6.33) and (6.34) into equation (6.31) yields

$$l = -nT\log(2\pi) - \frac{T}{2}\log\log\hat{\Sigma} - \frac{nT}{2}, \quad (6.35)$$

$$l_0 = -nT\log(2\pi) - \frac{T}{2}\log\left|\hat{\Sigma}_0\right| - \frac{nT}{2}. \quad (6.36)$$

The likelihood ratio test statistics is defined as

$$T_{ml} = -2(l_0 - l_1) = T\left(\log\left|\hat{\Sigma}_0\right| - \log\left|\hat{\Sigma}\right|\right),$$

which can be approximated by

$$T_{ml} = (T - c)\left(\log\left|\hat{\Sigma}_0\right| - \log\left|\hat{\Sigma}\right|\right), \quad (6.37)$$

where c is the number of parameters estimated in each equation of the unrestricted system.

Theorem 6.2: Likelihood Ratio Tests for Multivariate Granger Causality

1. The likelihood ratio statistics for testing the null hypothesis: $H_0^1 : A_{xy}(L) = 0$ is

$$T_{ml} = \left(T - (np+1)\right)\left(\log\left|\hat{\Sigma}_0\right| - \log\left|\hat{\Sigma}\right|\right), \quad (6.38)$$

which is asymptotically distributed as a central $\chi^2_{(n_1\times n_2\times p)}$ under the null hypothesis H_0^1.

2. The likelihood ratio statistics for testing the null hypothesis: $H_0^2 : A_{yx}(L)=0$ is

$$T_{ml}=\left(T-(np+1)\right)\left(\log\left|\hat{\sum}_0\right|-\log\left|\hat{\sum}\right|\right),$$

which is asymptotically distributed as a central $\chi^2_{(n_2\times n_1\times p)}$ under the null hypothesis H_0^2.

3. The likelihood ratio statistics for testing the null hypothesis: H_0^1 and $H_0^2 : A_{xy}(L)=0$ and $A_{yx}(L)=0$ is

$$T_{ml}=\left(T-(np+1)\right)\left(\log\left|\hat{\sum}_0\right|-\log\left|\hat{\sum}\right|\right),$$

which is asymptotically distributed as a central $\chi^2_{(2n_2\times n_1\times p)}$ under the hull hypothesis H_0^1 and H_0^2.

6.3.3 Nonstationary Time Series Granger Causal Analysis

6.3.3.1 Background

Basic tools in statistical analysis are the raw of large numbers and the central limit theorem. Applications of these tools usually assume that all moment functions are constant. When the moment functions of the time series vary over time, the raw of large numbers and the central limit theorem cannot be applied. In order to use basic probabilistic and statistical theories, the nonstationary time series must be transformed to stationary time series.

6.3.3.1.1 Stationarity

Definition 6.9: Stationary Time Series

The time series $Y, \ldots, Y_T$ are stationary if for any fixed p, the probability density function $P(Y_t, \ldots, Y_{t+p})$ does not change with time. Stationary time series include (1) The constant moments: $E[Y_t]=\mu$, $Var(Y_t)=\sigma^2$, and (2) the autocovariance between two time points only depends on the time between them and not the time of the start:

$$Cov(Y_t, Y_{t+s})=\gamma_s.$$

We are often interested in the first two moments of a time series. Therefore, we define the following weak stationary time series.

Definition 6.10: Weak Stationary Time Series

The time series $Y, \ldots, Y_T$ are weak stationary, if we have

$$E[Y_t]=\mu \text{ and } Cov(Y_t, Y_s) = Cov(Y_{t-s}, Y_0) \text{ for all } t>s.$$

Any time series which are not stationary and weak stationary are nonstationary time series. I introduce two types of nonstationary time series: nonstationary due to deterministic trends and nonstationary due to stochastic trends.

6.3.3.1.1.1 Deterministic Trends Consider the deterministic nonstationarity component of the time series and model the time series as

$$Y_t=\mu_t+\varepsilon_t, \quad (6.39)$$

where ε_t is a stationary noise term. If $\mu_t=\alpha+\beta t$ then the trend is a linear trend.

Three steps are used to model nonstationary time series with deterministic trend.

Step 1: Fit a model μ_t to Y_t.

Step 2: Calculate residual $\hat{\varepsilon}_t=Y_t-\mu_t$.

Step 3: Model stationary time series $\hat{\varepsilon}_t$.

6.3.3.1.1.2 Stochastic Trend One more realistic trend is a stochastic trend that models the trend changes from time to time as random. Consider a model for time series Y_t:

$$Y_t=\alpha+\beta t+u_t+\varepsilon_t, \quad (6.40)$$

where u_t is a random walk and ε_t is a stationary time series. Define

$$\mu_t=\alpha+\beta t+u_t$$

as the stochastic trend.

Example 6.1 Random Walk

Consider a model:

$$Y_t=Y_{t-1}+\varepsilon_t, \quad (6.41)$$

where ε_t is a white noise $N(0, \sigma^2)$.

Equation (6.41) can be rewritten as

$$Y_t = Y_{t-2} + \varepsilon_{t-1} + \varepsilon_t = \cdots = Y_0 + \sum_{\tau=0}^{t-1} \varepsilon_{t-\tau}, \qquad (6.42)$$

which implies that

$$Var(Y_t|Y_0) = Var\left(\sum_{\tau=0}^{t-1} \varepsilon_{t-\tau}\right) = t\sigma^2.$$

This shows that variance changes over time. Therefore, the time series defined in equation (6.41) is nonstationary.

Example 6.2 Random Walk with Drift

Consider a random walk with drift model:

$$Y_t = \alpha + Y_{t-1} + \varepsilon_t, \qquad (6.43)$$

where ε_t is a white noise $N(0, \sigma^2)$.

Again, the model (6.43) can be recursively rewritten as

$$Y_t = Y_0 + t\alpha + \sum_{\tau=0}^{t-1} \varepsilon_{t-\tau}.$$

Therefore, the conditional mean $E(Y_t \mid Y_0)$ and $Var(Y_t \mid Y_0)$ are

$E(Y_t|Y_0) = t\alpha$ and $Var(Y_t|Y_0) = t\sigma^2$, respectively.

The time series Y_t is nonstationary in terms of both conditional mean and conditional variance.

6.3.3.1.2 Unit Roots, Stationarity, and Differencing in AR Models First consider AR(1) model:

$$Y_t = \rho Y_{t-1} + \varepsilon_t,$$

where $\varepsilon_t \sim N(0, \sigma^2)$.

After some algebra, we can obtain

$$Y_t = \rho^t Y_0 + \sum_{\tau=0}^{t-1} \rho^\tau \varepsilon_{t-\tau}, \qquad (6.44)$$

which implies

$$Var(Y_t) = var\left(\sum_{\tau=0}^{t-1} \rho^\tau \varepsilon_{t-\tau}\right) = \sum_{\tau=0}^{t-1} \rho^{2\tau}\sigma^2 \to \sigma^2 \frac{1}{1-\rho^2}. \qquad (6.45)$$

It is clear that when $|\rho| < 1$, $Var(Y_t) \to \sigma^2 \frac{1}{1-\rho^2}$. In other words, when $|\rho| < 1$, the AR(1) is asymptotically stationary. Now we consider a general AR(p) time series.

Let $\theta(L) = 1 - \theta_1 L - \cdots - \theta_p L^p$. Consider AR($p$) model:

$$\theta(L) Y_t = Y_t - \theta_1 Y_{t-1} - \cdots - \theta_p Y_{t-p} = \varepsilon_t. \qquad (6.46)$$

For AR(1), we have $\theta(L) = 1 - \rho L$. The root of $\theta(L) = 1 - \rho L = 0$ is $L = \frac{1}{\rho}$. Equation (6.45) showed that when $|\rho| < 1$, i.e., $|L| > 1$, then the time series Y_t is stationary.

Consider the random walk model in equation (6.41):

$$Y_t = Y_{t-1} + \varepsilon_t,$$

which implies

$$Y_t - Y_{t-1} = (1 - L) Y_t = \varepsilon_t.$$

Therefore, for the random walk model, we have $\theta(L) = 1 - L$. The root of $\theta(L) = 0$ is 1. The above two examples show that the stationarity properties of time series Y_t are assessed by whether the roots of $\theta(L) = 0$ are outside the unit circle (stationary) or on it (nonstationary). Now we generalize this criterion to a general AR(p) process. Suppose that we have p roots: $\frac{1}{\theta_1}, \ldots, \frac{1}{\theta_p}$ and assume that $|\theta_i| \le 1$. Then, $\theta(L)$ can be factorized to $\theta(L) = (1 - \theta_1 L)\ldots(1 - \theta_p)$. We have the following theorem.

Theorem 6.3

For AR(p) process, if the roots of $\theta(L) = 0$ are outside the unit circle, then the time series Y_t is stationary and if the roots of $\theta(L) = 0$ are on the unit circle, the time series Y_t is nonstationary.

Now we discuss the decomposition of the nonstationary time series into stationary components and nonstationary components. Suppose that there are $k \le p$ unit roots and $p - k$ roots that are grater than one. $\theta(L)$

can be written as $\theta(L)=(1-\theta_1 L)\ldots(1-\theta_{p-k}L)(1-L)^k$. Equation (6.47) becomes

$$\begin{aligned}\theta(L)Y_t &= (1-\theta_1 L)\ \ldots\ (1-\theta_{p-k}L)(1-L)^k Y_t \\ &= (1-\theta_1 L)\ \ldots\ (1-\theta_{p-k}L)(\Delta^k Y_t) = \varepsilon_t,\end{aligned} \tag{6.47}$$

where $\Delta Y_t = (1-L)\,Y_t = Y_t - Y_{t-1},\ \ldots\ ,\Delta^k Y_t = (1-L)^k$ $Y_t = \Sigma_{j=0}^{k}\binom{k}{j}(-1)^{k-j}Y_{t-(k-j)}$.

Let $Z_t = \Delta^k Y_t$ and $\gamma(L)=(1-\theta_1 L)\ \ldots\ (1-\theta_{p-k}L)$. Then, the roots of $\gamma(L)=0$ are greater than one and outside the unit circle. Since $\gamma(L)Z_t = \varepsilon_t$, the time series Z_t, and hence $\Delta^k Y_t$ are stationary. Using differencing operation, we can change nonstationary time series to stationary time series.

Definition 6.11: Integrated of Order *k*

A series that has k unit roots is called integrated of order k time series. If Y_t is stationary, it is denoted by $I(0)$. If the first difference of Y_t is stationary, but the original time series Y_t is not, then Y_t is denoted by $I(1)$. If the first difference is nonstationary, but the second difference is stationary, then Y_t is denoted by $I(2)$. The integrated of order k time series is noted by $I(k)$.

The integrated of order k time series is stationary after being differenced k times. Recall that $Z_t = (1-L)^k Y_t = \Delta^k Y_t$. Inverse operation of difference operation $(1-L)$ is integration $(1-L)^{-1}$. The original time series can be obtained by

$$(1-L)^{-k} Z_t = (1-L)^{-k}(1-L)^k Y_t = Y_t.$$

The estimation of the parameters in the nonstationary time series by the OLS is biased. The association or correlation between two nonstationary time series may be spurious. Therefore, the first step for analyzing nonstationary time series is to convert nonstationary time series to stationary time series using difference operation.

6.3.3.1.3 Dickey-Fuller Tests for Unit Roots

6.3.3.1.3.1 Wiener Process A stochastic process W_t is a Wiener process if it satisfies the following four conditions:

1. $W_0 = 0$,
2. For $t \geq 0$ and $\Delta t \geq 0$, the increment $W_{t+\Delta t} - W_t$ is normally distributed as $N(0,\ \Delta t)$, which implies $W_t \sim N(0,\ t)$ and $W_1 \sim N(0,\ 1)$.
3. The increment $W_{s+\Delta s} - W_s$ and $W_{t+\Delta t} - W_t$ for any $s < s + \Delta s < t < t + \Delta t$ are independent,
4. With probability 1, the process W_t is continuous at t.

Example 6.3

Let $\{X_t, t = 1,2,\ldots, N\}$ be a sequence of independent and identically distributed discrete random variables such that $P(X_t = \mp 1) = \frac{1}{2}$. Then, we have

$$E[X_t] = 0 \text{ and } Var(X) = 1.$$

Define

$$S_0 = 0 \text{ and } S_N = \sum_{t=1}^{N} X_t.$$

Then, $E[S_N] = 0$ and $Var(S_N) = N$. Using central limit theorem, we obtain

$$\frac{S_N}{\sqrt{N}} \rightarrow N(0,1).$$

Rescale the process S_N to $S_{[Nt]}$. We define

$$W_t^N = \frac{S_{[Nt]}}{\sqrt{N}} = \sqrt{t}\,\frac{S_{[Nt]}}{\sqrt{Nt}} \sim N(0,\ t).$$

Therefore, when $N \rightarrow \infty$, $W_t^N \rightarrow W_t$.

Example 6.4 Random Walk

Consider the random walk model:

$$Y_t = Y_{t-1} + \varepsilon_t. \tag{6.48}$$

Equation (6.48) can be rewritten as

$$Y_T = Y_0 + \varepsilon_1 + \cdots + \varepsilon_T \sim N(0, T\sigma^2), \tag{6.49}$$

which implies

$$\frac{Y_T}{\sqrt{T}\sigma} \rightarrow N(0,1) = W(1), \tag{6.50}$$

where for a convenience of notation, W_t is written as $W(t)$.

In general, $\frac{Y_j}{\sqrt{T}\sigma} \to W(t)$, where $\frac{j-1}{T} \le t \le \frac{j}{T}$.

Example 6.5 Moments of Random Walk

Consider five moments of random walk:

$$S_{1T} = \frac{1}{T\sqrt{T}\sigma}\sum_{t=2}^{T} Y_{t-1}, \tag{6.51}$$

$$S_{2T} = \frac{1}{T^2\sigma^2}\sum_{j=1}^{T} Y_j^2, \tag{6.52}$$

$$S_{3T} = \frac{1}{T^2\sigma^2}\sum_{t=1}^{T}\left(Y_t - \bar{Y}\right)^2, \tag{6.53}$$

$$S_{4T} = T\left(\hat{\rho} - 1\right), \tag{6.54}$$

$$S_{5T} = \frac{\hat{\rho} - 1}{\hat{\sigma}\Big/\sqrt{\sum_{t=2}^{T} y_{t-1}^2}}, \tag{6.55}$$

where $\hat{\rho} = \frac{\sum_{t=2}^{T} Y_{t-1}Y_t}{\sum_{t=2}^{T} Y_{t-1}^2}$ and $\hat{\sigma}^2 = \frac{1}{T-1}\sum_{t=2}^{T}\left(Y_t - \hat{\rho}Y_{t-1}\right)^2$.

We will show that

$$\frac{1}{T\sqrt{T}\sigma}\sum_{t=2}^{T} Y_{t-1} \to \int_0^1 W(t)dt, \tag{6.56}$$

$$\frac{1}{T^2\sigma^2}\sum_{t=1}^{T} Y_t^2 \to \int_0^1 W^2(t)dt, \tag{6.57}$$

$$\frac{1}{T^2\sigma^2}\sum_{t=1}^{T}\left(Y_t - \bar{Y}\right)^2 \to \int_0^1\left(W(t) - \int_0^1 W(t)dt\right)^2 dt, \tag{6.58}$$

$$T\left(\hat{\rho} - 1\right) \to \frac{\frac{1}{2}\left(W^2(1) - 1\right)}{\int_0^1 W^2(t)dt} = \frac{\int_0^1 W(t)dW(t)}{\int_0^1 W^2(t)dt}, \tag{6.59}$$

$$\frac{\hat{\rho} - 1}{\hat{\sigma}\Big/\sqrt{\sum_{t=2}^{T} y_{t-1}^2}} \to \frac{\frac{1}{2}\left(W^2(1) - 1\right)}{\sqrt{\int_0^1 W^2(t)dt}} = \frac{\int_0^1 W(t)dW(t)}{\int_0^1 W^2(t)dt}. \tag{6.60}$$

Now I first show equation (6.56). Recall that

$$\frac{Y_{j-1}}{\sigma\sqrt{T}} \to W(t), \frac{j-1}{T} \le t_j \le \frac{j}{T}.$$

Then,

$$\begin{aligned} S_{1T} &= \frac{1}{T\sqrt{T}\sigma}\sum_{j=2}^{T} Y_{j-1} = \frac{1}{T}\sum_{j=2}^{T}\frac{Y_{j-1}}{\sqrt{T}\sigma} \\ &\to \frac{1}{T}\sum_{j=2}^{T} W\left(t_j\right) \to \int_0^1 W(t)dt. \end{aligned} \tag{6.61}$$

Next I show equation (6.57). It follows from equation (6.2) that

$$\begin{aligned} S_{2T} &= \frac{1}{T^2\sigma^2}\sum_{j=1}^{T} Y_j^2 = \frac{1}{T}\sum_{j=1}^{T}\left(\frac{Y_j}{\sqrt{T}\sigma}\right)^2 \\ &\to \frac{1}{T}\sum_{j=1}^{T} W^2\left(t_j\right) \to \int_0^1 W^2(t)dt. \end{aligned} \tag{6.62}$$

To show equation (6.58), I first reduce equation (6.53). It is clear that

$$\begin{aligned} S_{3T} &= \frac{1}{T^2\sigma^2}\sum_{j=1}^{T}\left(Y_j - \bar{Y}\right)^2 \\ &= \frac{1}{T^2\sigma^2}\sum_{j=1}^{T} Y_j^2 - \frac{1}{T^2\sigma^2} T\bar{Y}^2 \\ &= \frac{1}{T}\sum_{j=1}^{T}\left(\frac{Y_j}{\sqrt{T}\sigma}\right)^2 - \left(\frac{1}{T\sqrt{T}\sigma}\sum_{j=1}^{T} Y_j\right)^2. \end{aligned} \tag{6.63}$$

Using equations (6.61) and (6.62), we obtain

$$S_{3T}=\frac{1}{T}\sum_{j=1}^{T}\left(\frac{Y_j}{\sqrt{T}\sigma}\right)^2-\left(\frac{1}{T\sqrt{T}\sigma}\sum_{j=1}^{T}Y_j\right)^2$$

$$\to\int_0^1 W^2(t)dt-\left(\int_0^1 W(t)dt\right)^2$$

$$=\int_0^1\left(W(t)-\int_0^1 W(t)dt\right)^2 dt.$$

Now we show equation (6.54). Recall that the least square estimator of ρ is

$$\hat{\rho}=\frac{\sum_{j=2}^{T}Y_{j-1}Y_j}{\sum_{j=2}^{T}Y_{j-1}^2}\text{, which implies}$$

$$\hat{\rho}-1=\frac{\sum_{j=2}^{T}Y_{j-1}Y_j}{\sum_{j=2}^{T}Y_{j-1}^2}-1=\frac{\sum_{j=2}^{T}Y_{j-1}Y_j-\sum_{j=2}^{T}Y_{j-1}^2}{\sum_{j=2}^{T}Y_{j-1}^2}$$

$$=\frac{\sum_{j=2}^{T}Y_{j-1}e_j}{\sum_{j=2}^{T}Y_{j-1}^2}. \tag{6.64}$$

Let $U_T=\frac{1}{T\sigma^2}\sum_{j=2}^{T}Y_{j-1}e_j$ and $V_T=\frac{1}{T^2\sigma^2}\sum_{j=2}^{T}Y_{j-1}^2$. Note that $Y_{j-1}e_j=\frac{Y_j^2-Y_{j-1}^2-e_j^2}{2}$, which implies that

$$\sum_{j=2}^{T}Y_{j-1}e_j=\frac{1}{2}\sum_{j=2}^{T}\left(Y_j^2-Y_{j-1}^2-e_j^2\right). \tag{6.65}$$

Let $j'=j-1$. Then, we have

$$\sum_{j=2}^{T}Y_{j-1}^2=\sum_{j'=1}^{j'=T-1}Y_{j'}^2. \tag{6.66}$$

Substituting equation (6.66) into equation (6.65), we obtain

$$\sum_{j=2}^{T}Y_{j-1}e_j=\frac{1}{2}\left(Y_T^2-Y_1^2+e_1^2-\sum_{j=1}^{T}e_j^2\right)\text{, which implies that}$$

$$U_T=\frac{1}{2}\left(\left(\frac{Y_T}{\sqrt{T}\sigma}\right)^2-\frac{Y_1^2}{T\sigma^2}+\frac{e_1^2}{T\sigma^2}-\frac{1}{T\sigma^2}\sum_{j=1}^{T}e_j^2\right)$$

$$\to\frac{1}{2}\left(W^2(1)-1\right).$$

However,

$$\frac{1}{2}\left(W^2-1\right)=\int_0^1 W(t)dW(t)\text{, which implies}$$

$$U_T\to\int_0^1 W(t)dW(t). \tag{6.67}$$

Again, let $j'=j-1$. Then,

$$V_T=\frac{1}{T^2\sigma^2}\sum_{j=2}^{T}Y_{j-1}^2=\frac{1}{T^2\sigma^2}\sum_{j'=1}^{T-1}Y_{j'}^2$$

$$=\frac{1}{T^2\sigma^2}\left(\sum_{j=1}^{T}Y_j^2-Y_T^2\right)$$

$$=\frac{1}{T^2\sigma^2}\sum_{j=1}^{T}Y_j^2-\frac{1}{T}\left(\frac{Y_T}{\sqrt{T}\sigma}\right)^2. \tag{6.68}$$

Using equations (6.57) and (6.50), we obtain

$$V_T\to\int_0^1 W^2(t)dt. \tag{6.69}$$

It follows from equation (6.64) that

$$T(\hat{\rho}-1)=\frac{U_T}{V_T}\to\frac{\frac{1}{2}\left(W^2(1)-1\right)}{\int_0^1 W^2(t)dt}=\frac{\int_0^1 W(t)dW(t)}{\int_0^1 W^2(t)dt}. \tag{6.70}$$

Now we prove equation (6.60). S_{5T} in equation (6.55) can be rewritten in terms of U_T and V_T:

$$S_{5T}=\frac{1}{T}\frac{U_T}{V_T}\frac{\sqrt{\sum_{j=2}^{T}Y_{j-1}^2}}{\sigma}=\frac{1}{T}\frac{U_T}{V_T}\frac{T\sigma\sqrt{V_T}}{\sigma}$$

$$=\frac{U_T}{\sqrt{V_T}}\to\frac{\frac{1}{2}\left(W^2-1\right)}{\sqrt{\int_0^1 W^2(t)dt}}. \tag{6.71}$$

Again, consider the simple AR(1) model:

$$Y_t = \rho Y_{t-1} + \varepsilon_t.$$

The null and alternative hypotheses are

$$H_0 : \rho = 1, \left(Y_t \sim I(1)\right),$$

$$H_a : |\rho| < 1, \left(Y_t \sim I(0)\right).$$

The statistics for testing the null hypothesis H_0 is defined as

$$S_{st} = \frac{\hat{\rho} - 1}{\hat{\sigma} \Big/ \sqrt{\sum_{t=2}^{T} y_{t-1}^2}}, \tag{6.72}$$

where $\hat{\rho} = \frac{\sum_{t=2}^{T} Y_{t-1} Y_t}{\sum_{t=2}^{T} Y_{t-1}^2}$ and $\hat{\sigma}^2 = \frac{1}{T-1} \sum_{t=2}^{T} \left(Y_t - \hat{\rho} Y_{t-1}\right)^2$.

The distribution of the statistics S_{st} under the null hypothesis is Dickey-Fuller (DF) distribution as summarized in Theorem 6.4. The normalized bias $S_{nST} = T\left(\hat{\rho} - 1\right)$ has a well-defined asymptotical distribution that does not depend on nuisance parameters; therefore, it can also be used as a test statistic for the null hypothesis. Its distribution is also summarized in Theorem 6.4.

Theorem 6.4: Dickey-Fuller (DF) Distribution

The test statistics S_{st} under the null hypothesis is asymptotically distributed as DF distribution, which is defined by

$$S_{st} \sim \frac{\frac{1}{2}(W^2 - 1)}{\sqrt{\int_0^1 W^2(t)dt}}. \tag{6.73}$$

The asymptotical distribution of the normalized bias statistic is

$$S_{nST} = T\left(\hat{\rho} - 1\right) \sim \frac{\frac{1}{2}(W^2(1) - 1)}{\int_0^1 W^2(t)dt}. \tag{6.74}$$

6.3.3.1.3.2 Trend Cases The presence of trend under alternative hypothesis will determine both form of the test regression used and asymptotic distribution of the test statistics. We first consider the constant trend model:

$$Y_t = c + \rho Y_{t-1} + \varepsilon_t. \tag{6.75}$$

The hypotheses are

$H_0 : \rho = 1, c = 0$ implies that the time series Y_t is I(1) without drift,

$H_a : |\rho| < 1$ implies that the time series Y_t is I(0) with nonzero mean.

The estimator of constant c is mean

$$\bar{Y} = \frac{1}{T} \sum_{t=1}^{T} Y_t. \tag{6.76}$$

Using OLS method, we obtain the estimator $\hat{\rho}$:

$$\hat{\rho}_c = \frac{\sum_{j=2}^{T} \left(Y_j - \bar{Y}\right)\left(Y_{j-1} - \bar{Y}\right)}{\sum_{j=2}^{T} \left(Y_{j-1} - \bar{Y}\right)^2}. \tag{6.77}$$

Again, define the test statistics:

$$S_{stc} = \frac{\hat{\rho}_c - 1}{\hat{\sigma} \Big/ \sqrt{\sum_{t=2}^{T} \left(Y_{t-1} - \bar{Y}\right)^2}}, \tag{6.78}$$

$$S_{stnc} = T\left(\hat{\rho}_c - 1\right). \tag{6.79}$$

DF distribution and normalized bias distribution can be extended to time series AR(1) with constant trend.

Theorem 6.5: Dickey-Fuller (DF) Distribution for Constant Trend

The statistics S_{stc} under the null hypothesis is asymptotically distributed as DF distribution for the constant trend:

$$S_{stc} \to \frac{\int_0^1 \left(W(t) - \int_0^1 W(t)dt \right) dW(t)}{\sqrt{\int_0^1 \left(W(t) - \int_0^1 W(t)dt \right)^2 dt}}, \tag{6.80}$$

$$S_{stnc} \to \frac{\int_0^1 \left(W(t) - \int_0^1 W(t)dt \right) dW(t)}{\int_0^1 \left(W(t) - \int_0^1 W(t)dt \right)^2 dt}. \tag{6.81}$$

Now we first show equation (6.80). It follows from equation (6.77) that

$$\hat{\rho}_c - 1 = \frac{\sum_{j=2}^{T}(Y_j - \bar{Y})(Y_{j-1} - \bar{Y}) - \sum_{j=2}^{T}(Y_{j-1} - \bar{Y})^2}{\sum_{j=2}^{T}(Y_{j-1} - \bar{Y})^2}$$

$$= \frac{\sum_{j=2}^{T}(Y_j - \bar{Y})e_j}{\sum_{j=2}^{T}(Y_{j-1} - \bar{Y})^2}. \tag{6.82}$$

Let

$$U_{Tc} = \frac{1}{T\sigma^2}\sum_{j=2}^{T}(Y_j - \bar{Y})e_j \text{ and} \tag{6.83}$$

$$V_{Tc} = \frac{1}{T^2\sigma^2}\sum_{j=2}^{T}(Y_{j-1} - \bar{Y})^2. \tag{6.84}$$

Then, we have

$$U_{Tc} = \frac{1}{T\sigma^2}\left(\sum_{j=2}^{T} Y_{j-1}e_j - \bar{Y}\sum_{j=2}^{T} e_j\right).$$

It follows from equation (6.67) that

$$U_T = \frac{1}{T\sigma^2}\sum_{j=2}^{T} Y_{j-1}e_j \rightarrow \int_0^1 W(t)dW(t). \tag{6.85}$$

Recall that

$$\sum_{j=2}^{T} e_j = Y_T - Y_0 - e_1. \tag{6.86}$$

It follows from equation (6.50) that

$$\frac{Y_T}{\sqrt{T}\sigma} \rightarrow N(0,1) = W(1), \text{ which implies that}$$

$$\frac{1}{\sigma\sqrt{T}}\sum_{j=2}^{T} e_j = \frac{Y_T}{\sqrt{T}\sigma} - \frac{1}{\sqrt{T}\sigma}Y_0 - \frac{1}{\sqrt{T}\sigma}e_1 \rightarrow W(1). \tag{6.87}$$

Equation (6.61) showed that

$$\frac{\bar{Y}}{\sqrt{T}\sigma} = \frac{1}{T\sqrt{T}\sigma}\sum_{j=2}^{T} Y_{j-1} \rightarrow \int_0^1 W(t)dt. \tag{6.88}$$

Note that

$$U_{Tc} = U_T - \frac{\bar{Y}}{\sqrt{T}\sigma}\frac{1}{\sigma\sqrt{T}}\sum_{j=2}^{T} e_j. \tag{6.89}$$

Combining equations (6.85), (6.87)–(6.89), we obtain

$$U_{Tc} \rightarrow \int_0^1 W(t)dW(t) - \int_0^1 W(t)dt\ W(1). \tag{6.90}$$

Note that

$$\int_0^1 dW(t) = W(1) - W(0) = W(1). \tag{6.91}$$

Substituting equation (6.91) into equation (6.90) yields

$$U_{Tc} \rightarrow \int_0^1 \left(W(t) - \int_0^1 W(t)dt\right) dW(t). \tag{6.92}$$

Now we consider V_{Tc}. It follows from equation (6.58) that

$$V_{Tc} = \frac{1}{T^2\sigma^2}\sum_{j=2}^{T}(Y_{j-1} - \bar{Y})^2 \rightarrow \int_0^1 \left(W(t) - \int_0^1 W(t)dt\right)^2 dt. \tag{6.93}$$

Combining equations (6.82)–(6.84) yields

$$T(\hat{\rho}_c - 1) = \frac{U_{Tc}}{V_{Tc}} \rightarrow \frac{\int_0^1\left(W(t) - \int_0^1 W(t)dt\right)dW(t)}{\int_0^1\left(W(t) - \int_0^1 W(t)dt\right)^2 dt}. \tag{6.94}$$

Next we study S_{stc}. Combining equations (6.78), (6.82)–(6.84) yields

$$S_{stc} = \frac{U_{Tc}}{\sqrt{V_{Tc}}}. \tag{6.95}$$

It follows from equations (6.92), (6.93), and (6.95) that

$$S_{stc} \rightarrow \frac{\int_0^1\left(W(t) - \int_0^1 W(t)dt\right)dW(t)}{\sqrt{\int_0^1\left(W(t) - \int_0^1 W(t)dt\right)^2 dt}}. \tag{6.96}$$

Theorem 6.5 can be extended to the constant and time trend. Consider the test regression:

$$Y_t = c + \delta t + \rho Y_{t-1} + \varepsilon_t. \quad (6.97)$$

The hypothesis to be tested is

$H_0: \delta = 0,\ \rho = 1$, which implies that $Y_t \sim I(1)$ with drift,

$H_a: |\rho| < 1$, which implies that $Y_t \sim I(0)$ with deterministic time trend.

Using OLS to minimize

$$\min_{c,\delta,\rho} \sum_{t=1}^{T} \left(Y_t - c - \delta t - \rho Y_{t-1}\right)^2,$$

we obtain the estimator $\hat{\rho}$. Define the test statistics:

$$S_{tst} = \frac{\hat{\rho} - 1}{\sqrt{var(\hat{\rho})}}, \quad (6.98)$$

$$S_{tstn} = T(\hat{\rho} - 1). \quad (6.99)$$

Theorem 6.6: Dickey-Fuller (DF) Distribution for Deterministic Time Trend

The statistics S_{stc} under the null hypothesis is asymptotically distributed as DF distribution for the deterministic time trend:

$$S_{tst} \to \frac{\int_0^1 W^{\tau}(t)dW(t)}{\sqrt{\int_0^1 \left(W^{\tau}(t)\right)^2 dt}}, \quad (6.100)$$

$$S_{tstn} \to \frac{\int_0^1 W^{\tau}(t)dW(t)}{\int_0^1 \left(W^{\tau}(t)\right)^2 dt}. \quad (6.101)$$

6.3.3.1.4 Cointegration and Error Correction Model In some cases, two nonstationary time series "move together" so that their difference is stationary. Such series often follow a common stochastic trend.

Definition 6.12: Cointegration

Nonstationary time series which "move together" so that their difference is stationary are called cointegrated time series. Mathematically, cointegration is defined as follows. Suppose that there are a set of time series $Y_t = \left(Y_t^1, \ldots, Y_t^k\right)$, each time series being integrated of order d. If a linear combination of these time series $Z_t = \beta^T Y_t$ is integrated of order less than d, then this set of time series is called cointegrated with cointegration vector β.

Stationary attempts to pull series back to a fixed mean, while cointegration intends to pull two series back to (a fixed relationship with) each other. If two time series are cointegrated, the widely used OLS standard errors and t statistics cannot be used for analysis of such time series.

Cointegration is a powerful tool for analyzing multivariate time series. In practice, we often observe that each individual component of a multivariate time series may be nonstationary, but certain linear combinations of these components are stationary. The idea of cointegration comes from the notion of transformations which can make the data stationary.

6.3.3.1.4.1 Cointegration and the Error-Correction Mechanism (ECM) When both time series variables X_t and Y_t are nonstationary, we may obtain spurious results in a regression model (LIBRO Asterious_Applied-Econometrics 2020). Assume that X_t and Y_t are both I(1). Consider the regression:

$$Y_t = \beta_0 + \beta_1 X_t + \varepsilon_t. \quad (6.102)$$

The OLS estimators $\hat{\beta}_0$ and $\hat{\beta}_1$ are highly unreliable.

A natural way to solve this problem is to difference Y_t and X_t to convert the nonstationary time series variables to stationary time series variables. Consider the following regression model:

$$\Delta Y_t = \alpha_0 + \alpha_1 \Delta X_t + \Delta \varepsilon_t, \quad (6.103)$$

where Δ is difference operator. The regression (6.103) can solve the spurious regression problem and obtain the correct estimators $\hat{\alpha}_0$ and $\hat{\alpha}_1$. However, regression (6.103) only considers the short-run relationships between Y_t and X_t. Differences ΔY_t and ΔX_t provide no information about the long-run.

Cointegration and error of correction model (ECM) will be useful tools to incorporate the long-term dependence information into the regression models.

If we assume that Y_t and X_t are cointegrated and both are I(1), then there is a linear combination, e.g.,

$$\hat{\varepsilon}_t = Y_t - \hat{\beta}_0 - \hat{\beta}_1 X_t, \quad (6.104)$$

which is I(0)and stationary. The error $\hat{\varepsilon}_t$ can be taken as an ECM. We form the following regression model:

$$\Delta Y_t = \gamma_0 + \gamma_1 \Delta X_t - \pi \hat{\varepsilon}_{t-1} + e_t, \quad (6.105)$$

where γ_1 measures the short-term effect and π measures the long-term effect. The new regression model includes both short-run and long-run information. Now three time series variables are stationary. Since we assume that both Y_t and X_t are I(1), the differences ΔY_t and ΔX_t are stationary. We also assume that Y_t and X_t are cointegrated. Therefore, the residual $\hat{\varepsilon}_t$ in regression (6.104) is stationary. The ECM method measures the correction from disequilibrium of the previous period and adjusts for the stationary disequilibrium error term to prevent the errors in the long-run relationship from becoming larger and larger.

To further understand the error correction mechanism, we consider the simple linear autoregressive distributed lag (ARDL) model:

$$Y_t = \alpha_0 + \alpha_1 Y_{t-1} + \gamma_0 X_t + \gamma_1 X_{t-1} + \varepsilon_t, \quad (6.106)$$

where ε_t is i.i.d. variable with mean of zero and variance of σ^2.

The parameter γ_0 measures the short-term effect of the variable X_t on the variable Y_t. The long-term effect is to measure the reaction of Y_t after a change in X_t when the system is in equilibrium. In the equilibrium, we assume that

$$Y_t^* = Y_t = Y_{t-1} = \cdots = Y_{t-p} \text{ and } X_t^* = X_t = \cdots = X_{t-p}.$$

Under equilibrium, equation (6.106) is reduced to

$$Y_t^* = \alpha_0 + \alpha_1 Y_t^* + \gamma_0 X_t^* + \gamma_1 X_t^* + e_t. \quad (6.107)$$

Solving equation (6.107) for Y_t^*, we obtain

$$Y_t^* = \beta_0 + \beta_1 X_t^* + e_t, \quad (6.108)$$

where

$$\beta_0 = \frac{\alpha_0}{1-\alpha_1} \text{ and } \beta_1 = \frac{\gamma_0 + \gamma_1}{1-\alpha_1}.$$

Using definition of ΔY_t and ΔX_t, we obtain

$$Y_t = Y_{t-1} + \Delta Y_t \text{ and } X_t = X_{t-1} + \Delta X_t. \quad (6.109)$$

Substituting equations (6.108) and (6.109) into equation (6.106) yields

$$\Delta Y_t = \gamma_0 \Delta X_t - (1-a_1)(Y_{t-1} - \beta_0 - \beta_1 X_{t-1}) + \varepsilon_t, \text{ or} \quad (6.110)$$

$$\Delta Y_t = \gamma_0 \Delta X_t - \pi(Y_{t-1} - \beta_0 - \beta_1 X_{t-1}) + \varepsilon_t, \quad (6.111)$$

where $\pi = (1-a_1)$. The parameter π provides information about the speed of adjustment for disequilibrium.

The ECM method for solving nonstationary time series regression consists of two steps:

Step 1:

Pretest the nonstationary of the standard unit root DF testing to confirm that they are nonstationary. Use OLS to fit the regression model: $Y_t = \beta_0 + \beta_1 X_t + e_t$. Calculate the predicted residual: $\hat{e}_t = Y_{t-1} - \hat{\beta}_0 - \hat{\beta}_1 X_{t-1}$.

Step 2:

Fit the regression:

$$\Delta Y_t = \gamma_0 \Delta X_t - \pi \hat{e}_t + \varepsilon_t. \quad (6.112)$$

Now we extend the ECM to a more general model for large numbers of lagged terms:

$$Y_t = \mu + \sum_{i=1}^{p} \alpha_i Y_{t-i} + \sum_{i=0}^{q} \gamma_i X_{t-i} + \varepsilon_t. \quad (6.113)$$

Again, a long-term solution of the model is a point, where Y_t and X_t move to constant steady-state levels Y^* and X^*. Thus, assume that

$$Y^* = Y_t = \cdots = Y_{t-p} \text{ and } X^* = X_t = \cdots = X_{t-q}. \quad (6.114)$$

Substituting equation (6.114) into equation (6.113) leads to

$$Y^* = \delta_0 + \delta_1 X^*, \quad (6.115)$$

where

$$\delta_0 = \frac{\mu}{1-\sum_{i=1}^{p}\alpha_i} \text{ and } \delta_1 = \frac{\sum_{i=1}^{q}\gamma_i}{1-\sum_{i=1}^{p}\alpha_i}.$$

Equation (6.113) can be reduced to the following equation:

$$\Delta Y_t = \mu + \sum_{i=1}^{p-1} \alpha_i \Delta Y_{t-i} + \sum_{i=0}^{q-1} \gamma_i \Delta X_{t-i} + \varphi_1 Y_{t-1} + \theta_2 X_{t-1} + \varepsilon_t, \quad (6.116)$$

where

$\varphi_1 = -\left(1 - \sum_{i=1}^{p} \alpha_i\right)$ and $\theta_2 = \sum_{i=1}^{q} \gamma_i$.

Let $\theta_1 = \left(1 - \sum_{i=1}^{p} \alpha_i\right)$. Then, equation (6.116) can be reduced to

$$\Delta Y_t = \mu + \sum_{i=1}^{p-1} \alpha_i \Delta Y_{t-i} + \sum_{i=0}^{q-1} \gamma_i \Delta X_{t-i} - \theta_1 \left(Y_{t-1} - \frac{1}{\theta_1} - \frac{\theta_2}{\theta_1} X_{t-1} \right) + \varepsilon_t, \text{ or} \quad (6.117)$$

$$\Delta Y_t = \mu + \sum_{i=1}^{p-1} \alpha_i \Delta Y_{t-i} + \sum_{i=0}^{q-1} \gamma_i \Delta X_{t-i} - \theta_1 \left(Y_{t-1} - \beta_1 - \beta_2 X_{t-1} \right) + \varepsilon_t, \quad (6.118)$$

where $\beta_1 = 0$.

Fitting the stationary time regression:

$$Y_{t-1} - \beta_1 - \beta_2 X_{t-1} = e_{t-1},$$

we obtain

$$Y_{t-1} - \hat{\beta}_1 - \hat{\beta}_2 X_{t-1} = \hat{e}_{t-1}, \quad (6.119)$$

where e_{t-1} is an equilibrium error.

Substituting equation (6.119) into equation (6.118) yields

$$\Delta Y_t = \mu + \sum_{i=1}^{p-1} \alpha_i \Delta Y_{t-i} + \sum_{i=0}^{q-1} \gamma_i \Delta X_{t-i} - \pi \hat{e}_{t-1} + \varepsilon_t, \quad (6.120)$$

where π is the error-correction coefficient and measures how much of the equilibrium error is corrected.

Let

$$A(L) = a_0 + a_1 L + \cdots a_p L^p \text{ and}$$
$$B(L) = b_0 + b_1 L + \cdots + b_q L^q.$$

Equation (6.120) can be rewritten as

$$A(L)\Delta Y_t = \mu + B(L)\Delta X_t - \pi \hat{e}_{t-1} + \varepsilon_t. \quad (6.121)$$

The ECM method for solving general nonstationary time series regression consists of two steps:

Step 1:

Pre-test the nonstationary of the standard unit root DF testing to confirm that they are nonstationary. Use OLS to fit the regression model: $Y_{t-1} = \beta_0 + \beta_1 X_{t-1} + e_t$. Calculate the predicted residual: $\hat{e}_t = Y_{t-1} - \hat{\beta}_0 - \hat{\beta}_1 X_{t-1}$.

Step 2:

Fit the regression:

$$A(L)\Delta Y_t = \mu + B(L)\Delta X_t - \pi \hat{e}_{t-1} + \varepsilon_t.$$

6.3.3.1.5 Vector Autoregressions (VAR) Consider an m-variable VAR with p lags:

$$Y_t = \mu + \sum_{i=1}^{p} A_i Y_{t-i} + \varepsilon_t, \quad (6.122)$$

where Y_t is an m-dimensional vector, the A_i $(i = 1, \ldots, p)$ are $m \times m$ coefficient matrices and m-dimensional residual vector ε_t is assumed to have mean zero ($E[\varepsilon_t] = 0$, with no autocorrelation ($E[\varepsilon_t \varepsilon_{t-s}^T] = 0$) but can be correlated across equations ($E[\varepsilon_t \varepsilon_t^T] = \Sigma$).

Similar to ECM in equation (6.120), vector error correction model (VECM) consists of first differences of cointegrated $I(1)$ variables, their lags, and error correction terms:

$$\Delta Y_t = \mu + \prod Y_{t-1} + \sum_{i=1}^{p-1} \Phi_i \Delta Y_{t-i} + \varepsilon_t, \quad (6.123)$$

where matrixes $\prod$ and Φ_i $(i = 1, \ldots, p-1)$ are functions of matrices A_i $(i = 1, \ldots, p)$.

Equation (6.123) can be expanded to

$$\begin{aligned} Y_t - Y_{t-1} &= \mu + \prod Y_{t-1} + \sum_{i=1}^{p-1} \Phi_i Y_{t-i} - \sum_{i=1}^{p-1} \Phi_i Y_{t-i-1} + \varepsilon_t \\ &= \mu + \prod Y_{t-1} + \sum_{i=1}^{p-1} \Phi_i Y_{t-i} - \sum_{i=2}^{p} \Phi_{i-1} Y_{t-i} + \varepsilon_t \\ &= \mu + \left(\Phi_1 + \prod \right) Y_{t-1} + \sum_{i=2}^{p-1} (\Phi_i - \Phi_{i-1}) Y_{t-i} - \Phi_{p-1} Y_{t-p} + \varepsilon_t. \end{aligned} \quad (6.124)$$

Moving Y_{t-1} from the left sight to the right side, we obtain

$$Y_t = \mu + \left(I + \Phi_1 + \prod \right) Y_{t-1} + \sum_{i=2}^{p-1} (\Phi_i - \Phi_{i-1}) Y_{t-i}$$
$$- \Phi_{p-1} Y_{t-p} + \varepsilon_t . \quad (6.125)$$

Comparing equation (6.125) with equation (6.122), we obtain

$$I + \Phi_1 + \Pi = A_1, \quad (6.126)$$

$$\Phi_i - \Phi_{i-1} = A_i, i = 2, \ldots, p-1, \quad (6.127)$$

$$\Phi_{p-1} = -A_p . \quad (6.128)$$

Solving equations (6.126)–(6.128), we obtain

$$\Phi_j = -\sum_{i=j+1}^{p} A_i, \ j = 1, \ldots, p-1, \quad (6.129)$$

$$\prod = -\left(I - \sum_{i=1}^{p} A_i \right) = \Phi(1), \quad (6.130)$$

where $\Phi(Z) = I - \Phi_1 Z - \cdots - \Phi_p Z^p$ is the characteristic polynomial.

The coefficient matrix Π plays an important role. It is clear that if $\Pi = 0$, all time series terms in equation (6.123) are stationary. Equation (6.123) is reduced to

$$\Delta Y_t = \mu + \sum_{i=1}^{p-1} \Phi_i \Delta Y_{t-i} + \varepsilon_t , \quad (6.131)$$

If Π has a full rank, then we have

$$Y_{t-1} = \prod^{-1} \left(-\mu + \Delta Y_{t-1} - \sum_{i=1}^{p-1} \Phi_i \Delta Y_{t-i} + \varepsilon_t \right). \quad (6.132)$$

Since ΔY_{t-i} is stationary, time series Y_{t-1} is also stationary. Therefore, we use equation (6.122) to model the relationship among Y_t and do not use equation (6.123) to model their differences. There is no need the error correction.

Now consider Rank $(\Pi) = k, 0 < k < m$. In this case, we can factorize Π in the product of two matrices:

$$\prod = \alpha \beta^T , \quad (6.133)$$

where both α and β are $m \times k$ matrices and Rank $(\alpha) = \text{Rank}(\beta) = k$. Thus, we obtain

$$\prod Y_{t-1} = \alpha \left(\beta^T Y_{t-1} \right). \quad (6.134)$$

The vector $\beta^T Y_{t-1}$ is k stationary linear combinations and contains the error correction terms. The k columns of β are the cointegrating vectors. The elements of matrix α determine the size of the effects of the k error correction terms in the VECM. In this case, equation (6.123) is reduced to

$$\Delta Y_t = \mu + \alpha \left(\beta^T Y_{t-1} \right)$$
$$+ \sum_{i=1}^{p-1} \Phi_i \Delta Y_{t-i} + \varepsilon_t . \quad (6.135)$$

More general, we can consider a linear trend. Equation (6.135) can be changed into the following model incorporating the linear trend:

$$\Delta Y_t = \mu + \alpha \left(\beta^T Y_{t-1} + c_1 t \right)$$
$$+ \sum_{i=1}^{p-1} \Phi_i \Delta Y_{t-i} + \varepsilon_t . \quad (6.136)$$

6.3.3.2 *Multivariate Nonlinear Causality Test for Nonstationary Time Series*

6.3.3.2.1 ECM-VAR Model

When two nonstationary variables are cointegrated, the VAR model should be augmented with an error correction term for testing the Granger causality (Engle and Granger 1987).

Consider the ECM-VAR model:

$$\Delta Y_t = A_0 + A(L) \Delta Y_{t-1} + \alpha \left(ECM_{t-1} \right) + \varepsilon_t , \quad (6.137)$$

where

$$\Delta Y_t = \begin{bmatrix} \Delta Y_{1t} \\ \vdots \\ \Delta Y_{mt} \end{bmatrix}, A_0 = \begin{bmatrix} A_{10} \\ \vdots \\ A_{m0} \end{bmatrix}, A(L) = \begin{bmatrix} A_{11}(L) & \cdots & A_{1m}(L) \\ \vdots & \vdots & \vdots \\ A_{m1}(L) & \cdots & A_{mm(L)} \end{bmatrix},$$

$$\alpha = \begin{bmatrix} \alpha_{11} & \cdots & \alpha_{1k} \\ \vdots & \vdots & \vdots \\ \alpha_{m1} & \cdots & \alpha_{mk} \end{bmatrix}, ECM_{t-1} = \beta^T Y_{t-1} + Ct, Y_{t-1} = \begin{bmatrix} Y_{1,t-1} \\ \vdots \\ Y_{m,t-1} \end{bmatrix},$$

$$\beta = \begin{bmatrix} \beta_{11} & \cdots & \beta_{1k} \\ \vdots & \vdots & \vdots \\ \beta_{m1} & \cdots & \beta_{mk} \end{bmatrix}, C = \begin{bmatrix} c_1 \\ \vdots \\ c_m \end{bmatrix}.$$

Consider two nonstationary time series, $X_t = [X_{1t}, \ldots, X_{m,t}]^T$ and $Y_t = [Y_{1t}, \ldots, Y_{mt}]^T$. Let $m = m_1 + m_2$.

Suppose that X_t and Y_t are cointegrated with the residuals $VECM_t$. The ECM-VAR model for testing the Granger causality is given by

$$\begin{bmatrix} \Delta X_t \\ \Delta Y_t \end{bmatrix} = \begin{bmatrix} A_x \\ A_y \end{bmatrix} + \begin{bmatrix} A_{xx}(L) & A_{xy}(L) \\ A_{yx}(L) & A_{yy}(L) \end{bmatrix} \begin{bmatrix} \Delta X_{t-1} \\ \Delta Y_{t-1} \end{bmatrix} + \begin{bmatrix} \alpha_x \\ \alpha_y \end{bmatrix} ECM_{t-1} + \begin{bmatrix} \varepsilon_x \\ \varepsilon_y \end{bmatrix}, \tag{6.138}$$

where A_x and A_y are m_1 and m_2 dimensional vectors of intercept terms, respectively, $A_{xx}(L)$, $A_{xy}(L)$, $A_{yx}(L)$, and $A_{yy}(L)$ are $n_1 \times n_1$, $n_1 \times n_2$, $n_2 \times n_1$, and $n_2 \times n_2$ dimensional matrices of lag polynomials, respectively, α_x and α_y are n_1 and n_2 dimensional coefficient vectors for the error correction term ECM_{t-1}, respectively. The lag length was selected using the two-stage procedure (Abdalla and Murinde 1997).

6.3.3.2.2 Multivariate Nonlinear Causality Test

6.3.3.2.2.1 Introduction To identify existence of any nonlinear Granger causality relations between two vectors of time series consists of two steps (Bai et al. 2010).

Step 1:

Apply VAR model in (6.27) or (6.138) to two time series X_t and Y_t to identify their linear causal relationships and obtain their corresponding residuals $\hat{\varepsilon}_x$ and $\hat{\varepsilon}_y$.

Step 2:

Apply a nonlinear Granger causality test to the residual series instead of the original time series to identify the nonlinear Granger causality. For the convenience of presentation, let

$$X_t = \begin{bmatrix} \hat{\varepsilon}_{1x} \\ \vdots \\ \hat{\varepsilon}_{m_1 x} \end{bmatrix} \text{ and } Y_t = \begin{bmatrix} \hat{\varepsilon}_{1y} \\ \vdots \\ \hat{\varepsilon}_{m_2 y} \end{bmatrix}.$$

6.3.3.2.2.2 Multivariate Nonlinear Causality Hypothesis The null hypothesis for testing the causality relation between X_t and Y_t is

$$H_0: A_{xy}(L) = 0, \text{ or } A_{yx}(L) = 0.$$

Now we define the lead vector and lag vector of a time series. Define the m_{x_i}-length lead vector for $X_{i,t}$ and the m_{y_i}-length lead vector for $y_{i,t}$ as respectively.

$$X_{i,t}^{m_{x_i}} = \begin{bmatrix} X_{i,t} \\ \vdots \\ X_{i,t+m_{x_i}-1} \end{bmatrix}, m_{x_i} = 1,2,\ldots, Y_{i,t}^{m_{y_i}} = \begin{bmatrix} Y_{i,t} \\ \vdots \\ Y_{i,t+m_{y_i}-1} \end{bmatrix}, m_{y_i} = 1,2,\ldots, t = 1,2,\ldots, \tag{6.139}$$

The l_{x_i}-length lag vector for $X_{i,t}$ and l_{y_i}-length lag vector for y_{it} are defined, respectively, as

$$X_{i,t-l_{x_i}}^{l_{x_i}} = \begin{bmatrix} X_{i,t-l_{x_i}} \\ \vdots \\ X_{i,t-1} \end{bmatrix}, l_{x_i} = 1,2,\ldots, t = l_{x_i}+1, l_{x_i}+2, \ldots, \tag{6.140}$$

$$Y_{i,t-l_{y_i}}^{l_{y_i}} = \begin{bmatrix} Y_{i,t-l_{y_i}} \\ \vdots \\ Y_{i,t-1} \end{bmatrix}, l_{y_i} = 1,2,\ldots, t = l_{y_i}+1, l_{y_i}+2, \ldots, \tag{6.141}$$

We define $M_x = (m_{x_1}, \ldots, m_{x_{m_1}})$, $L_x = (l_{x_1}, \ldots, l_{x_{m_1}})$, $m_x = max(m_{x_1}, \ldots, m_{x_{m_1}})$, and $l_x = \max(l_{x_1}, \ldots, l_{x_{m_1}})$ for time series X_t. Similarly, we defined $M_y = \left(m_{y_1}, \ldots, m_{y_{m2}}\right)$, $l_y = \left(l_{y_1}, \ldots, l_{y_{m2}}\right)$, $m_y = max\left(m_{y_1}, \ldots, m_{y_{m2}}\right)$, and $l_y = \max\left(l_{y_1}, \ldots, l_{y_{m2}}\right)$ for Y_t.

After these definitions, we define the four events (Bai et al. 2010):

$$E_1 \equiv \left\{\left\|X_t^{M_x} - X_s^{M_x}\right\| < e\right\}$$

$$\equiv \left\{\left\|X_{i,t}^{m_{xi}} - X_{i,s}^{m_{xi}}\right\| < e, \text{ for all for all } i = 1, \ldots, m_1\right\};$$

$$E_2 \equiv \left\{\left\|X_{t-l_x}^{l_x} - X_{s-l_x}^{l_x}\right\| < e\right\}$$

$$\equiv \left\{\left\|X_{i,t-l_{x_i}}^{l_{xi}} - X_{i,s-l_{x_i}}^{l_{xi}}\right\| < e, \text{ for all } i = 1, \ldots, m_1\right\};$$

$$E_3 \equiv \left\{\left\|X_t^{M_y} - X_s^{M_y}\right\| < e\right\}$$

$$\equiv \left\{\left\|X_{i,t}^{m_{yi}} - X_{i,s}^{m_{yi}}\right\| < e, \text{ for all for all } i = 1, \ldots, m_2\right\};$$

$$E_4 \equiv \left\{\left\|X_{t-l_y}^{l_y} - X_{s-l_y}^{l_y}\right\| < e\right\}$$

$$\equiv \left\{\left\|X_{i,t-l_{y_i}}^{l_{yi}} - X_{i,s-l_{y_i}}^{l_{yi}}\right\| < e, \text{ for all } i = 1, \ldots, m_2\right\};$$

where $\|\cdot\|$ denotes the maximum norm.

Intuitively, if adding the vector series X_t does not improve the prediction of another vector series Y_t then, the vector series X_t is said not to strictly Granger cause another vector series Y_t. Formally, we have the following definition (Bai et al. 2010).

Definition 6.13: Granger Causality

The vector series X_t is said not to strictly Granger cause another vector series Y_t if

$$P\left(\left\|Y_t^{M_y} - Y_s^{M_y}\right\| < e \middle| \left\|Y_{t-l_y}^{l_y} - Y_{s-l_y}^{l_y}\right\| < e, \left\|X_{t-l_x}^{l_x} - X_{s-l_x}^{l_x}\right\| < e\right) = P\left(\left\|Y_t^{M_y} - Y_s^{M_y}\right\| < e \middle| \left\|Y_{t-l_y}^{l_y} - Y_{s-l_y}^{l_y}\right\| < e\right). \quad (6.142)$$

Now I introduce test statistic and its distribution (Bai et al. 2010).

6.3.3.2.2.3 Statistic for Testing Granger Causality and Its Asymptotic Distribution The test statistic for testing non-existence of nonlinear Granger causality can be extended from bivariate case to multivariate case.

Define

$$I(Z_1, Z_2, e) = \begin{cases} 1 & \|Z_1 - Z_2\| \le e \\ 0 & \|Z_1 - Z_2\| > e \end{cases}, \quad (6.143)$$

$$X_{t-l_{x_i}}^{m_{x_i}+l_{x_i}} = \left(X_{t-l_{x_i}}, \ldots, X_{t-1}, X_t, \ldots, X_{t+m_{x_i}-1}\right), \quad (6.144)$$

$$Y_{t-l_{y_i}}^{m_{y_i}+l_{y_i}} = \left(Y_{t-l_{y_i}}, \ldots, Y_{t-1}, Y_t, \ldots, Y_{t+m_{y_i}-1}\right). \quad (6.145)$$

$$C_1\left(M_y + l_y, l_x, e, n\right) = \frac{2}{n(n-1)} \sum\sum_{t<s} \prod_{i=1}^{m_1} I\left(Y_{i,t-l_{y_i}}^{m_{y_i}+l_{y_i}}, Y_{i,s-l_{y_i}}^{m_{y_i}+l_{y_i}}, e\right) \times \prod_{i=1}^{m_2} I\left(X_{i,t-l_{x_i}}^{l_{xi}}, X_{i,s-l_{x_i}}^{l_{xi}}, e\right), \quad (6.146)$$

$$C_2\left(l_x, l_y, e, n\right) = \frac{2}{n(n-1)} \sum\sum_{t<s} \prod_{i=1}^{m_1} I\left(X_{i,t-l_{x_i}}^{l_{xi}}, X_{i,s-l_{x_i}}^{l_{xi}}, e\right) \times \prod_{i=1}^{m_2} I\left(Y_{i,t-l_{y_i}}^{l_{yi}}, Y_{i,s-l_{y_i}}^{l_{yi}}, e\right). \quad (6.147)$$

$$C_3\left(M_y + l_y, e, n\right) = \frac{2}{n(n-1)} \sum\sum_{t<s} \prod_{i=1}^{m_2} I\left(X_{i,t-l_{y_i}}^{m_{y_i}+l_{y_i}}, X_{i,s-l_{y_i}}^{m_{y_i}+l_{y_i}}, e\right), \quad (6.148)$$

$$C_4\left(l_y, e, n\right) = \frac{2}{n(n-1)} \sum\sum_{t<s} \prod_{i=1}^{m_2} I\left(X_{i,t-l_{y_i}}^{l_{yi}}, X_{i,s-l_{y_i}}^{l_{yi}}, e\right), \quad (6.149)$$

where $t, s = \max\left(l_x, l_y\right) + 1, \ldots, T - m_y + 1, n = T + 1 - \max\left(l_x, l_y\right)$.

Define statistic T_{NG} for testing nonlinear Granger causality:

$$T_{NG} = \sqrt{n}\left(\frac{C_1\left(M_y + l_y, l_x, e, n\right)}{C_2\left(l_x, l_y, e, n\right)} - \frac{C_3\left(M_y + l_y, e, n\right)}{C_4\left(l_y, e, n\right)}\right). \quad (6.150)$$

Since the above defined statistics $C_j(*,n)$ in equations (6.146)–(6.149) are not U-statistic and the estimators of $C_j(*,n)$ are not consistent. Furthermore, the distribution of the statistic T_{NG} is not normal distribution, Bai et al. (2018) redefined $C_j(*,n)$ as follows.

For any given pair (t,s), define

$$C_1\left(M_y+l_y,l_x,e;t,s\right)$$

$$=P\left(\left\|Y_{t-l_y}^{M_y+l_y}-Y_{s-l_y}^{M_y+l_y}\right\|<e,\left\|X_{t-l_x}^{l_x}-X_{s-l_x}^{l_x}\right\|<e\right),$$

$$C_2\left(l_y,l_x,e:t,s\right)=P\left(\left\|Y_{t-l_y}^{l_y}-Y_{s-l_y}^{l_y}\right\|<e,\left\|X_{t-l_x}^{l_x}-X_{s-l_x}^{l_x}\right\|<e\right),$$

$$C_3\left(M_y+l_y,e;t,s\right)=P\left(\left\|Y_{t-l_y}^{M_y+l_y}-Y_{s-l_y}^{M_y+l_y}\right\|<e\right),$$

$$C_4\left(l_y,e;t,s\right)=P\left(\left\|Y_{t-l_y}^{l_y}-Y_{s-l_y}^{l_y}\right\|<e\right).$$

Since we assume that the time series X_t, Y_t are stationary, if $|t-s|=l|$, then the above formulas $C_j(*,t,s)$ can rewritten as $C_1\left(M_y+l_y,l_x,e;l\right), C_2\left(l_y,l_x,e:l\right)$, $C_3\left(M_y+l_y,e;l\right)$, and $C_4\left(l_y,e;l\right)$. The null hypothesis for testing the nonlinear Granger causality is

$$H_0: \frac{C_1\left(M_y+l_y,l_x,e;l\right)}{C_2\left(l_y,l_x,e:l\right)}=\frac{C_3\left(M_y+l_y,e;l\right)}{C_4\left(l_y,e;l\right)}. \quad (6.151)$$

Intuitively, the ratio $\frac{C_3\left(M_y+l_y,e;l\right)}{C_4\left(l_y,e;l\right)}$ measures how much information in time series $\left\{Y_{t-l_y},\ldots,Y_{t-1}\right\}$ the time series $\left\{Y_t,\ldots,Y_{t+M_y-1}\right\}$ contains. Similarly, the ratio $\frac{C_1\left(M_y+l_y,l_x,e;l\right)}{C_2\left(l_y,l_x,e:l\right)}$ measures summation of information in $\frac{C_3\left(M_y+l_y,e;l\right)}{C_4\left(l_y,e;l\right)}$ and additional information due to time series $\left\{X_{t-l_y},\ldots,X_{t-1}\right\}$. Equation (6.151) shows that under the null hypothesis, adding time series $\left\{X_{t-l_y},\ldots,X_{t-1}\right\}$ provides no additional information in predicting the time series $\left\{Y_{t-l_y},\ldots,Y_{t-1}\right\}$.

Let $L_{yx}=\max\left(l_y,l_x\right), n=T-L_{yx}-l-m_t+1$. Using frequency interpretation of the probability, we obtain the following consistent estimators of $C_1\left(M_y+l_y,l_x,e;l\right), C_2\left(l_y,l_x,e:l\right), C_3\left(M_y+l_y,e;l\right)$ and $C_4\left(l_y,e;l\right)$ (Bai et al. 2018):

$$\hat{C}_1\left(M_y+l_y,l_x,e;l\right)=\frac{1}{n}\sum_{t=l_{yx}+1}^{T-m_y+1}\prod_{i=1}^{m_2}I\left(Y_{i,t-l_{yi}}^{m_{yi}+l_{yi}},Y_{i,t+l-l_{yi}}^{m_{yi}+l_{yi}},e\right)\times\prod_{i=1}^{m_1}I\left(X_{i,t-l_{xi}}^{l_{xi}},X_{i,t+l-l_{xi}}^{l_{xi}},e\right), \quad (6.152)$$

$$\hat{C}_2\left(l_y,l_x,e;l\right)=\frac{1}{n}\sum_{t=l_{yx}+1}^{T-m_y+1}\prod_{i=1}^{m_2}I\left(Y_{i,t-l_{yi}}^{l_{yi}},Y_{i,t+l-l_{yi}}^{l_{yi}},e\right)\times\prod_{i=1}^{m_1}I\left(X_{i,t-l_{xi}}^{l_{xi}},X_{i,t+l-l_{xi}}^{l_{xi}},e\right). \quad (6.153)$$

$$\hat{C}_3\left(M_y+l_y,e;l\right)=\frac{1}{n}\sum_{t=l_{yx}+1}^{T-l-m_y+1}\prod_{i=1}^{m_2}I\left(Y_{i,t-l_{yi}}^{m_{yi}+l_{yi}},Y_{i,t+l-l_{yi}}^{m_{yi}+l_{yi}},e\right), \quad (6.154)$$

$$\hat{C}_4\left(l_y,e;l\right)=\frac{1}{n}\sum_{i=1}^{T-l-m_y+1}\prod_{i=1}^{m_2}I\left(X_{i,t-l_{yi}}^{l_{yi}},X_{i,t+l-l_{yi}}^{l_{yi}},e\right). \quad (6.155)$$

Now we define a new multivariate statistic for testing the nonlinear Granger causality (Bai et al. 2018):

$$T_{NNG}=\sqrt{n}\left(\frac{\hat{C}_1\left(M_y+l_y,l_x,e;l\right)}{\hat{C}_2\left(l_y,l_x,e;l\right)}-\frac{\hat{C}_3\left(M_y+l_y,e;l\right)}{\hat{C}_4\left(l_y,e;l\right)}\right). \quad (6.156)$$

Its asymptotic distribution under the null hypothesis is summarized in Theorem 6.7.

Theorem 6.7: Asymptotic Distribution of Nonlinear Multivariate Granger Causality Test

Assume that X_t and Y_t are stationary. Under hull hypothesis and some regular conditions (Bai et al. 2018), the statistic T_{NNG} is asymptotically distributed as a normal distribution: $N(0,\sigma^2\left(M_y,l_y,l_x,e,l\right))$, where

$$\sigma^2\left(M_y,l_y,l_x,e,l\right)=\nabla f^T\sum\nabla f,$$

$$\nabla f=\begin{bmatrix} \dfrac{1}{\hat{C}_2\left(l_y,l_x,e;l\right)} \\ -\dfrac{\hat{C}_1\left(M_y+l_y,l_x,e;l\right)}{\left(\hat{C}_2\left(l_y,l_x,e;l\right)\right)^2} \\ -\dfrac{1}{\hat{C}_4\left(l_y,e;l\right)} \\ \dfrac{\hat{C}_3\left(M_y+l_y,e;l\right)}{\left(\hat{C}_4\left(l_y,e;l\right)\right)^2} \end{bmatrix},$$

and $\sum$ is defined in Appendix 6A.

6.3.4 Granger Causal Networks

6.3.4.1 Introduction

In the previous sections, I introduced Grander causation between two time series and two sets of time series. This section will introduce Granger causal networks and use recurrent neural networks (RNNs) and generative models to infer Granger causal networks (Khanna and Tan 2019). Traditional causal graph learning methods include linear systems (VAR Granger analysis, regression) (Sommerlade et al. 2012), additive models (Sindhwani et al. 2013). However, additive models often miss nonlinear interaction information and fail to detect Granger causal connections. To overcome these limitations, we also introduce various neural networks for reconstruction of Granger causal networks (Tank et al. 2018).

The Grander causal networks have been applied to gene regulatory network mapping (Fujita et al. 2007) and the mapping of human brain connectome (Seth et al. (2015).

6.3.4.2 Architecture of Granger Causal Networks

Consider an n-dimensional vector of stationary time series $X_t=[X_{1,t,\ldots,}X_{n,t}]^T$. The Granger causal network is generalization of the VAR model. We assume that the samples $\{X_t,t=1,2,\ldots T\}$ are sequentially generated, following the nonlinear autoregressive model (Khanna and Tan 2019):

$$X_{i,t}=f_i\left(X_{t-1},\ldots,X_{t-p}\right)+\varepsilon_{i,t},i=1,2,\ldots,n. \quad (6.157)$$

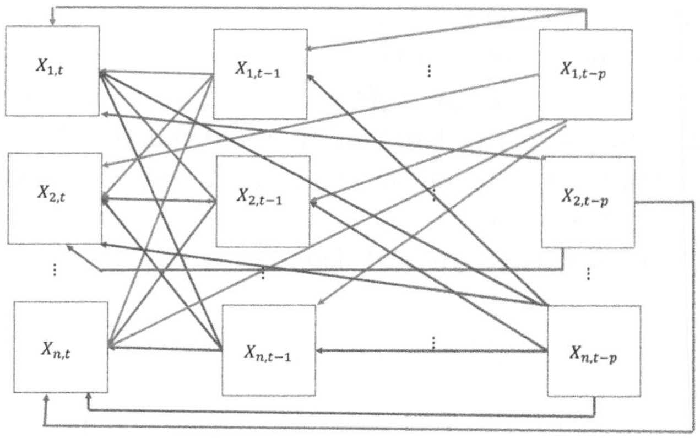

FIGURE 6.1 Architecture of Granger causal network.

where f_i is a component-wise nonlinear generating function which models all linear and nonlinear interactions between n time series up to time $t-1$. The architecture of Granger causal network underlying n time series is shown in Figure 6.1. The residual $\varepsilon_{i,t}$ summarizes the noises, the combined effect of all instantaneous and exogenous factors affecting the measurement of time series $X_{i,t}$ at time t.

Intuitively, if the past information of time series $X_{i,t}$ can improve the prediction of the current or future values of the time series $X_{j,t}$ then the time series $X_{i,t}$ Granger cause another time series $X_{j,t}$. Or, in terms of equation (6.157), if the time series $X_{j,t}$ does not Granger cause time series $X_{i,t}$, then the function f_i does not depend on the past measurements in time series $X_{j,t}$, i.e., for all $t\geq 1$ and all possible of distinct pair of time series $X_{j,t-p},\ldots,X_{j,t-1}$ and $X'_{j,t-p},\ldots,X'_{j,t-p},\ldots,X'_{j,t-1}$, the following equality which shows the function f_i does not depend on the past measurements in time series holds:

$$\begin{aligned} &f_i\left(X_{1,t-p},\ldots,X_{1,t-1},\ldots,X_{j,t-p},\ldots,X_{j,t-1},\ldots,\right.\\ &\left.\times X_{n,t-p},\ldots,X_{n,t-1}\right)\\ &=f_i\left(X_{1,t-p},\ldots,X_{1,t-1},\ldots,X'_{j,t-p},\ldots,X'_{j,t-1},\ldots,\right.\\ &\left.\times X_{n,t-p},\ldots,X_{n,t-1}.\right) \end{aligned} \quad (6.158)$$

Equation (6.158) implies that the detection of Granger non-causality is reduced to discover the components of the vector X_t of time series which are irrelevant to the individual function f_i. Equations (6.157) and (6.158) show that the principle for reconstruction of the Granger causal network is via pair-wise comparison. It compares two time series to infer the Granger causal relationships between a pair of time series. Let Θ_i denotes all parameters in the function f_i. If at least one parameter in Θ_i which are related to time series $X_{j,t-1}$ is significantly from zero, then $X_{j,t-1}$ Granger causes $X_{i,t}$ and arrow from $X_{j,t-1}$ to $X_{i,t}$ in the Granger causal network is present.

6.3.4.3 Component-Wise Multilayer Perceptron (cMPL) for Inferring Granger Causal Networks

To easily disentangle the effects from inputs to outputs, we model each component f_i with a separate MLP (Tank et al. 2018). The architecture of cMPL for inferring Granger causal networks is shown in Figure 6.2. Suppose that the number of hidden layers is L. Define

$$h_t^1 = \begin{bmatrix} h_t^{1,1} \\ \vdots \\ h_{,t}^{1,q_1} \end{bmatrix}, W^1 = \begin{bmatrix} W_1^1 & \cdots & W_P^1 \end{bmatrix},$$

$$W_p^1 = \begin{bmatrix} W_{1,p}^{1,1} & \cdots & W_{n,p}^{1,1} \\ \vdots & \vdots & \vdots \\ W_{1,p}^{1,q_1} & \cdots & W_{n,p}^{1,q_1} \end{bmatrix} \text{ and } X_{t-p} = \begin{bmatrix} X_{1,t-p} \\ \vdots \\ X_{n,t-p} \end{bmatrix},$$

where h_t^1 is the vector of values at the time t in the first hidden layer of nodes, $h_t^{1,j}$ is the value of the jth node in the first hidden layer at the time t, X_{t-p} is the vector of input time series at the time $t-p$, $X_{i,t-p}$ is the value of the ith time series at the time $t-p$, W^1 is the weight matrix connecting the input to the first hidden layer, W_p^1 is the vector weights connecting the input X_{t-p} to the nodes of the first hidden layer, and $W_{i,p}^{1,j}$ is the weight connecting the p lag of the ith time series to the jth node in the first hidden layer.

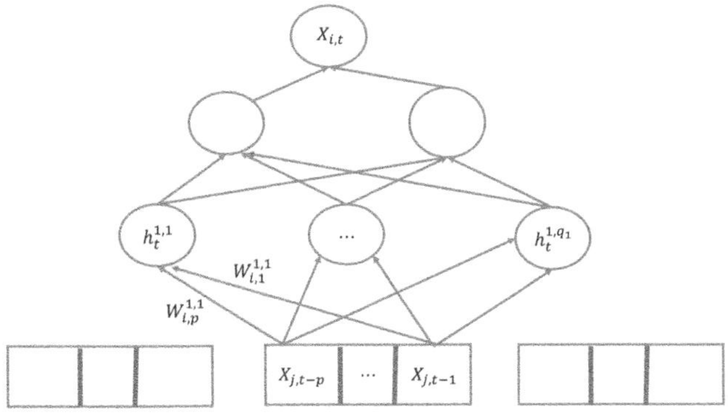

FIGURE 6.2 Architecture of cMPL.

Define the weight matrix W_i connecting the jth time series to all nodes in the first hidden layer:

$$W_j = \begin{bmatrix} W_{j,1}^1 & \cdots & W_{j,p}^1 \end{bmatrix} = \begin{bmatrix} W_{j,1}^{1,1} & \cdots & W_{j,P}^{1,1} \\ \vdots & \vdots & \vdots \\ W_{j,1}^{1,q_1} & \cdots & W_{j,P}^{1,q_1} \end{bmatrix}.$$

If all elements in the matrix W_j are zero, then the jth time series does not affect the values of the ith time series. As a consequence, the jth time series does not Granger causa the ith time series.

The values of the first hidden layer at the time t are given by

$$h_t^1 = \begin{bmatrix} h_t^{1,1} \\ \vdots \\ h_{,t}^{1,q_1} \end{bmatrix} = \begin{bmatrix} \sigma\left(\sum_{p=1}^{P} W_p^{1,1} X_{t-p}) + \sigma_1^1\right) \\ \vdots \\ \sigma\left(\sum_{p=1}^{P} W_p^{1,q_1} X_{t-p} + \sigma_{q_1}^1\right) \end{bmatrix} \tag{6.159}$$

$$= \sigma\left(\sum_{p=1}^{P} W_p^1 X_{t-p} + \sigma^1\right),$$

where σ is an element activation function and σ^1 is the bias vector at the first hidden layer.

The subsequent layers are fully connected and can be calculated by standard MLP as

$$h_t^l = \sigma\left(W^l h_t^{l-1} + b^l\right), \tag{6.160}$$

where

$$W^l = \begin{bmatrix} W_{11}^l & \cdots & W_{1q_{l-1}}^l \\ \vdots & \vdots & \vdots \\ W_{q_l 1}^l & \cdots & W_{q_l q_{l-1}}^l \end{bmatrix} \text{ and } b^l = \begin{bmatrix} b_1^l \\ \vdots \\ b_{q_1}^l \end{bmatrix}.$$

Finally, the output $X_{i,t}$ is given by

$$X_{i,t} = \sum_{j=1}^{q_L} W_{ij}^o h_t^{L,j} + b_o = W_o^T h_t^L + e_{it}, \tag{6.161}$$

where W_{ij}^{o} are the weights connecting the final hidden layer and output $X_{i,t}$, h_t^L is the final hidden vector from the final hidden layer L, and e_{it} is the error.

To make the Granger causality network sparse, we apply the group lasso penalty to the objective function (Tank et al. 2018). Therefore, the objective function for inferring sparse Granger causality network is given by

$$\min_{W} F\left(W, X_{t,}, \lambda\right) = \sum_{t=1}^{T}\left(X_{i,t} - W_o^T h_t^L\right)^2 + \lambda \left\|W_j\right\|_F, \quad (6.162)$$

where $W = \left[W^1, \ldots, W^L, W^o\right]$ and $\|.\|_F$ is the Froberius matrix norm.

The weight matrix W_j represents the relationships between input X_{jt} and output X_{it}. Its zero indicates the Granger non-causality between two time series X_{jt} and X_{it}.

To simultaneously select for both Granger causality and the lag order of the time series, we only need to take the lag P as a parameter to be optimized in objective function in equation (6.162):

$$\min_{W,P} F\left(W, X_{t,}, P, \lambda\right) = \sum_{t=1}^{T}\left(X_{i,t} - W_o^T h_t^L\right)^2 + \lambda \left\|W_j\right\|_F. \quad (6.163)$$

6.3.4.4 Component-Wise Recurrent Neural Networks (cRNNs) for Inferring Granger Causal Networks

RNNs compress the past time series into hidden states which allow to effectively extract nonlinear dependent information at quite long lags, and hence are a powerful tool for reconstruction of Granger causal networks (Tank et al. 2018). Similar to MLP, we develop cRNNs to model Granger causal relationships. We first introduce the standard cRNNs and then component-wise long short-term memory (cLSTM) for inferring Granger causal networks.

The architecture of cRNNs for construction of Granger networks is shown in Figure 6.3. Define

$$h_t = \begin{bmatrix} h_{1t} \\ \vdots \\ h_{mt} \end{bmatrix}, W_{hh} = \begin{bmatrix} W_{11}^{hh} & \cdots & W_{1m}^{hh} \\ \vdots & \vdots & \vdots \\ W_{m1}^{hh} & \cdots & W_{mm}^{hh} \end{bmatrix},$$

$$W_{xh} = \begin{bmatrix} W_{11}^{xh} & \cdots & W_{1j}^{xh} & \cdots & W_{1n}^{xh} \\ \vdots & & \vdots\;\vdots\;\vdots & & \vdots \\ W_{m1}^{xh} & \cdots & W_{mj}^{xh} & \cdots & W_{mn}^{xh} \end{bmatrix}.$$

The hidden state that contains information of the past time series is recursively updated by

$$h_t\left(W, X_t, h_{t-1}\right) = g\left(h_{t-1}, X_t\right)$$
$$= \tanh\left(W_{hh}h_{t-1} + W_{xh}X_t + b_h\right), \quad (6.164)$$

where b_h is a vector of bias.

The output X_{it} is modeled as

$$X_{it} = W_o^T h_t + e_{it}, \quad (6.165)$$

where $W_o = \begin{bmatrix} W_{1,o} & \cdots & W_{m,o} \end{bmatrix}^T$ and e_{it} is an error.

Similar to MLP, sufficient condition for Ganger non-causality of input time series $X_{j,t}$ on output time series $X_{i,t}$ is that all elements W_{lj}^{xh} of the jth column of W_{xh} are zero, $W_{lj}^{xh} = 0$, $l = 1, \ldots, m$.

Due to its recursive updating, hidden state $h_t\left(W, X_t, h_{t-1}\right)$ is a function $h_t\left(W, X_t, X_{t-1}, \ldots, X_1\right)$,

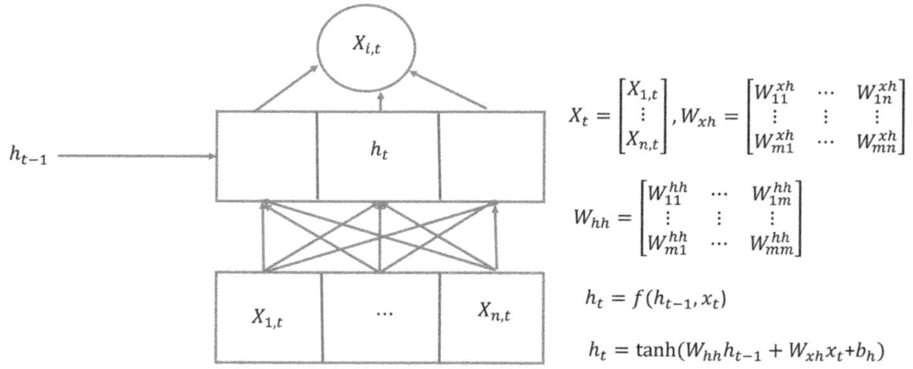

FIGURE 6.3 Architecture of cRNNS for construction of Granger causal networks.

which implies that $h_t(W, X_t, h_{t-1})$ contains information of all past time series $X_t, X_{t-1}, \ldots, X_1$.

Again, the objective function that includes group lasso penalty is

$$\min_{W_{xh}, W_{hh}, W_o, b_h} F\left(W_{xh}, W_{hh}, W_o, b_h, X_1, \ldots, X_t\right)$$
$$= \sum_{t=2}^{T} \left(X_{it} - W_o^T h_t\right)^2 + \lambda \sum_{j=1}^{n} \left\| W_{\cdot,j}^{xh} \right\|_2^2. \quad (6.166)$$

When λ increases, the number of zero weights will increase. The larger the parameter λ, the more sparse the Ganger causal network.

Next we introduce cLSTMs for inferring Granger causal networks. The LSTMs are a special kind of RNN, capable of learning long-term dependencies. The key to LSTMs is to introduce a cell state as the second hidden state. The cell state is like a conveyor belt. The LSTM does have the ability to remove or add information to the cell state, carefully regulated by structures called gates. The recursive equations for updating two hidden states are given by (Chapter 1)

$$f_t = \sigma\left(W_{xf} x_t + W_{hf} h_{t-1} + b_f\right),$$

$$\mathrm{i}_t = \tanh\left(\mathrm{W}_{xi} \mathrm{x}_t + \mathrm{W}_{hi} \mathrm{h}_{t-1} + \mathrm{b}_i\right),$$

$$g_t = \tanh\left(W_{xc} x_t + W_{hc} h_{t-1} + b_c\right),$$

$$C_t = f_t \circ C_{t-1} + i_t \circ g_t,$$

$$O_t = \sigma\left(W_{xo} x_t + W_{ho} h_{t-1} + b_o\right),$$

$$h_t = o_t \circ tanh\left(c_t\right),$$

$$X_{it} = W_o^T h_t + e_{it}.$$

Define a weight matrix that connects the forget gates, input gates, cell updates, output gate, and the update of the hidden state to the input time series:

$$W^x = \begin{bmatrix} W_{xf} \\ W_{xi} \\ W_{xc} \\ Wx_o \end{bmatrix} \text{ and } W^h = \begin{bmatrix} W_{hf} \\ W_{hi} \\ W_{hc} \\ W_{ho} \end{bmatrix}.$$

The jth column vector $W_{\cdot,j}^x$ of the weight matrix W^x measures the degree of influence of the time series $X_{j,t}$ on time series $X_{i,t}$. A sufficient condition for Granger non-causality of an input time series $X_{j,t}$ on an output time series $X_{i,t}$ is $W_{\cdot,j}^x = 0$. Similar to equation (6.166), the objective function for inferring sparse Granger causal network using cLSTM and group lasso is given by (Tank et al. 2018)

$$\min_{W^x, W^h W_o, b_h} F\left(W^x, W^h, W_o, b_h, X_1, \ldots, X_t\right)$$
$$= \sum_{t=2}^{T} \left(X_{it} - W_o^T h_t\right)^2 + \lambda \sum_{j=1}^{n} \left\| W_{\cdot,j}^x \right\|_2^2. \quad (6.167)$$

Again, the proximal gradient descent with line search can be used to solve the optimization problem in equation (6.167).

The remarkable feature of the cRNNs and cLSTMs is that they do not need to specify lags P of the time series.

6.3.4.5 Statistical Recurrent Units for Inferring Granger Causal Networks

In the previous section, we introduced cRNNs and cLSTM for inferring Granger causal networks. However, the highly nonlinear sigmoid function suffered from the vanishing/exploding gradient issue during training. To overcome this limitation, in this section, we introduce the statistical recurrent unit (SRU) that has no sigmoid gating functions and thus is less affected by the vanishing/exploding gradient issue during training for inferring Grander causal networks (Oliva et al. 2017; Khanna and Tan 2019). SRU can capture both short- and long-term temporal dependencies in a multivariate time series. This allows to take linear combinations of the summary statistics at different time scales for construction of predictive causal models with both highly component-specific and lag-specific features.

A key for the SRU to maintain long-term sequential dependencies is to use temporally aware summary statistics, which learn temporal order in a dataset. Recall that recursive function $h(X_t, h_{t-1})$ in the RNN which is not only as a function of X_t, but also as a function of the previous statistics h_{t-1} of X_{t-1}. The function $h(X_t, h_{t-1})$ is a summary statistics that can maintain temporal order information in the dataset. The function $h(X_t, h_{t-1})$ can recursively generate a sequence of summary statistics that keep temporal information:

$$h_1 = h(X_1, h_0),\ h_2 = h(X_2, h_1), \ldots, h_t = h(X_t, h_{t-1}), \ldots,$$

where h_0 is a vector of initial values.

Recurrent summary statistics at multiple scale can provide more temporal information. Define exponential moving averages:

$$\begin{aligned}\mu_t &= (1-\alpha)\mu_{t-1} + \alpha h_t \\ &= \cdots = (1-\alpha)^T \mu_0 + \alpha\left[(1-\alpha)^{T-1} h(X_1, h_0) \right. \\ &\quad \left. + \cdots + (1-\alpha)h(X_{t-1}, h_{t-2}) + h(X_t, h_{t-1})\right].\end{aligned}$$

Two features are combined to form a SRU. The ith SRU for inferring Granger causal network is shown in Figure 6.4 (Khanna and Tan 2019). The basic block of the SRU is recurrent statistic $\varnothing_t$:

$$\varnothing_{i,t} = h\left(W_{in}^i X_t + W_f^i r_{i,t-1} + b_{in}^i\right), \quad (6.168)$$

where $\varnothing_{i,t}$ is a $d_\varnothing$ dimensional vector of recurrent statistics, h is an element activation function such as ReLU, $r_{i,t-1}$ is a d_r dimensional feedback vector, b_{in}^i is a $d_\varnothing$ dimensional bias vector, W_{in}^i and W_f^i are two weight matrices and defined as

$$W_{in}^i = \begin{bmatrix} W_{11}^{i,in} & \cdots & W_{1j}^{i,in} & \cdots & W_{1n}^{i,in} \\ \vdots & & \vdots\ \vdots\ \vdots & & \vdots \\ W_{d_\varnothing n}^{i,in} & \cdots & W_{d_o j}^{i,in} & \cdots & W_{d_\varnothing n}^{i,in} \end{bmatrix},\ W_f^i = \begin{bmatrix} W_{11}^{i,f} & \cdots & W_{1d_r}^{i,f} \\ \vdots & \vdots & \vdots \\ W_{d_\varnothing 1}^{i,f} & \cdots & W_{d_\varnothing d_r}^{i,f} \end{bmatrix},$$

where $W_{\cdot,j}^{i,in} = \begin{bmatrix} W_{1j}^{i,in} & \cdots & W_{d_o j}^{i,in} \end{bmatrix}^T$ measures the influence of the jth time series $X_{j,t}$ on the time series $X_{i,t}$.

Feedback vector $r_{i,t-1}$ is generated by

$$r_{i,t} = g\left(W_r^i \mu_{i,t-1} + b_r^i\right), \quad (6.169)$$

where $\mu_{i,t-1}$ is an $md_\varnothing$ dimensional vector of multiscale summary statistics, b_r^i is a $d_\varnothing$ dimensional bias vector, W_r^i is a weight matrix defined as

$$W_r^i = \begin{bmatrix} W_{11}^{i,r} & \cdots & W_{1md_\varnothing}^{i,r} \\ \vdots & \vdots & \vdots \\ W_{d_r 1}^{i,r} & \cdots & W_{d_r md_\varnothing}^{i,r} \end{bmatrix}.$$

The SRU will take the following exponential moving averages to capture the temporal information in the dataset:

$$\mu_{i,t}^{\alpha_j} = \left(1-\alpha_j\right)\mu_{i,t-1}^{\alpha_j} + \alpha_j \varnothing_{i,t}, j = 1, \ldots, m, \text{ and} \quad (6.170)$$

$$\mu_{i,t} = \begin{bmatrix} \mu_{i,t}^{\alpha_1} \\ \vdots \\ \mu_{i,t}^{\alpha_m} \end{bmatrix}.$$

Output features are generated by multiscale summary statistics:

$$O_{i,t} = g\left(W_o^i \mu_{i,t} + b_o^i\right), \quad (6.171)$$

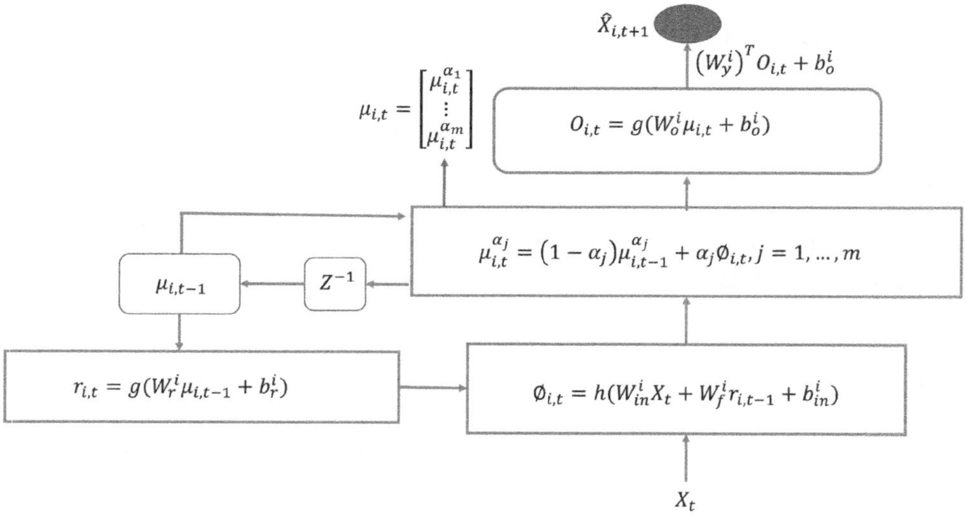

FIGURE 6.4 Outline of the ith SRU for inferring Granger causal networks.

where $O_{i,t}$ is a d_o dimensional vector of output features, b_o^i is a d_o dimensional bias vector, W_o^i is $d_o \times (md_\varnothing)$ dimensional weight matrix which is defined as

$$W_o^i = \begin{bmatrix} W_{1,1}^{i,o} & \cdots & W_{1,md_\varnothing}^{i,o} \\ \vdots & \vdots & \vdots \\ W_{d_o,1}^{i,o} & \cdots & W_{d_o,md_\varnothing}^{i,o} \end{bmatrix}. \quad (6.172)$$

Finally, the output $\hat{X}_{i,t+1}$ is given by a linear combination of nonlinear output features:

$$\hat{X}_{i,t+1} = \left(W_y^i\right)^T O_{i,t} + b_o^i \quad (6.173)$$

where $\hat{X}_{i,t+1}$ is the next-step prediction of the ith time series at the time $t+1$, b_o^i is a bias, W_y^i is a d_o dimensional vector of weights.

Recall that the jth column of vector $W_{.,j}^{i,in}$ of the weight matrix W_{in}^i in equation (6.168) quantifies the influence of the jth time series $X_{j,t}$ on the time series $X_{i,t}$. To learn the Granger causal network, we optimize the following penalized mean squared prediction error loss:

$$\min_{\Theta^i} F(\Theta^i, X_t) = \frac{1}{T-1} \sum_{t=1}^{T-1} \left(\hat{X}_{i,t+1} - X_{i,t+1}\right)^2 + \lambda \sum_{j=1}^{n} \left\| W_{.,j}^{i,in} \right\|_2, \quad (6.174)$$

where $\Theta^i = \left[W_y^i, b_o^i, W_o^i, b_o^i, W_r^i, b_r^i, W_{in}^i, W_f^i, b_{in}^i\right]$, $\hat{X}_{i,t+1}$ is calculated using equations (6.168)–(6.173), and λ is a penalty parameter. Optimization problem (6.174) is nonconvex which may have multiple local minima, and hence is difficult to find a global optimal solution. A first-order proximal gradient descent algorithm can be used to find a regularized solution of the SRU optimization problem (6.174) (Xiong 2018a; Khanna and Tan 2019).

The number of parameters in the SRU is much larger than the number of parameters in the standard RNN model. This will cause a serious overfitting problem. To overcome this limitation, Khanna and Tan (2019) used pseudo-autoencoder to reduce the dimensionality of the feedback operation in equation (6.169). A simple linear transformation is used as encoder to compress the multiscale md_ϕ dimensional summary statistics $\mu_{i,t}$ to a $d_{r'}$ dimensional vector $V_{i,t}$:

$$V_{i,t} = D_r^i \mu_{i,t}, \quad (6.175)$$

where $d_{r'} \ll md_\phi$. The elements of the transformation matrix D_r^i will be generated by independently sampling from a normal distribution $N\left(0, \frac{1}{d_{r'}}\right)$. The compressed vector $V_{i,t}$ can be thought as a low-dimensional projection of the multiscale summary statistics $\mu_{i,t}$. The compressed vector $V_{i,t}$ maintains most of the temporal information of the original data. The decoder consists of MLP, which maps the compressed summary statistics $V_{i,t}$ to the feedback vector $r_{i,t}$. Mathematically, we can write

$$r_{i,t} = \sigma\left(W_r^{*,i} V_{i,t} + b_r^{*,i}\right), \quad (6.176)$$

where $W_r^{*,i}$ is a $d_r \times d_{r'}$ dimensional weight matrix and $b_r^{*,i}$ is a d_r dimensional bias vector. The total number of parameters in the pseudo-autoencoder is $d_r(d_{r'}+1)$, which is much smaller than the number of parameters in generating feedback $r_{i,t}$ in the original SRU $(d_r(md_\varnothing+1))$.

Recall that the number of parameters in the matrix W_o^i is $md_\varnothing d_o$. It is clear that the weight matrix W_o^i often dominates the overall number of trainable parameters in the SRU. To alleviate the overfitting, we also need to reduce the number of parameters in the weight matrix W_o^i via penalization of the weights in the matrix W_o^i.

$$W_o^i = \begin{bmatrix} W_{1,1}^{i,o} & \cdots & W_{1,d_\varnothing}^{i,o} & \cdots & W_{1,(m-1)d_\varnothing+1}^{i,o} & \cdots & W_{1,md_\varnothing}^{i,o} \\ \vdots & \vdots & & & \vdots\vdots\vdots\vdots\vdots & & \\ W_{j,1}^{i,o} & \cdots & W_{j,d_\varnothing}^{i,o} & \cdots & W_{j,(m-1)d_\varnothing+1}^{i,o} & \cdots & W_{j,md_\varnothing}^{i,o} \\ \vdots & \vdots & & & \vdots\vdots\vdots\vdots\vdots & & \\ W_{d_o,1}^{i,o} & \cdots & W_{d_o,d_\varnothing}^{i,o} & \cdots & W_{d_o,(m-1)d_\varnothing+1}^{i,o} & \cdots & W_{1,md_\varnothing}^{i,o} \end{bmatrix}.$$

Let

$$W_{j,s(k)}^{i,o} = \begin{bmatrix} W_{j,k}^{i,o} \\ W_{j,d_\varnothing+k}^{i,o} \\ \vdots \\ W_{j,(m-1)d_\varnothing+k}^{i,o} \end{bmatrix}.$$

The group lasso for the regulation of the weights in the matrix W_o^i should be constructed by

$$G=\sum_{j=1}^{d_o}\sum_{k=1}^{d_\varnothing}\left\|W_{j,s(k)}^{i,o}\right\|_2. \quad (6.177)$$

Define the set of parameters $\hat{\Theta}_{eSRU}^i=\left(\hat{\Theta}^i \setminus W_r^i\right)\cup W_r^{*,i}$. Adding equation (6.177) into equation (6.174), we obtain the following objective function for inferring Granger causal network:

$$\min_{\Theta_{eSRU}^i} F\left(\Theta_{eSRU}^i, X_t\right)=\frac{1}{T-1}\sum_{t=1}^{T-1}\left(\hat{X}_{i,t+1}-X_{i,t+1}\right)^2$$

$$+\lambda_1\sum_{j=1}^{n}\left\|W_{.,j}^{i,in}\right\|_2+\lambda_2\sum_{j=1}^{d_o}\sum_{k=1}^{d_\varnothing}\left\|W_{j,s(k)}^{i,o}\right\|_2, \quad (6.178)$$

where λ_1 and λ_2 are two penalty parameters.

6.4 NONLINEAR STRUCTURAL EQUATION MODELS FOR CAUSAL INFERENCE ON MULTIVARIATE TIME SERIES

In this section, we introduce nonlinear structural equation models for causal inference on time series (Peters et al. 2013). First, we introduce several concepts.

Definition 6.14: Full Time Graph

Let $V=\{X_t^i, i=1,\ldots,n\}$ be a set of time series. If the infinite graph that contains each variable X_t^i as a node in the graph, then this graph is called full time graph.

Definition 6.15: Summary Time Graph

If the graph contains all components of the time series in the set V as nodes and a direction from X^i to X^j $(i\neq j)$ in the full time graph for some k is present, then such graph is called the summary time graph.

The Granger causal networks do not consider instantaneous causal relationships among time series. To overcome this limitation, we introduce restricted structural equation models for causal analysis in time series which is called **time series models with independent noise (TiMINo)** A remarkable feature of these models include nonlinear and instantaneous causal effects. The TiMINo is extension of the structural equation model from causal analysis for cross-section data to causal analysis for time series data. The TiMINo assumes that X_t is a function of all direct causes and some noise variable and that the collection of noises are jointly independent. Before formally defining the TiMINo, we introduce the notation of parent points of the time series variable $X_{i,t}$. Define $PA_0^i\subseteq X^{V\setminus i}$ as the set of points $X_{j,t}\rightarrow X_{i,t}$, $j=1,\ldots$ (Figure 6.5(a)), $PA_1^i\subseteq X^V$ as the set of pints $X_{j,t-1}\rightarrow X_{i,t-1}$, $j=1,\ldots$ (Figure 6.5(b)), and PA_k^i as the set of points $X_{j,t-k}\rightarrow X_{i,t-k}$, $j=1,\ldots$ (Figure 6.5(c)).

Formally, Definition 6.16 is given to precisely define the TiMINo (Peters et al. 2013).

Definition 6.16: TiMINo

Consider a multivariate time series $X_t=\left(X_{i,t}\right)_{i\in V}$. The time series satisfies a TiMINo if there is a $P>0$ and $\forall i\in V$, there are sets $PA_0^i\subseteq X^{V\setminus i}$, $PA_k^i\subseteq X^V$ such that

$$X_{i,t}=f_i\left(\left(PA_p^i\right)_{t-p},\ldots,\left(PA_1^i\right)_{t-1},\left(PA_0^i\right)_t, N_{i,t}\right), \quad (6.179)$$

with $N_{i,t}$ jointly independent over both i and t, and for each i, $N_{i,t}$ are identically distributed in t.

Equation (6.179) states that $X_{i,t}$ is a nonlinear function of its parents plus an additive noise term, with the

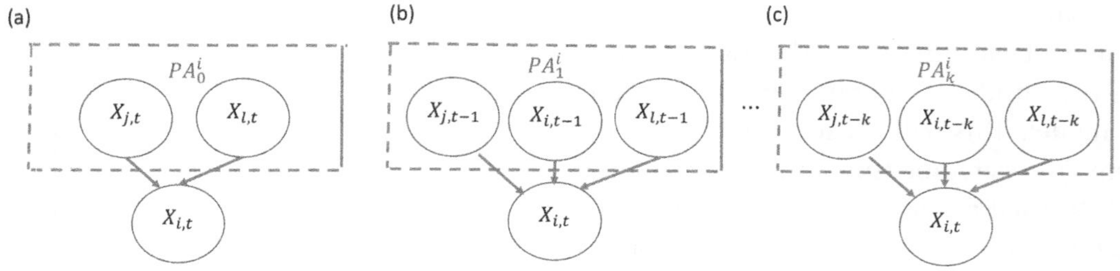

FIGURE 6.5 Scheme of parent points of $X_{i,t}$.

important condition that all noise terms are assumed to be jointly independent.

Equation (6.179) can be used to infer the full time graph. The full time graph can be generated by drawing arrows from any node that presents in the right-hand side of equation (6.179) to X_{it}.

In equation (6.179), $\left(PA_0^i\right)_t$ provides information about contemporaneous effect, while $\left(PA_p^i\right)_{t-p},\ldots,\left(PA_1^i\right)_{t-1}$ provide information about Granger causality.

Now we introduce the identifiability theorem of TiMINo (Peters et al. 2013).

Theorem 6.8: Identifiability of TiMINo

Assume that X_t can be represented as a TiMINo and that the parent set $PA\left(X_{it}\right)=\cup_0^p\left(PA_k^i\right)_{t-k}$ is direct cause of $X_{i,t}$. Further assume that one of the following conditions holds:

1. Equation (6.179) comes from either nonlinear functions f_i with additive normal noise $N_{i,t}$ or linear functions f_i with additive non-normal noise $N_{i,t}$.
2. Each component shows a time structure, i.e., $PA\left(X_{i,t}\right)$ contains at least one $X_{i,t-k}$, the joint distribution is faithful with respect to the full time graph, and the summary time graph is acyclic.

The proof is in the Appendix of the paper written by Peters et al. (2013).

Additive noise model for the cross-sectional data can be adapted to the TiMINo causal analysis. Algorithm 1 in Peter et al. (2013) for implementing adapted additive model for time series causal analysis is reintroduced as Algorithm 6.1.

Algorithm 6.1: Additive Noise Model for TiMINo Causality Analysis.

Step 1: Input:

Data sampled from a d-dimensional time series of length $T:\left(X_1,\ldots,X_T\right)$, maximal order p, $S\leftarrow(1,\ldots,d)$.

Step 2: Repeat

1. **for** k in S **do**
2. Fit TiMINo for X_t^K using $X_{t-p}^k,\ldots,X_{t-1}^K,X_{t-p}^i,\ldots,X_{t-1}^i,X_t^i$ for $i\in S\setminus\{k\}$.
3. Test the independence of residuals of $X^i, i\in S$.
   end for
4. Choose k^* to be the k with the weakest dependence, which implies that the additive model for k^* is satisfied (if there is no k with independence, break and output "I do not know. - bad model fit").
5. Find the parent set.

$$S\leftarrow S\setminus\left\{k^*\right\},\ pa\left(k^*\right)\leftarrow S.$$

   Until length $(S)=1$.
6. For all k, remove all parents that are not required to obtain independent residuals.
7. Output the parent set that determines the structure of causal network.

$$\left\{pa(1),\ldots,pa(d)\right\}.$$

The fitting TiMINo models are given as follows.

Linear Model:

$$f_i\left(p_1,\ldots,p_r,N_i\right)=a_{i,1}p_1+\cdots+a_rp_r+N_i, \quad (6.180)$$

Generalized Additive Model:

$$f_i\left(p_1,\ldots,p_r,N_i\right)=f_{i1}\left(p_1\right)+\cdots+f_{ir}\left(p_r\right)+N_i, \quad (6.181)$$

Nonlinear Regression:

$$f_i\left(p_1,\ldots,p_r,N_i\right)=f_i\left(p_1,\ldots,p_r\right)+N_i. \quad (6.182)$$

The order of the time series is determined by the AIC.

HSIC will be used to test for independence between two residual time series N_t^k and N_t^i. Specifically, the test statistic is defined as (Appendix 6B)

$$HSIC^2\left(\mathcal{F},\mathbb{Q},P_{XY}\right)=\frac{1}{n^2}\operatorname{tr}\left(KHLH\right), \quad (6.183)$$

where $H=I-\frac{1_n1_n^T}{n}$, $K=\left(k_{ij}\right)_{n\times n}$, and $L=\left(l_{ij}\right)_{n\times n}$, $k_{ij}=k\left(X_i,X_j\right)$, $l_{ij}=k\left(Y_i,Y_j\right)$, X and Y represent residual time series.

Some typical kernel functions (Bagnell 2008) are

Gaussian kernel: $k(x,y)=e^{-\frac{\|x-y\|^2}{2\sigma^2}}, \sigma>0$;

Laplace kernel: $k(x,y)=e^{-\frac{\|x-y\|}{\sigma}, \sigma>0}$;

Inverse multi-quadratics kernel: $k(x,y)=\frac{1}{\beta+\|x-y\|^{\alpha}}$, $\alpha,\beta>0$;

Fractional Brownian motion kernel: $k(x,y)=\frac{1}{2}\left(\|x\|^{2h}+\|y\|^{2h}-\|x-y\|^{2h}\right), 0<h<1$.

SOFTWARE PACKAGE

The bivariate linear Granger causality test is implemented in the *statsmodels* Python package [Python], in the *MSBVAR* package [R], the *lmtest* package [R][3], the *NlinTS* package [R], and the *vars* package [R].

APPENDIX 6A: TEST STATISTIC T_{NNG} ASYMPTOTICALLY FOLLOWS A NORMAL DISTRIBUTION

In this appendix, we completely stated the asymptotic normal distribution of the statistic T_{NNG}

Define

$$h_1(l_y,l_x,M_y,l.k)=I\left(Y^{l_y+M_y}_{l_{yx}+1+k-l_y}, Y^{l_y+M_y}_{l_{yx}+1+k+l-l_y}, e\right),$$

$$h_2(l_y,l_x,l,k)=I\left(X^{l_x}_{l_{yx}+1+k-l_x}, X^{l_x}_{l_{yx}+1+k+l-l_x}, e\right),$$

$$\begin{aligned}\sum_{11} &= E\Big[\big(h_1(l_y,l_x,M_y,l,0)h_2(l_y,l_x,l,o)\\ &\quad -C_1(M_y+l_y,l_x,e,l)^2\Big]\\ &+\sum_{k=1}^{n-1}2\left(1-\frac{k}{n}\right)E\Big[\big(h_1(l_y,l_x,M_y,l,0)h_2(l_y,l_x,l,o)\\ &\quad -C_1(M_y+l_y,l_x,e,l)\big)\big(h_1(l_y,l_x,M_y,l,k)h_2(l_y,l_x,l,k)\\ &\quad -C_1(M_y+l_y,l_x,e,l)\big)\Big],\end{aligned}$$

$$\begin{aligned}\sum_{12} &= E\Big[\big(h_1(l_y,l_x,M_y,l,0)h_2(l_y,l_x,l,o)\\ &\quad -C_1(M_y+l_y,l_x,e,l)\big)\big(h_1(l_y,l_x,0,l,0)h_2(l_y,l_x,l,o)\\ &\quad -C_2(l_y,l_x,e;l)\big)\Big]\\ &+\sum_{k=1}^{n-1}\left(1-\frac{k}{n}\right)E\Big[\big(h_1(l_y,l_x,M_y,l,0)h_2(l_y,l_x,l,o)\\ &\quad -C_1(M_y+l_y,l_x,e,l)\big)\\ &\quad \big(h_1(l_y,l_x,0,l,k)h_2(l_y,l_x,l,k)-C_2(l_y,l_x,e;l)\big)\Big]\\ &+\sum_{k=1}^{n-1}\left(1-\frac{k}{n}\right)E\Big[\big(h_1(l_y,l_x,M_y,l,k)h_2(l_y,l_x,l,k)\\ &\quad -C_1(M_y+l_y,l_x,e,l)\big)\big(h_1(l_y,l_x,0,l,k)h_2(l_y,l_x,l,k)\\ &\quad -C_2(l_y,l_x,e;l)\big)\Big],\end{aligned}$$

$$\begin{aligned}\sum_{13} &= E\Big[\big(h_1(l_y,l_x,M_y,l,0)h_2(l_y,l_x,l,o)\\ &\quad -C_1(M_y+l_y,l_x,e,l)\big)\big(h_1(l_y,l_x,M_y,l,0)\\ &\quad -C_3(M_y+l_y,e;l)\big)\Big]\\ &+\sum_{k=1}^{n-1}\left(1-\frac{k}{n}\right)E\Big[\big(h_1(l_y,l_x,M_y,l,0)h_2(l_y,l_x,l,o)\\ &\quad -C_1(M_y+l_y,l_x,e,l)\big)\big(h_1(l_y,l_x,M_y,l,k)\\ &\quad -C_3(M_y+l_y,e;l)\big)\Big]\\ &+\sum_{k=1}^{n-1}\left(1-\frac{k}{n}\right)E\Big[\big(h_1(l_y,l_x,M_y,l,k)h_2(l_y,l_x,l,k)\\ &\quad -C_1(M_y+l_y,l_x,e,l)\big)\big(h_1(l_y,l_x,M_y,l,0)\\ &\quad -C_3(M_y+l_y,e;l)\big)\Big],\end{aligned}$$

$$\begin{aligned}\sum_{14} &= E\Big[\big(h_1(l_y,l_x,M_y,l,0)h_2(l_y,l_x,l,o)\\ &\quad -C_1(M_y+l_y,l_x,e,l)\big)\big(h_1(l_y,l_x,0,l,0)\\ &\quad -C_4(l_y,e;l)\big)\Big]\\ &+\sum_{k=1}^{n-1}\left(1-\frac{k}{n}\right)E\Big[\big(h_1(l_y,l_x,M_y,l,0)h_2(l_y,l_x,l,o)\\ &\quad -C_1(M_y+l_y,l_x,e,l)\big)\big(h_1(l_y,l_x,0,l,k)\\ &\quad -C_4(l_y,e;l)\big)\Big]\\ &+\sum_{k=1}^{n-1}\left(1-\frac{k}{n}\right)E\Big[\big(h_1(l_y,l_x,M_y,l,k)h_2(l_y,l_x,l,k)\\ &\quad -C_1(M_y+l_y,l_x,e,l)\big)\big(h_1(l_y,l_x,0,l,0)\\ &\quad -C_4(l_y,e;l)\big)\Big],\end{aligned}$$

$$\sum_{22} = E\Big[\big(h_1(l_y,l_x,0,l,0)h_2(l_y,l_x,l,o) - C_2(l_y,l_x,e,l)\big)^2 + \sum_{k=1}^{n-1} 2\Big(1-\frac{k}{n}\Big)E\Big[\big(h_1(l_y,l_x,0,l,0)h_2(l_y,l_x,l,o) - C_2(l_y,l_x,e,l)\big)\big(h_1(l_y,l_x,0,l,k)h_2(l_y,l_x,l,k) - C_2(l_y,l_x,e,l)\big)\Big],$$

$$\sum_{23} = E\Big[\big(h_1(l_y,l_x,0,l,0)h_2(l_y,l_x,l,o) - C_2(l_y,l_x,e,l)\big)\big(h_1(l_y,l_x,M_y,l,0) - C_3(M_y+l_y,e;l)\big)\Big] + \sum_{k=1}^{n-1}\Big(1-\frac{k}{n}\Big)E\Big[\big(h_1(l_y,l_x,0,l,0)h_2(l_y,l_x,l,o) - C_2(l_y,l_x,e,l)\big)\big(h_1(l_y,l_x,M_y,l,k) - C_3(M_y+l_y,e;l)\big)\Big] + \sum_{k=1}^{n-1}\Big(1-\frac{k}{n}\Big)E\Big[\big(h_1(l_y,l_x,0,l,k)h_2(l_y,l_x,l,k) - C_2(l_y,l_x,e,l)\big)\big(h_1(l_y,l_x,M_y,l,0) - C_3(M_y+l_y,e;l)\big)\Big],$$

$$\sum_{24} = E\Big[\big(h_1(l_y,l_x,0,l,0)h_2(l_y,l_x,l,o) - C_2(l_y,l_x,e,l)\big)\big(h_1(l_y,l_x,0,l,0) - C_4(l_y,e;l)\big)\Big] + \sum_{k=1}^{n-1}\Big(1-\frac{k}{n}\Big)E\Big[\big(h_1(l_y,l_x,0,l,0)h_2(l_y,l_x,l,o) - C_2(l_y,l_x,e,l)\big)\big(h_1(l_y,l_x,0,l,k) - C_4(l_y,e;l)\big)\Big] + \sum_{k=1}^{n-1}\Big(1-\frac{k}{n}\Big)E\Big[\big(h_1(l_y,l_x,0,l,k)h_2(l_y,l_x,l,k) - C_2(l_y,l_x,e,l)\big)\big(h_1(l_y,l_x,0,l,0) - C_4(l_y,e;l)\big)\Big],$$

$$\sum_{33} = E\Big[\big(h_1(l_y,l_x,M_y,l,0) - C_3(M_y+l_y,e;l)\big)^2\Big] + \sum_{k=1}^{n-1} 2\Big(1-\frac{k}{n}\Big)E\Big[\big(h_1(l_y,l_x,M_y,l,0) - C_3(M_y+l_y,e,l)\big)\big(h_1(l_y,l_x,0,l,0) - C_4(l_y,e;l)\big)\Big],$$

:

$$\sum_{34} = E\Big[h_1(l_y,l_x,M_y,l,0) - C_3(M_y+l_y,e;l)\big(h_1(l_y,l_x,0,l,0) - C_4(l_y,e;l)\big)\Big] + \sum_{k=1}^{n-1}\Big(1-\frac{k}{n}\Big)E\Big[\big(h_1(l_y,l_x,M_y,l,k) - C_3(M_y+l_y,e;l)\big)\big(h_1(l_y,l_x,0,l,0) - C_4(l_y,e;l)\big)\Big],$$

$$\sum_{44} = E\Big[\big(h_1(l_y,l_x,0,l,0) - C_3(l_y,e;l)\big)^2\Big] + \sum_{k=1}^{n-1}\Big(1-\frac{k}{n}\Big)E\Big[\big(h_1(l_y,l_x,0,l,0) - C_4(l_y,e;l)\big) + \big(h_1(l_y,l_x,0,l,k) - C_4(l_y,e;l)\big)\Big].$$

Recall that

$$T_{NNG} = \sqrt{n}\left(\frac{\hat{C}_1((M_y+l_y,l_x,e;l)}{\hat{C}_2(l_y,l_x,e;l)} - \frac{\hat{C}_3(M_y+l_y,e;l)}{\hat{C}_4(l_y,e;l)}\right).$$

Let

$$f = \begin{bmatrix} C_1((M_y+l_y,l_x,e;l) \\ C_2(l_y,l_x,e;l) \\ C_3(M_y+l_y,e;l) \\ C_4(l_y,e;l) \end{bmatrix}.$$

Then,

$$\nabla f = \begin{bmatrix} \dfrac{1}{\hat{C}_2(l_y,l_x,e;l)} \\ -\dfrac{\hat{C}_1((M_y+l_y,l_x,e;l)}{\big(\hat{C}_2(l_y,l_x,e;l)\big)^2} \\ -\dfrac{1}{\hat{C}_4(l_y,e;l)} \\ \dfrac{\hat{C}_3(M_y+l_y,e;l)}{\big(\hat{C}_4(l_y,e;l)\big)^2} \end{bmatrix}.$$

Under the null hypothesis, applying the central limit theorem for the nonlinear function of statistic, we have

$$T_{NNG}=\sqrt{n}\left(\frac{\hat{C}_1\left(\left(M_y+l_y,l_x,e;l\right)\right.}{\hat{C}_2\left(l_y,l_x,e;l\right)}-\frac{\hat{C}_3\left(M_y+l_y,e;l\right)}{\hat{C}_4\left(l_y,e;l\right)}\right)$$

$$\xrightarrow{d}N\left(0,\sigma^2\left(M_y,l_y,l_x,e,l\right)\right),$$

where

$$\sigma^2\left(M_y,l_y,l_x,e,l\right)=\nabla f^T\sum\nabla f.$$

APPENDIX 6B: HSIC-BASED TESTS FOR INDEPENDENCE BETWEEN TWO STATIONARY MULTIVARIATE TIME SERIES

This appendix mainly briefly introduces some basic concepts on covariance operator and Hilbert-Schmidt Independence Criterion for testing independence between two time series (Gretton et al. 2005; Rubenstein et al. 2016; Wang et al. 2018).

6B1 Reproducing Kernel Hilbert Space

Definition 6B1: Inner Product

Let $\mathcal{H}$ be a vector space over $\mathbb{R}$. A function $\langle .,. \rangle_{\mathcal{H}}:\mathcal{H}\times\mathcal{H}\to\mathbb{R}$ is an inner product of two vectors on $\mathcal{H}$ if it satisfies the following three conditions:

1. Linear: $\langle \alpha f_1+\beta f_2,g\rangle_{\mathcal{H}}=\alpha\langle f_1,g\rangle_{\mathcal{H}}+\beta\langle f_2,g\rangle_{\mathcal{H}}$
2. Symmetric: $\langle f,g\rangle_{\mathcal{H}}=\langle g,f\rangle_{\mathcal{H}}$
3. Positivity: $\langle f,f\rangle_{\mathcal{H}}\geq 0$ and $\langle f,f\rangle_{\mathcal{H}}=0$ if and only if $f=0$.

The norm that is induced by the inner product is defined as $\|f\|_{\mathcal{H}}=\sqrt{<f,f>_{\mathcal{H}}}$.

Definition 6B2: Hilbert Space

Inner product space that contains Cauchy sequence limits is called Hilbert space.

Definition 6B3: Kernel

Let χ be a non-empty set. A function $k:\chi\times\chi\to\mathbb{R}$ is a kernel function if there exists an $\mathbb{R}$-Hilbert space and a feature map $\varnothing:\chi\to\mathcal{H}$ such that for all x,x', we have

$$k(x,x')=\langle\varnothing(x),\varnothing(x')\rangle_{\mathcal{H}}. \quad (6B1)$$

Example 6B1

Let $\chi=\mathbb{R}^2$ and $X=[X_1,X_2]^T$. Define feature map $\varnothing:\mathbb{R}^2\to\mathbb{R}^3$ as $\varnothing(X)=[X_1,X_2,X_1X_2]^T$. A kernel function is defined as

$$k(X,Y)=<\varnothing(X),\varnothing(Y)>_{\mathcal{H}}=X_1Y_1+X_2Y_2+X_1X_2Y_1Y_2.$$

We also can define $k(X,.)=\varnothing(X)$. Then, kernel function can be expressed as

$$k(X,Y)=<\varnothing(X),\varnothing(Y)>_{\mathcal{H}}=<k(X,.),k(Y,.>)_{\mathcal{H}}.$$

Define a linear function of the inputs X_1, X_2 and their product X_1X_2 as

$$f(X)=f_1X_1+f_2X_2+f_3X_1X_2.$$

Assume that equivalent representation of f is

$$f(.)=[f_1,f_2,f_3]^T,$$

where $f(.)$ refers to the function as a vector in $\mathbb{R}^3$ in this problem. The standard expression $f(X)$ is a function value evaluated at a data point and can be calculated by

$$f(X)=<f(.),\varnothing(X)>_{\mathcal{H}}=f(.)^T\varnothing(X). \quad (6B2)$$

Recall that $\varnothing(Y)=[Y_1,Y_2,Y_1Y_2]^T$ and $k(.,Y)=\varnothing(Y)$. The function $k(.,Y)$ can be evaluated at X by

$$\begin{aligned}<k(.,Y),\varnothing(X)>_{\mathcal{H}}&=k(X,Y)\\&=Y_1X_1+Y_2X_2+Y_1Y_2X_1X_2.\end{aligned} \quad (6B3)$$

Feature maps can also be written as

$$\varnothing(X)=k(.,X) \text{ and } \varnothing(Y)=k(.,Y).$$

Example 6B2: Gaussian Kernel

Consider Gaussian Kernel $k(X,Y)=exp\left(-\frac{\|X-Y\|^2}{2l^2}\right)$. We first define "physicists" Hermite polynomial: $H_n(x)=(-1)^n e^{x^2}\frac{d^n}{dx^n}\left(e^{-x^2}\right)$. Gaussian Kernel $k(X,Y)$

can be expressed in terms of Hermite polynomial $H_n(x)$. Hermite polynomials $H_n(x)$ are orthogonal with respect to the weight function $w(x)=e^{-x^2}$:

$$\int_{-\infty}^{\infty} H_m(x)H_n(x)e^{-x^2}dx=\begin{cases}\sqrt{\pi}2^n n! & m=n\\ 0 & m\neq n\end{cases}.$$

For example,

$$\int_{-\infty}^{\infty} H_1(x)H_0(x)e^{-x^2}dx=\int_{-\infty}^{\infty}(-1)e^{x2}\frac{d}{dx}\left(e^{-x^2}\right)e^{-x^2}dx$$

$$=-\int_{-\infty}^{\infty}\frac{d}{dx}\left(e^{-x^2}\right)dx=-e^{-x^2}\Big|_{-\infty}^{\infty}=0-0=0.$$

$$\int_{-\infty}^{\infty} H_1(x)H_1(x)e^{-x^2}dx$$

$$=\int_{-\infty}^{\infty}e^{x2}\frac{d}{dx}\left(e^{-x^2}\right)e^{x^2}\frac{d}{dx}\left(e^{-x^2}\right)e^{-x^2}dx$$

$$=\int_{-\infty}^{\infty}e^{x2}\frac{d}{dx}\left(e^{-x^2}\right)\frac{d}{dx}\left(e^{-x^2}\right)$$

$$=e^{-x^2}e^{x2}\frac{d}{dx}\left(e^{-x^2}\right)\Big|_{-\infty}^{\infty}-\int_{-\infty}^{\infty}e^{-x2}\frac{d}{dx}\left(e^{x^2}\frac{d}{dx}\left(e^{-x^2}\right)dx.\right. \quad \text{(6B4)}$$

Note that

$$e^{-x^2}e^{x^2}\frac{d}{dx}\left(e^{-x^2}\right)\Big|_{-\infty}^{\infty}=-2xe^{-x^2}\Big|_{-\infty}^{\infty}=0, \quad \text{(6B5)}$$

$$\int_{-\infty}^{\infty}e^{-x^2}\frac{d}{dx}\left(e^{x^2}\frac{d}{dx}\left(e^{-x^2}\right)dx\right.$$

$$=\int_{-\infty}^{\infty}e^{-x^2}\left(2xe^{x^2}\right)\frac{d}{dx}\left(e^{-x^2}\right)dx+\int_{-\infty}^{\infty}\frac{d}{dx}\left(\frac{d}{dx}\left(e^{-x^2}\right)\right)dx$$

$$=\int_{-\infty}^{\infty}2x\frac{d}{dx}\left(e^{-x^2}\right)dx+\int_{-\infty}^{\infty}\frac{d}{dx}\left(\frac{d}{dx}\left(e^{-x^2}\right)\right)dx. \quad \text{(6B6)}$$

Using calculus, we obtain

$\frac{d}{dx}\left(e^{-x^2}\right)=-2xe^{-x^2}$, which implies

$$\int_{-\infty}^{\infty}\frac{d}{dx}\left(\frac{d}{dx}\left(e^{-x^2}\right)\right)dx=\frac{d}{dx}\left(e^{-x^2}\right)\Big|_{-\infty}^{\infty}=-2xe^{-x^2}\Big|_{-\infty}^{\infty}=0. \quad \text{(6B7)}$$

The first term in equation (6B6) can be reduced to

$$\int_{-\infty}^{\infty}2x\frac{d}{dx}\left(e^{-x^2}\right)dx=2xe^{-x^2}\Big|_{-\infty}^{\infty}-2\int_{-\infty}^{\infty}e^{-x^2}dx=-2\int_{-\infty}^{\infty}e^{-x^2}dx=2\sqrt{\pi}. \quad \text{(6B8)}$$

Substituting equations (6B5)–(6B8) into equation (6B4) yields

$$\int_{-\infty}^{\infty}H_1(x)H_1(x)e^{-x^2}dx=2\sqrt{\pi}.$$

Let

$$\varnothing_n(x)=e^{-(c-a)x^2}H_n\left(\sqrt{2c}x\right).$$

Assume that orthogonality of functions in $L_2(\mathbb{R})$ is weighted by

$$p(x)=\sqrt{\frac{2a}{\pi}}e^{-ax^2}.$$

Then, $k(X,Y)=exp\left(-\frac{\|X-Y\|^2}{2l^2}\right)$ can be expanded as (Rasmussen and Williams 2006):

$$k(X,Y)=\sum_{n=1}^{\infty}\lambda_n\varnothing_n(X)\varnothing_n(Y),$$

where

$$\lambda_n=\sqrt{\frac{a}{B}}B^n,\ b=\frac{1}{2l^2},\ c=\sqrt{a^2+2ab},$$
$$A=a+b+c,\ B=\frac{b}{A}.$$

Define Infinite dimensional feature map:

$$\varnothing(X)=\left[\ldots,\sqrt{\lambda_n}\varnothing_n(X),\ldots\right]^T.$$

Then, kernel function $k(X,Y)$ is

$$k(X,Y)=exp\left(-\frac{\|X-Y\|^2}{2l^2}\right)=\varnothing(X)^T\varnothing(X)$$
$$=\sum_{n=1}^{\infty}\lambda_n\varnothing_n(X)\varnothing_n(Y).$$

Define $\mathcal{H}$ to be the space of functions: for $\{f_n\}_{n=1}^{\infty}\in l_2$. The function can be expressed as

$$f(X)=< f,\varnothing(X)>_{\mathcal{H}}=\sum_{n=1}^{\infty}f_n\sqrt{\lambda_n}\varnothing_n(X).$$

Definition 6B4: Reproducing Kernel Hilbert Space

Assume that $\mathcal{H}$ is a Hilbert space of $\mathbb{R}$-valued functions on a non-empty set of χ and a function $k(X,Y):\chi\times\chi\to\mathbb{R}$ is a kernel function. If

1. $\forall x\in\chi$, $k(.,X)\in\mathcal{H}$ and
2. $\forall x\in\chi$, $\forall f\in\mathcal{H}$, the reproducing property

$$< f(.), k(.,X)>_{\mathcal{H}}=f(X) \text{ holds,}$$

then $k(X,Y)$ is a reproducing kernel of $\mathcal{H}$ and $\mathcal{H}$ is a reproducing kernel Hilbert space (RKHS).

Next we use RKHS to introduce the mean embedding and covariance operator (Wang et al. 2018). Let X be a random variable with distribution P_x.

Definition 6B5: Mean Embedding

Mean embedding $\mu[P_x]$ is defined as

$$\mu[P_x]=E_x[k(X,.)]. \quad (6B9)$$

Its sampling formula is

$$\mu[X]=\frac{1}{m}\sum_{i=1}^{m}k(X_i,.). \quad (6B10)$$

Consider a nonlinear function of a random variable $f(X)$. Then, the expectation of random function is

$$E_x[f(X)]=E_x[< f, k(X,.)>_{\mathcal{H}}=< f,E_x[k(X,.)]>_{\mathcal{H}}$$
$$=< f,\mu[P_x]>_{\mathcal{H}}. \quad (6B11)$$

Combining equations (6C9)–(6C11), we obtain the sampling formula for $E_x[f(X)]$:

$$S_x[f(X)]=< f,\mu[X]>_{\mathcal{H}}$$
$$=< f,\frac{1}{m}\sum_{i=1}^{m}k(X_i,.)>_{\mathcal{H}}=\frac{1}{m}\sum_{i=1}^{m}< f,k(X_i,.)>_{\mathcal{H}}$$
$$=\frac{1}{m}\sum_{i=1}^{m}f(X_i). \quad (6B12)$$

Before introducing covariance operator, we first study the Hilbert space of Hilbert-Schmidt operators (Gretton 2020). A key concept of the Hilbert-Schmidt operator is the Hilbert-Schmidt norm of the operator which is defined in terms of bases in the space.

Let $\mathcal{F}$ and $\mathbb{Q}$ be separable Hilbert spaces. Define two sets of orthonormal bases $(e_i)_{i\in I}$ and $(u_j)_{j\in J}$ for $\mathcal{F}$ and $\mathbb{Q}$, respectively. Define two compact linear operators $L:\mathbb{Q}\to\mathcal{F}$ and $M:\mathbb{Q}\to\mathcal{F}$. Then, we have

$$Lu_j=\sum_i\beta_ie_i, \text{ which implies}$$

$$\beta_i=\langle Lu_j,e_i\rangle_{\mathcal{F}}.$$

The Hilbert-Schmidt norm of the operator can be defined as follows.

Definition 6B6: Hilbert-Schmidt Norm

The Hilbert-Schmidt norm of the operators L is defined as

$$\|L\|_{HS}^2=\sum_{j\in J}\|Lu_j\|_{HS}^2=\sum_{i\in I}\sum_{j\in J}|< Lu_j,e_i>_{\mathcal{F}}|^2. \quad (6B13)$$

Next we calculate the inner product of two operators L and M. Since both operators map $\mathbb{Q}\to\mathcal{F}$, the inner product of two operators should be defined by the inner product in the Hilbert space as follows:

$$< L, M >_{HS} = \sum_{j \in J} < Lu_j, Mu_j >_{\mathcal{F}}$$
$$= \sum_{i \in I} \sum_{j \in J} < Lu_j, e_i >_{\mathcal{F}} < Mu_j, e_i >_{\mathcal{F}} . \quad (6B14)$$

6B2 Tensor Product

A matrix can be generated by the following product of specific vectors:

$$T = ba^T . \quad (6B15)$$

The product of the matrix T and a vector f is

$$Tf = ba^T f =< a, f >_{\mathcal{F}} b. \quad (6B16)$$

The product in equation (6B15) is a special case of tensor product and is written as

$$T = b \otimes a. \quad (6B17)$$

Equation (6B16) can then be written as

$$Tf = (b \otimes a) f =< f, a >_{\mathcal{F}} b. \quad (6B18)$$

Definition 6B7: Tensor Product

Let $b \in \mathbb{Q}$ and $a \in \mathcal{F}$. $b \otimes a$ is called tensor product if it satisfies

$$(b \otimes a) f =< f, a >_{\mathcal{F}} b.$$

The tensor product $b \otimes a$ is also called a rank-one operator from $\mathbb{Q}$ to $\mathcal{F}$. Now we calculate the Hilbert-Schmidt norm as follows:

$$\|b \otimes a\|_{HS}^2 = \sum_{j \in J} \left\|(b \otimes a) f_j\right\|_{\mathcal{F}}^2$$
$$= \sum_{j \in J} \left\|b < a, f_j >_{\mathcal{F}}\right\|_{\mathcal{F}}^2 \quad (6B19)$$
$$= \|b\|_{\mathbb{Q}}^2 \sum_{j \in J} \left|< a, f_j >_{\mathcal{F}}\right|^2 = \|b\|_{\mathbb{Q}}^2 \|a\|_{\mathcal{F}}^2 .$$

Therefore, the rank-one operator is Hilbert-Schmidt. Now we introduce a lemma useful for calculation of inner product in Hilbert-Schmidt space (HS) (Gretton 2020).

Lemma 6B.1

Let $L \in HS(\mathbb{Q}, \mathcal{F})$ be a Hilbert-Schmidt operator. Then, we have

$$< L, b \otimes a >_{HS} = < b, La >_{\mathcal{F}} . \quad (6B20)$$

Proof

Let u_j be orthonormal basis in $\mathcal{F}$, using equation (6B14), we obtain

$$< L, b \otimes a >_{HS} = \sum_{j \in J} < Lu_j, b \otimes au_j >_{\mathcal{F}} . \quad (6B21)$$

Using equation (6B18), we have

$$b \otimes au_j = b < a, u_j >_{\mathcal{F}} . \quad (6B22)$$

Substituting equation (6B22) into equation (6B21) yields

$$< L, b \otimes a >_{HS} = \sum_{j \in J} < Lu_j, b < a, u_j >_{\mathcal{F}} >_{\mathcal{F}}$$
$$= \sum_{j \in J} < Lu_j, b >_{\mathcal{F}} < a, u_j >_{\mathcal{F}} . \quad (6B23)$$

Since $a \in \mathcal{F}$, expanding a in terms of u_j leads to

$$a = \sum_{j \in J} < a, u_j >_{\mathcal{F}} u_j, \text{ which implies}$$
$$La = \sum_{j \in J} < a, u_j >_{\mathcal{F}} Lu_j \text{ and}$$
$$< b, La >_{\mathcal{F}} = \sum_{j \in J} < a, u_j >_{\mathcal{F}} < Lu_j, b >_{\mathcal{F}} . \quad (6B24)$$

Combining equations (6B23) and (6B24), we prove that

$$< L, b \otimes a >_{HS} = < b, La >_{\mathcal{F}} .$$

Let $L = c \otimes d$. Lemma 6B.1 implies the following corollary.

Corollary 6B.1

$$<c\otimes d,b\otimes a>_{HS}=<c,b>_{\mathcal{F}}<d,a>_{\mathbb{Q}}. \quad (6B25)$$

Proof

It follows from Lemma 6B.1 that

$$<c\otimes d,b\otimes a>_{HS}=<b,(c\otimes d)a>_{\mathcal{F}}. \quad (6B26)$$

However, using equation (6B18), we obtain

$$(c\otimes d)a=<d,a>_{\mathbb{Q}}c. \quad (6B27)$$

Substituting equation (6B27) into equation (6B26), we obtain

$$<c\otimes d,b\otimes a>_{HS}=<b,<d,a>_{\mathbb{Q}}c>_{\mathcal{F}}$$
$$=<c,b>_{\mathcal{F}}<d,a>_{\mathbb{Q}}.$$

6B3 Cross-Covariance Operator

Cross-covariance operator is a generalization of $n\times n$ dimensional covariance matrix to infinite dimensional feature space. Let X and Y be n-dimensional random vectors. The covariance matrix between X and Y is defined as

$$Cov(X,Y)=E_{xy}\left[(X-\mu_X)(Y-\mu_Y)^T\right]$$
$$=E_{xy}\left[XY^T\right]-\mu_X\mu_Y^T \quad (6B28)$$
$$=E_{XY}\left[X\otimes Y\right]-\mu_X\mu_Y^T$$

Define two vectors of nonlinear functions $f(X)$ and $g(Y)$. Then, we can define

$$Cov\left(f(X),g(Y)\right)=E_{XY}\left[f(X)g(Y)\right]$$
$$-E_X\left[f(X)\right]E_Y\left[g(Y)\right]. \quad (6B29)$$

The first term in equation (6B29) can be further reduced to

$$E_{XY}\left[f(X)g(Y)\right]$$
$$=E_{XY}\left[<f(.),k(X,.)>_{\mathcal{F}}<g(.),k(Y,.)>_{\mathbb{Q}}\right]$$
$$=E_{XY}\left[<f(.)\otimes g(.),k(X,.)\otimes k(Y,.)>_{HS}\right]$$

$$=<f(.)\otimes g(.),E_{XY}\left[k(X,.)\otimes k(Y,.)\right]>_{HS}$$
$$=<f(.)\otimes g(.),\bar{C}(X,Y)>_{HS} \quad (6B30)$$
$$=<f,\tilde{C}(X,Y)g>_{\mathcal{F}},$$

where

$$\tilde{C}(X,Y)=E_{XY}\left[k(X,.)\otimes k(Y,.)\right]. \quad (6B31)$$

The second term in equation (6B29) can be reduced to

$$E_X\left[f(X)\right]E_Y\left[g(Y)\right]$$
$$=E_X\left[<f(.),k(X,.)>_{\mathcal{F}}\right]E_Y\left[<g(.),k(Y,.)>_{\mathbb{Q}}\right]$$
$$=<f(.),E_X\left[k(X,.)\right]>_{\mathcal{F}}<g(.),E_Y\left[k(Y,.)\right]>_{\mathbb{Q}}$$
$$=<f(.)\otimes g(.),E_X\left[k(X,.)\right]\otimes E_Y\left[k(Y,.)\right]>_{HS}. \quad (6B32)$$

Combining equations (6B29), (6B30), and (6B32), we obtain

$$Cov\left(f(X),g(Y)\right)$$
$$=<f(.)\otimes g(.),E_{XY}\left[k(X,.)\otimes k(Y,.)\right]$$
$$-E_X\left[k(X,.)\right]\otimes E_Y\left[k(Y,.)\right]>_{HS}$$
$$=<f(.)\otimes g(.),C(X,Y)>_{HS} \quad (6B33)$$
$$=<f,C(X,Y)g>_{\mathcal{F}},$$

where

$$C(X,Y)=E_{XY}\left[k(X,.)\otimes k(Y,.)\right]$$
$$-E_X\left[k(X,.)\right]\otimes E_Y\left[k(Y,.)\right]. \quad (6B34)$$

Definition 6B8: Cross-Covariance Operator

A covariance operator without center is defined as

$$\tilde{C}(X,Y)=E_{XY}\left[k(X,.)\otimes k(Y,.)\right]. \quad (6B35)$$

A covariance operator with center is defined as

$$C(X,Y)=E_{XY}\left[k(X,.)\otimes k(Y,.)\right]-E_X\left[k(X,.)\right]$$
$$\otimes E_Y\left[k(Y,.)\right]. \quad (6B36)$$

Sampling formula for the cross-covariance operator with center is given by

$$C(X,Y)=\frac{1}{n}\sum_{i=1}^{n}k(X_i,.)\otimes k(Y_i,.)$$

$$-\left[\frac{1}{n}\sum_{i=1}^{n}k(X,.)\right]\otimes\left[\frac{1}{n}\sum_{i=1}^{n}k(Y,.)\right]. \quad (6B37)$$

6B4 The Hilbert-Schmidt Independence Criterion

Covariance operator measures dependence between two random variables. The Hilbert-Schmidt norm of the covariance operator can be used as criterion for assessing independence between two random variables. The Hilbert-Schmidt norm of the centered covariance operator is defined as

$$HSIC^2(\mathcal{F},\mathbb{Q},P_{XY})$$

$$=\|C_{XY}\|_{HS}^2=\big\|E_{XY}[k(X,.)\otimes k(Y,.)] - E_X[k(X,.)]\otimes E_Y[k(Y,.)]\big\|_{HS}^2$$

$$=E_{XY}E_{X'F'}[<k(X,.)\otimes l(Y,.),k(X',.)\otimes l(Y',.)>_{HS}]$$

$$-2<E_{XY}[k(X,.)\otimes l(Y,.)],E_{X'}[k(X',.)]\otimes E_{Y'}[l(Y',.)]>_{HS} \quad (6B38)$$

$$+<E_X[k(X,.)]\otimes E_Y[k(Y,.)],E_{X'}[k(X',.)]\otimes E_{Y'}[k(Y',.)]>_{HS}.$$

Equation (6B27) implies that the first term in equation can be reduced to

$$E_{XY}E_{X'F'}[<k(X,.)\otimes l(Y,.),k(X',.)\otimes l(Y',.)>_{HS}]$$

$$=E_{XY}E_{X'F'}[k(X,X')l(Y,Y')]. \quad (6B39)$$

Again, equation (6B27) implies that the second term in equation can be reduced to

$$<E_{XY}[k(X,.)\otimes l(Y,.)],E_{X'}[k(X',.)]$$

$$\otimes E_{Y'}[l(Y',.)]>_{HS}=E_{XY}[E_{X'}[k(X.X')]E_{Y'}[l(Y,Y')]]. \quad (6B40)$$

Using the similar argument, the third term in equation (6B38) can be reduced to

$$<E_X[k(X,.)]\otimes E_Y[k(Y,.)],E_{X'}[k(X',.)]$$

$$\otimes E_{Y'}[k(Y',.)]>_{HS}=E_{XX'}[k(X,X')]E_{YY'}[l(Y,Y')]. \quad (6B41)$$

Combining equations (6B38)–(6B41), we obtain

$$HSIC^2(\mathcal{F},\mathbb{Q},P_{XY})=E_{XY}E_{X'F'}[k(X,X')l(Y,Y')] -2E_{XY}[E_{X'}[k(X.X')] \times E_{Y'}[l(Y,Y')]]+E_XE_{X'} \times[k(X,X')]E_YE_{Y'}[l(Y,Y')]. \quad (6B42)$$

$HSIC^2(\mathcal{F},\mathbb{Q},P_{XY})$ can be approximated by its sampling formula:

$$HSIC^2(\mathcal{F},\mathbb{Q},P_{XY})=\frac{1}{n^2}\sum_{i=1}^{n}\sum_{j=1}^{n}k_{ij}l_{ij}-\frac{2}{n^3}\sum_{i=1}^{n}\sum_{j=1}^{n}\sum_{r=1}^{n}k_{ij}l_{ir}+\frac{1}{n^4}\sum_{i=1}^{n}\sum_{j=1}^{n}\sum_{q=1}^{n}\sum_{r=1}^{n}k_{ij}l_{qr}, \quad (6B43)$$

where $k_{ij}=k(X_i,X_j)$, $l_{ij}=k(Y_i,Y_j)$.

Let $K=(k_{ij})_{n\times n}$ and $L=(l_{ij})_{n\times n}$ be $n\times n$ dimensional matrices. Then, equation (6B43) can be further reduced to

$$HSIC^2(\mathcal{F},\mathbb{Q},P_{XY})=\frac{1}{n^2}\mathrm{tr}\,(KL)-\frac{2}{n^3}\mathbf{1}_n^TKL\mathbf{1}_n+\frac{1}{n^4}\mathbf{1}_n^TK\mathbf{1}_n\mathbf{1}_n^TL\mathbf{1}_n. \quad (6B44)$$

Let

$$A=\frac{1}{n^2}\mathrm{tr}\,(KL)-\frac{2}{n^3}\mathbf{1}_n^TKL\mathbf{1}_n+\frac{1}{n^4}\mathbf{1}_n^TK\mathbf{1}_n\mathbf{1}_n^TL\mathbf{1}_n. \quad (6B45)$$

Note that

$$\frac{2}{n^3}\mathbf{1}_n^TKL\mathbf{1}_n=\frac{1}{n^2}\left[\frac{1}{n}\mathbf{1}_n^TKL\mathbf{1}_n+\frac{1}{n}\mathbf{1}_n^TLK\mathbf{1}_n\right] \text{ and}$$

$$\frac{2}{n^3}\mathbf{1}_n^TKL\mathbf{1}_n=\frac{1}{n^2}\mathrm{tr}\left(\frac{1}{n}\mathbf{1}_n^TKL\mathbf{1}_n+\frac{1}{n}\mathbf{1}_n^TLK\mathbf{1}_n\right) \quad (6B46)$$

$$=\frac{1}{n^2}\mathrm{tr}\left(\frac{\mathbf{1}_n\mathbf{1}_n^T}{n}KL+\frac{\mathbf{1}_n\mathbf{1}_n^T}{n}LK\right).$$

Again, $\frac{1}{n^4}\mathbf{1}_n^T K\mathbf{1}_n\mathbf{1}_n^T L\mathbf{1}_n$ can be reduced to

$$\frac{1}{n^4}\mathbf{1}_n^T K\mathbf{1}_n\mathbf{1}_n^T L\mathbf{1}_n = \frac{1}{n^2}\operatorname{tr}\left(\frac{\mathbf{1}_n\mathbf{1}_n^T}{n^2}K\mathbf{1}_n\mathbf{1}_n^T L\right). \quad (6B47)$$

Substituting equations (6B46) and (6B47) into equation (6B45) yields

$$\begin{aligned} A &= \frac{1}{n^2}\operatorname{tr}\left(KL - \frac{\mathbf{1}_n\mathbf{1}_n^T}{n}KL - \frac{\mathbf{1}_n\mathbf{1}_n^T}{n}LK + \frac{\mathbf{1}_n\mathbf{1}_n^T}{n}K\frac{\mathbf{1}_n\mathbf{1}_n^T}{n}L\right) \\ &= \frac{1}{n^2}\operatorname{tr}\left(HKL - HK\frac{\mathbf{1}_n\mathbf{1}_n^T}{n}L\right) \\ &= \frac{1}{n^2}\operatorname{tr}\left(HK\left(I - \frac{\mathbf{1}_n\mathbf{1}_n^T}{n}\right)L\right) \quad (6B48) \\ &= \frac{1}{n^2}\operatorname{tr}(HKHL) \\ &= \frac{1}{n^2}\operatorname{tr}(KHLH), \end{aligned}$$

where $H = I - \frac{\mathbf{1}_n\mathbf{1}_n^T}{n}$.

Some typical kernel functions are

Gaussian kernel: $k(x,y) = e^{-\frac{\|x-y\|^2}{2\sigma^2}}, \sigma > 0$;

Laplace kernel: $k(x,y) = e^{-\frac{\|x-y\|}{\sigma}, \sigma>0}$;

Inverse multi-quadratics kernel: $k(x,y) = \frac{1}{\beta+\|x-y\|^\alpha}$, $\alpha, \beta > 0$;

Fractional Brownian motion kernel: $k(x,y) = \frac{1}{2}\left(\|x\|^{2h} + \|y\|^{2h} - \|x-y\|^{2h}\right), 0 < h < 1$.

EXERCISES

EXERCISE 6.1

Show that the maximum likelihood estimators of Σ and Σ_0 are, respectively, given by

$$\hat{\Sigma} = \begin{pmatrix} e_x \\ e_y \end{pmatrix}\begin{pmatrix} e_x^T & e_y^T \end{pmatrix} \text{ and}$$

$$\hat{\Sigma}_0 = \begin{pmatrix} e_{x_0} \\ e_{y_0} \end{pmatrix}\begin{pmatrix} e_{x_0}^T & e_{y_0}^T \end{pmatrix}.$$

EXERCISE 6.2

Show that

$$\begin{pmatrix} e_x^T & e_y^T \end{pmatrix}\hat{\Sigma}^{-1}\begin{pmatrix} e_x \\ e_y \end{pmatrix} = n \text{ and } \begin{pmatrix} e_{x_0}^T & e_{y_0}^T \end{pmatrix}\hat{\Sigma}_0^{-1}\begin{pmatrix} e_{x_0} \\ e_{y_0} \end{pmatrix} = n.$$

EXERCISE 6.3

Let $H_n(x) = (-1)^n e^{x^2}\frac{d^n}{dx^n}\left(e^{-x^2}\right)$. Show that

$$\int_{-\infty}^{\infty} H_m(x)H_n(x)e^{-x^2} = \begin{cases} \sqrt{\pi}\,2^n n! & m=n \\ 0 & m \neq n \end{cases}.$$

EXERCISE 6.4

Please show

$$\begin{aligned} &HSIC^2(\mathcal{F}, \mathbb{Q}, P_{XY}) \\ &= \frac{1}{n^2}\operatorname{tr}(KL) - \frac{2}{n^3}\mathbf{1}_n^T KL\mathbf{1}_n + \frac{1}{n^4}\mathbf{1}_n^T K\mathbf{1}_n\mathbf{1}_n^T L\mathbf{1}_n. \end{aligned}$$

CHAPTER 7

Deep Learning for Counterfactual Inference and Treatment Effect Estimation

7.1 INTRODUCTION

Causal inference and causal discovery have two formal frameworks: structural causal models (SCMs) and the potential outcome framework (counterfactual inference) (Guo et al. 2020). In Chapters 3, 4, and 5, we discussed applications of the variational autoencoder (VAE) and generative adversarial network (GAN) to inferring SCMs. In this chapter, we will introduce application of the VAE and GAN to counterfactual inference and treatment effect estimation.

An essential issue in causal inference is that the counterfactual outcomes are unobservable (Hassanpour and Greiner 2021). This makes estimating causal effects a great challenge. There are two basic deep learning methods for counterfactual inference: discriminative methods and generative methods. Discriminative methods directly predict the potential outcome via modeling the conditional distribution of the outcome, given the treatment and feature variables. The recent developed deep learning-based discriminative methods include the learning representation for counterfactual inference (Johansson et al. 2016; Hassanpour and Greiner 2021), counterfactual regression (Colombo et al. 2019), counterfactual survival analysis (Chapfuwa et al. 2020), and counterfactual risk prediction (Pfohl et al. 2019). The generative methods for counterfactual inferences can also divided into (1) VAE for counterfactual analysis and treatment effect estimation, and (2) GAN for counterfactual outcome generation and individualized treatment effect (ITE) estimation. The VAE-based counterfactual inference includes hierarchical conditional VAE for counterfactual distribution estimation (Vercheval and Pizurica 2021), VAE for counterfactual generation (Popescu et al. 2021), VAE for counterfactual generative network (Sauer and Geiger 2021), targeted VAE and targeted learning (Vowels et al. 2020), counterfactual VAE for treatment effect estimation (Wu and Fukumizu 2021), conditional subspace VAE (Downs et al. 2020), and disentangled causal effect VAE (Kim et al. 2020). Recently developed GANs for counterfactual inference and ITE estimation include counterfactual generative networks (Sauer and Geiger 2021), conditional counterfactual generative networks (Looveren et al. 2021), residual GAN for generating counterfactual (Nemirovsky et al. 2020), GAN and conditional GAN for ITE estimation (Yoon et al. 2018; Bica et al. 2020; Ge et al. 2020), generative models for scientific discovery in molecular biology (Lopez et al. 2020), and integrated VAE and GAN for single sell data analysis (Yu and Welch 2021).

7.1.1 Potential Outcome Framework and Counterfactual Causal Inference

In history, structural equation models and potential outcome framework and counterfactuals developed

DOI: 10.1201/9781003028543-7

relatively independently in different fields, but they can be unified using interventional queries with *do*-calculus. This allows methods and algorithms developed within one framework to be easily applied to one another, and also allows predictions about the consequences of intervening upon (rather than merely observing) the variables, and provides a method of evaluating counterfactual claims. The SCMs and counterfactual causal inference can be unified by *do*-calculus (Dablander 2020).

In Chapter 5, we introduced the SCMs and application of VAE and GAN to the SCMs. In this chapter, we will introduce the counterfactual causal inference and applications of the VAE and GAN to the counterfactuals and treatment effect estimation.

The observed outcome is called factual. Contrary to fact is called counterfactual. A counterfactual is a potential outcome. The counterfactual is an unobserved alternative to the observed factual. Counterfactual is an imagined world. We define the variable to be manipulated as a treatment which is denoted by T. The variable that responds to the treatment is defined as the outcome and denoted by Y. In counterfactual terminology of cause and effect, the event A causes the even B can be viewed as "the even B would not have occurred if it were not for the even A", which uses the subjunctive mood. The cause-effect description is based on the imaged and unobserved counterfactual.

The Potential Outcome Framework which is also known as the Neyman-Rubin Potential Outcomes, or the Rubin Causal Model, is widely used causal tool alternative to the SCMs. The treatment can be binary, discrete, and continuous. Similarly, the outcome can also be binary, discrete and continuous. There are total of nine combinations of treatment and outcome. The description of the potential outcome framework will focus on the binary treatment and outcome in this section.

We first introduce definition of potential outcome (Rosenbaum and Rubin 1983; Guo et al. 2020).

Definition 7.1: Neyman-Rubin Potential Outcome

Given the treatment T and outcome Y, the potential outcome Y_i^t of individual i, is the outcome that would have been observed if the individual i had received treatment $T = t$.

Potential outcome is originally defined for binary treat. They assume that selected individuals are divided into two groups: treatment group and non-treatment groups. Each individual in both groups has potential outcomes in two states: one potential outcome is observed and alternative potential outcome is not observed. Let Y_i^1 and Y_i^0 denote the potential outcomes in the treated and untreated states, respectively. The measured outcome Y_i under the Neyman-Rubin potential outcome framework can be expressed as

$$Y_i = T_i Y_i^1 + (1 - T_i) Y_i^0, \tag{7.1}$$

where $T_i = 1$ denotes the receipt of treatment and $T_i = 0$ denotes nonreceipt.

7.1.2 Assumptions and Average Treatment Effect

Equation (7.1) indicates which of two outcomes would be observed in the real data. The observed outcome Y_i involves two potential outcomes Y_i^0 and y_i^1. Unlike SCM which directly tests causation of the treatment variable T with the observed outcome Y_i, counterfactual framework for testing causation must simultaneously check both potential outcomes Y_i^0 and Y_i^1. The counterfactual framework cannot test causation using only observed outcome Y. In other words, the counterfactual framework must compare difference between potential outcomes of a certain individual under two different treatments. Individualized causal effects cannot be directly observed. Below we give definition of treatment effect. The counterfactual framework tests for causation via estimating treatment effect.

Definition 7.2: Individual Treatment Effect and Average Treatment Effect

Consider a binary treatment. Given individual i and potential outcome Y_i^t, the individual treatment effect is defined as $\tau_i = Y_i^1 - Y_i^0$ and the average treatment effect (ATE) is defined as

$$\tau = E[\tau_i] = E\left[Y_i^1 - Y_i^0\right]. \tag{7.2}$$

The fundamental problem of using equation (7.2) to estimate the treatment effect is that we can only observe one of these two outcomes Y_i^1, Y_i^0, since a given person can only be treated or not treated. To solve this problem, we can imagine that there are units that can be assumed to be identical or almost identical. We can use the outcomes of these matched units in the population to replace the counterfactuals.

Instead of viewing the potential outcome as a point value, we assume the existence of the distribution of the potential outcome (Lattimore and Ongv 2018). The distribution of Y_i^1 is the distribution of Y, if everyone was treated. Similarly, the distribution of Y_i^0 is the distribution of Y, if everyone is not treated. Let $P(Y^1)$ be the probability of recovery, across the population, if everyone was treated and $P(Y^0)$ be the probability of recovery, across the population, if everyone was not treated.

Half quantities in the counterfactual framework are missing values. Without assumptions, the counterfactual framework provides no basis for causal analysis.

Assumptions 7.1

The following assumptions are essential for potential outcome framework as basis of causal inference:

1. The ignorable treatment assignment assumption. The assignment of the study individuals to binary treatment is independent of the outcome of the treatment Y^1 and the outcome of the non-treatment Y^0, i.e.,

$$(Y^1, Y^0) \perp\!\!\!\perp T.$$

2. No interference: Outcomes only depend on the treatment applied to the unit, and not affected by treatments applied to other units.
3. Single version of treatment. The treatments for all individuals are exactly the same version (the same brand, dosage, etc.).
   Assumptions 7.2 and 7.3 are together often known as the Stable Unit Treatment Value Assumption, or SUTVA.
   Consider $N(i=1,\ldots,N)$ units (individuals), $T(t=1,\ldots,T)$ treatments, and outcome variable Y which take values Y_i^t for the ith individual assigned treatment t. SUTVA states that the value Y_i^t will be the same regardless what mechanism is used to assign treatment t to unit i and regardless what treatments the other units receive and this holds for all $i=1\ldots,N$ and all $t=1,\ldots,T$.

Counterfactuals can be interpreted by *do*-calculus. The counterfactual questions: "what would have been the distribution of Y had $T=t$?" can read as $p(Y \mid do(x))$ by *do*-calculus. For binary treatment, "what would the distribution of outcomes look like if everyone was treated" can be interpreted as $p(Y \mid do\,(T=1)$ and "what would the distribution of outcomes look like if no one was treated" can be interpreted as $p(Y \mid do(T=0))$ by *do*-calculus.

Since potential outcomes Y_i^1 and Y_i^0 are unobservable, even ATEs cannot be directly estimated. We need to make transformation of equation (7.2). From equation (7.1), the potential outcomes Y_i^1 and Y_i^0 can be estimated by

$$Y_i T_i = T_i^2 Y_i^1 + T_i(1-T_i)Y_i^0. \tag{7.3}$$

Taking conditional expectation on both sides of equation (7.3) over the population, we obtain

$$E\left[Y_i T_i \mid T_i = 1\right] = E\left[T_i^2 Y_i^1 + T_i(1-T_i)Y_i^0 \mid T_i = 1\right], \tag{7.4}$$

which implies that

$$E\left[Y_i \mid T_i = 1\right] = E\left[Y_i^1 \mid T_i = 1\right]. \tag{7.5}$$

By the ignorable treatment assignment assumption, we have

$$E\left[Y_i^1 \mid T_i = 1\right] = E\left[Y_i^1\right]. \tag{7.6}$$

Substituting equation (7.6) into equation (7.5), we obtain

$$E\left[Y_i^1\right] = E\left[Y_i \mid T_i = 1\right]. \tag{7.7}$$

Similarly, we have

$$E\left[Y_i^0\right] = E\left[Y_i \mid T_i = 0\right]. \tag{7.8}$$

Substituting equations (7.7) and (7.8) into equation (7.2), we obtain

$$\tau = E\left[Y_i \mid T_i = 1\right] - E\left[Y_i \mid T_i = 0\right], \text{ or}$$
$$\tau = E\left[Y \mid T = 1\right] - E\left[Y \mid T = 0\right]. \tag{7.9}$$

Definition 7.3: ATE

Under the ignorable treatment assignment assumption, the ATE is defined as

$$ATE = \tau = E[Y \mid T = 1] - E[Y \mid T = 0]. \quad (7.10)$$

In the observational studies, the ignorability assumption, in general, is difficult to be satisfied. Therefore, we make further assumptions to extend the ignorability to conditional ignorability:

$$(Y_i^1, Y_i^0) \perp\!\!\!\perp T \mid X, \quad (7.11)$$

where X is a set of variables.

Conditional ignorability in equation (7.11) assumes that the potential outcomes, Y_i^1 and Y_i^0 are jointly independent of treatment assignment conditional on the groups defined by the value of X.

After introducing a set of variables X, assumptions (7.1) should be changed to assumptions (7.2):

Assumptions 7.2

The following assumptions are essential for potential outcome framework as basis of causal inference:

1. The modified ignorable treatment assignment assumption.
   Given the set of variables X, the assignment of the study individuals to binary treatment is conditionally independent of the outcome of the treatment Y^1 and the outcome of the nontreatment Y^0, i.e.,
   $$(Y^1, Y^0) \perp\!\!\!\perp T \mid X.$$
2. No interference. Outcomes only depend on the treatment applied to the unit, and not affected by treatments applied to other units.
3. Single version of treatment. The treatments for all individuals are exactly the same version (the same brand, dosage, etc.).
4. Overlap. For all x and all $i \in \{1, 2, \ldots, t\}$,
   $$0 < p(T_i = 1 \mid X = x) < 1.$$
   This assumption ensures that given any set of variables x, the conditional probability of being given T_i for every i is positive.

After introducing X, we can define the conditional average treatment effect (CATE) as follows.

Definition 7.4: Under modified ignorable treatment assignment assumption, the CATE is defined as

$$CATE = \tau(x) = E[Y \mid T = 1, X = x] - E[Y \mid T = 0, X = x]. \quad (7.12)$$

We also can define the ATE on the treated or on the untreated.

Definition 7.5: The ATE on the treated is defined as

$$ATT = E[Y^1 - Y^0 \mid T = 1], \quad (7.13)$$

and the ATE on the untreated is defined as

$$ATC = E[Y^1 - Y^0 \mid T = 0]. \quad (7.14)$$

The ATT measures measure the marginal treatment effect in the subpopulation that received the treatment, and the ATC measures measure the marginal treatment effect in the subpopulation that did not receive the treatment.

In practice, ATE, ATT, and ATC can be estimated by their conditional partners:

$$ATE = \sum_x \begin{pmatrix} E[Y \mid T = 1, X = x] \\ -E[Y \mid T = 0, X = x] \end{pmatrix} P(X = x), \quad (7.15)$$

$$ATT = E[Y \mid T = 1] - \sum_x E[Y \mid T = 0, X = x] P(X = x \mid T = 1), \quad (7.16)$$

$$ATC = \sum_x \big(E[Y \mid T = 1, X = x] P(X = x \mid T = 0) - E[Y \mid T = 0]. \quad (7.17)$$

Now we introduce traditional methods for treatment estimation. We mainly used the results from the review (Guo et al. 2020).

7.1.3 Traditional Methods without Unobserved Confounders

In this section, we assume that the observed covariates include all confounders. Adjustment for a subset of covariates X will be used to remove the effects of the observed confounders. There are three major adjustment methods: (1) regression, (2) propensity score, and (3) covariate balancing methods.

7.1.3.1 Regression Adjustment

Consider a dataset $\{Y_i, t_i, x_i\}_{i=1}^n$. Let $Y_i^{t_i}$ be the factual outcome and $Y_i^{1-t_i}$ be the counterfactual outcome. There are two regression models to fit the data: (1) fit a single function to the data and (2) fit a model for each potential outcome.

1. A single function model:

$$Y = \mu(T, X) + \varepsilon, \tag{7.18}$$

where $\mu(T,X) = E[Y \mid T, X]$.

2. Two models with each potential outcome having one:

$$Y = \mu(1, X) + \varepsilon_1, \tag{7.19}$$

where $\mu(1,X) = E[Y \mid T = 1, X]$,

$$Y = \mu(0, X) + \varepsilon_2, \tag{7.20}$$

where $\mu(0,X) = E[Y \mid T = 0, X]$.

The potential outcome Y_i^t is estimated by

$$\hat{Y}_i^t = \mu(t, X_i). \tag{7.21}$$

Therefore, we can use the estimated potential outcomes to estimate the ATE:

$$ATE = \frac{1}{n}\sum_{i=1}^{n}\left(\hat{Y}_i^1 - \hat{Y}_i^0\right). \tag{7.22}$$

7.1.3.2 Propensity Score Methods

We first define the propensity score. The propensity score is defined as the conditional probability $p(t \mid X)$ of treatment assignment given observed baseline features X. The propensity score methods try to reducing the effects of confounding in observational studies by mimicking some of the particular features of a randomized controlled trial (RCT). Conditional on the propensity score, the distribution of observed features will be similar between treated and control groups (Austin 2011). Unlike randomized control trials, in the observational studies, the treatment selection is often affected by individual features, which results in systematic differences in baseline features between the treated and control groups. Therefore, unbiased estimation of the treatment effects must consider the difference in baseline features between treated and untreated individuals. The propensity score methods try to reduce the effects of confounding on the estimation. In observational studies, the true propensity score is, in general, unknown, and hence needs to be estimated from the observed data. Traditionally, the logistic regression where the treatment is regressed on the observed baseline features is often used to estimate the propensity score.

There are four types of propensity score methods: propensity score matching (PSM), propensity score stratification, inverse probability of treatment weighting (IPTW), and adjustment based on propensity score. In this section, we focus on the PSM and IPTW. Other two methods can be similarly derived (Beygelzimer et al. 2019; Guo et al. 2020).

7.1.3.2.1 Propensity Score Matching The most common implementation of PSM is to match a treated individual to a set of untreated individuals with similar propensity scores or simply to make one-to-one matching, where a treated individual matches an untreated individual with similar propensity score (Schwab et al. 2018). After a pair of matched samples has been formed, the treatment effect can be estimated by directly comparing difference in outcomes between treated and untreated individuals in the matched sample. The ATE can be estimated by

$$ATE = \frac{1}{n}\left[\sum_{i:T_i=1}\left(Y_i - Y_j\right) + \sum_{i:T_i=0}\left(Y_j - Y_i\right)\right]. \tag{7.23}$$

7.1.3.2.2 Inverse Probability of Treatment Weighting (IPTW) IPTW synthesizes a "variate balancing pseudo-population" by weighting samples with the propensity score, which resulting in removing confounding. The sample assigned treatment may not be quite representative of the broader population. The goal is to make the sample look more like the population. To achieve this, we can assign a larger weight to the individuals who are underrepresented in the sample and a lower weight to those who are over-represented.

A typical way to defined weight is given by

$$W_i = \left(\frac{T_i}{p(T_i|X_i)}, \frac{1-T_i}{1-p(T_i|X_i)} \right).$$

Thus, the weighted ATE is given by

$$ATE = \frac{1}{n}\sum_{i=1}^{n} \frac{T_i Y_i}{p(T_i \mid X_i)} - \frac{1}{n}\sum_{i=1}^{n} \frac{(1-T_i)Y_i}{1-p(T_i \mid X_i)}. \quad (7.24)$$

Similarly, the weights for the treated population and the untreated population are, respectively, given by

$$\left(1, \frac{p(T_i|X_i)}{1-p(T_i|X_i)}\right) \text{ and } \left(\frac{1-p(T_i|X_i)}{p(T_i|X_i)}, 1-T_i\right). \quad (7.25)$$

The weighted ATT and ATC can be estimated, respectively, by

$$ATT = \frac{1}{n}\sum_{i=1}^{n} T_i Y_i - \frac{1}{n}\sum_{i=1}^{n} \frac{p(T_i|X_i)}{1-p(T_i|X_i)}(1-T_i)Y_i, \quad (7.26)$$

and

$$ATC = \frac{1}{n}\sum_{i=1}^{n} \frac{1-p(T_i|X_i)}{p(T_i|X_i)} T_i Y_i - \frac{1}{n}\sum_{i=1}^{n}(1-T_i)Y_i. \quad (7.27)$$

7.1.3.3 Doubly Robust Estimation (DRE) and G-Methods

We first introduce back-door criterion which is useful in DRE before studying General (G)-methods and DRE.

Definition 7.6: Back-Door Path

Given a pair of treatment and outcome variables (T,Y), a path connecting T and Y is a back-door path for (T,Y), if and only if the path is not a directed path and the path is not blocked (it has no collider).

Definition 7.7: Back-Door Criterion

Given a treatment-outcome pair (T,Y), a set of features X satisfies the back-door criterion of (T,Y), if and only if conditioning on X, all back-door paths of (T,Y) are blocked (Figure 7.1).

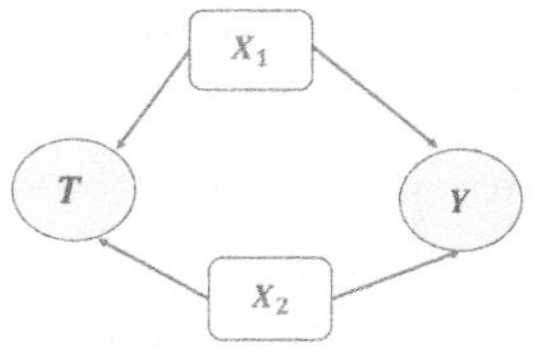

FIGURE 7.1 Back-door and conditional DAG.

G-method, which was first introduced by Robins (1986), is a very useful tool for recovering unbiased estimates of treatment effect. G-methods use "reweighting" and/or "standardizing" a population to synthesize a new "pseudo-population" in which, conditioning on a set of features, the treatment and outcome are d-separated. G-methods include Standardization via the parametric/non-parametric g-formula and IPW (Rothman 2020). In this section, we will focus on the doubly robust estimator of the treatment effect, which combines Standardization and IPW.

7.1.3.3.1 Outcome Model and Standardization Consider a DAG shown in Figure 7.1, where Y represents a continuous potential outcome, T represents a binary treat, and X_1 and X_2 are binary covariates. Conditional on covariates X_1 and X_2, all back-door paths from the treatment T to outcome Y are blocked. Therefore, conditional on X_1 and X_2, the outcome Y and the treatment T are independent. In other words, we have

$$Y \perp\!\!\!\perp T \mid X_1, X_2. \quad (7.28)$$

The outcome model is given by

$$E[Y|T, X_1, X_2] = \beta_0 + \beta_1 T + \beta_2 X_1 + \beta_3 X_2, \quad (7.29)$$

$$\hat{Y} \mid T, X_1, X_2 = \hat{\beta}_0 + \hat{\beta}_1 T + \hat{\beta}_2 X_1 + \hat{\beta}_3 X_2. \quad (7.30)$$

We can show that the parameter estimate $\hat{\beta}_1$ is unbiased estimate of the average effect difference of the treatment T on the outcome Y. In fact, it follows from equation (7.30) that

$$\begin{aligned}\hat{\beta}_1 &= \left(\hat{Y} \mid T=1, X_1, X_2\right) - \left(\hat{Y} \mid T=0, X_1, X_2\right) \\ &= E[Y|T=1, X_1, X_2] - E[Y|T=0, X_1, X_2] \\ &= E[Y^1|X_1, X_2] - E[Y^0 \mid X_1, X_2]. \end{aligned} \quad (7.31)$$

Using the outcome model, we can recover the marginal counterfactual estimates of the mean outcome under $T = 1$ and $T = 0$:

$$E[Y^T] = \sum_{x_1}\sum_{x_2} E[Y^T|X_1 = x_1, X_2 = x_2] \, P(X_1 = x_1, X_2 = x_2). \tag{7.32}$$

Equation (7.32) is often called Standardization with the parametric g-formula (Rothman 2020).

7.1.3.3.2 Doubly Robust Estimation (DRE) Intervention model is introduced in Section 7.1.3.2. In summary, with Standardization, we assume we correctly specify the true Outcome Model, and with IPW, we assume we correctly specify the true Intervention Model (Rothman 2020):

Outcome Model: $E[Y \mid T, X_1, X_2]$,

Intervention Model: $P(T = 1 \mid X_1, X_2]$.

The correct models include both the correct variables and the correct function forms. Incorrect models may cause the estimations of the treatment effects biased.

In practice, only Standardization or IPW is correct, but not both. However, we do not know which model is correct. In this section, we introduce Double Robust Estimator that tries to recover estimates of the mean counterfactual quantities of interest using both Standardization and IPW approaches, and then combine the results of the two analyses together (Rothman 2020). The remarkable feature of the double robust estimator is that as long as at least one model is correct (the Outcome Model or Intervention Model) then the combined sampling estimator generates unbiased estimates of the mean counterfactual quantities of interest.

The DRE for the outcomes Y^1, Y^0, and ATE are, respectively, given by (Rothman 2020):

$$\widehat{DRE}_{T=1} = \frac{1}{n}\sum_{i=1}^{n}\left(\frac{T_i\left(Y_i - \hat{Y}_i^1\right)}{p\left[T_i = 1 \mid X_i\right]} + \hat{Y}_i^1\right), \tag{7.33}$$

$$\widehat{DRE}_{T=0} = \frac{1}{n}\sum_{i=1}^{n}\left(\frac{(1-T_i)\left(Y_i - \hat{Y}_i^0\right)}{1 - p\left(T_i = 1 \mid X_i\right)} + \hat{Y}_i^0\right), \tag{7.34}$$

$$\widehat{DRE}_{T=1} - \widehat{DRE}_{T=0} = \frac{1}{n}\sum_{i=1}^{n}\left(\frac{T_i\left(Y_i - \hat{Y}_i^1\right)}{p\left(T_i = 1 | X_i\right)} + \hat{Y}_i^1\right) - \frac{1}{n}\sum_{i=1}^{n}\left(\frac{(1-T_i)\left(Y_i - \hat{Y}_i^0\right)}{1 - p(T_i = 1 \mid X_i)} + \hat{Y}_i^0\right). \tag{7.35}$$

Now we show that

$$E\left[\widehat{DRE}_{T=1}\right] = E[Y^1], \tag{7.36}$$

$$E\left[\widehat{DRE}_{T=0}\right] = E[Y^0], \tag{7.37}$$

$$E\left[\widehat{DRE}_{T=1} - \widehat{DRE}_{T=0}\right] = E[Y^1] - E[Y^0]. \tag{7.38}$$

We first consider that he Outcome Model is correctly specified, but the Intervention Model is misspecified. Under this assumption, large number of theory ensures that

$$\frac{1}{n}\sum_{i=1}^{n}\hat{Y}_i^1 = E[Y^1]. \tag{7.39}$$

Next we show that

$$E\left[\frac{T_i\left(Y_i - \hat{Y}_i^1\right)}{p\left(T_i = 1 \mid X_i\right.}\right] = 0. \tag{7.40}$$

In fact,

$$E\left[\frac{T_i\left(Y_i - \hat{Y}_i^1\right)}{p\left(T_i = 1 \mid X_i\right.}\right] = \frac{1}{p\left(T_i = 1 \mid X_i\right)} E\left[T_i\left(Y_i - \hat{Y}_i^1\right)\right]. \tag{7.41}$$

But,

$$\begin{aligned} E\left[T_i\left(Y_i - \hat{Y}_i^1\right)\right] &= E\left[I(T_i = 1)\left(Y_i - \hat{Y}_i^1\right)\right] \\ &= p(T_i = 1)E\left[\left(Y_i - \hat{Y}_i^1\right) \mid T_i = 1\right] \\ &= p(T_i = 1)\left(E\left[Y_i \mid T_i = 1\right] - E\left[\hat{Y}_i^1 \mid T_i = 1\right]\right) \\ &= p(T_i = 1)\left(\hat{Y}_i^1 - \hat{Y}_i^1\right) = 0. \end{aligned}$$

This shows that equations (7.40) and (7.41) hold.

Combining equations (7.39) and (7.40), we obtain

$$\widehat{E[DRE}_{T=1}] = E\left[\frac{1}{n}\sum_{i=1}^{n}\left(\frac{T_i\left(Y_i - \hat{Y}_i^1\right)}{p\left(T_i = 1 \mid X_i\right)} + \hat{Y}_i^1\right)\right]$$

$$= \frac{1}{n}\sum_{i=1}^{n} E\left[\left(\frac{T_i\left(Y_i - \hat{Y}_i^1\right)}{p\left(T_i = 1 \mid X_i\right)} + \hat{Y}_i^1\right)\right]$$

$$= \frac{1}{n}\sum_{i=1}^{n} E\left[\frac{T_i\left(Y_i - \hat{Y}_i^1\right)}{p\left(T_i = 1 \mid X_i\right)}\right] + \frac{1}{n}\sum_{i=1}^{n} E\left[\hat{Y}_i^1\right]$$

$$= \frac{\sum_{i=1}^{n} E\left[\hat{Y}_i^1\right]}{n} = E\left[E\left[\hat{Y}_i^1\right]\right] = E\left[Y^1\right]. \tag{7.42}$$

Similarly, we can show

$$E\left[\widehat{DRE}_{T=0}\right] = E\left[Y^0\right], \tag{7.43}$$

$$E\left[\widehat{DRE}_{T=1} - \widehat{DRE}_{T=0}\right] = E\left[Y^1\right] - E\left[Y^0\right]. \tag{7.44}$$

Now we consider the second case: incorrect outcome model and correct intervention model. In Exercise 7.2, we show that under correct intervention model, we have

$$\frac{1}{n}\sum_{i=1}^{n}\frac{E\left[Y_i \mid T_i = 1\right]}{p\left(T_i = 1 \mid X_i\right)} = E\left[Y^1\right], \tag{7.45}$$

$$\frac{1}{n}\sum_{i=1}^{n}\frac{E\left[Y_i \mid T_i = 0\right]}{p\left(T_i = 0 \mid X_i\right)} = E\left[Y^0\right]. \tag{7.46}$$

Under correct intervention model and incorrect outcome model, we define

$$\widehat{DRE}_{T=1} = \frac{1}{n}\sum_{i=1}^{n}\left(\frac{T_i Y_i}{p\left(T_i \mid X_i\right)} - \frac{\hat{Y}_i^1\left(T_i - p\left(T_i \mid X_i\right)\right)}{p\left(T_i \mid X_i\right)}\right), \tag{7.47}$$

$$\widehat{DRE}_{T=0}$$

$$= \frac{1}{n}\sum_{i=1}^{n}\left(\frac{\left(1 - T_i\right)Y_i}{1 - p\left(T_i \mid X_i\right)} - \frac{\hat{Y}_i^0\left(1 - T_i - \left(1 - p\left(T_i \mid X_i\right)\right)\right)}{1 - p\left(T_i \mid X_i\right)}\right), \tag{7.48}$$

$$\widehat{DRE}_{T=1} - \widehat{DRE}_{T=0}$$

$$= \frac{1}{n}\sum_{i=1}^{n}\left(\left(\frac{T_i Y_i}{p\left(T_i \mid X_i\right)} - \frac{\hat{Y}_i^1\left(T_i - p\left(T_i \mid X_i\right)\right)}{p\left(T_i \mid X_i\right)}\right) - \left(\frac{\left(1 - T_i\right)Y_i}{1 - p\left(T_i \mid X_i\right)} - \frac{\hat{Y}_i^0\left(1 - T_i - \left(1 - p\left(T_i \mid X_i\right)\right)\right)}{1 - p\left(T_i \mid X_i\right)}\right)\right). \tag{7.49}$$

Now we show that under the correct intervention model, the estimator $\widehat{DRE}_{T=1}$ is unbiased estimator:

$$E\left[\widehat{DRE}_{T=1}\right] = E\left[Y^1\right]. \tag{7.50}$$

In fact, it follows from equation (7.47) that

$$E\left[\widehat{DRE}_{T=1}\right] = \frac{1}{n}E\left[\sum_{i=1}^{n}\left(\frac{T_i Y_i}{p\left(T_i \mid X_i\right)} - \frac{\hat{Y}_i^1\left(T_i - p\left(T_i \mid X_i\right)\right)}{p\left(T_i \mid X_i\right)}\right)\right]$$
$$= \frac{1}{n}\sum_{i=1}^{n}\left(E\left[\frac{T_i Y_i}{p\left(T_i \mid X_i\right)}\right] - E\left[\frac{\hat{Y}_i^1\left(T_i - p\left(T_i \mid X_i\right)\right)}{p\left(T_i \mid X_i\right)}\right]\right). \tag{7.51}$$

But,

$$E\left[\frac{\hat{Y}_i^1\left(T_i - p\left(T_i \mid X_i\right)\right)}{p\left(T_i \mid X_i\right)}\right]$$

$$= E\left[E\left[\frac{\hat{Y}_i^1\left(T_i - p\left(T_i \mid X_i\right)\right)}{p\left(T_i \mid X_i\right)} \mid X_i\right]\right]$$

$$= E\left[\frac{\hat{Y}_i^1}{p\left(T_i \mid X_i\right)} E\left[\left(T_i - p\left(T_i \mid X_i\right)\right) \mid X_i\right]\right]$$

$$= E\left[\frac{\hat{Y}_i^1}{p\left(T_i \mid X_i\right)}\left(E\left[T_i \mid X_i\right] - p\left(T_i \mid X_i\right)\right)\right] = 0. \tag{7.52}$$

Substituting equation (7.52) into equation (7.51), we obtain

$$E\left[\widehat{DRE}_{T=1}\right] = \frac{1}{n}\sum_{i=1}^{n} E\left[\frac{T_i Y_i}{p\left(T_i \mid X_i\right)}\right]. \tag{7.53}$$

Combining equations (7.45) and (7.53), we obtain

$$E\left[\widehat{DRE}_{T=1}\right] = E\left[Y^1\right]. \tag{7.54}$$

Similarly, we can show

$$E\left[\widehat{DRE}_{T=0}\right] = E\left[Y^0\right], \tag{7.55}$$

$$E\left[\widehat{DRE}_{T=1}-\widehat{DRE}_{T=0}\right]=E\left[Y^1\right]-E\left[Y^0\right]. \quad (7.56)$$

All the above results can be summarized in Lemma 7.1.

Lemma 7.1

Assume that either of the outcome model or the intervention model or both the outcome and intervention models are correct. Then, the DRE defined as

$$\widehat{DRE}_{T=1}=\frac{1}{n}\sum_{i=1}^{n}\left(\frac{T_iY_i}{p(T_i\mid X_i)}-\frac{\tilde{Y}_i^1\left(T_i-p(T_i\mid X_i)\right)}{p(T_i\mid X_i)}\right), \quad (7.57)$$

$$\widehat{DRE}_{T=0}=\frac{1}{n}\sum_{i=1}^{n}\left(\frac{(1-T_i)Y_i}{1-p(T_i\mid X_i)}-\frac{\tilde{Y}_i^0\left(1-T_i-\left(1-p(T_i|X_i)\right)\right)}{1-p(T_i\mid X_i)}\right), \quad (7.58)$$

$$\widehat{DRE}_{T=1}-\widehat{DRE}_{T=0}=\frac{1}{n}\sum_{i=1}^{n}\left(\left(\frac{T_iY_i}{p(T_i\mid X_i)}-\frac{\tilde{Y}_i^1\left(T_i-p(T_i\mid X_i)\right)}{p(T_i\mid X_i)}\right)\right. \quad (7.59)$$
$$\left.-\left(\frac{(1-T_i)Y_i}{1-p(T_i\mid X_i)}-\frac{\tilde{Y}_i^0(1-T_i-\left(1-p(T_i|X_i)\right))}{1-p(T_i\mid X_i)}\right)\right).$$

where $\tilde{Y}_i^{T_i}$ is the estimated potential outcomes of the individuals with regression adjustment $E[Y\mid T,X]$ is unbiased estimators of Y^1, Y^0, and Y^1-Y^0.

7.1.3.4 Targeted Maximum Likelihood Estimator (TMLE)

Targeted maximum likelihood estimation is a semiparametric double-robust method and is asymptotically efficient. The TMLE consists of five steps (Luque-Fernandez et al. 2018).

Step 1: Initialization. Fit the outcome model to initialize the estimator of the outcome:

$$\bar{Q}^0(T,X)=E[Y\mid T,X].$$

Let $\text{logit}(w)=\log\frac{w}{1-w}$. The outcome model can be fitted using a standard logistic regression:

$$\text{logit}\left(E[Y|T,X]\right)=\text{logit}\left(p(Y=1|T,X)\right)=\beta_0+\beta_1T+\beta_2^TX. \quad (7.60)$$

Let expit (w) denote the inverse logit function:

$$\text{expit}(w)=\frac{e^w}{1+e^w}.$$

The initial probability for the outcome $E[Y\mid T,X]$ can be estimated by

$$\bar{Q}^0(0,X)=\text{expit}\left(\hat{\beta}_0+\hat{\beta}_2^TX\right), \quad (7.61)$$

$$\bar{Q}^0(1,X)=\text{expit}\left(\hat{\beta}_0+\hat{\beta}_1+\hat{\beta}_2^TX\right). \quad (7.62)$$

Step 2: Prediction of the propensity score $\hat{p}(T,X)$. A logistic regression can be used to model the propensity score:

$$\text{logit}\left(p(T=1|X)\right)=\alpha_0+\alpha_1^TX. \quad (7.63)$$

After the logistic regression model is fitted, the propensity score is then estimated using

$$\hat{p}(T=1|X)=expit\left(\hat{\alpha}_0+\hat{\alpha}_1^TX\right). \quad (7.64)$$

Step 3: The new covariate and estimation of residual ε.
Define a new covariate:

$$H(1,X_i)=\frac{T_i}{\hat{p}(T_i=1\mid X_i)} \text{ and } H(0,X_i)=\frac{1-T_i}{\hat{p}(T_i=0\mid X_i)}, \quad (7.65)$$

where the new covariate is a very similar to inverse probability of treatment weights.

We assume that the outcome $E[Y\mid T,X](\varepsilon)$ is a function of two fluctuation parameters $\varepsilon=(\varepsilon_0,\varepsilon_1)$.

Consider a model with the observed outcome (Y) as dependent variable and the logit of the initial prediction of $\bar{Q}^0(T,X)$, two new covariates $H(1,X)$ and $H(0,X)$ as independent predictors:

$$E[Y=1|T,X](\varepsilon)=\frac{1}{1+\exp\left(-\log\frac{\bar{Q}^0(T,X)}{1-\bar{Q}^0(T,X)}-\varepsilon_0H(0,X)-\varepsilon_1H(1,X)\right)}. \quad (7.66)$$

Step 4: Update $\bar{Q}^0(T,X)$ to $\bar{Q}^1(T,X)$

Update the estimate of outcome $E[Y\mid T,X]$ from $\bar{Q}^0(T,X)$ to $\bar{Q}^1(T,X)$:

$$\bar{Q}^1(0,X)=\text{expit}\left[\text{logit}\left(\bar{Q}^0(0,X)\right)++\frac{\hat{\varepsilon}_0}{\hat{p}(T=0\mid X)}\right], \tag{7.67}$$

$$\bar{Q}^1(1,X)=\text{expit}\left[\text{logit}\left(\bar{Q}^0(1,X)\right)++\frac{\hat{\varepsilon}_1}{\hat{p}(T=1\mid X)}\right]. \tag{7.68}$$

Step 5: Targeted estimate of the ATE

The targeted estimate of the ATE is given by

$$\widehat{ATE}_{TMLE}=\frac{1}{n}\sum_{i=1}^{n}\left(\bar{Q}^1(1,X)-\bar{Q}^1(0,X)\right). \tag{7.69}$$

7.2 COMBINE DEEP LEARNING WITH CLASSICAL TREATMENT EFFECT ESTIMATION METHODS

There are increasing interests in application of deep learning to treatment effect estimation (Johansson et al. 2016; Shalit et al. 2016; Alaa and van der Schaar 2017; Louizos et al. 2017; Farrell et al. 2018; Künzel et al. 2019; Schwab et al. 2019; Shi et al. 2019; Nie et al. 2021). Widely used deep learning methods for treatment effect estimation are two stage methods. First stage is to fit the outcome model and propensity score using deep neural networks. Second stage is to plug these fitted models into a downstream classical treatment effect estimator.

Although neural networks are a powerful tool to predict the outcome Y from the treatment and covariates, and another to predict the treatment from the covariates, neural networks focus on performance of their predictions and are less concerned with the quality of the downstream treatment effect estimation (Shi et al. 2019). In this section, we will introduce recent works on modifying design and training of neural networks to increase the accuracy of treatment effect estimation.

7.2.1 Adaptive Learning for Treatment Effect Estimation

7.2.1.1 Problem Formulation

We consider a binary treatment T, a set of covariates X and outcome Y. Assume that the observed covariates X include all common causes of the treatment and outcome, i.e., there are no unmeasured confounders. The ATE is defined as

$$\tau=E\left[E[Y|T=1,X]-E[Y|T=0,X]\right]. \tag{7.70}$$

Let $\hat{Q}$ be an estimate of the conditional outcome $Q(T,X)=E[Y\mid T,X]$. Then, the sampling approximation of the ATE is given by

$$\hat{\tau}=\frac{1}{n}\sum_{i=1}^{n}\left(\hat{Q}(1,X_i)-\hat{Q}(0,X_i)\right). \tag{7.71}$$

7.2.2 Architecture of Neural Networks

Let $g(X)=(T\mid X)$ be the propensity score. It showed that the propensity score $g(X)$ is a sufficient statistic for the treatment effect variable τ. We cite the Theorem 2.1 in Shi et al. (2019) as Theorem 7.1 here.

Theorem 7.1: Sufficiency of Propensity Score

If the ATE is identifiable from observational data by adjusting for X, i.e., $\tau=E\left[E[Y|T=1,X]-E[Y|T=0,X]\right]$, *then adjusting for the propensity score also suffices:*

$$\tau=E\left[E[Y|T=1,g(X)]-E[Y|T=0,g(X)]\right]. \tag{7.72}$$

Theorem 7.1 tells us that the propensity score is a sufficient statistic for predicting the outcome. It suffices to adjust for only information in X contained in the treatment. The part of X that is only useful for predicting the outcome but not the treatment can be viewed as noise and should be discarded in estimation of the treatment effect (Shi et al. 2019).

Figure 7.2 shows the architecture of deep learning for predicting propensity score and conditional outcome from covariates and treatment information (Shi et al. 2019). The deep learning consists of two blocks.

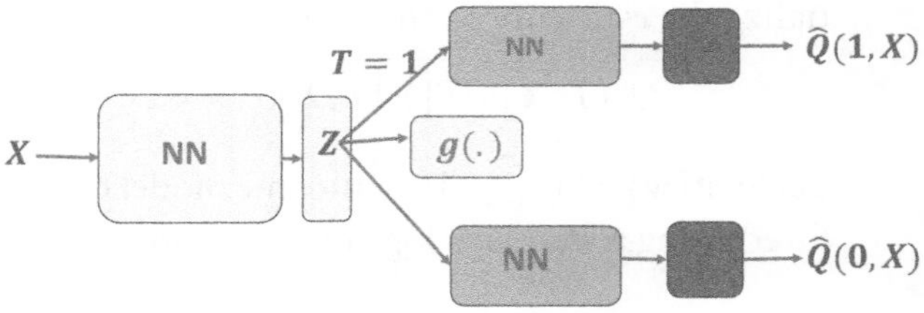

FIGURE 7.2 Architecture of deep learning for ATE estimation.

The first block is a neural network. It maps the input X to a representation space (layer) $Z \in R^p$. The second block also consists of neural networks. It predicts both the treatment (propensity score) $g(X)$and potential outcomes $Q(T,X)$ from this shared representation Z. Two separated neural networks predict the outcomes $\hat{Q}(0,X): R^p \to R$ and $\hat{Q}(1,X): R^p \to R$. One neural network with a simple linear map and a sigmoid activation function predicts the propensity score $\hat{g}(X)$.

Assume that n individuals are sampled. The loss function for training deep learning networks is defined as

$$\mathcal{L}(X,T,\lambda,\theta) = \frac{1}{n}\sum_{i=1}^{n}\Big[\big(Q(T_i, X_i;\theta) - Y_i\big)^2 + \lambda\big(T_i \log g(X_i;\theta) + (1-T_i)\log\big(1-g(X_i;\theta)\big)\big)\Big], \tag{7.73}$$

where λ is a penalty parameter.

We use backpropagation algorithm to solve the following optimization problem:

$$\hat{\theta} = \underset{\theta}{\operatorname{argmin}}\, \mathcal{L}(X,T,\lambda,\theta). \tag{7.74}$$

After the optimization problem (7.74) is solved, we can estimate the counterfactual outcome:

$$\hat{Q}(T_i, X_i) = Q\big(T_i, X_i; \hat{\theta}\big). \tag{7.75}$$

As a result, we can estimate the ATE:

$$\hat{\tau} = \frac{1}{n}\sum_{i=1}^{n}\big[\hat{Q}(1,X_i) - \hat{Q}(0,X_i)\big]. \tag{7.76}$$

7.2.3 Targeted Regularization

To guarantee desirable asymptotic properties of the ATE estimator, we need to modify the objective function $\mathcal{L}(X,T,\lambda,\theta)$. This modification depends on nonparametric estimation theory (Shi et al. 2019).

Define the efficient influence curve of τ:

$$\varphi(Y,T,X;Q,g,\tau) = Q(1,X) - Q(0,X) + \left(\frac{T}{g(X)} - \frac{1-T}{1-g(X)}\right) \times \big(Y - Q(T,X)\big) - \tau. \tag{7.77}$$

If the estimators of the conditional outcome $\hat{Q}(T,X)$ and propensity score $\hat{g}(X)$ are consistent, and the tuple $(\hat{Q}(T,X), \hat{g}(X), \hat{\tau})$ satisfies the following nonparametric estimating equation:

$$\frac{1}{n}\sum_{i=1}^{n}\varphi\big(Y,T,X;\hat{Q},\hat{g},\hat{\tau}\big) = 0, \tag{7.78}$$

then the estimator $\hat{\tau}$ of the ATE will have asymptotical nice properties such as robustness and efficiency (Chernozhukov et al. 2017; Shi et al. 2019).

To develop targeted regularization, we require that $Q(T,X)$ and $g(X)$ should be modeled by a neural network with output heads $Q(T_i,X_i;\theta)$ and $g(X_i;\theta)$. We assume that the neural network is trained by minimizing a differentiable objective function $\mathcal{L}(X,T,\lambda,\theta)$. Define a regularization term $\gamma(Y,T,X;\theta,\varepsilon)$ for the objective function $\mathcal{L}(X,T,\lambda,\theta)$:

$$\tilde{Q}(T_i, X_i;\theta,\varepsilon) = Q(T_i,X_i;\theta) + \varepsilon\left[\frac{T_i}{g(X_i;\theta)} - \frac{1-T_i}{1-g(X_i;\theta)}\right], \tag{7.79}$$

$$\gamma(Y_i, T_i, X_i;\theta,\varepsilon) = \big(Y_i - \tilde{Q}(T_i, X_i;\theta,\varepsilon)\big)^2. \tag{7.80}$$

Define the regularized objective function $\tilde{\mathcal{L}}(X,T;\lambda,\beta,\theta,\varepsilon)$:

$$\tilde{\mathcal{L}}(X,T;\lambda,\beta,\theta,\varepsilon) = \hat{\mathcal{L}}(X,T,\lambda,\theta) + \beta\frac{1}{n}\sum_{i=1}^{n}\gamma(Y_i, T_i, X_i;\theta,\varepsilon), \tag{7.81}$$

where β is a penalty parameter.

Minimizing the regularized objective function, we obtain the estimators of the weights θ in the neural networks and ε:

$$\big[\hat{\theta},\hat{\varepsilon}\big] = \underset{\theta,\varepsilon}{\operatorname{argmin}}\, \tilde{\mathcal{L}}(X,T;\lambda,\beta,\theta,\varepsilon). \tag{7.82}$$

Let

$$\hat{Q}^{treg}(T, X_i) = \tilde{Q}\big(T, X_i;\hat{\theta},\hat{\varepsilon}\big).$$

We then define a target regularization-based ATE estimator $\hat{\tau}^{treg}$ as

$$\hat{\tau}^{treg} = \frac{1}{n}\sum_{i=1}^{n}\left(\hat{Q}^{treg}(1, X_i) - \hat{Q}^{treg}(0, X_i)\right). \quad (7.83)$$

The optimal solutions $\hat{\theta}, \hat{\varepsilon}$ of the target regularized objective function must satisfy

$$\left.\frac{\partial \tilde{\mathcal{L}}(X,T;\lambda,\beta,\theta,\varepsilon)}{\partial \varepsilon}\right|_{\varepsilon} = 0. \quad (7.84)$$

However,

$$\frac{\partial \tilde{\mathcal{L}}(X,T;\lambda,\beta,\theta,\varepsilon)}{\partial \varepsilon} = -2\beta\frac{1}{n}\sum_{i=1}^{n}\left(Y_i - \tilde{Q}(T_i, X_i;\theta,\varepsilon)\right)\frac{\partial \tilde{Q}(T_i, X_i;\theta,\varepsilon)}{\partial \varepsilon}$$
$$= -2\beta\frac{1}{n}\sum_{i=1}^{n}\left(Y_i - \tilde{Q}(T_i, X_i;\theta,\varepsilon)\right)\begin{bmatrix}\frac{T_i}{g(X_i;\theta)} \\ -\frac{1-T_i}{1-g(X_i;\theta)}\end{bmatrix}. \quad (7.85)$$

Combining equations (7.79), (7.84), and (7.85), we obtain

$$\left.\frac{\partial \tilde{\mathcal{L}}(X,T;\lambda,\beta,\theta,\varepsilon)}{\partial \varepsilon}\right|_{\hat{\varepsilon}} = -2\beta\frac{1}{n}\sum_{i=1}^{n}\left(Y_i - Q(T_i, X_i;\theta)\right)\left[\frac{T_i}{g(X_i;\theta)} - \frac{1-T_i}{1-g(X_i;\theta)}\right]. \quad (7.86)$$

Combining equations (7.77) and (7.83), we obtain

$$\frac{1}{n}\sum_{i=1}^{n}\varphi\begin{pmatrix}Y_i, T_i, X_i, \\ \hat{Q}^{treg}, \hat{g}, \hat{\tau}^{treg}\end{pmatrix} = \frac{1}{n}\sum_{i=1}^{n}\left[\hat{Q}^{treg}(1,x) - \hat{Q}^{treg}(1,x) - \hat{\tau}^{treg} + \left(Y_i - \hat{Q}^{treg}(T_i, X_i, \hat{\theta})\right)\right]\left[\frac{T_i}{g(X_i;\theta)} - \frac{1-T_i}{1-g(X_i;\theta)}\right]. \quad (7.87)$$

Substituting equation (7.83) into equation (7.87), we obtain

$$\frac{1}{n}\sum_{i=1}^{n}\varphi\left(Y_i, T_i, X_i, \hat{Q}^{treg}, \hat{g}, \hat{\tau}^{treg}\right) = \frac{1}{n}\sum_{i=1}^{n}\left(Y_i - \hat{Q}^{treg}(T_i, X_i, \hat{\theta})\right)\left[\frac{T_i}{g(X_i;\theta)} - \frac{1-T_i}{1-g(X_i;\theta)}\right]. \quad (7.88)$$

Combining equations (7.86) and (7.88), we obtain

$$\left.\frac{\partial \tilde{\mathcal{L}}(X,T;\lambda,\beta,\theta,\varepsilon)}{\partial \varepsilon}\right|_{\hat{\varepsilon}} = -2\beta\frac{1}{n}\sum_{i=1}^{n}\varphi\left(Y_i, T_i, X_i, \hat{Q}^{treg}, \hat{g}, \hat{\tau}^{treg}\right). \quad (7.89)$$

Therefore, $\left.\frac{\partial \tilde{\mathcal{L}}(X,T;\lambda,\beta,\theta,\varepsilon)}{\partial \varepsilon}\right|_{\hat{\varepsilon}} = 0$ implies the non-parametric estimating equation (7.78). This shows that the presented target regularized deep learning for ATE estimation is robust and asymptotically efficient.

7.3 COUNTERFACTUAL VARIATIONAL AUTOENCODER

7.3.1 Introduction

In this section, we introduce extension of VAE to estimation of ITE in the presence of unobserved confounding (Wu and Fukumizu 2021). In the previous discussion, we assume that a large number of covariates are collected and all confounding are included in the collection of covariates, and hence can be adjusted in the analysis. However, in practice, this assumption may be violated. Some confounding variables may be unobserved.

In causal inference, several methods, including instrumental variables (Kuang et al. 2020), proxy (or surrogate) variables (Miao et al. 2018), network structure (Ogburn et al. 2020), and multiple causes (Wang and Blei 2019), for dealing with unobserved confounding have been developed. The counterfactual VAE (CFVAE) is a remarkable work of application of artificial intelligence (AI) to estimation of ITE in the presence of unobserved confounding ((Wu and Fukumizu 2021). The CFVAE has several nice properties. First, the CFVAE model for the treatment estimation is identifiable (Khemakhem et al. 2019, Roeder et al. 2020). Second, the results of the CFVAE are interpretable. Thirdly, the CFVAE is able to estimate ITE.

7.3.2 Variational Autoencoders

Before reviewing autoencoders, we first introduce balancing score (Wu and Fukumizu 2021). Balancing score including propensity score, as its special case, is often used for unbiased assigning individuals to the treatment and control groups in observational data to remove confounding bias in estimation of causal effect (Wijayatunga 2015). Let function $b(z)$ be a balancing score. Balancing score plays an essential role in estimating the treatment effect. Therefore, here we cite theorem proposed by Rosenbaum and Rubin (1983) (Wu and Fukumizu 2021).

Theorem 7.2: Balancing Score

Let $b(Z)$ be a function of random variable z. Then $b(Z)$ is a balancing score, i.e., $T \perp\!\!\!\perp Z \mid b(Z)$, if and only if $f(b(Z)) = p(T=1 \mid Z)$ for some function f, where $e(Z) = p(T=1 \mid Z)$ is a propensity score. Assume further that z satisfies strong ignorability (i.e., $(Y^0, Y^1) \perp\!\!\!\perp T \mid Z,\ 0 < p(T=1|Z=z) < 1$), then so does $b(z)$.

Now we introduce several types of VAEs (Wu and Fukumizu 2021). The VAE is a class of latent variable model. The VAE consists of encoder and decoder. The encoder maps the observed outcome, the treatment t and the covariates x to the latent space with the latent variables z. The decoder generates the outcome y with the probability $p(y \mid z,t)$.

In Chapters 3 and 5, we showed that the evidence of lower bound (ELBO) of the log-likelihood is given by

$$\log p(y) \geq \log p(y) - KL(q_\varnothing(z \mid y) \,\|\, p_\theta(z|y)) = E_{z\sim q_\varnothing(z|y)} \log p_\theta(y|z) - KL(q_\varnothing(z|y) \,\|\, p_\theta(z)). \tag{7.90}$$

Let

$$ELBO(\theta,\varnothing, y) = \mathcal{L}(\theta,\varnothing, y) = E_{z\sim q_\varnothing(z|y)} \log p_\theta(y|z) - KL(q_\varnothing(z|y) \,\|\, p_\theta(z)). \tag{7.91}$$

Define $q_\varnothing(z|y)$ as encoder which approximates the posterior distribution $p_\theta(z|y)$, and $p_\theta(y|z)$ as decoder which generates the outcome y.

7.3.2.1 CVAE

By adding conditional c, we define conditional VAE (CVAE). The ELBO for the CVAE is given by

$$\mathcal{L}_{CVAE}(\theta,\varnothing, y,c) = E_{z\sim q_\varnothing(z|y,c)} \log p_\theta(y|z,c) - KL(q_\varnothing(z|y,c) \| p_\theta(z|c)). \tag{7.92}$$

7.3.2.2 iVAE

By introducing auxiliary variable u, Khemakhem et al. (2019), proposed the identifiable VAE (iVAE). The iVAE assumes $y \perp\!\!\!\perp u \mid z$, i.e., $p(y|z,u) = p(y \mid z)$. Under this assumption, the VAE is identifiable. The ELBO for the iVAE is

$$\log p(y|u) \geq \mathcal{L}_{iVAE}(\theta,\varnothing, y,u),$$

where

$$\mathcal{L}_{iVAE}(\theta,\varnothing, y,u) = E_{z\sim q_\varnothing(z|y,u)} \log p_f(y|z) - KL(q_\varnothing(z \mid y,u) \,\|\, p_{T,\lambda}(z,u)), \tag{7.93}$$

where y satisfies the following equation

$y = f(z) + \varepsilon$, ε is additive noise, and distribution of z is exponential family with sufficient statistics T and the parameter λ. The identifiability indicates that the functional parameters (f, T, λ) are uniquely determined (up to a simple transformation).

7.3.3 Architecture of CFVAE

Next we introduce architecture of CFVAE developed in Wu and Fukumizu (2021). Assume strong ignorability, the treatment effect can be estimated by

$$\mu_t(x) = E[Y^t|X=x] = E[E[Y^t|X=x,Z]] = \int p(y|z,x,t)\, y dy p(z|x) dz. \tag{7.94}$$

The estimation of the treatment effect involves the probability $p(y,z \mid x,t)$. The probability $p(y,z \mid x,t)$ can be written as

$$p(y,z|x,t) = p(y|z,x,t)\, p(z \mid x,t). \tag{7.95}$$

We assume that the covariates x are proxy variables and

$$y \perp\!\!\!\perp x|z,t. \tag{7.96}$$

Then, we have

$$p(y|z,x,t) = p(y|z,t). \tag{7.97}$$

Substituting equation (7.97) into equation (7.95) yields

$$p(y,z|x,t) = p(y|z,t)\,p(z|x,t). \tag{7.98}$$

The VAE should be designed to model the joint distribution in equation (7.98). The VAE learns to recover the causal representation of the latent variables z, which can be used to estimate the treatment effect in the presence of confounding. Theorem 7.2 shows that to unbiasedly estimate the treatment effect, we do not need to recover the true confounders z. We only need to recover the causal representation $b(z)$ which is a function of the propensity score $e(z)$, the part of z that contain information for balancing the treatment assignment.

A key for designing the VAE is to make the designed VAE identifiable. The essential assumption to ensure that iVAE is identifiable, is that

$$y \perp\!\!\!\perp u \,|\, z. \tag{7.99}$$

Comparing equation (7.96) with equation (7.99) motives us to take covariates x as auxiliary variable u in iVAE. As results, the joint distribution in equation (7.98) requires that CFVAE should be an iVAE. The joint distribution in equation (7.98) is a conditional distribution, given the treatment T. The CFVAE also should be CVAE. Therefore, the CFVAE should be designed as a combination of iVAE and CVAE with treatment T and covariates X as conditioning and auxiliary variable, respectively. The architecture of the CFVAE is shown in Figure 7.3.

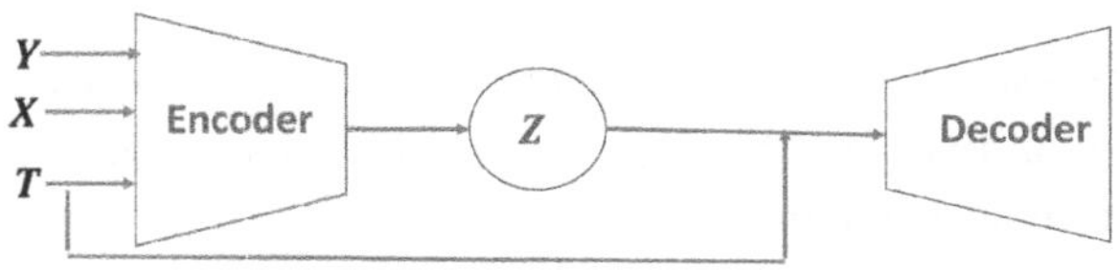

FIGURE 7.3 Architecture of CFVAE.

7.3.4 ELBO

The ELBO of the log-likelihood $\log p(y\,|\,x,t)$ is given by (Wu and Fukumizu 2021):

$$\log p(y|x,t) \geq \log p(y|x,t) - KL(q(z|x,y,t)\|$$
$$p(z|x,y,t)) = E_{z\sim q_{\varnothing}(z|x,y,t)} \log p_{\theta}(y|z,t)$$
$$-KL(q_{\varnothing}(z\,|\,x,y,t)\,\|\,p_{\theta}(z\,|\,x,t)) = \mathcal{L}_{CFVAE}(\varnothing,\theta,x,y,t). \tag{7.100}$$

7.3.4.1 Encoder

The encoder is defined as $q_{\varnothing}(z\,|\,x,y,t)$, where $\varnothing = (r,s)$. The encoder maps the observed data (X,Y,T) to the latent space Z. Define the transformation function:

$$Z^{(i,l)} = \mu_r^{(i)} + \sigma_s^{(i)} \odot \varepsilon^{(l)},\ \varepsilon^{(l)} \sim N(0,I), \tag{7.101}$$

where

$$\left[\mu_r^{(i)}\,|\log\sigma_s^{(i)}\right] = MLP(x^{(i)}, y^{(i)}, t^{(i)}, W_z), \tag{7.102}$$

MLP denotes multilayered perceptron. Two layer MLP is defined as

$$MLP(x^{(i)}, y^{(i)}, t^{(i)}, W_z) = \text{ReLU}\left(\left(x^{(i)}\ y^{(i)}\ t^{(i)}\right)W_z^1\right)W_z^2. \tag{7.103}$$

The distribution of the encoder is given by

$$q_{r,s}(z|x,y,t) = \prod_{i=1}^{n} N\left(Z^{(i)};\, \mu_r^{(i)}, \sigma_s^{(i)}\right). \tag{7.104}$$

7.3.4.2 Decoder

The decoder is defined as $p_{f,g}(y|z,t)$. The distribution function $p_{f,g}(y|x,t)$ is assumed to be

$$p_{f,g}\left(y^{(j)}|z^{(j,l)}, t^{(j)}\right) = N\left(Y^{(j)};\mu_f^{(j,l)}, \sigma_g^{(j,l)}\right), \tag{7.105}$$

where

$$\left[\mu_f^{(j,l)}\,|\,\sigma_g^{(j,l)}\right] = MLP(z^{(j,l)}, t^{(j)}, W_y). \tag{7.106}$$

Two layer MLP is defined as

$$MLP(z^{(j,l)}, t^j, W_y^1, W_y^2) = \text{ReLU}\left(\left(z^{(j,l)}\ t^{(j)}\right)W_y^1\right)W_y^2.$$

7.3.4.3 Computation of the KL Distance

Now we calculate $-KL(q(z\,|\,x,y,t)\,\|\,p(z\,|\,x,t))$ in equation (7.100). Under some assumptions, the KL distance can often be integrated analytically. We assume that the dimensionality of the latent vector z is M. Let μ_h^m and σ_h^m denote the mth element of the mean vector $\boldsymbol{\mu}_h$ and standard deviation vector $\boldsymbol{\sigma}_h$ of the latent vector $\boldsymbol{z}$. Assume that $p(z \mid x,t)$ is a Gaussian distribution:

$$p(z|\boldsymbol{x},t)=\prod_{m=1}^{M}\frac{1}{\sqrt{2\pi\sigma_h^m}}\exp\left\{-\frac{\left(z^m-\mu_h^m\right)^2}{2\left(\sigma_h^m\right)^2}\right\}. \quad (7.107)$$

We assume that the distribution $q_{r,s}(\boldsymbol{z} \mid \boldsymbol{x},\boldsymbol{y},t)$ of encoder is given by

$$q_{r,s}(z|\boldsymbol{x},\boldsymbol{y},t)=\prod_{m=1}^{M}\frac{1}{\sqrt{2\pi\sigma_s^m}}\exp\left\{-\frac{\left(z^m-\mu_r^m\right)^2}{2\left(\sigma_s^m\right)^2}\right\}. \quad (7.108)$$

We can show (Exercise 7.4) that KL distance is given by

$$-KL\left(q(z \mid x,y,t)\,||\,p(z|x,t)\right)$$

$$=\frac{1}{2}\left\{\sum_{m=1}^{M}\left(\log\left(\sigma_s^m\right)^2+1\right)-\sum_{m=1}^{M}\left[\log\left(\sigma_h^m\right)^2+\frac{\left(\sigma_s^m\right)^2}{\left(\sigma_h^m\right)^2}+\frac{\left(\mu_r^m-\mu_h^m\right)^2}{\left(\sigma_h^m\right)^2}\right]\right\}. \quad (7.109)$$

7.3.4.4 Calculation of ELBO

Now we calculate the first term in equation (7.100). It follows from equation (7.105) that

$$\log p\left(y^{(j)}|z^{(j,l)},t^{(j)}\right)$$

$$=-\frac{1}{2}\left[\log(2\pi)+\log\sigma^{j,l}+\frac{\left(y^{(j)}-\mu^{(j,l)}\right)^2}{\left(\sigma^{j,l}\right)^2}\right]. \quad (7.110)$$

The first term in equation (7.100) can be approximated by

$$\frac{1}{nL}\sum_{j=1}^{n}\sum_{l=1}^{L}\log p\left(y^{(j)}|z^{(j,l)},t^{(j)}\right)$$

$$=-\frac{1}{2nL}\sum_{j=1}^{n}\sum_{l=1}^{L}\left[\log(2\pi)+\log\sigma^{j,l}+\frac{\left(y^{(j)}-\mu^{(j,l)}\right)^2}{\left(\sigma^{j,l}\right)^2}\right]. \quad (7.111)$$

Combining equations (7.109) and (7.111), we obtain the ELBO:

$$\mathcal{L}_{CFVAE}(\varnothing,\boldsymbol{\theta},x,y,t)$$

$$=-\frac{1}{2nL}\sum_{j=1}^{n}\sum_{l=1}^{L}\left[\log(2\pi)+\log\sigma^{j,l}+\frac{\left(y^{(j)}-\mu^{(j,l)}\right)^2}{\left(\sigma^{j,l}\right)^2}\right]$$

$$+\frac{1}{2}\left\{\sum_{m=1}^{M}\left(\log\left(\sigma_s^m\right)^2+1\right)-\sum_{m=1}^{M}\left[\log\left(\sigma_h^m\right)^2+\frac{\left(\sigma_s^m\right)^2}{\left(\sigma_h^m\right)^2}+\frac{\left(\mu_r^m-\mu_h^m\right)^2}{\left(\sigma_h^m\right)^2}\right]\right\}. \quad (7.112)$$

7.4 VARIATIONAL AUTOENCODER FOR SURVIVAL ANALYSIS

7.4.1 Introduction

Causal survival analysis is interested in the estimation and prediction of causal effect of a given intervention or treatment on survival time (Paidamoyo et al. 2020). Classical causal survival analysis is performed by a RCT, where the treatment is randomly assigned to individuals. However, the RCTs are usually unethical, expensive, and infeasible. Alternatively, observational data such as electronic health records (EHRs) can be used to assess the causal effect of the treatment on the survival (Häyrinen et al. 2008; Lu et al. 2018).

Causal survival analysis from observational data raises two great challenges (Chapfuwa et al. 2020). The first challenge is presence of confounders that affect both the treatment and survival time (Halloran and Hudgens 2012). The second challenge is the censoring problem where we only know that an event has not occurred up to a certain point in time and do not know the exact time when time-to-event takes place (Li et al. 2018; Paidamoyo et al. 2020). Classical statistical causal survival-analysis approaches often use the Cox proportional hazards (CoxPH) model (Saha-Chaudhuri and Juwara 2021) and the accelerated failure time (AFT) model (Pang et al. 2021) with proper weighting for each individual to account for confounding bias to estimate the effect of the treatment or covariates (Díaz 2019; Hernán and Robins 2020). These models assume a linear relationship between the covariates and survival distribution and are difficult to deal with high-dimensional data and to capture complex nonlinear interactions. Furthermore, these models lack a counterfactual prediction mechanism, and hence are not able to estimate ITE.

Recently, tree-based machine learning methods such as Random Survival Forest (RSF) (Nasejje et al. 2017) and Bayesian Additive regression trees (BART) (Tan and Roy 2019) have been extended to causal survival analysis (Hu et al. 2020). Alternatively, the neural network-based causal methods use the learned representation to balance distributions across treatment and control groups. Therefore, the deep Cox models and deep latent models for causal survival analysis which can account for the confounding bias have been recently developed (Beaulac et al. 2018; Ching et al. 2018; Katzman et al. 2018; Chapfuwa et al. 2020; Huang et al. 2020). In this section, we introduce the VAE-based causal survival analysis methods which utilize the balanced (latent) representation learning to predict counterfactual survival outcomes and estimate ITE in observational studies (Chapfuwa et al. 2020).

7.4.2 Notations and Problem Formulation

Consider N_1 treated individuals and N_0 untreated individuals. Thus, the total number of sampled individuals is $N = N_0 + N_1$. Let X be a set of covariates which can be a mixture of categorical and continuous covariates. Let A be a treatment assignment indicator, where $A = 1$ for the treated and $A = 0$ for the control. Let T be outcome (survival time), T_1 be potential survival time if the individual would receive the treatment, and T_0 be potential survival time if the individual would receive no treatment. Let T_A be the factual survival time and T_{1-A} be the counterfactual survival time. In practice we only observe the factual survival time T_A. In survival analysis, we need to introduce a (right) censoring which is most likely due to the loss of follow-up. Let C be a censoring time and $\delta \in \{0,1\}$ be a censoring indictor. The observed time is $Y = \min(C_A, T_A)$, i.e., if $T_A < C_A$ then the non-censored time T_A is observed and set $\delta = 1$.

ITE is defined as

$$ITE = E\left[T_1 - T_0 \mid X\right]. \tag{7.113}$$

We are also interested in estimation of the conditional distributions $P(T_1 \mid X)$ and $P(T_0 \mid X)$ of the survival time, given the covariates. In summary, we observe the dataset $D = \{y_i, x_i, a_i, \delta_i, c_i\}_{i=1}^{N}$.

7.4.3 Classical Survival Analysis Theory

We review some basic concepts and theory in survival analysis (Chapfuwa et al. 2020). Let $F(t \mid X)$ be the cumulative distribution function $p(T \le t|X)$ of the event (delth) time t, given a set of the covariates X. Define conditional survival function $S(t|X) = p(T > t|X) = 1 - F(t \mid X)$, which measures the probability of survival up to time t. The hazard function $\lambda(t)$ is defined as the instantaneous probability of the event occurring between $\{t, t + \Delta t\}$, given $T > t$ and $\Delta t \to 0$:

$$\lambda(t) = \lim_{\Delta t \to 0} \frac{p(t < T \langle t + \Delta t | T.t)}{\Delta t}. \tag{7.114}$$

We can show that (Exercise 7.5)

$$\begin{aligned} \lambda(t \mid X) &= \lim_{\Delta t \to 0} \frac{p(t < T < t + dt | X = x)}{p(T > t | X = x) dt} \\ &= -\frac{d \log S(t|X)}{dt} = \frac{f(t \mid X)}{S(t \mid X)}, \end{aligned} \tag{7.115}$$

where

$$f(t|X) = \frac{dF(t \mid X)}{dt} = P(T = t \mid X = x). \tag{7.116}$$

We can easily see from equation (7.115) that

$$f(t|X) = \lambda(t \mid X) S(t \mid X). \tag{7.117}$$

Let $S_A(t \mid X)$ and $\lambda_A(t \mid X)$ be the conditional survival and hazard functions for the potential outcomes T_A, i.e., T_1 and T_0, given the set of covariates X, respectively. To further measure potential outcomes for the treatment, we give two definitions.

Definition 7.8: Difference in Expected Lifetime

Difference in expected lifetime is defined as

$$\begin{aligned} ITE_{life}(t_{max}, x) &= E\left[T_1 - T_0 | X\right] \\ &= \int_0^{t_{max}} \left(p(T_1 > t|X] - p(T_0 > t \mid X)\right) dt \\ &= \int_0^{t_{max}} \left(S_1(t|x) - S_0(t|x)\right) dt. \end{aligned} \tag{7.118}$$

Definition 7.9: Difference in Survival Function

Difference in survival function is defined as

$$ITE(t,x) = S_1(t|x) - S_0(t|x). \quad (7.119)$$

Definition 7.10: Hazard Ratio

Hazard ratio is defined as

$$HR(t,x) = \frac{\lambda_1(t|x)}{\lambda_0(t|x)}. \quad (7.120)$$

Now we extend the ignorability and overlap assumptions from cross-sectional data to survival data.

Assumptions 7.3

1. Ignorability

$$\{T_1, T_2\} \perp\!\!\!\perp A|X\ . \quad (7.121)$$

2. Overlap
In the covariate support, almost surely if $p(X=x)>0$ then we have

$$0 < p(A=1|X=x) < 1. \quad (7.122)$$

3. Informative censoring:

$$T \perp\!\!\!\perp C \mid X, A, \text{or} \quad (7.123)$$

4. Non-informative censoring:

$$T \perp\!\!\!\perp C. \quad (7.124)$$

Let C_1 and C_0 be potential censoring time. Define $p(C_1|X)$ and $p(C_0|X)$.

7.4.4 Potential Outcome (Survival Time) and Censoring Time Distributions

We now define distribution function of the potential outcomes. Let t_a be potential outcome. Assume that t_a is distributed as

$$t_a \sim p_{h,\Phi}(T \mid X=x, A=a), \quad (7.125)$$

where h, Φ are parameters of the distribution.

By the strong ignorability assumption, i.e., $\{T_1, T_0\} \perp\!\!\!\perp A \mid X$, outcome T is conditionally independent of the treatment assignment A, given the covariates X. Therefore, $p_{h,\Phi}(T \mid X=x, A=a)$ can be written as $p_{h,\Phi}(T \mid X)$. Therefore, equation (7.125) can be rewritten as

$$t_a \sim p_{h,\Phi}(T_a \mid X=x). \quad (7.126)$$

Let C_A be a censoring time of individual receiving treatment A. Similarly, censoring time C_A is distributed as

$$C_A \sim p_{v,\Phi}(C_A \mid X=x), \quad (7.127)$$

where v and Φ are parameters of the distribution. Let $f_{h,\Phi}(t_a \mid x)$ and $F_{h,\Phi}(t_a \mid x)$ be, respectively, conditional density and cumulative function of the potential survival time t_a, given covariates x.Let $S_{h,\Phi}(t_a \mid x)$ be conditional survival function of t_a, given covariates x. Let $e_{v,\Phi}(c_a \mid x)$ and $g_{v,\Phi}(c_a \mid x)$ be conditional density and cumulative function of the censoring time c_a, given x. Let $G_{v,\Phi}(c_a \mid x)$ be conditional survival function of c_a, given x. We assume $T \perp\!\!\!\perp C \mid X, A$, i.e., informative censoring.

Now we study how to calculate $p_{h,\Phi}(T_a \mid X=x)$ and $p_{v,\Phi}(C_A \mid X=x)$. If $\delta=1$, then T_a is uncensored and the subject contributes $f_{h,\Phi}(t_a \mid x)$ to the likelihood. If $\delta=0$, then T_a is censored, then the subject contributes $p(T_a > t_a|X=x) = S_{h,\Phi}(t_a \mid x)$ to the likelihood. In summary, we obtain

$$p_{h,\Phi}(T_a|X=x) = f_{h,\Phi}(t_a \mid x)^{\delta} S_{h,\Phi}(t_a \mid x)^{1-\delta}. \quad (7.128)$$

Now we jointly consider the potential survival time T_a and censoring time C_a. The likelihood contributions for the two types of observations are summarized in Table 7.1.

TABLE 7.1 Likelihood Contribution of Survival and Censoring Time

Event	Expressed as	Likelihood Contribution
$T_a = t_a, \delta = 1, C_a \ge c_a$	$[T_a = t_a, C_a \ge c_a]$	$f_{h,\Phi}(t_a \mid x) G_{v,\Phi}(c_a \mid x)$
$T_a > t_a, \delta = 0, C_a = c_a$	$[T_a \ge t_a, C_a = c_a]$	$S_{h,\Phi}(t_a \mid x) e_{v,\Phi}(c_a \mid x)$

It follows from Table 7.1 that the likelihood function $p_{h,\Phi,v}(t_a, c_a \mid x)$ is given by

$$p_{h,\Phi,v}(t_a, c_a|x) = \left[f_{h,\Phi}(t_a \mid x) G_{v,\Phi}(c_a \mid x)\right]^{\delta} \left[S_{h,\Phi}(t_a \mid x) e_{v,\Phi}(c_a \mid x)\right]^{1-\delta}$$
$$= \left[f_{h,\Phi}(t_a \mid x)\right]^{\delta}\left[S_{h,\Phi}(t_a \mid x)\right]^{1-\delta} \left[e_{v,\Phi}(c_a \mid x)\right]^{1-\delta}\left[G_{v,\Phi}(c_a \mid x)\right]^{\delta}$$
$$= p_{h,\Phi}(T_a|X = x) p_{v,\Phi}(C_A \mid X = x), \quad (7.129)$$

where

$$p_{h,\Phi}(T_a|X = x) = \left[f_{h,\Phi}(t_a \mid x)\right]^{\delta}\left[S_{h,\Phi}(t_a \mid x)\right]^{1-\delta}, \quad (7.130)$$

$$p_{v,\Phi}(C_A|X = x) = \left[e_{v,\Phi}(c_a \mid x)\right]^{1-\delta}\left[G_{v,\Phi}(c_a \mid x)\right]^{\delta}. \quad (7.131)$$

7.4.5 VAE Causal Survival Analysis

The doctors often select treatments T, depending on the observed covariates X, which characterize the true patient status. This leads to the selection bias. The true patient status is unknown and hidden. In addition, the unobserved confounding that affects both the potential outcomes and treatments. The confounding variables are also hidden. The hidden variables include both selection bias variables and confounding variables. To overcome these problems, the deep latent models can be used to model the latent variables. Variational inference can then be used or inference on latent variable models (Beaulac et al. 2018).

7.4.5.1 Deep Latent Model

The deep latent model maps the observed survival time T, the treatment A, and covariates x to the latent variables z. The set of latent variables z include the true patient health status and confounding variables. The covariates x collected as proxy of the true health status and confounders. The distribution for the response T is based on the patient health status, confounders, and the treatment selected A.

We denote $T^a = T(a)$. From equation (7.113) we can define

$$\mu_a(x) = E\left[T(a) \mid X = x\right]. \quad (7.132)$$

Then, using computing expectation by conditioning, we obtain

$$\mu_a(x) = E\left[T \mid A = a, X = x\right]$$
$$= E\left[E\left[T \mid A = a, X = x\right] \mid Z = z\right]$$
$$= \int tp(t|a,x,z) p(z|x) dtdz. \quad (7.133)$$

Equation (7.133) involves distribution $p(t,z \mid a,x)$. We assume the conditional independence $T \perp\!\!\!\perp X \mid Z, A$. Thus, distribution $p(t,z \mid a,x)$ can be factorized to

$$p(t,z|a,x) = p(t|z,x,a) p(z \mid x,a)$$
$$= p(t|z,a) p(z \mid x,a). \quad (7.134)$$

As we discussed in Section 7.3, a combination of CVAE and iVAE, with treatment a and covariate x as conditioning and auxiliary variable, respectively, can be used to design an architecture of a VAE (Figure 7.4). Similar to equation (7.130), taking censored data into account, we obtain the log-likelihood for the conditional survival function as

$$\log p_\theta(t|a,z) = \delta \log f_\theta(t|a,z) + (1-\delta)\log S_\theta(t \mid a,z), \quad (7.135)$$

where δ is a censoring indicator variable and defined as in Section 7.3, $f_\theta(t|a,z)$ is the conditional density function and can be modeled as Weibull distribution, and $S_\theta(t \mid a,z)$ is the conditional survival function.

7.4.5.2 ELBO

The maximum likelihood estimation of the parameters can be transformed to maximize the ELBO. We can show that the ELBO ($\mathcal{L}_{SVAE}$) for the survival VAE (SVAE) is given by (Appendix 7A)

$$\log p(t \mid x,a) \ge \mathcal{L}_{SVAE} = -KL\left(q_\varnothing(z \mid x,t,a) \,\|\, p_\theta(z|x,a)\right) + E_{q_\varnothing(z|x,t,a)} \log p_\theta(t \mid z,a). \quad (7.136)$$

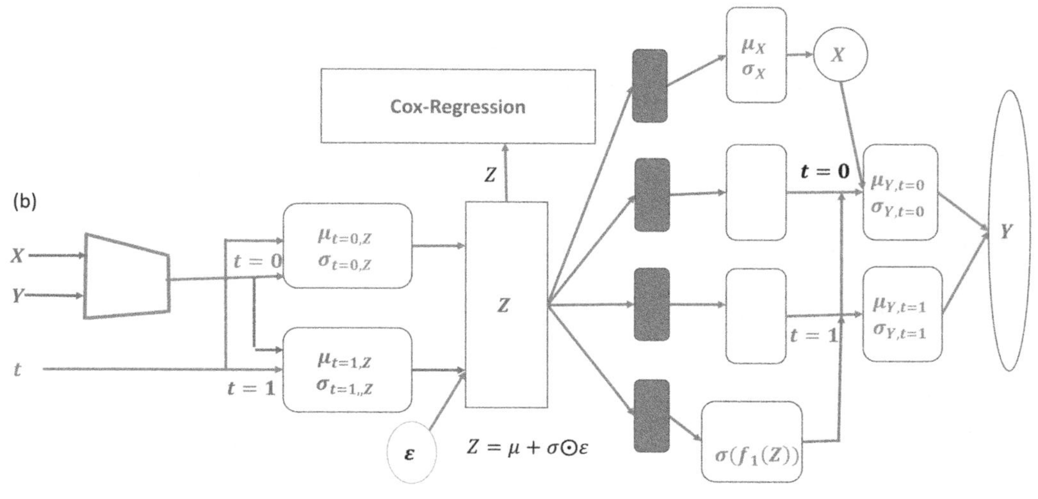

FIGURE 7.4 Architecture of SVAE.

In order to make the SVAE identifiable, the second term in the ELBO drops the dependence on x.

7.4.5.3 Encoder

The encoder is defined as $q_{\varnothing}(z \mid x,t,a)$, where $\varnothing = (r,s)$. The encoder maps the observed data (x,t,a) to the latent space Z. Define the transformation function:

$$Z^{(i,l)} = \mu_r^{(i)} + \sigma_s^{(i)} \odot \varepsilon^{(l)},\ \varepsilon^{(l)} \sim N(0,I), \quad (7.137)$$

where

$$\left[\mu_r^{(i)} \mid \log \sigma_s^{(i)}\right] = MLP\left(x^{(i)}, a^{(i)}, t^{(i)}, W_z\right), \quad (7.138)$$

MLP denotes multilayered perceptron. Two layer MLP is defined as

$$MLP\left(x^{(i)}, a^{(i)}, t^{(i)}, W_z\right) = \text{ReLU}\left(\left(x^{(i)}, a^{(i)}, t^{(i)}\right)W_z^1\right)W_z^2. \quad (7.139)$$

The distribution of the encoder is given by

$$q_{r,s}(z|x,t,a) = \prod_{i=1}^{m} N\left(Z^{(i)}; \mu_r^{(i)}, \sigma_s^{(i)}\right), \quad (7.140)$$

where m is the dimension of the latent space.

7.4.5.4 Decoder

The decoder is defined as $p_\theta\left(t^{(i)}|z^{(i,l)}, a^{(i)}\right)$. The distribution function $p_\theta\left(t^{(i)}|z^{(i,l)}, a^{(i)}\right)$ is calculated by

$$\log p_\theta\left(t^{(i)}|z^{(i,l)}, a^{(i)}\right) = \delta^{(i)} \log f_\theta\left(t^{(i)}|z^{(i,l)}, a^{(i)}\right) + \left(1-\delta^{(i)}\right)\log S_\theta\left(t^{(i)} \mid z^{(i,l)}, a^{(i)}\right), \quad (7.141)$$

where $f_\theta(t|z,a)$ is assumed to follow a Weibull distribution (λ, k):

$$f_\theta\left(t^{(i)}\right) = \frac{k}{\lambda}\left(\frac{t^{(i)}}{\lambda}\right)^{k-1} e^{-\left(\frac{t^{(i)}}{\lambda}\right)^k},\ t \geq 0, \quad (7.142)$$

where

$$\left[\lambda^{(i)}, k^{(i)}\right] = MLP\left(z^{(i,l)}, a^{(i)}, W_t\right), \quad (7.143)$$

Two layer MLP is defined as

$$MLP\left(z^{(i,l)}, a^{(i)}, W_t\right) = \text{ReLU}\left(\left(z^{(i,l)}, a^{(i)}\right)W_t^1\right)W_t^2. \quad (7.144)$$

The survival function of the Weibull distribution is given by (Exercise 7.7)

$$S_\theta\left(t^{(i)}|z^{(i,l)}, a^{(i)}\right) = \exp\left\{-\left(\frac{t^{(i)}}{\lambda^{(i)}}\right)^{k^{(i)}}\right\}. \quad (7.145)$$

7.4.5.5 Computation of the KL Distance

Now we calculate $-KL\left(q(z \mid x,t.a) \| p(z \mid x,a)\right)$ in equation (7.136). Under some assumptions, the KL distance can often be integrated analytically. We assume that the dimensionality of the latent vector $\boldsymbol{z}$ is M. Let μ_h^m and σ_h^m denote the mth element of the mean vector $\boldsymbol{\mu_h}$ and standard deviation vector $\boldsymbol{\sigma_h}$ of the latent vector $\boldsymbol{z}$. Assume that $p(z \mid x,a)$ is a Gaussian distribution $N(z;\ \mu_h, \sigma_h)$:

$$p(z|\boldsymbol{x},a) = \prod_{m=1}^{M} \frac{1}{\sqrt{2\pi}\sigma_h^m} \exp\left\{-\frac{\left(z^m - \mu_h^m\right)^2}{2\left(\sigma_h^m\right)^2}\right\}, \quad (7.146)$$

where

$$[\mu_h \mid \log \sigma_h] = MLP\left(x^{(i)}, a^{(i)}, W_{pr}\right). \quad (7.147)$$

Two layer MLP is defined as

$$MLP\left(x^{(i)}, a^{(i)}, W_{pr}\right) = \text{ReLU}\left(\left(x^{(i)}, a^{(i)}\right)W_{pr}^1\right)W_{pr}^2. \quad (7.148)$$

We assume that the distribution $q_{r,s}(z \mid x, t, a)$ of encoder is given by

$$q_{r,s}(z|x,t,a) = \prod_{m=1}^{M} \frac{1}{\sqrt{2\pi}\sigma_s^m} \exp\left\{-\frac{\left(z^m - \mu_r^m\right)^2}{2\left(\sigma_s^m\right)^2}\right\}. \quad (7.149)$$

We can show (Exercise 7.4) that KL distance is given by

$$-KL\left(q(z \mid x,t,a) \| p(z|x,a)\right) = \frac{1}{2}\left\{\sum_{m=1}^{M}\left(\log\left(\sigma_s^m\right)^2 + 1\right) - \sum_{m=1}^{M}\left[\log\left(\sigma_h^m\right)^2 + \frac{\left(\sigma_s^m\right)^2}{\left(\sigma_h^m\right)^2} + \frac{\left(\mu_r^m - \mu_h^m\right)^2}{\left(\sigma_h^m\right)^2}\right]\right\}. \quad (7.150)$$

7.4.5.6 Calculation of ELBO

Now we calculate the second term in equation (7.136). It follows from equation (7.141) that

$$\log p_\theta\left(t^{(i)}|z^{(i,l)}, a^{(i)}\right) = \delta^{(i)} \log f_\theta\left(t^{(i)}|z^{(i,l)}, a^{(i)}\right) + \left(1 - \delta^{(i)}\right)\log S_\theta\left(t^{(i)} \mid z^{(i,l)}, a^{(i)}\right)$$

The second term in equation (7.136) can be approximated by

$$\frac{1}{nL}\sum_{i=1}^{n}\sum_{l=1}^{L} \log p_\theta\left(t^{(i)}|z^{(i,l)}, a^{(i)}\right) = \frac{1}{nL}\sum_{i=1}^{n}\sum_{l=1}^{L}\left[\delta^{(i)} \log f_\theta\left(t^{(i)}|z^{(i,l)}, a^{(i)}\right) + \left(1 - \delta^{(i)}\right)\log S_\theta\left(t^{(i)} \mid z^{(i,l)}, a^{(i)}\right)\right]. \quad (7.151)$$

Combining equations (7.150) and (7.151), we obtain the ELBO:

$$\mathcal{L}_{SVAE}(\varnothing, \theta, x, t, a) = \frac{1}{2}\left\{\sum_{m=1}^{M}\left(\log\left(\sigma_s^m\right)^2 + 1\right) - \sum_{m=1}^{M}\left[\log\left(\sigma_h^m\right)^2 + \frac{\left(\sigma_s^m\right)^2}{\left(\sigma_h^m\right)^2} + \frac{\left(\mu_r^m - \mu_h^m\right)^2}{\left(\sigma_h^m\right)^2}\right]\right\} + \frac{1}{nL}\sum_{i=1}^{n}\sum_{l=1}^{L}\left[\begin{array}{l}\delta^{(i)} \log f_\theta\left(t^{(i)}|z^{(i,l)}, a^{(i)}\right) \\ +\left(1 - \delta^{(i)}\right)\log S_\theta\left(t^{(i)} \mid z^{(i,l)}, a^{(i)}\right)\end{array}\right]. \quad (7.152)$$

We use backpropagation algorithm and the ADAM Optimizer to solve the following maximization problem (Kingma and Ba 2014):

$$\max_{\varnothing, \theta} \mathcal{L}_{SVAE}(\varnothing, \theta, x, t, a). \quad (7.153)$$

7.4.5.7 Prediction

We can use the fitted SVAE model to predict survival time for a new patient with characteristics x and a given treatment t. The prediction distribution $p(t \mid x, a)$ can be calculated by

$$p(t|x,a) = \int p(t|a,z)p(z|a,x)dz = \frac{1}{L}\sum_{l=1}^{L} p(z^l|a,x)p(t|a, z^l) = \sum_{l=1}^{L} w_l p(t|a, z^l), \quad (7.154)$$

where $w_l = \frac{1}{L} p(z^l \mid a, x)$.

By the similar argument, survival function $p(T > t \mid a, x)$ can be calculated by

$$p(T > t|a,x) = \sum_{l=1}^{L} w_l p(T > t \mid a, z^l). \quad (7.155)$$

Define

$$d = p(T > t|a_1, x) - p(T > t \mid a_0, x). \quad (7.156)$$

The quantity d can be used to select the treatment: If $d > 0$ then select the treatment a_1, otherwise, select the treatment a_0.

7.4.6 VAE-Cox Model for Survival Analysis

There have been increasing interests in combining autoencoder or VAE with a Cox model for survival analysis. The widely used approach to combine the autoencoder (AE) or VAE with the Cox model is to use AE or VAE as the data reduction tools for pre-data processing and then use the reduced latent variables as the input to the Cox model (Kim et al. 2020). Some authors also integrate the AE with the Cox model for survival analysis (Huang et al. 2020). In this section, we integrate the SVAE with the Cox model.

7.4.6.1 Cox Model

The CoxPH model (Huang et al. 2020) assumes that the hazard function at time t for patient i which is defined as the event rate at time t can be factored as

$$\lambda(t|z_i) = \lambda_0(t)\exp(\beta^T z_i), \tag{7.157}$$

where $\lambda_0(t)$ is the baseline hazard function—common to all individuals in the population and dependent on only time t, and reflects the underlying hazard for subjects with all covariates $z_1, \ldots, z_p$ equal to 0, $\beta = (\beta_1, \ldots, \beta_p)$ is the vector of parameters of the model, z_i is the vector of the latent variable for the patient i.

It follows from equation (7.120) that the hazard ratio is equal to

$$HR(t \mid z_i) = \exp(\beta^T z_i). \tag{7.158}$$

Then, the log of the hazard ratio for the ith individual to the baseline is given by

$$\log HR(t \mid z_i) = \log \frac{\lambda(t|z_i)}{\lambda_0(t)} = \beta^T z_i. \tag{7.159}$$

Equation (7.159) is another version of the Cox model. It states that the CoxPH model is a linear model for the log of the hazard ratio. The advantage of the Cox model is that we can estimate the parameters β without having to estimate $\lambda_0(t)$.

7.4.6.2 Likelihood Estimation for the Cox Model

Let (t_i, δ_i, z_i) be the observed dataset for the individual i, where

t_i is a possibly censored failure (survival) time random variable,

δ_i is the failure (survival)/censoring indicator ($\delta_i = 1$ for fail, $\delta_i = 0$ for censor),

z_i is a set of covariates or latent variables as defined earlier.

Assume that there are K distinct failure times. Let $\tau_1 < \tau_2 < \cdots < \tau_K$ be the K ordered, distinct failure times.

Let $R(t_i) = \{j : t_j \geq t_i\}$ be a risk set for the time t_i, which denotes the set of individuals who are "at risk" for failure at time t_i. We define the partial likelihood as a product of conditional probabilities of occurrence of a failure event at the time t, given the risk set at that time and that one failure is to happen. Let t_i be the failure time associated with individual i and $L_i(\beta)$ be the partial likelihood for individual i. Statistically, the partial likelihood $L_i(\beta)$ is expressed as

$$\begin{aligned} L_i(\beta) &= p(\text{individual } i \text{ fails at time } t_i \mid one \\ &\qquad \text{failyre from the risk set } R(t_i)) \\ &= \frac{p(\text{individual } i \text{ fails} \mid \text{at risk at } t_i)}{\sum_{l \in R(t_i)} p(\text{individual } l \text{ fails} \mid \text{at risk at } t_i)} \\ &= \frac{\lambda(t_i \mid z_i)}{\sum_{l \in R(t_i)} \lambda(t_i \mid z_l)}. \end{aligned} \tag{7.160}$$

Therefore, the partial likelihood $L(\beta)$ for the entire sample size is given by

$$L(\beta) = \prod_{i=1}^{K} L_i(\beta) = \prod_{i=1}^{K} \frac{\lambda(t_i \mid z_i)}{\sum_{l \in R(t_i)} \lambda(t_i \mid z_l)}. \tag{7.161}$$

Substituting equation (7.157) into equation (7.161) yields

$$\begin{aligned} L(\beta) &= \prod_{i=1}^{K} \frac{\lambda_0(t)\exp(\beta^T z_i)}{\sum_{l \in R(t_i)} \lambda_0(t)\exp(\beta^T z_l)} \\ &= \prod_{i=1}^{K} \frac{\exp(\beta^T z_i)}{\sum_{l \in R(t_i)} \exp(\beta^T z_l)}. \end{aligned} \tag{7.162}$$

7.4.6.3 A Censored-Data Likelihood

Consider the censored data, where we may not observe the failure (survival) times $T_1, \ldots, T_K$. Define two variables:

$$U_i = \min(T_i, C_i), \tag{7.163}$$

$$\delta_i = I(T_i \le C_i), \tag{7.164}$$

where C_i is the (fixed or random) potential censoring time which is defined as a censored observation at time τ if $C_i = \tau$ and $T_i > \tau$.

A continuous observed variable U_i and a binary censoring indicator variable δ_i can take the forms:

$(U_i, \delta_i) = (\tau_i, 1)$, if T_i is uncensored at time τ_i,

$(U_i, \delta_i) = (\tau_i, 0)$, if T_i is censored at time τ_i.

Now we introduce the censored data likelihood. Consider two cases.

Case 1: C_i are known constants

$$L_i(\beta) = \begin{cases} f(\tau_i) & \delta_i = 1 \\ 1 - F(\tau_i) & \delta_i = 0 \end{cases} \tag{7.165}$$

which implies

$$L_i(\beta) = [f(\tau_i)]^{\delta_i}[1 - F(\tau_i)]^{1-\delta_i} = [f(\tau_i)]^{\delta_i}[S(\tau_i)]^{1-\delta_i}. \tag{7.166}$$

Substituting equation (7.117) into equation (7.166), we obtain

$$L_i(\beta) = [\lambda(\tau_i|z_i)]^{\delta_i} S(\tau_i \mid z_i). \tag{7.167}$$

The partial likelihood for the entire sample is given by

$$L(\beta) = \prod_{i=1}^{K} L_i(\beta) = \prod_{i=1}^{K} [\lambda(\tau_i|z_i)]^{\delta_i} S(\tau_i \mid z_i) = \prod_{i=1}^{K} \left[\frac{\lambda(\tau_i|z_i)}{\sum_{l \in R(t_i)} \lambda(t_i|z_l)}\right]^{\delta_i} \left[\sum_{l \in R(t_i)} \lambda(t_i|z_l)\right]^{\delta_i} S(\tau_i \mid z_i). \tag{7.168}$$

The first term in equation (7.168) contains the majority of the information about the parameters β, while the last two terms involve mainly the baseline hazard $\lambda_0(t)$. To simplify the analysis, the partial likelihood often ignores the last two terms and is reduced to

$$L(\beta) = \prod_{i=1}^{K} \left[\frac{\lambda(\tau_i|z_i)}{\sum_{l \in R(t_i)} \lambda(\tau_i|z_l)}\right]^{\delta_i} = \prod_{i=1}^{K} \left[\frac{\lambda_0(\tau)\exp(\beta^T z_i)}{\sum_{l \in R(t_i)} \lambda_0(\tau)\exp(\beta^T z_l)}\right]^{\delta_i} = \prod_{i=1}^{K} \left[\frac{\exp(\beta^T z_i)}{\sum_{l \in R(t_i)} \exp(\beta^T z_l)}\right]^{\delta_i}. \tag{7.169}$$

This is the partial likelihood that does not depend on the baseline hazard function $\lambda_0(\tau)$.

Case 2: C_i are independent and identically distributed as continuous G distribution with density function g.

In equation (7.129), we show that

$$L(\beta) = \prod_{i=1}^{K} [f(\tau_i)]^{\delta_i} [1 - F(\delta_i)]^{1-\delta_i} \prod_{i=1}^{K} [1 - G(\tau_i)]^{\delta_i} [g(\tau_i)]^{1-\delta_i}. \tag{7.170}$$

Equation (7.170) implies that likelihood functions for survival time and censoring time can be calculated separately, if we assume that distributions F and G are functionally independent. In other words, $F = F_\theta$ and $G = G_\phi$, $\theta \in \Theta, \phi \in \Phi$. This provides a very nice property. When we maximize the likelihood with the parameters θ, we can view the likelihood involving G as constants. The parameters θ and ϕ can be estimated separately. In Case 2 when censoring times are random variables, we still can use equation (7.169) in Case 1 as likelihood function for the survival time. Therefore, the likelihood functions for survival time in Cases 1 and 2 can take the same form.

The log-likelihood is

$$l(\beta) = \log L(\beta) = \sum_{i=1}^{K} l_i(\beta), \tag{7.171}$$

where

$$l_i(\beta) = \delta_i \left[\beta^T z_i - \log\left\{ \sum_{l \in R(\tau_i)} e^{\beta^T z_l} \right\} \right]. \quad (7.172)$$

7.4.6.4 Object Function for VAE-Cox Model

Combining equations (7.152) and (7.172), we obtain the objective function for VAE-Cox model:

$$\mathcal{L}_{VAE-Cox}(\theta, \beta) = \mathcal{L}_{SVAE}(\varnothing, \theta, x, t, a) + \gamma l(\beta), \quad (7.173)$$

where

$$\mathcal{L}_{SVAE}(\varnothing, \theta, x, t, a)$$
$$= \frac{1}{2}\left\{ \sum_{m=1}^{M} \left(\log(\sigma_s^m)^2 + 1 \right) - \sum_{m=1}^{M} \left[\log(\sigma_h^m)^2 + \frac{(\sigma_s^m)^2}{(\sigma_h^m)^2} + \frac{(\mu_r^m - \mu_h^m)^2}{(\sigma_h^m)^2} \right] \right\}$$
$$+ \frac{1}{nL}\sum_{i=1}^{n}\sum_{l=1}^{L} \left[\begin{array}{l} \delta^{(i)} \log f_\theta\left(t^{(i,l)} | z^{(i,l)}, a^{(i)}\right) \\ +\left(1-\delta^{(i)}\right) \log S_\theta\left(t^{(i)} \mid z^{(i,l)}, a^{(i)}\right) \end{array} \right],$$

$$\left[\mu_r^{(i)} \mid \log \sigma_s^{(i)}\right] = MLP\left(x^{(i)}, a^{(i)}, t^{(i)}, W_z\right),$$

$$Z^{(i,l)} = \mu_r^{(i)} + \sigma_s^{(i)} \odot \varepsilon^{(l)}, \ \varepsilon^{(l)} \sim N(0, I),$$

$$\left[\mu_h \mid \log \sigma_h\right] = MLP\left(x^{(i)}, a^{(i)}, W_{pr}\right),$$

$$l(\beta) = \sum_{i=1}^{K} \delta_i \left[\beta^T z_i - \log\left\{ \sum_{l \in R(\tau_i)} e^{\beta^T z_l} \right\} \right],$$

7.5 TIME SERIES CAUSAL SURVIVAL ANALYSIS

7.5.1 Introduction

The most survival models consider the binary case: non-fatal and fatal state and model the survival probability of transition from non-fatal to fatal state, given a set of individual covariates (Groha et al. 2020). Some extensions of the CoxPH models, including relaxing the linear and proportional hazards assumptions, have been developed. These extensions model $\lambda(t) = \lambda_0(t)\exp(f_\theta(x))$ with a deep neural network for $f_\theta(x)$ (Lee et al. 2018) and $\lambda(t) = \lambda_0(t)\exp(f_\theta(t,x))$ with continuous time models (Giunchiglia et al. 2018; Katzman et al. 2018; Ren et al. 2018). The survival analysis methods introduced in the previous sections, including classical Cox model, study the occurrence of single fatal events. However, in the real world, multiple states take places, for example, the progress of COVID-19 might be modeled as not infected, infected with non-symptom, infected with symptom, sever COVID-19, hospitalized, and dead. We often observe the individuals with significantly different features and quite different disease trajectories.

To meet these challenges, we need to extend the survival analysis from the single state to multiple states and model multiple non-fatal states and complicated transitions among all states (Gerstung et al. 2017; Grinfeld et al. 2018; Lee et al. 2018; Rueda et al. 2019; Groha et al. 2020; Nicora et al. 2020). In this section, we introduce neural ordinary differential equations which were discussed in Chapter 1 to model the Kolomogorov forward equation of the underlying transitions among the multiple states for multi-state causal survival analysis and producing desired outcomes (Groha et al. 2020).

7.5.2 Multi-State Survival Models

7.5.2.1 Notations and Basic Concepts

We first introduce some basic concepts and notations. The state space is defined as the set of all the possible system states. We number the states by integers from 0 to r. The state space is denoted by

$$Y_s = \{0, 1, 2, \ldots, r\}.$$

Let $p_i(t) = p(Y(t) = i)$ be the probability that the system is in state i at time t. We denote the state probability distribution by $\boldsymbol{P}(t) = \left[p_0(t), p_1(t), \ldots, p_r(t)\right]^T$. Assume that the process is in state i at time S. The stochastic process $Y(t)$ is Markov if

$$\begin{aligned} p\left(Y(s+t) = j | Y(s) = i, Y(u) = y(u), 0 \le u < t\right) \\ = p\left(Y(s+t) = j | Y(s) = i\right) \end{aligned} \quad (7.174)$$

for all possible $Y(u), 0 \le u < t$.

We consider continuous time, finite state space Markov processes. Let $p_{ij}(s,t)$ be the transition probability from state i at time s to the state j at time t, i.e.,

$$p_{ij}(s,t) = p\left(Y(t) = j \mid Y(s) = i\right). \quad (7.175)$$

The Markov process is called a time-homogeneous process, if for all i, j in the state space it satisfies

$$p\left(Y(t+s)=j \mid Y(s)=i\right)=P\left(Y(t)=j \mid Y(s)=i\right), \tag{7.176}$$

for all $s,t \geq 0$.

For the time-homogeneous Markov process, the transition probability $p_{ij}(t)=p\left(Y(t)=j \mid Y(0)=i\right)$ may be arranged as a matrix:

$$P(t)=\begin{bmatrix} p_{00}(t) & \cdots & p_{0r}(t) \\ \vdots & \vdots & \vdots \\ p_{r0}(t) & \cdots & p_{rr}(t) \end{bmatrix}. \tag{7.177}$$

The elements in the transition probability matrix must satisfy

$$\sum_{j=0}^{r} p_{ij}(t)=1. \tag{7.178}$$

A trajectory of the Markov process is shown in Figure 7.5, where $0=S_0 \leq S_1 \leq S_2 \leq \ldots$ be the times at which transitions occur. Let $\tilde{T}_i$ be the sojourn time spent in state i during a visit to state i. The sojourn times $\tilde{T}_1, \tilde{T}_2, \ldots$ are independent and exponentially distributed.

7.5.3 Multi-State Survival Models

In this section, we will introduce works of Groha et al. (2020) on multi-state survival models. Multi-state models are defined as Markov jump processes $\{Y(t); 0 \leq t \leq T\}$, which take values in a finite state space $Y_s=\{1, 2, \ldots, r\}$ and movements are discrete. The discrete movements are called jumps. The Markov jump processes are often modeled as a simple or compound Poisson process.

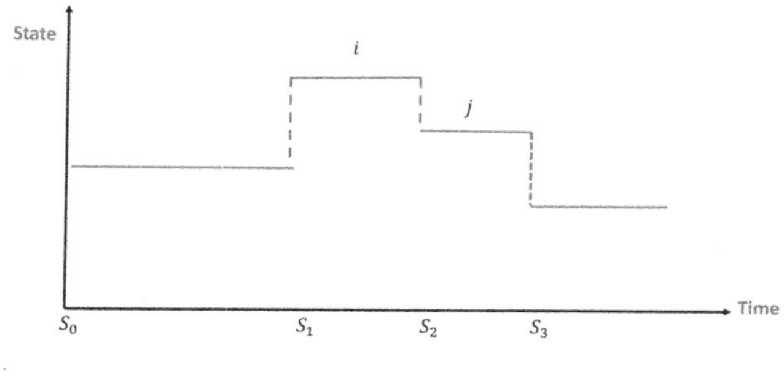

FIGURE 7.5 A trajectory of the Markov process.

7.5.3.1 Transition Probabilities, the Kolmogorov Forward Equations and Likelihood Function

The transition probabilities $p_{ij}(s,t)$ satisfy the Kolmogorov forward equations (Appendix 7B):

$$\frac{dp_{ij}(s,t)}{dt}=\sum_{k} p_{ik}(s,t)\lambda_{kj}(t). \tag{7.179}$$

For each individual, we can observe the Markov process $Y(t)$ with discrete jumps. Consider m time-indexed states $y(t_1), \ldots, y(t_m)$ for a single observation. The likelihood for the single observation is given by

$$\begin{aligned} &p\left(y(t_1), \ldots, y(t_m), \theta\right) \\ &= p\left(y(t_1)\right)\prod_{j=2}^{m} p_{y(t_{j-1})y(t_{j-1})}\left(t_{j-1}, t_j, \theta\right)\lambda_{y(t_{j-1})y(t_j)}\left(t_j \mid \theta\right), \end{aligned} \tag{7.180}$$

where

$p_{y(s)y(s)}\left(t_{j-1}, t_j, \theta\right)$ is the probability that the system is in state $y(s)$ from time t_{j-1} to time t_j, $\lambda_{y(t_{j-1})y(t_j)}$ is the transition rate from state $y\left(t_{j-1}\right)$ to state $y(t_j)$ at time t_j and $\Pi_{j=2}^{1}=1$.

Let $t_1^i, \ldots, t_{m_i}^i$ be the m_i time points for the individual i, $y_i=\left\{y_i\left(t_1^i\right), \ldots, y_i\left(t_{m_i}^i\right)\right\}$ be a set of states, and $Y=\left\{y_1, \ldots, y_n\right\}$. The log-likelihood for the observations of n individuals is defined as

$$l(\theta, Y)=\sum_{i=1}^{n} \log p\left(y_i\left(t_1^i\right), \ldots, y_i\left(t_{m_i}^i\right); \theta\right), \tag{7.181}$$

where the transition probability $p\left(y_i\left(t_1^i\right), \ldots, y_i\left(t_{m_i}^i\right); \theta\right)$ is defined in equation (7.180).

Let $p_{y(s)y(t)}(s,t)=T\left(y(s), y(t)\right)$. Then, under Markov assumption, the transition probability $p_{y(s)y(t)}$ is determined by the Kolmogorov forward equations:

$$\frac{dp_{y(s)y(t)}(s,t)}{dt}=\sum_{k} p_{y(s)k}(s,t)\lambda_{ky(t)}(t), \tag{7.182}$$

where $s=t_{j-1}, t=t_j$ and the transition rate $\lambda_{ky(t)}(t)$ only depends on the time t.

7.5.3.2 Likelihood Function with Interval Censoring

Assume the right censoring. Let $D=\{x_i, y_i, \delta_i, i=1,\ldots,n, j=1,\ldots,m_i\}$ be a set of n observations, where x_i are the covariates for the individual i, y_i are the states or the states at time of last contact (censoring time) for the individual i, and δ_i is a censoring indicator variable defined as

$$\delta_i = \begin{cases} 0 & \text{censored} \\ 1 & \text{observed} \end{cases}.$$

The log-likelihood for a set of n censoring data is given by

$$\begin{aligned} l(\theta,Y) = \sum_{i=1}^{n} \Bigg\{ & \log p\left(y_i(t_1^i)\right) \sum_{j=2}^{m_i-1} \Big[\log p_{y_i(t_{j-1}^i) y_i(t_{j-1}^i)} \\ & \left(t_{j-1}^i, t_j^i; \theta\right) + \log \lambda_{y_i(t_{j-1}^i) y_i(t_j^i)}\left(t_j^i; \theta\right) \Big] \\ & + \log p_{y_i(t_{m_i-1}^i) y_i(t_{m_i-1}^i)}\left(t_{m_i-1}^i, t_{m_i}^i; \theta\right) \\ & + \delta_i \log \lambda_{y_i(t_{m_i-1}^i) y_i(t_{m_i}^i)}\left(t_{m_i}^i; \theta\right) \Bigg\}. \end{aligned} \tag{7.183}$$

7.5.3.3 Ordinary Differential Equations (NODE) for Multi-State Survival Models

A transition rate matrix can be expressed as three forms:

$$Q(s,t) = \begin{bmatrix} \lambda_{00}(s,t) & \cdots & \lambda_{0r}(s,t) \\ \vdots & \vdots & \vdots \\ \lambda_{r0}(s,t) & \cdots & \lambda_{rr}(s,t) \end{bmatrix}, \tag{7.184}$$

$$Q(t) = \begin{bmatrix} \lambda_{00}(t) & \cdots & \lambda_{0r}(t) \\ \vdots & \vdots & \vdots \\ \lambda_{r0}(t) & \cdots & \lambda_{rr}(t) \end{bmatrix}, \tag{7.185}$$

$$Q = \begin{bmatrix} \lambda_{00} & \cdots & \lambda_{0r} \\ \vdots & \vdots & \vdots \\ \lambda_{r0} & \cdots & \lambda_{rr} \end{bmatrix}. \tag{7.186}$$

Equation (7.184) is a general form of the transition matrix. The elements in the transition rate matrix should satisfy three conditions:

1. $0 \le -\lambda_{ii}(s,t) < \infty$,
2. $\lambda_{ij}(s,t) \ge 0, i \ne j$,
3. $\sum_j \lambda_{ij}(s,t) = 0$ for all i or

$$-\lambda_{ii}(s,t) = \sum_{j \ne i} \lambda_{ij}(s,t). \tag{7.187}$$

The transition rate matrix can be modeled by a neural network. To ensure positive transition rates, a softplus activation will be used on the last layer of the neural network. Since homogeneous Markov process is widely used, in the below introduction we will focus on the transition matrix $Q(t)$.

Transition rates also depend on the covariates. We need to incorporate the covariates and the history of the evolution into the model. To achieve this, auxiliary memory states $m(t) = [m_1(t), \ldots, m_n(t)]^T$ are introduced. The memory state equation is given by

$$\frac{dm_i}{dt} = M_i(t, \mathbf{P}(t), m(t)), \tag{7.188}$$

where, $\mathbf{P}(t) = [p_0(t), p_1(t), \ldots, p_r(t)]^T$ is the vector of the state probability distributions. The covariates of the individuals are encoded in the initial conditions of ODE (7.188):

$$m(0) = f(x),$$

where $f(x)$ is a vector of nonlinear functions and modeled by a neural network.

To incorporate the covariates into equation (7.182), we need to model the transition rates as a function of covariates:

$$\lambda_{kj}(t, P(0,t), m(t), x), \tag{7.189}$$

where $P(s,t)$ is a transition probability matrix. Let $s=0$. The Kolmogorov forward and backward equations incorporating covariates are given by

$$\frac{dp_{ij}(0,t)}{dt} = \sum_k p_{ik}(0,t) \lambda_{kj}(t, P(0,t), m(t), x) \tag{7.190}$$

$$\frac{dp_{ij}(s,0)}{ds} = -\sum_k \lambda_{ik}(s, P(0,s), m(s), x) p_{kj}(s,0). \quad (7.191)$$

Let

$$P(0,t) = \begin{bmatrix} p_{00}(0,t) & \cdots & p_{0r}(0,t) \\ \vdots & \vdots & \vdots \\ p_{r0}(0,t) & \cdots & p_{rr}(0,t) \end{bmatrix},$$

$$Q_{ij}(t,x) = \lambda_{ij}(t, P(0,t), m(t), x),$$

$$Q(t,x) = \begin{bmatrix} Q_{00}(t,x) & \cdots & Q_{0r}(t,x) \\ \vdots & \vdots & \vdots \\ Q_{r0}(t,x) & \cdots & Q_{rr}(t,x) \end{bmatrix},$$

$$M(t) = \begin{bmatrix} M_0(t, P(t), m(t)) \\ \vdots \\ M_r(t, P(t), m(t)) \end{bmatrix}.$$

Then, equations (7.188), (7.190), and (7.191) can be rewritten in a vector or matrix form:

$$\frac{dm(t)}{dt} = M(t), \quad (7.192)$$

$$\frac{dP(0,t)}{dt} = P(0,t)Q(t,x), \quad (7.193)$$

$$\frac{dP(s,0)}{ds} = -Q(s,x)P(s,0). \quad (7.194)$$

Now we introduce Algorithm 7.1 for solving ODE using PyTorch that can automatically compute gradients of models specified by the users and solve ODE (Paszke et al. 2019; Groha et al. 2021). Before introducing the algorithm, we recall that the softplus activation function is defined as $f(x) = \log(1 + e^x)$.

Algorithm 7.1: Solve ODE for $p_{ij}(s,t)$ and $\lambda_{ij}(t)$

Input: Covariates x and time interval (s,t).

Step 1: Initialization.

Set $m(0) = f_\theta(x)$, $\boldsymbol{P}(0,0) = 1$ and

$$S_0 = (\boldsymbol{P}(0,0), \boldsymbol{P}(0,0), \boldsymbol{m}(0)).$$

Step 2: Define Kolmogorov forward and backward equation.

Function KFE_KBE (**P**(0,t), **P**(t,0), **m**(t), t)

$\lambda_{ij}(t), M_i(t) = g_\phi(P(0,t), m(t), x, t)$ {g_ϕ implemented by NN with softplus for λ}

$\lambda_{ii}(t) = -\sum_k \lambda_{ik}(t)$ {transition rate equality}

$\frac{dP_{ij}(0,t)}{dt} = \sum_k p_{ik}(0,t)\lambda_{kj}(t)$ {calculate gradient for Kolmogorov forward equation}

$\frac{dp_{ij}(t,0)}{dt} = -\sum_k \lambda_{ik}(t) p_{kj}(t,0)$ {calculate gradient for Kolmogorov backward equation}

$\frac{dm_i(t)}{dt} = M_i(t)$ {calculate gradient for augmented equation}

return $\left[\frac{dp_{ij}(0,t)}{dt}, \frac{dp_{ij}(t,0)}{dt}, \frac{dm_i(t)}{dt}\right]$ {calculate gradients}

$$\boldsymbol{P}(0,t), \boldsymbol{P}(s,0), \boldsymbol{m}(t)$$
$$= ODESolve(S_0, KFE_{KBE}, (0,t), save_{at} = (s,t))$$

$\boldsymbol{P}(s,t) = \boldsymbol{P}(s,0)\boldsymbol{P}(0,t)$ {calculate the transition probability matrix}

$$p_{ij}(s,t), \lambda_{ij}(t)$$

7.6 NEURAL ORDINARY DIFFERENTIAL EQUATION APPROACH TO TREATMENT EFFECT ESTIMATION AND INTERVENTION ANALYSIS

7.6.1 Introduction

A key issue for neural networks is the instability of training neural networks, which is due to gradient descent optimization algorithm (Qin et al. 2020). Suppose that we want to minimize the objective function $L(x)$. The popular discrete dynamic approach to unconstrained optimization is gradient methods. Specifically, the optimization problem

$$\min_x L(x) \quad (7.195)$$

can be solved by iterative discrete gradient methods:

$$X_{t+1} = X_t - \nabla_x L(X_t). \quad (7.196)$$

Discretization error of iterative gradient may cause instability of the neural network training. To overcome this problem, we take continuous approach. Equation (7.196) can be transformed to (Lin 1991):

$$\frac{dX}{dt} = -\nabla_x L(X) \quad (7.197)$$

with the initial condition

$$X(0)=X_0.$$

When $t\to\infty$, then $X(t)$ will converge to X_* and $\nabla_x L(X_*)=0$, which is the necessary condition that X_* is a minimizer of the objective function $L(X)$. The idea that optimization problem can be solved by ordinary differential equation can be applied to neural network learning. The ordinary differential equation that is derived from neural learning is called neural ordinary differential equation (NODE) (Chen et al. 2018).

Consider an optimization problem:

$$\min_{\theta} l(z(t),\theta) \tag{7.198}$$

Subject to

$$\frac{dz(t)}{dt}=f(z(t),t,\theta) \tag{7.199}$$

$$z(t_0)=z_0. \tag{7.200}$$

The dynamics of residual networks, recurrent neural network decoders, and normalizing flows are often described by the following difference equation:

$$z_{t+1}=z_t+f(z_t,\theta), \tag{7.201}$$

where z_t is an input to the neural network, z_{t+1} is the output of the neural network, and $f(z_t,\theta)$ is a neural network transformation function. The solution to ODE (7.199) can be written as

$$z(t)=z(t_0)+\int_{t_0}^{t} f(z(\tau),\tau,\theta)d\tau. \tag{7.202}$$

Viewing θ as control variables and $J(\theta)=l(z(t),t,\theta)$ be the terminal cost function, we can take the optimization problem (7.198)–(7.200) as an optimal control problem. Let $\psi(t)$ be the adjoint and $H(z(t),\theta(t),\psi(t),t)=\psi^T(t)f(z(t),\theta,t)$ be the Hamiltonian function. Then, applying equations (7D11 and 7D12) in Appendix 7D (the Pontryagin's maximum principle), we obtain

$$\psi(t)=-\frac{\partial l(z(t),\theta)}{\partial z} \tag{7.203}$$

$$\dot{\psi}(t)=-\psi^T(t)\frac{\partial f(z(t),t,\theta)}{\partial z}. \tag{7.204}$$

Substituting equation (7.202) into $l(z(t),\theta)$ we obtain

$$l(z(t),\theta)=l\left(z(t_0)+\int_{t_0}^{t} f(z(\tau),\tau,\theta)d\tau\right). \tag{7.205}$$

Now we calculate the gradient of the cost function. Taking derivative with respect to θ on both sides of equation (7.205), we obtain

$$\frac{\partial l}{\partial\theta}=\left(\frac{\partial l}{\partial z}\right)^T\int_{t_0}^{t}\frac{\partial f(z(\tau),\tau,\theta)}{\partial\theta}d\tau. \tag{7.206}$$

Substituting equation (7.203) into equation (7.206) yields

$$\begin{aligned}\frac{\partial l}{\partial\theta}&=-\psi^T(t)\int_{t_0}^{t}\frac{\partial f(z(\tau),\tau,\theta)}{\partial\theta}d\tau\\&=-\int_{t_0}^{t}\psi^T(t)\frac{\partial f(z(\tau),\tau,\theta)}{\partial\theta}d\tau\\&=\int_{t}^{t_0}\psi^T(t)\frac{\partial f(z(\tau),\tau,\theta)}{\partial\theta}d\tau.\end{aligned} \tag{7.207}$$

Equations (7.203), (7.204), and (7.207) form NODE for neural learning. Next we introduce several applications of NODE to neural learning.

7.6.2 Latent NODE for Irregularly-Sampled Time Series

In Chapter 1, we introduce RNN as a power tool for modeling regularly-sampled time series data. However, the RNN is less efficient to fit for irregularly-sampled time series data. The irregularly-sampled time series can be modeled by a continuous time latent trajectory (Chen et al. 2018).

Each trajectory is determined by the following dynamic system with a local state $z_{t_0}^i$ and a global function $f(z(t),t,\theta_1)$ across all n time series:

$$\begin{aligned}\frac{dz}{dt}&=f(z(t),t,\theta_1),\\ z^i(t_0)&=z_{t_0}^i,\ i=1,\ldots,n.\end{aligned} \tag{7.208}$$

Following a standard VAE approach, time series $x_{t_0,\ldots,}x_{t_N}$ is mapped to a latent space and a variational posterior distribution $q\left(z_{t_0} \mid x_{t_0,\dots,}x_{t_N}\right)$ will be determined by an RNN. The RNN is modeled by

$$h\left(t_j\right)=g\left(h\left(t_{j-1}\right), x\left(t_j\right)\right). \quad (7.209)$$

We assume that the variational posterior distribution $q\left(z_{t_0} \mid x_{t_0,\dots,}x_{t_N}\right)$ follows a D dimensional Gaussian distribution:

$$q\left(z_{t_0}|x_{t_0,\dots,}x_{t_N}\right)=N\left(z_{t_0}; \mu_{z_0}\left(h_{t_N}\right), \sigma_{z_0}^2\left(h\left(t_N\right)\right)\right),$$

where

$$\left[\mu_{z_0}\left(h_{t_N}\right)\|\log\sigma_{z_0}\left(h\left(t_N\right)\right)\right]=MLP\left(h\left(t_N\right),W\right), \quad (7.210)$$

MLP denotes a multiple perceptron and W denotes weight matrices.

Define

$$z^l\left(t_0\right)=\mu_{z_0}\left(h_{t_N}\right)+\sigma_{z_0}\left(h\left(t_N\right)\right)\varepsilon^l,$$
$$\varepsilon^l \sim N(0,1), l=1,\dots, L. \quad (7.211)$$

Let $p\left(z_{t_0}\right) \sim N(0,I)$ and $\theta=\left(\theta_1, \theta_2\right)$. Calculate the ELBO as follows:

$$J(\theta, z(t))=\frac{1}{L}\sum_{i=1}^{L}\sum_{j=1}^{M}\log p\left(x\left(t_j\right)\|z^i\left(t_j\right), \theta_1\right)$$
$$-KL\left(q\left(z_{t_0}|x_{t_0,\dots,}x_{t_N},\theta_2\right)\|p\left(z_{t_0}\right)\right). \quad (7.212)$$

Define

$$p\left(x\left(t_j\right)\|z^i\left(t_j\right), \theta_1\right)=N(x\left(t_j\right);$$
$\mu\left(z^i\left(t_j\right)\right), \sigma^2\left(z^i\left(t_j\right)\right)$, which implies
$$\log p\left(x\left(t_j\right)\|z^i\left(t_j\right), \theta_1\right)$$
$$=-\frac{1}{2}\{m\log(2\pi)$$
$$+\sum_{g=1}^{m}\left[\log\sigma_g^2\left(z^i\left(t_j\right)\right)-\frac{\left(x_g\left(t_j\right)-\mu_g\left(z^i\left(t_j\right)\right)\right)^2}{\sigma_g^2\left(z^i\left(t_j\right)\right)}\right]. \quad (7.213)$$

We can show (Exercise 7.9) that KL distance $KL\left(q\left(z_{t_0}|x_{t_0,\dots,}x_{t_N}, \theta_2\right)\|p\left(z_{t_0}\right)\right)$ is given by

$$-KL\left(q\left(z_{t_0}|x_{t_0},\dots, x_{t_N}, \theta_2\right)p\left(z_{t_0}\right)\right)$$
$$=\frac{1}{2}\sum_{d=1}^{D}\left[\log\sigma_{z_0d}^2\left(h\left(t_N\right)\right)+1\right.$$
$$\left.-\sigma_{z_0d}^2\left(h\left(t_N\right)\right)-\mu_{z_0d}^2\left(h\left(t_N\right)\right)\right]. \quad (7.214)$$

Substituting equations (7.213) and (7.214) into equation (7.212), we obtain

$$J(\theta,z(t))=-\frac{1}{L}\sum_{i=1}^{L}\sum_{j=1}^{M}$$
$$\times\left\{m\log(2\pi)+\sum_{g=1}^{m}\left[\log\sigma_g^2\left(z^i\left(t_j\right)\right)-\frac{\left(x_g\left(t_j\right)-\mu_g\left(z^i\left(t_j\right)\right)\right)^2}{\sigma_g^2\left(z^i\left(t_j\right)\right)}\right]\right\}$$
$$+\frac{1}{2}\sum_{d=1}^{D}\left[\log\sigma_{z_0d}^2\left(h\left(t_N\right)\right)+1-\sigma_{z_0d}^2\left(h\left(t_N\right)\right)\right.$$
$$\left.-\mu_{z_0d}^2\left(h\left(t_N\right)\right)\right]. \quad (7.215)$$

NODEs are given by

$$\psi(t)=-\frac{\partial J(\theta,z(t))}{\partial z} \quad (7.216)$$

$$\dot{\psi}(t)=-\psi^T(t)\frac{\partial f(z(t),t,\theta)}{\partial z} \quad (7.217)$$

$$\frac{\partial J(\theta,z(t))}{\partial\theta}=\int_t^{t_0}\psi^T(\tau)\frac{\partial f(z(\tau),\tau,\theta)}{\partial\theta}d\tau \quad (7.218)$$

$$\frac{dz}{dt}=f\left(z(t),t, \theta_1\right) \quad (7.219)$$

$$z^i\left(t_0\right)=z_{t_0}^i, i=1,\dots, n.$$

Algorithm for solving NODEs (7.216)–(7.219) is left as exercise.

7.6.3 Augmented Counterfactual ODE for Effect Estimation of Time Series Interventions with Confounders

We consider time series data where both response and intervention are time varying. Time series causal analysis has wide applications. For example, EHRs collect individual longitudinal data with the treatment records and the patients' response. Similar to the treatment effect estimation for cross-sectional data, a general framework for treatment effect estimation with time series (longitudinal) data is also counterfactual inference. Treatment effects can be estimated as the difference between the factual and counterfactual outcomes.

The methods for time series counterfactual inference without unobserved confounders have been developed, including modeling the intervention and response as linear time-invariant (LTI) dynamic system (Soleimani et al. 2017), using Counterfactual Gaussian Process (CGP) to model the counterfactual future outcomes under sequences of future interventions (Schulam and Saria 2017), and recurrent marginal structural networks for learning dynamic treatment responses over time (Lim 2018). Li et al. (2021) recently proposed to use augmented counterfactual ordinary differential equations (ACODEs) for time series counterfactual inference with hidden confounders. This section we focus on introducing application of NODE to estimation of dynamic treatment effect with hidden confounders (Li et al. 2021).

7.6.3.1 Potential Outcome Framework for Estimation of Effect of Time Series Interventions

Consider a multivariate time series outcomes $Y(t)$, a multivariate time series covariates $X(t)$, and continuous-time time-dependent interventions $A(t)$. The observational data consists of multiple realizations of time series outcomes, covariates, and interventions. Since the realization are independent to each other, for the simplicity in the below discussion, we only consider one realization. Suppose that up to time t, N time series outcomes and covariates $D_N(t)=\left\{y_{i(t)},x_i(t)\right\}_{i=1}^{N}$ and continuous-time interventions $\left\{a(s),s\le t\right\}$ are observed. Let $a_{>t}$ denote $\{a(s):s>t\}$. Time series potential outcome framework is defined as learning the conditional distribution of the potential outcomes $Y(a_{>t})$ under future sequence of interventions $a_{>t}$, given the historical data $\left\{y_i,x_i\right\}_{i=1}^{N}$:

$$p\left(Y\left(a_{>t}\right)|\,a_{\le t},\left\{y_i,x_i,t_i\right\}_{i=1}^{N}\right). \tag{7.220}$$

We cannot directly estimate the distribution of the potential outcome. However, under some assumptions, we can fit the regression model to the observational data to estimate the distribution (Li et al. 2021)

$$p\left(Y(s>t)\,|\,a_{>t},a_{\le t},\left\{y_i,x_i,t_i\right\}_{i=1}^{N}\right), \tag{7.221}$$

which can approximate $p\left(Y\left(a_{>t}\right)|a_{\le t},\left\{y_i,x_i,t_i\right\}_{i=1}^{N}\right)$.

Now we state three essential assumptions for the potential outcome framework for causal inference.

Assumptions 7.4

We introduce assumptions in the Neyman-Rubin model (Bica et al. 2020; Li et al. 2021).

Assumption 1. (Consistency). The potential outcome for receiving an intervention plan $a_{\ge t}$ is the same as the observed outcome $Y\left(a_{\ge t}\right)=Y\,|\,a_{\ge t}$.

Assumption 2. (Overlap). For all $\left(a_1,x_1,\ldots,a_{t-1},x_{t-1}\right)$, we have

$$\begin{aligned}0<p\big(&A_t=a_t|A_1=a_1,X_1=x_1,\ldots,\\&A_{t-1}=a_{t-1},X_{t-1}=x_{t-1}\big)<1.\end{aligned} \tag{7.222}$$

Assumption 3. (Sequential strong ignorability). Conditional on $a_{<t},x_{<t},y_{<t}$, the potential outcomes $Y\left(a_{\ge t}\right)$ are independent of $a(t)$,

$$Y\left(a_{\ge t}\right)\perp\!\!\!\perp a(t)\,|\,a_{<t},x_{<t},y_{<t}. \tag{7.223}$$

Assumption 3 implies that there are no confounders, which affect both outcomes and interventions. In practice, the assumptions cannot be tested. To mitigate the effect of hidden confounders, we augment the space that includes the true temporal dynamics as its subspace. Working on the time series in the augmented space with high dimensions may make it easier to approximate the true temporal dynamics in the original space with low dimensions. Next we introduce augmented ODE that is proposed by Li et al. (2021).

7.6.3.2 Augmented Counterfactual Ordinary Differential Equations

We first introduce k auxiliary time series $U(t)\in R^k$ along with the original time series $Y(t)\in R^d$ and

$X(t) \in R^m$ to generate the augmented time series $G(t) = \begin{bmatrix} Y(t) \\ X(t) \\ U(t) \end{bmatrix} \in R^{d+m+k}$. In the augmented space, we assume

$$p\left(y(a_{>t})|a_{\le t}, \{u(t_i), t_i, x(t_i)\}_{i=1}^n\right) = p\left(y_{>t} \mid a_{>t}, a_{\le t}, \{u(t_i), t_i, x(t_i)\}_{i=1}^n\right), \quad (7.224)$$

where in general, auxiliary variables $U(t) = u(t)$ are initialized with all zero vectors.

VAE is used to model training and counterfactual inference process (Rubanova et al. 2019; Li et al. 2021). We first introduce latent counterfactual differential equation (CDE), which is a basic component of both encoder and decoder.

7.6.3.2.1 Latent Counterfactual Differential Equation We assume that the latent dynamic system is an additive model where forces drive the dynamics of the latent system is the state space alone $f_z(z(t), \theta_z)$ and a combination of states and interventions $f_a(z(t), a(t), \theta_a)$. The CDF is given by

$$\frac{dz(t)}{dt} = f_z(z(t), \theta_z) + f_a(z(t), a(t), \theta_a) \quad (7.225)$$

$$z(t_0) = z_0$$

$$G(t) \sim p(G(t)|z(t)), \text{ or } Y(t) \sim p(Y(t)|z(t))$$

$$\text{and } X(t) \sim p(X(t)|z(t)).$$

where both $f_z(z(t), \theta_z)$ and $f_a(z(t), a(t), \theta_a)$ are parameterized by neural networks:

$$f_z(z(t), \theta_z) = \text{ReLU}(z(t)W_z^1)W_z^2 \quad (7.226)$$

$$f_a(z(t), a(t), \theta_a) = \text{ReLU}([z(t), a(t)]W_a^1)W_a^2, \quad (7.227)$$

and

$Y(t)$ and $X(t)$ are Gaussian distributions:

$$Y(t) \sim N(Y(t); \mu_Y(t), \Sigma_Y(t)),$$
$$X(t) \sim N(X(t); \mu_X(t), \Sigma_X(t)). \quad (7.228)$$

The means $\mu_Y(t)$, $\mu_X(t)$, and covariance matrices $\Sigma_Y(t)$, $\Sigma_X(t)$ are parameterized by neural networks. For example, two layer MLPs are defined as

$$[\mu_Y(t)|\Sigma_Y(t)] = \text{ReLU}(z(t)W_Y^1)W_Y^2, \quad (7.229)$$

$$[\mu_X(t)|\Sigma_X(t)] = \text{ReLU}(z(t)W_X^1)W_X^2. \quad (7.230)$$

We modified algorithm for solving the latent CDE in Li et al. (2021) to Algorithm 7.2, including covariates $x(t)$.

Algorithm 7.2: Solve the Latent CDE

Input: A distribution of initial state z_0 of the latent variable $p(z_0)$, time points $\{t_i\}_{i=1}^n$, continuous-time intervention process $a_{\le t}$.

Output: Time series observations $\{(y_i, x_i, t_i\}_{i=1}^n$, corresponding latent states $\{z_i, t_i\}_{i=1}^n$.

Step 1: Sampling z_0 from $p(z_0)$.

Step 2: Define functions f_z and f_a in equations (7.226) and (7.227).

Step 3: Compute latent states $z_1, \ldots, z_N = \text{ODESolve}\left(f_z, f_a, z_0, a_{\le t}, \{t_i\}_{i=1}^n\right)$.

Step 4: Iteration. **for** $i = 1, \ldots, N$ **do**

a. Compute

$$[\mu_Y(t_i)|\Sigma_Y(t_i) = \text{ReLU}(z(t_i)W_Y^1)W_Y^2$$

$$[\mu_X(t_i)|\Sigma_X(t_i)] = \text{ReLU}(z(t_i)W_X^1)W_X^2$$

b. Sample

$$y_i \sim N(\mu_Y(t_i), \Sigma_Y(t_i)), x_i \sim N(\mu_X(t_i), \Sigma_X(t_i)).$$

end for

Step 5: return $\{y_i, x_i, z_i\}_{i=1}^N$.

Now we introduce encoder (Li et al. 2021). The encoder maps the observed data $(y(t_i), x(t_i), u(t_i))$ to the hidden state in RNN. Unlike the vanilla RNN where each hidden state has only one variable, each hidden state in the new RNN is decomposed to two hidden substates (h_i, h_i'), where h_i' is an additional state and determined by ODE. The encoder consists of two components: CDE and

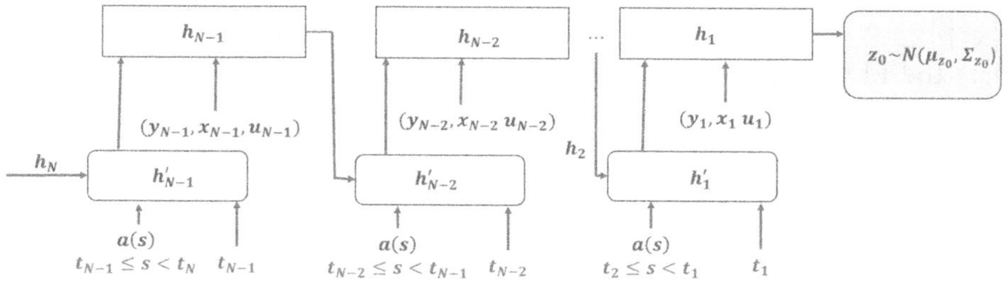

FIGURE 7.6 Architecture of the CDE-RNN encoder.

RNN. Thus, the encoder with two components CDE and RNN is called the CDE-RNN encoder. We define two neural network parameterized functions for the CDE.

$$g_h(h(t)) = \text{ReLU}(h(t)W_h^1)W_h^2 \tag{7.231}$$

$$g_a(h(t)) = \text{ReLU}([h(t), a(t)]W_{ha}^1)W_{ha}^2. \tag{7.232}$$

The architecture of the CDE-RNN encoder is shown in Figure 7.6 and its computational procedure is summarized in Algorithm 7.3 (Li et al. 2021).

Algorithm 7.3: The CDE-RNN Encoder

Input: Time series observations: $\{y_i, x_i, t_i\}_{i=1}^n$, the corresponding time intervention process $a_{\leq t}$.

Output: Hidden states of the RNN $\{h_i, t_i\}_{i=1}^n$.

Step 1: Initialization. Set $h_N = 0, u_i = 0, i = 1, \ldots, N$.

Step 2: Define functions.

$$g_h(h(t)) = \text{ReLU}(h(t)W_h^1)W_h^2, g_a(h(t)) = \text{ReLU}([h(t), a(t)]W_{ha}^1)W_{ha}^2.$$

Step 3: for $i = N-1, \ldots, 1$ **do**

Update $h'_i = \text{ODESolve}\left(g_h, g_a, (t_i, t_{i+1}), h_{i+1}, \{a(s), t_i \leq s < t_i\}\right)$

$$h_i = RNN\ (h'_i, g_i), g_i = \begin{bmatrix} y(t_i) \\ x(t_i) \\ u(t_i) \end{bmatrix}.$$

end for

Step 4: return $\{h_i\}_{i=N-1}^0$.

The posterior $q\left(z_0 \mid \{y_i, x_i, t_i\}_{i=1}^N, a_{\leq t}\right)$ is defined as

$$q\left(z_0 | \{y_i, x_i, t_i\}_{i=1}^N, a_{\leq t}\right) = N\left(\mu_{z_0}, \Sigma_{z_0}\right), \tag{7.233}$$

$$\left[\mu_{z_0} | \Sigma_{z_0}\right] = \text{ReLU}\left(h_1 W_{z_0}^1\right)W_{z_0}^2, \tag{7.234}$$

where $W_{z_0}^1$ and $W_{z_0}^2$ are weight matrices of two layer MLP.

Next we introduce decoder. Decoder is summarized in Algorithm 7.4.

Algorithm 7.4: CDE Decoder

Input: μ_{z_0}, Σ_{z_0}, the time intervention process $a_{\leq t}$.

Output: The reconstruct time series $\{y_i, x_i, t_i\}_{i=1}^N$

Step 1: Initialization.

Sample $z_0 \sim N(\mu_{z_0}, \Sigma_{z_0})$.

Step 2: Define functions.

$$f_z(z(t), \theta_z) = \text{ReLU}(z(t)W_z^1)W_z^2$$

$$f_a(z(t), a(t), \theta_a) = \text{ReLU}([z(t), a(t)]W_a^1)W_a^2$$

Step 3: Compute

$$z_1 \ldots z_N = \text{ODESolve}\left(f_z, f_a, z_0, a_{\leq t}, \{t_i\}_{i=1}^N\right)$$

Step 4: Iteration.

for $i = 1, \ldots, N$ **do**

Compute $\left[\mu_{g_i}(z_i) | \Sigma_{g_i}(z_i)\right] = \text{ReLU}\ (z_i W_g^1)W_g^2$

Sample $g_i \sim N\left(\mu_{g_i}(z_i), \Sigma_{g_i}(z_i)\right)$

end for

Step 5: return $\{g_i\}_{i=1}^N$

Now we calculate the ELBO. Assume $z_0 \in R^J$. We can show (Exercise 7.11) that

$$-KL\left(q\left(z_0|\{y_i, x_i, t_i\}_{i=1}^N, a_{\le t}\right)||p(z_0)\right) = \frac{1}{2}\left[\log|\Sigma_{z_0}| + J - \mathrm{tr}\left(\Sigma_{z_0}\right) - \mu_{z_0}^T \mu_{z_0}\right]. \quad (7.235)$$

Next we calculate $E_{z_0 \sim q_\varnothing(z_0|\{g_i\}_{i=1}^N, a_{\le t})}\left[\log p_\theta(g_1,\ldots, g_N | z_1,\ldots, z_n)\right]$. Sampling $z_0^l \sim N\left(\mu_{z_0}, \Sigma_{z_0}\right), l = 1,\ldots, L$ and using ODESolve $\left(f_z, f_a, z_0, a_{\le t}, \{t_i\}_{i=1}^N\right)$, we obtain the sampled sequences of latent variables $z_1^l,\ldots, z_N^l, l = 1,\ldots, L$ and sampling formula:

$$E_{z_0 \sim q_\varnothing(z_0|\{g_i\}_{i=1}^N, a_{\le t})}\left[\log p_\theta(g_1,\ldots,g_N|z_1,\ldots, z_n)\right] = -\frac{1}{2L}\sum_{l=1}^{L}\sum_{i=1}^{N}\left[(d+k+m)\log(2\pi) + \log\left|\sum_{g_i}^{l}\right| + \left(g_i^l - \mu_{g_i}^l\right)^T\left(\sum_{g_i}^{l}\right)^{-1}\left(g_i^l - \mu_{g_i}^l\right)\right]. \quad (7.236)$$

Combining equations (7.235) and (7.236), we obtain the ELBO:

$$\mathcal{L}_{ELBO} = -\frac{1}{2L}\sum_{l=1}^{L}\sum_{i=1}^{N}\left[(d+k+m)\log(2\pi) + \log\left|\sum_{g_i}^{l}\right| + \left(g_i^l - \mu_{g_i}^l\right)^T\left(\sum_{g_i}^{l}\right)^{-1}\left(g_i^l - \mu_{g_i}^l\right)\right] + \frac{1}{2}\left[\log\left|\sum_{z_0}\right| + J - \mathrm{tr}\left(\sum_{z_0}\right) - \mu_{z_0}^T\mu_{z_0}\right]. \quad (7.237)$$

The overall learning procedure of the augmented counterfactual ODE for estimation of effect of time series intervention is summarized in Algorithm 7.5 (Li et al. 2021).

Algorithm 7.5: Overall Procedure of the Augmented Counterfactual ODE

Input: Time series $\{y_i, x_i\}_{i=1}^n$, continuous-time intervention $a_{\le t}$. Initial values of parameters $(\theta, \varnothing)$.

While not converged **do**

Step 1: Select a time series $\{(y_i, x_i)\}_{i=1}^N$ and intervention process $a_{\le t}$.

Initialize auxiliary variable u_i with all zero vectors for $i = 1,\ldots, N$.

Augment time series $g_i = \begin{bmatrix} y_i \\ x_i \\ u_i \end{bmatrix}$ for $i = 1,\ldots, N$.

Initialize hidden state vector h_N with zeros.

Step 2: for $i = N-1,\ldots, 0$ **do**

Update $h_i' = \text{ODESolve}\left(g_h, g_a, (t_i, t_{i+1}), h_{i+1}, \{a(s): t_i \le s < t_{i+1}\}\right)$

Update $h_i = RNN\left(h_i', g_i\right)$

end for

Step 3: Compute $\left[\mu_{z_0}|\Sigma_{z_0}\right] = \text{ReLU}\left(h_1 W_{z_0}^1\right)W_{z_0}^2$.

Sample $z_0 \sim N\left(\mu_{z_0}, \Sigma_{z_0}\right)$.

Step 4: Compute $z_1,\ldots, z_n = \text{ODESolve}\left(f_z, f_a, z_0, a_{>t}, \{t_i\}_{i=1}^n\right)$

Step 5: for $i = 1,\ldots, N$ **do**

Compute $\left[\mu_{g_i}(z_i)|\Sigma_{g_i}(z_i)\right] = \text{ReLU}\left(z_i W_g^1\right)W_g^2$

Sample $g_i \sim N\left(\mu_{g_i}(z_i), \Sigma_{g_i}(z_i)\right)$

end for

Step 6: Compute the gradient $\nabla_{\theta,\varnothing}\mathcal{L}_{ELBO}(\theta,\varnothing)$ in equation (7.237).

Iterate $(\theta, \varnothing)$ with Adam optimizer.

end while

7.7 GENERATIVE ADVERSARIAL NETWORKS FOR COUNTERFACTUAL AND TREATMENT EFFECT ESTIMATION

Treatment response is heterogeneous. However, the classical methods treat the treatment response as homogeneous and estimate the ATEs. Therapy should be offered personally to ensure that the right therapy is offered to "the right patient at the right time" (Subbiah and Kurzrock 2018; Ge et al. 2020). Consequently, alternative to calculating the average effect of an intervention over a population, we should estimate ITEs. Methods for estimation of ITEs using observational data largely differ from standard statistical estimation methods. A key to estimating treatment effect is to infer counterfactual response of the treatments. Since counterfactual outcomes can never be observed, the counterfactual outcome estimation is an essentially missing value problem. GANs are a powerful tool for imputing missing data (Goodfellow et al. 2014; Yoon et al. 2018). The GAN-based models for unbiasedly estimating ITE in the absence of unobserved confounding, including vanilla GAN (Yoon et al. 2018; Ge et al. 2020), the Counterfactual-GAN (cGAN) (Averitt et al. 2020), Residual GAN (Nemirovsky et al. 2020), and significantly modified GAN model for estimating effects of the continuous-valued treatments have been recently developed. To solve the problem of unmeasured confounders, the integration models of VAE and GAN for estimating ITE in the presence of unmeasured confounders have been proposed, including one GAN and two VAEs with measurable proxy for the latent confounders (Lee et al. 2018) and GAN with Adversarial Balancing-based representation for ITE estimation (Du et al. 2019).

This section will focus on basic principles for using GAN for ITE estimation with discrete and continuous-valued treatment, and in the absence and presence of unmeasured confounders.

7.7.1 A General GAN Model for Estimation of ITE with Discrete Outcome and Any Type of Treatment

7.7.1.1 Potential Framework

The Rubin causal model for estimation of treatment effects is assumed and the GAN model is modified to estimate ITE (Yoon et al. 2018; Ge et al. 2020). Consider K treatments. Let T_k be the kth treatment variable that can be binary, categorical, or continuous, and $T=\left[T_1,\dots,T_K\right]^T$ be the treatment vector. We assume that there is precisely one non-zero component of the treatment vector T, which is denoted by T_η, where η is the index of this component. Each sample has one and only one assigned treatment T_η. To extend the binary treatment to include categorical and continuous treatments, we define the treatment assignment indicator vector $M=\left[M_1,\dots,M_k,\dots,M_K\right]^T$ as

$$M_k=\begin{cases}1 & k=\eta\\ 0 & \text{otherwise}\end{cases},$$

where $\sum_{k=1}^K M_k=1$.

For example, if

$$T=\begin{bmatrix}0\\ T_2\\ 0\end{bmatrix},$$

then $\eta=2$ and

$$M=\begin{bmatrix}0\\ 1\\ 0\end{bmatrix}.$$

In special case where we consider only treated and untreated cases, we assume $K=2$. Let T_1 denote the treatment then $T_1=1$ denotes the presence of treatment and $T_1=0$ represents others. We use T_2 to denote absence of treatment where $T_2=1$ indicates no treatment. For the sample with the treatment, we have

$$T=\begin{bmatrix}T_1\\ 0\end{bmatrix} \text{ and } M=\begin{bmatrix}1\\ 0\end{bmatrix}.$$

For the sample with no treatment, we have

$$T=\begin{bmatrix}0\\ T_2\end{bmatrix} \text{ and } M=\begin{bmatrix}0\\ 1\end{bmatrix}.$$

Define the vector of potential outcome $Y(T)=\left[Y(T_1),\dots,Y(T_K)\right]^T$, where $Y(T_k)$ is the potential outcome of the sample under the treatment T_k. The outcomes can take binary, categorical and continuous values. When $K=2$, the potential outcome $Y(T_1)$ corresponds to the widely used notation for one treatment Y^1, the potential outcome of the treated sample, while the potential outcome $Y(T_2)$ corresponds to Y^0, the potential outcome of the untreated sample. Only one of the potential outcomes can be observed. The observed outcome that corresponds to the potential outcome of the individual receiving the treatment T_η is denoted by $Y(T_\eta)$. The observed outcome is called the factual outcome and the unobserved potential outcomes are called counterfactual outcomes, or simply counterfactuals. For the convenience of notation, the factual outcome is also denoted by Y_f and the counterfactuals are denoted by Y_{cf}.

The observed outcome Y_f can be expressed as

$$Y_f=Y_\eta=\sum_{k=1}^{K}M_kY(T_k).$$

When $K=2$, we have $M_2=1-M_1$. The above equation becomes

$$Y_f=M_1Y(T_1)+(1-M_1)Y(T_2)=M_1Y^1+(1-M_1)Y^0,$$

which coincides with the standard expression of the observed outcome for one treatment.

Let $X=\left[X_1,\ldots,X_q\right]^T$ be the q-dimensional feature vector such as age, sex, ethnicity, genomic variables, imaging signals, etc. Assume that n individuals are sampled. Let $T^{(i)}=\left[T_1^{(i)},\ldots,T_K^{(i)}\right]^T$, $Y^{(i)}=\left[Y^{(i)}\left(T_1^{(i)}\right),\ldots,Y^{(i)}\left(T_K^{(i)}\right)\right]^T$ and $X^{(i)}=\left[X_1^{(i)},\ldots,X_q^{(i)}\right]^T$, $i=1,\ldots,n$, be the treatment vector, the vector of potential outcomes, and feature vector of the ith individual, respectively. The vector of potential outcomes includes one observed outcome. The remaining outcomes in the vector of potential outcomes are counterfactuals.

The most widely used measure of the treatment effect for the multiple treatment is the pairwise treatment effect. The individual effect $\xi_{jk}^{(i)}$ between the pairwise treatments: T_j and T_k is defined as $\xi_{jk}^{(i)}=Y^{(i)}\left(T_j^{(i)}\right)-Y^{(i)}\left(T_k^{i}\right)$, the average pairwise treatment effect

$\tau_{jk}=E\left[\xi_{jk}^{(i)}\right]$. The CATE $\tau_{jk|T_j}$, given the patients treated with T_j is defined as $\tau_{jk|T_j}=E\left[\xi_{jk}^{(i)}\mid T_j\right]$.

Our goal is to estimate the conditional distribution of treatment effect, given the feature vector X. Let $F_{Y|X}(T_k)$ be the conditional distribution of the potential outcome $Y(T_k)$ under the treatment T_k, given the feature vector X, and $F_{Y|X}(T)$ be the conditional joint distribution of the potential outcome vector $Y(T)$ under the K treatment T, given the feature vector X. Assume that n individuals are sampled. For the ith individual, T_η treatment ($M_\eta=1$) is assigned. Let $X^{(i)}$ and $Y_\eta^{(i)}\left(T_\eta^{(i)}\right)=Y_f^{(i)}$ be the observed feature vectors and the observed potential outcomes of the ith individual. Therefore, the observed dataset is given by $D=\left(X^{(i)},T^{(i)},Y_\eta^{(i)},i=1,\ldots,n\right)$. The factual and counterfactual outcomes of the ith individual are denoted by $y_f^{(i)}$ and $y_{cf}^{(i)}$, respectively.

To estimate the treatment effects, we often make the following three assumptions (Rubin 1974; Yoon et al. 2018; Ge et al. 2020):

Assumption 1. (Ignorability Assumption). Conditional on X, the potential outcomes $Y(T)$ and the treatment T are independent,

$$Y(T)=\left(Y(T_1),\ldots,Y(T_K)\right)\perp\!\!\!\perp T\mid X. \quad (7.238)$$

This assumption requires no unmeasured confounding variables.

Assumption 2. (Common Support). For the feature vector X and all treatments,

$$0<P\left(T_k=t_k|X\right)<1. \quad (7.239)$$

Assumption 3. (Stable Unit Treatment Value Assumption). No interference (units do not interfere with each other).

7.7.1.2 Conditional GAN as a General Framework for Estimation of ITE

The key issue for the estimation of ITE is unbiased counterfactual estimation. Counterfactuals will never be observed and cannot be tested by data. The true counterfactuals are unknown. Recently developed GANs that are a perfect tool for missing data imputation started a revolution in deep learning (Luo and Zhu 2017). An incredible potential of GANs is to accurately generate the hidden (missing) data distribution given some of the features in the data. Therefore, we can use GANs to generate counterfactual outcomes.

GANs consist of two parts: the "generative" part that is called the generator and "adversarial" part that is called the discriminator. They are realized by neural networks. Typically, a K-dimensional noise vector is input into the generator that converts the noise vector to a new fake data instance. Then the generated new data instance is input into the discriminator to evaluate them for authenticity. The generator constantly learns to generate better fake data instances, while the discriminator constantly obtains both real data and fake data and improves accuracy of evaluation for authenticity.

7.7.1.2.1 Architecture of Conditional Generative Adversarial Networks (CGANs) for Generating Potential Outcomes Features provide essential information for estimation of counterfactual outcomes. Therefore, we use CGANs (Mirza and Osindero 2014) as a general framework for ITE estimation. The CGANs for ITE estimation consist of two blocks. The first imputation block is to impute the counterfactual outcomes. The second ITE block is to estimate distribution of the treatment effects using the complete dataset that is generated in the imputation block. The architecture of CGANs is shown in Figure 7.7. Both the generator and discriminator are implemented by feedforward neural networks.

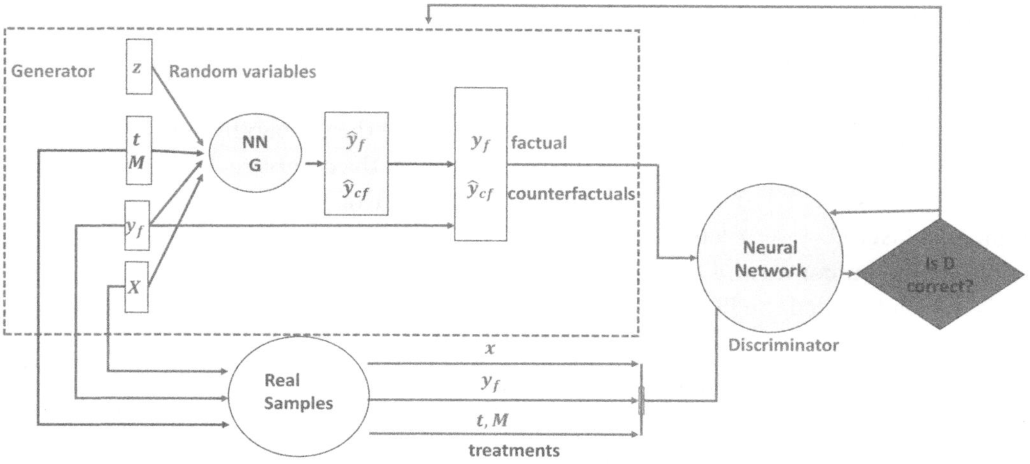

FIGURE 7.7 The architecture of CGAN for ITE estimation.

7.7.1.2.2 Imputation Block A counterfactual generator in the imputation block is a nonlinear function of the feature vector, treatment vector T, treatment assignment indicator vector M, observed factual outcome y_f, and K dimensional random vector z_G with uniform distribution $z_G \sim U\left((-1,1)^K\right)$, where $Y_f = Y_\eta$. The generator is denoted by

$$\tilde{Y} = G\left(X, Y_f, T \odot M, (1-M) \odot z_G, \theta_G\right), \quad (7.240)$$

where output $\tilde{Y}$ represents a sample of G. It can take binary values, categorical values, or continuous values. 1 is a vector of 1, $\odot$ denotes element-wise multiplication, and θ_G is the parameters in the generator. We use $\bar{Y}$ to denote the complete dataset that is obtained by replacing $\tilde{Y}_\eta$ with Y_f.

The distribution of $\tilde{Y}$ depends on the determinant of the Jacobian matrix of the transformation function $G\left(X, Y_f, T, M, z_G, \theta_G\right)$. Changing the transformation function can change the distribution of the generated counterfactual outcomes. Let $P_{Y|x,t,m,y_f}(y)$ be the conditional distribution of the potential outcomes, given $X = x$, $T = t$, $M = m$, $Y_f = y_f$. The goal of the generator is to learn the neural network G such that $G\left(x, y_f, t, m, z_G, \theta_G\right) \sim P_{Y|x,t,m,y_f}(y)$.

Unlike the discriminator in the standard CGANs where the discriminator evaluates the input data for their authenticity (real or fake data), the counterfactual discriminator D_G that maps pairs $(x, \bar{y})$ to vectors in $[0,1]^k$ attempts to distinguish the factual component from the counterfactual components. The output of the counterfactual discriminator D_G is a vector of probabilities that the component represents the factual outcome. Let $D_G\left(x, \tilde{y}, t, m, \theta_d\right)_i$ represent the probability that the ith component of $\tilde{y}$ is the factual outcome, i.e., $i = \eta$, where θ_d denotes the parameters in the discriminator. The goal of the counterfactual discriminator is to maximize the probability $D_G(x, \tilde{y}, t, m, \theta_d)_i$ for correctly identifying the factual component η via changing the parameters in the discriminator neural network D_G.

7.7.1.2.3 Loss Function The imputation block in CGAN attempts to impute counterfactual outcomes by extending the loss function of the binary treatment in GANITE (Yoon et al. 2018) to all types of treatments: binary, categorical, or continuous. We define the loss function $V(D_G, G)$ as

$$E_{(x,t,m,y_f) \sim P_{data}(x,t,m,y_f)} E_{z_G \sim u((-1,1)^K)}$$
$$\left[M^T \log D_G\left(X, \tilde{Y}, T, M, \theta_d\right)\right.$$
$$\left. + (\mathbf{1} - M)^T \log\left(\mathbf{1} - D_G\left(X, \tilde{Y}, T, M, \theta_d\right)\right)\right],$$

where log is an element-wise operation. The goal of the imputation block is to maximize the counterfactual discriminator D_G and then minimize the counterfactual generator G:

$$\min_G \max_{D_G} V\left(D_G, G, \theta_d\right). \quad (7.241)$$

In other words, we train the counterfactual discriminator D_G to maximize the probability of correctly identifying the assigned treatment M_η and the quantity of the treatment T_η or the observed factual outcome $Y_f\left(Y_\eta\right)$, and then train the counterfactual generator G to minimize the probability of correctly identifying M_η and T_η. After the imputation block is performed, the counterfactual generator G produces the complete dataset $\bar{D}=\left\{X,\bar{Y}\right\}$. Next, we use the imputed complete dataset $\bar{D}=\left\{X,\bar{Y}\right\}$ to generate the distribution of potential outcomes and to estimate the ITE via CGANs, which is called the ITE block.

7.7.1.2.4 ITE Block The CGANs for ITE block consist of three parts: generator, discriminator, and loss function, which are summarized as follows (Yoon et al. 2018; Ge et al. 2020).

7.7.1.2.4.1 ITE Generator Unlike the ITE in GANITE where the ITE generator is a nonlinear transform function of only X and Z_I, the ITE generator G_I introduced here (Ge et al. 2020) is a nonlinear transform function of X, T, and Z_I:

$$\hat{Y}=G_I\left(X,T,Z_I,\theta_{gI}\right), \tag{7.242}$$

where $\hat{Y}$ is the generated K-dimensional vector of potential outcomes, X is a feature vector, T is a treatment vector, and Z_I is a K-dimensional vector of random variables and follows the uniform distribution $Z_I\sim u\left((-1,1)^K\right)$. The ITE generator attempts to find the transformation $\hat{Y}=G_I\left(X,T,Z_{I,}\theta_{gI}\right)$ such that $\hat{Y}\sim P_{Y|X,T}(y)$.

7.7.1.2.4.2 ITE Discriminator Following the CGANs, we define a discriminator D_I as a nonlinear classifier with $\left(X,T,Y^*=\bar{Y}\right)$ or $\left(X,T,Y^*=\hat{Y}\right)$ as input and a scalar that outputs the probability of Y^* being from the complete dataset $\bar{D}$.

7.7.1.2.4.3 Loss Function Again, unlike the loss function in GANITE where the decision function is $D_I\left(X,Y^*\right)$, a decision function here (Ge et al. 2020) is defined as $D\left(X,T,Y^*\right)$. The loss function for the ITE block is then defined as

$$V_I\left(D_I,G_I\right)=E_{X,T\sim P(x,T)}\left[E_{Y^*\sim P_{Y|X,T}}(y)\left[\log D_I\left(X,T,Y^*\right)\right]\right.$$
$$\left.+E_{Z_I\sim u\left((-1,1)^K\right)}\left[\log\left(1-D_I\left(X,T,Y^*\right)\right)\right]\right], \tag{7.243}$$

where $D_I\left(X,T,Y^*\right)$ is the nonlinear classifier that determines whether Y^* is from the complete dataset $\bar{D}$ or from the generator G_I. The goal of the ITE block is to maximize the probability of correctly identifying that Y^* is from the complete dataset $\bar{D}$ and to minimize the probability of a correct classification. Mathematically, the ITE attempts

$$\min_{G_I}\max_{D_I} V_I\left(D_I,G_I\right). \tag{7.244}$$

The algorithms for numerically solving the optimization problems (7.242) and (7.244) are summarized in the Appendices 7E and 7F.

7.7.2 Adversarial Variational Autoencoder-Generative Adversarial Network (AVAE-GAN) for Estimation in the Presence of Unmeasured Confounders

In Section 7.7.1, we assume that confounders do not exist or observed variables contain exactly all the Confounders. However, these assumptions in the real world may be violated. In this section, we introduce a VAE approach that first simultaneously infers latent factors from the observed data, then disentangles the inferred latent factors into three disjoint sets corresponding to the instrumental, confounding, and risk factors, and finally incorporates the confounders into analysis (Zhang et al. 2020). The observed variables, in general, can be categorized into three clusters. The first cluster includes instrumental variables that are only involved in the treatment, but has no effect on the outcome. The second cluster includes the risk factors that only have effects on the outcome, but not on the treatment. The third cluster includes confounding variables that have effects on both the treatment and the outcome.

To stabilize the training process of the VAE, we further modify the training of the VAE with dual objectives—a classical VAE reconstruction error criterion, and an adversarial training criterion to match the aggregated posterior distribution of the latent representation of the VAE to an arbitrary prior distribution (Makhzani et al. 2015; Lin et al. 2020). After the VAE is trained, the latent variables will be used as covariates to input to the CGAN that was discussed in Section 7.7.1 to estimate the ITE. In other words, we introduce AVAE-GAN for ITE estimation where two GANs (one GAN for VAE training and another GAN for ITE estimation) are used.

7.7.2.1 Architecture of AVAE-GAN

Architecture of AVAE-GAN for ITE estimation is shown in Figure 7.8. The AVAE-GAN consists of adversarial variational autoencoder (AVAE) which maps the feature vector x to the latent variables z_x containing confounders, and GAN which replaces covariates x by z_x and uses the algorithm presented in Section 7.7.1 for ITE estimation. The AVAE combines the concepts of adversarial learning and VAE (Liu et al. 2020; Dai et al. 2017). Let $q(z_x \mid x)$ be an approximate posterior distribution of the latent variables z_x, given x and $p_{data}(x)$ be the data distribution. Defines an aggregated posterior distribution of $q(z_x)$ on the latent variables z_x as

$$q(z_x) = \int q(z_x|x) p_{data}(x) dx.$$

The AVAE is regularized by matching the aggregated posterior $q(z_x)$, to an arbitrary prior, $p(z_x)$. The encoder of the VAE $q(z_x)$ is the generator of the GAN. The AVAE has due loss functions. One loss function is the reconstruction error of the VAE. Second loss function is the discrimination accuracy of the GAN.

Second GAN mainly plays a role in ITE estimation. The VAE compresses the original data x into the latent variables z_x that contain confounders. The latent variables z_x are connected to the generator and discriminator of the second GAN (CGAN) as input conditional covariates, which might mitigate the effects of the confounders. The CGAN that was introduced in Section 7.7.1 or other GNAs dedicated to the treatment effect estimation can be used as the second GAN in the AVAE-GAN.

The VAE that was introduced in Section 5.5.2 can be used as the VAE in the AVAE-GAN. Next we will introduce another VAE with the disentangled latent factors as the VAE in the AVAE-GAN.

7.7.2.2 VAE with Disentangled Latent Factors

7.7.2.2.1 The Model for Proxy Variables, Treatments, Outcomes, and Latent Variables We assume that the observed features or covariates can be taken as proxy variables of the confounders. The data do not provide information about whether the observed set of features is large enough to include all unmeasured confounders. Usually, we hope that we can include as many features as possible. For examples, in cancer research, we can include somatic mutations, gene expression, miRNA, methylation, protein expression, and images as the set of covariates x. Some of them might be related with the confounders, while others might have nothing to do with the confounders. The VAE may disentangle them.

Assume that the mapped latent variables can be partitioned into three disjoint sets of factors $Z = (Z_t, Z_y, Z_c)$, where $Z_t \in R^{d_t}$ are the instrument factors influencing only the treatment $T \in R^K$, $Z_y \in R^{d_y}$ are the risk factors influencing only the outcome $Y \in R^K$, and $Z_c \in R^{d_c}$ are the confounding factors influencing both the treatment and outcome. We also assume that the observed feature variables $X \in R^{d_x}$ are generated by the latent factors.

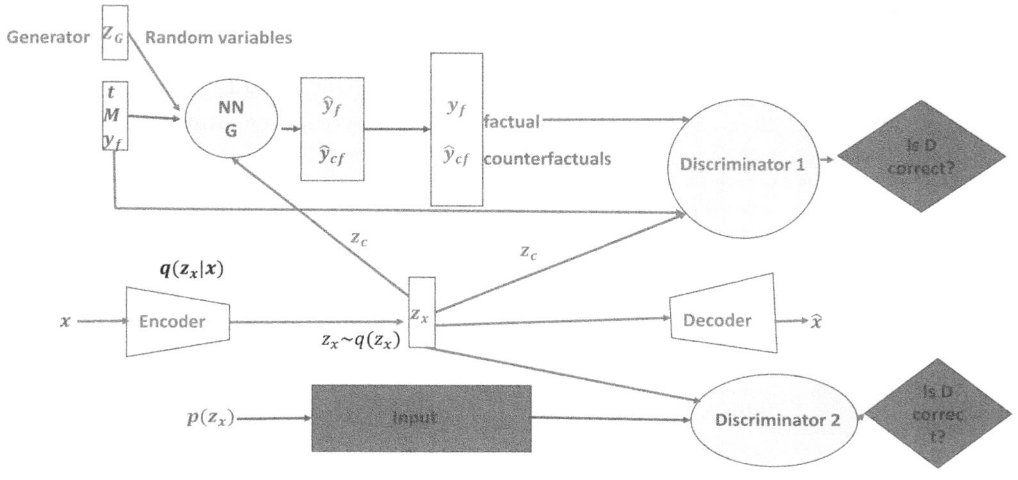

FIGURE 7.8 Architecture of AVAE-GAN for ITE estimation.

This model implies that the outcome is determined by the treatments and confounders. Recall that the treatment effect is determined by

$$p(Y|do(T=t)) = \sum_{Z_t}\sum_{Z_c} p(Y, Z_t, Z_c \,|\, do(T=t)). \tag{7.245}$$

Next we convert the do-probability to the ordinary statistical probability. The variables Y, T, Z_t and Z_c form a Bayesian network (Figure 7.9). Recall that in the Bayesian network, when calculating $p(Y, Z_t, Z_c \,|\, do(T=t)$, we need to disconnect all the nodes that are directed to the node T (dashed lines in Figure 7.9 (b)). Therefore, we obtain

$$\left(Y, Z_t, Z_c \,|\, do(T=t) = p(Z_t)p(Z_c)p(Y \,|(T=t), Z_c\right). \tag{7.246}$$

Substituting equation (7.246) into equation (7.245), we obtain

$$\begin{aligned} p(Y|do(T=t)) &= \sum_{Z_t}\sum_{Z_c} p(Z_t)p(Z_c)p(Y \,|(T=t), Z_c) \\ &= \sum_{Z_c} p(Z_c)p(Y \,|(T=t), Z_c)\sum_{Z_t} p(Z_t). \\ &= \sum_{Z_t} p(Y \,|(T=t), Z_c)p(Z_c). \end{aligned} \tag{7.247}$$

Next we can show that the conditional average effect of the treatment on the outcome, given the feature variables can be unbiasedly estimated if we can recover the confounding factors and risk factors. Recall that the conditional probability of the outcome, given the treatment and feature variables is defined as

$$p(Y|do(T=t), X). \tag{7.248}$$

Since we assume that X is generated by Z_T, Z_Y, and Z_c, then equation (7.248) can be reduced to

$$p(Y|do(T=t), X) = p(Y \,|\, do(T=t), Z_T, Z_y, Z_c). \tag{7.249}$$

Since given Z_Y and Z_c, every path from the node Z_T to Y in Figure 7.9(b) is blocked, then we have $Y \perp\!\!\!\perp Z_T \,|\, T, Z_Y, Z_c$ in $G_{\bar{T}}$, where $G_{\bar{T}}$ indicates that all incoming edges of T are deleted. Applying the Pear's first role of *Do*-calculus in Theorem 5.1 (equations (5.4) and (5.5) in Chapter 5, $Z = Z_T$, $W = (Z_Y, Z_c)$, we obtain

$$p(Y|do(T=t), Z_T, Z_Y, Z_c) = p(Y \,|\, do(T=t), Z_Y, Z_c). \tag{7.250}$$

Let $G_{\underline{T}}$ denote the graph by deleting all outgoing edges of the node T (Figure 7.9(c)). Since every path from treatment T to outcome Y is blocked by the nodes (Z_c, Z_Y) in $G_{\underline{T}}$, then we have $Y \perp\!\!\!\perp T \,|\, Z_c, Z_Y$.

Let $X = \varnothing$, $Z = T$, $W = (Z_c, Z_Y)$. Applying the Pear's second role of *Do*-calculus in Theorem 5.1 (equation (5.6)), we obtain

$$p(Y|do(T=t), Z_Y, Z_c) = P(Y \,|\, T=t, Z_Y, Z_c). \tag{7.251}$$

Substituting equation (7.251) into equation (7.250), we obtain

$$p(Y|do(T=t), Z_T, Z_Y, Z_c) = P(Y \,|\, T, Z_Y, Z_c). \tag{7.252}$$

Equation (7.252) implies that the CATE of the treatment T on the outcome Y, given features X can be unbiasedly estimated if we can discover the confounding factors Z_c and Z_Y from their proxy variables X. These results also imply that detangling the latent variables and identify the appropriate

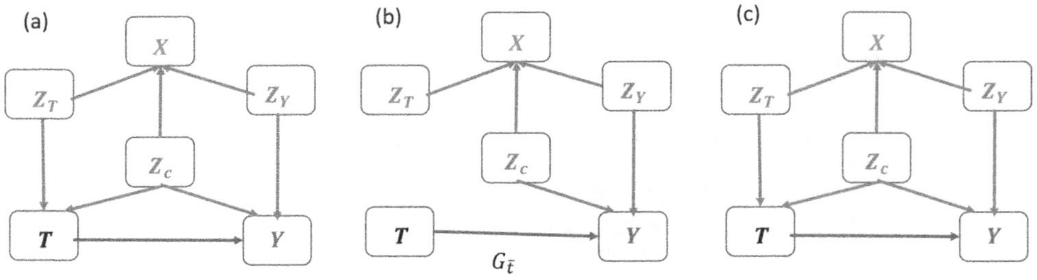

FIGURE 7.9 Graph notation for latent variables and their relations with the treatment and outcome.

confounding factors allow us to select treatment effect related information and remove irrelevant information. This is essential for ITE estimation. Next we discuss how to use VAE to map and disentangle the latent factors.

7.7.2.2.2 Disentangled Representation of Latent Factors The available data are observed features X. We use AVAE to map the observed variables X into the disjoint latent subspace to learn the confounding and risk factors. Recall that the first step of the VAE is to use the variational distribution $q_{\varnothing}(Z_T, Z_Y, Z_c \mid X)$ to approximate the posterior distribution $p_\theta(X \mid Z_T, Z_Y, Z_c)$. The variational distribution $q_{\varnothing}(Z_T, Z_Y, Z_c \mid X)$ is called encoders. To disentangle the latent factors, we decompose the encoders into three separate encoders $q_{\varnothing_T}(Z_T|X)$, $q_{\varnothing_Y}(Z_Y \mid X)$, and $q_{\varnothing_c}(Z_c \mid X)$. Correspondingly, the prior distribution $p(Z_T, Z_Y, Z_c)$ are assumed to be the product of three independent normal distributions $p(Z_T)$, $p(Z_Y)$, and $p(Z_c)$. In other words, we assume that these independent prior distributions are given by

$$p(Z_T)=\prod_{j=1}^{d_c} N(Z_{tj}|0,1),\ p(Z_Y)$$

$$=\prod_{j=1}^{d_y} N(Z_{yj}\mid 0,1)\text{and }(Z_c)=\prod_{j=1}^{d_c} N(Z_{cj}\mid 0,1).$$

7.7.2.2.3 Encoder The encoder is defined as

$$\begin{aligned} &q_{\varnothing}(Z_T, Z_{Y,} Z_c|X,T,Y) \\ &= q_{\varnothing}(Z_T, Z_{Y,} Z_c|X) \\ &= q_{\varnothing_t}(Z_T|X) q_{\varnothing_y}(Z_Y|X) q_{\varnothing_c}(Z_c|X). \end{aligned} \tag{7.253}$$

$$q_{\varnothing_t}(Z_T|X)=\prod_{j=1}^{d_t} N(Z_{tj};\mu_{tj},\sigma_{tj}^2),\ q_{\varnothing_c}(Z_c|X)$$

$$=\prod_{j=1}^{d_c} N(Z_{cj};\mu_{cj},\sigma_{cj}^2),$$

$$q_{\varnothing_y}(Z_Y|X)=\prod_{j=1}^{d_y} N(Z_{yj};\mu_{yj},\sigma_{yj}^2),$$

where the means and variances of the Gaussian distributions are given by the neural networks, for examples, the following two layers MLP:

$$[\mu_t \mid \log\sigma_t^2] = \text{ReLU}(XW_t^1)W_t^2$$

$$[\mu_c \mid \log\sigma_c^2] = \text{ReLU}(XW_c^1)W_c^2$$

$$[\mu_y \mid \log\sigma_y^2] = \text{ReLU}(XW_y^1)W_y^2.$$

The KL distance between variational posterior distribution and prior distribution are given by

$$\begin{aligned} KL(q_{\varnothing}(Z|X)\|P(Z)) &= KL(q_{\varnothing_t}(Z_t|X)\|p(Z_t)) \\ &\quad + KL(q_{\varnothing_y}(Z_y|X)\|p(Z_y)) \\ &\quad + KL(q_{\varnothing_c}(Z_c \mid X)\| p(Z_c)), \end{aligned} \tag{7.254}$$

where

$$-KL(q_{\varnothing_t}(Z_t|X)\|p(Z_t)) = \frac{1}{2}\sum_{j=1}^{d_t}[1+\log\sigma_{tj}^2-\mu_{tj}^2-\sigma_{tj}^2] \tag{7.255}$$

$$-KL(q_{\varnothing_y}(Z_y|X)\|p(Z_y)) = \frac{1}{2}\sum_{j=1}^{d_y}[1+\log\sigma_{yj}^2-\mu_{yj}^2-\sigma_{yj}^2] \tag{7.256}$$

$$-KL(q_{\varnothing_c}(Z_c \mid X)\| p(Z_c)) = \frac{1}{2}\sum_{j=1}^{d_c}[1+\log\sigma_{cj}^2-\mu_{cj}^2-\sigma_{cj}^2]. \tag{7.257}$$

7.7.2.2.4 Decoder Decoder is defined as

$$p_\theta(X,Y,T \mid Z_T, Z_Y, Z_c). \tag{7.258}$$

It follows from Figure 7.9 that

$$\begin{aligned} &p_\theta(X,Y,T|Z_T, Z_Y, Z_c) \\ &= p_\theta(X|Z_T, Z_Y, Z_c)p_\theta(T|Z_T, Z_c)p_\theta(Y|T, Z_Y, Z_c). \end{aligned} \tag{7.259}$$

We assume that $p_\theta(T|Z_T, Z_c)$ is given by

$$p_\theta(T|Z_T, Z_c)=\prod_{j=1}^{K} Bern(f_{tj}(Z_T, Z_c)), \tag{7.260}$$

where *Bern* denotes a Bernoulli distribution with the parameter $p_{tj} = f_{tj}(Z_T, Z_c)$ and

$$f_{tj}(Z_T, Z_c) = \text{ReLU}\left(\left(Z_T, Z_c\right)W_{tj}^1\right)W_{tj}^2,\ j = 1,\ldots, K. \tag{7.261}$$

If outcomes are multiple binary treats, the probability $p_\theta(Y \mid T, Z_Y, Z_c)$ can be similarly defined as that for binary treatment. If outcomes are continuous variable, then the probability $p_\theta(Y \mid T, Z_Y, Z_c)$ is defined as follows:

$$p_\theta(Y|T, Z_Y, Z_c) = \prod_{j=1}^{K} N\left(Y_j; \mu_{Y_j}, \sigma_{Yj}^2\right), \tag{7.262}$$

where

$$\mu_{Y_j} = \sum_{i=1}^{K} t_i f_{ij}(Z_Y, Z_c), \tag{7.263}$$

$$\sigma_{Yj}^2 = \sum_{i=1}^{k} t_i g_{ij}(Z_Y, Z_c), \tag{7.264}$$

$$f_{ij}(Z_{Yj}, Z_{cj}) = \text{ReLU}\left(\left(Z_{Yj}, Z_{cj}\right)W_{ij}^1\right)W_{ij}^2, \tag{7.265}$$

$$g_{ij}(Z_{Yj}, Z_{cj}) = \text{ReLU}\left(\left(Z_{Yj}, Z_{cj}\right)W_{ij}^1\right)W_{ij}^4,\ j = 1,\ldots, K, \tag{7.266}$$

functions $f_{ij}(Z_{Yj}, Z_{cj})$ and $g_{ij}(Z_{Yj}, Z_{cj})$ are modeled by two layer MLP.

The probability $p_\theta(X \mid Z_T, Z_Y, Z_c)$ is defined as

$$p_\theta(X|Z_T, Z_Y, Z_c) = \prod_{j=1}^{d_x} p_\theta(X_j \mid Z_T, Z_Y, Z_c). \tag{7.267}$$

Case 1: Continuous variables

If X_j is continuous variable, then $p_\theta(X_j \mid Z_T, Z_Y, Z_c)$ is assumed to be Gaussian distribution:

$$p_\theta(X_j|Z_T, Z_Y, Z_c) = N\left(X_j; \mu_{xj}, \sigma_{xj}^2\right), \tag{7.268}$$

where

$$\mu_{xj} = ReLU\left(\left(Z_{tj}, Z_{yj}, Z_{cj}\right)W_{xj}^1\right)W_{xj}^2, \tag{7.269}$$

$$\sigma_{xj}^2 = ReLU\left(\left(Z_{tj}, Z_{yj}, Z_{cj}\right)W_{xj}^3\right)W_{xj}^4. \tag{7.270}$$

Case 2: Discrete variables

If the variable X_j is discrete, then $p_\theta(X_j \mid Z_T, Z_Y, Z_c)$ can be modeled as logistic regression:

$$p_\theta(X_j|Z_T, Z_Y, Z_c) = \sum_{i=1}^{m}\left[x_{ji}\log p_{xji} + (1 - x_{ji})\log(1 - p_{xji})\right], \tag{7.271}$$

where

$$p_{x_j} = ReLU\left(\left(Z_{tj}, Z_{yj}, Z_{cj}\right)W_{xj}^5\right)W_{xj}^6. \tag{7.272}$$

7.7.2.2.5 ELBO The ELBO is defined as

$$\begin{aligned}\mathcal{L}_{AVAE}(\theta) &= \log p_\theta(X, Y, T|Z_T, Z_Y, Z_c) \\ &\quad - KL\left(q_\varnothing(Z|X, Y, T)\|p(Z)\right) \\ &= \frac{1}{L}\sum_{l=1}^{L}\log p_\theta(X, Y, T|Z_T^l, Z_Y^l, Z_c^l) \\ &\quad - KL\left(q_\varnothing(Z|X, Y, T)\|p(Z)\right),\end{aligned} \tag{7.273}$$

where

$$Z_T^l = \mu_T + \sigma_T \odot \varepsilon_T^l,\ \varepsilon_T^l \sim N\left(0, I_{d_T}\right) \tag{7.274}$$

$$Z_Y^l = \mu_Y + \sigma_Y \odot \varepsilon_Y^l,\ \varepsilon_Y^l \sim N\left(0, I_{d_Y}\right) \tag{7.275}$$

$$Z_c^l = \mu_c + \sigma_c \odot \varepsilon_c^l,\ \varepsilon_Y^l \sim N\left(0, I_{d_c}\right). \tag{7.276}$$

It follows from equation (7.259) that $\log p_\theta(X, Y, T|Z_T^l, Z_Y^l, Z_c^l)$ can be calculated by

$$\begin{aligned}\log p_\theta(X, Y, T|Z_T^l, Z_Y^l, Z_c^l) &= \log p_\theta(X|Z_T, Z_Y, Z_c) \\ &\quad + \log p_\theta(T|Z_T, Z_c) \\ &\quad + \log p_\theta(Y \mid T, Z_Y, Z_c).\end{aligned}$$

Again, equations (7.253) and (7.254) imply that KL distance $KL\left(q_\varnothing(Z|X, Y, T)\|p(Z)\right)$ can be calculated by

$$\begin{aligned}&KL\left(q_\varnothing(Z|X, Y, T)\|p(Z)\right) \\ &= KL\left(q_{\varnothing_t}(Z_t|X)\|p(Z_t)\right) + KL\left(q_{\varnothing_y}(Z_y|X)\|p(Z_y)\right) \\ &\quad + KL\left(q_{\varnothing_c}(Z_c \mid X)\| p(Z_c)\right).\end{aligned}$$

During training, backpropagation algorithm can be used to maximize $\mathcal{L}_{AVAE}(\theta)$ for parameter estimation, which leads to estimate Z_c. The latent variables Z_c are taken as input to the second GAN for ITE estimation.

SOFTWARE PACKAGE

Code for "counterfactual generative network" was posted in https://github.com/autonomousvision/counterfactual_generative_networks

Code for "Adapting Neural Networks for the Estimation of Treatment Effects" is available at github.com/claudiashi57/dragonet

Code for "Deephit: A deep learning approach to survival analysis with competing risks" is posted at https://github.com/chl8856/DeepHit

APPENDIX 7A: DERIVE EVIDENCE OF LOWER BOUND

Since integrand $\log p(t\mid x,a)$ is a constant with respect to expectation $E_{q(z|x,a)}[.]$, we obtain

$$\log p(t\mid x,a) = E_{q(z|x,t,a)}\left[\log p(t\mid x,a)\right], \quad (7A1)$$

which implies

$$\begin{aligned} & E_{q(z|x,a)}\left[\log p(t|x,a)\right] \\ & = E_{q(z|x,t,a)}\left[\log \frac{p(t,z|x,a)}{p(z|x,a,t)}\right] \\ & = E_{q(z|x,t,a)}\left[\log \frac{p(t,z|x,a)}{q(z\mid x,t,a)}\frac{q(z\mid x,t,a)}{p(z|x,a,t)}\right] \\ & = E_{q(z|x,t,a)}\left[\log \frac{q(z\mid x,t,a)}{p(z|x,a,t)}\right] \\ & \quad + E_{q(z|x,t,a)}\left[\log \frac{p(t,z|x,a)}{q(z\mid x,t,a)}\right] \\ & = KL\left(q(z|x,t,a)\|p(z|x,a,t)\right) \\ & \quad + E_{q(z|x,t,a)}\left[\log \frac{p(t,z|x,a)}{q(z|x,t,a)}\right]. \end{aligned} \quad (7A2)$$

Since $KL\left(q(z|x,t,a)\|p(z|x,a,t)\right) \geq 0$, equation (7A2) can be reduced to

$$E_{q(z|x,t,a)}\left[\log p(t|x,a)\right] \geq E_{q(z|x,t,a)}\left[\log \frac{p(t,z|x,a)}{q(z\mid x,t,a)}\right]. \quad (7A3)$$

Substituting equation (7.134) into equation (7A3) yields

$$\begin{aligned} & E_{q(z|x,t,a)} \\ & \left[\log \frac{p(t,z|x,a)}{q(z\mid x,t,a)}\right] = E_{q(z|x,t,a)}\left[\frac{p(t|z,a)p(z\mid x,a)}{q(z\mid x,t,a)}\right] \\ & = E_{q(z|x,t,a)}\left[\log \frac{p(z\mid x,a)}{q(z\mid x,t,a)}\right] \\ & \quad + E_{q(z|x,t,a)}\left[p(t|z,a)\right] \\ & = -KL\left(q(z|x,t,a)\|p(z|x,a)\right) \\ & \quad + E_{q(z|x,t,a)}\left[p(t|z,a)\right]. \end{aligned} \quad (7A4)$$

Substituting equation (7A4) into equation (7A3), we obtain the ELBO:

$$\begin{aligned} E_{q(z|x,t,a)}\left[\log p(t|x,a)\right] \geq & -KL\left(q(z|x,t,a)\|p(z|x,a)\right) \\ & + E_{q(z|x,t,a)}\left[p(t|z,a)\right]. \end{aligned} \quad (7A5)$$

This proves the ELBO in equation (7.136).

APPENDIX 7B: DERIVATION OF KOLMOGOROV FORWARD EQUATIONS

Let $\tilde{T}_i(t)$ be the process spends in state i before making a transition at time t into a different state is exponentially distributed with rate $\lambda_i(t)$. The rate $\lambda_i(t)$ can be interpreted as the rate at which the process leaves state i at time t. Its expectation is then given by

$$E\left[\tilde{T}_i(t)\right] = \frac{1}{\lambda_i(t)}. \quad (7B1)$$

Recall that $p_{ij}(s,t)$ is the probability that the process goes to state j from state i at time t. Let $\lambda_{ij}(t)$ be the transition rate from state i to state j at time t. Define the transition rate $\lambda_{ij}(t)$ as

$$\lambda_{ij}(t) = \lambda_i(t) p_{ij}(s,t). \quad (7B2)$$

Since $\sum_{j=0, j\neq i}^{r} p_{ij}(s,t) = 1$, then making summation on both sides of equation (7B2) yields

$$\lambda_i(t) = \sum_{j=0,\ j\neq i}^{r} \lambda_{ij}(t). \quad (7B3)$$

Let $T_{ij}(t)$ be the time the process spends in state i before entering into state j $(j \neq i)$ at time t. The time $T_{ij}(t)$ is exponentially distributed with rate $\lambda_{ij}(t)$.

Let Δt be a short time interval. Since $\tilde{T}_i(t)$ and $T_{ij}(t)$ are exponentially distributed, then we obtain

$$p_{ii}(t, t+\Delta t) = p\left(\tilde{T}_i(t) > \Delta t\right) = e^{-\lambda_i(t)\Delta t} \approx 1 - \lambda_i(t)\Delta t, \quad (7B4)$$

$$p_{ij}(t,t+\Delta t) = p\left(T_{ij}(t) \leq \Delta t\right) = 1 - e^{-\lambda_{ij}(t)\Delta t} \approx \lambda_{ij}(t)\Delta t, \quad (7B5)$$

which implies

$$\lim_{\Delta t \to 0} \frac{1 - p_{ii}(t, t+\Delta t)}{\Delta t} = \lambda_i(t), \quad (7B6)$$

$$\lim_{\Delta t \to 0} \frac{p_{ij}(t, t+\Delta t)}{\Delta t} = \lambda_{ij}(t). \quad (7B7)$$

Using the law of total probability, we obtain the Chapman-Kolmogorov equations:

$$p_{ij}(s,t+\Delta t) = \sum_{k=0}^{r} p_{ik}(s,t) p_{kj}(t,t+\Delta t). \quad (7B8)$$

Subtracting $p_{ij}(s,t)$ from equation (7B8), we obtain

$$\begin{aligned} &p_{ij}(s,t+\Delta t) - p_{ij}(s,t) \\ &= \sum_{k=0}^{r} p_{ik}(s,t) p_{kj}(t,t+\Delta t) - p_{ij}(s,t) \\ &= \sum_{k=0,k\neq j}^{r} p_{ik}(s,t) p_{kj}(t,t+\Delta t) - \left(1 - p_{jj}(t, t+\Delta t)\right) p_{ij}(s,t). \end{aligned} \quad (7B9)$$

Dividing by Δt on both sides of equation (7B9) and then taking limit, we obtain

$$\begin{aligned} &\lim_{\Delta t \to 0} \frac{p_{ij}(s,t+\Delta t) - p_{ij}(s,t)}{\Delta t} \\ &= \lim_{\Delta t \to 0} \frac{\sum_{k=0,k\neq j}^{r} p_{ik}(s,t) p_{kj}(t,t+\Delta t)}{\Delta t} \\ &\quad - \lim_{\Delta t \to 0} \frac{\left(1 - p_{jj}(t, t+\Delta t)\right) p_{ij}(s,t)}{\Delta t}. \end{aligned} \quad (7B10)$$

Substituting equations (7B6) and (7B7) into equation (7B10) yields

$$\frac{dp_{ij}(s,t)}{dt} = \sum_{k=0,\ k\neq j}^{r} p_{ik}(s,t)\lambda_{kj}(t) - \lambda_i(t) p_{ij}(t). \quad (7B11)$$

We can show (Exercise 7.8) that

$$-\lambda_j(t) = \lambda_{jj}(t). \quad (7B12)$$

Substituting equation (7B12) into equation (7B11), we prove the Kolmogorov forward equations:

$$\frac{dp_{ij}(s,t)}{dt} = \sum_{k} p_{ik}(s,t)\lambda_{kj}(t).$$

Similarly, we can get the Kolmogorov backward equation:

$$\frac{dp_{ij}(s,t)}{ds} = \sum_{k} \lambda_{ik}(s) p_{kj}(s,t). \quad (7B13)$$

APPENDIX 7C: INVERSE RELATIONSHIP OF THE KOLMOGOROV BACKWARD EQUATION

Assume that $s > 0$. Let $P(0,s)$ be the solution to the Kolmogorov forward equation (7.193) and $P(s,0)$ be the solution to backward equation (7.194). We will show that $P(s,0) = P^{-1}(0,s)$. Recall that

$$P(0,s)P^{-1}(0,s) = I. \quad (7C1)$$

Taking derivative with respect to s on both sides of equation (7C1), we obtain

$$\begin{aligned} &\frac{d(P(0,s)P^{-1}(0,s))}{ds} \\ &= \frac{dP(0,s)}{ds} P^{-1}(0,s) + P(0,s)\frac{d\ P^{-1}(0,s)}{ds} = 0. \end{aligned} \quad (7C2)$$

It follows from equation (7.193) that

$$\frac{dP(0,s)}{ds}=P(0,s)Q(0,x). \quad (7C3)$$

Substituting equation (7C3) into equation (7C2) yields

$$P(0,s)Q(0,x)P^{-1}(0,s)+P(0,s)\frac{d\,P^{-1}(0,s)}{ds}=0. \quad (7C4)$$

Multiplying by $P^{-1}(0,s)$ on both sides of equation (7C4), we obtain

$$Q(0,x)P^{-1}(0,s)+\frac{d\,P^{-1}(0,s)}{ds}=0. \quad (7C5)$$

Comparing equation (7C5) with equation (7.194), we obtain

$$P^{-1}(0,s)=P(s,0). \quad (7C6)$$

APPENDIX 7D: INTRODUCTION TO PONTRYAGIN'S MAXIMUM PRINCIPLE

In this appendix, we briefly introduce Pontryagin's maximum principle (Vinter 2020). Define an optimal control problem with unrestricted control:

$$\dot{x}=f(x,u,t), \quad (7D1)$$
$$x(t_0)=x_{0,}\ t\in[t_0\ t_f],$$

The objective function is

$$J(u(t))=K(x(t_f),t_f)+\int_{t_0}^{t_f}L(x(t),u(t),t)dt. \quad (7D2)$$

Let $u^*(t)$ be the minimizer of function $J(u(t))$ under constraints (7D1). Then, we have

$$J(u^*(t))\le J(u(t)),\ \text{for all } u(t)\in U_{[t_0,t_f]}. \quad (7D3)$$

Let $\psi(t)$ be Lagrange multiplier function, which is also called the adjoint variable. Then, the constrained optimal control problem (7D2) can be transformed to unconstrained optimal control problem using the Lagrange multiplier function:

$$J_1(u(t),x(t))=K(x(t_f),t_f)+\int_{t_0}^{t_f}\{L(x(t),u(t),t)dt$$
$$+\psi^T(t)[\dot{x}-f(x,u,t)]dt$$
$$=K(x(t_f),t_f)$$
$$+\psi^T(t_f)x(t_f)-\psi^T(t_0)x(t_0)$$
$$-\int_{t_0}^{t_f}[\dot{\psi}^T(t)x(t)+H(x(t),$$
$$u(t),\psi(t),t)]dt, \quad (7D4)$$

where

$$H(x(t),u(t),\psi(t),t)$$
$$=\psi^T(t)f(x,u,t)-L(x(t),u(t),t)$$

is Hamiltonian function.

Now we use variation calculus to derive the necessary of the optimal control. Consider

$$u(t)=u^*(t)+\varepsilon\delta u(t), \quad (7D5)$$
$$x(t)=x^*(t)+\varepsilon\delta x(t), \quad (7D6)$$

where $\delta u(t)$ and $\delta x(t)$ are differential near the optimal solution $(u^*(t),x^*(t))$. Now we calculate $J_1(u^*(t)+\varepsilon\delta u(t),x^*(t)+\varepsilon\delta x(t))$:

$$J_1(u^*(t)+\varepsilon\delta u(t),x^*(t)+\varepsilon\delta x(t))$$
$$=K(x(t_f)+\varepsilon\delta x(t_f)+\psi^T(t_f)(x(t_f)+\varepsilon\delta x(t_f))$$
$$-\psi^T(t_0)x(t_0)-\int_{t_0}^{t_f}[\dot{\psi}^T(t)(x(t)+\varepsilon\delta x(t))$$
$$++H(x(t)+\varepsilon\delta x(t),u(t)+\varepsilon\delta u(t),\psi(t),t)]dt. \quad (7D7)$$

The first order differential of $J_1(u^*(t),x^*(t))$ can be calculated by

$$\left.\frac{\partial J_1(u^*(t)+\varepsilon\delta u(t),x^*(t)+\varepsilon\delta x(t))}{\partial\varepsilon}\right|_{\varepsilon=0}$$
$$=\frac{\partial K^T(x(t_f))}{\partial x}\delta x(t_f)+\psi^T(t_f)\delta x(t_f)$$
$$-\int_{t_0}^{t_f}\left\{\dot{\psi}^T(t)\delta x(t)+\left(\frac{\partial H(x(t),u(t),\psi(t),t)}{\partial x}\right)^T\delta x(t)\right.$$
$$\left.+\left(\frac{\partial H(x(t),u(t),\psi(t),t)}{\partial u}\right)^T\delta u(t)\right\}dt. \quad (7D8)$$

Let

$$\delta x(t_f)=\frac{\partial K(x(t_f))}{\partial x}+\psi(t_f),\ \delta x(t)=\dot{\psi}(t)$$
$$+\frac{\partial H(x(t),u(t),\psi(t),t)}{\partial x}\ \text{and}\ \delta u(t)$$
$$=\frac{\partial H(x(t),u(t),\psi(t),t)}{\partial u}. \tag{7D9}$$

Substituting equation (7D9) into equation (7D8), we obtain

$$\left.\frac{\partial J_1(u^*(t)+\varepsilon\delta u(t),x^*(t)+\varepsilon\delta x(t))}{\partial\varepsilon}\right|_{\varepsilon=0}$$
$$=\left\|\frac{\partial K(x(t_f))}{\partial x}+\psi(t_f)\right\|_2^2$$
$$+\int_{t_0}^{t_f}\left\{\left\|\dot{\psi}(t)+\frac{\partial H(x(t),u(t),\psi(t),t)}{\partial x}\right\|_2^2+\left\|\frac{\partial H(x(t),u(t),\psi(t),t)}{\partial u}\right\|_2^2\right\}dt. \tag{7D10}$$

Setting $\left.\frac{\partial J_1(u^*(t)+\varepsilon\delta u(t),x^*(t)+\varepsilon\delta x(t))}{\partial\varepsilon}\right|_{\varepsilon=0}=0$, we obtain the necessary conditions of the optimal control:

$$\frac{\partial K(x(t_f))}{\partial x}+\psi(t_f)=0,$$
$$\dot{\psi}(t)+\frac{\partial H(x(t),u(t),\psi(t),t)}{\partial x}=0,$$
$$\frac{\partial H(x(t),u(t),\psi(t),t)}{\partial u}=0\ \text{or} \tag{7D11}$$

$$\psi(t_f)=-\frac{\partial K(x(t_f))}{\partial x}$$
$$,\ \dot{\psi}(t)=-\frac{\partial H(x(t),u(t),\psi(t),t)}{\partial x} \tag{7D12}$$
$$\frac{\partial H(x(t),u(t),\psi(t),t)}{\partial u}=0. \tag{7D13}$$

Equation (7D1) can be rewritten as

$$\dot{x}=\frac{\partial H(x(t),u(t),\psi(t),t)}{\partial\psi}. \tag{7D14}$$

Equations (7D11)–(7D14) form the Pontryagin's maximum principle.

APPENDIX 7E: ALGORITHM FOR ITE BLOCK OPTIMIZATION

Algorithms

To solve the optimization problems (7.242) and (7.244), we use sampling formulas to approximate the expectations. We first discuss implementation of the imputation block.

Imputation block optimization

Assume that n individuals are sampled. Sampling approximation of $V(D_G, G)$ is given by

$$\hat{V}(D_G,G)$$
$$\approx\frac{1}{n}\sum_{i=1}^{n}\Big[M_i^T\log\big(D_G\big(X^{(i)},T^{(i)},\tilde{Y}^{(i)},M_i,\theta_d\big)\big)$$
$$+(1-M_i)^T\log\big(1-D_G\big(X^{(i)},T^{(i)},\tilde{Y}^{(i)},M_i,\theta_d\big)\big)\Big], \tag{7E1}$$

where $\tilde{Y}=G(X,Y_f,T\odot M,(1-M)\odot Z_G,\theta_G)$.

To enforce that the estimated factual outcome $\tilde{Y}_\eta^{(i)}$should be as close to the observed factual outcome $Y_f^{(i)}$ as possible, we post the following restriction:

$$l(G)=\frac{1}{n}\sum_{i=1}^{n}\big(Y_f^{(i)}-\tilde{Y}_\eta^{(i)}\big)^2. \tag{7E2}$$

The optimization problem (7.242) can be implemented by

$$\min_{D_G}-\hat{V}(D_G,G), \tag{7E3}$$

$$\min_{G}\hat{V}(D_G,G)+\lambda l(G). \tag{7E4}$$

Optimization problems (7E3) and (7E4) can be solved by backpropagation (stochastic gradient decent) algorithms. The details for the algorithms are given in Appendix 7F.

ITE block optimization

ITE block intends to estimate the counterfactual outcomes using the observed outcomes and imputed counterfactual outcomes. Its performance metrics are defined for $K = 2$ (binary treatments):

$$L(G_I) = \frac{1}{n}\sum_{i=1}^{n}\left[\left(\left(\bar{y}_1^{(i)} - \bar{y}_0^{(i)}\right) - \left(\hat{y}_1^{(i)} - \hat{y}_0^{(i)}\right)\right)^2\right], \quad (7E5)$$

for $K > 2$:

$$L(G_I) = \frac{1}{n}\sum_{i=1}^{n} \| \bar{\boldsymbol{y}}^{(i)} - \hat{\boldsymbol{y}}^{(i)} \|_2^2 . \quad (7E6)$$

Sampling formula for $V_I(D_I, G_I)$ is

$$\hat{V}_I(D_I, G_I) = \frac{1}{n}\sum_{i=1}^{n}\left[\log D_I\left(x^{(i)}, (y^*)^{(i)}\right) + \log\left(1 - D_I\left(x^{(i)}, (y^*)^{(i)}\right)\right)\right]. \quad (7E7)$$

The optimization problem (7E7) for ITE can be reformulated as

$$\min_{D_I} -\hat{V}_I(D_I, G_I), \quad (7E8)$$

$$\min_{G_I} \hat{V}_I(D_I, G_I) + \gamma L(G_I). \quad (7E9)$$

Again, stochastic gradient descent methods can be used to solve optimization problems (7E8) and (7E9). Algorithms for their numerical implementation are similar to algorithms for the imputation block.

APPENDIX 7F: ALGORITHMS FOR IMPLEMENTING STOCHASTIC GRADIENT DECENT

This Algorithm is a modified version of Algorithms in (Goodfellow et al. 2014).

Let l be the number of steps to apply to the discriminator.

for number of training iterations **do**

for l steps **do**

1. Sampling minibatch of m noise vectors $\{z^{(1)},\ldots,z^{(m)}\}$ from prior $P_g(z)$.
2. Sampling minibatch of m samples of feature, observed outcomes, treatment, and treatment assignment indicator data $\{X^{(i)}, T^{(i)}, M^{(i)}, y_f^{(i)}, i = 1,\ldots,m\}$ from data generating distributions.
3. Update the parameters in discriminator by descending its stochastic gradient:

$$\nabla_{\theta_d} \hat{V}(D_G, G, \theta_d). \quad (7F1)$$

end if

a. Sampling minibatch of m noise vectors $\{z^{(1)},\ldots,z^{(m)}\}$ from prior $P_g(z)$.

b. Update the parameters in generator by descending its stochastic gradient:

$$\nabla_{\theta_g} \hat{V}\left(D_G, G(\theta_g)\right) + \lambda l\left(G(\theta_g)\right). \quad (7F2)$$

end if

EXERCISES

EXERCISE 7.1
Show that

$$\tau = E[Y|T=1) - E[Y\,|\,T=0].$$

EXERCISE 7.2
Show that

$$E\left[\frac{YT}{p(T|X)}\right] = E[Y^1] \text{and}$$
$$E\left[\frac{Y(1-T)}{p(T|X)}\right] = E[Y^0].$$

EXERCISE 7.3
Show that

$$\log p(y) - KL\left(q_\varnothing(z|y) \| p_\theta(z|y)\right) = E_{z\sim q_\varnothing(z|y)} \log p_\theta(y|z) - KL\left(q_\varnothing(z|y) \| p_\theta(z)\right).$$

EXERCISE 7.4
Show that

$$-KL\left(q(z\,|\,x,y,t) \,\|\, p(z|x,t)\right) = \frac{1}{2}\left\{\sum_{m=1}^{M}\left(\log(\sigma_s^m)^2 + 1\right) - \sum_{m=1}^{M}\left[\log(\sigma_h^m)^2 + \frac{(\sigma_s^m)^2}{(\sigma_h^m)^2} + \frac{(\mu_r^m - \mu_h^m)^2}{(\sigma_h^m)^2}\right]\right\}.$$

EXERCISE 7.5
Show that

$$\lambda(t\mid X)=\frac{f(t\mid X)}{S(t\mid X)}.$$

EXERCISE 7.6
Show that

$$E[T|X=x]=\int_0^{t_{max}} S(t|x)dt.$$

EXERCISE 7.7
Show that

The survival function of the Weibull distribution is given by

$$S_\theta\left(t^{(i)}|z^{(i,l)},a^{(i)}\right)=\exp\left\{-\left(\frac{t^{(i)}}{\lambda^{(i)}}\right)^{k(i)}\right\}.$$

EXERCISE 7. 8
Show that

$$-\lambda_j(t)=\lambda_{jj}(t).$$

EXERCISE 7.9
Show that

$$-KL\left(q\left(z_{t_0}|x_{t_0},\ldots,x_{t_N},\theta_2\right)||p\left(z_{t_0}\right)\right)$$
$$=\frac{1}{2}\sum_{d=1}^{D}\left[\log\sigma^2_{z_0d}\left(h(t_N)\right)+1\right.$$
$$\left.-\sigma^2_{z_0d}\left(h(t_N)\right)-\mu^2_{z_0d}\left(h(t_N)\right)\right].$$

EXERCISE 7.10
Develop algorithm for solving NODEs (7.216)–(7.219).

EXERCISE 7.11
Show that

$$-KL\left(q\left(z_0|\{y_i,x_i,t_i\}_{i=1}^N,a_{\le t}\right)||p(z_0)\right)$$
$$=\frac{1}{2}\left[\log\left|\Sigma_{z_0}\right|+J-\text{tr}\left(\Sigma_{z_0}\right)-\mu_{z_0}^T\mu_{z_0}\right].$$

EXERCISE 7.12
Show that

$$KL\left(q_\varnothing(Z\mid X)||P(Z)\right)=KL\left(q_{\varnothing_t}(Z_t\mid X)||p(Z_t)\right)$$
$$+KL\left(q_{\varnothing_y}(Z_y\mid X)||p(Z_y)\right)$$
$$+KL\left(q_{\varnothing_c}(Z_c\mid X)||\,p(Z_c)\right).$$

EXERCISE 7.13
Show that

$$p_\theta\left(X,Y,T|Z_T,Z_Y,Z_c\right)$$
$$=p_\theta\left(X|Z_T,Z_Y,Z_c\right)p_\theta\left(T|Z_T,Z_c\right)p_\theta\left(Y|T,Z_Y,Z_c\right).$$

CHAPTER 8

Reinforcement Learning and Causal Inference

8.1 INTRODUCTION

Fundamental challenge in science and engineering is learning to make good decisions under uncertainty. Reinforcement learning (RL) is a goal oriented machine learning and a powerful tool for making decision. RL mainly deal with a sequential decision-making where the agent or decision maker learns an optimal policy from the past history and environment by trial-and-error experience to achieve a long-term highest reward (François-Lavet et al. 2018; Kiumarsi et al. 2018). The RL agent or decision maker learns the state of the dynamic system from the environment and selects the actions that generate the most reward. A great challenge which the RL faces is that the selected and implemented actions affect not only the immediate reward, but also the future states of the system, which in turn affects all future rewards (Varghese and Mahmoud 2020). The RL can be classified into off-line learning and online learning. The off-line learning uses acquired experiences and data as a batch for learning, and the online learning uses sequentially available data to progressively update the decision of the agent. The RL attempts to learn the optimal policy and value function for a highly uncertain system. Therefore, the RL is often called approximate dynamic programming (ADP).

The RL has been successfully applied to robotics (Malik 2020), self-driving cars (Grigorescu et al. 2019), economics and financing (Charpentier et al. 2020), health care (Yu et al. 2019).

In this chapter we introduce the formulation of the RL problem as an agent for making decision which will cover a wide variety of tasks and capture many essential features of the RL. We will also introduce several widely used approaches to learning sequential decision-making tasks and how RL is applied to science, engineering, and health care areas.

8.2 BASIC REINFORCEMENT LEARNING THEORY

8.2.1 Formalization of the Problem

8.2.1.1 Markov Decision Process and Notation

RL is to learn actions or to make decisions. The RL can be formulated as a problem of finite Markov decision processes (MDPs) (Sutton and Barto 2018). The RL consists of state $S_t \in S$ in the system, action $A_t \in \mathcal{A}$ which in turn will change the system to a new state S_{t+1}, observation $\omega_{t+1} \in \Omega$, and gives a reward $R_t \in \mathcal{R}$ (Figure 8.1). The state of a system is a parameter or a set of parameters that determines the dynamics of a system. The MDP assumes the Markov property: the future is independent of the past given the present. In the RL language, we consider an agent and an environment. The environment is evolved as a dynamic system. The learner or decision maker is called the agent and everything outside the agent is called the environment. We assume that at each of a sequence of time points $t = 0, 1, 2, 3, 4, \ldots$, the agent interacts with the environment to generate a sequence of data:

$$S_0, A_0, R_1, S_1, A_1, R_2, S_2, A_2, R_3, S_3, A_3, R_4, \ldots.$$

DOI: 10.1201/9781003028543-8

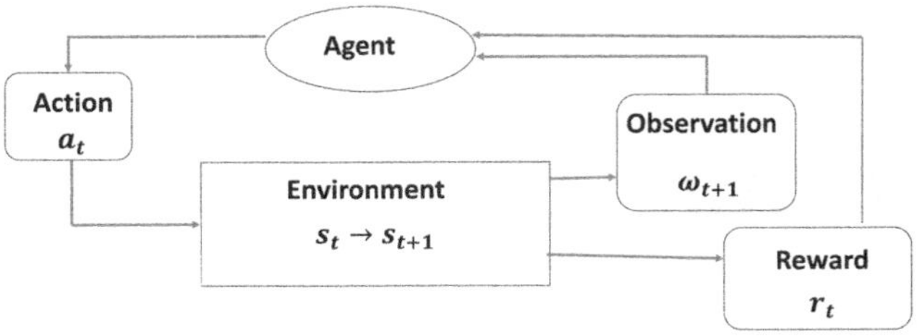

FIGURE 8.1 Architecture of reinforcement learning.

Now we consider system dynamics. In the deterministic, we assume the transition equation:

$$S_{t+1} = f(S_t, A_t), \tag{8.1}$$

where f is a nonlinear function.

In the stochastic case, the system transition can be described by the following transition probability:

$$P(S_{t+1} = s' | S_t = s, A_t = a). \tag{8.2}$$

In a finite MDP, we assume that the numbers of elements in the sets of states, actions, and rewards are finite. Let $R_{t+1}(s_t, a_t, s_{t+1})$ be a reward function, which indicates that starting at state $S_t = s_t$, taking action $A_t = a_t$, the system transits to the state $S_{t+1} = s_{t+1}$, we obtain the reward $R_{t+1}(s_t, a_t, s_{t+1})$.

If we incorporate the reward into equation (8.2), we have

$$P(s', r | s, a) = P(S_{t+1} = s', R_t = r | S_t = s, A_t = a). \tag{8.3}$$

The transition probability function which maps $P: \mathcal{S} \times \mathcal{R} \times \mathcal{S} \times \mathcal{R} \rightarrow [0,1]$ defines the dynamics of the MDP. Equation (8.3) completely determines the state-transition probability $P(s'|s, a)$ by the following equation:

$$P(s'|s, a) = \sum_{r \in \mathcal{R}} P(s', r | s, a). \tag{8.4}$$

Give dynamic equation (8.3), we can compute the expected reward:

$$\begin{aligned} r_t(s, a) &= E\left[R_{t+1}(s_t, a_t, s_{t+1}) | S_t = s, A_t = a\right] \\ &= \sum_{r \in \mathcal{R}} r \sum_{s' \in \mathcal{S}} P(s', r | s, a). \end{aligned} \tag{8.5}$$

Suppose that $r_{t+1}, r_{t+2}, r_{t+3}$ is a sequence of rewards received after time t. Then,

$$r_{t+1} = E_a\left[R_{t+1}(s_t, a, s_{t+1})\right]. \tag{8.6}$$

Let G_t be the total rewards received after time t. Then, G_t is given by

$$G_t = \sum_{\tau=1}^{T-t} r_{t+\tau}.$$

The total reward G_t after time t will go to infinite when time goes to infinite. To overcome this problem, we introduce a discounter factor $0 < \gamma < 1$. Then, equation (8.6) is modified to

$$G_t = \sum_{k=0}^{\infty} \gamma^k r_{t+k+1}. \tag{8.7}$$

The reward G_t can also be recursively computed as follows:

$$G_t = r_{t+1} + \gamma G_{t+1}. \tag{8.8}$$

An illustration of an MDP is shown in Figure 8.2. At time t, the agent takes action a_t that changes the states from S_t to S_{t+1} in the environment (the system). The environment offers a reward $R_{t+1}(S_t, a_t, S_{t+1})$.

8.2.1.2 State-Value Function and Policy

The RL attempts to reach the goal with the smallest cost or gains the best reward. The reward depends on what state when the agent interacts with the environment. The value is a function of state and denoted by $V(s)$.

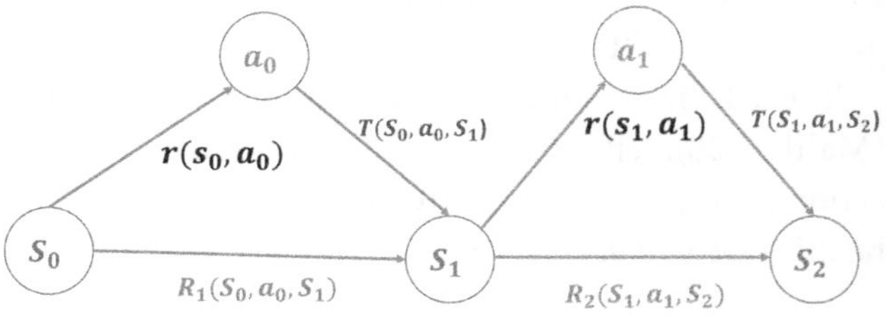

FIGURE 8.2 Scheme of an MDP.

Definition 8.1: State-Action Value Function

A state-action value function $Q(s,a)$ is defined as the expected return (or future reward), given the agent to be in the state s and take action a.

How the agent selects action? We need a rule to select action. A rule for action selection is called a policy. Policy can be grouped into deterministic and stochastic classes. Formally, in the deterministic case, a policy is defined as a mapping from the state to the action $\pi(s): \mathcal{S} \to \mathcal{A}$. In the stochastic case, a policy is defined as a mapping from the state to probabilities of selecting each possible action. If the agent is in the state $S_t = s$ at the time point t, then a policy is a mapping from the state $S_t = s$ to a conditional probability of selecting action $A_t = a$, given state s, and is denoted by

$$\pi(a|s) = P(A_t = a|S_t = s). \tag{8.9}$$

Next we define state-value function for policy π, which is denoted by $V_\pi(s)$.

Definition 8.2: State-Value Function for Policy

State-value function under policy π is defined as

$$V_\pi(s) = E_\pi[G_t|S_t = s] \text{ for all } s \in \mathcal{S}. \tag{8.10}$$

Substituting equation (8.7) into equation (8.10), we obtain

$$V_\pi(s) = E_\pi\left[\sum_{k=0}^{\infty} \gamma^k r_{t+k+1}|S_t = s\right], \tag{8.11}$$

where

$$r_{t+1} = R_{t+1}(S_t, a, S_{t+1}),$$

$$E_\pi[r_{t+1}] = E_{a\sim\pi(S_t,.)}[R_{t+1}(S_t, a, S_{t+1})],$$

transition probability $p(S_{t+1}|S_t, a_t)$ with $a_t \sim \pi(S_t)$.

Equation (8.11) can also be written as

$$V_\pi(S_t) = E_\pi\left[\sum_{k=0}^{\infty} \gamma^k r_{t+k+1}|S_t\right]. \tag{8.12}$$

Now we introduce Bellman's equation to recursively calculate state-value function for policy.

Theorem 8.1: Bellman Equation

$$V_\pi(s) = \sum_{a\in\mathcal{A}} \pi(a|s) \sum_{s'\in S} p(s'|s,a)[R(s,a,s') + \gamma V_\pi(s')]. \tag{8.13}$$

Diagram of Bellman equation is shown in Figure 8.4.

Proof

By definition of $V_\pi(s)$, we obtain

$$V_\pi(s) = E_\pi\left[\sum_{k=0}^{\infty} \gamma^k r_{t+k+1}|S_t = s\right]$$

$$= E_\pi\left[r_{t+1} + \gamma \sum_{k=0}^{\infty} \gamma^k r_{t+1+k+1}|S_t = s\right]$$

$$= \sum_{a\in\mathcal{A}} \pi(a_t = a|S_t = s) \sum_{s'\in S} p(S_{t+1} = s'|S_t = s,\ a_t = a)$$

$$[R(s,a,s') + \gamma \sum_{k=0}^{\infty} \gamma^k r_{t+1+k+1}|S_t = s,$$

$$a_t = a,\ S_{t+1} = s')]$$

$$= \sum_{a\in\mathcal{A}} \pi(a|s) \sum_{s'\in S} p(s'|s,\ a)[R(s,a,s') + \gamma V_\pi(s')].$$

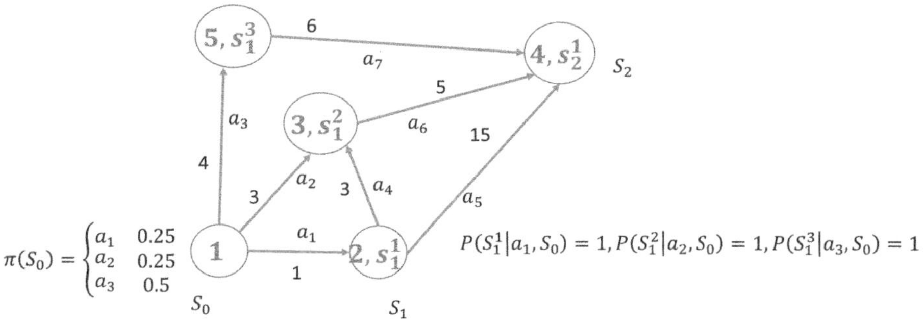

FIGURE 8.3 path and MDP.

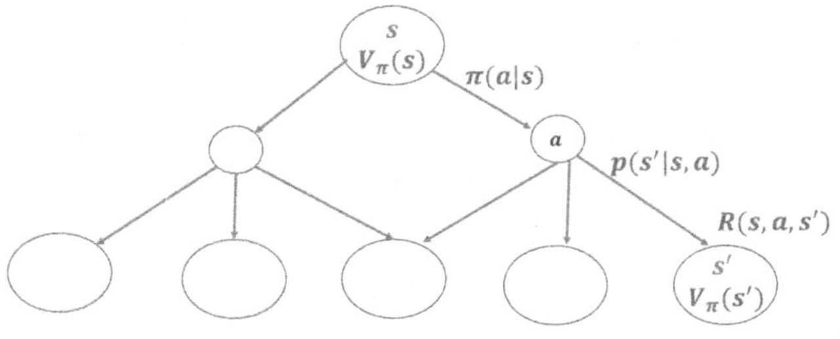

FIGURE 8.4 Diagram of Bellman equation.

From the above proof process, we can also obtain the alternative recursive formula for state-value function:

$$V_\pi(S_t) = E_\pi\left[r_{t+1} + \gamma V_\pi(S_{t+1}) | S_t = s, a_t = a\right]. \quad (8.14)$$

Definition 8.3: State-Value Function

A value function $V(s)$ at the state s is defined as the optimal expected return (or future reward), given the agent to be in the state s.

In other words, the value function $V(s)$ estimates how good it is when the agent is in a given state s.

Mathematically, the state-value function $V(s)$ is defined as

$$V(s) = \max_{\pi \in \Pi} V_\pi(s), \quad (8.15)$$

where Π is a set of policies.

Similarly, state-action value function can be extended to policy. Let $Q_\pi(s,a)$ be state-action value function under policy π. Then, we define state-action value function under the policy as follows.

Definition 8.4: State-Action Value Function for Policy

State-action value function for policy π is defined as the expected return, given the agent in state $S_t = s$ and taking action $A_t = a$, i.e.,

$$Q_\pi(s,a) = E_\pi\left[G_t | S_t = s, A_t = a\right]$$
$$= E\left[\sum_{k=0}^{\infty} \gamma^k r_{t+k+1} | S_t = s, A_t = a, \pi\right], \quad (8.16)$$

where

$$r_{t+1} = R_{t+1}(S_t, a, S_{t+1}).$$

State-action value function is also called Q action-value function. State value function and state-action value function are shown in Figure 8.5. The relationship between Bellman equation for state-action value function is given in Corollary 8.1.

Corollary 8.1: Bellman Equation for Sate-Action Value Function

$$Q_\pi(s,a) = \sum_{s' \in S} p(s'|(s,a)$$
$$\left[R(s,a,s') + \gamma \sum_{a' \in \mathcal{A}} \pi(a'|s') Q_\pi(s',a')\right]. \quad (8.17)$$

State-value function for policy $V_\pi(s)$ and state-action value function for policy have close relationship. We can show (Exercise 8.2) that

$$Q_\pi(s,a) = \sum_{s' \in S} p(s'|s, a)\left[R(s, a, s') + \gamma V_\pi(s')\right], \quad (8.18)$$

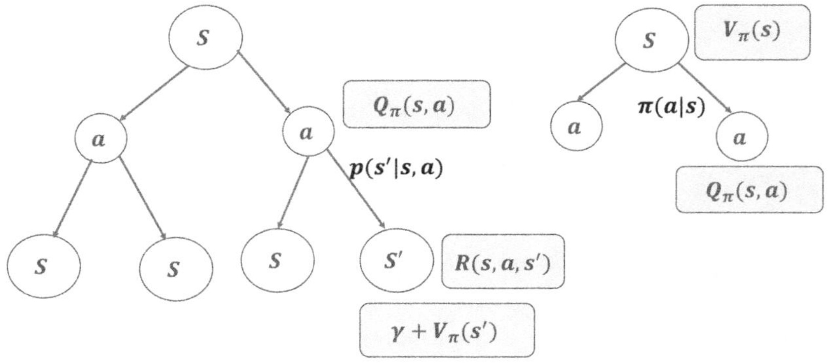

FIGURE 8.5 Relation between state-value function and state-action value function.

and

$$V_\pi(s)=\sum_{a\in\mathcal{A}}\pi(a|s)Q_\pi(s,a). \quad (8.19)$$

8.2.1.3 Optimal Value Functions and Policies

Next we introduce optimal value function and optimal policies (Sutton and Barto 2018). The RL intends to find optimal policies that reach the optimal reward in a long run. Before defining optimal value function and optimal policies, we first define a partial ordering over policies. A partial ordering $\pi_1 \geq \pi_2$ between policies is defined as if and only if

$$V_{\pi_1}(s)\geq V_{\pi_2}(s) \text{ for all } s. \quad (8.20)$$

Equation (8.20) implies that there is at least one policy that is better than or equal to all other policies. Therefore, we have the following optimal policy and optimal state-value function definition.

Definition 8.5: Optimal State-Value Function and Optimal Policy

The optimal state-value function $V_*(s)$ is defined as

$$V_*(s)=\max_\pi V_\pi(s), \quad (8.21)$$

where $V_*(s)$ shows the maximum expected reward that one can achieve from state s.

An optimal policy π_* is defined as a policy that is better than or equal to all other policies, i.e.,

$$V_{\pi_*}(s)=\max_\pi V_\pi(s), \forall s\in S. \quad (8.22)$$

We can also define an optimal action-value function.

Definition 8.6: Optimal Action-Value Function

An optimal action-value function $Q_*(s, a$ is defined as

$$Q_*(s,a)=\max_\pi Q_\pi(s,a), \forall s\in S, a\in\mathcal{A}. \quad (8.23)$$

Equation (8.23) indicates that this function gives the optimal expected return for an optimal policy π_*, with initial action a in state s.

The optimal action-value function and optimal value function have close relationship. The following Theorem 8.2 summarizes this important property.

Theorem 8.2: Optimal Value Function and Optimal Action-Value Function

The optimal action-value function can be expressed in terms of optimal value function:

$$Q_*(s,a)=E\left[r_{t+1}+\gamma V_*(S_{t+1})|S_t=s, a_t=a\right] \quad (8.24)$$

$$=\sum_{s'\in S}p(s'|s,a)\left[R(s,a,s')+\gamma V_*(s')\right] \quad (8.25)$$

$$V_*(s)=\max_{a\in\mathcal{A}}Q_*(s,a). \quad (8.26)$$

Proof

It follows from equation (8.18) that

$$Q_\pi(s,a)=\sum_{s'\in S}p(s'|s,a)\left[R(s,a,s')+\gamma V_\pi(s')\right]. \quad (8.27)$$

Taking optimal with respect to policy, we obtain

$$Q_*(s,a)=\sum_{s'\in S}p(s'|s,a)\left[R(s,a,s')+\gamma V_*(s')\right].$$

Similarly, $Q_\pi(s,a)$ can be expressed as

$$Q_\pi(s,a)=E\left[r_{t+1}+\gamma V_\pi(S_{t+1})|S_t=s, a_t=a\right]. \quad (8.28)$$

Taking optimal with respect to policy, we obtain

$$Q_*(s,a)=E\left[r_{t+1}+\gamma V_*(S_{t+1})|S_t=s, a_t=a\right].$$

Recall from equation (8.19), we obtain

$$V_\pi(s)=\sum_{a\in\mathcal{A}}\pi(a|s)Q_\pi(s,a),$$

which implies (Exercise 8.3) that

$$V_*(s)=\max_{a\in\mathcal{A}}Q_*(s,a).$$

Now we introduce the existence theorem of the optimal policy without proof.

Theorem 8.3: The Existence of the Optimal Policy

For any MDPs:

1. there exists an optimal policy π_* that is at least as good as all other policies:

$$\pi_* \geq \pi, \forall \pi;$$

2. there can be many optimal policies, but all optimal policies achieve the optimal value function:

$$V_{\pi_*}(s) = V_*(s), \forall s;$$

3. all optimal policies achieve the optimal action-value functions,

$$Q_{\pi_*}(s, a) = Q_*(s, a), \forall s, a;$$

4. there is always a deterministic optimal policy for any MDP.

Example 8.1

Diagram for Example 8.1 is shown in Figure 8.6, where the number in red color denotes the reward and the number in dark color denotes the action index. We assume

1. three states: A, B, and C;
2. two possible actions in each state, denoted by 1 and 2;
3. $V_\pi(C) = 10$;
4. the following transition probabilities:

$$p(s'|s, a) = \begin{cases} 0.9 & \text{right direction} \\ 0.1 & \text{wrong direction} \end{cases}$$

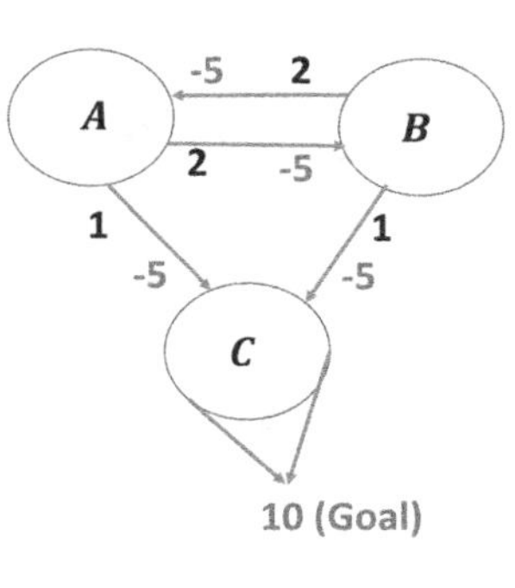

FIGURE 8.6 Diagram for Example 8.1.

5. $\pi(1|A) = 1, \pi(2|A) = 0$

$$\pi(1|B) = 1, \pi(2|B) = 0$$

$$\pi(1|C) = 1, \pi(2|C) = 0.$$

Equations can be derived by the following Bellman recursive formula:

$$V_\pi(s) = \sum_{a \in A} \pi(a|s) \sum_{s' \in S} p(s'|s, a) \times \left[R(s, a, s') + \gamma V_\pi(s')\right], \text{ which implies that}$$

$$V_\pi(B) = 1 * 0.9 * (-5 + 10) + 1 * 0.1(-5 + V_\pi(A)) = 4 + 0.1 V_\pi(A), \quad (8.29)$$

$$V_\pi(A) = 1 * 0.9(-5 + V_\pi(C)) + 1 : 0.1(-5 + V_\pi(B)) = 4 + 0.1\ V_\pi(B). \quad (8.30)$$

Solving equations (8.29) and (8.30), we obtain

$$V_\pi(A) = 4.444 \text{ and } V_\pi(B) = 4.44.$$

8.2.1.4 Bellman Optimality Equation

Equations (8.24)–(8.26) allow us to easily extend Bellman equations to Bellman optimality equations.

Theorem 8.4: Bellman Optimality Equations

Bellman optimality equation for V_* is

$$V_*(s) = \max_{a \in A} \sum_{s' \in S} p(s'|s, a)\left[R(s, a, s') + \gamma V_*(s')\right]. \quad (8.31)$$

Bellman optimality equation for Q_* is

$$Q_*(s,a) = E\left[r_{t+1} + \gamma \max_{a'} Q_*(S_{t+1}, a') | S_t = s, a_t = a\right] \quad (8.32)$$

$$= \sum_{s' \in S} p(s'|s, a)\left[R(s, a, s') + \gamma \max_{a'} Q_*(s', a')\right]. \quad (8.33)$$

Proof

Recall equation (8.26) that

$$V_*(s) = \max_{a \in A} Q_*(s, a). \quad (8.34)$$

Substituting equation (8.25) into equation (8.34) yields

$$V_*(s) = \max_{a \in \mathcal{A}} \sum_{s' \in S} p(s'|s, a)\left[R(s, a, s') + \gamma V_*(s')\right].$$

which proves the equation (8.31).

Substituting equation (8.25) into equation (8.24), we obtain

$$Q_*(s,a) = E\left[r_{t+1} + \gamma \max_{a'} Q_*(S_{t+1}, a') | S_t = s, a_t = a)\right],$$

which proves equation (8.32).

Similarly, substituting equation (8.26) into equation (8.25), we prove equation (8.33).

The backup diagram on the left (Figure 8.7(a)) graphically represents the Bellman optimality equation for $V_*(s)$, and the backup diagram on the right (Figure 8.7(b)) represents the Bellman optimality equation for $Q_*(s, a)$.

It follows from equation (8.31) that one state specifies one Bellman optimality equation. If the system has k states then there must have k equations. If the transition probabilities are known, in principle, we can solve the system of Bellman optimality equations. Suppose that we have k states and m actions, then we will have mk Bellman optimality equation for $Q_*(s, a)$.

Once the optimal value function $V_*(s)$ is found, next step is to find the optimal policy. One candidate for the optimal policy is the greedy policy. The word "greedy" represents any search or decision procedure that selects alternative action based only on local or immediate considerations, without considering long-term consequence (Sutton and Barto 2018).

Definition 8.7: Greedy Policy

Greedy policy for a given $Q(s,a)$ function is defined as

$$\pi(a|s) = \begin{cases} 1 \text{ if } a = \operatorname{argmax}_a Q(s,a) \\ 0 \qquad \text{otheraisse} \end{cases}. \tag{8.35}$$

Greedy policy for a given value $V(s)$ function is defined as

$$\pi(a|s) = \begin{cases} 1 \text{ if } a = \operatorname{argmax}_a p(s'|s,a)\left(R(s,a,s') + \gamma V(s')\right) \\ 0 \qquad \text{otheraise} \end{cases}. \tag{8.36}$$

Definition 8.8: Optimal Policy

Optimal policy π_* is defined as

$$\pi_* = \operatorname{argmax}_{\pi} V_\pi(s), \forall s \in S. \tag{8.37}$$

It is clear from equation (8.36) that a greedy policy that evaluates the short-term consequences of actions—specifically, the one-step consequences—is actually optimal in the long-term sense because V_* already takes into account the reward consequences of all possible future system states and transition actions. Therefore, we have the following theorem.

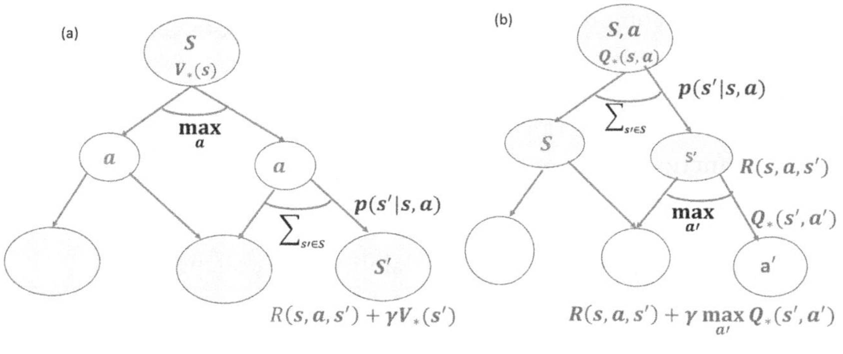

FIGURE 8.7 (a) Backup diagram for V_*, (b) backup diagram for Q_*.

Theorem 8.5: Equivalence between Greedy Policy and Optimal Policy

A greedy policy for V_* is an optimal policy π_*, which can be expressed as

$$\pi_*(s) = \arg\max_{a \in \mathcal{A}} \sum_{s' \in S} p(s'|s,a)\left[R(s,a,s') + \gamma V_*(s')\right]. \quad (8.38)$$

Theorem 8.5 states that a one-step-ahead search leads to the long-term optimal actions.

8.2.2 Dynamic Programming

RL is also named as approximate dynamic programming, and neurodynamic programming. Although exact dynamic programming algorithms are of limited utility in RL due to their perfect model assumption and heavy computation, dynamic programming provides a foundation for RL (Seinfeld and Lapidus 1968; Bertsekas 2019). The RL has the same goal as the dynamic programming, but with less assumptions and computations. This section introduces basic concepts and algorithms of dynamic programming (Sutton and Barto 2018).

We assume that the environment is an MDP. The MDP is assumed to have finite number of states, finite number of actions, and finite reward set. The dynamics of the MDP are determined by a set of transition probabilities $p(s', r|s, a)$ for all $s \in \mathcal{S}$, $a \in \mathcal{A}(s)$, $r \in \mathcal{R}$ and $s' \in S^+$, where S^+ denotes S plus a terminate state.

Introduction of dynamic programming starts with calculation of value function and policy using Bellman equation. Illustration of dynamic programming via shortest path is shown in Figure 8.3.

8.2.2.1 Policy Evaluation

We first introduce policy evaluation with Bellman operator. Policy evaluation is to compute the state-value function V_π for an arbitrary policy π. The algorithms for computing value function are an iterative algorithm, where the computed value function finally converges to the true value function as the number of iterations goes infinity. In mathematics, this is a fixed-Point problem (Rao and Jelvis 2021). In other words, the iteration of computed value function converges to the fix-point (the true value function). Therefore, introducing the Bellman operator starts with the fixed-point problem.

Definition 8.9: Fixed-Point

Let $f(x)$ be a function. The fixed-point of the function is a point x that satisfies the equation:

$$x = f(x). \quad (8.39)$$

The fixed-points may be multiple or may not exist. The dynamic programming has a unique fixed-point problem.

Definition 8.10: Contraction Map

Let $\mathcal{X}$ be a metric space and $f(x): \mathcal{X} \to \mathcal{X}$ be a map. If there exists some constant $0 < k < 1$ such that

$$d(f(x), f(x^*)) \leq k d(x, x^*) \text{ for all } x, x^* \in \mathcal{X}. \quad (8.40)$$

Theorem 8.6: Banach's Fixed Point Theorem

Let $\mathcal{X}$ be a complete matric space, and $f(x)$ be a contraction map on $\mathcal{X}$. Then

1. there exists a unique fixed point x_* such that

$$f(x_*) = x_*$$

2. For any initial value $x_0 \in \mathcal{X}$ and any sequences $\{x_i, i = 0, 1, \ldots\}$ defined as $x_{i+1} = f(x_i)$, we have

$$\lim_{i \to \infty} x_i = x_* \quad (8.41)$$

3. $$d(x_*, x_i) \leq \frac{k^i}{1-k} d(x_1, x_0). \quad (8.42)$$

Next we introduce the Bellman policy operator and policy evaluation algorithm (Rao and Jelvis 2021). Assume that the states of the MDP are $S = \{s_1, \ldots, s_n\}$ and $\mathcal{N} = \{s_1, \ldots, s_m\}$ are the non-terminal states. Let $R(s,a)$ be the expected reward upon action a in state s. In other words,

$$R(s,a) = \sum_{s' \in S} p(s'|s,a) R(s,a,s'). \quad (8.43)$$

Define the reward function $R_\pi(s)$ and transition probability for a policy $\pi(a|s)$

$$R_\pi(s) = \sum_{a \in \mathcal{A}} \pi(a|s) R(s,a) \quad (8.44)$$

$$p_\pi(s,s') = \sum_{a \in A} p(s,a,s')\pi(a|s). \quad (8.45)$$

Let

$$R_\pi = \begin{bmatrix} R_\pi(s_1) \\ \vdots \\ R_\pi(s_n) \end{bmatrix} \text{ and}$$

$$P_\pi = \begin{bmatrix} p_\pi(s_1, s_1') & \cdots & p_\pi(s_1, s_n') \\ \vdots & \vdots & \vdots \\ p_n(s_n, s_1') & \cdots & p_n(s_n, s_n') \end{bmatrix}.$$

Definition 8.11: Bellman Policy Operator

Bellman policy operator is defined as

$$B_\pi V(s) = \sum_{a \in A} \pi(a|s) \sum_{s' \in S} p(s'|s,a) \times [R(s,a,s') + \gamma V(s')]. \quad (8.46)$$

Bellman policy operator in vector and matrix form is defined as

$$B_\pi V = R_\pi + \gamma P_\pi V. \quad (8.47)$$

Equations (8.46) and (8.47) are equivalent. It follows from equation (8.47) that

$$(B_\pi V)(s) = R_\pi(s) + (\gamma P_\pi V)(s). \quad (8.48)$$

Using equations (8.43) and (8.44), we obtain

$$R_\pi(s) = \sum_{a \in A} \pi(a|s) \sum_{s' \in S} p(s'|s,a) R(s,a,s'). \quad (8.49)$$

Using definition of $(P_\pi V)(s)$ and equation (8.45), we obtain

$$+(\gamma P_\pi V)(s) = \gamma \sum_{a \in A} \pi(a|s) \sum_{s' \in S} p(s'|s,a) V(s'). \quad (8.50)$$

Combining equations (8.48)–(8.50) yields

$$B_\pi V(s) = \sum_{a \in A} \pi(a|s) \sum_{s' \in S} p(s'|s,a)[R(s,a,s') + \gamma V(s')].$$

This shows that equations (8.46) and (8.47) are equivalent.

Recall that Bellman equation is given by

$$V_\pi(s) = \max_{a \in A} \sum_{s' \in S} p(s'|s,a)[R(s,a,s') + \gamma V_\pi(s')]. \quad (8.51)$$

It follows from equation (8.46) that

$$B_\pi V_\pi(s) = \sum_{a \in A} \pi(a|s) \sum_{s' \in S} p(s'|s,a) \times [R(s,a,s') + \gamma V_\pi(s')]. \quad (8.52)$$

Equations (8.51) and (8.52) imply that

$$B_\pi V_\pi(s) = V_\pi(s). \quad (8.53)$$

Equation (8.53) shows that the Bellman policy operator is a linear operator with fixed point $V_\pi(s)$.

Definition 8.12: Bellman Optimality Operator

Bellman optimality operator is defined as

$$(B_* V)(s) = \max_\pi (B_\pi V)(s) = \max_a \sum_{s' \in S} p(s'|s,a)[R(s,a,s') + \gamma V(s')]. \quad (8.54)$$

We can easily show that Bellman optimality operator B_* is a nonlinear operator with fixed point V_*, i.e., (Exercise 8.4)

$$B_* V_* = V_*. \quad (8.55)$$

Definition 8.13: Greedy Policy

Greedy policy $G(V)$ is defined as

$$G(V)(s) = \operatorname{argmax}_a \sum_{s' \in S} p(s'|s,a) \times [R(s,a,s') + \gamma V(s')]. \quad (8.56)$$

Equation (8.56) implies

$$B_{G(V)} V = B_* V. \quad (8.57)$$

Using iterative applications of the Bellman policy operator B_π, we can evaluate the value function. To use Banach Fixed-Point Theorem for proving the convergence of iterated sequence, we need to develop a metric in which the Bellman policy operator B_π is a contraction map. L_∞ norm is such matric.

Definition 8.14: L_∞ Norm of the Value Function

Let $V \in R^d$ and $W \in R^d$ be value functions, i.e.,

$$V(s)=\begin{bmatrix} V_1(s) \\ \vdots \\ V_d(s) \end{bmatrix}, W(s)=\begin{bmatrix} W_1(s) \\ \vdots \\ W_d(s) \end{bmatrix}.$$

The L_∞ norm of the vector is defined as

$$\|V(s) - W(s)\|_\infty$$

$$= \max\left(|V_1(s)-W_1(s)|,\ldots,|V_d(s)-W_d(s)|\right). \quad (8.58)$$

The L_∞ norm of the value function is defined as

$$\|V-W\|_\infty = \max_s\left(\|V(s_1)-W(s_1)\|_\infty,\ldots,\|V(s_m)-W(s_m)\|_\infty\right) \quad (8.59)$$

$$= \max_{s\in\mathcal{N}}\left(\|(V-W)(s)\|_\infty\right). \quad (8.60)$$

Now we prove that both Bellman policy operator and Bellman optimality operator are γ contraction operators in L_∞ norm.

Theorem 8.7: Contraction Property of Bellman Policy Operator and Bellman Optimality Operator

Both Bellman policy operator and Bellman optimality operator are γ contraction operators in L_∞ norm, i.e., for any value functions V and W, we have

$$\|B_\pi V - B_\pi W\|_\infty \le \|\gamma V - W\|_\infty \quad (8.61)$$

$$\|B_* V - B_* W\|_\infty \le \|\gamma V - W\|_\infty. \quad (8.62)$$

Proof

$$\|B_\pi V - B_\pi W\|_\infty = \max_{s\in\mathcal{N}}\|(B_\pi V - B_\pi W)(s)\|_\infty. \quad (8.63)$$

Substituting equation (8.47) into equation (8.63) yields

$$\|B_\pi V - B_\pi W\|_\infty = \max_{s\in\mathcal{N}}\|(R_\pi + \gamma P_\pi V - R_\pi - \gamma P_\pi W)(s)\|_\infty$$

$$= \gamma \max_{s\in\mathcal{N}}\|(P_\pi V - P_\pi W)(s)\|_\infty$$

$$\le \gamma \max_{s\in\mathcal{N}}\|(V-W)(s)\|_\infty$$

$$= \gamma\|V-W\|_\infty.$$

Similarly, we can prove equation (8.62) (Exercise 8.5).

Next we will show the monotonicity of the Bellman policy operator and Bellman optimality operator. If for any value functions V_1 and V_2 that satisfy the inequality $V_1(s) \le V_2(s)$, $\forall s \in S$, then value function V monotonically increases and is denoted by $V_1 \le V_2$. We can show Bellman policy operator and Bellman optimality operator are monotonic operators.

Theorem 8.8: Monotonic Bellman Policy Operator and Bellman Optimality Operator

If $V_1 \le V_2$ then

$$B_\pi V_1 \le B_\pi V_2 \quad (8.64)$$

$$B_* V_1 \le B_* V_2. \quad (8.65)$$

Proof

Recall (equation (8.47)) that the Bellman policy operator is defined as

$$B_\pi V = R_\pi + \gamma P_\pi V.$$

Since every element in the matrix P_π and vector value functions V_1 and V_2 is positive, then we have

$$\gamma P_\pi V_1 \le \gamma P_\pi V_2, \quad (8.66)$$

which implies

$$R_\pi + \gamma P_\pi V_1 \le R_\pi + \gamma P_\pi V_2. \quad (8.67)$$

However,

$$B_\pi V_1 = R_\pi + \gamma P_\pi V_1, B_\pi V_2 = R_\pi + \gamma P_\pi V_2. \quad (8.68)$$

It follows from equations (8.67) and (8.68) that

$$B_\pi V_1 \le B_\pi V_2.$$

We can similarly show (Exercise 8.6) that

$$B_* V_1 \le B_* V_2.$$

Theorem 8.7 implies the following policy evaluation convergence theorem (Rao and Jelvis 2021).

Theorem 8.9: Policy Evaluation Convergence Theorem

For a finite MDP with m non-terminal states and $\gamma < 1$, if the value function V_π is evaluated with a fixed policy π, then value function V_π is the unique fixed-point of the Bellman policy operator $B_\pi : R^m \to R^m$ and any iterative sequence of $V_{k+1} = B_\pi V_k$, starting with any initial value function V_0, will converge to the fixed point V_π, i.e.,

$$\lim_{k\to\infty} V_k = V_\pi \text{ or} \tag{8.69}$$

$$\lim_{k\to\infty} (B_\pi)^k (V_0) = V_\pi. \tag{8.70}$$

Theorem 8.9 provides foundation for the following policy evaluation algorithm (Rao and Jelvis 2021).

Algorithm 8.1: Policy Evaluation Algorithm

Step 1: Input any initial value function $V_0 \in R^m$
Step 2: While not converge **do**

$$V_{k+1} = B_\pi (V_k) = R_\pi + \gamma P_\pi V_k. \tag{8.71}$$

Step 3: Check convergence

$$\max_{s\in\mathcal{N}} \left\| (V_{k+1} - V_k)(s) \right\|_\infty < \varepsilon$$

End return V_{k+1}.

8.2.2.2 Value Function and Policy Improvement

Next we will introduce dynamic programming for solving the control problem. Improving value function and policy provides a foundation for solving the control problem. Recall Definition 8.7 for greedy policy. The greedy policy is defined as a function that maps a value function V to a deterministic policy $\pi'_D : \mathcal{N} \sim \mathcal{A}$:

$$G(V)(s) = \pi'_D(s) = \operatorname*{argmax}_{a\in\mathcal{A}} Q(s,a), \tag{8.72}$$

where

$$Q(s,a) = R(s,a) + \gamma \sum_{s\in\mathcal{N}} p(s,a,s')V(s').$$

Greedy policy can also be defined as

$$G(V)(s) = \operatorname*{argmax}_{a\in\mathcal{A}} \left\{ \sum_{s'\in S} \sum_{r\in D} p(s,a,r,s')(r + \gamma W(s')) \right\}, \quad \forall s \in \mathcal{N}, \tag{8.73}$$

where

$$W(s') = \begin{cases} V(s') & \text{if } s' \in \mathcal{N} \\ 0 & \text{if } s' \in T = S - \mathcal{N} \end{cases}.$$

Greedy policy implies that applying the greedy policy function G on value function V_π can improve the policy. The mapped value function $V_{\pi'_D}$ is "greater" than the original value function V_π.

Now we give the definition for value function comparison (Sutton and Barto 2018; Rao and Jelvis 2021).

Definition 8.15: Value Function Comparison

Let $V_1, V_2 : \mathcal{N} \to R$ be value functions of an MDP. Value function inequality $V_1 \ge V_2$ is defined as

$$V_1(s) \ge V_2(s), \ \forall s \in \mathcal{N}. \tag{8.74}$$

Thus, improved value function, or better value function means that the better value function is *no worse for each of the states.*

Computing the value function for a policy can help us to find better policies. The value function for a policy provides information to assess how good it is if we follow the current policy from the current state. The value function also provides a tool to improve the whole policy by one step improving value function. In other words, we can select an action a in the current state s and thereafter following the existing policy π. We can formally state the following improvement theorem.

Theorem 8.10: Policy Improvement

Consider two deterministic policies π and π' where π' is identical to, except that $\pi'(s)=a\neq\pi(s)$. If

$$Q_\pi(s,\pi'(s))\geq V_\pi(s), \tag{8.75}$$

then

$$V_{\pi'}\geq V_\pi \text{ and } \pi'\geq\pi, \tag{8.76}$$

i.e., the he changed policy π' is indeed better than.

Proof

First we show that

$$Q_\pi(s,\pi'(s))=E\left[r_{t+1}+\gamma V_\pi(S_{t+1})|S_t=s, A_t=\pi'(s)\right]. \tag{8.77}$$

Using definition of $Q_\pi(s,\pi'(s))$, we obtain

$$Q_\pi(s,\pi'(s))=E\left[\sum_{k=0}^{\infty}\gamma^k r_{t+k+1}|S_t=s, A_t=\pi'(s)\right]$$

$$=E\left[r_{t+1}+\sum_{k=1}^{\infty}\gamma^k r_{t+k+1}|S_t=s, A_t=\pi'(s)\right]$$

$$=E\left[r_{t+1}+\gamma\sum_{k=1}^{\infty}\gamma^{k-1} r_{t+k+1}|S_t=s, A_t=\pi'(s)\right]. \tag{8.78}$$

Substituting $k'=k-1$ into equation (8.78), we obtain

$$Q_\pi(s,\pi'(s))=E\left[r_{t+1}+\gamma\sum_{k'=0}^{\infty}\gamma^{k'} r_{t+1+k'+1}|S_t=s, A_t=\pi'(s)\right]. \tag{8.79}$$

However, by assumption that π' is identical to π after time t and definition of the value function for a policy V_π, we obtain

$$V_\pi(S_{t+1})=E\left[\sum_{k'=0}^{\infty}\gamma^{k'} r_{t+1+k'+1}|S_t=s, A_t=\pi'(s)\right]. \tag{8.80}$$

Substituting equation (8.80) into equation (8.79) yields

$$Q_\pi(s,\pi'(s))=E\left[r_{t+1}+\gamma V_\pi(S_{t+1})|S_t=s,\right.$$
$$\left.A_t=\pi'(s)\right]. \tag{8.81}$$

Since $A_t=\pi'(s)$, equation (8.81) implies that

$$Q_\pi(s,\pi'(s))=E_{\pi'}\left[r_{t+1}+\gamma V_\pi(S_{t+1})|S_t=s\right]. \tag{8.82}$$

Using assumption (8.75), we obtain

$$Q_\pi(s,\pi'(s))\leq E_{\pi'}\left[r_{t+1}+\gamma Q_\pi(S_{t+1},\pi'(S_{t+1}))|S_t=s\right]. \tag{8.83}$$

Using equation (8.81), we obtain

$$E_{\pi'}\left[r_{t+1}+\gamma Q_\pi(S_{t+1},\pi'(S_{t+1}))|S_t=s\right]$$
$$=E_{\pi'}\left[r_{t+1}+\gamma E\left[r_{t+2}+\gamma V_\pi(S_{t+2)})|S_{t+1}=S,\right.\right.$$
$$\left.\left.A_{t+1}=\pi'(S_{t+1})\right]|S_t=s\right]$$
$$=E_{\pi'}\left[r_{t+1}+\gamma r_{t+2}+\gamma^2 V_\pi(S_{t+2})|S_t=s\right]$$
$$\leq E_{\pi'}\left[r_{t+1}+\gamma r_{t+2}+\gamma^2 r_{t+3}+\gamma^3 V_\pi(S_{t+3})|S_t=s\right]$$
$$\vdots$$
$$\leq E_{\pi'}\left[r_{t+1}+\gamma r_{t+2}+\gamma^2 r_{t+3}+\gamma^3 r_{t+4}+\ldots|S_t=s\right]$$
$$=V_{\pi'}(s). \tag{8.84}$$

Combining equations (8.75), (8.83), and (8.84), we obtain

$$V_\pi(\mathrm{s})\leq V_{\pi'}(s).$$

This proves the Theorem 8.10.

Theorem 8.10 shows that given a state-value function for a policy, we can use state-action value function to find a better policy. We can extend Theorem 8.10 to more general case that selecting at each state the action that appears best according to state-action value function $Q_\pi(s,a)$.

Next we show that greedy policy can improve value function (Sutton and Barto 2018; Rao and Jelvis 2021).

Theorem 8.11: Greedy Policy Theorem

For a finite MDP and any policy π, a greedy policy, defined as

$$G(V_\pi)(s) = \pi'_D(s) = \underset{a\in\mathcal{A}}{\operatorname{argmax}}\, Q(s,a)$$
$$= \underset{a}{\operatorname{argmax}}\, E\left[r_{t+1} + \gamma V_\pi(S_{t+1}) \mid S_t = s,\, A_t = a\right],$$

Then greedy policy $G(V_\pi)(s)$ is better than π, i.e.,

$$V_{\pi'_D} = V_{G(V_\pi)} \geq V_\pi. \tag{8.85}$$

Proof

Theorem 8.9 implies that repeatedly applying the Bellman policy operator $B_{\pi'_D}$, starting with the value function V_π, the resulting sequence $\left(B_{\pi'_D}\right)^k (V_\pi)$ will converge to the value function $V_{\pi'_D}$, i.e.,

$$\lim_{k\to\infty} \left(B_{\pi'_D}\right)^k (V_\pi) = V_{\pi'_D}. \tag{8.86}$$

Theorem 8.8 shows that the sequence of the Bellman policy operator $\left(B_{\pi'_D}\right)^k (V_\pi)$ monotonically increases, i.e.,

$$V_{\pi'_D} \geq \left(B_{\pi'_D}\right)^k (V_\pi) \geq \left(B_{\pi'_D}\right)^{k-1}$$
$$(V_\pi) \geq \ldots \geq \left(B_{\pi'_D}\right)^{(0)} (V_\pi) = V_\pi.$$

This completes the proof of theorem.

Algorithms for RL are to train an agent to make decisions that will maximize rewards over time. Methods for finding the optimal policy consist of model-based approaches and model-free approaches. Model-based approaches can be classified as policy iteration approaches and value iteration approaches. Model-free approaches that do not need to know the models can be further decomposed into two types of methods: Monte Carlo methods and temporal difference (TD) methods. Now we first introduce policy iteration methods (Sutton and Barto 2018; Rao and Jelvis 2021).

8.2.2.3 Policy Iteration

Once a policy π_k has been improved using $G(V_\pi)$, we can infer a better policy π_{k+1}. Then, we compute the value function $V_{\pi_{k+1}}$ and apply greedy policy again to obtain a better policy π_{k+2}. Repeating above process, we obtain a sequence of monotonically increasing value functions and policies:

$$\pi_0 \to V_{\pi_0} \to G\left(V_{\pi_0}\right) = \pi_1 \to V_{\pi_1} \to G\left(V_{\pi_1}\right)$$
$$= \pi_2 \to \ldots V_{\pi_k} \to G\left(V_{\pi_k}\right)$$
$$= \pi_{k+1} \to \ldots \to \pi_* \to V_{\pi_*}.$$

The generated sequence of greedy policies is strictly monotonically increasing. Since an MDP has only a number of finite policies, the sequence of greedy policies will finally converge to the optimal policy. Now we formally summarize the algorithms for policy iteration and value iteration (Sutton and Barto 2018; Rao and Jelvis 2021).

Algorithm 8.2: Policy Iteration

Step 1: Initialization.
$V(s) \in R$, $\pi(s) \in \mathcal{A}(s)$, $\forall s \in S$, where $\pi(s)$ is a deterministic policy.
$\delta > 0$ is a small threshold parameter.

Step 2: Compute Value Function.
Repeat

$$\Delta \leftarrow 0$$

for all $s \in S$ **do**

$$V \leftarrow V(s)$$

$$a \leftarrow \pi(s)$$

$$V(s) \leftarrow \sum_{s'\in S} p(s'|s,a)\left[R(s,a,s') + \gamma V(s')\right]$$

$$\Delta \leftarrow \max\left(\Delta, \left|V - V(s)\right|\right)$$

end for
until $\Delta < \delta$

Step 3: Policy Improvement
Poly Stable $\leftarrow$ True
for all $s \in S$ **do**

$$b \leftarrow \pi(s)$$

$$\pi(s) \leftarrow \underset{a \in \mathcal{A}(s)}{\operatorname{argmax}} \sum_{s \in S} p(s'|s,a)\left[R(s,a,s') + \gamma V(s')\right]$$

If $b \neq \pi(s)$ then
Policy stable $\leftarrow$ false
end if
end for
If policy stable then
Stop and return $V_* \leftarrow V$, $\pi_* \leftarrow \pi$
else
Go to Step 2 (Compute Value Function)
endif

Limitation of policy iteration is that each of its iterations involves policy evaluation, which substantially increases computations. Many practical examples show that policy iteration only needs a few policy evaluations. To overcome the heavy computation limitation, the policy evaluation step of policy iteration can be truncated and value function evaluation and truncated policy evaluation steps can be combined to form Algorithm 8.3.

Algorithm 8.3: Value Iteration

Step 1: Initialization

$$k = 0,\ V_0(s) \in R,\ \forall s \in S.$$

Step 2: Repeat
for all $s \in S$ **do**

$$V_{k+1}(s) \leftarrow \max_{a \in \mathcal{A}(s)} \sum_{s' \in S} p(s'|s,a)\left[R(s,a,s') + \gamma V_k(s')\right]$$

end for
until $V_k(,)$ **converged**
Step 3: Output π_*, V_*.

The previous approaches assume the model: transition probability $p(s,a,s')$ and reward function $R(s,a,s')$. However, in practice, the model is often unknown. We need to directly access the physical data from the environment or dynamic system (Weinan 2017). What we observe is individual experiences of next state and reward, rather than the actual probabilities of occurrence of next states and rewards. Next we introduce the model-free methods for finding optimal policy and optimal value function (Sutton and Barto 2018; Rao and Jelvis 2021).

8.2.2.4 Monte Carlo Policy Evaluation

Monte Carlo methods do not assume the model and complete knowledge of the environments or dynamic systems. Only observed data from experience, i.e., sample sequences of states, actions, and rewards from actual or simulated interaction with an environment are available for Monte Carlo methods. The experience may be generated from online or from simulation. Although simulating experience also requires a model, the model need only generate sample transitions, not the complete probability distributions of all possible transitions. Simulation of experience may be much easier than obtaining the distribution in an explicit form.

Monte Carlo methods are only defined for episodic tasks. Suppose we want to estimate the value function $V_\pi(s)$ of a state s under a policy π, given a set of episodes generated by following the policy π and passing through the state s. In an episode, the state s may be visited multiple times. Each occurrence of state s in an episode is referred to as a visit to s. Within a given episode, the first time when s is visited is referred to as the first visit to s. The first visit Monte Carlo estimate $V_\pi(s)$ is defined as the average of the rewards following first visit. We also can define the every-visit Monte Carlo estimate $V_\pi(s)$ as the average of the rewards following all visits to s.

Assume that N episodes are generated by the policy π. The ith episode is denoted as

$$S_0^i, A_0^i, R_1^i, S_1^i, A_1^i, \ldots, R_{T-1}^i, S_{T-1}^i, A_{T-1}^i, R_T^i.$$

Let $R^i(s)$ be the total reward following the ith visit to s. Recall that the value $V_\pi(s)$ is defined as

$$V_\pi(s) = E_\pi\left[\sum_{k=0}^{\infty} \gamma^k r_{t+k+1} | S_t = s\right],$$

which can be approximated by

$$V_\pi(s) \approx \frac{1}{N}\sum_{i=1}^{N} R^i(s). \tag{8.87}$$

Now we introduce Algorithm 8.4 for first-visit Monte Carlo estimation of $V_\pi(s)$ (Sutton and Barto 2018; Rao and Jelvis 2021).

Algorithm 8.4: First-Visit Monte Carlo Estimation of $V_\pi(s)$

Step 1: Initialization.
$V(s) \in R$, arbitrarily, $\forall s \in S$
Returns $(s) \leftarrow$ an empty list, $\forall s \in S$.
Step 2: for all $s \in S$ **do**
Step 3: While $V(s)$ not converges **do**

1. Generate an episode using π: $S_0, A_0, R_1, S_1, A_1, R_2, \ldots, S_{T-1}, A_{T-1}, R_T$
2. $G \leftarrow 0$
3. **for** $t = T-1, T-2, \ldots, 1, 0$ **do**

$$G \leftarrow \gamma G + R_{t+1}$$

If s appears in $S_0, S_1, \ldots, S_{t-1}$
then Append G to *Returns* (s)
end if

$$V(s) \leftarrow average(Returns(s))$$

end while

8.2.2.5 Temporal-Difference Learning

To combine dynamic programming and Monte Carlo methods, we introduce TD learning (Sutton and Barto 2018; Rao and Jelvis 2021). TD learning is also model-free methods. It can learn value functions and policies directly from individual experience without a model of the system dynamics. TD learning attempts to predict the combination of immediate reward and the reward prediction at the next moment in time. In this section, we focus on learning value function V_π under a policy π.

Recall rom equation (8.87) that $V_\pi(s)$ is defined as

$$V_\pi(s) = E_\pi\left[\sum_{k=0}^{\infty} \gamma^k r_{t+k+1} | S_t = s\right], \tag{8.88}$$

which can be approximated by

$$V_\pi(s) \approx \frac{1}{k}\sum_{i=1}^{k} R_i(s). \tag{8.89}$$

Equation (8.89) can be rewritten as

$$V_\pi(s) \approx \hat{V}_k(s) = \frac{1}{k}\sum_{i=1}^{k} R_i(s). \tag{8.90}$$

But, equation (8.90) can be further reduced to

$$\hat{V}_k(s) = \frac{\sum_{i=1}^{k-1} R_i(s) + R_k(s)}{k}$$

$$= \frac{k-1}{k}\frac{\sum_{i=1}^{k-1} R_i(s) + R_k(s)}{k-1}$$

$$= \left(1 - \frac{1}{k}\right)\left[\frac{1}{k-1}\sum_{i=1}^{k-1} R_i(s) + \frac{1}{k}R_k(s)\right]. \tag{8.91}$$

Substituting equation (8.90) into equation (8.91) yields

$$\hat{V}_k(s) = \hat{V}_{k-1}(s) + \frac{1}{k}\left[R_k(s) - \hat{V}_{k-1}(s)\right]$$

$$= \hat{V}_{k-1}(s) + \alpha_k\left[R_k(s) - \hat{V}_{k-1}(s)\right]. \tag{8.92}$$

It follows from equations (8.88) and (8.89) that

$$R_k(s) = \sum_{i=0}^{\infty} \gamma^i r_{t+i+1}, \tag{8.93}$$

and

$$V_k(s) \approx R_k(s). \tag{8.94}$$

Equation (8.93) can be further reduced to

$$R_k(S_t) = r_{t+1} + \gamma\sum_{i=0}^{\infty} \gamma^i r_{t+1+i+1}$$

$$= r_{t+1} + \gamma R_{k-1}(S_{t+1}). \tag{8.95}$$

Combining equations (8.94) and (8.95), we obtain

$$R_k(S_t) = r_{t+1} + \gamma V_{k-1}(S_{t+1}). \tag{8.96}$$

Substituting equation (8.96) into equation (8.92) leads to

$$\hat{V}_k(S_t) = \hat{V}_{k-1}(S_t) + \alpha_k\left[r_{t+1} + \gamma V_{k-1}(S_{t+1}) - \hat{V}_{k-1}(S_t)\right], \tag{8.97}$$

where $r_{t+1} + \gamma V_{k-1}(S_{t+1}) - \hat{V}_{k-1}(S_t)$ is called TD.

One-step TD or TD(0) method is summarized in Algorithm 8.5 (Sutton and Barto 2018; Rao and Jelvis 2021).

Algorithm 8.5: TD(0) for V_π Estimation

Input: The policy π to be evaluated, step size $\alpha \in (0, 1]$.

Step 1: Initialization.
Given arbitrary $V(s)$ with constraints $(\text{terminal}) = 0$, $\forall s \in S^+$.
Step 2: for each episode **do**

a. Initialize S

b. for each step of episode

$$A \leftarrow \text{action } \pi(s)$$

c. Take action A, observe R, S'

d. $V(S) \leftarrow V(S) + \alpha\left[R + \gamma V(S') - V(S)\right]$

$$S \leftarrow S'$$

e. Until S is terminal

8.2.2.6 Comparisons: Dynamic Programming, Monte Carlo Methods, and Temporal Difference Methods

Now we summarize and compare dynamic programming, Monte Carlo methods, and TD methods (Sutton and Barto 2018; Rao and Jelvis 2021). Recall that Monte Carlo methods use $V_\pi(s) = E_\pi\left[G_t | S_t = s\right]$ as a target. However, the model and expected value are unknown, expectation is approximated with random samples (sample returns) $V_k(S_t) \approx R_t(S_t)$ and needs to wait for the end of the episodes.

The target of the dynamic programming is

$$V_\pi(s) = E_\pi\left[R_{t+1} + \gamma V_\pi(S_{t+1}) | S_t = s\right]. \quad (8.98)$$

The dynamic programming needs to know the model. Since $V_\pi(S_{t+1})$ is unknown, the expectation needs to be estimated using the Bellman equation:

$$V_k(S_t) \approx E_\pi\left[R_{t+1} + \gamma V_{k-1}(S_{t+1}) | S_t\right],$$

where $V_{k-1}(S_{t+1})$ replaces $V_\pi(S_{t+1})$.

The TD samples the expected values in equation (8.98) and uses the current estimate $V_{k-1}(S_{t+1})$ to replace $V_\pi(S_{t+1})$:

$$V_k(S_t) \approx \left(R_{t+1} + \gamma V_{k-1}(S_{t+1})\right).$$

It utilizes the advantage of both sampling of the Montel Carlo methods and boot strapping of the dynamic programming. The TD methods do not need to wait for the end of the episodes.

8.3 APPROXIMATE FUNCTION AND APPROXIMATE DYNAMIC PROGRAMMING

8.3.1 Introduction

The exact dynamic programming needs to calculate value functions and policy for all states at each iteration. The value functions in the dynamic programming are specified for each state and represented by a lookup table. Every state s has a value function $V(s)$ and every state-action pair has a state-action value function $Q(s, a)$. When the state space is large, exact dynamic programming requires prohibitive storage and computations. Therefore, when the state space is large, the value function needs to be approximated (Rao and Jelvis 2021). Value function approximation is a fundamental problem in RL (Asadi et al. 2021).

A framework for value function approximation is to represent a value function $V(s)$ and a state-action value function $Q(s,a)$ by parameterized functions $V(s;w)$ and $Q(s,a;w)$, instead of a table, where w are parameters. Many function approximation methods, including linear combination of features, neural networks, decision trees, nearest neighbors, and Fourier/Wavelet bases, have been developed. We will focus on two widely used differentiable function approximation methods: linear feature representations and neural networks.

Let x be the state and y be the value function for that state. We wish to estimate the conditional probability distribution function of random variable y, given x, from the data. The conditional probability of y, given x, is denoted as $f(x;w)(y)$. Suppose that a sequence of data $\{x_i, y_i\}_{i=1}^n$ is given. The likelihood or log-likelihood function is

$$L = \prod_{i=1}^{n} f(x_i, w)(y_i), \quad (8.99)$$

$$l(x,y,w) = \log L = \sum_{i=1}^{n} \log f(x_i, w)(y_i). \quad (8.100)$$

The maximum likelihood estimation is defined as

$$w^* = \underset{w}{\operatorname{argmax}} \left\{ \sum_{i=1}^{n} \log f(x_i; w)(y_i) \right\}. \quad (8.101)$$

We assume that the function f specifies the model probability distribution M of $y|x$ and the data $\{x_i, y_i\}_{i=1}^{n}$ specifies the empirical probability distribution D of $y|x$. The log-likelihood function can also be written as

$$\log M = \log f.$$

Equation (8.101) can be approximated by

$$E_D[\log M] \approx \frac{1}{n} \sum_{i=1}^{n} \log f(x_i; w)(y_i). \quad (8.102)$$

The maximum likelihood estimation can be interpreted as minimization of a loss function defined as the cross-entropy

$$J(w) = -E_D[\log M]. \quad (8.103)$$

The model can specify the conditional expected value of y, given x:

$$E_M[y|x] = E_{f(x;w)}[y] = \int y f(x;w)(y)dy. \quad (8.104)$$

Minimization of the loss function $J(w)$ is again usually solved by gradient methods:

$$w^{k+1} = w^k - \eta \nabla_w J(w), \quad (8.105)$$

where η is a learning rate.

8.3.2 Linear Function Approximation

Define a sequence of feature map function:

$$\varnothing_j(x): \mathcal{X} \to R, \ j = 1, 2, \ldots, m$$

and a vector of feature functions $\Phi(x): \mathcal{X} \to R^m$ and a vector of weights W:

$$\Phi(x) = \begin{bmatrix} \phi_1(x) \\ \vdots \\ \phi_m(x) \end{bmatrix}, W = \begin{bmatrix} w_1 \\ \vdots \\ w_m \end{bmatrix}.$$

Assume that the conditional density function for $y|x$ is a Gaussian distribution with mean $\Phi(x)^T W$ and variance σ^2:

$$p(y|x) = f(x;w)(y)$$
$$= \frac{1}{\sqrt{2\pi\sigma}} \exp\left(-\frac{\left(y - \Phi(x)^T W\right)^2}{2\sigma^2} \right). \quad (8.106)$$

The cross-entropy loss function for the function $f(x;w)(y)$ in equation (8.106) is given by

$$J(W) = -E_D[\log M]$$
$$= -E_D\left[\log\left(f(x;w)(y)\right)\right]$$
$$= -\frac{1}{n} \sum_{i=1}^{n} \log\left(f(x_i;w)(y_i)\right). \quad (8.107)$$

Substituting equation (8.106) into equation (8.107) and ignoring constant terms, we obtain

$$J(W) = \frac{1}{2n} \sum_{i=1}^{n} \left(y_i - \Phi(x_i)^T W\right)^2, \quad (8.108)$$

which is identical to the mean-squared-error of the linear regression.

The regularized loss function incorporating L^2 regularization with λ as the regularization parameter is given by

$$J(W) = \frac{1}{2n} \sum_{i=1}^{n} \left(y_i - \Phi(x_i)^T W\right)^2 + \lambda \|W\|_2^2. \quad (8.109)$$

The gradient of $J(W)$ with respect to W is

$$\nabla_W J(W) = \frac{1}{n} \sum_{i=1}^{n} \Phi(x_i)$$
$$\times \left(y_i - \Phi(x_i)^T W\right) + \lambda \|W\|_2. \quad (8.110)$$

8.3.3 Neural Network Approximation

In this section we introduce neural network approximation (Rao and Jelvis 2021). Consider a neural network with L hidden layers $l=0,1,\ldots,L-1$. Let $l=0$ be input layer with input $I_0=\Phi(x)$. The output layer is the Lth layer with the output $I_L=y$. Let I_l be the input vector to layer l and O_l be the output to layer l. We assume $I_{l+1}=O_l$ for all $l=0,1,\ldots,L-1$. Define

$$O_L=E_M\left[y|x\right].$$

In summary, we define

$$\begin{aligned}&I_0=\Phi(x),\ I_{l+1}=O_l,\ O_L\\&=E_M\left[y\,|\,x\right] \text{ for all } l=0,1,\ldots,L-1.\end{aligned} \tag{8.111}$$

Let the matrix W_l be the parameters for layer l. Let w_{ij}^l be the weight of the neuron j in layer $l-1$ connected to the neuron i in layer l, n_{l-1} and n_l be the number of neurons in the layers $l-1$ and l, respectively. Then, the matrix W^l is

$$W^l=\begin{bmatrix} w_{11}^l & \cdots & w_{1n_{l-1}}^l \\ \vdots & \vdots & \vdots \\ w_{n_l1}^l & \cdots & w_{n_ln_{l-1}}^l \end{bmatrix}.$$

The neurons in the layer l receive signals S^l that define a linear transformation from layer input I_l to the signals S^l. Therefore, we obtain

$$S^l=W^lI_l,\ l=0,1,\ldots,L. \tag{8.112}$$

Define the output function O_l as

$$O_l=g_l\left(S^l\right) \text{ for all } l=0,1,\ldots,L, \tag{8.113}$$

where g_l is the vector of activation function of layer l and is element-wise operation.

Equations (8.111) and (8.112) define the forward-propagation of the neural network that is used to calculate the value function in the RL.

To learn the weights in the neural network, we will use the gradient methods to minimize the cross entropy loss function. Thus, we first calculate the gradient of the cross entropy loss function with respect to W^l for $l=0,1,\ldots,L$.

Let

$$W_{i.}^l=\begin{bmatrix} w_{i1}^l & \cdots & w_{in_{l-1}}^l \end{bmatrix}. \tag{8.114}$$

Then,

$$W^l=\begin{bmatrix} W_{1.}^l \\ \vdots \\ W_{n_l.}^l \end{bmatrix}, \text{ and} \tag{8.115}$$

$$\nabla_{W^l}J(W)=\begin{bmatrix} \dfrac{\partial J(W)}{\partial W_{1.}^l} \\ \vdots \\ \dfrac{\partial J(W)}{\partial W_{n_l.}^l} \end{bmatrix}. \tag{8.116}$$

Using chain rule, we obtain

$$\frac{\partial J(W)}{\partial W_{i.}^l}=\frac{\partial J(W)}{\partial S_i^l}\frac{\partial S_i^l}{\partial W_{i.}^l}. \tag{8.117}$$

It follows from equation (8.112) that

$$S_i^l=\left(I_l\right)^T\left(W_{i.}^l\right)^T. \tag{8.118}$$

Using equation (8.118) and definition of gradient of the scale variable, we obtain

$$\frac{\partial S_i^l}{\partial W_{i.}^l}=\left(I_l\right)^T. \tag{8.119}$$

Substituting equation (8.119) into equation (8.117), we obtain

$$\begin{aligned}\frac{\partial J(W)}{\partial W_{i.}^l}&=\frac{\partial J(W)}{\partial S_i^l}\left(I_l\right)^T\\&=\frac{\partial J(W)}{\partial S_i^l}\nabla_{W^l}J(W),\end{aligned} \tag{8.120}$$

Combining equations (8.116) and (8.120), we obtain

$$\nabla_{W^l}J(W)=P_l\left(I_l\right)^T, \tag{8.121}$$

where $P_l=\dfrac{\partial J(W)}{\partial S^l}$.

If the loss function $J(W)$ incorporates L^2 regularization, then equation (8.121) is reduced to

$$\nabla_{W^l} J(W) = P_l (I_l)^T + \lambda W^l. \quad (8.122)$$

Finally, we introduce the formula for the analytical calculation of the vector P_l that has a recursive formulation and is an essential component of the backpropagation algorithm (Rao and Jelvis 2021).

Theorem 8.12: Recursive Formula for P_l Calculation

$$P_l = \left((W^{l+1})^T P_{l+1}\right) \odot g_l'(S_l), \; l = 0, 1, \ldots, L-1, \quad (8.123)$$

where g_l' is derivative of g_l and $\odot$ denotes element-wise multiplication.

Proof

Using definition of P_l, we have

$$P_l = \frac{\partial J(S^{l+1})}{\partial S^l} = \begin{bmatrix} \frac{\partial S_1^{l+1}}{\partial S_1^l} & \cdots & \frac{\partial S_{n_{l+1}}^{l+1}}{\partial S_1^l} \\ \vdots & \vdots & \vdots \\ \frac{\partial S_1^{l+1}}{\partial S_{n_l}^l} & \cdots & \frac{\partial S_{n_{l+1}}^{l+1}}{\partial S_{n_l}^l} \end{bmatrix} \begin{bmatrix} \frac{\partial J(S^{l+1})}{\partial S_1^{l+1}} \\ \vdots \\ \frac{\partial J(S^{l+1})}{\partial S_{n_{l+1}}^{l+1}} \end{bmatrix} \quad (8.124)$$

$$= \frac{\partial (S_{l+1})^T}{\partial S_l} P_{l+1}.$$

It follows from equation (8.112) that

$$S^{l+1} = W^{l+1} I_{l+1}. \quad (8.125)$$

Equation (8.111) defines

$$I_{l+1} = O_l. \quad (8.126)$$

Substituting equation (8.113) into equation (8.126), we obtain

$$I_{l+1} = g_l(S^l). \quad (8.127)$$

Again substituting equation (8.127) into equation (8.125) yields

$$S^{l+1} = W^{l+1} g_l(S^l). \quad (8.128)$$

Taking partial derivatives $\frac{\partial S^{l+1}}{\partial (S^l)^T}$ on equation (8.128), we obtain

$$\frac{\partial S^{l+1}}{\partial (S^l)^T} = W^{l+1} diag(g_l'(S^l)). \quad (8.129)$$

Substituting equation (8.129) into equation (8.124) yields

$$P_l = \left[diag(g_1'(S^l))\right]^T (W^{l+1})^T P_{l+1}$$
$$= g_1'(S^l) \odot (W^{l+1})^T P_{l+1}$$
$$= \left((W^{l+1})^T P_{l+1}\right) \odot g_l'(S_l).$$

This proves the theorem.

Finally we need to calculate $P_L = \frac{\partial J}{\partial S^L}$. Recall that the cross-entropy function is defined in terms of $P(y|S^L)$. To calculate $P_L = \frac{\partial J}{\partial S^L}$, we need to specify a functional form for $P(y|S^L)$. We assume that $P(y|S^L)$ follows a generic exponential family:

$$P(y|S^L, \tau) = h(y, \tau) e^{\frac{S^L y - A(S^L)}{d(\tau)}}. \quad (8.130)$$

Recall that $O_L = g_L(S^L) = E_p[Y|S^L]$. It is well known that

$$E_p[Y|S^L] = A'(S^L). \quad (8.131)$$

Thus, we require that

$$O_L = g_L(S^L) = A'(S^L). \quad (8.132)$$

In other words, the output layer activation function $g_L(S^L)$ should be equal to the derivative of the $A(S^L)$ function.

Assume that expectation with real data is approximated by a single data point (x, y). Then, the cross-Entropy loss function is

$$J = -\log P(y|S^L)$$
$$= -\log h(y, \tau) + \frac{A(S^L) - S^L y}{d(\tau)}. \quad (8.133)$$

It follows from equation (8.133) that

$$P_L = \frac{\partial J}{\partial S^L} = \frac{A'(S^L) - y}{d(\tau)}. \quad (8.134)$$

Substituting equation (8.132) into equation (8.134), we obtain

$$P_L = \frac{\partial J}{\partial S^L} = \frac{O_L - y}{d(\tau)}. \quad (8.135)$$

The above results can be summarized in Theorem 8.13.

Theorem 8.13:

$$P_L = \frac{\partial J}{\partial S^L} = \frac{O_L - y}{d(\tau)}.$$

Example 8.2: Normal Distribution

Normal density function is

$$p(Y|S^L,\tau) = \frac{1}{\sqrt{2\pi}\sigma}\exp\left\{-\frac{(y-\mu)^2}{2\sigma^2}\right\}$$

$$= \frac{1}{\sqrt{2\pi}\sigma}\exp\left\{-\frac{y^2}{2\sigma^2}\right\}\exp\left\{\frac{y\mu - \frac{1}{2}\mu^2}{\sigma^2}\right\}. \quad (8.136)$$

Set $S^L = \mu$, $\tau = \sigma, h(y,\tau)$

$$= \frac{1}{\sqrt{2\pi}\tau}\exp\left\{-\frac{y^2}{2\tau^2}\right\},$$

$$d(\tau) = \tau,\ A(S^L) = \frac{(S^L)^2}{2}.$$

It follows from equation (8.132) that

$$O_L = g_L(S^L) = A'(S^L) = S^L, \quad (8.137)$$

which implies that activation function in the output layer is the identity function.

For the normal distribution, we have

$$P_L = \frac{S^L - y}{\tau}. \quad (8.138)$$

Example 8.3: Bernoulli Distribution

Density function for Bernoulli distribution is given by

$$p(Y|S^L,\tau) = \exp\left\{y\log\frac{p}{1-p} + \log(1-p)\right\}$$

$$= \exp\left\{yS^L + \log\left(1+e^{S^L}\right)\right\}.$$

Set

$$S^L = \log\frac{p}{1-p},\ \tau = 1,\ h(y,\tau) = 1,$$

$$d(\tau) = 1,\ A(S^L) = \log\left(1+e^{S^L}\right) \text{ and}$$

$$O_L = g_L(S^L) = E\left[y|S^L\right] = \frac{1}{1+e^{-S^L}}.$$

Thus,

$$P_L = \frac{1}{1+e^{-S^L}} - y, \quad (8.139)$$

and output layer activation function is the logistic function.

8.3.4 Value-Based Methods

The value-based methods for RL first learn a value function and then define a policy to optimize the value functions (François-Lavet et al. 2018). We will introduce a simple and popular value-based algorithm, the Q-learning algorithm, the fitted Q-learning algorithm, and the deep Q-network (DQN) algorithm (Rahman et al. 2018; Ohnishi et al. 2019). To further get into the insights of the Q-network algorithm, we will also review various improvements of the DQN algorithm and provide the link between the value-based algorithms and policy-based algorithms.

8.3.4.1 Q-Learning

The Q-learning is an off policy RL algorithm that seeks to find the best action to take given the current state. One way to learn the best action is to find the optimal Q-value function by looking up the table of values $Q(s,a)$. In other words, the Q-learning uses the Bellman equation to learn the unique optimal $Q^*(s,a)$ function:

$$(BQ^*)(s,a) = Q^*(s,a), \quad (8.140)$$

where the Bellman operator is defined as

$$(BQ)(s,a)=\sum_{s'\in S} p(s,a,s')\left(R(s,a,s')+\gamma \max_{a'\in \mathcal{A}} Q(s',a')\right). \tag{8.141}$$

This approach is simple and easy to implement. However, when the state-action space is large, both memory size and computational time are huge. The exact Q-learning is infeasible in practice. Below we introduce the fitted Q-learning where the parameterized state-action value function $Q(s,a,\theta)$ will be used. We will focus on DQN where neural networks are used to approximate the action-value function (Mnih et al. 2013; Mnih et al. 2015), but we also briefly introduce double DQN (Hasselt et al. 2016) and deep recurrent Ql-learning (Romac and Béraud 2019).

8.3.4.2 Deep Q-Network

DQN attempts to train an intelligent agent and learn optimal actions while interacting with an environment (Wang et al. 2021). Training DQN has some remarkable features that are different from supervised learning. First, unlike supervised learning where a predefined set of data is used for training, the DQN learns from the agent's dynamically changing experiences. Second, it is difficult to understand the behavior of a DQN agent, whether it is intentional or it is just a random choice. Finally, it is crucial to the training to specify a random rate for input. The random inputs allow the agent to flexibly explore the unknown part of the environment.

The DQN learning process generates a sequence of states, actions, and rewards:

$$s_0, a_0, r_1, s_1, a_1, r_2, \ldots, s_{n-1}, a_{n-1}, r_n, s_n.$$

The total reward for one episode (i.e., from start to end) is defined as

$$R=r_1+r_2+\ldots r_n.$$

Assume that the agent achieves the optimal reward at time t and wishes to choose a sequence of appropriate actions to optimize its future rewards defined as

$$R_t=r_t+r_{t+1}+\ldots+r_n,$$

which can be extended to the discounted future reward to take random environment into account:

$$R_t=r_t+\gamma r_{t+1}+\ldots+\gamma^{n-1}r_n=r_t+\gamma R_{t+1}, \tag{8.142}$$

where γ is a discount constant.

Equation (8.142) implies that the optimal reward from the time t onward is equal to the summation of the currently achieved reward and the optimal discounted future reward.

Recall that similar to equation (8.142), Bellman equation defines

$$Q(s,a)=r+\gamma \max_{a'} Q(s'a'), \tag{8.143}$$

where (s',a') is the state-action pair after pair (s,a). If the number of states and number of action are large, exactly solving Bellman equation is computationally prohibitive. Therefore, neural networks are used to approximate state-action value function $Q(s,a)$.

The DQN takes states as input and expected reward $Q(s,a)$ as output (Wang and Ueda 2021).

Let $Q(s,a,\theta)$ be neural network-based approximation of $Q(s,a)$ with θ representing the parameters (the weights of neural networks). The Q-network takes the current state s as input and outputs the predicted action and reward for the action. The action is determined as

$$a'=\underset{a}{\operatorname{argmax}}\, Q(s,a,\theta_i), \tag{8.144}$$

and the predicted q-value is given by

$$q=\max_a Q(s,a,\theta_i). \tag{8.145}$$

The objective function in Q-network at the ith iteration is defined as

$$L_i(\theta_i)=E_{(s,a,r,s')\sim U(D)}\left[\left(r+\gamma \max_{a'} \hat{Q}(s',a';\theta_i^-)-Q(s,a;\theta_i)\right)^2\right], \tag{8.146}$$

where $r+\gamma \max_{a'} Q(s',a';\theta_i^-)$ is the target for the ith iteration, θ^- are parameters in the target network and are updated every C time steps. A very important issue in objective function (8.146) is the distribution $U(D)$. First, we define D. Sampling in DQN is based on the agent's experience. Each iteration stores experience sequence $e_t = (s_t, a_t, r_t, s_{t+1})$. Define experience replay data space as $D_t = \{e_1, e_2, \ldots, e_t\}$. Randomly draw samples of experience $(s,a,r,s') \sim U(D)$ in calculation of expectation $E_{(s,a,r,s') \sim U(D)}$ and update Q-function by sampling from experience in minibatch fashion.

To update the parameters θ_i, we need to calculate the gradient $\nabla_{\theta_i} L_i(\theta_i)$:

$$\nabla_{\theta_i} L_i(\theta_i) = -2E_{(s,a,r,s') \sim U(D)} \times \left[\left(r + \gamma \max_{a'} \hat{Q}(s', a'; \theta_i^-) - Q(s,a;\theta_i) \right) \nabla_{\theta_i} Q(s,a;\theta_i) \right]. \quad (8.147)$$

Define

$$v(\theta_i, t) = \gamma v(\theta_i, t-1) + (1-\gamma) \left(\nabla_{\theta_i} Q(s,a;\theta_i) \right)^2. \quad (8.148)$$

The parameters θ are updated by

$$\theta_i \leftarrow \theta_i - \frac{\eta}{\sqrt{v(\theta_i, t)}} \nabla_{\theta_i} Q(s,a;\theta_i). \quad (8.149)$$

The above solution procedures are summarized as Algorithm 8.6 (Mnih et al. 2013).

Algorithm 8.6: Deep Q-Networks

Step 1: Initialization.

Initialize replay memory D to capacity $\boldsymbol{N}$.

Randomly assign weights θ to action-value function Q as initial values.

Randomly assign weights θ^- to action-value function $\hat{Q}$ as initial values.

Step 2: Iteration (episode)

for episode =1, M **do**

Initialize sequence $s_1 = \{x_1\}$ (x_1: image or other input) and preprocessed sequence for state $\varnothing_1 = \varnothing(s_1)$.

Step 3: Iteration (time, fixed episode)

for t=1,T **do**

with probability ε select a random action a_t, otherwise select $a_t = \arg\max_a Q(\varnothing(s_t), a; \theta)$

Execute action a_t and observe reward r_t and image x_{t+1} (or other quantity)

Set $s_{t+1} = (s_t, a_t, x_{t+1})$ and preprocess $\varnothing_{t+1} = \varnothing(s_{t+1})$

Store transition $(\varnothing_t, a_t, r_t, \varnothing_{t+1})$ in D

Sample random minibatch of transitions $(\varnothing_j, a_j, r_j, \varnothing_{j+1}\}$ from D

Set

$$y_j = \begin{cases} r_j & \text{if episode terminates at step } j+1 \\ r + \gamma \max_{a'} \hat{Q}(\varnothing_{j+1}, a'; \theta^-) & \text{otherwise} \end{cases}$$

Perform a gradient descent step on $\left(y_j - Q(\varnothing_j, a_j; \theta)\right)^2$ with respect to the network parameters θ

Every c steps set $\hat{Q} = Q$

end for

end for

8.4 POLICY GRADIENT METHODS

8.4.1 Introduction

The previously introduced methods for RL are action-value methods. The action-value methods learn the values of actions and select actions based on the estimated values of actions. Next we introduce policy-based methods, which learn policy without calculating a value function. The most popular policy-based methods are policy gradient methods. Policy gradient methods can be classified model-based and model-free policy gradient methods (D'Oro et al. 2019; Lan et al. 2021).

A policy can be parameterized by a policy vector parameter $\theta \in R^d$. A policy can be written as

$$\pi(a|s,\theta) = p(A_t = a | S_t = s, \theta_t = \theta). \quad (8.150)$$

Let $J(\theta)$ be a scalar performance measure. We seek to maximize performance measure $J(\theta)$ using gradient ascent. The parameter θ_{t+1} at time $t+1$ is updated as

$$\theta_{t+1} = \theta_t + \alpha \widehat{\nabla_\theta J(\theta_t)}, \quad (8.151)$$

where $\widehat{\nabla_\theta J(\theta_t)}$ is an estimator of the gradient $\nabla_\theta J(\theta)$. All methods that follow parameter update equation (8.151) are referred to as policy gradient methods. Methods that learn both value and policy approximation functions are referred to as actor-critic methods, where "actor" represents the learned policy and "critic" represents the learned value function, usually a state-value function.

8.4.2 Policy Approximation

Let $\pi(a|s,\theta)$ be parametrized policy. The parameterized policy function $\pi(a|s,\theta)$ should be differentiable with respect to parameters θ. We assume that gradient $\nabla_\theta\, \pi(a|s,\theta)$ exists and is finite for all $s \in S$, $a \in \mathcal{A}(s)$, and $\theta \in R^d$ (Sutton and Barto 2018).

To implement exploration, the policy $\pi(a|s,\theta)$ is assumed random and is often defined as

$$\pi(a|s,\theta)=\frac{e^{h(s,a,\theta)}}{\sum_b e^{h(s,b,\theta)}}, \quad (8.152)$$

where $h(s,a,\theta)\in R$ for each state-action pair is often called action preference and is used to define an exponential soft-max distribution. Again, the action preferences can also be parameterized by a neural network or simply a linear function of features:

$$h(s,a,\theta)=\theta^T x(s,a), \quad (8.153)$$

where $x(s,a)\in R^d$ is a feature vector.

Next we introduce the policy gradient theorem to show that policy-gradient methods have stronger convergence feature than action-value methods (Sutton and Barto 2018). We first define the performance measure $J(\theta)$ and the value of the start state of the episode:

$$J(\theta)=V_{\pi_\theta}(s_0), \quad (8.154)$$

where s_0 is the start state of the episode, the policy is determined by θ and V_{π_θ} is the true value function for π_θ.

Next we calculate the gradient of performance measure $\nabla_\theta J(\theta)$. Recall (equation (8.19)) that state-value function under the policy can be written as

$$V_{\pi_\theta}(s)=\sum_{a\in\mathcal{A}}\pi_\theta(a|s)Q_{\pi_\theta}(s,a). \quad (8.155)$$

Taking gradient on both sides of equation (8.155), we obtain

$$\nabla_\theta V_{\pi_\theta}(s)=\sum_{a\in\mathcal{A}}\big[\nabla_\theta\pi_\theta(a|s)Q_{\pi_\theta}(s,a) + \pi_\theta(a|s)\nabla_\theta Q_{\pi_\theta}(s,a)\big]. \quad (8.156)$$

By definition, $Q_{\pi_\theta}(s,a)$ is equal to the summation of the current reward r and the future state value $V_\pi(s')$:

$$Q_{\pi_\theta}(s,a)=\sum_{s'}\sum_r p(s',r|s,a)\big(r+V_\pi(s')\big). \quad (8.157)$$

Substituting equation (8.157) into equation (8.156) yields

$$\begin{aligned}\nabla_\theta V_{\pi_\theta}(s) &=\sum_{a\in\mathcal{A}}\Big[\nabla_\theta\pi_\theta(a|s)Q_{\pi_\theta}(s,a)+\pi_\theta(a|s)\nabla_\theta\sum_{s'}\sum_r p(s',r|s,a)\big(\mathrm{r}+V_\pi(s')\big)\Big]\\ &=\sum_{a\in\mathcal{A}}\Big[\nabla_\theta\pi_\theta(a|s)Q_{\pi_\theta}(s,a)+\pi_\theta(a|s)\sum_{s'}\sum_r p(s',r|s,a)\nabla_\theta V_\pi(s')\Big].\end{aligned} \quad (8.158)$$

Recall (equation 8.4) that

$$P(s'|s,a)=\sum_{r\in\mathcal{R}}P(s',r|s,a). \quad (8.159)$$

Substituting equation (8.159) into equation (8.158), we obtain

$$\nabla_\theta V_{\pi_\theta}(s)=\sum_{a\in\mathcal{A}}\big[\nabla_\theta\pi_\theta(a|s)Q_{\pi_\theta}(s,a) + \pi_\theta(a|s)\sum_{s'}P(s'|s,a)\nabla_\theta V_\pi(s')\big] \quad (8.160)$$

Let

$$\varnothing(s)=\sum_{a\in\mathcal{A}}\nabla_\theta\pi_\theta(a|s)Q_{\pi_\theta}(s,a). \quad (8.161)$$

Substituting equation (8.161) into equation (8.160), we obtain

$$\begin{aligned}\nabla_\theta V_{\pi_\theta}(s)&=\varnothing(s)+\sum_{a\in\mathcal{A}}\pi_\theta(a|s)\sum_{s'}P(s'|s,a)\nabla_\theta V_\pi(s')\\ &=\varnothing(s)+\sum_{s'}\rho_\pi(s\to s',1)\,\nabla_\theta V_\pi(s'),\end{aligned} \quad (8.162)$$

where

$$\rho_\pi(s\to s',1)=\sum_{a\in\mathcal{A}}\pi_\theta(a|s)P(s'|s,a). \quad (8.163)$$

Repeating equation (8.162) one time, we obtain

$$\nabla_\theta V_{\pi_\theta}(s)=\varnothing(s)+\sum_{s'}\rho_\pi(s\to s',1) \times\Big[\varnothing(s')+\sum_{s''}\rho_\pi(s'\to s'',1)\nabla_\theta V_\pi(s'')\Big]$$

$$=\varnothing(s)+\sum_{s'}\rho_\pi(s\to s',1)\varnothing(s')$$

$$+\sum_{s''}\sum_{s'}\rho_\pi(s\to s',1)\rho_\pi(s'\to s'',1)\nabla_\theta V_\pi(s'')$$

$$=\varnothing(s)+\sum_{s'}\rho_\pi(s\to s',1)\varnothing(s')$$

$$+\sum_{s''}\rho_\pi(s'\to s'',2)\,\nabla_\theta V_\pi(s''), \quad (8.164)$$

where

$$\rho_\pi(s\to s'',2)$$

$$=\sum_{s'}\rho_\pi(s\to s',1)\rho_\pi(s'\to s'',1). \quad (8.165)$$

Define a recursive formula:

$$\rho_\pi(s\to x,k+1)$$

$$=\sum_{s'}\rho_\pi(s\to s',\,k)\rho_\pi(s'\to x,1). \quad (8.166)$$

Repeat equation (8.164) many times and using equation (8.166), we obtain

$$\nabla_\theta V_{\pi_\theta}(s)=\varnothing(s)+\sum_{s'}\rho_\pi(s\to s',1)\varnothing(s')$$

$$+\sum_{s''}\rho_\pi(s'\to s'',2)\varnothing(s'')$$

$$+\sum_{s'''}\rho_\pi(s''\to s''',3)\nabla_\theta V_\pi(s''')$$

$$=\dots$$

$$=\sum_{x\in s}\sum_{k=0}^{\infty}\rho_\pi(s\to x,k)\varnothing(x). \quad (8.167)$$

In equation (8.167), the derivative of Q-value function, $\nabla_\theta Q_\pi(s,a)$ is removed. By plugging equation (8.167) into the objective function $J(\theta)$ (equation (8.154)), we obtain

$$\nabla_\theta J(\theta)=\nabla_\theta V_\pi(s)$$

$$=\sum_{s}\sum_{k=0}^{\infty}\rho_\pi(s_0\to s,k)\varnothing(s). \quad (8.168)$$

Let $\eta(s)=\sum_{k=0}^{\infty}\rho_\pi(s_0\to s,k)$. Then, equation (8.168) can be rewritten as

$$\nabla_\theta J(\theta)=\sum_{s}\eta(s)\varnothing(s)$$

$$=\left(\sum_{s}\eta(s)\right)\sum_{s}\frac{\eta(s)}{\sum_{s}\eta(s)}\varnothing(s)$$

$$\propto\sum_{s}\frac{\eta(s)}{\sum_{s}\eta(s)}\varnothing(s)$$

$$=\sum_{s}d_\pi(s)\varnothing(s), \quad (8.169)$$

where

$$d_\pi(s)=\frac{\eta(s)}{\sum_{s}\eta(s)}. \quad (8.170)$$

Substituting equation (8.161) into equation (8.169) yields

$$\nabla_\theta J(\theta)\propto\sum_{s}d_\pi(s)\sum_{a\in\mathcal{A}}\nabla_\theta\pi_\theta(a|s)Q_{\pi_\theta}(s,a). \quad (8.171)$$

$$=\sum_{s}d_\pi(s)\sum_{a\in\mathcal{A}}Q_{\pi_\theta}(s,a)\nabla_\theta\pi_\theta(a|s)$$

$$=\sum_{s}d_\pi(s)\sum_{a\in\mathcal{A}}\pi_\theta(a|s)Q_{\pi_\theta}(s,a)\frac{\nabla_\theta\pi_\theta(a|s)}{\pi_\theta(a|s)}. \quad (8.172)$$

Note that

$$\nabla_\theta\log\pi_\theta(a|s)=\frac{\nabla_\theta\pi_\theta(a|s)}{\pi_\theta(a|s)}. \quad (8.173)$$

Substituting equation (8.173) into equation (8.172), we obtain

$$\nabla_\theta J(\theta)\propto\sum_{s}d_\pi(s)\sum_{a\in\mathcal{A}}\pi_\theta(a|s)Q_{\pi_\theta}(s,a)\nabla_\theta\log\pi_\theta(a|s).$$

$$=E_{s\sim d_\pi,a\sim\pi_\theta}\left[Q_{\pi_\theta}(s,a)\nabla_\theta\log\pi_\theta(a|s)\right). \quad (8.174)$$

This proves the policy gradient theorem (Sutton and Barto 2018).

Theorem 8.14: The Policy Gradient Theorem

Assume $\gamma = 1$ and Episode case. The gradient of performance measure $\nabla_\theta J(\theta)$ is proportional to

$$\nabla_\theta J(\theta) \propto \sum_s d_\pi(s) \sum_{a \in A} \pi_\theta(a|s) Q_{\pi_\theta}(s,a) \nabla_\theta \log \pi_\theta(a|s)$$

$$= E_{s \sim d_\pi, a \sim \pi_\theta}\left[Q_{\pi_\theta}(s,a) \nabla_\theta \log \pi_\theta(a|s)\right],$$

where $d_\pi(s)$ is the on-policy distribution under the policy π.

8.4.3 REINFORCE: Monte Carlo Policy Gradient

The policy gradient theorem shows that its right-hand side is a sum over states weighted by how often the states are visited by the episode under the target policy π. Therefore, equation (8.174) can be written as

$$\nabla J(\theta) \propto E_\pi\left[\sum_a Q_\pi(S_t, a) \nabla_\theta \log \pi_\theta(a|S_t)\right]. \quad (8.175)$$

The parameters can be updated by

$$\theta_{t+1} = \theta_t + \alpha \sum_a \hat{Q}(S_t, a) \nabla_\theta \log \pi_\theta(a|S_t), \quad (8.176)$$

where $\hat{Q}(S_t, a)$ is an approximation to $Q_\pi(S_t, a)$. However, this algorithm involves all actions. To simplify algorithm, we use the one action A_t actually taken at time t, instead of all actions (Williams 1992; Sutton and Barto 2018). Therefore, replacing all actions by sampling $A_t \sim \pi$, equation (8.175) can be written as

$$\nabla J(\theta) \propto E_\pi\left[\hat{Q}(S_t, A_t) \nabla_\theta \log \pi_\theta(A_t|S_t)\right]. \quad (8.177)$$

It follows from equation (8.16) that

$$Q_\pi(S_t, A_t) = E_\pi\left[G_t | S_t, A_t\right]. \quad (8.178)$$

Using equations (8.171) and (8.178), we obtain,

$$\nabla J(\theta) \propto E_\pi\left[G_t \nabla_\theta \log \pi_\theta(A_t|S_t)\right], \quad (8.179)$$

which implies that we can measure return G_t from real sample trajectory and use it to update the parameters (Weng 2021):

$$\theta_{t+1} = \theta_t + \alpha G_t \nabla_\theta \log \pi_\theta(A_t|S_t). \quad (8.180)$$

The process for updating the parameters is summarized in Algorithm 8.7 (Sutton and Barto 2018).

Algorithm 8.7: REINFORCE (Monte-Carlo Policy Gradient)

Step 1: Initialization.
Randomly initialize the policy parameter θ. Input step size $\alpha > 0$.

Step 2: Loop forever (for each episode)
Generate one trajectory on policy $\pi_\theta : S_0, A_0, R_1, S_1, A_1, R_2, \ldots, S_{T-1}$.

Step 3: For each step of the episode: $t = 0, 1, 2, \ldots, T-1$ **do**

1. Estimate the return $G_t = \sum_{k=t+1}^{T} \gamma^{k-t-1} R_k$.
2. Update the policy parameter

$$\theta \leftarrow \theta + \alpha \gamma^t G_t \nabla_\theta \log \pi_\theta(A_t|S_t).$$

end for
end for

8.4.4 REINFORCE with Baseline

An arbitrary baseline $b(s)$ can be subtracted from the action value in equation (8.171) to generalize the policy gradient theorem:

$$\nabla_\theta J(\theta) \propto \sum_s d_\pi(s) \sum_{a \in A} \left(Q_{\pi_\theta}(s,a) - b(s)\right) \nabla_\theta \pi_\theta(a|s). \quad (8.181)$$

The baseline can be any function, but cannot vary with action a. The rule for updating the parameter is given by

$$\theta_{t+1} = \theta_t + \alpha\left(G_t - b(S_t)\right) \nabla_\theta \log \pi_\theta(A_t|S_t). \quad (8.182)$$

The role of baseline is to reduce the variance of the parameter estimator. We can choose an estimate of the state value $\hat{V}(S_t, W)$ as a baseline. A Monte Carlo method is often used to estimate the state-value weights W. REINFORCE with baseline method is summarized in Algorithm 8.8 (Sutton and Barto 2018).

Algorithm 8.8: REINFORCE with Baseline

Step 1: Initialization.

Input a differentiable and parameterized policy function $\pi_\theta(a|s)$.

Input a differentiable and parameterized state-value function $\hat{V}(s,w)$.

Input step sizes $\alpha_\theta > 0$, $\alpha_w > 0$.

Initialize policy parameter θ and state-value weights W.

Step 2: Loop forever (for each episode)

Generate one trajectory on policy $\pi_\theta : S_0, A_0, R_1, S_1, A_1, R_2, \ldots, S_{T-1}$, following $\pi_\theta(.|.)$.

Step 3: For each step of the episode: $t = 0, 1, 2, \ldots, T-1$ **do**

1. Estimate the return $G \leftarrow \sum_{k=t+1}^{T} \gamma^{k-t-1} R_k$
2. Calculate difference $\delta \leftarrow G - \hat{V}(S_t, w)$
3. Update the state-value weights

$$w \leftarrow w + \alpha_w \delta \nabla_w \hat{V}(S_t, w)$$

4. Update the policy parameter

$$\theta \leftarrow \theta + \alpha_\theta \gamma^t \delta \nabla_\theta \log \pi_\theta(A_t|S_t).$$

end for
end for

8.4.5 Actor–Critic Methods

Each state transition involves two states. REINFORCE with baseline uses the estimated value of the first state of the transition as a baseline for the subsequent return prior to the transition action. Therefore, the value function of the first state cannot be used to assess the performance of the action. One step return $G_{t:t+1}$ that is summation of the reward and the value-function of the second state. The one step return is an estimated of the actual return and can be used to assess the action (Sutton and Barto 2018).

Because the state-value function is used to assess actions, the value function is referred to as a critic, the policy is referred to as an actor. Either action-value $Q_w(a|s)$ or state-value $V_w(s)$ can be used as critic.

Replacing full return by one-step return and baseline by estimated state-value function in equation (8.182), we obtain

$$\begin{aligned}\theta_{t+1} &= \theta_t + \alpha\left(G_{t:t+1} - \hat{V}(S_t, w)\right)\nabla_\theta \log \pi_\theta(A_t \mid S_t) \\ &= \theta_t + \alpha\Big(R_{t+1} + \gamma \hat{V}(S_{t+1}, w) \\ &\quad - \hat{V}(S_t, w)\Big)\nabla_\theta \log \pi_\theta(A_t|S_t) \\ &= \theta_t + \alpha \delta_t \nabla_\theta \log \pi_\theta(A_t|S_t),\end{aligned} \tag{8.183}$$

where

$$\delta_t = R_{t+1} + \gamma \hat{V}(S_{t+1}, w) - \hat{V}(S_t, w).$$

One-step actor-critic method for estimating policy π_θ is summarized in Algorithm 8.9 (Sutton and Barto 2018).

Algorithm 8.9: One-Step Actor-Critic (Episodic) for Estimating Policy π_θ

Step 1: Initialization.

Input a differentiable and parameterized policy function $\pi_\theta(a|s)$.

Input a differentiable and parameterized state-value function $\hat{V}(s,w)$.

Input step sizes $\alpha_\theta > 0$, $\alpha_w > 0$.

Initialize policy parameter θ and state-value weights W.

Step 2: Loop forever (for each episode)

Initialize S (first state of episode)

$$I \leftarrow 1$$

Step 3: Loop while S is not terminal (for each time step)

1. $A \sim \pi_\theta(.|s)$
2. Take action A, observe S', R
3. $\delta \leftarrow R + \gamma \hat{V}(S', w) - \hat{V}(S, w)$ (If S' is terminal, then $\hat{V}(S', w) = 0$)
4. $w \leftarrow w + \alpha_w \delta \nabla_w \hat{V}(S, w)$
5. $\theta \leftarrow \theta + \alpha_\theta I \delta \nabla_\theta \log \pi_\theta(A|S)$

6. $I \leftarrow \gamma I$

7. $S \leftarrow S'$

end loop
end loop

8.4.6 *n*–Step Temporal Difference (TD)

Methods that extend TD over n steps are called n–step TD methods (Sutton and Barto 2018). The n–step TD methods generalize both MC methods and one-step TD methods and are in the middle between the MC methods and one-step TD methods. We will introduce n–step TD methods for prediction and control (Sarsa) in this section.

*8.4.6.1 **n**– Step Prediction*

Consider a sequence of $S_t, R_{t+1}, S_{t+1}, R_{t+2}, \ldots, R_T, S_T$ generated by an episode. For updating the estimate of the value function $V_\pi(S_t)$, the MC methods use the complete return:

$$G_t = R_{t+1} + \gamma R_{t+2} + \gamma^2 R_{t+3} + \ldots + \gamma^{T-t-1} R_T, \quad (8.184)$$

while one-step TD methods use the one-step return:

$$G_{t:t+1} = R_{t+1} + \gamma V_t(S_{t+1}),$$

where T is the last time step of the episode and $V_t : S \to R$ is the estimate at time t of V_π, i.e.,

$$V_t(S_{t+1}) = R_{t+2} + \gamma R_{t+3} + \ldots + \gamma^{T-t-2} R_T.$$

Similarly, two-step update is given by

$$G_{t:t+2} = R_{t+1} + \gamma \mathrm{R}_{t+2} + \gamma^2 V_t(S_{t+2}),$$

where $\gamma^2 V_{t+1}(S_{t+2})$ accounts for the missing of the terms $\gamma^2 R_{t+3} + \ldots + \gamma^{T-t-1} R_T$. In general, the n–step update is defined as

$$G_{t:t+n} = R_{t+1} + \gamma R_{t+2} + \gamma^2 R_{t+3} + \ldots + \gamma^{n-1} R_{t+n} + \gamma^n V_t(S_{t+n}), \quad (8.185)$$

where $n \geq 1, 0 \leq t < T - n$. If $t + n \geq T$, then $G_{t:t+n} = G_t$, n–step return is equal to the full return. $G_{t:t+n}$ can be considered approximate returns, truncated after n steps and then corrected for the remaining missing terms, in equation (8.186) by $V_t(S_{t+n})$. The general n–step return can be defined as

$$G_{t:t+n}(c) = R_{t+1} + \gamma R_{t+2} + \gamma^2 R_{t+3} + \ldots + \gamma^{n-1} R_h + \gamma^n c, \quad (8.186)$$

where $c \in R$ is a scalar correction. The time $h = t + n$ is referred to as the horizon of the n–step return.

Therefore, we obtain

$$G_{t:t+n}\left(V_t(S_{t+n})\right) = R_{t+1} + \gamma R_{t+2} + \gamma^2 R_{t+3} + \ldots + \gamma^{n-1} R_{t+n} + \gamma^n V_t(S_{t+n}),$$

Computation of n–step return requires estimate of V_{t+n-1}, which is computed only after R_{t+n} is available. Therefore, the state-value learning algorithm using n–step return is

$$V_{t+n}(S_t) = V_{t+n-1}(S_t) + \alpha\left[G_{t:t+n}\left(V_t(S_{t+n})\right) - V_{t+n-1}(S_t)\right], 0 \leq t < T. \quad (8.187)$$

It should know that the values of all other states except for S_t remain unchanged, i.e., $V_{t+n}(s) = V_{t+n-1}(s)$ for all $s \neq S_t$. During the first $n - 1$ steps of each episode we do not make any changes. We summarize here in Algorithm 8.10 (Sutton and Barto 2018).

Algorithm 8.10: *n*– Step TD for Estimating *V*

Step 1: Initialization

Input: A policy π, step size $\alpha \in (0,1]$, a positive integer n

Initialize value function $V(s)$ arbitrarily, for all $s \in S$

All store and access operations (for S_t and R_t) can take their index mod $n+1$

Step 2: Loop for each episode

Initialize and store $S_0 \neq$ terminal

$$T \leftarrow \infty$$

Step 3: Loop for $t = 0, 1, 2, \ldots$

If $t < T$ then

Take an action according to $\pi(.\mid S_t)$

Observe and store the next reward as R_{t+1} and the next state as S_{t+1}

If S_{t+1} is terminal then

$$T \leftarrow t+1$$

end if

$\tau \leftarrow t-n+1$ (τ is the time whose state's estimation is being updated)

If $\tau \geq 0$

$$G \leftarrow \sum_{i=\tau+1}^{\min(\tau+n,T)} \gamma^{t-i-1} R_i$$

If $\tau + n < T$ then

$$G \leftarrow G + \gamma^n V(S_{\tau+n})$$

endif

$$V(S_\tau) \leftarrow V(S_\tau) + \alpha\left[G - V(S_\tau)\right]$$

Endif

Until $\tau = T-1$

endif

Now we introduce increment. The $n-$step backup at time t can produce the increment $\Delta V_t(S_t)$ which is defined as

$$\Delta V_t(S_t) = \alpha\left[G_{t:(t+n)}\left(V_t(S_{t+n})\right) - V_t(S_t)\right], \quad (8.188)$$

where α is a positive step-size parameter.

If $s \neq S_t$ then $\Delta V_t(s)$ is defined as zero.

In online updating, the update is performed at each time step by

$$V_{t+1}(s) = V_t(s) + \Delta V_t(s). \quad (8.189)$$

However, in off-line updating, the update is only made at the end of the episode by summing the increments during the episode:

$$V_{t+1}(s) = V_t(s), \ \forall t < T$$

$$V_T(s) = V_{T-1}(s) + \sum_{t=0}^{T-1} \Delta V_t(s). \quad (8.190)$$

The $n-$step TD has a nice error reduction property. Suppose that the $n-$step TD starts with any value function $V(s)$. Define the error of the $n-$step TD estimate of $V_\pi(s)$ as

$$|E_\pi\left[G_{t:t+n}\left(V(S_{t+n})\right)|S_t = s\right] - V_\pi(s)|. \quad (8.191)$$

Then, the worse error of the $n-$step TD estimate of $V_\pi(s)$ is less than or equal to γ^n times the worst error between $V_\pi(s)$ and its estimate $V(s)$, i.e.,

$$\max_S |E_\pi\left[G_{t:t+n}\left(V(S_{t+n})\right)|S_t = s\right] - V_\pi(s)| \leq \gamma^n \max_S \left|V(s) - V_\pi(S)\right|. \quad (8.192)$$

8.4.7 *TD*(λ) Methods

Next we introduce $TD(\lambda)$ algorithm (Sutton and Barto 2018). Any average of $n-$step returns, where the sum of the weights for the average should be equal to 1, can be used to develop new algorithms with the average of $n-$step returns as their target returns. Consider a sequence of weights $(1-\lambda), (1-\lambda)\lambda, (1-\lambda)\lambda^2, \ldots, (1-\lambda)\lambda^n, \ldots$ and define $TD(\lambda)$ algorithm as

$$TD(\lambda) = (1-\lambda)\sum_{n=1}^{\infty} \lambda^{n-1} TD_n, \quad (8.193)$$

where TD_n is the $n-$step TD algorithm and $(1-\lambda)\sum_{n=1}^{\infty} \lambda^{n-1} = (1-\lambda)\frac{1}{(1-\lambda)} = 1$.

Since the return of $TD(\lambda)$ is given by $G_{t:t+n}\left(V_t(S_{t+n})\right)$, the return of $TD(\lambda)$, which is called the λ-return, is given by

$$L_t = (1-\lambda)\sum_{n=1}^{\infty} \lambda^{n-1} G_{t:t+n}\left(V_t(S_{t+n})\right). \quad (8.194)$$

Equation (8.194) implies that when $n \geq T-t$, the n-step TD passes the terminal state, then the return is $G_{t:t+n} = G_{t:T} = G_t$. Equation (8.194) is reduced to

$$L_t = (1-\lambda)\sum_{n=1}^{T-t-1} \lambda^{n-1} G_{t:t+n}\left(V_t(S_{t+n})\right) + \lambda^{T-t-1} G_t. \quad (8.195)$$

When $\lambda = 1$, $L_t = G_t$, $TD(\lambda)$ is reduced to Monte Carlo algorithm. When $\lambda = 0$, $L_t = G_{t:t+1}\left(V_t(S_{t+1})\right)$, the one-step return. Thus, for $\lambda = 0$, TD(0) is the one-step TD algorithm.

We also can compute an increment $\Delta V_t(S)$:

$$\Delta V_t(S) = \begin{cases} \alpha\left(L_t - V_t(S_t)\right) & S = S_t \\ 0 & S \neq S_t \end{cases}. \quad (8.196)$$

As in equation (8.189) or equation (8.190), the updating can be either online or off-line.

For each state visited, we update the value function by looking forward to all the future rewards and taking their best combination. This approach is referred to as forward view of $TD(\lambda)$.

Next we consider backward view of $TD(\lambda)$ (Sutton and Barto 2018). To look back, we need a variable to record past visited states. This additional memory random variable is called eligibility trace, which is denoted $E_t(s) \in R^t$. The eligibility trace variable is recursively updated:

$$E_t(s) = \begin{cases} \gamma\lambda E_{t-1}(s) & \forall s \in S, s \neq S_t \\ \gamma\lambda E_{t-1}(S_t) + 1 & s = S_t \end{cases}, \quad (8.197)$$

where S_t is the state visited at time t, γ is the discounter rate and λ is the parameter in $TD(\lambda)$. Equation (8.197) implies that each time the state is visited, the eligibility trace is accumulated, and then it is reduced when the states are not visited. Eligibility trace records the recently visited states and measures the degree with which each state is eligible for learning changes when a reinforcing event takes place.

There are two alternative variations of eligibility trace. The first one is the replacing trace. Consider a visited state. Suppose that the trace due to the first visit has completely decayed to zero before the second revisit. In this case, the replacing trace methods reset the trace variable to 1:

$$E_t(S_t) = 1. \quad (8.198)$$

The second trace variation is the Dutch trace. The Dutch trace is between the accumulating and replacing traces and is defined as

$$E_t(S_t) = (1-\alpha)\gamma\lambda E_{t-1}(S_t) + 1, \quad (8.199)$$

where α approaches to zero, the Dutch trace becomes accumulating trace, and if $\alpha = 1$, the Dutch trace is reduced to the replacing trace.

When the global TD errors occur, the value functions of recently visited states are updated by

$$\Delta V_t(s) = \alpha\delta_t E_t(s), \forall s \in S, \quad (8.200)$$

where

$$\delta_t = R_{t+1} + \gamma V_t(S_{t+1}) - V_t(S_T). \quad (8.201)$$

The backward view or mechanic online algorithm of $TD(\lambda)$ is summarized in Algorithm 8.11.

Algorithm 8.11

Step 1: Initialization

If s is the terminate state, set $V(s) = 0$. Otherwise, randomly assign arbitrary value to $V(s)$.

Step 2: Repeat for each episode:

Initialization for each episode

Initialize $E(s) = 0, \forall s \ni \mathcal{S}$.

Initialize S

Step 3: Repeat for each step of episode

1. $A \leftarrow$ action given by $\pi(.|S)$ for S
2. Take action A, observe reward R, and next state S'.
3. Calculate

$$\delta \leftarrow R + \gamma V(S') - V(S)$$

4. $E(S) \leftarrow E(S) + 1$ (accumulating traces)
   or $E(S) \leftarrow (1-\alpha)E(S) + 1$ (Dutch traces)
   or $E(S) \leftarrow 1$
5. For all $s \in S$

$$V(s) \leftarrow V(s) + \alpha\delta E(s)$$

$$E(s) \leftarrow \gamma\lambda E(s)$$

$$S \leftarrow S'$$

Until S is terminal

Next we introduce the true online $TD(\lambda)$. The value function update for the true online $TD(\lambda)$ is defined as

$$V_{t+1}(s) = V_t(s) + \alpha\left[\delta_t + V_t(S_t) - V_{t-1}(S_t)\right]E_t(s) \\ - \alpha I_{sS_t}\left[V_t(S_t) - V_{t-1}(S_t)\right], \forall s \in \mathcal{S}, \quad (8.202)$$

where

$$I_{sS_t} = \begin{cases} 1 & s = S_t \\ 0 & s \neq S_t \end{cases}.$$

An efficient implementation is summarized in Algorithm 8.12 (Sutton and Barto 2018).

Algorithm 8.12

Step 1: Initialization

If s is terminal then $V(s)=0$, otherwise initialize $V(s)$ arbitrarily

$$V_{old} \leftarrow 0$$

Step 2: Repeat for each episode

Initialize $E(s)=0, \forall s \in \mathcal{S}$

Initialize S

Step 3: Repeat for each step of episode:

$A \leftarrow$ action given by $\pi(.|S)$

Take action A, observe reward R, and next state S'

$$\Delta \leftarrow V(S) - V_{old}$$

$$V_{old} \leftarrow V(S')$$

$$\delta \leftarrow R + \gamma V(S') - V(S)$$

$$E(S) \leftarrow (1-\alpha)E(S) + 1$$

Step 4: For all $s \in \mathcal{S}$

$$V(s) \leftarrow V(s) + \alpha(\delta + \Delta)E(s)$$

$$E(s) \leftarrow \gamma\lambda E(s)$$

$$V(S) \leftarrow V(S) - \alpha\Delta$$

$$S \leftarrow S'$$

Until S is terminal

8.4.8 Sarsa and Sarsa (λ)

In the previous section, we discuss the value function prediction. In this section, we discuss the action control and application of eligibility trace to action control. Action control attempts to learn state-action function (Q-function). Sarsa algorithm is a slight variation of the popular Q-learning algorithm. The previously introduced Q-learning technique is an Off-Policy technique. Sarsa that stands for State Action Reward State Action, on the other hand, is an On-line Policy, which uses the action induced by the current policy to learn the Q-value (Sutton and Barto 2018).

Recall that Q-learning algorithm updates the $Q(s_t, a_t)$ function by

$$Q(s_t, a_t) = Q(s_t, a_t) + \alpha\left(R_{t+1} + \gamma \max_a Q(s_{t+1}, a) - Q(s_t, a_t)\right). \quad (8.203)$$

The update equation for Sarsa which involves the current state, current action, their induced reward, the next state, and next action is given by

$$Q(s_t, a_t) = Q(s_t, a_t) + \alpha(R_{t+1}, + \gamma Q(s_{t+1}, a_{t+1}) - Q(s_t, a_t)). \quad (8.204)$$

Using update equation (8.204), we can design Algorithm 8.13 for Sarsa.

Algorithm 8.13: Sarsa: An On-Policy TD Control

Step 1: Initialize $Q(s,a)$ arbitrarily

Step 2: Repeat for each episode

Initialize s

Choose a from s using policy derived from $Q(.|s)$ (e.g., ε-greedy)

Step 3: Repeat for each step of episode

1. Take action a, observe R, next state s'
2. Choose a' from s', using policy derived from $Q(.|s')$ (e.g., ε-greedy)
3. Update

$$Q(s,a) \leftarrow Q(s,a) + \alpha(R + \gamma Q(s',a') - Q(s,a))$$

4. $s \leftarrow s', a \leftarrow a'$

Until s is a terminal state

Next we introduce the eligibility trace version of Sarsa, which is referred to as Sarsa (λ) (Sutton and Barto 2018). This is application of $TD(\lambda)$ prediction methods to state-action pair. The trace for the state should be extended to state-action pair s, a. The update should be triggered by visiting the state-action pair and are implemented by the identity-indicator notation. The eligibility trace functions for state-pair are defined as

accumulating: $E_t(s,a) = \gamma\lambda E_{t-1}(s,a) + I_{sS_t} I_{aA_t}$
Dutch: $E_t(s,a) = (1-\alpha)\gamma\lambda E_{t-1}(s,a) + I_{sS_t} I_{aA_t}$
replacing: $E_t(s,a) = \left(1 - I_{sS_t} I_{aA_t}\right)\gamma\lambda E_{t-1}(s,a) + I_{sS_t} I_{aA_t}$,
for all $s \in \mathcal{S}$, $a \in \mathcal{A}$.

Substituting state-action variable $Q_t(s,a)$ for state variable $V_t(s)$ and trace variable $E_t(s,a)$ for $E_t(s)$ in the update equation (8.200) for $TD(\lambda)$, we obtain the update equation for Sarsa (λ):

$$Q_{t+1}(s,a) = Q_t(s,a) + \alpha\delta_t E_t(s,a), \text{ for all } s,a, \quad (8.205)$$

where

$$\delta_t = R_{t+1} + \gamma Q_t(S_{t+1}, A_{t+1}) - Q_t(S_t, A_t).$$

The complete Sarsa (λ) algorithm is summarized as Algorithm 8.14 (Sutton and Barto 2018).

Algorithm 8.14: Sarsa (λ)

Step 1: Initialize $Q(s,a)$ arbitrarily, $\forall s \in \mathcal{S}$, $a \in \mathcal{A}$

Step 2: Repeat for each episode
Initialization

$$E(s,a) = 0, \forall s \in \mathcal{S}, \ a \in \mathcal{A}$$

Initialize S, A

Step 3: Repeat for each step of episode
Take action A, observe R, S'
Choose action A', using policy $\pi(A'|S')$ derived from Q (e.g., ε-greedy)
Calculate update

$$\delta \leftarrow R + \gamma Q(S',A') - Q(S,A)$$

Update traces
$E(S,A) \leftarrow E(S,A) + 1$ (accumulating traces)
or $E(S,A) \leftarrow (1-\alpha)E(S,A) + 1$ (Dutch traces)
or $E(S,A) \leftarrow 1$ (replacing traces)
For all $s \in \mathcal{S}$, $a \in \mathcal{A}(s)$

$$Q(s,a) \leftarrow Q(s,a) + \alpha\delta E(s,a)$$

$$E(s,a) \leftarrow \gamma\lambda E(s,a)$$

$$S \leftarrow S', \ A \leftarrow A'$$

Until S is a terminal state

8.4.9 Watkin's $\mathbf{Q}(\lambda)$

The Watkin's $Q(\lambda)$ algorithm is an off-policy method (Sutton and Barto 2018). $Q(\lambda)$ algorithm also can combine the Q-learning with eligibility traces. However, unlike $TD(\lambda)$ or Sarsa (λ), $Q(\lambda)$ algorithm stops at the first non-greedy action (exploratory) taken. In general, if A_{t+n} is the first exploratory action, then the longest backup is

$$R_{t+1} + \gamma R_{t+2} + \ldots + \gamma^{n-1} R_{t+n} + \gamma^n \max_a Q_t(S_{t+n}, a),$$

where updating is off-line.

The eligibility traces are updated in two steps. When an exploratory action was taken, the eligibility traces are set to zero for all state-action pairs. Otherwise, the eligibility traces for all state-action pairs decayed by $\gamma\lambda$. In the second step, the accumulative trace value for the current state-action pair is incremental by 1 and the replacing trace value for the current state-action pair is set to 1.

In summary, the eligibility trace value is updated by

$$E_t(s,a) = \begin{cases} \gamma\lambda E_{t-1}(s,a) + I_{sS_t} I_{aA_t} & \text{if } Q_{t-1}(S_t, A_t) = \max_a Q_{t-1}(S_t, a) \\ I_{sS_t} I_{aA_t} & \text{otherwise} \end{cases}. \quad (8.206)$$

The state-action variable is updated by

$$Q_{t+1}(s,a) = Q_t(s,a) + \alpha\delta_t E_t(s,a), \quad \forall s \in \mathcal{S}, \ a \in \mathcal{A}(s), \quad (8.207)$$

where

$$\delta_t = R_{t+1} + \gamma \max_{a'} Q_t(S_{t+1}, a') - Q_t(S_t, A_t).$$

The complete Watkin's $Q(\lambda)$ algorithm is summarized in Algorithm 8.15.

Algorithm 8.15: Watkin's $Q(\lambda)$ Algorithm

Step 1: Initialize $Q(s,a)$ arbitrarily, $\forall s \in \mathcal{S},\ a \in \mathcal{A}$

Step 2:. Repeat for each episode
Initialization

$$E(s,a)=0,\ \forall s \in \mathcal{S},\ a \in \mathcal{A}$$

Initialize S, A

Step 3: Repeat for each step of episode
Take action A, observe R, S'
Choose action A', using policy $\pi(A'|S')$ derived from Q (e.g., ε-greedy)
$A^* \leftarrow \max_a Q(S',a)$ (if A' ties for the max, then $A^* \leftarrow A'$)

$$\delta \leftarrow R+\gamma Q(S',A^*)-Q(S,A)$$

Update traces
$E(S,A) \leftarrow E(S,A)+1$ (accumulating traces)
or $E(S,A) \leftarrow (1-\alpha)E(S,A)+1$ (Dutch traces)
or $E(S,A) \leftarrow 1$ (replacing traces)
For all $s \in \mathcal{S},\ a \in \mathcal{A}(s)$

$$Q(s,a) \leftarrow Q(s,a)+\alpha\delta E(s,a)$$

$$\text{If } A' = A^* \text{ then } E(s,a) \leftarrow \gamma\lambda E(s,a)$$

Else $E(s,a) \leftarrow 0$

$$S \leftarrow S',\ A \leftarrow A'$$

Until S is a terminal state

8.4.10 Actor-Critic and Eligibility Trace

Now we extend the actor-critic methods to incorporate the eligibility traces (Grondman et al. 2012). The received reward often involves the results of several steps. Eligibility trace provides information on the earlier visited several steps. Both actor part and critic part need to use eligibility traces for each state and each state-action pair. Let $E_t(s)$ and $E_t(s,a)$ be the eligibility trace at the time t for the state s and state-action pair (s,a), respectively. The updating equation for $E_t(s)$ is given by (Sutton and Barto 2018)

$$E_t(s)=\begin{cases} \gamma\lambda E_{t-1}(s) & \forall s \in \mathcal{S},\ s \neq S_t \\ \gamma\lambda E_{t-1}(s)+1 & s = S_t \end{cases}, \quad (8.208)$$

and the updating equation for $E_t(s,a)$ is given by

$$E_t(s,a)=\begin{cases} \gamma\lambda E_{t-1}(s,a)+1-\pi_t(S_t, A_t) & \text{if } s=S_t, a=A_t \\ \gamma\lambda E_{t-1}(s,a) & \text{otherwise} \end{cases}. \quad (8.209)$$

The actor-critic updating equations with the use of eligibility traces are given by

$$W_{t+1}=W_t+\alpha_{c,t}\delta_t E_t(s) \quad (8.210)$$

$$\theta_{t+1}=\theta_t+\alpha_{a,t}\delta_t E_t(s,a)\nabla_\theta \log\pi_\theta(A_t|S_t), \quad (8.211)$$

where

$$\delta_t = R_{t+1}+\gamma\hat{V}(S_{t+1},w)-\hat{V}(S_t,w).$$

8.5 CAUSAL INFERENCE AND REINFORCEMENT LEARNING

Causal inference and RL are two different, but closely related concepts and disciplines. Combining causal inference and RL involves two basic topics: (1) application of causal inference to RL and (2) application of RL to causal inference. Application of causal inference to RL is referred to as causal RL (Bareinboim et al. 2021; Lu et al. 2021). In the previous chapters, we introduced two basic approaches to causal inference: (1) structural model and (2) counterfactual framework. Current application of structural model to RL is to use structural causal model to deconfound RL for observational data (Bareinboim et al. 2015; Lu et al. 2018; de Haan et al. 2019).

Counterfactual can be used as an alternative causal inference framework which is incorporated into RL to remove spurious return (Bottou et al. 2013; Buesing et al. 2018; Pitis et al. 2020). RL can also be used to causal discovery (Madumal et al. 2019; Zhu et al. 2020). In this section, we introduce some typical papers in these two topics.

8.5.1 Deconfounding Reinforcement Learning

In this section we introduce a general framework for RL with unmeasured confounding (Lu et al. 2018).

8.5.1.1 *Adjust for Measured Confounders*

We first consider the observed confounders and then study the unobserved confounders. Let C denote a set of observed confounders across all times. Let a_t denote action and r_t denote a reward received at time t. Assume that the confounder C affects both the action a_t and reward r_t. Without loss of generality, we further assume that the common confounders C are time-independent for each individual or for each procedure.

Theoretic foundation for adjusting the observed confounders to remove spurious association and retain true causation is *do*-calculus, which was introduced in Section 5.2. The effect r_t caused by action a_t can be expressed as

$$p\left(r_t | do\left(a_t\right)\right). \tag{8.212}$$

We are unable to calculate the probability $p\left(r_t | do\left(a_t\right)\right)$ from observational data. To overcome this problem, in Section 5.2, we introduced three inference rules of *do*-calculus to map interventional distributions to observational probability distributions to allow us to estimate the interventional distribution from observational data alone (Pearl 2012). In this section, we will use back-door criterion and front-door criterion to convert the interventional distribution to classical statistical probability distribution.

Definition 8.16: Back-Door Path

When we estimate the effect of X on Y, a back-door path is an undirected path between X and Y with an arrow into X.

Definition 8.17: Back-Door Criterion

A set of conditioning variables S satisfies the back-door criterion if (i) the set S blocks every back-door path between X and Y and (ii) the set S does not contain a descendent of X. We need to block all non-causal paths (Figure 8.8(a)).

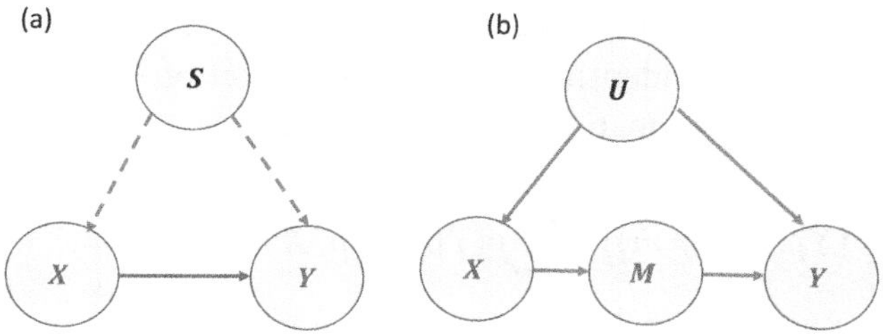

FIGURE 8.8 Illustration of the back-door and front door criteria.

When S meets the back-door criterion, the intervention distribution can be mapped to a statistical distribution as follows.

$$p\left(Y | do\left(X=x\right)\right)=\sum_s p\left(Y | X=x, S=s\right)p\left(S=s\right). \tag{8.213}$$

Definition 8.18: Front-Door Criterion

A set of variables M satisfies the front-door criterion if (i) M blocks all directed paths from X to Y, (ii) there are no unblocked back-door paths from X to M, and (iii) X blocks all back-door paths from M to Y (Figure 8.8(b)).

Again, intervention distribution can be calculated by the statistical distributions:

$$p\left(Y | do\left(X=x\right)\right)=\sum_m p\left(M=m | X=x\right)\sum_{x'} p\left(Y | X=x', M=m\right)p\left(X=x'\right). \tag{8.214}$$

Equation (8.214) can be further explained as follows. By condition (i), M blocks all directed paths from X to Y, any causal dependence of Y on X must be mediated via M, which implies

$$p\left(Y | do\left(X=x\right)\right) = \sum_m p\left(Y | do\left(M=m\right)\right)p\left(M=m | Do\left(X=x\right)\right). \tag{8.215}$$

Condition (ii) implies that by back-door criterion, we can obtain the causal effect of X on M directly,

$$p\left(M=m | do\left(X=x\right)\right)=p\left(M=m | X=x\right), \tag{8.216}$$

Condition (iii) states that X satisfies the back-door criterion for identifying the causal effect of M on Y. Then, using equation (8.213), we obtain

$$p(Y|do(M=m))=\sum_{x'}p(Y|M=m,\ X=x')P(X=x'). \tag{8.217}$$

Substituting equations (8.216) and (8.217) into equation (8.215), we obtain equation (8.214).

Under the assumption that the confounders are measured, using back-door criterion (equation (8.213), where $Y=r_t$, $X=a_t$ and $C=c$), we obtain

$$p(r_t|do(a_t=a))=\sum_{c}p(r_t|a_t=a,\ C=c)p(C=c). \tag{8.218}$$

Equation (8.218) provides formula for confounder adjustment. Back-door and front-door offer a powerful tool to map intervention distributions to observational distributions, and hence are very useful for causal analysis.

8.5.1.2 Proxy Variable Approximation to Unobserved Confounding

In Section 8.5.1.1, we discussed about adjustment of observed confounding. In real problems, many confounders are unobserved. There have been no methods which can directly adjust for unobserved confounders. One solution is to use proxy variables of the confounders for identifying unobserved hidden confounders. However, it is unknown what variables are proxy variables of the confounders. Therefore, this solution consists of two steps. First step is to collect many variables such as genomics, geographic, lab test results, clinical data, images, and language variables. The second step is to use dimension reduction methods to map the collected data into the latent variables. Then, we develop a causal model to infer the causal relations among the latent variables, action and reward variables. From the inferred causal model, we identify the latent variables that simultaneously affect both action and reward variables. Such identified latent variables that affect both action and reward variables can be used as proxy variables of confounders (Lu et al. 2018).

8.5.1.3 Deep Latent Model for Identifying the Proxy Variables of Confounders

In this section, we assume that the observed covariates at different time are independent. Then, we can use VAE to infer the latent model for each time point t, separately. Let $x=(x_{1,\dots,}x_T)$, $x_t\in R^{D_x}$ be a sequence of observed covariates, $a=(a_1,\dots,a_{T-1})$, $a_t\in R^{D_a}$ be a sequence of action data and $r=(r_2,\dots,r_{T+1})$, $r_t\in R^{D_r}$ be a sequence of rewards. We assume a sequence of observational covariates, actions, and rewards, where actions and rewards are confounded by one or several unknown factors. Let c be a set of common unobserved confounders. The VAE deep latent model is shown in Figure 8.9. The model in Figure 8.9 defines the following distributions:

$$p(z_t)=\prod_{j=1}^{D_z}N(z_{ptj}|0,1),\ p(c)=\prod_{j=1}^{D_c}N(c_j|0,1) \tag{8.219}$$

$$p(x_t|z_t,c)=N\left(x_t|\mu_t^x,(\sigma_t^x)^2\right),\ \mu_t^x=MLP(z_t,c,w_\mu^x),$$
$$(\sigma_t^x)^2=MLP(z_t,c,w_\sigma^x) \tag{8.220}$$

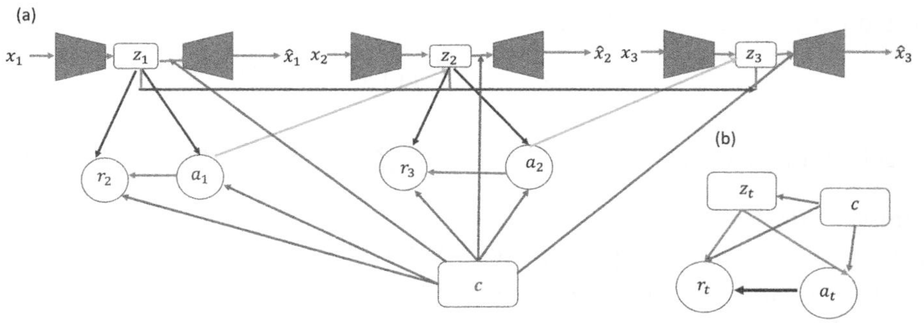

FIGURE 8.9 VAE latent model for deconfounding.

$$p(a_t|z_t,c) = N\left(a_t|\mu_t^a, \left(\sigma_t^a\right)^2\right), \mu_t^a = MLP\left(z_t, c, w_\mu^a\right),$$
$$\left(\sigma_t^a\right)^2 = MLP\left(z_t, c, w_\sigma^a\right) \tag{8.221}$$

$$p(r_{t+1}|z_t, a_t, c) = N\left(r_{t+1}|\mu_t^r, \left(\sigma_t^r\right)^2\right),$$
$$\mu_t^a = MLP\left(z_t, a_t, c, w_\mu^r\right), \left(\sigma_t^r\right)^2 = MLP\left(z_t, a_t, c, w_\sigma^r\right) \tag{8.222}$$

$$p(z_t|z_{t-1}, a_{t-1}) = N\left(z_t|\mu_{pt}^z, \left(\sigma_{pt}^z\right)^2\right),$$
$$\mu_{pt}^z = MLP\left(z_{t-1}, a_{t-1}, w_{p\mu}^z\right),$$
$$\left(\sigma_{pt}^z\right)^2 = MLP\left(z_{t-1}, a_{t-1}, w_{p\sigma}^z\right). \tag{8.223}$$

where *MLP* represents a multilayer perceptron.

8.5.1.4 Reward and Causal Effect Estimation

There are two types of confounders: the time-independent global confounder c and time-dependent confounders $\{z_t\}_{t=1}^T$ (Lu et al. 2018). The time-dependent confounder z_t plays a state role in RL. The reward r_t is the immediate reward when the action a_t in the state z_t is taken. The goal of RL is to find an optimal policy $\pi_*(a_t|z_t)$, which determines what action a_t is taken, given the state z_t. Therefore, the time-dependent confounder z_t can be viewed as the state rather than a confounder of a_t and r_t. Therefore, we do not need to adjust for z_t.

The causal effect r_t of action a_t, given the state z_t is defined as

$$p\left(r_t|z_t, do\left(a_t = a\right)\right). \tag{8.224}$$

If we view the pair $\left(r_t, z_t\right)$ in Figure 8.9(b) as an effect, then c satisfies the back-door criterion in the graph Figure 8.9(b). The interventional distribution $p\left(r_t|z_t, do\left(a_t = a\right)\right)$ can be converted to statistical distribution:

$$p\left(r_t|z_t, do\left(a_t = a\right)\right) = \int p\left(r_t|z_t, a_t = a, c\right)p(c)dc. \tag{8.225}$$

Equation (8.225) shows that $p\left(r_t|z_t, do\left(a_t = a\right)\right)$ can be identified from the joint probability distribution $p(c,z,x,a,r)$. Since VAE can map the observed x to the latent variables c and z, the joint probability distribution $p(c,z,x,a,r)$ can be recovered from the observed data (x,a,r).

8.5.1.5 Variational Autoencoder for Reinforcement Learning

The RL data consist of the observed covariates $x = \left[x_1, \ldots, x_T\right]$, set of actions $a = \left[a_1, \ldots, a_T\right]$ and set of rewards $r = \left[r_2, \ldots, r_{T+1}\right]$. The log-likelihood function, denoted by $\log p_\theta(x,a,r)$, is a nonlinear function and parameterized by neural networks. The explicit log-likelihood function is intractable. VAE that is introduced in Chapter 3 should be used to approximate the log-likelihood function. The VAE lower bound for $\log p_\theta(x,a,r)$ is given by

$$\begin{aligned}\log p_\theta(x,a,r) &\geq \mathcal{L}(x,a,r,\theta,\varnothing)\\ &= E_{q_\varnothing(z,c|x,a,r)}\left[\log p_\theta(x,a,r|z,c)\right]\\ &\quad - KL\left(q_\varnothing(z,c|x,a,r)||p_\theta(z,c)\right).\end{aligned} \tag{8.226}$$

Using the Markov property of the proposed model (Figure 8.9), the joint distribution can be factorized as follows:

$$\begin{aligned}p_\theta(x,a,r,z,c) = &\prod_{t=1}^T p(x_t|z_t,c)p(a_t|z_t,c)p(r_{t+1}|z_t, a_t, c)\\ &p(c)p(z_1)\prod_{t=2}^T p(z_t|z_{t-1}, a_{t-1}),\end{aligned} \tag{8.227}$$

which implies that

$$\begin{aligned}p_\theta(x,a,r|z,c) &= \frac{p_\theta(x,a,r,z,c)}{p(z,c)}\\ &= \prod_{t=1}^T p(x_t|z_t,c)p(a_t|z_t,c)p(r_{t+1}|z_t, a_t, c).\end{aligned} \tag{8.228}$$

Similarly, we obtain

$$\begin{aligned}p(z,c|x,a,r) &= p(c|x,a,r)p(z|x,a,r,c)\\ &= p(c|x,a,r)p(z_1|x,a,r)\prod_{t=2}^T p(z_t|z_{t-1},x,a,r).\end{aligned} \tag{8.229}$$

Equation (8.229) implies that the approximation of the posterior distribution $q(z,c|x,a,r)$ can be given by

$$q(z,c|x,a,r) = q(c|x,a,r)q(z_1|x,a,r)\prod_{t=2}^{T} q(z_t|z_{t-1},x,a,r). \quad (8.230)$$

Using equations (8.228) and (8.230), we can reduce equation (8.226) to

$$\begin{aligned}\log p_\theta(x,a,r) &\geq \mathcal{L}(x,a,r,\theta,\varnothing) \\ &= \sum_{t=1}^{T} E_{c\sim q(c|x,a,r),\, z_t\sim q(z_t|z_{t-1},x,a,r)}\big[\log p(x_t|z_t,c) \\ &\quad +\log p(a_t|z_t,c)+\log p(r_{t+1}|z_t,a_t,c)\big]. \\ &\quad -KL(q(c|x,a,r)\|p(c)) - KL(q(z_1|x,a,r)\|p(z_1)) \\ &\quad -\sum_{t=2}^{T} E_{z_{t-1}\sim q(z_{t-1}|z_{t-2},x,a,r)} \\ &\quad \times\big[KL(q(z_t|z_{t-1},x,a,r)\|p(z_t|z_{t-1},x,a,r))\big],\end{aligned} \quad (8.231)$$

where for simplicity of notations, we omit subscripts θ and $\varnothing$.

8.5.1.6 Encoder

Encoder maps the observed data (x,a,r) to the latent space. Architecture of encoding network is shown in Figure 8.10. There are two types of encoding functions: $q(c|x,a,r)$ and $q(z|x,a,r)$. We assume that both of two types of encoding functions follow Gaussian distributions:

$$q(c|x,a,r) = N(c|\mu_c,\sigma_c^2),\ \mu_c = MLP(x,a,r,w_\mu^c),$$
$$\sigma_c^2 = MLP(x,a,r,w_\sigma^c), \quad (8.232)$$

$$q(z_t|x,a,r) = N\left(z|\mu_t^z,(\sigma_t^z)^2\right), \quad (8.233)$$

where MLP represents a multilayer perceptron, w_μ^c, w_σ^c are weights in MLPs.

The mean μ_t^z and variance $(\sigma_t^z)^2$ of the encoding function are respectively given by (Appendix 8A)

$$\mu_t^z = w_\mu h_{combined} + b_\mu, \quad (8.234)$$

$$(\sigma_t^z)^2 = \text{softplus}\,(w_\sigma h_{combined} + b_\sigma), \quad (8.235)$$

where softplus function is defined as

softplus$(x) = \log(1+e^x)$ and $h_{combined}$ is defined by equations (8A1–8A3) in Appendix 8A.

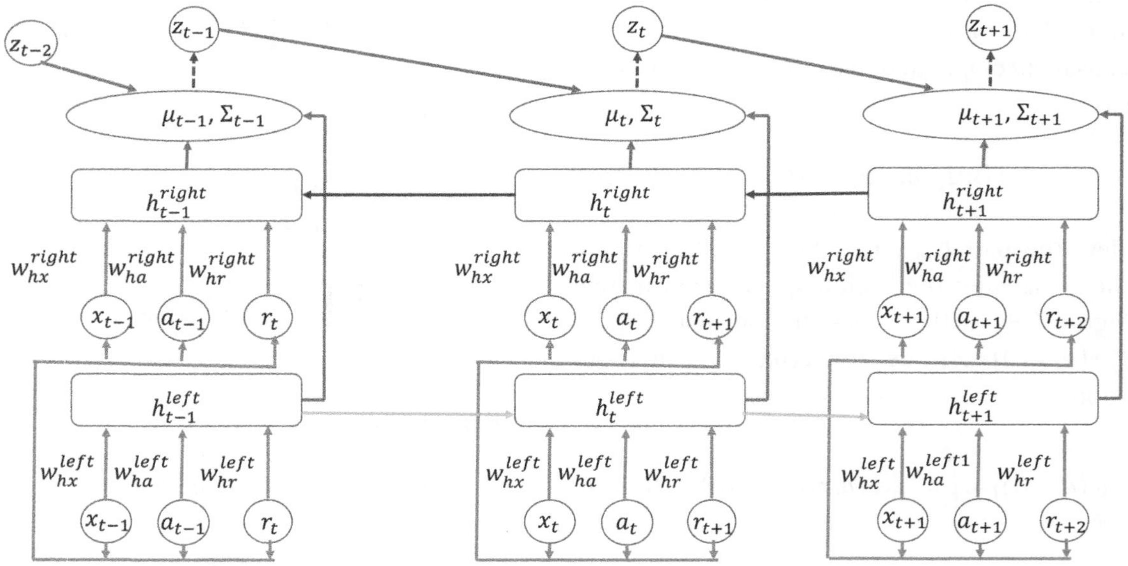

FIGURE 8.10 Architecture of encoding network.

Counterfactual reasoning requires that at any time step t, given a new x_t, we predict a_t and r_{t+1}, which in turn, are mapped to the latent variable z_t. Therefore, we need to estimate two conditional probabilities:

$$q(a_t|x_t) = N\left(\mu = \hat{\mu}_t^a, \sigma^2 = \left(\hat{\sigma}_t^a\right)^2\right),$$
$$\hat{\mu}_t^a = MLP\left(x_t, w_\mu^a\right), \left(\hat{\sigma}_t^a\right)^2 = MLP\left(x_t, w_\sigma^a\right) \tag{8.236}$$

$$q(r_{t+1}|x_t, \hat{a}_t) = N\left(\left(\mu = \hat{\mu}_t^r, \sigma^2 = \left(\hat{\sigma}_t^r\right)^2\right),\right.$$
$$\hat{\mu}_t^r = MLP\left(x_t, a_t, w_\mu^r\right), \left(\hat{\sigma}_t^r\right)^2 = MLP\left(x_t, a_t, w_\sigma^r\right) \tag{8.237}$$

We incorporate probabilities $q(a_t|x_t)$ and $q(r_{t+1}|x_t, \hat{a}_t)$ into the evidence of lower bound to obtain the modified objective function:

$$\mathcal{L}_{DRL} = \mathcal{L}(x,a,r,\theta,\varnothing) + \sum_{t=1}^{T}\left[\log q(a_t|x_t) + \log q(r_{t+1}|x_t, \hat{a}_t)\right]. \tag{8.238}$$

The KL divergence is computed in Appendix 8B and its final result is

$$-KL\left(q_\varnothing(z,c|x,a,r)\|p_\theta(z,c)\right)$$
$$= \frac{1}{2}\sum_{j=1}^{D_c}\left[\log\left(\sigma_{tj}^c\right)^2 + 1 - \left(\sigma_{tj}^c\right)^2 - \left(\mu_{tj}^c\right)^2\right.$$
$$+ \frac{1}{2}\sum_{j=1}^{D_z}\left[\log\left(\sigma_{1j}^z\right)^2 + 1 - \left(\sigma_{1j}^z\right)^2 - \left(\mu_{1j}^z\right)^2\right]$$
$$+ \frac{1}{2}\sum_{t=2}^{T} E_{z_{t-1}\sim q(z_{t-1}|z_{t-2},x,a,r)}\sum_{j=1}^{D_z}\left[\log\left(\frac{\sigma_{tj}^z}{\sigma_{ptj}^z}\right)^2\right.$$
$$\left. + 1 - \left(\frac{\sigma_{tj}^z}{\sigma_{ptj}^z}\right)^2 - \left(\frac{\mu_{tj}^z - \mu_{ptj}^z}{\sigma_{ptj}^z}\right)^2\right]. \tag{8.239}$$

8.5.1.7 Decoder and ELBO

It follows from equation (8.226) that decoding function is defined as

$$p_\theta(x,a,r|z,c) = \prod_{t=1}^{T} p(x_t|z_t,c)\,p(a_t|z_t,c)\,p(r_{t+1}|z_t, a_t,c), \tag{8.240}$$

where $p(x_t|z_t,c)$, $p(a_t|z_t,c)$, $p(r_{t+1}|z_t, a_t,c)$ are respectively defined in equations (8.220), (8.221), and (8.222).

Monte Carlo method can be used to calculate the first term $E_{q_\varnothing(z,c|x,a,r)}\left[\log p_\theta(x,a,r\,|\,z,c)\right]$ of ELBO:

$$E_{q_\varnothing(z,c|x,a,r)}\left[\log p_\theta(x,a,r|z,c)\right] \approx \frac{1}{L}\sum_{l=1}^{L}\log p_\theta\left(x,a,r|z^l,c^l\right). \tag{8.241}$$

Recall from equation (8.230) that

$$q(z,c|x,a,r) = q(c|x,a,r)\,q(z_1|x,a,r)\prod_{t=2}^{T} q(z_t|z_{t-1},x,a,r). \tag{8.242}$$

$q(c|x,a,r)$ follows a multivariate normal distribution. Its mean and variance are defined by neural networks (equation (8.232)).

Define

$$C^l = \mu_c + \sigma_c \odot \varepsilon^l, \ \varepsilon^l \sim N(0,I), l = 1,\ldots, L. \tag{8.243}$$

Sampling from the joint distribution $q(z_1|x,a,r)\prod_{t=2}^{T} q(z_t|z_{t-1},x,a,r)$ by sampling recursively from $q(z_t|z_{t-1},x,a,r)$ (equations (8.233)–(8.235)) from $t = 1$ to T. Define

$$Z_t^l = \mu_t^z + \sigma_t^z \odot \varepsilon^l, \ \varepsilon^l \sim N(0,I), l = 1,\ldots, L, \tag{8.244}$$

where when $t = 1$, set

$$h_{combined} = \frac{1}{2}\left(h_t^{left} + h_t^{right}\right).$$

Note that

$$\log q(a_t|x_t) = -\frac{D_a}{2}\log(2\pi) - \frac{1}{2}\sum_{j=1}^{D_a}\left[\log\left(\sigma_{tj}^a\right)^2 + \left(\frac{a_{tj} - \mu_{tj}^a}{\sigma_{tj}^a}\right)^2\right] \tag{8.245}$$

$$\log q(r_{t+1}|x_t, a_t) = -\frac{D_r}{2}\log(2\pi) - \sum_{j=1}^{D_r}\left[\log\left(\sigma_{tj}^r\right)^2 + \left(\frac{r_{tj} - \mu_{tj}^r}{\sigma_{tj}^r}\right)^2\right] \tag{8.246}$$

The ELBO can be reduced to

$$\mathcal{L}_{DRL} \approx \frac{1}{L}\sum_{l=1}^{L}\Big[\log p_\theta\left(x,a,r|z^l,c^l\right)$$
$$-KL\left(q_\varnothing\left(z^l,c^l|x,a,r\right)||p_\theta\left(z^l,c^l\right)\right)$$
$$-\frac{1}{2}\sum_{t=1}^{T}\left[\frac{D_a}{2}\log(2\pi)+\frac{D_r}{2}\log(2\pi)\right.$$
$$+\frac{1}{2}\sum_{j=1}^{D_a}\left[\log\left(\sigma_{tj}^a\right)^2+\left(\frac{a_{tj}-\mu_{tj}^a}{\sigma_{tj}^a}\right)^2\right.$$
$$\left.\left.+\log\left(\sigma_{tj}^r\right)^2+\left(\frac{r_{tj}-\mu_{tj}^r}{\sigma_{tj}^r}\right)^2\right]\right]. \qquad (8.247)$$

8.5.1.8 Deconfounding Causal Effect Estimation and Actor-Critic Methods

Using equation (8.225), we obtain Monte Carlo method for calculating the interventional distribution $p\left(r_t|z_t, do\left(a_t=a\right)\right)$ as follows:

$$p\left(r_t|z_t, do\left(a_t=a\right)\right)$$
$$\approx \frac{1}{N}\sum_{l=1}^{N}p\left(r_t|z_t=z,\, a_t=a,\, c^l\right), \qquad (8.248)$$
$$c^l \sim \prod_{j=1}^{D_c}N\left(c_j^l|0.1\right),$$

where equation (8.243) can also be used to sample c^l.

Recall from equation (8.183) that the gradient of Actor-Critic loss function is given by

$$\nabla_\theta J(\theta)=E_\pi\Big[\left(r_{t+1}+\gamma\hat{V}\left(z_{t+1},\theta_v\right)-\hat{V}\left(z_t,\theta_v\right)\right)$$
$$\times\nabla_\theta\log\pi_\theta\left(a_t|z_t\right)\Big], \qquad (8.249)$$

where r_{t+1} follows distribution $p\left(r_t|z_t, do\left(a_t=a\right)\right)$, which is approximated by equation (8.248).

The parameter θ for the policy $\pi_\theta\left(a_t|z_t\right)$ is updated by

$$\theta^{k+1}=\theta^k+\alpha\nabla_\theta J(\theta). \qquad (8.250)$$

8.5.2 Counterfactuals and Reinforcement Learning

The set of actions or interventions for control of dynamic system is often limited. The environments that determine the transition dynamics may change rapidly over time. The future environments of the dynamic system may be substantially different from the previous one. The actions or interventions cannot be only inferred from the historical data. To fully design optimal actions or interventions in the RL may not be feasible. In addition, it is difficult to collect data for RL. RL data are often sparse (Feinberg et al. 2018). Therefore, we need to design alternative actions and predict counterfactual rewards or outcomes in the future under a sequence of alterative actions in the RL (Buesing et al. 2018; Ge et al. 2020; Lu et al. 2020; Kuremoto et al. 2021). In this section, we introduce data augmentation by counterfactual reasoning for RL (Lu et al. 2020).

8.5.2.1 Structural Causal Model for Counterfactual Inference

Causal network models encode scientific assumptions and counterfactuals emerge as a general framework for causal inference (Chen and Pearl 2015). If the model is identified, a counterfactual query will be raised to predict values of the variables in the model using the available information. These predicted values that are unobserved can be used as counterfactuals.

Let $Y=\left[Y_1,\ldots, Y_n\right]$ be an n-dimensional vector of observed variables. A structural causal model that defines the causal relationships among n variables is given by a set of functional equations:

$$Y_i=f_i\left(pa\left(Y_i\right), U_i\right), i=1,\ldots, n \qquad (8.251)$$

where $pa\left(Y_i\right)$ denotes the set of endogenous parents of Y_i, which directly causes Y_i, U_i denotes a noise variable for Y_i. We assume that U_i are jointly independent.

Example 8.4

A structural causal model for a causal network in Figure 8.11 is

$$Y_1=f_1\left(U_1\right)$$
$$Y_2=f_2\left(Y_1, Y_4, U_2\right)$$
$$Y_3=f_3\left(Y_1, U_3\right)$$
$$Y_4=f_4\left(U_4\right) \qquad (8.252)$$

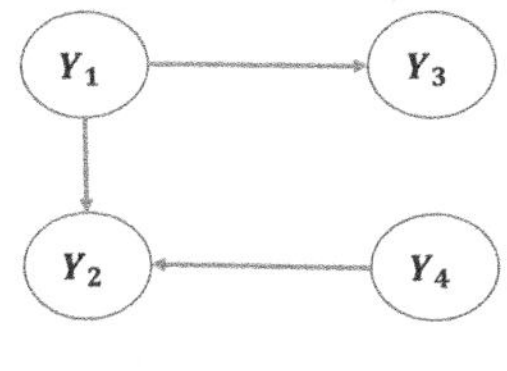

FIGURE 8.11 An example of structural causal model.

Once the structural causal model (8.251) is identified, we can infer how the effect distributions response to the changes of cause distributions. Now we discuss how to use the structural causal model for making counterfactual reasoning. Assume that the structural causal model M in (8.251) is given and we observe $Y_i = y_i$ if the parents of Y_i take values $pa(y_i)$. The procedures that show how to estimate counterfactual Y_i', if values of the parents of Y_i are changed to $pa(y_i')$ (Lu et al. 2020):

1. Use observations $(Y_i = y_i, pa(Y_i) = pa(y_i))$ to determine the noise $\hat{U}_i$.
2. Replace the structural equations for the variables in $pa(Y_i)$ with the function $pa(Y_i) = pa(y_i')$ to obtain the modified model $M_{y'}$.
3. Use the modified model $M_{y'}$ and the value of $\hat{U}_i$ to estimate the counterfactual value Y_i'.

The counterfactual value Y_i' is often denoted by $Y_{pa(y')} | Y = y, pa(Y) = pa(Y')$.

8.5.2.2 Bidirectional Conditional GAN (BiCoGAN) for Estimation of Causal Mechanism

8.5.2.2.1 Counterfactual Reinforcement Learning for a General Policy Define the structural causal model for the RL:

$$S_{t+1} = f(S_t, A_t, U_{t+1}), \tag{8.253}$$

where S_t and A_t are the state and action at time t, S_{t+1} is the sate at time $t+1$ and U_{t+1} is a noise term. A core of inferring counterfactuals is to use BiCoGAN (Jaiswal et al. 2017; Carrara et al. 2020, Section 4.3.2) for estimation of causal mechanism f which can be any functional class of causal models, e.g., linear causal model, nonlinear additive model, and post-nonlinear causal model. Given state S_t, action A_t, noise U_{t+1} at the time step t, and function f, the system will change to the state S_{t+1}. To estimate f, we can take the generated $\hat{S}_{t+1}$ as the fake data and observed S_{t+1} as real data. Using GAN, we can make distribution of $\hat{S}_{t+1}$ as close to the true distribution of S_{t+1} as possible. The resulting generator G can be taken as the approximation of the function f. To simultaneously estimate both the function f and noise term U_{t+1}, we use BiCoGAN. Specifically, the generator learns a mapping $G(S_t, A_t, U_{t+1})$ from the distribution $p_z(S_t, A_t, U_{t+1})$ to the distribution p_G with the goal of making p_G as close as possible to $p_{data}(S_{t+1})$, where $Z_t = (S_t, A_t, U_{t+1})$. Since the generator maps (S_t, A_t, U_{t+1}) to S_{t+1}. Thus, the generator is also called decoder. The output of the decoder is $(S_t, A_t, U_{t+1}, \hat{S}_{t+1})$. The distribution of the output of the decoder can be written as

$$p(S_t, A_t, U_{t+1}, \hat{S}_{t+1}) = p(S_t, A_t, U_{t+1}) p(\hat{S}_{t+1} | S_t, A_t, U_{t+1}). \tag{8.254}$$

The BiCoGAN also performs the inverse map from S_{t+1} to (S_t, A_t, U_{t+1}), i.e., it encoders information of S_{t+1} into (S_t, A_t, U_{t+1}). Therefore, the second map is called encoder (Figure 8.12(a)). The output of the encoder is $(\hat{S}_t, \hat{A}_t, \hat{U}_{t+1}, S_{t+1})$. Its distribution can be written as

$$p(\hat{S}_t, \hat{A}_t, \hat{U}_{t+1}, S_{t+1}) = p(S_{t+1}) p(\hat{S}_t, \hat{A}_t, \hat{U}_{t+1} | S_{t+1}). \tag{8.255}$$

The discriminator is trained such that the samples from the decoder distribution in equation (8.254) and encoder distribution in equation (8.255) cannot be distinguishable. Therefore, we obtain

$$\begin{aligned} &p(S_t, A_t, U_{t+1}) p(\hat{S}_{t+1} | S_t, A_t, U_{t+1}) \\ &= p(S_{t+1}) p(\hat{S}_t, \hat{A}_t, \hat{U}_{t+1} | S_{t+1}), \end{aligned} \tag{8.256}$$

which implies that

$$S_{t+1} \approx \hat{S}_{t+1} = G(S_t, A_t, U_{t+1}) \tag{8.257}$$

and

$$U_{t+1} \approx \hat{U}_{t+1}.$$

BiCoGAN improves the modeling of the latent space Z_t by exposing Z_t to the discriminator together with the images $G(Z_t)$ generated from Z_t. An encoder

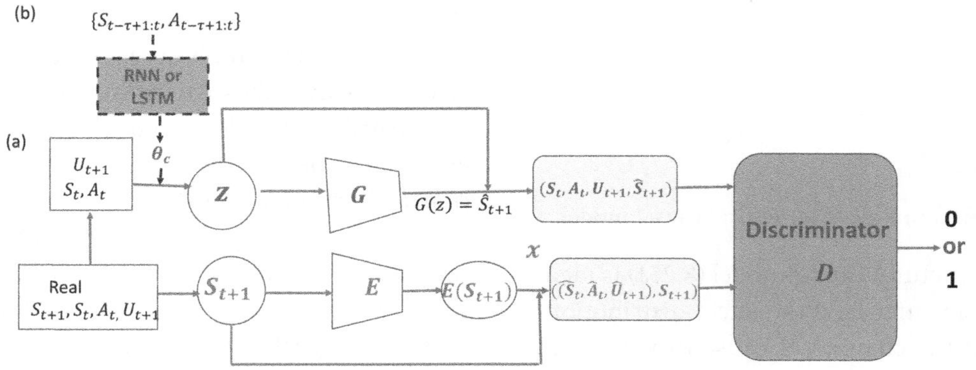

FIGURE 8.12 (a) Architecture of BiCoGAN, (b) RNN for personalized policy.

$E(S_{t+1})$ is introduced to map real samples S_{t+1} to the corresponding latent space Z_t. The encoder module $E(S_{t+1})$ is trained together with the decoder $G(Z_t)$. The discriminator $D(S_{t+1}, E(S_{t+1})) \in [0,1]$ is trained to assess whether the couple $(S_{t+1}, E(S_{t+1}))$ comes from a real $(S_{t+1}, E(S_{t+1}))$ or generated image $(G(Z_t), Z_t)$. Fooling discriminator D leads G and E to minimize the difference between $(G(Z_t), Z_t)$ and $(S_{t+1}, E(S_{t+1}))$ couples. The minimax problem for the BiCoGAN is

$$\min_{G,E}\max_{D} V(D,G,E) = E_{S_{t+1}\sim p_{data}(S_{t+1})}\left[\log D\left(S_{t+1}, E(S_{t+1})\right)\right] + E_{Z_t\sim p(Z_t)}\log\left(1 - D\left(G(Z_t), Z_t\right)\right) \quad (8.258)$$

Latent space consists of the intrinsic factors $z = U_{t+1}$ and the extrinsic factors $c = (S_t, A_t)$. While the intrinsic factors are sampled randomly from a simple latent distribution such as a uniform or a normal distribution, the more specialized extrinsic factors are more difficult to model (Jaiswal et al. 2017). To match the intrinsic and extrinsic factors and stabilize the encoder training process, we introduce the extrinsic factor loss $R(c, E_c(S_{t+1}))$, which is defined as the distance between the intrinsic factors c and the extrinsic factors $E_c(S_{t+1}) = (\hat{S}_t, \hat{A}_t)$ to stabilize the training process for the encoder and prevent overfitting the estimation error $R(c, E_c(S_{t+1}))$ of (S_t, A_t). Therefore, the extrinsic factor loss $R(c, E_c(S_{t+1}))$ is added to the objective function to regularize the training process:

$$\min_{G,E}\max_{D} V'(D,G,E) = V(D,G,E) + \gamma E_{(S_t, A_t, S_{t+1})\sim p_{data}(S_t, A_t, S_{t+1})}\left[R\left((S_t, A_t), E_c(S_{t+1})\right)\right], \quad (8.259)$$

where γ is the penalty parameter.

After learning the structural causal model: $G = \hat{f}$ and $\hat{U}_{t+1}$, for $t = 1,\ldots, T$, via the BiCoGAN, we can infer the counterfactual S'_{t+1} (Lu et al. 2020). Suppose at time $t+1$, we observe $(S_t = s_t, A_t = a, S_{t+1} = s_{t+1})$. We want to predict what would have been the next state S'_{t+1}, if we had taken the action a'. To solve make prediction for counterfactual under action a', we input $s_t, a', \hat{u}_{t+1}$ to the generator. The output $G(s_t, a', \hat{u}_{t+1}) = S'_{t+1}$ is the inferred counterfactual under alternative action a'. This provides a powerful tool for imputing the missing data or augmenting the data via inferring the counterfactual states under a set of alternative actions. The actions can be either discrete or continuous. Assume that the support of the action space $\mathcal{A}$ is $[b^-, b^+]$. We uniformly sample action a' from $[b^-, b^+]$ and use identified structural causal model to infer the counterfactual outcome S'_{t+1}. The original data and inferred counterfactual data are to form augmented data $\tilde{D}$ for RL.

8.5.2.2.2 Counterfactual Reinforcement Learning of Personalized Policies In practice, we often observe the different responses of different individuals to the identical action. To deal with this problem, we introduce personalized counterfactual inference (Lu et al. 2020). We modify the structural causal model (8.253) to

$$S_{t+1} = f(S_t, A_t, U_{t+1}, \theta_c), \quad (8.260)$$

where θ_c represents the hidden factor of the individual c. To estimate θ_c, for each individual at time t, we include the data sequence with size τ: $\{S_{t-i+1}, A_{t-i+1}, S_{t-i+2}\}_{i=1}^{\tau}$ into the dataset.

Taking the data sequence $\{S_{t-\tau+1:t}, A_{t-\tau+1:t}\}$ as an input to a vanilla recurrent neural network (RNN) or long short-term memory (LSTM) (Jaquesn et al. 2017) (Section 1.1.3), we output $\hat{\theta}_c$ from the RNN or LSTM, which, in turn, is taken as a new input to the generator G (Figure 8.12(b)). The output of the BiCoGAN is

$$\hat{S}_{t+1}^{c} = G\left(S_t, A_t, U_{t+1}, \theta_c\right), \tag{8.261}$$

where $\hat{S}_{t+1}^{c}$ is the estimated state of the individual c at time $t+1$. The latent variable θ_c that is learned from the additional sequence of data $\{S_{t-\tau+1:t}, A_{t-\tau+1:t}\}$ provides information on personalized counterfactual response.

The estimated $\hat{\theta}_c$ can be used to group individuals and perform group RL. Specifically, the k-means algorithm is used to cluster $\hat{\theta}_c$ into several groups. The estimated centroid from k-means is a new $\bar{\theta}_c$ to represent the group and is taken as a new input for the group. Repeating counterfactual reasoning on each group of individuals, we obtain the augmented dataset $\bar{D}_i$ for the ith group.

8.5.2.3 Dueling Double-Deep Q-Networks and Augmented Counterfactual Data for Reinforcement Learning

Dueling double deep Q-network (D3QN) (Wang et al. 2015; Raghu et al. 2017; Lu et al. 2020) can be applied to the counterfactually augmented dataset $\tilde{D}$. The D3QN is a state-of-the-art deep RL method, which has found many successful application to complex sequential decision-making problems (Lu et al. 2020). The remarkable feature of dueling architecture is to more quickly identify the correct action during policy evaluation (Wang et al. 2015).

8.5.2.3.1 Deep Q-Networks Q-value functions may be high dimensional. A DQN $Q(S,A,\theta)$ with parameters θ in the neural network can be used to approximate Q-value functions (Wang et al. 2015). The loss function of Q-Network is defined as

$$L_i(\theta_i) = E_{s,a,r,s'}\left[\left(y_i^{DQN} - Q(S,A,\theta_i)\right)^2\right], \tag{8.262}$$

where

$$y_i^{DQN} = r + \gamma \max_{a'} Q(S', A', \theta^-), \tag{8.263}$$

θ^- represents the parameters of a fixed and separate target network. Stochastic gradient descent method can be used to update θ:

$$\theta_{i+1} = \theta_i + \alpha\left[y_i^{DQN} - Q(S_i, A_i, \theta_i)\right]\nabla_{\theta_i} Q(S_i, A_i, \theta_i). \tag{8.264}$$

8.5.2.3.2 Double Deep Q-Network The max operator in the DQN uses the same values for both selecting and evaluating an action, which may cause to select overestimated values and lead to overoptimistic value estimates (Wang et al. 2015). To overcome this limitation, double deep Q-Network (DDQN) is developed. It defines the following target network and loss function:

$$y_i^{DDQN} = r + \gamma Q\left(S, \operatorname*{argmax}_{A'} Q\left((S', A', \theta_i); \theta^-\right)\right), \tag{8.265}$$

$$L_i(\theta_i) = E_{S,A,r,S'}\left[\left(y_i^{DDQN} - Q(S,A,\theta_i)\right)^2\right]. \tag{8.266}$$

8.5.2.3.3 Dueling Double Deep Q-Network In practice, we often observe that it is unnecessary to select all actions and estimate the value of each action. Wang et al. (2021) proposed the D3QN for RL. Before studying D3QN, we introduce an advantage function which is defined as

$$A_\pi(s,a) = Q_\pi(s,a) - V_\pi(s). \tag{8.267}$$

The architecture of the D3QN is given as follows. The lower layers of the D3QN are convolutional to convert high-dimensional image signals to the state, followed the lower layers are two sequences (streams) of fully connected layers. One stream estimates (scalar) state-value function $V(s, \theta, \beta)$ and second stream estimates the advantage function for each action $A(s,a,\theta,\alpha)$. Their estimations are separated. The final module combines two streams to generate a single output Q function:

$$\begin{aligned} &Q(s,a,\theta,\alpha,\beta) \\ &= V(s,\theta,\beta) + \left(A(s,a,\theta,\alpha) - \frac{1}{|A|}\sum_{a'} A(s,a',\theta,\alpha)\right), \end{aligned} \tag{8.268}$$

where $|A|$ denotes the number of actions in the set A.

The entire procedure of RL with counterfactual-based data augmentation is given in Algorithm 8.16 (Lu et al. 2020).

Algorithm 8.16

Step 1: Input: observed triplet $\left(S_t, A_t, S_{t+1}\right)$, $t=1,\ldots,T$ from all individuals.

Step 2: Estimation of a general policy

1. Estimate the structural causal model in (8.262) with BiCoGAN.
2. Use the estimated structural causal model to generate counterfactual states and noises for a set of alternative actions. Such counterfactually augmented data are denoted by $\tilde{D}$.
3. Perform D3QN on $\tilde{D}$ and output the general policy.

Step 3: Estimation of personalized policies

4. Use BiCoGAN to estimate the structural causal model in (8.260) and the individual factor θ_c.
5. For each individual, use the estimated structural causal model to generate counterfactual data for a set of alternative actions. Denote the counterfactually augmented dataset for the individual i by $\tilde{D}_i$.
6. Run D3QN on the data $\tilde{D}_i$ and output the personalized policy for the individual i.

8.6 REINFORCEMENT LEARNING FOR INFERRING CAUSAL NETWORKS

8.6.1 Instruction

The widely used traditional methods for construction of causal networks are based on score functions. The causal networks are formulated as a directed acyclic graph (DAG). For each DAG, a score is assigned. Inferring causal network is reduced to a combinatorial optimization problem. Various algorithms have been developed to search for the DAG with the best score. Finding the optimal combinatorial problem solution is NP-hard (Chickering et al. 2004). Zheng et al. (2018) and Yu et al. (2019) formulated acyclic constraint in term of smooth continuous function and transformed the combinatorial optimization problem to a continuous optimization problem. However, Zheng et al. (2018) can only deal with linear causal network analysis and Yu et al. (2019) can only perform quasi-linear causal network analysis, although they used VAE and matrix normal to extend analysis to the DAG with multiple features of the nodes. In the past several years, a combination of RL and graph encoding-decoding algorithms is used to search for the best solution to combinatorial optimization problem (Vaswani et al. 2017; Kool et al. 2019; Kwon et al. 2020; Mazyavkina et al. 2020). Following this path, Zhu et al. (2020) applied RL to causal network discovery. Searching best solution to combinatorial optimization problem is then transformed to search optimization policy which is a continuous optimization problem. This opens a new way for construction of causal networks. However, their structural causal models are still not fully nonlinear. In this section, we introduce (1) the recent development in attention networks, (2) application of RL to solving combinatorial optimization problems, and (3) a general framework for causal discovery with RL which can completely solve nonlinear causal network problems.

8.6.2 Mathematic Formulation of Inferring Causal Networks Using Bidirectional Conditional GAN

Consider a network with m_1 endogenous variables $\left(Y_1,\ldots,Y_{m_1}\right)$ and m_2 exogenous variables $(X_1,\ldots,X_{m_2})$ with n samples, where $Y_i \in R^n$, $X_i \in R^n$. The set of variables $\left(Y_1,\ldots,Y_{m_1}\right)$ defines a DAG $\mathcal{G}=(V,E)$, where V is a set of nodes and E is the set of edges. Each node $i \in V$ corresponds to exactly one variable Y_i. Therefore, the node and variable can be exchangeable used. Define a functional causal model:

$$Y_i = f_i\left(Y_{pa(i)}, X_{pa(i)}, e_i\right), i=1,\ldots,m_1, \quad (8.269)$$

where $Y_{pa(i)}$ are the endogenous variables of the parents of the endogenous variable Y_i and $X_{pa(i)}$ are the exogenous variables of the parents of the endogenous variable Y_i, e_i, i=1,…, m_1 are noise variables and are jointly independent. If the network $\mathcal{G}$ is given, then $Y_{pa(i)}, X_{pa(i)}$ are also given.

In Section 8.5.2.2.1, we introduced BiCoGAN method ((Jaiswal et al. 2017; Carrara et al. 2020) for

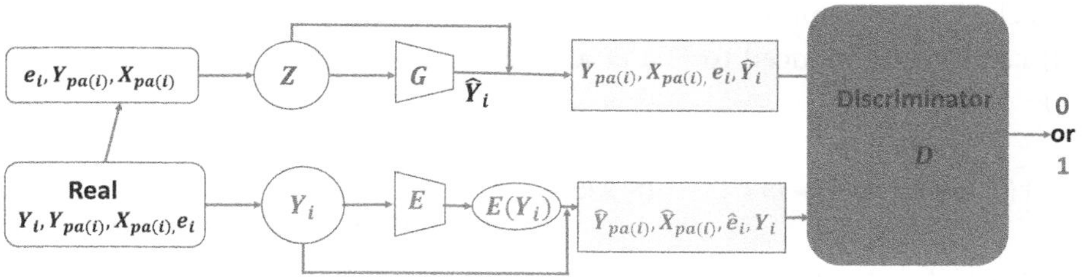

FIGURE 8.13 Architecture of BiCoGAN for functional causal model.

estimating structural causal model in RL. Now we extend the BiCoGAN to inferring the functional causal models (8.269). The BiCoGAN is shown in Figure 8.13. The BiCoGAN is applied to each function causal model of the endogenous variable Y_i. Therefore, m_1 BiCoGANs fit m_1 functional causal model in parallel.

BiCoGAN improves the modeling $\left(Y_{pa(i)}, X_{pa(i)}, e_i\right)$ by exposing $\left(Y_{pa(i)}, X_{pa(i)}, e_i\right)$ to the discriminator together with the images $f_i\left(Y_{pa(i)}, X_{pa(i)}, e_i\right)$ generated from $\left(Y_{pa(i)}, X_{pa(i)}, e_i\right)$. An encoder $E(Y_i)$ is introduced to map real samples Y_i to the corresponding encoding space $\left(Y_{pa(i)}, X_{pa(i)}, e_i\right)$. The encoder module $E(Y_i)$ is trained together with the decoder $f_i\left(Y_{pa(i)}, X_{pa(i)}, e_i\right)$. The discriminator $D(Y_i, E(Y_i)) \in [0,1]$ is trained to assess whether the couple $(Y_i, E(Y_i))$ comes from a real $(Y_i, E(Y_i))$ or generated image $\left(f_i\left(Y_{pa(i)}, X_{pa(i)}, e_i\right), Y_{pa(i)}, X_{pa(i)}, e_i\right)$. Fooling discriminator D leads f_i and E to minimize the difference between $\left(f_i\left(Y_{pa(i)}, X_{pa(i)}, e_i\right), Y_{pa(i)}, X_{pa(i)}, e_i\right)$ and $(Y_i, E(Y_i))$ couples. The minimax problem for the BiCoGAN is

$$\begin{aligned}&\min_{G,E}\max_{D} V(D,G,E)\\&= E_{Y_i \sim p_{data}(Y_i)}\left[\log D(Y_i, E(Y_i))\right]\\&\quad + E_{Z \sim p(Z)} \log\left(1 - D(f_i(Z), Z)\right),\end{aligned} \tag{8.270}$$

where $Z = \left(Y_{pa(i)}, X_{pa(i)}, e_i\right)$.

Assume that n samples are available. Then, we obtain

$$\hat{Y}_i^j = \hat{f}_i\left(Y_{pa(i)}^j, X_{pa(i)}^j, \hat{e}_i\right), \; j = 1, \ldots, n. \tag{8.271}$$

Define

$$\hat{Y}_i = \begin{bmatrix} \hat{Y}_i^1 \\ \vdots \\ \hat{Y}_i^n \end{bmatrix}, \hat{e}_i = \begin{bmatrix} \hat{e}_i^1 \\ \vdots \\ \hat{e}_i^n \end{bmatrix}, Y_i = \begin{bmatrix} Y_i^1 \\ \vdots \\ Y_i^n \end{bmatrix},$$

$$Y_{pa(i)} = \begin{bmatrix} Y_{pa(i)}^1 \\ \vdots \\ Y_{pa(i)}^n \end{bmatrix}, X_{pa(i)} = \begin{bmatrix} X_{pa(i)}^1 \\ \vdots \\ X_{pa(i)}^n \end{bmatrix}.$$

For each node, we define the score as

$$score_i = \|Y_i - \hat{Y}_i\|_2^2, \tag{8.272}$$

and the Bayesian Information Criterion (BIC) score for the entire network as

$$S_{BIC}(\mathcal{G}) = \sum_{i=1}^{m_1} n \log \frac{\|Y_i - \hat{Y}_i\|_2^2}{n} + m_{edge} \log n, \tag{8.273}$$

where m_{edge} denotes the number of edges.

Next we define constraints to ensure the acyclicity of the graph. The acyclicity constraints are often specified by combinatorial methods. The optimization problem is in nature, a combinatorial optimization problem, and hence is an NP-hard problem (Chickering et al. 2004).

To meet this computational challenge, Zhang et al. (2018) proposed to transform these combinatorial optimization problems into continuous optimization problem using a new smooth acyclicity constraint.

An adjacency matrix A is a DAG, if and only if (Section 5.2.2)

$$\mathrm{h}(A) = \mathrm{Tr}\left(e^{A \circ A}\right) - m_1 = 0, \tag{8.274}$$

where $A \circ A$ denotes element-wise multiplication. The constrained (8.274) can also be reduced to (Yu et al. 2019; Section 5.3.5):

$$h(A) = \mathrm{Tr}\left[(I + \alpha A \circ A)^m\right] - m = 0. \quad (8.275)$$

The score function $S_{BIC}(\mathcal{G})$ depends on the topology of the graph. Construction of the causal network is finally formulated as a combinatorial optimization problem:

$$\min_{\mathcal{G}} S_{BIC}(\mathcal{G}) \quad (8.276)$$

$$\text{s.t. } h(A) = 0,$$

where A is the adjacency matrix of the DAG $\mathcal{G}$.

The constrained combinatorial optimization problem (8.276) can be solved by a sequence of unconstrained combinatorial optimization problem using Lagrange multiplier method:

$$\max_{\lambda} D(\lambda) = \min_{\mathcal{G}} S(\mathcal{G}, \lambda), \quad (8.277)$$

where

$$S(\mathcal{G}, \lambda) = S_{BIC}(\mathcal{G}) + \lambda\, h(A). \quad (8.278)$$

To improve the convergence rate, we add the second constraint to the objective function:

$$S(\mathcal{G}, \lambda_1, \lambda_2) = S_{BIC}(\mathcal{G}) + \lambda_1 h(A) + \lambda_2 I(\mathcal{G} \notin DAGs), \quad (8.279)$$

where I is an indicator function.

Therefore, finally we need to solve the problem:

$$\max_{\lambda_1, \lambda_2} D(\lambda_1, \lambda_2) = \min_{\mathcal{G}} S(\mathcal{G}, \lambda_1, \lambda_2). \quad (8.280)$$

Gradient method can also be used to solve the optimization problem (8.280). Update formula is

$$\lambda_1^{k+1} = \lambda_1^k + \eta_1 h(A_*) \quad (8.281)$$

$$\lambda_2^{k+1} = \lambda_2^k + \eta_2\, I(\mathcal{G} \notin DAGs), \quad (8.282)$$

where A_* is the adjacency matrix that corresponds to the graph $\mathcal{G}_* = \underset{\mathcal{G}}{\mathrm{argmin}}\, S(\mathcal{G}, \lambda_1, \lambda_2)$.

For fixed DAG and penalty parameters λ_1, λ_2, we define the reward as:

$$R = -S(\mathcal{G}, \lambda_1, \lambda_2). \quad (8.283)$$

8.6.3 Framework of Reinforcement Learning for Combinatorial Optimization

RL for combinatorial optimization can be defined as the following MDP (Mazyavkina et al. 2020).

Definition 8.19: MDP of RL for Combinatorial Optimization

The MDP of RL for combinatorial optimization consists of state space $\mathcal{S}$, action space $\mathcal{A}$, reward R, transition probability P, scalar discount factor γ, and the length of episode T.

1. State $s_t \in \mathcal{S}$. A state s is either defined as a partial solution to the problem (e.g., a partially constructed DAG) or suboptimal solution (e.g., suboptimal DAG, or simply a DAG).
2. Action $a_t \in \mathcal{A}$ is defined as either as an addition to a partial solution (edges to partial DAG) or changing complete solution (complete DAG).
3. Reward function R is a mapping from states and actions into real numbers $R: \mathcal{S} \times \mathcal{A} \to R$. It can be defined as the changes in the cost after taking action. The reward function depends on the current state s_t and current action a_t. Thus, the reward function is often denoted as $R(s_t, a_t)$.
4. Transition probability $P(S_{t+1} = s_{t+1} | S_t = s_t, A_t = a_t)$ is defined as a conditional probability that the dynamic system in state s_t at time t moves to the state s_{t+1} at time $t+1$ after taking action a_t.
5. Scalar discount factor $0 < \gamma \le 1$ is defined as the factor that accounts for the amount of reward reduced when the system moves away from the current state.
6. T, the length of the episode, where an episode is defined as a sequence $\{s_t, a_t, r_{t+1}, s_{t+1}, a_{t+1}, r_{t+2}, s_{t+2}, \ldots s_T\}$.

The goal of RL is to find a policy $\pi(a|s)$ that maps the states into actions (Mazyavkina et al. 2020). The RL attempts to find the optimal policy that optimizes the expected discounted sum of cost (reward):

$$\pi_* = \underset{\pi}{\operatorname{argmax}} E_\pi \left[\sum_{t=0}^{T} \gamma^t R(s_t, a_t) \right]. \quad (8.284)$$

As we discussed before, there are two types of RL algorithms: value-based algorithms and policy-based algorithms. No matter what type of algorithm is used, all RL algorithms include two important elements: state and action. State in our problem represents the given graph, while actions are often represented by numbers. Therefore, an RL for solving combinatorial optimization problem needs to encode a state to a number.

The pipeline for solving causal network estimation problem with RL is presented as follows. BiCoGAN is first used to estimate the functional causal model. Then, inferring causal network based on the estimated functional causal model is formulated as a constrained combinatorial problem, which in turn is reformulated as a sequence of unconstrained combinatorial optimization problem using the Lagrange multiplier. The negative objective function for the unconstrained combinatorial optimization problem is used as a reward function. We define the graph as the state and the action as the graph generating process. A key step is to encode the input state into a numerical vector (Q-values or probability of each action). Next, the RL selects an action, the environment moves to a new state (changes to a new graph), and the agent receives a reward for the taken action. Repeat the process until the suboptimal graph for the fixed penalty parameters is reached. Finally, update the penalty parameters and repeat the iteration process again until reach the global optimal DAG that defines the causal network.

8.6.4 Graph Encoder and Decoder

Both encoding and decoding graph-structured data are essential to application of RL to causal inference. The classical methods for encoding graphs use graph embedding or fixed feature selection to map graphs into real vector spaces, while the recent methods for graph encoding directly use graph neural networks. Similarly, the classical techniques for decoding graphs use the rule-based, graph-driving methods, while recent methods learn the decoding directly via deep learning (Hamilton et al. 2018; Paolis Kaluza et al. 2018; Xu 2020).

Graph encoding methods can be classified into three categories (Cui et al. 2018): matrix factorization-based methods, random walk-based methods, and neural network-based methods (Cao et al. 2015; Ou et al. 2016). In this section, I will mainly introduce the neural network-based methods.

8.6.4.1 Mathematic Formulation of Graph Embedding

Consider a graph $G(V,E)$ wth associated adjacency matrix A, where $V = \{v_1, \ldots, v_m\}$ is a set of nodes and E is a set of edges. An edge e_{ij} can be represented by a pair of (v_i, v_j), $v_i, v_j \in V$. If the edge e_{ij} is directed edge, then (v_i, v_j) represents direction of $v_i \to v_j$. The nodes v_i and v_j are called adjacent nodes or neighboring nodes. Let $X \in R^{d \times m}$ be the real-valued matrix of node features, where d is the number of features for each node and m is the number of nodes. The goal of graph embedding is to use the adjacency matrix A and node feature matrix X to map each node or a subgraph to a vector $Z \in R^l$, $l \ll m$ in the low dimensional space.

8.6.4.2 Node Embedding

We first discuss embedding nodes. Node embedding maps nodes to a low-dimensional latent space where their graph position and their neighborhood structure are encoded (Figure 8.13) (Hamilton et al. 2018). Node embedding consists of two parts: encoder and decoder. The encoder maps each node to a low-dimensional latent vector and the decoder recovers graph structural information from the mapped latent vectors. Now we mathematically define the encoder and decoder.

Definition 8.20: Encoder and Decoder

Mathematically, the encoder is defined as a function,

$$ENC: V \to R^d, \quad (8.285)$$

i.e., maps a node $v_i \in V$ to the latent vector $z_i \in R^d$ (z_i is called the embedding of the node v_i). The decoder is also defined as a function of two variables,

$$DEC: R^d \times R^d \to R^+, \quad (8.286)$$

i.e., maps a pair of node embedding to a positive similarity measure in the original graph.

Consider two nodes v_i and v_j, and their embedding (z_i, z_j). Let $S_G(v_i, v_j)$ be the similarity measure between v_i and v_j in the original graph. The quality of reconstruction of the encoder-decoder is quantified by the similarity. The similarity of the reconstruction of encoder-decoder is measured by $DEC(ENC(v_i), ENC(v_j))$. The goal of the encoder-decoder is to make the similarity of the reconstruction as close to the similarity measure $S_G(v_i, v_j)$ between the pair of nodes v_i and v_j as possible, i.e.,

$$DEC(ENC(v_i), ENC(v_j)) \approx S_G(v_i, v_j). \quad (8.287)$$

Let $l: R \times R \to R$ be a loss function which measures the difference between the reconstruction similarity measure $DEC(z_i, z_j)$ and similarity measure $S_G(v_i, v_j)$ in the original graph. The encoder-decoder attempts to minimize the loss $\mathcal{L}$ over a set of training node pairs D:

$$\mathcal{L} = \sum_{(v_i, v_j) \in D} l\left(DEC(z_i, z_j), S_G(v_i, v_j)\right). \quad (8.288)$$

8.6.4.3 Shallow Embedding Approaches

Shallow embedding is a key component of the node embedding algorithms (Hamilton et al. 2018). The encoder function for shallow embedding is defined as

$$ENC(v_i) = \mathbf{Z}\mathbf{v}_i, \quad (8.289)$$

where $\mathbf{Z} \in R^{d \times m}$ is a matrix consisting of the embedding vectors for all nodes in the graph and $\mathbf{v}_i \in I_V$ is a one-hot indicator vector indicating that the ith column of $\mathbf{Z}$ corresponds to node v_i.

Specifically,

$$\mathbf{Z}\mathbf{v}_i = \begin{bmatrix} z_{11} & \cdots & z_{1i} & \cdots & z_{1m} \\ \vdots & \vdots & & \vdots\vdots\vdots & \\ z_{d1} & \cdots & z_{di} & \cdots & z_{dm} \end{bmatrix} \begin{bmatrix} 0 \\ \vdots \\ 1 \\ \vdots \\ 0 \end{bmatrix} = \begin{bmatrix} z_{1i} \\ \vdots \\ z_{di} \end{bmatrix} = \mathbf{Z}_i.$$

8.6.4.3.1 Factorization-Based Approaches The encoder function for the factorization-based approach is defined in equation (8.289), while the decoder function and loss function in the factorization-based approaches for the Laplacian eigenmaps are respectively defined as

$$DEC(z_i, z_j) = \left\| z_i - z_j \right\|_2^2 \quad (8.290)$$

and

$$\mathcal{L} = \sum_{(v_i, v_j) \in D} DEC(z_i, z_j) S_G(v_i, v_j). \quad (8.291)$$

The decoder function and loss function for the inner-product methods are respectively, defined as

$$DEC(z_i, z_j) = z_i^T z_j \quad (8.292)$$

and

$$\mathcal{L} = \sum_{(v_i, v_j) \in D} \left\| DEC(z_i, z_j) - S_G(v_i, v_j) \right\|_2^2. \quad (8.293)$$

Several graph factorization methods use the adjacency matrix to define the loss function. The Graph Factorization algorithm defines the loss function as the element of the adjacency matrix, i.e., $S_G(v_i, v_j) = A_{ij}$ and the GraRep use the power of the adjacency matrix to define the loss function, i.e., $S_G(v_i, v_j) = A_{ij}^2$.

For the inner product decoder function, using equation (8.292), we obtain

$$\sum_{i=1}^{d} \sum_{j=1}^{d} DEC(z_i, z_j) = \sum_{i=1}^{d} \sum_{j=1}^{d} z_i^T z_j = Z^T Z, \quad (8.294)$$

where $Z = [z_1, \ldots, z_d]$.

Define the similarity matrix of the graph as

$$S = \left(S_G(v_i, v_j)\right)_{d \times d}. \quad (8.295)$$

Then, the similarity matrix can be factorized as

$$S \approx Z^T Z. \quad (8.296)$$

The loss function for matrix-factorization approaches is

$$\mathcal{L} = \left\| Z^T Z - S \right\|_F^2, \quad (8.297)$$

where F denotes the Frobenius norm of the matrix. Equation (8.297) implies that learning embedding for each node by matrix-factorization approaches is implemented by making the inner product between the learned embedding vectors to approximate the deterministic measure of node similarity.

8.6.4.3.2 Random Walk Approaches Random walk approach is also another popular class of shallow embedding methods. For random walks, the similarity of nodes is measured by their c-occurrence on short random walks over the graph. Unlike the matrix-factorization approaches where deterministic node similarity measures are used, random walk approaches use statistics of random walks to define the similarity measures. Let $p_{G,T}(v_j \mid v_i)$ be the probability of visiting v_j on a length-T ($T \in [2,\ldots, l_T]$) random walk starting v_i.

The decoder function for DeepWalk (Peruzzi et al. 2014) and node2vec (Grover and Leskovec 2016) is defined as

$$Dec(z_i, z_j) = \frac{\exp(z_i^T z_j)}{\sum_{v_k \in V} \exp(z_i^T z_k)} \tag{8.298}$$

We mainly introduce the DeepWalk and briefly introduce node2vec. The DepWalk algorithm include two major parts: a random walk generator and a representation update procedure (Peruzzi et al. 2014)

8.6.4.3.2.1 Random Walk Generator The random walk generator takes a graph G as an input, and uniformly samples a random node v_i as the root of the random walk which is denoted by W_{v_i} where v_i indicates that the random walk W_{v_i} starts at the root node v_i. The random walk W_{v_i} repeatedly samples uniformly from the neighbors of the last node visited until the maximum length t is reached. At each node, we perform γ random walks of length t. The random work procedures are summarized in Algorithm 8.17 (Algorithm 1 in (Peruzzi et al. 2014)).

Algorithm 8.17: Random Walk

Input:

Graph $G(V, E)$
Window Size: ω
Number of walk per node: γ
Walk length t

Output:

Matrix of node representations $\Phi \in R^{|V| \times d}$

Step 1: Initialization. Sample Φ from $\mathcal{U}^{|V| \times d}$

Step 2: Construct a binary tree T from the set of nodes V

Step 3: for $i = 0$ to γ do (iteration of random walk generation)

Step 4: $\mathcal{O} = \text{shuffle}(V)$

Step 5: for each $v_i \in \mathcal{O}$ do (perform random walks for all nodes)

$W_{v_i} = RandomWalk(G, v_i, \omega)$, generate a random walk with length t, starting at v_i

SkipGram (Φ, W_{v_i}, ω)

end for

end for

Now we introduce the SkipGram to update representations (Peruzzi et al. 2014). The purpose of SkipGram is to learn the latent representation $\Phi : v \in V \rightarrow R^{|V| \times d}$. The representation function Φ is often parameterized by a $|V| \times d$ matrix. Our goal is to use the representations $\Phi(v_1), \ldots, \Phi(v_{i-1})$ of the visited nodes $v_1, \ldots, v_{i-1}$ to estimate the conditional probability of observing v_i of next visit, given the representations of the past visited nodes $v_1, \ldots, v_{i-1}$:

$$p(v_i | \Phi(v_1), \ldots, \Phi(v_{i-1})). \tag{8.299}$$

The problem for this calculation is that the computation of this conditional probability quickly becomes intractable when the walk length increases. To overcome this limitation, we make three changes. The first change is to replace prediction of a missing word using its context with prediction of the context using the word. The second change is to redefine the context as the composed of the words presenting on the right and left of the given word. The third change is to remove the ordering constraints in mathematical formulation of the problem. After these changes, the estimation problem (8.299) becomes

$$\min_{\Phi} - \log p(\{v_{i-1}, \ldots, v_{i+w}\} \setminus v_i | \Phi(v_i)). \tag{8.300}$$

This formulation has several remarkable features. First, this model assumes the order independence which allows to develop small models as one node is given at time. Secondly, the models capture the shared similarity in local graph structure between the nodes. Finally, by combining both truncated random walks and language models, the DeepWalk generalizes representations of social network, which encode the latent forms of community membership to the general networks.

Using independent assumption, we can reduce the conditional probability in equation (8.300) to

$$p(\{v_{i-\omega},\ldots,v_{i+\omega}\}\backslash v_i|\Phi(v_i))=\prod_{j=i-w,\ j\neq i}^{i+\omega}p(v_j|\Phi(v_i)). \quad (8.301)$$

The core element in equation (8.301) is the conditional probability $p(v_j|\Phi(v_i))$. A logistic regression can be used to compute $p(v_j|\Phi(v_i))$. However, modeling the conditional probability using logistic regression would result in a large number of labels $(|V|)$ (Peruzzi et al. 2014). To overcome this limitation, we use hierarchical softmax (Ruder et al. 2016).

Recall that the softmax is defined as

$$p(v_i|c)=\frac{\exp(h^T\Phi(v_i))}{\sum_{v_j\in V}\exp(h^T\Phi(v_j))}, \quad (8.302)$$

where h is the output vector of the penultimate network layer. Since the denominator involves computing the inner product between h and outpu embedding of every node in the network, computation of softmax is very expansive. To overcome this limitation, hierarchical softmax can be used as an approximation to the softmax using binary tree (Morin and Bengio 2005). We first build a binary tree. We assign the nodes to the leaves of the binary tree. Let b_0 be the root of the tree and $b_{[\log|V|]}=u_k$. The inner nodes between the root and leaves are indexed by $b_1, b_2, \ldots$ (Figure 8.14). Define the conditional probability $p(b_l\,|\,\Phi(v_i))$ as

$$p(b_l|\Phi(v_i))=\frac{1}{1+\exp(-\Phi^T(v_j)\Psi(b_l))}, \quad (8.303)$$

where $\Psi(b_l)\in R^d$ is the representation assigned to tree node b_l. For example, $p(v_3|\Phi(v_2))$ is equal to the product of the probabilities along the path $b_0\to b_1\to b_2\to b_3=v_3$ in Figure 8.14, i.e.,

$$\begin{aligned}p(v_3|\Phi(v_2))=&\frac{1}{1+\exp(-\Phi^T(v_2)\Psi(b_0))}\\&\times\frac{1}{1+\exp(-\Phi^T(v_2)\Psi(b_1))}\\&\times\frac{1}{1+\exp(-\Phi^T(v_2)\Psi(b_2))}\\&\times\frac{1}{1+\exp(-\Phi^T(v_2)\Psi(v_3))}.\end{aligned}$$

8.6.4.4 *Attention and Transformer for Combinatorial Optimization and Construction of Directed Acyclic Graph*

8.6.4.4.1 Notations and Definitions We introduce DAG-RNN with most materials from Amizadeh et al. (2019), and attention and transformer for

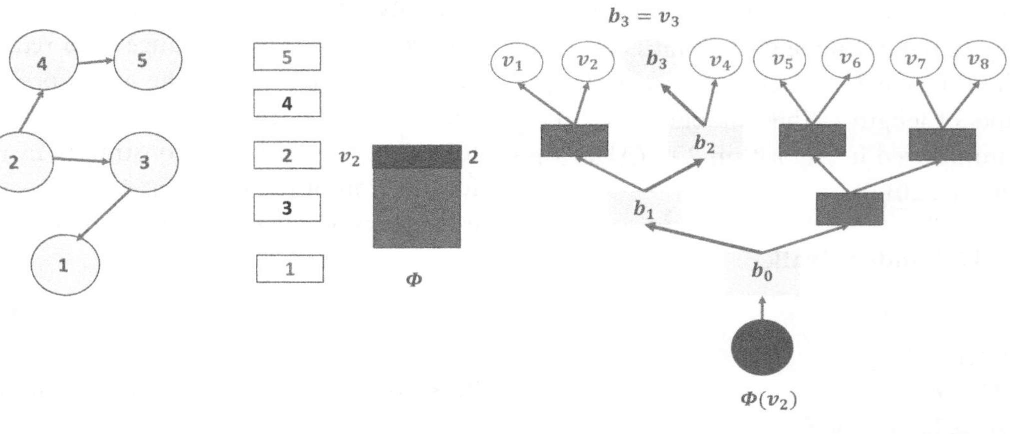

FIGURE 8.14 Illustration of DEEP WALK.

combinatorial optimization and construction of DAG using RL with most materials from Kool et al. (2019).

Consider a DAG $G=(V,E)$, where $V=\{v_i\}_{i=1}^n$ and $E=\{e_{ij}\}$ are the sets of direct edges with e_{ij} denoting an edge from v_i to v_j. Assume that the topological sort of the DAG determines the order of the nodes in the DGA. For any node $v \in V$, $\pi_{pa}(v)$ denotes the set of parents of v in G. The reverse DAG, G^r is defined as the DAG with the same set of nodes in G, but reversed edges. The nodes of G^r appear in the reverse order of those of G. For each node v in a given G, define $\mu_G(v): V \rightarrow R^d$ (a d-dimensional vector function) as a DAG function (Figure 8.15). All possible d-dimensional functions $\mu_G(v)$ form a space, denoted by G^d. Let $F_\theta(\mu_G): G^d \rightarrow \mathcal{O}$ be a parametric function that maps a DAG function μ_G in G^d to some output space $\mathcal{O}$.

8.6.4.4.2 DAG Embedding Layer The embedding of DAG is defined as a mapping from a DAG to the continuously changed vector space, which is often called latent space. A DAG consists of nodes and structure of the graph. Therefore, the embedding of DAG includes mapping each node in the input graph to a latent vector that is called the node state. The structure information of the graph is included in the embedding of DAG via the interactive update of the node states, which takes the graph structure into account. The DAG recurrent networks (DAG-RNNs) (Shuai et al. 2016) in a junction with the gated graph sequential neural networks (GGS-NN) (Li et al. 2015) are used for the embedding of DAG.

The input node feature and mapped node state vectors are defined as a DAG function. Specifically, the d-dimensional DAG function $\mu_G(v)$ is used to define the node feature vector $x_v = \mu_G(v)$, while some unknown q-dimensional DAG function $\delta_G(v): V \rightarrow R^q$ is used to define the node state vector $h_v = \delta_G(v)$. Unlike the classical RNN where the hidden state h_{v_i} of the node v_i is a nonlinear function of the input feature vector Y_{v_i} or X_{v_i} of the current node v_i and the hidden state vector $h_{v_{i-1}}$ of the previous node v_{i-1}, the state vector h_{v_i} of the node v_i is defined as

$$h_{v_i} = GRU\left(Y_{v_i}, h'_{v_i}\right), \tag{8.304}$$

where $h'_{v_i} = \mathcal{A}\left(\left\{h_u \mid u \in \pi_{pa}(v_i)\right\}\right)$, GRU is the gated recurrent unit, $\mathcal{A}$ is a deep set aggregator function (Zaheer et al. 2017). The deep set function is invariant to the permutation of its input. Equation (8.304) states that the state vector is generated by the GRU with the node feature vector and the aggregated state of its parent nodes.

To complete one layer (forward) embedding of the input DAG feature μ_G, we apply equation (8.304) sequentially to the nodes of the DAG in the topological set order to obtain the state vector h_v for all nodes

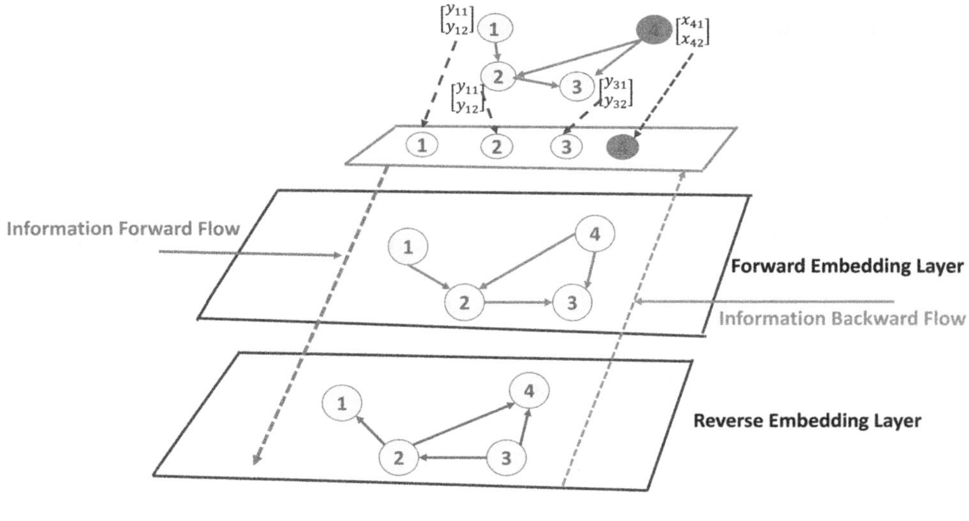

FIGURE 8.15 Illustration of DAG encoding.

of G. Let $\varepsilon_\beta : G^d \to G^q$ be the embedding function that maps the input d-dimensional DAG function to a q-dimensional DAG function space. Then, we have $\varepsilon_\beta(\mu_G) = \delta_G$.

After completing one layer (forward) embedding of the input DAG feature function which is processed in the reversed order. Let $\varepsilon^r : G^d \to G^q$ be the reverse embedding function that maps the input d-dimensional DAG function to a q-dimensional DAG function space in the reverse order. Then, we have $\varepsilon^r(\mu_G) = \varepsilon(\mu_{G^r})$, where G^r denotes the reversed G of the DAG G. Forward embedding layer takes information from its ancestor nodes, while reverse embedding layer considers information flowing from its descendant nodes.

8.6.4.4.3 Attention Layer

8.6.4.4.3.1 Single Attention The embedding are updated using N attention layers (Vaswani et al. 2017; Kool et al. 2019; Choromanski et al. 2021). Each attention layer consists of two sublayers: multi-head attention (MHA) layer and fully connected feed-forward (FF) layer. Each sublayer is followed by batch normalization (Ioffe and Szegedy 2015).

Attention is measured by the similarity between the target node and its neighbors and used as weight. If two nodes are chosen, they are more similar, the more attention should be paid to these two nodes. Let $q_i \in R^{d_k}$ be a vector of query, $k_i \in R^{d_k}$ be a vector of key, $v_i \in R^{d_v}$ be a vector of values, and h_i be the state vector of the node i. Furthermore, let $W^Q \in R^{d_k \times d_h}$, $W^K \in R^{d_k \times d_h}$ and $W^V \in R^{d_v \times d_h}$ be the query, key, and value matrices. Define the transformations:

$$q_i = W^Q h_i,\ k_i = W^K h_i,\ v_i = W^V h_i. \tag{8.305}$$

The similarity measure $u_{ij} \in R$ between node i and node j is defined as

$$u_{ij} = \begin{cases} \dfrac{q_i^T k_j}{\sqrt{d_k}} & \text{if } i \text{ adjacent to } j \\ -\infty & \text{otherwise} \end{cases} \tag{8.306}$$

Using a softmax, we define the attention weights $\alpha_{ij} \in [0.1]$ between node i and node j:

$$\alpha_{ij} = \frac{e^{u_{ij}}}{\sum_l e^{u_{il}}}. \tag{8.307}$$

The state vector of the node i is updated as the convex combination of the value v_j from the neighborhood nodes:

$$h_i' = \sum_j \alpha_{ij} v_j. \tag{8.308}$$

8.6.4.4.3.2 Multi-Head Attention T allow nodes to receive different types of messages from different neighbor nodes, we define MHA. Let M be the number of heads, h_{im}', $m \in [1, \ldots, M]$ be the updated state vector of he node i using the mth attention mechanism, and $W_m^O \in R^{d_h \times d_v}$ be the weight matrix. The final MHA state vector of values for node i is a function of the states $h_1, \ldots, h_n$ and defined as

$$MHA_i(h_1, \ldots, h_n) = \sum_{m=1}^{M} W_m^O h_{im}'. \tag{8.309}$$

8.6.4.4.3.3 Batch Normalization Let n be the batch size, $\bar{h} = \frac{1}{n}\sum_{i=1}^n h_i$, $\sigma_i^2 = \frac{1}{n}\sum_{i=1}^n (h_i - \bar{h})^2$, and $\hat{h}_i = \frac{h_i - \bar{h}}{\sqrt{\sigma_i^2 + \varepsilon}}$, ε is small value to ensure $\sigma_i^2 + \varepsilon > 0$, W^{b_n} and b^{b_n} be the learnable d_h-dimensional affine parameters. The batch normalization is defined as

$$BN(h_i) = W^{b_n} \odot \hat{h}_i + b^{b_n}, \tag{8.310}$$

where $\odot$ denotes the element-wise multiplication.

8.6.4.4.3.4 Feed-Forward Sublayer Let $W^{ff,0} \in R^{d_{ff} \times d_h}$, $W^{ff,1} \in R^{d_{ff} \times d_{ff}}$ be weight matrices, $b^{ff,0} \in R^{d_{hh}}$ and $b^{ff,1} \in d^{ff}$ be bias vectors. The output of FF sublayer is defined as

$$FF(\hat{h}_i) = W^{ff,1} ReLu\left(W^{ff,0}\hat{h}_i + b^{ff,0}\right) + b^{ff,1}. \tag{8.311}$$

8.6.4.4.3.5 Attention Layer There are N attention layers. Consider the lth attention layer, which consists of an MHA layer and FF layer, each followed by batch normalization. Summarizing the above discussions, we obtain the output of the lth attention layer:

$$\hat{h}_i = BN^{(l)}\left(h_i^{(l-1)} + MHA_i^{(l)}\left(h_1^{(l-1)}, \ldots, h_n^{(l-1)}\right)\right), \tag{8.312}$$

$$h_i^{(l)} = BN^{(l)}\left(\hat{h}_i + FF^{(l)}\left(\hat{h}_i\right)\right). \tag{8.313}$$

One of the final output is an aggregated embedding $\bar{h}^{(N)}$, defined as

$$\bar{h}^{(N)} = \frac{1}{N}\sum_{i=1}^{n} h_i^{(N)}. \tag{8.314}$$

Both the node embedding $h_i^{(N)}$ and graph embedding $\bar{h}^{(N)}$ are the output of the attention layer.

8.6.4.4.4 Decoder The decoder outputs the adjacency matrix, given the graph embedding $\bar{h}^{(N)}$ and the node embedding $h_i^{(N)}$ $(i=1,\ldots,n)$ (Figure 8.16). To improve the decoding and utilize the context information, a special context node is added to the graph. The decoder will take the output of an attention (sub)layer on the top of the encoder as its input. A single head attention mechanism (Bernoulli probability sampling) will be used to compute the adjacency matrix of the DAG.

8.6.4.4.4.1 Context Embedding The context of the decoder for the node i consists of the graph embedding $\bar{h}^{(N)}$ and the node embedding $h_i^{(N)}$ of the node i. The context of the decoder for the node i is given by

$$h_{c(i)}^{N} = \left[\, \bar{h}^{(N)} \ h_i^{(N)} \,\right], i=1,\ldots,n, \tag{8.315}$$

where $\left[\, \bar{h}^{(N)} \ h_i^{(N)} \,\right]$ is the horizontal concatenation operator.

Let $N+1$ index the node embedding layer of the decoder. Next we use the MHA mechanism to calculate the context node embedding $h_{c(i)}^{N+1}$. Let W_m^Q, W_m^K and W_m^V be the query, key, and value weight matrices for the mth head. Define

$$q_{c(i)}^{m} = W_m^Q h_{c(i)},\ k_i^m = W_m^K h_i \text{ and } v_i^m = W_m^V h_i. \tag{8.316}$$

Calculate the compatibility $u_{c(i)j}^{m} \in R$ of the query $q_{c(i)}^{m}$ of the context $c(i)$ with the key k_j^m of the node j:

$$u_{c(i)j}^{m} = \frac{\left(q_{c(i)}^{m}\right)^T k_j^m}{\sqrt{d_k}},\ d_k = \frac{d_h}{M} \tag{8.317}$$

and attention weights $\alpha_{c(i)j}^{m} \in [0,1]$ for the mth head:

$$\alpha_{c(i)j}^{m} = \frac{\exp\left(u_{c(i)j}^{m}\right)}{\sum_l \exp\left(u_{c(i)l}^{m}\right)}. \tag{8.318}$$

Then, we calculate the state vector for the context node:

$$h_{c(i)}^{m} = \sum_j \alpha_{c(i)j}^{m} v_j^m. \tag{8.319}$$

Let $W_m^O \in R^{d_h \times d_v}$ be the parameter matrices. Then, the final MHA value for the context $c(i)$ is defined as

$$h_{c(i)}^{N+1} = \sum_{m=1}^{M} W_m^O h_{c(i)}^{m}. \tag{8.320}$$

8.6.4.4.4.2 Compute the Adjacency Matrix for the Decoded DAG To calculate the adjacency matrix, we add one final decoder layer with a single attention head. We define

$$q_{c(i)} = W^Q h_{c(i)}^{N+1},\ k_j = W^K h_j^N,\ v_j = W^V h_j^N. \tag{8.321}$$

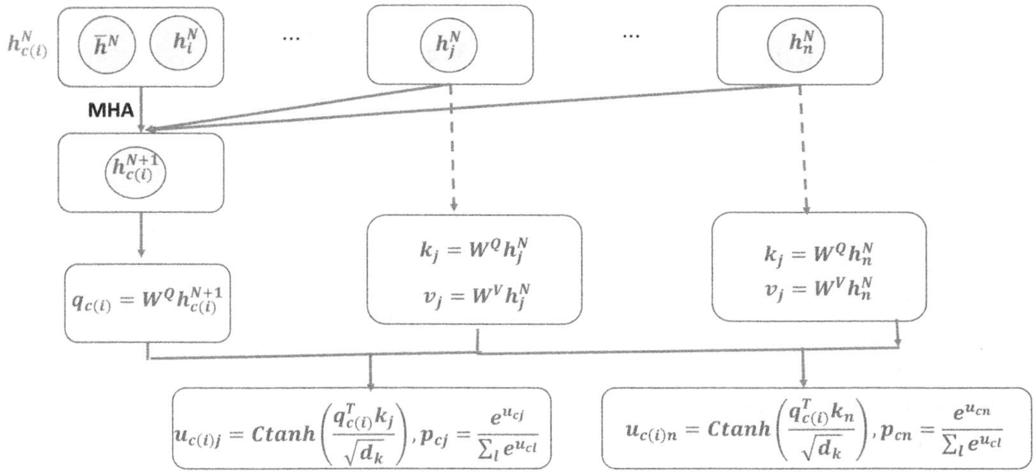

FIGURE 8.16 Decoder for DAG.

We calculate the compatibility $u_{c(i)j}$

$$u_{c(i)j} = C\tanh\left(\frac{q_{c(i)}^T k_j}{\sqrt{d_k}}\right), d_k = d_h \tag{8.322}$$

and probability $p_{ij} = p_\theta\left(A_{ij}|G_s\right)$

$$p_{ij} = \frac{e^{u_{c(i)j}}}{\sum_l e^{u_{c(i)l}}}, j = 1,\ldots, n, j \neq i, \tag{8.323}$$

where $C \in [-10,10]$.

To generate a binary adjacency matrix, the element A_{ij} of the adjacency matrix, indicating that the presence of an edge from the node i to the node j is sampled according to the Bernoulli distribution with the probability p_{ij}. Taking $\pi = A$, we denote the policy by $p_\theta\left(\pi|G_s\right)$.

8.6.4.4.5 Reinforce Learning with Greedy Rollout Baseline RL with greedy rollout baseline is used to generate policy, which produces improved graphs (Kool et al. 2019). In the previous section, we studied how to obtain a probability distribution $p_\theta\left(\pi|G_s\right)$. Sampling from $p_\theta\left(\pi|G_s\right)$, we obtain a directed graph. For a fixed π, the score of its corresponding graph G is $S\left(G, \lambda_1, \lambda_2\right)$. Define the cost of π (or the return) as

$$L(\pi) = S\left(G, \lambda_1, \lambda_2\right) \tag{8.324}$$

and the loss

$$J\left(\theta|G_s\right) = E_{p_\theta(\pi|G_s)}\left[L(\pi)\right]. \tag{8.325}$$

REINFORCE with baseline (Section 8.4.4, Kool et al. 2019) can be used to update the parameter $\boldsymbol{\theta}$. The gradient of the loss function can be approximated by

$$8.4.4\nabla J\left(\theta|G_s\right) = E_{p_\theta(\pi|G_s)}\left[L(\pi) - b(s)\right]\nabla \log p_\theta\left(\pi|G_s\right). \tag{8.326}$$

We have some freedom of choice to select a baseline for reducing the variance of the gradient $\nabla J\left(\theta|G_s\right)$. In general, there are two choices: sample rollout and greedy rollout baselines (Kwon et al. 2020). Since greedy-rollout baseline can reduce more variance, we will mainly introduce greedy-rollout baseline. The baseline policy is denoted by $p_{\theta_{BL}}\left(\pi|G_s\right)$. To stabilize the baseline and reduce the oscillation, we freeze the greedy-rollout policy $p_{\theta_{BL}}\left(\pi|G_s\right)$ for a fixed number of steps of each epoch. At the end of each epoch, we compare the current training policy $p_\theta\left(\pi|G_s\right)$ ith the baseline policy $p_{\theta_{BL}}\left(\pi|G_s\right)$, if significant improvement is observed (by some test, for example, a paired one-sided t-test with significance level $\alpha = 0.05$ on 10,000 separate instances), then the current policy will be used as baseline policy (i.e., $\theta \rightarrow \theta_{BL}$) to update the baseline policy and sample new evaluation instance.

If the sampled policy improves the performance, $L(\pi)$ will be less than the baseline and the gradient $\nabla J\left(\theta|G_s\right)$ will be negative. The parameter θ will be updated, which, in turn, causes to change actions and reduce the loss. Algorithm 1 in Kool et al. (2019) is adapted to Algorithm 8.18.

Algorithm 8.18: REINFORCE with Greedy Rollout Baseline for DAG

Step 1: Input

Training set $S(Y,X)$. Penalty parameters λ_1, λ_2. Score function $S\left(G, \lambda_1, \lambda_2\right)$, number of epochs E, Steps T, per epoch, batch size B, significance level α.

Step 2: Initialization

$$\theta, \theta_{BL} \leftarrow \theta_0.$$

Step 3: Iteration

for epoch = 1, …, E **do**

for step = 1, …, T **do**

1. $S_i \leftarrow$ *Sample Input* $(S), \forall i \in \{1,\ldots, B\}$ (Sample a training set)
2. $\pi_i \leftarrow$ *SampleRollout* $\left(S_i, p_\theta\left(\pi|G_s\right)\right), \forall i \in \{1,\ldots, B\}$ (sample A)
3. $\pi_i^{(BL)} \leftarrow$ *GrredyRollout* $\left(S_i, p_{\theta_{BL}}\left(\pi|G_s\right)\right), \forall i \in \{1,\ldots, B\}$ (Select a A_*)
4. $\nabla J \leftarrow \sum_{i=1}^B S\left(\pi_i = G_i, \lambda_1, \lambda_2\right) - S\left(\pi_i^{BL} = G_i^{BL}, \lambda_1, \lambda_2\right) \times \nabla_\theta \log p_\theta\left(\pi_i|G_s\right)$
5. $\theta \leftarrow$ *Adam* $(\theta, \nabla J)$

end for

If OneSidedPairedTtest $\left(p_\theta, p_{\theta_{BL}}\right) < \alpha$ then

$$\theta_{BL} \leftarrow \theta$$

end if

end for

SOFTWARE PACKAGE

Code for "Explainable reinforcement learning through a causal lens" is posted in https://github.com/EthicalML/xai

Code for "Deconfounding Reinforcement Learning (DRL)" is posted in https://github.com/CausalRL/DRL

Code for "Structured Inference Networks for Nonlinear State Space Models" is posted in https://github.com/clinicalml/structuredinferencelearned

Deep Q-learning and four of its most important supplements: Double DQN, Dueling DQN, Noisy DQN and DQN are linked to the code: https://github.com/Parsa33033/Deep-Reinforcement-Learning-DQN

Software for the paper "node2vec: Scalable feature learning for networks" is posted on the github: https://github.com/aditya-grover/node2vec

The paper "Attention, learn to solve routing problems!" is linked to the github: https://github.com/wouterkool/attention-learn-to-route

This github "https://github.com/huawei-noah/trustworthyAI" includes the software for the paper "Causal discovery with reinforcement learning".

The code for BERT and pre-trained models are available at https://github.com/googleresearch/bert

APPENDIX 8A: BIDIRECTIONAL RNN FOR ENCODING

A major component of the encoder is a bidirectional RNN. The bidirectional RNN consists of a forward RNN and a backward RNN (Krishnan et al. 2017). The forward RNN passes information in a forward direction and the backward RNN passes information in a backward direction.

For any time step t, given input x_t, a_t and r_{t+1}. Let the hidden layer activation function be $\varnothing_h$. The forward and backward hidden state updates are given as follows:

$$h_t^{left} = \varnothing_h\left(W_{hh}^{left} h_{t-1}^{left} + W_{hx}^{left} x_t + w_{ha}^{left} a_t + w_{hr}^{left} r_{t+1} + b_t^{left}\right), \quad (8A1)$$

$$h_t^{right} = \varnothing_h\left(w_{hh}^{right} h_{t+1}^{right} + w_{hx}^{right} x_t + w_{ha}^{right} a_t + w_{hr}^{right} r_{t+1} + b_t^{right}\right), \quad (8A2)$$

where $W_{hh}^{left}, W_{hx}^{left}, w_{ha}^{left}, w_{hr}^{left}, w_{hh}^{right}, w_{hx}^{right}, w_{ha}^{right}, w_{hr}^{right}$ are the weight matrices and b_t^{left}, b_t^{right} are the bias vectors in the RNNs.

The outputs of Bi-RNN are mean μ_t and variance σ_t^2 of the encoding function $q(z_t \,|, z_{t-1}, x, a, r)$. There are four sources that influence μ_t and σ_t^2: (1) the value z_{t-1}, (2) the action a_{t-1}, (3) the hidden state h_t^{left} of the left RNN, and (4) the hidden state h_t^{right} of the right RNN. The function for combining four sources is then given by (Lu et al. 2018)

$$h_{combined} = \frac{1}{4}\left(\tanh\left(w_z z_{t-1} + b_z\right) + \tanh\left(w_a a_{t-1} + b_a\right) + h_t^{left} + h_t^{right}\right) \quad (8A3)$$

Therefore, the mean μ_t and variance σ_t^2 of the encoding function are respectively given by

$$\mu_t = w_\mu h_{combined} + b_\mu, \quad (8A4)$$

$$\sigma_t^2 = \text{softplus}\left(w_\sigma\, h_{combined} + b_\sigma\right), \quad (8A5)$$

where softplus function is defined as

$$\text{softplus}(x) = \log(1 + e^x). \quad (8A6)$$

APPENDIX 8B: CALCULATION OF KL DIVERGENCE

Now we calculate KL divergence. We first calculate $-KL\left(q(c \mid x, a, r) \| p(c)\right)$. It follows from equations (8.219) and (8.232) that

$$\log q(c|x,a,r) = -\frac{D_c}{2}\log(2\pi) - \frac{1}{2}\sum_{j=1}^{D_c}\left[\log\left(\sigma_{tj}^c\right)^2 + \frac{\left(c_j - \mu_{tj}^c\right)^2}{\left(\sigma_{tj}^c\right)^2}\right], \quad (8B1)$$

$$\log p(c) = -\frac{D_c}{2}\log(2\pi) - \frac{1}{2}\sum_{j=1}^{D_c} c_j^2. \quad (8B2)$$

It follows from equation (8B1) that

$$E_{q(c|x,a,r)}\left[\log q(c|x,a,r)\right] = -\frac{D_c}{2}\log(2\pi) - \frac{1}{2}\sum_{j=1}^{D_c}\left[\log\left(\sigma_{tj}^c\right)^2 + 1\right], \quad (8B3)$$

$$E_{q(c|x,a,r)}[\log p(c)] = -\frac{D_c}{2}\log(2\pi) - \frac{1}{2}\sum_{j=1}^{D_c}\left[(\sigma_{tj}^c)^2 + (\mu_{tj}^c)^2\right]. \tag{8B4}$$

Combining equations (8B3) and (8B4), we obtain

$$-KL(q(c|x,a,r)||p(c)) = \frac{1}{2}\sum_{j=1}^{D_c}\left[\log(\sigma_{tj}^c)^2 + 1 - (\sigma_{tj}^c)^2 - (\mu_{tj}^c)^2\right]. \tag{8B5}$$

Recall from equation (8.219) and equation (8.233) that

$$\log p(z_1) = -\frac{D_z}{2}\log(2\pi) - \frac{1}{2}\sum_{j=1}^{D_z} z_{p1j}^2$$

$$\log q(z_1) = -\frac{D_z}{2}\log(2\pi) - \frac{1}{2}\sum_{j=1}^{D_z}\log(\sigma_{1j}^z)^2 - \frac{1}{2}\sum_{j=1}^{D_z}\frac{(z_{1j} - \mu_{1j}^z)^2}{(\sigma_{1j}^z)^2}.$$

Again, by the similar argument, we obtain

$$-KL(q(z_1|x,a,r)||p(z_1)) = \frac{1}{2}\sum_{j=1}^{D_z}\left[\log(\sigma_{1j}^z)^2 + 1 - (\sigma_{1j}^z)^2 - (\mu_{1j}^z)^2\right]. \tag{8B6}$$

Next we calculate $KL(q(z_t|z_{t-1}, x,a,r)||p(z_t|z_{t-1},x,a,r))$. It follows from equations (8.223) and (8.233) that

$$\log p(z_t|z_{t-1}, x,a,r) = -\frac{D_z}{2}\log(2\pi) - \frac{1}{2}\sum_{j=1}^{D_z}\left[\log(\sigma_{ptj}^z)^2 + \frac{(z_{ptj} - \mu_{ptj}^z)^2}{(\sigma_{ptj}^z)^2}\right] \tag{8B7}$$

$$\log q(z_t|z_{t-1},x,a,r) = -\frac{D_z}{2}\log(2\pi) - \frac{1}{2}\sum_{j=1}^{D_z}\left[\log(\sigma_{tj}^z)^2 + \frac{(z_{ptj} - \mu_{tj}^z)^2}{(\sigma_{tj}^z)^2}\right]. \tag{8B8}$$

Then, we obtain

$$E_{q(z_t|z_{t-1},x,a,r)}[\log q(z_t|z_{t-1},x,a,r)] = -\frac{D_z}{2}\log(2\pi) - \frac{1}{2}\sum_{j=1}^{D_z}\left[\log(\sigma_{tj}^z)^2 + 1\right] \tag{8B9}$$

$$E_{q(z_t|z_{t-1},x,a,r)}[\log p(z_t|z_{t-1}, x,a,r)] = -\frac{D_z}{2}\log(2\pi) - \frac{1}{2}\sum_{j=1}^{D_z}\left[\log(\sigma_{ptj}^z)^2 + \frac{(\sigma_{tj}^z)^2}{(\sigma_{ptj}^z)^2} + \frac{(\mu_{tj}^z - \mu_{ptj}^z)^2}{(\sigma_{ptj}^z)^2}\right]. \tag{8B10}$$

Using equations (8B9) and (8B10), we obtain

$$-KL(q(z_t|z_{t-1}, x,a,r)||p(z_t|z_{t-1},x,a,r)) = \frac{1}{2}\sum_{j=1}^{D_z}\left[\log\left(\frac{\sigma_{tj}^z}{\sigma_{ptj}^z}\right)^2 + 1 - \left(\frac{\sigma_{tj}^z}{\sigma_{ptj}^z}\right)^2 - \left(\frac{\mu_{tj}^z - \mu_{ptj}^z}{\sigma_{ptj}^z}\right)^2\right], \tag{8B11}$$

which implies that

$$-\sum_{t=2}^{T} E_{z_{t-1}\sim q(z_{t-1}|z_{t-2},x,a,r)} \times\left[KL(q(z_t|z_{t-1}, x,a,r)||p(z_t|z_{t-1},x,a,r))\right] = \frac{1}{2}\sum_{t=2}^{T} E_{z_{t-1}\sim q(z_{t-1}|z_{t-2},x,a,r)}\sum_{j=1}^{D_z}\left[\log\left(\frac{\sigma_{tj}^z}{\sigma_{ptj}^z}\right)^2 + 1 - \left(\frac{\sigma_{tj}^z}{\sigma_{ptj}^z}\right)^2 - \left(\frac{\mu_{tj}^z - \mu_{ptj}^z}{\sigma_{ptj}^z}\right)^2\right]. \tag{8B12}$$

Combining equations (8B5), (8B6), and (8B12), we obtain

$$
\begin{aligned}
&-KL\left(q_{\varnothing}(z,u|x,a,r)\|p_{\theta}(z,c)\right)\\
&=\frac{1}{2}\sum_{j=1}^{D_c}\left[\log\left(\sigma_{tj}^{c}\right)^2+1-\left(\sigma_{tj}^{c}\right)^2-\left(\mu_{tj}^{c}\right)^2\right]\\
&+\frac{1}{2}\sum_{j=1}^{D_z}\left[\log\left(\sigma_{1j}^{z}\right)^2+1-\left(\sigma_{1j}^{z}\right)^2-\left(\mu_{1j}^{z}\right)^2\right] \qquad (8B13)\\
&+\frac{1}{2}\sum_{t=2}^{T}E_{z_{t-1}\sim q(z_{t-1}|z_{t-2},x,a,r)}\sum_{j=1}^{D_z}\left[\log\left(\frac{\sigma_{tj}^{z}}{\sigma_{ptj}^{z}}\right)^2\right.\\
&\left.+1-\left(\frac{\sigma_{tj}^{z}}{\sigma_{ptj}^{z}}\right)^2-\left(\frac{\mu_{tj}^{z}-\mu_{ptj}^{z}}{\sigma_{ptj}^{z}}\right)^2\right].
\end{aligned}
$$

EXERCISES

EXERCISE 8.1
Show that

$$G_t = R_{t+1} + \gamma G_{t+1} = \sum_{k=0}^{\infty}\gamma^k R_{t+k+1}.$$

EXERCISE 8.2
Show that

$$Q_\pi(s,a)=\sum_{s'\in S}p(s'|s,a)\left[R(s,a,s')+\gamma V_\pi(s')\right],$$

and

$$V_\pi(s)=\sum_{a\in\mathcal{A}}\pi(a|s)Q_\pi(s,a).$$

EXERCISE 8.3
Show that

$$V_*(s)=\max_{a\in\mathcal{A}}Q_*(s,a).$$

EXERCISE 8.4
Show that Bellman optimality operator B_* is a nonlinear operator with fixed point V_*, i.e.,

$$B_*V_* = V_*.$$

EXERCISE 8.5
Show that

$$||B_*V - B_*W||_\infty \le \gamma||V-W||_\infty.$$

EXERCISE 8.6
Show that

$$B_*V_1 \le B_*V_2.$$

EXERCISE 8.7
Show that for Poison distribution, we have

$$P_L = e^{S^L} - y.$$

EXERCISE 8.8
Write loss function for BiCoGAN in Figure 8.13.

EXERCISE 8.9
Find optimal solutions for the loss function in Exercise 8.8.

EXERCISE 8.10
Calculate $P(v_4|v_2)$ in Figure 8.14.

EXERCISE 8.11
For the Figure 8.17, calculate the exponential constraint.

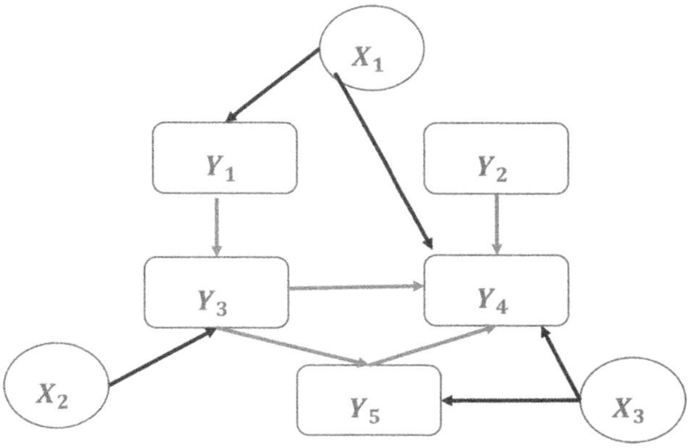

FIGURE 8.17 Architecture of causal network.

EXERCISE 8.12
If $h_{c(i)}^N$ in Figure 8.16 is replaced by $h_{c(i,j)}^N = \left[\ \bar{h}^N\ h_i^N\ h_j^N\ \right]$, please calculate $u_{c(i,j),n}$.

References

Abdalla, I. and Murinde, V. (1997). Exchange rate and stock price interactions in emerging financial markets: Evidence on India, Korea, Pakistan and the Philippines, Applied Financial Economics 7: 25–35.

Aggarwal, K., Kirchmeyer, M., Yadav, P., Keerthi, S. S. and Gallinari, P. (2020). Benchmarking regression methods: A comparison with CGAN. arXiv:1905.12868.

Alaa, A. and van der Schaar, M. (2017). Bayesian inference of individualized treatment effects using multi-task gaussian processes. arXiv e-prints arXiv:1704.02801.

Altché, F. and Fortelle, A. D. L. (2017). An LSTM network for highway trajectory prediction, Proceedings of the IEEE 20th International Conference on Intelligent Transportation Systems.

Amari, S. (1982). Differential geometry of curved exponential families-curvatures and information loss, The Annals of Statistics 10(2): 357–385.

Amizadeh, S., Matusevych, S. and Weimer, M. (2019). Learning to solve circuit-SAT: An unsupervised differentiable approach ICLR (Poster) 2019.

Amornbunchornvej, C., Zheleva, E. and Berger-Wolf, T. Y. (2019). Variable-lag Granger causality for time series analysis. 2019 IEEE International Conference on Data Science and Advanced Analytics (DSAA). DOI: 10.1109/DSAA.2019.00016.

Aragam, B. (2020). DYNOTEARS: Structure learning from time-series data. arXiv:2002.00498.

Aragam, B., Amini, A. A. and Zhou, Q. (2016). Learning directed acyclic graphs with penalized neighbourhood regression. arXiv:1511.08963.

Arjovsky, M. and Bottou, L. (2017). Towards principled methods for training generative adversarial networks. arXiv:1701.04862.

Arjovsky, M., Chintala, S. and Bottou, L. (2017). Wasserstein gan. arXiv preprint arXiv:1701.07875.

Asadi, K., Parikh, N., Parr, R. E, Konidaris, G. D. and Littman, M. L. (2021). Deep radial-basis value functions for continuous control. arXiv:2002.01883.

Athey, S. and Imbens, G. (2016). The state of applied econometrics - causality and policy evaluation. arXiv:1607. 00699.

Austin, P. C. (2011). An introduction to propensity score methods for reducing the effects of confounding in observational studies. Multivariate Behavioral Research 46: 399–424.

Averitt, A. J, Vanitchanant, N., Ranganath, R. and Perotte, A. J. (2020). The counterfactual χ-GAN. arXiv:2001. 03115.

Ayed, I., de Bézenac, E., Pajot, A., Brajard, J. and Gallinari, P. (2019). Learning dynamical systems from partial observations. arXiv:1902.11136.

Bach, F., Jenatton, R., Mairal, J. and Obozinski, G. (2012). Optimization with sparsity-inducing penalties, Foundations and Trends in Machine Learning 4: 1–106.

Bagnell, D. (2008). Kernel methods/functional gradient descent. Statistical Techniques in Robotics (16–831, F08). http://www.cs.cmu.edu/~16831-f14/notes/F08/lecture23/16831_lecture23.dmunoz.pdf

Bai, Z., Hui, Y., Jiang, D., Lv, Z., Wong, W. K. and Zheng, S. (2018). A new test of multivariate nonlinear causality, PLoS One 13: e0185155.

Bai, Z., Wong, W. K. and Zhang, B. (2010). Multivariate linear and nonlinear causality tests, Mathematics and Computers in Simulation 81: 5–17.

Baiocchi, M., Cheng, J. and Small, D. S. (2014). Instrumental variable methods for causal inference, Statistics in Medicine 33: 2297–2340.

Ban, Y., Alameda-Pineda, X., Girin, L. and Horaud, R. (2018). Variational Bayesian inference for audio-visual tracking of multiple speakers. arXiv:1809.10961.

Bao, J., Chen, D., Wen, F., Li, H. and Hua, G. (2017). CVAE-GAN: Fine-grained image generation through asymmetric training. 2017 IEEE International Conference on Computer Vision (ICCV), Venice, 2017, 2764–2773. DOI: 10.1109/ICCV.2017.299.

Bareinboim, E., Correa, J., Ibeling, D. and Icard, T. (2021). On Pearl's hierarchy and the foundations of causal inference, Technical Report, R-60, Columbia University, March, 2021.

Bareinboim, E., Forney, A. and Pearl, J. (2015). Bandits with unobserved confounders: A causal approach, Advances in Neural Information Processing Systems 28(NIPS 2015): 1342–1350.

Bartels, S. and Hennig, P. (2019). Conjugate gradients for kernel machines. arXiv:1911.06048.

Bartlett, M. and Cussens, J (2017). Integer linear programming for the Bayesian network structure learning problem, Artificial Intelligence 244: 258–271.

Beaulac, C., Rosenthal, J. S. and Hodgson, D. (2018). A deep latent-variable model application to select treatment intensity in survival analysis, Proceedings of the Machine Learning for Health (ML4H) Workshop at NeurIPS 2018.

Beaulieu-Jones, B. K., Wu, Z. S., Williams, C., Lee, R., Bhawani, S. P., Byrd, J. B. and Greene, C. S. (2017). Privacy-preserving generative deep neural networks support clinical data sharing, Cardiovascular Quality and Outcomes 12: e005122.

Bellot, A. and van der Schaar, M. (2019). Conditional independence testing using generative adversarial networks. arXiv:1907.04068.

Bennett, A., Kallus, N. and Schnabel, T. (2019). Deep generalized method of moments for instrumental variable analysis. arXiv:1905.12495.

Bertsekas, D. P. (1996). Constrained Optimization and Lagrange Multiplier Methods. Athena Scientific, Belmont, MA.

Bertsekas, D. P. (2019). Reinforcement Learning and Optimal Control. Athena Scientific, Belmont, MA.

Beygelzimer, A., d'Alch'e Buc, F., Fox, E. and Garnett, R. (eds.), Equitable stable matchings in quadratic time, Advances in Neural Information Processing Systems 32(NIPS 2019): 455–465.

Bica, I., Alaa, A. M. and van der Schaar, M. (2020). Time series deconfounder: Estimating treatment effects over time in the presence of hidden confounders. In Daum'e, Hal, III and Aarti Singh (eds.), Proceedings of the International Conference on Machine Learning: 3944–3954.

Bica, I., Jordon, J. and van der Schaar, M. (2020). Estimating the effects of continuous-valued interventions using generative adversarial networks, Neural Information Processing Systems (NeurIPS) 2020: 455–465.

Bishop, C. (2006). Pattern Recognition and Machine Learning. Springer, New York, NY.

Blei, D. M., Kucukelbir, A. and McAuliffe, J. D. (2018). Variational inference: A review for statisticians. arXiv:1601.00670.

Blöbaum, P., Janzing, D., Washio, T., Shimizu, S. and Schölkopf, B. (2018). Analysis of cause-effect inference by comparing regression errors. arXiv: 1802.06698.

Bloebaum, P., Janzing, D., Washio, T., Shimizu, S. and Schoelkopf, B. (2018). Cause-effect inference by comparing regression errors, Proceedings of the Twenty-First International Conference on Artificial Intelligence and Statistics, PMLR 84: 900–909.

Blum, M. G. B., Valeri, L., François, O., Cadiou, S., Siroux, V., Lepeule, J. and Slama, R. (2020). Challenges raised by mediation analysis in a high-dimension setting, Environmental Health Perspectives 128: 55001.

Borovykh, A. (2019). Analytic expressions for the output evolution of a deep neural network. arXiv: 1912.08526.

Bottou, L., Peters, J., Quiñonero-Candela, J., Charles, D. X., Chickering, M., Portugaly, E., Ray, D., Simard, P. and Snelson, E. (2013). Counterfactual reasoning and learning systems: The example of computational advertising, Journal of Machine Learning Research 14: 3207–3260.

Bourlard, H. A. and Morgan, N. (2012). Connectionist speech recognition: A hybrid approach, Springer Science & Business Media 247: 27–58.

Britz, D. (2015). Understanding convolutional neural networks for NLP. http://www. wildml. com/2015/11/understanding-convolutional-neuralnetworks-for-nlp/(visited on 11/07/2015)

Brock, A., Donahue, J. and Simonyan, K. (2018). Large scale GAN training for high fidelity natural image synthesis. arXiv:1809.11096.

Bruna, J., Zaremba, W., Szlam, A. and LeCun, Y. (2013). Spectral networks and locally connected networks on graphs. arXiv:1312.6203.

Buesing, L., Weber, T., Zwols, Y., Racaniere, S., Guez, A., Lespiau, J-B. and Heess N. (2018). Woulda, coulda, shoulda: Counterfactually-guided policy search. arXiv: 1811.06272.

Bühlmann, P., Peters, J. and Ernest, J. (2014). CAM: Causal additive models, high-dimensional order search and penalized regression, Annals of Statistics 42: 2526–2556.

Bun, M. J. G. and Harrison, T. D. (2019). OLS and IV estimation of regression models including endogenous interaction terms, Econometric Reviews 38: 814–827.

Burda, Y., Grosse, R. and Salakhutdinov, R. (2015). Importance weighted autoencoders. arXiv:1509. 00519.

Byrd, R. H., Nocedal, J. and Schnabel, R. B. (1994). Representations of quasi-Newton matrices and their use in limited memory methods, Mathematical Programming 63: 129–156.

Byrd, R. H., Nocedal, J. and Yuan, Y. (1987). Global convergence of a class of quasi-Newton methods on convex problems, SIAM Journal on Numerical Analysis 24: 1171–1190.

Cai, D., Chen, T., Wu, P. and Zhang, H. (2013). Introduction to mediation analysis with structural equation modeling, Shanghai Arch Psychiatry 25: 390–394.

Cai, R., Qiao, J., Zhang, K., Zhang, Z. and Hao, Z. (2019). Causal discovery with cascade nonlinear additive noise models. arXiv:1905.09442.

Cao, S., Lu, W. and Xu, Q. (2015). Grarep: Learning graph representations with global structural information, Proceedings of the 24th ACM International on Conference on Information and Knowledge Management: 891–900.

Cao, S., Lu, W., Xu, Q. (2016). Deep neural networks for learning graph representations, AAAI'16: Proceedings of the Thirtieth AAAI Conference on Artificial Intelligence: 1145–1152.

Carrara, F., Amato, G., Brombin, L., Falchi, F. and Gennaro, C. (2020). Combining GANs and autoencoders for efficient anomaly detection. arXiv:2011.08102.

Cashin, A. G., McAuley, J. H., Lamb, S. E., Hopewell, S., Kamper, S. J., Williams, C. M., Henschke, N. and Lee, H. (2020). Development of a guideline for reporting mediation analyses (AGReMA), BMC Med Res Methodol 20: 19.

Chan, L. W. and Szeto, C. C. (1999). Training recurrent network with blockdiagonal approximated levenberg-marquardt algorithm. International Joint Conference on Neural Networks, IJCNN'99 3: 1521–1526.

Chang, B., Meng, L., Haber, E., Ruthotto, L., Begert, D. and Holtham, E. (2017). Reversible architectures for arbitrarily deep residual neural networks. arXiv:1709.03698.

Chang, B., Meng, L., Haber, E., Tung, F. and Begert, D. (2018). Multi-level residual networks from dynamical systems view. arXiv:1710.10348.

Chapfuwa, P., Assaad, S., Zeng, S., Pencina, M., Carin, L. and Henao, R. (2020). Survival analysis meets counterfactual inference. arXiv:2006.07756.

Chapfuwa, P., Tao, C., Li, C., Page, C., Goldstein, B., Carin, L. and Henao, R. (2018). Adversarial time-to-event modeling. arXiv:1804.03184.

Charpentier, A., Elie, R. and Remlinger, C. (2020). Reinforcement learning in economics and finance. arXiv:2003.10014.

Chen, B. and Pearl, J. (2015). Graphical tools for linear structural equation modeling, UCLA Cognitive Systems Laboratory, Technical Report (R-432).

Chen, R. T. Q., Rubanova, Y., Bettencourt, J. and Duvenaud, D. (2018). Neural ordinary differential equations. arXiv:1806.07366.

Chernozhukov, V., Chetverikov, D., Demirer, M., Duflo, E., Hansen, C. and Newey, W. (2017). Double/debiased/neyman machine learning of treatment effects, American Economic Review 5: 261–265.

Chickering, D. M., Heckerman, D. and Meek, C. (2004). Large-sample learning of Bayesian networks is NP-hard, Journal of Machine Learning Research 5: 1287–1330.

Ching, T., Zhu, X., Garmire, L. X. (2018). Cox-nnet: An artificial neural network method for prognosis prediction of high-throughput omics data, PLoS Comput Biol 14: e1006076.

Cho, Y. and Saul, L. (2009). Kernel methods for deep learning. Advances in Neural Information Processing Systems 22 (NIPS 2009).

Choi, E., Bahadori, M. T., Schuetz, A., Stewart, W. F. and Sun, J. (2016). Doctor AI: Predicting clinical events via recurrent neural networks, Proceedings of the 1st Machine Learning for Healthcare Conference 56: 301–308.

Choromanski, K., Likhosherstov, V., Dohan, D., Song, X., Gane, A., Sarlos, T., Hawkins, P., Davis, J., Mohiuddin, A., Kaiser, L., Belanger, D., Colwell, L. and Weller, A. (2021). Rethinking attention with performers, ICLR 2021.

Chrysos, G. G, Kossaifi, J. and Zafeiriou, S. (2020). RoCGAN: Robust conditional GAN, Int J Comput Vis 128: 2665–2683.

Colombo, N., Silva, R., Kang, S. M. and Gretton, A. (2019). Counterfactual distribution regression for structured inference. arXiv:1908.07193.

Cui, P., Wang, X., Pei, J. and Zhu, W. (2018). A survey on network embedding, IEEE Transactions on Knowledge and Data Engineering 31: 833–852.

Dablander, F. (2020). An introduction to causal inference. PsyArXiv Preprint.

Dai, Q., Li, Q., Tang, J. and Wang, D. (2017). Adversarial network embedding. arXiv:1711.07838.

Dandekar, R. and Barbastathis, G. (2020). Quantifying the effect of quarantine control in Covid-19 infectious spread using machine learning. medRxiv preprint. DOI: https://doi.org/10.1101/2020.04.03.20052084

de Haan, P., Jayaraman, D. and Levine, S. (2019). Causal confusion in imitation learning. arXiv:1905.11979.

Defferrard, M., Bresson, X. and Vandergheynst, P. (2016). Convolutional neural networks on graphs with fast localized spectral filtering, Proceedings of NIPS: 3844–3852.

Delasalles, E., Ali, Ziat., Denoyer, L. and Gallinari, P. (2019). Spatio-temporal neural networks for space-time series forecasting and relations discovery, Knowledge and Information Systems 61(3): 1241–1267.

Devlin, J., Chang, M.-W., Lee, K. and Toutanova, K. Bert. (2018). Pre-training of deep bidirectional transformers for language understanding. arXiv:1810.04805.

Dey, R. and Salem, F. M. (2017). Gate-variants of gated recurrent unit (GRU) neural networks. arXiv:1701.05923.

Díaz, I. (2019). Statistical inference for data-adaptive doubly robust estimators with survival outcomes, Statistics in Medicine 38: 2735–2748.

Dickson, B. (2020). How do you measure trust in deep learning? https://bdtechtalks.com/2020/11/23/deep-learning-trust-metrics/

Ding, S. and Cook, R. D. (2014). Dimension folding PCA and PFC for matrix-valued predictors. Supplementary Material, Statistica Sinica 24: 463–492.

Dinga, R., Schmaal, L., Penninx, B. W. J. H., Veltman, D. J. and Marquand, A. F. (2020). Controlling for effects of confounding variables on machine learning predictions. bioRxiv preprint, DOI: https://doi.org/10.1101/2020.08.17.255034

DiPrete, T. A., Burik, C. A. P. and Koellinger, P. D. (2019). Genetic instrumental variable regression: Explaining socioeconomic and health outcomes in nonexperimental data, Proceedings of the National Academy of Sciences of the United States of America 115: E4970–E4979.

Donahue, J., Krähenbühl, P. and Darrell, T. (2017). Adversarial feature learning, Arxiv: 1605.09782.

D'Oro, P., Metelli, A. M., Tirinzoni, A., Papini, M. and Restelli, M. (2019). Gradient-aware model-based policy search. arXiv:1909.04115.

Downs, M., Chu, J., Yacoby, Y., Doshi-Velez, F. and WeiWei, P. (2020). CRUDS: Counterfactual recourse using disentangled subspaces, ICML Workshop on Human Interpretability in Machine Learning 1–23.

Du, X., Sun, L., Duivesteijn, W., Nikolaev, A. and Pechenizkiy, M. (2019). Adversarial balancing-based representation learning for causal effect inference with observational data. arXiv:1904.13335.

Dupont, E., Doucet, A., and The, Y. W. (2019). Augmented neural odes. arXiv:1904.01681.

Duvenaud, D. (2020). The kernel cookbook: Advice on covariance functions. https://www.cs.toronto.edu/~duvenaud/cookbook/

Eaton, D. and Murphy, K. (2007). Exact Bayesian structure learning from uncertain interventions, Artificial Intelligence and Statistics 2:107–114.

Eichler, M. (2013). Causal inference with multiple time series: Principles and problems, Philosophical Transactions of the Royal Society A 371: 20110613. http://dx.doi.org/10.1098/rsta.2011.0613

Emami, H., Dong, M., Nejad-Davarani, S. P. and Glide-Hurst, C. K. (2018). Generating synthetic CTs from magnetic resonance images using generative adversarial networks, Medical Physics 10.1002/mp.13047. DOI: 10.1002/mp.13047

Engle, R. F. and Granger, C. W. J. (1987). Cointegration and error correction: Representation, estimation, and testing, Econometrica 55: 251–276.

Faez, F., Ommi, Y., Baghshah, M. S. and Rabiee, H. R. (2020). Deep graph generators: A survey. arXiv:2012.15544.

Farrell, M. H., Liang, T. and Misra, S. (2018). Deep neural networks for estimation and inference. arXiv:1809.09953.

Fathoby, R. and Goela, N. (2018). Discrete wasserstein generative adversarial networks (DWGAN). ICLR 2018 Conference.

Feinberg, V., Wan, A., Stoica, I., Jordan, M. I., Gonzalez, J. and Levine, S. (2018). Model-based value estimation for efficient model-free reinforcement learning. arXiv:1803.00101.

François-Lavet, V., Henderson, P., Islam, R., Bellemare, M. G. and Pineau, J. (2018). An introduction to deep reinforcement learning, Foundations and Trends in Machine Learning 11: 3–4. DOI: 10.1561/2200000071.

Fujita, A., Sato, J. R., Garay, M., Yamaguchi, R., Miyano, S., Sogayar, M. and Ferreira, C. E. (2007). Modeling gene expression regulatory networks with the sparse vector autoregressive model, BMC Systems Biology 1: 39.

Galbusera, F., Niemeyer, F., Seyfried, M., Bassani, T., Casaroli, G., Kienle, A., Wilke, H. J. (2018). Exploring the potential of generative adversarial networks for synthesizing radiological images of the spine to be used in In Silico Trials, Frontiers in Bioengineering and Biotechnology 6: 53.

Gallicchio, C. and Micheli, A. (2010). Graph echo state networks in IJCNN, IEEE, 2010: 1–8.

Gasulla, D. G. (2015). Link prediction in large directed graphs. Ph.D. thesis.

Ge, Q., Hu, Z., Zhang, K., Li, S., Lin, W., Jin, L. and Xiong, M. M. (2020). Recurrent neural reinforcement learning for counterfactual evaluation of public health interventions on the spread of Covid-19 in the world. DOI: https://doi.org/10.1101/2020.07.08.20149146

Ge, Q., Huang, X., Fang, S., Guo, S., Liu, Y., Lin, W. and Xiong, M. M. (2020). Conditional generative adversarial networks for individualized treatment effect estimation and treatment selection, Frontiers in Genetics 11: 585804.

Geoffrion, A. (1971). Duality in nonlinear programming: A simplified applications-oriented development, Siam Review 13: 1–37.

Gerstung, M., Papaemmanuil, E., Martincorena, I., Bullinger, L., Gaidzik, V. I., Paschka, P., Heuser, M., Thol, F., Bolli, N., Ganly, P., Ganser, A., McDermott, U., Döhner, K., Schlenk, R. F., Döhner, H. and Campbell, P. J. (2017). Precision oncology for acute myeloid leukemia using a knowledge bank approach, Nature Genetics 49: 332–340.

Ghosh, A., Kulharia, V., Namboodiri, V. P., Torr, P. H. and Dokania, P. K. (2018). Multi-agent diverse generative adversarial networks, Proceedings of the IEEE Conference on Computer Vision and Pattern Recognition 1: 8513–8521.

Gilmer, J., Schoenholz, S. S., Riley, P. F., Vinyals, O. and Dahl, G. E. (2017). Neural message passing for quantum chemistry. arXiv:1704.01212.

Girin, L., Leglaive, S., Bie, X., Diard, J., Hueber, T. and Alameda-Pineda, X. (2020). Dynamical variational autoencoders: A comprehensive review. arXiv:2008.12595.

Giunchiglia, E., Nemchenko, A. and van der Schaar, M. (2018). RNN-SURV: A deep recurrent model for survival analysis. In: Kůrkovám V., Manolopoulosm Y., Hammerm B., Iliadism L. and Maglogiannis, I. (eds.), Artificial Neural Networks and Machine Learning – ICANN 2018. ICANN 2018. Lecture Notes in Computer Science, vol 11141. Springer, Cham.

Glymour, C., Zhang, K. and Spirtes, P. (2019). Review of causal discovery methods based on graphical models, Frontiers in Genetics 10.

Goodfellow, I. J., Pouget-Abadie, J., Mirza, M., Xu, B. and Warde-Farley, D. (2014). Generative Adversarial Networks, Advances in Neural Information Processing Systems 3: 2672–2680.

Gori, M., Monfardini, G. and Scarselli, F. (2005). A new model for learning in graph domains, Proceeding of IJCNN 2: 729–734.

Goudet, O., Kalainathan, D., Caillou, P., Guyon, I., Lopez-Paz, D., Guyon, I., Sebag, M., Tritas, A. and Tubaro, P. (2017). Learning functional causal models with generative neural networks. arXiv:1709.05321.

Goudet, O., Kalainathan, D., Caillou, P., Guyon, I., Lopez-Paz, D. and Sebag, M. (2018). Causal generative neural networks. arXiv:1711.08936.

Granger, C. W. J. (1969). Investigating causal relations by econometric models and cross-spectral graph theory, Applied and Computational Harmonic Analysis 30: 129–150.

Gretton, A., Borgwardt, K. M., Rasch, M. J., Schölkopf, B. and Smola, A. J. (2012). A kernel two-sample test, Journal of Machine Learning Research 13:723–773.

Gretton, A., Bousquet, O., Smola, A., Schölkopf, B. (2005). Measuring statistical dependence with HilbertSchmidt norms, Algorithmic Learning Theory: 16th International Conference (ALT 2005), 63–78.

Gretton, A. (2020). Notes on mean embeddings and covariance Operators. https://www.gatsby.ucl.ac.uk/~gretton/coursefiles/lecture5_covarianceOperator.pdf

Grigorescu, S., Trasnea, B., Cocias, T. and Macesanu, G. (2019). A survey of deep learning techniques for autonomous driving. arXiv:1910.07738.

Grinfeld, J., Nangalia, J., Baxter, E. J., Wedge, D. C., Angelopoulos, N., Cantrill, R., Godfrey, A. L., Papaemmanuil, E., Gundem, G., MacLean, C., Cook, J., O'Neil, L., O'Meara, S., Teague J. W., Butler, A. P, Massie, C. E, Williams, N., Nice, F. L, Andersen, C. L., Hasselbalch, H. C., Guglielmelli, P., McMullin, M. F., Vannucchi, A. M., Harrison, C. N., Gerstung, M., Green, A. R. and Campbell, P. J. (2018). Classification and personalized prognosis in myeloproliferative neoplasms, New England Journal of Medicine 379: 1416–1430.

Groha, S., Schmon, S. M. and Gusev, A. (2020). A general framework for survival analysis and multi-state modelling. arXiv:2006.04893.

Grondman, I., Busoniu, L., Lopes, G. A. D. and Babuska, R. (2012). A survey of actor-critic reinforcement learning: Standard and natural policy gradients, Systems Man and Cybernetics Part C: Applications and Reviews IEEE Transactions on 42: 1291–1307.

Grover, A. and Leskovec, J. (2016). node2vec: Scalable feature learning for networks, Knowledge Discovery and Data Mining.

Grover, A., Dhar, M. and Ermon, S. (2018). Flow-GAN: Combining maximum likelihood and adversarial learning in generative models. arXiv:1705.08868.

Gulrajani, I., Ahmed, F., Arjovsky, M., Dumoulin, V. and Courville, A. (2017). Improved training of wasserstein gans. arXiv:1704.00028.

Gunzler, D., Chen, T., Wu, P. and Zhang, H. (2013). Introduction to mediation analysis with structural equation modeling. Shanghai Archives of Psychiatry 25: 390–394.

Guo, R., Cheng, L., Li, J. P., Hahn, R. and Liu, H. (2020). A survey of learning causality with data: Problems and methods. arXiv:1809.09337.

Haber, E. and Ruthotto, L. (2017). Stable architectures for deep neural networks. arXiv:1705.03341.

Haber, E., Ruthotto, L. and Holtham, E. (2017). Learning across scales—A multiscale method for convolution neural networks. arXiv:1703.02009.

Hagmayer, Y., Sloman, S., Lagnado, D. and Waldmann, M. R. (2007). Causal reasoning through intervention. In Gopnik, A. and Schulz, L. (eds.), Causal Learning: Psychology, Philosophy, and Computation (pp.86–100). Oxford University Press, Oxford, England. https://doi.org/10.1093/acprof:oso/9780195176803.003.0007

Halloran, M. E. and Hudgens, M. G. (2012). Causal inference for vaccine effects on infectiousness, The International Journal of Biostatistics 8: 10.2202/1557-4679.1354/j/ijb.2012.8.issue-2/1557-4679.1354/1557-4679.1354.xml.

Hamilton, W. L., Ying, R. and Leskovec, J. (2018). Representation learning on graphs: Methods and applications. arXiv:1709.05584.

Hammond, D. K., Vandergheynst, P. and Gribonval, R. (2011). Wavelets on graphs via spectral graph theory, Applied and Computational Harmonic Analysis 30: 129–150.

Hartford, J., Lewis, G., Leyton-Brown, K. and Taddy, M. (2016). Counterfactual prediction with deep instrumental variables networks. arXiv:1612.09596.

Hartford, J., Lewis, G., Leyton-Brown, K. and Taddy, M. (2017). Deep IV: A flexible approach for counterfactual prediction, Proceedings of the 34th International Conference on Machine Learning, PMLR 70: 1414–1423.

Hassanpour, N. and Greiner, R. (2021). Variational autoencoder architectures that excel at causal inference, International Conference on Learning Representations, 2021.

Häyrinen, K., Saranto, K. and Nykänen, P. (2008). Definition, structure, content, use and impacts of electronic health records: A review of the research literature, International Journal of Medical Informatics 77: 291–304.

He, K., Zhang, X., Ren, S. and Sun, J. (2015). Deep residual learning for image recognition. arXiv:1512.03385.

He, T. and Droppo, J. (2016). Exploiting LSTM structure in deep neural networks for speech recognition, Proceedings of the IEEE International Conference on Acoustics, Speech and Signal Processing: 5445–5449.

He, K., Zhang, X., Ren, S. and Sun, J. (2016). Identity mappings in deep residual networks. arXiv:1603.05027.

Henaff, M., Bruna, J. and LeCun, Y. (2015). Deep convolutional networks on graph-structured data. arXiv:1506.05163.

Hernán, M. A. and Robins, J. M. (2020). Causal Inference: What If. Chapman & Hill/CRC, Boca Raton, FL.

Herrmann, V. (2021). Wasserstein GAN and the Kantorovich-Rubinstein duality. https://vincentherrmann.github.io/blog/wasserstein/

Hochreiter, S. and Schmidhuber, J. (1997). Long short-term memory, Neural Computation 9: 1735–1780.

Hoffman, M. D., Blei, D. M, Wang, C. and Paisley, J. (2013). Stochastic variational inference, Journal of Machine Learning Research 14: 1303–1347.

Hornik, K. (1991). Approximation capabilities of muitilayer feedforward networks. Neural Networks 4: 251–257.

Hu, Z., Jiao, R., Wang, P., Zhu, Y., Zhao, J., De Jager, P., Bennett, D. A., Jin, L. and Xiong, M. (2020). Shared causal paths underlying Alzheimer's dementia and Type 2 diabetes, Scientific Reports 10: 4107.

Huang, Z., Johnson, T. S., Han, Z., Helm, B., Cao, S., Zhang, C., Salama, P., Rizkalla, M., Yu, C.Y., Cheng, J., Xiang, S., Zhan, X., Zhang, J. and Huang, K. (2020). Deep learning-based cancer survival prognosis from RNA-seq data: Approaches and evaluations, BMC Medical Genomics 13(Suppl 5): 41.

Ioffe, S. and Szegedy, C. (2015). Batch normalization: Accelerating deep network training by reducing internal covariate shift, In International Conference on Machine Learning 448–456.

Jaiswal, A., AbdAlmageed, W., Wu, Y. and Natarajan, P. (2017). Bidirectional conditional generative adversarial networks. arXiv:1711.07461.

Janzing, D, Mooij, J., Zhang, K., Lemeire, J., Zscheischler, J., Daniusis, P., Steudel, B. and Schölkopf, B. (2012). Information-geometric approach to inferring causal directions, Artificial Intelligence 182: 1–31.

Jaquesn, N., Gu, S., Turner, R. and Eck, D. (2017). Tuning recurrent neural networks with reinforcement learning, International Conference on Learning Representations (ICLR) workshop, Toulon, France.

Jitkrittum, W., Szabo, Z., Chwialkowski, K. and Gretton A. (2016). Interpretable distribution features with maximum testing power, NeurIPS Proceedings 2016.

Johansson, F. D., Shalit, U. and Sontag, D. (2016). Learning representations for counterfactual inference. arXiv:1605.03661.

Johansson, F. D., Shalit, U. and Sontag, D. (2016). Learning representations for counterfactual inference, ICML'16: Proceedings of the 33rd International Conference on Machine Learning 48: 3020–3029.

Kalainathan, D., Goudet, O., Guyon, I., Lopez-Paz, D. and Sebag, M. (2018). SAM: Structural agnostic model, causal discovery and penalized adversarial learning. arXiv:1803.04929.

Kallus, N., Mao, X. and Udell, M. (2018). Causal inference with noisy and missing covariates via matrix factorization, Advances in Neural Information Processing Systems: 6921–6932.

Karras, T., Laine, S. and Aila, T. (2019). A style-based generator architecture for generative adversarial networks. arXiv:1812.04948.

Katzman, J. L., Shaham, U., Cloninger, A., Bates, J., Jiang, T. and Kluger, Y. (2018). DeepSurv: Personalized treatment recommender system using a Cox proportional hazards deep neural network, BMC Medical Research Methodology 18: 24.

Keren, G. and Schuller, B. (2016). Convolutional RNN: An enhanced model for extracting features from sequential data. arXiv:1602.05875.

Khanna, S. and Tan, V. Y. F. (2019). Economy statistical recurrent units for inferring nonlinear Granger causality. arXiv:1911.09879.

Khemakhem, I., Kingma, D. P. and Hyvärinen, A. (2019). Variational autoencoders and nonlinear ICA: A unifying framework. CoRR, abs/1907.04809.

Kim, H., Shin, S., Jang, J., Song, K., Joo, W., Kang, W. and Moon IC. (2020). Counterfactual fairness with disentangled causal effect variational autoencoder. arXiv:2011.11878.

Kim, S., Kim, K., Choe, J., Lee, I. and Kang, J. (2020). Improved survival analysis by learning shared genomic information from pan-cancer data, Bioinformatics 36(Supplement 1): i389–i398.

Kingma, D. P. and Ba, J. (2014). Adam: A method for stochastic optimization. arXiv:1412.6980.

Kingma, D. P. and Welling, M. (2013). Auto-encoding variational Bayes. arXiv:1312.6114.

Kingma, D. P. and Welling, M. (2019). An introduction to variational autoencoders. arXiv:1906.02691.

Kipf, T. N. and Welling, M. (2016). Semi-supervised classification with graph convolutional networks. arXiv:1609.02907.

Kipf T. N. and Welling M. (2017). Semi-supervised classification with graph convolutional networks. arXiv:1609.02907.

Knoblauch J., Jewson, J. and Damoulas, T. (2019). Generalized variational inference. arXiv:1904.02063.

Kocaoglu, M., Snyder, C., Dimakis, A. G. and Vishwanath, S. (2017). CausalGAN: Learning causal implicit generative models with adversarial training. arXiv:1709.02023.

Kochmann, D. M. (2017). Computational solid mechanics (151-0519-00L). https://ethz.ch/content/dam/ethz/special-interest/mavt/mechanical-systems/mm-dam/documents/Notes/TensorNotes1.pdf

Koivisto, M. (2012). Advances in exact Bayesian structure discovery in Bayesian networks. arXiv:1206.6828.

Kong, J., Kim, J. and Bae, J. (2020). HiFi-GAN: Generative adversarial networks for efficient and high fidelity speech synthesis. arXiv:2010.05646.

Kool, W., van Hoof, H. and Welling M. (2019). Attention, learn to solve routing problems! arXiv:1803.08475.

Krishnan, R. G., Shalit, U. and Sontag D. (2017). Structured inference networks for nonlinear state space models, Association for the Advancement of Artificial Intelligence 2101–2109.

Künzel, S. R., Sekhon, J. S., Bickel, P. J. and Yu, B. (2019). Metalearners for estimating heterogeneous treatment effects using machine learning, Proceeding of the National Academy of Sciences 116: 4156–4165.

Kuang, Z., Sala, F. and Sohoni, N. (2020). Ivy: Instrumental variable synthesis for causal inference, Proceedings of the Twenty Third International Conference on Artificial Intelligence and Statistics, PMLR 108:398–410.

Kuremoto, T., Hirata, T., Obayashi, M., Mabu, S. and Kobayashi, K. (2021). Training deep neural networks with reinforcement learning for time series forecasting. https://cdn.intechopen.com/pdfs/66346.pdf

Kwon, Y. D., Choo, J., Kim, B., Yoon, I., Gwon, Y. and Min, S. (2020). POMO: Policy optimization with multiple optima for reinforcement learning. arXiv:2010.16011.

Lachapelle, S., Brouillard, P., Deleu, T. and Lacoste-Julien, S. (2019). Gradient-based neural DAG learning. arXiv:1906.02226.

Lan, Q., Tosatto, S., Farrahi, H., and Mahmood, A. R. (2021). Model-free Policy Learning with Reward Gradients. arXiv:2103.05147.

Lattimore, F. and Ong, C. S. (2018). A Primer on causal analysis. arXiv:1806.01488.

Lee, C., Mastronarde, N. and van der Schaar, M. (2018). Estimation of individual treatment effect in latent confounder models via adversarial learning. arXiv:1811.08943.

Lee, C., Zame, W. R., Yoon, J. and van der Schaar, M. (2018). Deephit: A deep learning approach to survival analysis with competing risks, Thirty-Second AAAI Conference on Artificial Intelligence: 2314–2321.

Lee, J., Bahri, Y., Novak, R., Schoenholz, S., Pennington, J. and Sohl-dickstein, J. (2018). Deep neural networks as Gaussian processes, arXiv:1711.00165.

Lee, J., Xiao, L., Schoenholz, S. S., Bahri, Y., Novak, R., Sohl-Dickstein, J. and Pennington, J. (2019). Wide neural networks of any depth evolve as linear models under gradient descent. arXiv:1902.06720.

Leeb, F., Annadani, Y., Bauer, S. and Schölkopf, B. (2020). Structural autoencoders improve representations for generation and transfer. arXiv:2006.07796.

Li, C., Welling, M., Zhu, J. and Zhang, B. (2018). Graphical generative adversarial networks. arXiv:1804.03429.

Li, F., Zhu, Z., Zhang, X., Cheng, J. and Zhao, Y. (2019). Diffusion induced graph representation learning, Neurocomputing 360: 220–229.

Li, G., Chen, J., Assefa, S. A. and Liu, Y. (2021). Time series counterfactual inference with hidden confounders, ICLR 2021 Conference. https://openreview.net/forum?id=JVs1OrQgR3A

Li, H., Xiao, Q. and Tian, J. (2020). Supervised whole DAG causal discovery. arXiv:2006.04697.

Li, J., Luo, X. and Qiao, M. (2019). On generalization error bounds of noisy gradient methods for non-convex earning. arXiv:1902.00621.

Li, L., Wu, C. H., Ning, J., Huang, X., TinaShih, Y. C. and Shen, Y. (2018). Semiparametric estimation of longitudinal medical cost trajectory, Journal of the American Statistical Association 113: 582–592.

Li, M., Lin, J., Ding, Y., Liu, Z., Zhu, J-Y. and Han, S. (2020). GAN compression: Efficient architectures for interactive conditional GANs. arXiv:2003.08936.

Li, Q. and Hao, S. (2018). An optimal control approach to deep learning and applications to discrete-weight neural networks. arXiv:1803.01299.

Li, X., Zhang, H. and Zhang, R. (2020). Embedding graph auto-encoder with joint clustering via adjacency sharing. arXiv:2002.08643.

Li, Y., Tarlow, D., Brockschmidt, M. and Zemel, R. (2015). Gated graph sequence neural networks. arXiv:1511.05493.

LIBRO (2020). Cointegration and the error-correction mechanism (ECM): a general approach. Time Series Econometrics. 356-363. http://www.ecostat.unical.it/Algieri/Didattica/Financial%20Markets/Tutorials/LIBRO%20Asterious_Applied-Econometrics-387-392.pdf

Lim, B. (2018). Forecasting treatment responses over time using recurrent marginal structural networks. In Bengio, S., Wallach, H., Larochelle, H., Grauman, K., Cesa-Bianchi, N. and Garnett, R. (eds.), Advances in Neural Information Processing Systems 31: 7483–7493. Curran Associates.

Lim, J., Ryu, S., Kim, J. W. and Kim, W. Y. (2018). Molecular generative model based on conditional variational autoencoder for de novo molecular design, Journal of Cheminformatics 10: 31.

Lin, E., Mukherjee, S. and Kannan, S. (2020). A deep adversarial variational autoencoder model for dimensionality reduction in single-cell RNA sequencing analysis, BMC Bioinformatics 21: 64.

Lin, T. W. (1991). Solving optimization problems using ordinary differential equations, Journal of the Chinese Institute of Engineers 14: 391–395.

Lin, W., Khan, M. E. and Schmidt, M. (2019). Fast and simple natural-gradient variational inference with mixture of exponential-family approximations. arXiv:1906.02914.

Liu, B., Tan, C., Li, S., He, J. and Wang, H. (2020). A data augmentation method based on generative adversarial networks for grape leaf disease identification, IEEE Access 8: 102188–102198.

Liu, C. (2014). Optimal algorithms and the BFGS updating techniques for solving unconstrained nonlinear minimization problems, Journal of Applied Mathematics 2014, Article ID 324181, 14 pages. https://doi.org/10.1155/2014/324181

Liu, G.-H. and Theodorou, E. A. (2019). Deep learning theory review: An optimal control and dynamical systems perspective. arXiv:1908.10920.

Liu, H. T. (1991). Qualitative theory of differential equations. China Science and Technology Press Pub: Beijing.

Liu, Q., Allamanis, M., Brockschmidt, M. and Gaunt, A. L. (2018). Constrained graph variational autoencoders for molecule design. arXiv:1805.09076.

Locatello, F., Bauer, S., Lucic, M., Rätsch, G., Gelly, S., Schölkopf, B. and Bachem, O. (2018). Challenging common assumptions in the unsupervised learning of disentangled representations. arXiv:1811.12359.

Loh, P. L. and Buhlmann, P. (2014). High-dimensional learning of linear causal networks via inverse covariance estimation, Journal of Machine Learning Research 15: 3065–3105.

Long, Z., Lu, Y., Ma, X. and Dong, B. (2017). Pde-net: Learning pdes from data. arXiv preprint arXiv: 1710.09668.

Looveren, A. V., Klaise, J., Vacanti, G. and Cobb, O. (2021). Conditional generative models for counterfactual explanations. arXiv:2101.10123.

Lopez, R., Gayoso, A. and Yosef, N. (2020). Enhancing scientific discoveries in molecular biology with deep generative models, Molecular Systems Biology 16: e9198.

Lopez-Paz, D. and Oquab, M. (2017). Revisiting classifier two-sample tests, International Conference on Learning Representations (ICLR) 2017: 1–15.

Louizos, C., Shalit, U., Mooij, J., Sontag, D., Zemel, R. and Welling, M. (2017). Causal effect inference with deep latent-variable models. arXiv:1705.08821.

Lousdal, M. L. (2018). An introduction to instrumental variable assumptions, validation and estimation, Lousdal Emerg Themes Epidemiol 15: 1.

Lu, B., Cai, D. and Tong, X. (2018). Testing causal effects in observational survival data using propensity score matching design, Statistics in Medicine 37: 1846–1858.

Lu, C. (2021). Causal reinforcement learning in healthcare and medicine. https://causallu.files.wordpress.com/2021/01/talkpekinguniv13jan2021.pdf

Lu, C., Huang, B., Wang, K., Hernández-Lobato, J. M., Zhang, K. and Schölkopf B. (2020). Sample-efficient reinforcement learning via counterfactual-based data augmentation. arXiv:2012.09092.

Lu, C., Schölkopf, B. and Hernández-Lobato, J. M. (2018). Deconfounding reinforcement learning in observational settings. arXiv:1812.10576.

Lu, M. Y., Shahn, Z., Sow, D., Doshi-Velez, F. and Lehman, L. H. (2020). Is deep reinforcement learning ready for practical applications in healthcare? A sensitivity analysis of Duel-DDQN for hemodynamic management in sepsis patients. arXiv:2005.04301.

Lu, Y. and Lu, J. (2020). A universal approximation theorem of deep neural networks for expressing distributions. arXiv:2004.08867.

Lu, Y., Zhong, A., Li, Q. and Dong, B. (2017). Beyond finite layer neural networks: Bridging deep architectures and numerical differential equations. arXiv preprint arXiv:1710.10121.

Luo, W. and Zhu, Y. (2017). Matching using sufficient dimension reduction for causal inference. arXiv:1702.00444.

Luque-Fernandez, M. A., Schomaker, M., Rachet, B. and Schnitzer, M. E. (2018). Targeted maximum likelihood estimation for a binary treatment: A tutorial, Statistics in Medicine 37: 2530–2546.

Ma, S. and Ji, C. (1998). A unified approach on fast training of feedforward and recurrent networks using em algorithm, IEEE Transactions on Signal Processing 46: 2270–2274.

Madumal, P., Miller, T., Sonenberg, L. and Vetere, F. (2019). Explainable reinforcement learning through a causal lens. arXiv:1905.10958.

Makhzani, A., Shlens, J., Jaitly, N., Goodfellow, I. and Frey, B. (2016). Adversarial autoencoders, International Conference on Learning Representations 2016.

Makhzani, A., Shlens, J., Jaitly, N., Goodfellow, I. and Frey, B. (2015). Adversarial autoencoders. arXiv:151105644 2015.

Malik, A. A. (2020). Robots and COVID-19: Challenges in integrating robots for collaborative automation. arXiv:2006.15975.

Mallinar, N. and Rosset, C. (2018). Deep canonically correlated LSTMs. arXiv:1801.05407.

Manzour, H., Küçükyavuz, S. and Shojaie, A. (2019). Integer programming for learning directed acyclic graphs from continuous data. arXiv:1904.10574.

Manzour, H., Küçükyavuz, S., Wu, H. and Shojaie, A. (2020). Integer programming for learning directed acyclic graphs from continuous data, INFORMS Journal on Optimization Published online in Articles in Advance 03 Nov 2020. https://doi.org/10.1287/ijoo.2019.0040

Marx, A. and Vreeken, J. (2017). Telling cause from effect using MDL-based local and global regression, In 2017 IEEE International Conference on Data Mining (ICDM) 307–316.

Matthews, A. G. d. G., Hron, J., Rowland, M., Turner, R. E. and Ghahramani, Z. (2018). Gaussian process behaviour in wide deep neural networks, arXiv:1804.11271.

Mattia, F. D., Galeone, P., Simoni, M. D. and Ghelfi, E. (2019). A survey on GANs for anomaly detection. arXiv:1906.11632.

Mazyavkina, N., Sviridov, S., Ivanov, S. and Burnaev, E. (2020). Reinforcement learning for combinatorial optimization: A survey. arXiv:2003.03600.

Mazzarisi, P., Zaoli, S., Campajola, C. and Lillo, F. (2020). Tail Granger causalities and where to find them: Extreme risk spillovers vs. spurious linkages. arXiv:2005.01160.

Mehrotra, A. and Dukkipati, A. (2017). Generative adversarial residual pairwise networks for one shot learning. arXiv:1703.08033.

Miao, W., Geng, Z. and Tchetgen Tchetgen, E. J. (2018). Identifying causal effects with proxy variables of an unmeasured confounder, Biometrika 105: 987–993.

Micheli, A. (2009). Neural network for graphs: A contextual constructive approach, IEEE Transactions on Neural Networks 20: 498–511.

Miersemann, E. (2012). Partial differential equations, Lecture Notes. https://www.math.uni-leipzig.de/~miersemann/pdebook.pdf

Mikolov, T., Deoras, A., Kombrink, S., Burget, L. and Cernock, J. (2011). Empirical evaluation and combination of advanced language modeling techniques, Twelfth Annual Conference of the International Speech Communication Association: 605–608.

Mirza, M. and Osindero, S. (2014). Conditional generative adversarial nets. arXiv 1411.1784.

Miyato, T., Kataoka, T., Koyama, M. and Yoshida, Y. (2018). Spectral normalization for generative adversarial networks. CoRR, abs/1802.05957.

Mnih, V., Kavukcuoglu, K., Silver, D., Graves, A., Antonoglou, I., Wierstra, D. and Riedmiller, M. (2013). Playing Atari with Deep Reinforcement Learning. arXiv:1312.5602.

Mnih, V., Kavukcuoglu, K., Silver, D. et al. (2015). Human-level control through deep reinforcement learning. Nature 518: 529–533.

Mohamed, S. and Lakshminarayanan, B. (2017). Learning in implicit generative models. arXiv:1610.03483.

Mooij, J. J. P., Janzing, D., Zscheischler, J. and Schölkopf, B. (2016). Distinguishing cause from effect using observational data: Methods and benchmarks, Journal of Machine Learning Research 17: 1–102.

Morin, F. and Bengio, Y. (2005). Hierarchical probabilistic neural network language model, Proceedings of the Tenth International Workshop on Artificial Intelligence and Statistics, PMLR R5: 246–252.

Nasejje, J. B., Mwambi, H., Dheda, K. and Lesosky, M. (2017). A comparison of the conditional inference survival forest model to random survival forests based on a simulation study as well as on two applications with time-to-event data, BMC Medical Research Methodology 17: 115.

Nauata, N., Chang, K. H., Cheng, C. Y., Mori, G. and Furukawa, Y. (2020). House-GAN: Relational generative adversarial networks for graph-constrained house layout generation. arXiv:2003.06988.

Nemirovsky, D., Thiebaut, N., Xu, Y. and Gupta, A. (2020). CounteRGAN: Generating realistic counterfactuals with residual generative adversarial nets. arXiv:2009.05199.

Neyman, J. (1923/1990). Sur les applications de la theorie des probabilites aux experiences agricoles: Essai desprincipes (1923). English translations excerpts by D. Dabrowska and T. Speed, Statistical Science 4(5): 465–480.

Nguyen, C. V., Yingzhen, L., Bui, T. D. and Turner, R. E. (2017). Variational continual learning. arXiv:1710.10628.

Nguyen, D. Q., Nguyen, T. D. and Phung, D. (2020). Universal graph transformer self-attention networks. arXiv:1909.11855.

Nicora, G., Moretti, F., Sauta, E., Porta, M. D., Malcovati, L., Cazzola, M., Quaglini, S. and Bellazzi, R. (2020). A continuous-time Markov model approach for modeling myelodysplastic syndromes progression from cross-sectional data, Journal of Biomedical Informatics 104: 103398.

Nie, L., Ye, M., Liu, Q. and Nicolae, D. (2021). Varying coefficient neural network with functional targeted regularization for estimating continuous treatment effects, ICLR 2021.

Niepert, M., Ahmed, M. and Kutzkov, K. (2016). Learning convolutional neural networks for graphs. arXiv:1605.05273.

Novak, R., Xiao, L., Lee, J., Bahri, Y., Yang, G., Hron, J., Abolafia, D. A., Pennington, J. and Sohl-Dickstein, J. (2019). Bayesian deep convolutional networks with many channels are gaussian processes, arXiv:1810.05148.

Nowozin, S., Cseke, B. and Tomioka, R. (2016). f-GAN: Training generative neural samplers using variational divergence minimization. arXiv:1606.00709.

Nugroho, H., Susanty, M., Irawan, A., Koyimatu, M. and Yunita, A. (2020). Fully convolutional variational auto-encoder for feature extraction of fire detection system, Journal of Computer Science and Information 13: 9–15. DOI: http://dx:doi:org/10:21609/jiki:v13i1:761

Ogburn, E. L., Sofrygin, O., Diaz, I. and van der Laan, M. J. (2020). Causal inference for social network data. arXiv:1705.08527.

Ohnishi, S., Uchibe, E., Yamaguchi, Y., Nakanishi, K., Yasui, Y. and Ishii, S. (2019). Constrained deep Q-learning gradually approaching ordinary Q-learning, Frontiers in Neurorobotics 13: 103.

Oliva, J. B., P´oczos, B. and Schneider, J. (2017). The statistical recurrent unit, Proceedings of the 34th International Conference on Machine Learning 70: 2671–2680.

O'Shaughnessy, M., Canal, G., Connor, M., Davenport, M. and Rozell, C. (2020). Generative causal explanations of black-box classifiers. arXiv:2006.13913.

Ou, M., Cui, P., Pei, J., Zhang, Z. and Zhu, W. (2016). Asymmetric transitivity preserving graph embedding, Proceedings of the 22nd ACM SIGKDD International Conference on Knowledge discovery and Data Mining: 1105–1114.

Paidamoyo, C., Serge, A., Shuxi, Z., Michael, P., Lawrence, C. and Ricardo, H. (2020). Survival analysis meets counterfactual inference. arXiv:2006.07756.

Pamfil, R., Sriwattanaworachai, N., Desai, S., Pilgerstorfer, P., Beaumont, P., Georgatzis, K. and Aragam, B. (2020). DYNOTEARS: Structure learning from time-series data. arXiv:2002.00498.

Pan, S., Hu, R., Long, G., Jiang, J., Yao, L. and Zhang, C. (2018). Adversarially regularized graphn autoencoder for graph embedding. arXiv:1802.04407.

Pang, M., Platt, R. W., Schuster, T. and Abrahamowicz, M. (2021). Spline-based accelerated failure time model, Statistics in Medicine 40: 481–497.

Paolis Kaluza, M. C. D. P., Amizadeh, S. and Yu, R. (2018). A neural framework for learning DAG to DAG translation, NeurIPS'2018 Workshop, 2018.

Parikh, N. and Boyd, S. (2013). Proximal algorithms, Foundations and Trends in Optimization 1: 123–231.

Park, J., Lee, M., Chang, H. J., Lee, K. and Choi, J. Y. (2019). Symmetric graph convolutional autoencoder for unsupervised graph representation learning, Proceedings of the IEEE International Conference on Computer Vision, 6519–6528.

Paszke, A., Gross, S., Massa, F., Lerer, A., Bradbury, J., Chanan, G. and Killeen, T. (2019). Pytorch: An imperative style, high-performance deep learning library. arXiv:1912.01703.

Pearl, J. (2009). Causality: Models, Reasoning and Inference. Second Edition, Cambridge University Press, New York, NY.

Pearl, J. (1995). Causal diagrams for empirical research. Biometrika 82: 669–710.

Pearl, J. (2012). The do-calculus revisited. In Nando de Freitas and Kevin Murphy (eds.), Proceedings of the Twenty-Eighth Conference on Uncertainty in Artificial Intelligence. AUAI Press: Corvallis, OR, 4–11.

Pearl, J., Glymour, M. and Jewell, N. P. (2016). Causal Inference in Statistics: A Primer. Wiley, Hoboken, NJ.

Peruzzi, B., Al-Rfou, R. and Skiena, S. (2014). DeepWalk: Online learning of social representations. arXiv:1403.6652.

Peters, J., Janzing, D. and Schölkopf, B. (2011). Causal inference on discrete data using additive noise models, IEEE Transaction on Pattern Analysis and Machine Intelligence 33: 2350–2436.

Peters, J., Janzing, D. and Schölkopf, B. (2013). Causal inference on time series using restricted structural equation models, NIPS'13: Proceedings of the 26th International Conference on Neural Information Processing Systems 1: 154–162.

Peters, J., Janzing, D. and Schölkopf, B. (2017). Elements of Causal Inference – Foundations and Learning Algorithms. MIT Press, Cambridge, MA. http://www.math.ku.dk/~peters/

Petneházi, G. (2019). Recurrent neural networks for time series forecasting. arXiv:1901.00069.

Pfohl, S. R., Duan, T., Ding, D. Y. and Shah, N. H. (2019). Counterfactual reasoning for fair clinical risk prediction, Proceedings of the 4th Machine Learning for Healthcare Conference, PMLR 106: 325–358.

Pidhorskyi, S., Adjeroh, D. and Doretto G. (2020). Adversarial atent autoencoders. arXiv:2004.04467.

Pitis, S., Creager, E. and Garg, A. (2020). Counterfactual data augmentation using locally factored dynamics. arXiv:2007.02863.

Pontryagin, L. S. (1987). Mathematical Theory of Optimal Processes. CRC Press, Boca Raton, FL.

Popescu, O. I., Shadaydeh, M. and Denzler, J. (2021). Counterfactual generation with Knockoffs. arXiv: 2102.00951.

Puskorius, G. V. and Feldkamp, L. A. (1994). Neurocontrol of nonlinear dynamical systems with kalman filter trained recurrent networks, IEEE Transactions on Neural Networks 5: 279–297.

Qin, C., Wu, Y., Springenberg, J. T., Brock, A., Donahue, J., Lillicrap, T. and Kohli P. (2020). Training generative adversarial networks by solving ordinary differential equations. Larochelle, H., Ranzato, M., Hadsell, R., Balcan, M. F. and Lin, H. (eds.), Advances in Neural Information Processing Systems 33: 5599–5609 (NeurIPS 2020).

Quaglino, A., Gallieri, M., Masci, J. and Koutn'ık, J. (2019). Accelerating neural odes with spectral elements. arXiv:1906.07038.

Radford, A., Metz, L. and Chintala, A. (2015). Unsupervised representation learning with deep convolutional generative adversarial networks, International Conference on Learning Representations 2015.

Raghu, A., Komorowski, M., Ahmed, I., Celi, L., Szolovits, P. and Ghassemi, M. (2017). Deep reinforcement learning for sepsis treatment. arXiv:1711.09602.

Rahman, M. M., Rashid, S. M. H. and Hossain, M. M. (2018). Implementation of Q learning and deep Q network for controlling a self balancing robot model, Robotics and Biomimetics 5: 8.

Ramachandra, V. (2018). Deep learning for causal inference. arXiv:1803.00149.

Ranganath, R., Tran, D. and Blei, D. M. (2016). Hierarchical variational models. Proceedings of the 33rd International Conference on Machine Learning 2568–2577.

Rao, A. and Jelvis, T. (2021). Foundations of reinforcement learning with applications in finance. https://stanford.edu/~ashlearn/RLForFinanceBook/book.pdf

Rasmussen, C. E. and Williams C. (2006). Gaussian Processes for Machine Learning. MIT Press, Cambridge, MA (online version at http://www.gaussianprocess.org/gpml/).

Ren, K., Qin, J., Zheng, L., Yang, Z., Zhang, W., Qiu, L. and Yu, Y. (2018). Deep recurrent survival analysis. arXiv:1809.02403.

Rezende, D. J., Mohamed, S. and Wierstra, D. (2014). Stochastic backpropagation and approximate inference in deep generative models, International Conference on Machine Learning 1278–1286.

Rijnhart, J. J. M., Valente, M. J., MacKinnon, D. P., Twisk, J. W. R. and Heymans, M. W. (2020). The use of traditional and causal estimators for mediation models with a binary outcome and exposure-mediator interaction, Structural Equation Modeling. DOI: 10.1080/10705511.2020.1811709

Robins, J. (1986). A new approach to causal inference in mortality studies with a sustained exposure period—application to control of the healthy worker survivor effect, Math Model. 7: 1393–1512.

Roda, W. C., Varughese, M. B., Han, D. and Li, M. Y. (2020). Why is it difficult to accurately predict the COVID-19 epidemic?, Infectious Disease Modelling 5: 271–281.

Roeder, F., Meldolesi, E., Gerum, S., Valentini, V. and Rödel, C. (2020). Recent advances in (chemo-)radiation

therapy for rectal cancer: A comprehensive review, Radiation Oncology 15: 262.
Romac, C. and Béraud, V. (2019). Deep recurrent Q-Learning vs deep Q-Learning on a simple partially observable Markov decision process with Minecraft. arXiv:1903.04311.
Rosenbaum, P. R. and Rubin, D. B. (1983). The central role of the propensity score in observational studies for causal effects, Biometrika 70:41–55.
Rossi, A., Firmani, D., Matinata, A., Merialdo, P. and Barbosa, D. (2020). Knowledge graph embedding for link prediction: A comparative analysis. arXiv: 2002.00819.
Rothman, A. (2020). Causal inference in data science: Doubly robust estimation of g-methods. Hands-On Tutorials. https://towardsdatascience.com/doubly-robust-estimators-for-causal-inference-in-statistical-estimation-3c00847e9db
Rubanova, Y., Chen, R. T. Q. and Duvenaud, D. K. (2019). Latent ordinary differential equations for irregularly-sampled time series. In Wallach, H., Larochelle, H., Beygelzimer, A., d'Alch´e Buc, F., Fox, E. and Garnett, R. (eds.), Advances in Neural Information Processing Systems 32: 5320–5330. Curran Associates.
Rubenstein, P. K., Bongers, S., Schoelkopf, B. and Mooij, J. M. (2016). From deterministic ODEs to dynamic structural causal models. arXiv:1608.08028.
Rubin, D. B. (1974). Estimating causal effects of treatments in randomized and nonrandomized studies, Journal of Educational Psychology 66: 688–701.
Ruder, S. (2016). On word embeddings - Part 2: Approximating the Softmax. https://ruder.io/word-embeddings-softmax/
Rueda, O. R., Sammut, S. J., Seoane, J. A., Chin, S. F., Caswell-Jin, J. L., et al. (2019). Dynamics of breast-cancer relapse reveal late-recurring ER-positive genomic subgroups, Nature 567: 399–404.
Russin, J., Fernandez, R., Palangi, H., Rosen, E., Jojic, N., Smolensky, P. and Gao, J. (2021). Compositional processing emerges in neural networks solving math problems. arXiv:2105.08961.
Sagan, H. (1969). Introduction to the Calculus of Variation. Dover Publications Inc., New York, NY.
Saha-Chaudhuri, P. and Juwara, L. (2021). Survival analysis under the Cox proportional hazards model with pooled covariates, Statistics in Medicine 40: 998–1020.
Salehi, P., Chalechale, A. and Taghizadeh, M. (2020). Generative adversarial networks (GANs): An overview of theoretical model, evaluation metrics, and recent developments. arXiv:2005.13178.
Salehinejad, H., Sankar, S., Barfett, J., Colak, E. and Valaee, S. (2018). Recent advances in recurrent neural networks. arXiv:1801.01078.
Salha, G., Hennequin, R. and Vazirgiannis, M. (2020). Simple and effective graph autoencoders with one-hop linear models. arXiv:2001.07614.
Salha, G., Limnios, S., Hennequin, R., Tran, V. A. and Vazirgiannis, M. (2019). Gravity-inspired graph autoencoders for directed link prediction. arXiv:1905.09570.
Särkkä, S. (2012) Applied Stochastic Differential Equations. https://users.aalto.fi/~ssarkka/course_s2012/pdf/sde_course_booklet_2012.pdf
Sastry, S. (1999). Lyapunov stability theory. Nonlinear Systems. 182–234, Springer: Berlin, Germany.
Sauer, A. and Geiger, A. (2021). Counterfactual generative networks. arXiv:2101.06046.
Scarselli, F., Gori, M., Tsoi, A. C., Hagenbuchner, M. and Monfardini, G. (2009). The graph neural network model, IEEE Transactions on Neural Networks 20: 61–80.
Schulam, P. and Saria, S. (2017). Reliable decision support using counterfactual models. In Guyon, I., Luxburg, U. V., Bengio, S., Wallach, H., Fergus, R., Vishwanathan, S. and Garnett, R. (eds.), Advances in Neural Information Processing Systems 30: 1697–1708. Curran Associates.
Schulz, E., Speekenbrink, M. and Krause, A. (2018). A tutorial on Gaussian process regression: Modelling, exploring, and exploiting functions, Journal of Mathematical Psychology 85: 1–16.
Schwab, P., Linhardt, L. and Karlen, W. (2018). Perfect match: A simple method for learning representations for counterfactual inference with neural networks. arXiv:1810.00656.
Schwab, P., Linhardt, L., Bauer, S., Buhmann, J. M. and Karlen, W. (2019). Learning counterfactual representations for estimating individual dose-response curves. arXiv:1902.00981.
Seinfeld, J. H. and Lapidus, L. (1968). Aspects of forward dynamic programming algorithm, Industrial & Engineering Chemistry Process Design and Development 7: 475–478.
Seth, A. K., Barrett, A. B. and Barnett, L. (2015). Granger causality analysis in neuroscience and neuroimaging, Journal of Neuroscience 35: 3293–3297.
Sgouritsa, E., Janzing, D., Hennig, P. and Scholkopf, B. (2015). Inference of cause and effect with unsupervised inverse regression, Proceedings of the 18th International Conference on Artificial Intelligence and Statistics (AISTATS) 38: 847–855.
Shalit, U., Johansson, F. D. and Sontag, D. (2016). Estimating individual treatment effect: Generalization bounds and algorithms. arXiv:1606.03976.
Shamsolmoali, P., Zareapoor, M., Zhou, H., Wang, R. and Yang, J. (2020). Road segmentation for remote sensing images using adversarial spatial pyramid networks. arXiv:2008.04021.
Shen, W. and Liu, R. (2017). Learning residual images for face attribute manipulation. arXiv:1612.05363.
Shen, X., Liu, F., Dong, H., Lian, Q., Chen, Z. and Zhang, T. (2020). Disentangled generative causal representation learning. arXiv:2010.02637.

Shrestha, A. and Mahmood, A. (2019). Review of deep learning algorithms and architectures, IEEE Access 99: 1–1.

Sherstinsky, A. (2018). Fundamentals of recurrent neural network (RNN) and long short-term memory (LSTM) network. arXiv:1808.03314.

Shi, C., Blei, D. and Veitch, V. (2019). Adapting neural networks for the estimation of treatment effects, Advances in Neural Information Processing Systems 32: 2503–2513.

Shimizu, S., Hoyer, P., Hyvarinen, A. and Kerminen, A. (2006). A linear non-Gaussian acyclic model for causal discovery, Journal of Machine Learning Research 7: 2003–2030.

Shuai, B., Zuo, Z., Wang, G. and Wang, B. (2015). DAG-recurrent neural networks for scene labeling. arXiv:1509.00552.

Simmons, G. F. (2016). Differential Equations with Applications and Historical Notes. CRC, Press, Boca Raton, FL.

Sims, C. A. (1972). Money, income and causality, The American Economic Review 62: 540–552.

Sindhwani, V., Minh, H. Q. and Lozano, A. C. (2013). Scalable matrix-valued kernel learning for high-dimensional nonlinear multivariate regression and Granger causality, Proceedings of the Twenty-Ninth Conference on Uncertainty in Artificial Intelligence 2013, 586–595.

Smith, K. E. and Smith, A. O. (2020). Conditional GAN for time series generation. arXiv:2006.16477.

Soleimani, H., Subbaswamy, A. and Saria, S. (2017). Treatment-response models for counterfactual reasoning with continuous-time, continuous-valued interventions, Proceedings of the Conference on Uncertainty in Artificial Intelligence 2017: 266.

Sommerlade, L., Thiel, M., Platt, B., Plano, A., Riedel, G., Grebogi, C., Timmer, J. and Schelter, B. (2012). Inference of Granger causal time-dependent influences in noisy multivariate time series, Journal of Neuroscience Methods 203: 173–185.

Spirtes, P., Glymour, C. and Scheines, R. (2000). Causation, Prediction, and Search. MIT press, Cambridge, MA.

Subbiah, V. and Kurzrock, R. (2018). Challenging standard-of-care paradigms in the precision oncology era, Trends Cancer 4: 101–109.

Suganthan, P. N. (2018). Letter: On non-iterative learning algorithms with closed-form solution, Applied Soft Computing 70: 1078–1082.

Sutton, R. S. and Barto, A. G. (2018). Reinforcement Learning: An Introduction. Second Edition, MIT Press, Cambridge, MA.

Suttorp, M. M, Siegerink, B., Jager, K. J., Zoccali, C. and Dekker, F. W. (2015). Graphical presentation of confounding in directed acyclic graphs, Nephrology, Dialysis, Transplantation 30: 1418–1423.

Tan, Y. V. and Roy, J. (2019). Bayesian additive regression trees and the General BART model, Statistics in Medicine 38: 5048–5069.

Tang, D., Liang, D., Jebara, T., Ruozzi, N. (2019). Correlated variational auto-encoders. arXiv:1905.05335.

Tank, A., Covert, I., Foti, N., Shojaie, A. and Fox, E. (2018). Neural Granger causality for nonlinear time series. arXiv:1802.05842.

Teumer, A. (2018). Common methods for performing Mendelian randomization, Frontiers in Cardiovascular Medicine 5: 51.

Toderici, G., Vincent, D., Johnston, N., Hwang, S. J., Minnen, D., Shor, J. and Covell, M. (2017). Full resolution image compression with recurrent neural networks. Proceedings of the IEEE Conference on Computer Vision and Pattern Recognition. 5306–5314.

van Hasselt, H., Guez, A. and Silver, D. (2016). Deep reinforcement learning with double Q-learning, AAAI'16: Proceedings of the Thirtieth AAAI Conference on Artificial Intelligence: 2094–2100.

Vanderweele, T. J., Tchetgen Tchetgen, E. J., Cornelis, M. and Kraft, P. (2014). Methodological challenges in Mendelian randomization, Epidemiology 25: 427–435.

Varghese, N. V. and Mahmoud, Q. (2020). A survey of multi-task deep reinforcement learning. Electronics 9: 1363.

Vaswani, A., Shazeer, N., Parmar, N., Uszkoreit, J., Jones, L., Gomez, A. N., Kaiser, L. and Polosukhin, I. (2017). Attention is all you need, Advances in Neural Information Processing Systems 30: 5998–6008.

Veitch, V., Wang, Y. and Blei, D. M. (2019). Using embeddings to correct for unobserved confounding in networks. arXiv:1902.04114.

Venu, S. K. and Ravula, S. (2021). Evaluation of Deep Convolutional Generative Adversarial Networks for data augmentation of chest X-ray images, Future Internet 13(1): 8.

Venugopalan, S., Xu, H., Donahue, J., Rohrbach, M., Mooney, R. and Saenko, K. (2014). Translating videos to natural language using deep recurrent neural networks. arXiv preprint arXiv:1412.4729.

Vercheval, N. and Pizurica, A. (2021). Hierarchical variational autoencoder for visual counterfactuals. arXiv:2102.00854.

Vinter, R. (2020). Optimal control and Pontryagin's maximum principle. In: Baillieul, J. and Samad, T. (eds.), Encyclopedia of Systems and Control. Springer, London. https://doi.org/10.1007/978-1-4471-5102-9_200-2.

Viola, J., Chen, Y. Q. and Wang, J. (2021). FaultFace: Deep Convolutional Generative Adversarial Network (DCGAN) based Ball-Bearing failure detection method. Information Sciences 542: 195–211.

Vowels, M. J., Camgoz, N. C. and Bowden, R. (2020). Targeted VAE: Structured inference and targeted learning for causal parameter estimation. arXiv:2009.13472.

Wainwright, M. J. and Jordan, M. I. (2008). Graphical models, exponential families, and variational inference,

Foundations and Trends in Machine Learning 1: 1–305.

Wang, C., Pan, S., Long, G., Zhu, X. and Jiang, J. (2017). MGAE: Marginalized graph autoencoder for graph clustering, Proceedings of the 2017 ACM on Conference on Information and Knowledge Management: 889–898.

Wang, D., Cui, P. and Zhu, W. (2016). Structural deep network embedding, Proceedings of the 22nd ACM SIGKDD International Conference on Knowledge Discovery and Data Mining: 1225–1234.

Wang, G., Li, W. K. and Zhu, K. (2018). New HSIC-based tests for independence between two stationary multivariate time series. arXiv:1804.09866.

Wang, H., Wang, J., Wang, J., Zhao, M., Zhang, W., Zhang, F., Xie, X. and Guo, M. (2018). Graphgan: Graph representation learning with generative adversarial nets, Thirty-Second AAAI Conference on Artificial Intelligence 2018: 2508–2515.

Wang, L., Zhang, W., He, X. and Zha, H. (2018). Supervised reinforcement learning with recurrent neural network for dynamic treatment recommendation. arXiv:1807.01473.

Wang, X., Tan, K., Du, Q., Chen, Y. and Du, P. J. (2020). CVA2E: A conditional variational autoencoder with an adversarial training process for hyperspectral imagery classification. IEEE Transactions on Geoscience and Remote Sensing 58: 5676–5692.

Wang, Y. and Blei, D. M. (2019). Multiple causes: A causal graphical view. arXiv:1805.06826.

Wang, Z., Schaul, T., Hessel, M., van Hasselt, H., Lanctot, M. and de Freitas, N. (2015). Dueling network architectures for deep reinforcement learning. arXiv:1511.06581.

Wang, Z. T. and Ueda, M. (2021). A convergent and efficient deep Q Network algorithm. arXiv:2106.15419.

Weng, L. (2021). Controllable neural text generation. https://lilianweng.github.io/lil-log/2021/01/02/controllable-neural-text-generation.html

Wei, X., Gong, B., Liu, Z., Lu, W. and Wang, L. (2018). Improving the improved training of Wasserstein GANs: A consistency term and its dual effect. arXiv:1803.01541.

Weinan, W. (2017). A proposal on machine learning via dynamical systems, Communications in Mathematics and Statistics 5: 1–11.

Wen, Y., Singh, R. and Raj, B. (2019). Reconstructing faces from voices. arXiv:1905.10604.

White, H. and Lu, X. (2010). Granger causality and dynamic structural systems, Journal of Financial Econometrics 8: 193–243.

Wijayatunga, P. (2015). Probabilistic analysis of balancing scores for causal inference, Journal of Mathematics Research 7(2): 2015.

Williams, R. J. (1992). Simple statistical gradient-following algorithms for connectionist reinforcement learning, Machine Learning 8: 229–256.

Wright, S. (1921). Correlation and causation, Journal of Agricultural Research 20: 557–580.

Wright, S. J. (2015). Coordinate descent algorithms, Mathematical Programming 151: 3–34.

Wu, P. and Fukumizu, K. (2020). Causal mosaic: Cause-effect inference via nonlinear ICA and ensemble method, Proceedings of Machine Learning Research 108: 1157–1167.

Wu, P. and Fukumizu, K. (2021). Identifying treatment effects under unobserved confounding by causal representation learning. arXiv:2101.06662.

Wu, Z., Pan, S., Chen, F., Long, G., Zhang, C. and Yu, P. S. (2019). A comprehensive survey on graph neural networks. arXiv:1901.00596.

Xie, F., Cai, R., Huang, B., Glymour, C., Hao, Z. and Zhang, K. (2020). Generalized independent noise condition for estimating latent variable causal graphs. arXiv:2010.04917.

Xiong, M. M. (2018a). Big Data in Omics and Imaging: Integrated Analysis and Causal Inference. CRC Press, Boca Raton, Florida.

Xiong, M. M. (2018b). Big Data in Omics and Imaging: Association Analysis. CRC Press, Boca Raton, Florida.

Xiong, M. M. and Wang, P. (1992). Learning potential functions and differential inclusion, Proceedings of IJCNN'92 I: 401–406.

Xu, L., Chen, Y., Srinivasan, S., de Freitas, N., Doucet, A. and Gretton, A. (2020). Learning deep features in instrumental variable regression. arXiv:2010.07154.

Xu, M. (2020). Understanding graph embedding methods and their applications. arXiv:2012.08019.

Yang, C., Zhuang, P., Shi, W., Luu, A. and Li, P. (2019). Conditional structure generation through graph variational generative adversarial nets. NeurIPS 2019: 1338–1349.

Yang, L., Cheung, N., Li, J. and Fang, J. (2019). Deep clustering by Gaussian mixture variational autoencoders with graph embedding, 2019 IEEE/CVF International Conference on Computer Vision (ICCV). DOI: 10.1109/ICCV.2019.00654.

Yang, M., Liu, F., Chen, Z., Shen, X., Hao, J. and Wang, J. (2020). CausalVAE: Disentangled representation learning via neural structural causal models. arXiv:2004.08697.

Yellapragada, M. S. and Konkimalla, C. P. (2019). Variational Bayes: A report on approaches and applications. arXiv:1905.10744.

Yoon, J., Jarrett, D. and van der Schaar, M. (2019). Time-series generative adversarial networks, Advances in Neural Information Processing Systems 32 (NeurIPS 2019).

Yoon, J., Jordon, J. and van der Schaar, M. (2018). Estimation of individualized treatment effects using generative adversarial nets, ICLR 2018.

Yu, C., Liu, J. and Nemati, S. (2019). Reinforcement learning in healthcare: A survey. arXiv:1908.08796.

Yu, H. and Welch, J. D. (2021). MichiGAN: Sampling from disentangled representations of single-cell data using generative adversarial networks, Genome Biology 22: 158 (2021).

Yu, Y., Chen, J., Gao, T. and Yu M. (2019). DAG-GNN: DAG structure learning with graph neural networks. arXiv:1904.10098.

Yuan, C. and Malone, B. (2013). Learning optimal Bayesian networks: A shortest path perspective, The Journal of Artificial Intelligence Research 48: 23–65.

Zagoruyko, S. and Komodakis, N. (2017). Wide residual networks. arXiv:1605.07146.

Zaheer, M., Kottur, S., Ravanbakhsh, S., Poczos, B., Salakhutdinov, R. R. and Smola, A. J. (2017). Deep sets, Advances in Neural Information Processing Systems 30: 3394–3404.

Zaremba, W., Sutskever, I. and Vinyals, O. (2014). Recurrent neural network regularization. arXiv:1409.2329.

Zenati, H., Foo, C. S., Lecouat, B., Manek, G. and Chandrasekhar, V. R. (2018). Efficient GAN-based anomaly detection. arXiv:1802.06222.

Zhang, C., Butepage, J., Kjellstrom, H. and Mandt, S. (2019). Advances in variational inference, IEEE Transactions on Pattern Analysis and Machine Intelligence 41: 2008–2026.

Zhang, C., Zhang, K. and Li, Y. (2020). A causal view on robustness of neural networks. arXiv:2005.01095.

Zhang, K., Wang, Z., Zhang, J. and Schölkopf, B. (2015). On estimation of functional causal models: General results and application to the postnonlinear causal model, ACM Transactions on Intelligent Systems and Technology 7: 13:1–13:22.

Zhang, R., Zhang, Y. and Li, X. (2020). Graph convolutional auto-encoder with bi-decoder and adaptive-sharing adjacency. arXiv:2003.04508.

Zhang, W., Liu. L. and Li, J. (2020). Treatment effect estimation with disentangled latent factors. arXiv: 2001.10652.

Zhang, X., Li, Z., Loy, C. Chen. and Lin, D. (2017). Polynet: A pursuit of structural diversity in very deep networks. arXiv:1611.05725.

Zhang, X., Liu, H., Li, Q. and Wu, X. M. (2019). Attributed graph clustering via adaptive graph convolution. arXiv:1906.01210.

Zhao, B., Chang, B., Jie, Z. and Sigal, L. (2018). Modular generative adversarial networks. arXiv:1804.03343.

Zhao, Y. and Luo, X. (2019). Granger mediation analysis of multiple time series with an application to functional magnetic resonance imaging, Biometrics 75: 788–798.

Zheng, S., Zhu, Z., Zhang, X., Liu, Z., Cheng, J. and Zhao, Y. (2020). Distribution-induced bidirectional generative adversarial network for graph representation learning. arXiv:1912.01899.

Zheng, X., Aragam, B., Ravikumar, P. and Xing, E. P. (2018). AGs with NO TEARS: Continuous otimization for structure learning. arXiv:1803.01422.

Zhu J-Y, Park, T., Isola, P. and Efros, A. A. (2017). Unpaired image-to-image translation using cycle-consistent adversarial networks, Proceedings of the IEEE International Conference on Computer Vision, 2223–2232.

Zhu, S., Ng, I. and Chen, Z. (2020). Causal discovery with reinforcement learning. arXiv:1906.04477.

Zhuo, C., Kyle, M., Elmer, S., Gopal, G. and Lakshman, T. (2016). A physician advisory system for chronic heart failure management based on knowledge patterns, Theory and Practice of Logic Programming 16: 604–618.

Index

Note: Locators in *italics* represent figures and **bold** indicate tables in the text.

O

P

Q

R

S

T

For Product Safety Concerns and Information please contact our EU
representative GPSR@taylorandfrancis.com Taylor & Francis Verlag GmbH,
Kaufingerstraße 24, 80331 München, Germany

Printed and bound by CPI Group (UK) Ltd, Croydon, CR0 4YY
08/05/2025
01864539-0001